Carbon Incentives, Plans, and Policies in Agriculture

Carbon Incentives, Plans, and Policies in Agriculture

Edited by

Ram Swaroop Meena
Department of Agronomy, Institute of Agricultural Sciences,
Banaras Hindu University, Varanasi, Uttar Pradesh, India

Manoj Kumar Jhariya
Department of Farm Forestry, Sant Gahira Guru Vishwavidyalaya,
Ambikapur, Chhattisgarh, India

Arnab Banerjee
School of Studies in Environmental Science, Pt. Ravishankar Shukla University,
Raipur, Chhattisgarh, India

Sandeep Kumar
ICAR-Indian Agricultural Research Institute, Regional Station,
Karnal, Haryana, India

ELSEVIER

ACADEMIC PRESS
An imprint of Elsevier

Academic Press is an imprint of Elsevier
125 London Wall, London EC2Y 5AS, United Kingdom
50 Hampshire Street, 5th Floor, Cambridge, MA 02139, United States

ISBN: 978-0-443-33557-0

For Information on all Academic Press publications
visit our website at https://www.elsevier.com/books-and-journals

Publisher: Andre Wolff
Acquisitions Editor: Chiara Giglio
Editorial Project Manager: Ritika Dagar
Production Project Manager: M. Balakrishnan
Cover Designer: Raman Kumar

Typeset by MPS Limited, Chennai, India

Contents

16. Carbon credit estimation in croplands

Victor V. Ngaiza, Emmanuely Z. Nungula, Norbert J. Ngowi, Luciana R. Chappa, Rogerio P. Soratto, and Harun I. Gitari

17. Carbon credit estimation in agroforestry systems

Anjana J. Atapattu, Tharindu D. Nuwarapaksha, Nuwandhya S. Dissanayaka, and Shashi S. Udumann

Section V
Plans and policies for carbon incentives

23. Research, development, and future directives toward carbon financing

Donald Mlambo

24. Policy and legal frameworks for the successful implementation of green credits and financing

Arnab Banerjee, Manoj Kumar Jhariya, Ram Swaroop Meena, Abhishek Raj, Sharda Prasad Tripathi, Tamoghna Achharya, and Srimanta Gupta

27. Action plans toward net-zero emissions

Krishna Kant Mina, Suchismita Satapathy, Satish Kumar Singh, Ramawatar Meena, Shashi Shekhar, Gourisankar Pradhan, Sanskriti, Sandeep Kumar, Manoj Kumar Jhariya, Arnab Banerjee, and Sanjeev Kumar

List of contributors

Abey M. Abdi Central Highlands Eco-Region Research Programme, Kenya Forestry Research Institute, Nairobi, Kenya

Muhammed Abdulazeez Department of Microbiology, School of Sciences and Information Technology, Skyline University Nigeria, Kano, Nigeria

Abdulhakeem Idris Abdulhakeem Department of Fisheries and Aquaculture, Federal University Dutsin-Ma, Dutsin-Ma, Katsina, Nigeria

Mustapha Abdulsalam Department of Microbiology, School of Sciences and Information Technology, Skyline University Nigeria, Kano, Nigeria

Tamoghna Achharya Department of Marine Science, Berhampore University, Brahmapur, Odisha, India

S.B. Agrawal Department of Forestry, Jawaharlal Nehru Agricultural University, Jabalpur, Madhya Pradesh, India

Joshua Oluwatobi Akinola Department of Natural Sciences, University of Maryland, College Park, MD, United States

Rizal Akbar Aldyan National Research and Innovation Agency (BRIN), Research Center for Social Welfare, Villages and Connectivity, Jakarta, Indonesia

Keivan Arastou Department of Civil Engineering, University of Isfahan, Isfahan, Iran

Anjana J. Atapattu Agronomy Division, Coconut Research Institute, Lunuwila, Sri Lanka

R. Bajpai Department of Forestry, Jawaharlal Nehru Agricultural University, Jabalpur, Madhya Pradesh, India

Sunday Zeal Bala Department of Education, University of Maryland Eastern Shore, Princess Anne, MD, United States

Arnab Banerjee School of Studies in Environmental Science, Pt. Ravishankar Shukla University, Raipur, Chhattisgarh, India

Tyas Mutiara Basuki Research Center for Ecology, National Research and Innovation Agency (BRIN), Cibinong, Indonesia

Amna Bouri Higher Institute of Water Sciences and Techniques, University of Gabès, Gabès, Tunisia

Salem Bouri Water, Energy, and Environment Laboratory, National Engineering School of Sfax, Sfax, Tunisia

Jan Buleca Faculty of Economics, Technical University of Kosice, Kosice, Slovak Republic

Sigit Andy Cahyono Research Center for Ecology, National Research and Innovation Agency (BRIN), Cibinong, Indonesia

Chanderkant Dr Yashwant Singh Parmar University of Horticulture and Forestry, Nauni, Solan, Himachal Pradesh, India

Luciana R. Chappa Department of Agricultural Sciences, College of Agriculture, Mwalimu Nyerere University of Agriculture and Technology, Butiama, Tanzania

Suddhasuchi Das Malda Krishi Vigyan Kendra, Uttar Banga Krishi Vishwavidyalaya, Ratua, Malda, West Bengal, India

Nafisa Dayyabu Faculty of Earth and Environmental Sciences, Bayero University, Kano, Nigeria

Annpurna Devi Department of Farm Forestry, Sant Gahira Guru Vishwavidyalaya, Ambikapur, Chhattisgarh, India

D.M.N.S. Dissanayaka Agronomy Division, Coconut Research Institute, Lunuwila, Sri Lanka

Shehu-Alimi Elelu Department of Chemistry, Howard University, Washington, DC, United States

Mary Gathara Central Highlands Eco-Region Research Programme, Kenya Forestry Research Institute, Nairobi, Kenya

Harun I. Gitari Department of Agricultural Science and Technology, School of Agriculture and Environmental Sciences, Kenyatta University, Nairobi, Kenya

Srimanta Gupta Department of Environmental Science, The University of Burdwan, Burdwan, West Bengal, India

Zied Haj-Amor Water, Energy, and Environment Laboratory, National Engineering School of Sfax, Sfax, Tunisia

Madinat Hassan Foresight Institute of Research and Translation (FIRAT), Kigali, Rwanda

Ganiyat Omotayo Ibrahim Department of Chemistry, Nottingham Trent University, Nottingham, United Kingdom

Husna Farouq Ibrahim Department of Microbiology, Bayero University, Kano, Nigeria

Ijah U.J.J. Department of Microbiology, Federal University of Technology Minna, Minna, Nigeria

Paisteras H. Ilomo Department Engineering Management Studies, Mzumbe University, Morogoro, Tanzania

Musa Ojeba Innocent Department of Microbiology, School of Sciences and Information Technology, Skyline University Nigeria, Kano, Nigeria

Shervin Jamshidi Department of Civil Engineering, University of Isfahan, Isfahan, Iran

Manoj Kumar Jhariya Department of Farm Forestry, Sant Gahira Guru Vishwavidyalaya, Ambikapur, Chhattisgarh, India

Abdulsalam Khadija Department of Agricultural Sciences, University of Ilorin, Ilorin, Kwara, Nigeria

Bhimappa Honnappa Kittur Faculty of Technical Forestry, Indian Institute of Forest Management, Bhopal, Madhya Pradesh, India

Inna Koblianska Department of International Economic Relations, Sumy State University, Sumy, Ukraine

Sandeep Kumar ICAR-Indian Agricultural Research Institute, Regional Station, Karnal, Haryana, India

Sanjeev Kumar Department of Agronomy, Institute of Agricultural Sciences, Banaras Hindu University, Varanasi, Uttar Pradesh, India

Vandana Kumari Department of Soil Science, Advance Centre on Sericulture, Dr. Kalam Agricultural College, Bihar Agricultural University, Kishanganj, Bihar, India

Qiang Li School of Energy and Environment, City University of Hong Kong, Hong Kong, P.R. China

Miracle Uwa Livinus Department of Biochemistry, School of Sciences and Information Technology, Skyline University Nigeria, Kano, Nigeria

Leo Rio Ependi Malau Research Center for Behavioral and Circular Economics, National Research and Innovation Agency (BRIN), Jakarta Selatan, Indonesia

Asmau M. Maude Department of Microbiology, Federal University of Technology Minna, Minna, Nigeria

Ram Swaroop Meena Department of Agronomy, Institute of Agricultural Sciences, Banaras Hindu University, Varanasi, Uttar Pradesh, India

Ramawatar Meena Department of Soil Science and Agricultural Chemistry, Institute of Agricultural Sciences, Banaras Hindu University, Varanasi, Uttar Pradesh, India

Krishna Kant Mina Department of Soil Science and Agricultural Chemistry, Institute of Agricultural Sciences, Banaras Hindu University, Varanasi, Uttar Pradesh, India

Donald Mlambo Department of Forest Resources and Wildlife Management, National University of Science and Technology, Bulawayo, Zimbabwe

Niladri Sekhar Mondal Department of Environmental Science, School of Sciences, Netaji Subhas Open University, Kolkata, West Bengal, India

Saikat Mondal Raghunathpur College, Raghunathpur, Purulia, West Bengal, India

Muhammad Rifqi Tirta Mudhofir Research Center for Behavioral and Circular Economics, National Research and Innovation Agency (BRIN), Jakarta Selatan, Indonesia

Edward Mufandaedza Department of Forest Resources and Wildlife Management, National University of Science and Technology, Bulawayo, Zimbabwe

Adams Mustapha Department of Microbiology, Federal University of Technology Minna, Minna, Nigeria

Riziki Mwadalu Central Highlands Eco-Region Research Programme, Kenya Forestry Research Institute, Nairobi, Kenya

Stephen Mwakabonga Department of Environment and Sustainable Development, Mzumbe University, Morogoro, Tanzania

Debabrata Nath International Rice Research Institute (IRRI)-South Asia Regional Centre (ISARC), Varanasi, Uttar Pradesh, India

Victor V. Ngaiza School of Life Sciences and Bio-engineering, The Nelson Mandela African Institution of Sciences and Technology, Arusha, Tanzania

Norbert J. Ngowi Department of Environment and Sustainable Development, Mzumbe University, Morogoro, Tanzania

Niharika Dr Yashwant Singh Parmar University of Horticulture and Forestry, Nauni, Solan, Himachal Pradesh, India

Emmanuely Z. Nungula Department of Environment and Sustainable Development, Mzumbe University, Morogoro, Tanzania

Tharindu D. Nuwarapaksha Agronomy Division, Coconut Research Institute, Lunuwila, Sri Lanka

Kesia A. Nyambele Faculty of Agribiology, Food and Natural Resources, Czech University of Life Sciences, Prague, Czech Republic

Debnath Palit Durgapur Government College, Durgapur, West Bengal, India

Pankaj Department of Farm Forestry, Sant Gahira Guru Vishwavidyalaya, Ambikapur, Chhattisgarh, India

Abioye O. Peter Department of Microbiology, Federal University of Technology Minna, Minna, Nigeria

Gourisankar Pradhan Department of Agronomy, Institute of Agricultural Sciences, Banaras Hindu University, Varanasi, Uttar Pradesh, India

Abhishek Raj Pt. Deendayal Upadhyay College of Horticulture and Forestry, Dr. Rajendra Prasad Central Agricultural University, Samastipur, Bihar, India

Janusz Ruszkowski Institute of Political Science and Security Studies, University of Szczecin, Szczecin, Poland

Mercy Rutto Central Highlands Eco-Region Research Programme, Kenya Forestry Research Institute, Nairobi, Kenya

Priyanka Saha ICAR Krishi Vigyan Kendra, West Garo Hills, ICAR RC NEH Region, Umiam, Meghalaya, India

Job Oloruntoba Samuel Department of Microbiology, Federal University of Technology Minna, Minna, Nigeria

Sanskriti Department of Agronomy, Institute of Agricultural Sciences, Banaras Hindu University, Varanasi, Uttar Pradesh, India

S. Sarvade Department of Forestry, Jawaharlal Nehru Agricultural University, Jabalpur, Madhya Pradesh, India

Suchismita Satapathy Department of Agronomy, Institute of Agricultural Sciences, Banaras Hindu University, Varanasi, Uttar Pradesh, India

Hari Harjanto Setiawan National Research and Innovation Agency (BRIN), Research Center for Social Welfare, Villages and Connectivity, Jakarta, Indonesia

Shashi Shekhar Department of Agricultural Engineering, Institute of Agricultural Sciences, Banaras Hindu University, Varanasi, Uttar Pradesh, India

Satish Kumar Singh Department of Soil Science and Agricultural Chemistry, Institute of Agricultural Sciences, Banaras Hindu University, Varanasi, Uttar Pradesh, India

Vishnu K. Solanki Department of Forestry, Jawaharlal Nehru Agricultural University, Jabalpur, Madhya Pradesh, India

Rogerio P. Soratto Department of Crop Science, College of Agricultural Sciences, São Paulo State University (UNESP), Botucatu, Brazil

Muhammed Lawal Subair Department of Accounting and Finance, Kwara State University, Malete, Kwara, Nigeria

Sumino Institute for Rural Technology Development (LPTP), Jl. Raya Palur, Karanganyar, Indonesia

Tijani Abiola Tajudeen Department of Soil Science, Federal University of Technology, Minna, Nigeria

Maryna Tatar Faculty of Economics, Technical University of Kosice, Kosice, Slovak Republic

Bata Timothy Department of Geology, Skyline University Nigeria, Kano, Nigeria

Sharda Prasad Tripathi Sant Gahira Guru Vishwavidyalya, Surguja, Ambikapur, Chhattisgarh, India

Shashi S. Udumann Agronomy Division, Coconut Research Institute, Lunuwila, Sri Lanka

Katimu Yusuf Department of Microbiology, Federal University Gashua, Gashua, Yobe State, Nigeria

Lin Zhang School of Energy and Environment, City University of Hong Kong, Hong Kong, P.R. China

About the editors

Dr. Ram Swaroop Meena was born into a farming family and is working in the Department of Agronomy at the Institute of Agricultural Sciences (IAS), Banaras Hindu University (BHU), Varanasi, Uttar Pradesh, India. With an excellent academic background, he has been elected a fellow of several academies, including the National Academy of Sciences, India (NASI), the National Academy of Agricultural Sciences (NAAS), the National Academy of Biological Sciences (NABS), the Indian Society of Agronomy (ISA), and the Society for Rapeseed-Mustard Research (SRMR). The Ministry of Education has also awarded him the Raman Fellowship for the USA under World Food Prize Laureate Prof. Rattan Lal, and he was an INSA visiting scientist at ISARC-IRRI (IRRI South Asia Regional Center-The International Rice Research Institute). Additionally, he has been listed among the world's top 2% of scientists for his research impact. He has guided PhD and master's students and completed 10 externally funded projects from IRRI, The 'Institute of Eminence' (IoE)-Banaras Hindu University (BHU), Science and Engineering Research Board (SERB)-Department of Science and Technology (DST), The Indian Council of Agricultural Research (ICAR), Ministry of Education (MOE), The Government of India (GOI). He has published books, at an international level (Springer, Elsevier, etc.). He has made significant research contributions to natural resource management and to farm and industrial waste recycling for soil health. Dr. Meena's research evaluates energy flow, atmospheric CO_2 capture, carbon credit and auditing, and eco-friendly techniques for a sustainable food system and soil regeneration, as evidenced by high-impact publications and extension work at the farmer level. His work is crucial for understanding the long-term stability of soil carbon, which enhances soil microbial and nutrient dynamics in rice-based diversified cropping systems. He trained farmers from the Deep Forest Tribe to adopt climate-smart practices to secure their livelihoods under the DST-SEED grant project. Dr. Meena has also contributed to agricultural extension activities in villages designated under the Hon'ble Prime Minister of India's initiative. His future work aims to reduce soil organic carbon oxidation and enhance its stability in agroecosystems, enabling farmers to benefit from carbon credits and generate additional income.

Dr. Manoj Kumar Jhariya is a senior assistant professor in the Department of Farm Forestry at Sant Gahira Guru Vishwavidyalaya, Sarguja, Ambikapur, Chhattisgarh, India. His research focuses on forest ecology, agroforestry, natural resource management and sustainability, and carbon sequestration. Dr. Jhariya was listed among the world's top 2% scientists by Stanford University and Elsevier in 2023, 2024, and 2025, thanks to his extensive publishing activities (Springer, Elsevier, Wiley, Apple Academic Press, Taylor & Francis, CRC, etc.). Dr. Jhariya holds a BSc in Agriculture, an MSc in Forestry, and a PhD in Forestry from Indira Gandhi Krishi Vishwavidyalaya, Raipur, Chhattisgarh, India. He serves on the editorial boards of several journals and is a life member of numerous professional societies, including the Indian Science Congress Association, the Applied and Natural Science Foundation, the Society for Advancement of Human and Nature, the Medicinal and Aromatic Plants Association of India, the ISDS Society, the International Journal of Development and Sustainability, the Asia Society of Researchers, the Agriculture Forum for Technical Education of Farming Society, the Society of Green Development (India), the National Academy of Biological Sciences (India), the Society for Science and Climate Change and Sustainable Environment, and the Indian Society of Agroforestry. Currently, Dr. Jhariya is working on carbon incentives and green economy, agroforestry-based land restoration, quality planting materials in forestry, non-timber forest products' utilization and value addition, and the application of underutilized plants in land restoration.

Dr. Arnab Banerjee is an associate professor in the School of Studies in Environmental Science, Pt. Ravishankar Shukla University, Raipur, Chhattisgarh, India. He completed an MSc and a PhD (environmental science) from Burdwan University, Purba Bardhaman, West Bengal, India, and MPhil (environmental science) from Kalyani University, Kalyani, West Bengal, India. He was awarded the University Gold Medal for securing first place with first-class honors in the MSc examination. He has also received the Young Scientist Award for best oral presentation at an international conference held at the University of Burdwan. He was a project fellow under a UGC-sponsored major research project. He is a life member of the Academy of Environmental Biology. He has supervised postgraduate students and is engaged in postgraduate teaching and research, alongside publishing books, book chapters, and research articles in reputed national and international journals.

Dr. Sandeep Kumar was born into a farming family in the village of Jaswantpura, Rajgarh (Churu), Rajasthan, India. He joined the Agricultural Research Services (ARS) in 2020 at the ICAR-Indian Institute of Pulses Research (IIPR), Regional Centre, Bhopal, Madhya Pradesh, India. Currently, he is a scientist (agronomy) at the ICAR-Indian Agricultural Research Institute (IARI), Regional Station, Karnal, Haryana, India. Dr. Kumar completed his schooling in his native village and earned his bachelor's degree in agriculture (Hons.) in 2013 from the College of Agriculture, Lalsot, SKRAU, Bikaner, Rajasthan, India. He completed his master's degree from BHU, Varanasi, Uttar Pradesh, India, and his PhD in agronomy from Chaudhary Charan Singh Haryana Agricultural University, Hisar, Haryana, India, with a specialization in nutrient management in grain legumes. Throughout his academic career, he secured first division and qualified for the ICAR-NET three times, in addition to receiving prestigious SRF fellowships from the Indian Council of Agricultural Research. His areas of expertise include improving resource-use efficiency in conservation agriculture-based rice–wheat systems, cropping system diversification, carbon sequestration, and carbon credits for sustainable crop production. A prolific author and editor, Dr. Kumar has received several accolades, including the Land 2020 Best Paper Award from MDPI and Wiley's Top Cited Article Award for 2022–2023. He was also honored with the Best Young Scientist Award (Regional Station) for 2024–2025 by ICAR-IARI, New Delhi. Dr. Kumar is actively engaged in teaching postgraduate and PhD students at the IARI–IIWBR Mega Hub, Karnal, Haryana, India. He also plays a key role in institutional development and resource generation initiatives. His future research goals focus on soil carbon sequestration, climate change adaptation, carbon credit generation, regenerative agriculture, and low-emission technologies in a systems-based approach.

Foreword

In the era of climate change, the world must shift toward a green economy as a crucial step in climate action. Agriculture is one of the most promising sectors that need transformation to address the current challenges, starting with maintaining yields while adopting climate-smart practices. Achieving net-negative emissions in agriculture requires changing farming methods to reduce agroecological footprints and leverage natural processes to absorb or offset emissions. Managing the use of natural resources in agriculture will require coordinated efforts among farmers, governments, researchers, and consumers. It will involve a combination of agro-technological innovations for carbon financing and a shift toward more sustainable diets and consumption patterns for long-term ecosystem sustainability.

Vision 2050 is an ambitious plan to achieve net-negative emissions in agriculture and support the successful decarbonization of agricultural systems. Various methods, such as conservation agriculture, the carbon credit mechanism, precision farming, carbon-negative agriculture, natural farming, agroforestry, and producing more from less, are among the policy issues that need to be addressed to transform agriculture. Therefore, this book offers a timely approach to identify potential areas, roadmaps, strategies, solutions, and policy arenas toward a greener economy. It will also provide in-depth insights into the concepts of the carbon market, incentive models, mechanisms of carbon financing, and technological advancements. Additionally, it will help estimate the potential of carbon financing as an additional source of income for farmers' communities. The present book will serve as a guideline for the academic and administrative sectors to establish the link between policy and science, assess the effectiveness of carbon financing, and harmonize the economy and agroecology.

Rattan Lal
Distinguished University Professor of Soil Science,
SENR Director, CFAES Rattan Lal Center for Carbon Management and Sequestration
The Ohio State University, Columbus, OH, United States

Preface

Agriculture is a major economic activity for human beings that interlinks livelihood security, climate response, and food security. The sector plays a dual role in greenhouse gas (GHG) emissions and carbon sequestration (C_{seq}). As we move toward the green economy through various forms of agricultural transformation, economic incentives in the form of carbon (C) credits could be a suitable option globally. In this context, various policies—such as carbon taxes and subsidy programs, as well as an emphasis on carbon sinks across various sectors—help achieve the Sustainable Development Goals (SDGs) for long-term sustainability. Economic gain mechanisms such as green accounting, sustainable accounting, credit systems, ecosystem service payments, and carbon trading have now gained worldwide attention. Green credit is the most recent approach to greening the economic system by fostering eco-sensitive behavior among individuals, industries, corporate sectors, academic institutions, and other allied sectors.

Agriculture is one of the major sectors worldwide that contributes significantly to the gross domestic product (GDP) of developing nations. This book therefore covers the accounting system and the reforms needed in the economy through proper carbon incentives and policy formulation to increase net economic returns in agro-based industries. This would also help boost the socioeconomic development of the farming community and fulfill emission reduction and other environmental obligations. Technological interventions—such as organic waste recycling, crop rotation involving legume and nonlegume systems, agroforestry, zero-tillage agriculture, the adoption of biofertilizers, and integrated nutrient management practices—are showing significant promise in fulfilling the SDGs and the overall sustainability of the soil ecosystem. Such activities support conservation, sustainable production, and fertilizer management and also have the potential to generate subsidies through the carbon credit mechanism. Economic incentives should be provided in the form of green credits to promote activities that enhance the soil carbon pool and soil fertility. Agroforestry has shown significant promise in this direction and can therefore be subsidized to encourage its wider adoption. Suitable policy formulation for carbon restoration would also have a positive impact on agricultural production and contribute to a greener economy.

Moving toward a green economy requires the adoption of green credit mechanisms across various sectors to promote innovative approaches that support sustainable agriculture and environmental conservation. Such approaches would help reduce environmental costs, along with lower pollution and energy consumption across various enterprises. The key challenges for the effective implementation of the green credit mechanism across various sectors include developing a proper methodology, ensuring transparency in the process, implementing regulatory control, and maintaining a stable economic system. Addressing these challenges would help fulfill three major SDG 2030 goals—SDG-12, SDG-13, and SDG-15—which aim toward sustainable consumption and production, climate resilience, and life on land, respectively. The United Nations (UN) has worked to mediate financial support for achieving SDG 2030 and has initiated a resource efficiency program through which it promotes green credit and financing mechanisms across various sectors to achieve the SDGs.

This book explores carbon incentive options in the agricultural sector to achieve the target of a green economy. It engages different players in the circular economy to help global readers understand the mechanisms for monetizing agronomic incentives. This book presents an in-depth compilation of plans and incentives related to the carbon economy, carbon financing, monetization, green accounting, ecological economics, environmental economics, and carbon credit and trading, as well as other sustainability issues within the agroecosystem, including carbon management and the green economy for the benefit of the farming community across the globe. This approach would encourage young researchers in sustainability science to develop smart solutions for sustainable agroecosystems and their services and to work toward the successful implementation of carbon financing. It would also help create a conceptual framework for future policy and planning perspectives on developing a carbon-based green economy in agroecosystems.

Ram Swaroop Meena
Manoj Kumar Jhariya
Arnab Banerjee
Sandeep Kumar

Section I

Introduction

Chapter 1

Agriculture for capturing atmospheric CO_2 and monetizing carbon credits in the green economy: policies and actions

Ram Swaroop Meena[1], Manoj Kumar Jhariya[2], Arnab Banerjee[3], and Sandeep Kumar[4]

[1]*Department of Agronomy, Institute of Agricultural Sciences, Banaras Hindu University, Varanasi, Uttar Pradesh, India,* [2]*Department of Farm Forestry, Sant Gahira Guru Vishwavidyalaya, Ambikapur, Chhattisgarh, India,* [3]*School of Studies in Environmental Science, Pt. Ravishankar Shukla University, Raipur, Chhattisgarh, India,* [4]*ICAR-Indian Agricultural Research Institute, Regional Station, Karnal, Haryana, India*

Highlights

- Showcase the agroecosystem's potential as a carbon sink.
- Monetization of stored carbon in an agroecosystem.
- Carbon credit generation and flow to the farming community.
- Approaches for carbon monetization in agroecosystems.
- The exploration and feasibility analysis of carbon monetization in an agroecosystem.

1.1 Introduction

Globally, agriculture and agroecosystems are the key components for growth, development, and the economy. Hence, in an agricultural system, people, or the farming community, stress on better yields and productivity (Kumar et al., 2024). They also aim to ensure that the cost-benefit ratio is on the positive side. Currently, major events such as climate change, natural disasters, environmental degradation, and global warming are severely impacting the agricultural system, leading to an imbalance in the sustainability of the agroecosystem. The symptoms include loss of soil quality and fertility, land degradation, reduced yield and productivity, and alterations in the carbon (C) cycle across different agricultural cropping systems (Meena et al., 2022). It has now become a debatable issue whether combating climate change through climate-resilient agroecosystems is an efficient and suitable mechanism, such as green credit or C credit-like policies, for maintaining the sustainability of agroecosystems (Raj et al., 2025; Meena et al., 2025). Additionally, various reports have found that atmospheric CO2 concentrations have risen significantly over the past few decades, leading to an increase in Earth's temperature, commonly referred to as global warming. Now, this has to be taken into account in the field of agricultural production, as many crops have shown significant promise in capturing atmospheric CO_2 and utilizing it in cellular metabolism to give better yield and productivity (Meena et al., 2024a, 2024b, 2024c). This has opened up the areas of C and green credit for exploration by both agricultural scientists and other allied scientific branches. Developing appropriate financing for agricultural components that would increase income for the farming community is important, as in the developing world most farmers are small landholders or cultivate on a small scale. Hence, in this context, the C credit system would provide additional income opportunities for the farming community.

The Intergovernmental Panel on Climate Change (IPCC) has recommended, in all its meetings, reducing atmospheric CO_2 concentration to bring the Earth's average temperature below 1.5°C. According to the Paris Agreement and the 2030 Agenda, emission reductions should exceed 40% to achieve the desired targets (Lee et al., 2023). To achieve the target of emission reduction, the development of appropriate markets for C through green economics has proven a suitable alternative and an effective strategy to promote C emission reduction in agro-based industries and practices. Associated

Carbon Incentives, Plans, and Policies in Agriculture. DOI: https://doi.org/10.1016/B978-0-443-33557-0.00021-5

with this, the concept of C neutrality was also introduced to balance every ton of CO_2 emitted and captured, thereby maintaining C balance. The development of the C market emerged under the CDM (Clean Development Mechanism) framework. Interestingly, the mechanism of sharing C credits and lowering emissions was achieved by economically stronger nations by investing in C-reduction projects in developing countries. Such developments in the agriculture sector have led to the origin of the C farming era. Under these, economic incentives were prioritized for the farming community that would demonstrate a commitment to C emission reduction and CO_2 capture through innovative, eco-friendly technologies in the agroecosystem (Prasad et al., 2021). The European Union has already formulated and is working to build a robust credit system, which is generating significant economic gains across both corporate and noncorporate sectors. Furthermore, the EU aims to achieve net-zero emissions by 2050 through the effective implementation of the carbon farming mechanism. Hence, all the policies and strategies were aimed at sustainable agricultural practices that would reduce carbon and ecological footprints in sustainable production and yield (Banerjee et al., 2021). Under such a scheme, the EU is attempting to reduce fertilizer use by 50%, which would lead to less nutrient loss, which would enhance the deterioration of soil fertility on one hand and reduce fertilizer use up to 20% by 2030 on the other. To achieve these mechanisms, such as promoting organic farming on 1/4 of the arable land by 2030, efforts were made to reduce the agricultural carbon footprint. To this end, C farming has emphasized a management system that helps utilize atmospheric C in various components of the soil–plant system in the form of sequestered C and C stocks (Jhariya et al., 2024a). The overall objective was to reduce atmospheric emissions and increase the flow of C into the soil–plant system to gradually build up the soil C pool (Rodias et al., 2021; Banerjee et al., 2023; Meena et al., 2024b).

This chapter aims to provide a global perspective on greenhouse gas (GHG) emissions in the agricultural sector, followed by an examination of the associated C dynamics within the agroecosystem. Furthermore, the chapter sheds light on the monetization mechanism of C through C credit worldwide. Finally, the chapter discusses policies, strategies, decision-making, and R&D activities in this new field of monetizing C credits within the agroecosystem to move toward a green economy.

1.2 Agriculture and GHG emission scenarios

Among anthropogenic activities, agriculture occupies 50% of the Earth's surface, accounting for one-third of global GHG emissions. It has been further reported that agricultural activities account for 31% of GHG emissions (Rodríguez-Espinosa et al., 2023). Hence, agriculture has become one of the priority sources for environmental impact after housing and transportation activities (The Circularity Gap Report, 2019). From the Indian perspective, GHG emissions account for 14% of emissions. Combinations of food industries and agricultural activities result in a carbon footprint value of approximately 10 billion tons of CO_2 equivalent. Among the various farming activities, more than half (54%) of the GHG contribution comes from livestock, followed by the application of nitrogenous fertilizers, which account for about 20% of GHG emissions (Food and Agriculture Organization of the United Nations (FAO), 2018). Rice is the most staple food crop under tropical climatic conditions and hence contributes one-third of methane emissions on a global basis, with an 11% contribution in terms of nitrous oxide emissions (Food and Agriculture Organization of the United Nations (FAO), 2018).

Various reports have shown that GHG concentrations have increased significantly over the past few decades, leading to major global climate change events (Gopalakrishnan et al., 2019). According to Outhwaite et al. (2022), GHG emissions have increased by two-thirds over the last three decades. Among the GHGs, CO_2, which makes the largest contribution (76%), is followed by CH_4 at 16% and N_2O at 6% (Crippa et al., 2021). Such a rise in GHG emissions has reflected severe socio-economic and ecological vulnerabilities at both regional and global levels. GHG emissions should be reduced by limiting the C emissions to about 570 gigatonnes, along with other GHGs, with specific emission limits (Robinson, 2020). Now, when we consider agricultural activity, it appears that agriculture is a potential source of GHGs such as N2O and CH4, which have a greater potential to cause global warming (Robinson, 2020; Meena et al., 2024a).

GHG emissions have become the primary drivers of the global climate crisis. Various sectors of economic development are contributing significantly to GHG emissions, as are sectors of the economy, in both developed and developing nations (Banerjee et al., 2024). The significant contribution from AFOLU (Agriculture, Forestry, and Other Land Use Change) accounts for 1/4 of global emissions and up to 18% in the Indian context. Soil is expected to be a major carbon sink, accounting for 89% of the C sequestration (Cseq) potential through agricultural activities. Therefore, climate-resilient practices in agricultural systems may significantly improve the Cseq potential of the soil ecosystem for agriculture-based countries like India (Jhariya et al., 2018; Meena and Pradhan, 2023).

1.3 Agroecosystems and carbon dynamics

In the era of industrialization, the pressure is on natural resources and their unsustainable exploitation. The extensive use of fossil fuels, land-use changes, and a consumptive, unsustainable lifestyle have all contributed to increased GHG emissions, leading to the global climate change crisis. Hence, moving toward C neutrality by 2050 is the major task facing humanity. According to some studies, by 2050, the global population would increase by up to 9.9 billion people, consuming an additional 80% of energy resources and 70% of food resources, leading to increasingly unsustainable forms of behaviour and habitat (IISD, 2020). According to one estimate in the last two decades, 30% of GHG emissions have increased from agricultural sources. This may be attributed to increased production activities, both in crop and animal husbandry (Frank et al., 2019). Across various agricultural systems, CO_2 emissions range from 3.3 to 4.2 Gt CO2eq annually due to crop-livestock interactions (Poore and Nemecek, 2018).

Given the agricultural sector's contribution to GHG emissions, alternative systems are needed to reframe the agroecosystem not as an emitter but as a carbon sink, thereby reducing GHG emissions (Frank et al., 2019; Raj et al., 2024). Various strategies, such as altering dietary intake from animal-based foods to plant-based foods and optimizing the use of fertilizer and water resources, may bring some benefit to reducing GHG emissions from agroecosystems (Kumar et al., 2020; Shang et al., 2021). Nutrient management through the development of compost and vermicompost may help reduce C and nitrogen losses, thereby reducing their subsequent emissions and increasing atmospheric concentrations or load (Harindintwali et al., 2021; Meena et al., 2023).

Considering agricultural C dynamics, the C economy within the agroecosystem is an essential aspect for the effective implementation of C monetization for the farming community. Further, in C dynamics, C flow, flux, and stock are the important components that regulate the distribution and stocking of C in different components of the ecosystems (Yadav et al., 2022; Jhariya et al., 2024b; Baghel et al., 2025; Meena et al., 2024c) (Fig. 1.1). For effective monetization, creating an artificial market that would have proper demand for C credits is also an important aspect of the C economy in agroecosystems (Painkra et al., 2025). In this connection, the development of eco-friendly technology that would promote increased C sink for the agricultural ecosystem would also help achieve the target of emission reduction, followed by the monetization of stored C within the agroecosystem and its components.

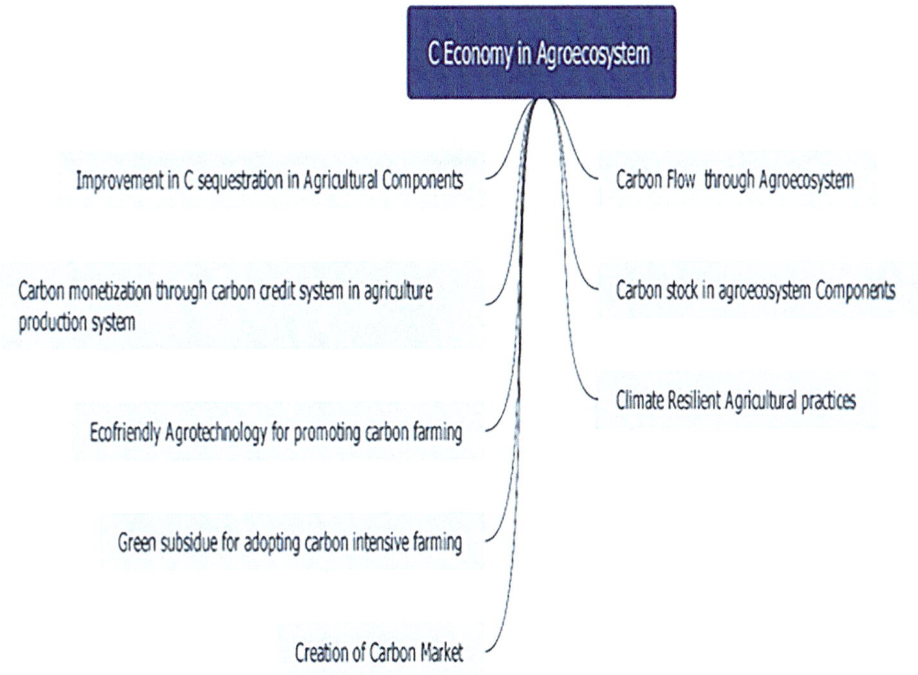

FIGURE 1.1 Carbon economy in the agroecosystem and its components.

1.4 Carbon credits in the agricultural system: carbon capture through agricultural practices

Agricultural intervention in terms of diversifying practices for C capture is the need of the hour for effective implementation of C monetization for the benefit of agriculture (Fig. 1.2).

Such approaches would provide additional benefits and socioeconomic uplift for the farming communities of developing countries as a whole. This would also fulfill the target of doubling the income of farming communities under the banner of C trading, sequestration, emission reduction, and achieving net-zero emissions or C neutrality. To incentivize the C credit in the agricultural system, the proper development of the market is essential to facilitate the purchase mechanism of C credits among buyers and sellers. It would also benefit the C offset programs worldwide that contribute to the emission-reduction mechanism.

Globally, implementing C credits requires effective incentives for the farming community to transition from conventional agricultural practices to climate-smart agroecosystem development through best management practices (Raj et al., 2023). It requires the creation of voluntary C credits, followed by a C offset mechanism, to reduce anthropogenic emissions and combat climate change (Pradhan and Meena, 2023b). In the USA, two corporate houses, *viz.*, Indigo and Nori, are working in this perspective of offsetting C from agricultural soil through the voluntary market mechanism. The initiation of the C market formulation by these corporate houses began in 2019 and 2017. A comparative study was conducted to examine the willingness and nonwillingness of the farming community to adopt climate-resilient practices in terms of soil carbon management within the agroecosystem. The main mechanism behind the C credits system within the agricultural sector involves screening activities that promote Cseq and monetizing them in terms of C credits. This would promote greater involvement in and participation in climate-resilient practices, which would be impossible without such a financial mechanism. However, it has some inherent problems, including the proper assessment of Cseq through on-farm activities. The amount of C_{seq} under this process is also an essential aspect of the proper evaluation of the amount of C credit generated for the particular agroecosystem and its actual benefit for the farming communities.

Development of proper protocols by the nonprofit organization for accurate estimation and verification of the amount of C_{seq} from the agricultural system (Zelikova et al., 2021) (). A mechanism involving the screening of activities that provide a GHG surplus through proper standardized protocols (such as the Soil Enrichment Protocol and Climate Action Reserve Protocol) is essential for determining the eligibility of a particular activity for consideration under C credit evaluation (Soil Enrichment Protocol, 2020). Another process would involve identifying farming communities that have adopted climate-resilient practices, whether or not they received financial subsidies or other incentives. However, those who have not

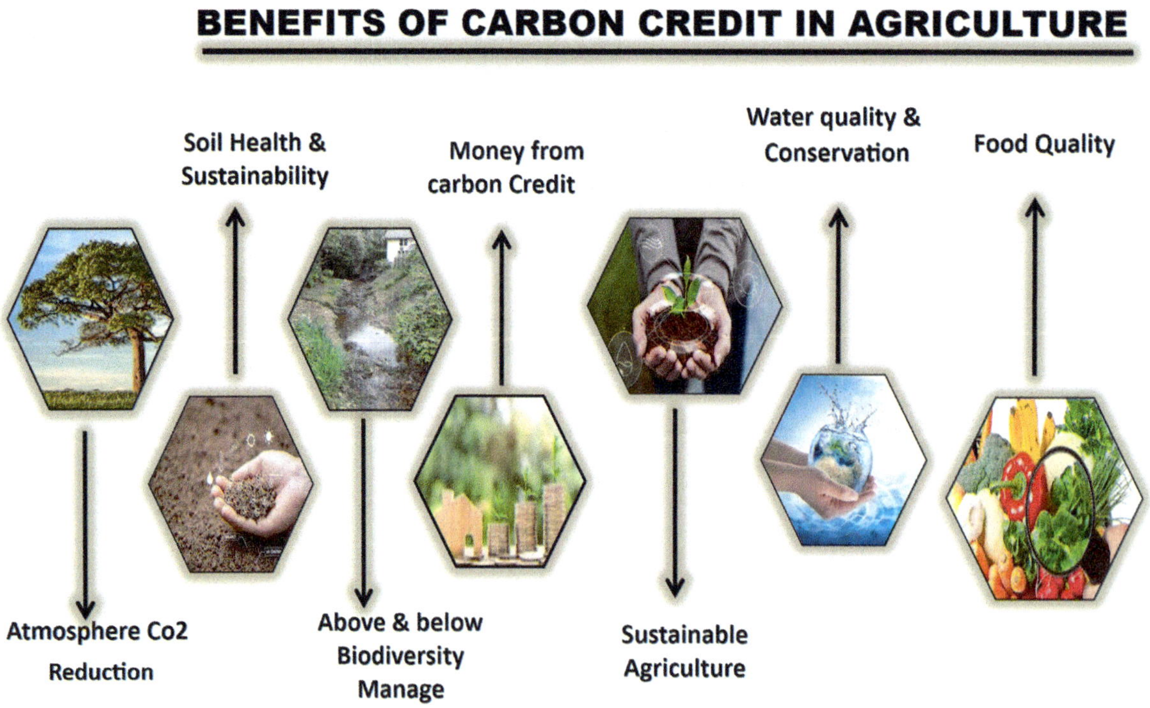

FIGURE 1.2 Benefits of carbon credit in agriculture.

adopted under such a mechanism may also be brought under the scheme for future promotion of the C incentive mechanism.

The EU C trading system is the leading example of the effective implementation of C financing and credit systems across various sectors. They have also emphasized the implementation of C farming as part of a green business model that would help reduce emissions and lead to financial benefits for the farming community (Wolf, 2022).

Extensive tillage breaks down soil organic matter and other essential soil components. Hence, a no-tillage approach was undertaken for soil aggregate stabilization, followed by a lesser emission of GHGs. The no-till approach has also shown significant promise in soil Cseq, leading to the buildup of the soil organic carbon (SOC) pool (Derpsch et al., 2010). Biochar application to soil has also shown significant promise for CO2 capture in soil systems through SOC (Gross et al., 2021). Mulching is another suitable approach that inhibits soil erosion, thereby enhancing topsoil organic carbon (OC) content. Apart from that, the use of crop residues, compost, and green manure is an alternative that enhances the soil Cseq through microbial decomposition and build-up of SCP (soil carbon pool) (Chen et al., 2018). Focusing on cultivating indigenous and traditional varieties also leads to a higher rate of Cseq. It has been reported that such approaches improve soil quality and promote vigorous root growth, which helps bind and aggregate soil. Maintaining soil biota, which includes soil macro- and microbiomes, is another important step that supports proper C and other nutrient cycling, leading to better soil health. This, on the other hand, helps reduce the use of fertilizer, water, and GHG emissions, making it the best choice. Proper crop rotation with polyculture practices helps maintain soil biodiversity and, by indirectly promoting soil microbial biomass and intensive root development, leads to better SOC sequestration (Abagandura et al., 2020). Cultivation of perennial grasses helps develop an intensive root system, which improves soil structure and enhances the soil's ability to act as a carbon sink. It has been reported that they maintain suitable conditions for the agroecosystem to continue on a long-term basis with reduced inputs of various agrochemicals and reduced GHG emissions (Sarkar et al., 2020; Pradhan and Meena, 2023a). Planting tree belts is another option that helps prevent soil erosion and maintain soil temperature. Plantation for harvesting is another suitable option that increases C input into the soil during vegetative growth while also serving as an erosion-preventive structure. Retention of slash and stubble mulches is another source of C sink, as they provide soil protection and contribute to SCP through proper decomposition by soil microbes. This also enhances the soil biota and their capacity to increase the Cseq (Liu et al., 2009).

1.5 Monetizing carbon credits in developed and developing nations

C credit is a concept introduced under the Kyoto Protocol in 1997 to address GHG emission reduction (Xu et al., 2023). At present, the demand for C credit is expanding rapidly in both developed and developing nations. It has been estimated that the C credit requirement worldwide will increase by 15-fold by 2030 to fulfill the SDGs. It would be highly beneficial for the farming community as it would provide additional economic benefits from C_{seq} on the agricultural lands (Morizet-Davis et al., 2023). It has been estimated that farmers can earn up to Rs. 800 per C credit from the carbon market. However, it may be extended up to Rs. 2000 per C credit for large agro-based corporations that are capable of purchasing C credits directly from the farming community (Blaufelder et al., 2021). It has been found that farmers can obtain up to 4 C credits per acre through agricultural practices. However, it is associated with factors such as water-holding capacity, higher water infiltration rate, better nutrient availability, and lower soil surface temperature (Babu et al., 2023).

The value of C credit varies significantly depending on the market economy, the nature and type of project activities, and the topography and geography of the observed site. Hence, the participation of the farming community in the effective implementation of the C credit mechanism and the involvement of the government body in various forms of agricultural activities are essential (Leifeld, 2023). Carbon credits play a central role in the formulation of the green economy, which emphasizes net-zero emissions, followed by carbon monetization to achieve the target of carbon neutrality and carbon offset within the agroecosystem, as presented in Fig. 1.3

In the EU, the European Green Deal initiative was initiated, which led to the development of the EU trading system, providing a solid framework for monetizing C credits in terms of technology transfer or financial benefits. It also helped develop the country's image by taking the initiative to reduce C emissions across various sectors for the green economy, as shown in Fig. 1.4

The EU trading system is the largest market in terms of monetizing and regulating the C economy across the globe (Green et al., 2023). Another important step initiated by the European Green Deal, under the Farm to Fork concept, aimed to formulate an eco-friendly and sustainable climate-resilient agroecosystem that would enrich the SCP while maximizing yield and productivity (European Commission, 2019). This has also become part of the major action plan, titled "Decade of Action," to achieve the SDGs (United Nations, 2021). However, there are greater challenges in terms of agronomy, economics, and environmental considerations for the effective implementation of such a trade mechanism in the agricultural sector and its practical benefits in the agroecosystem.

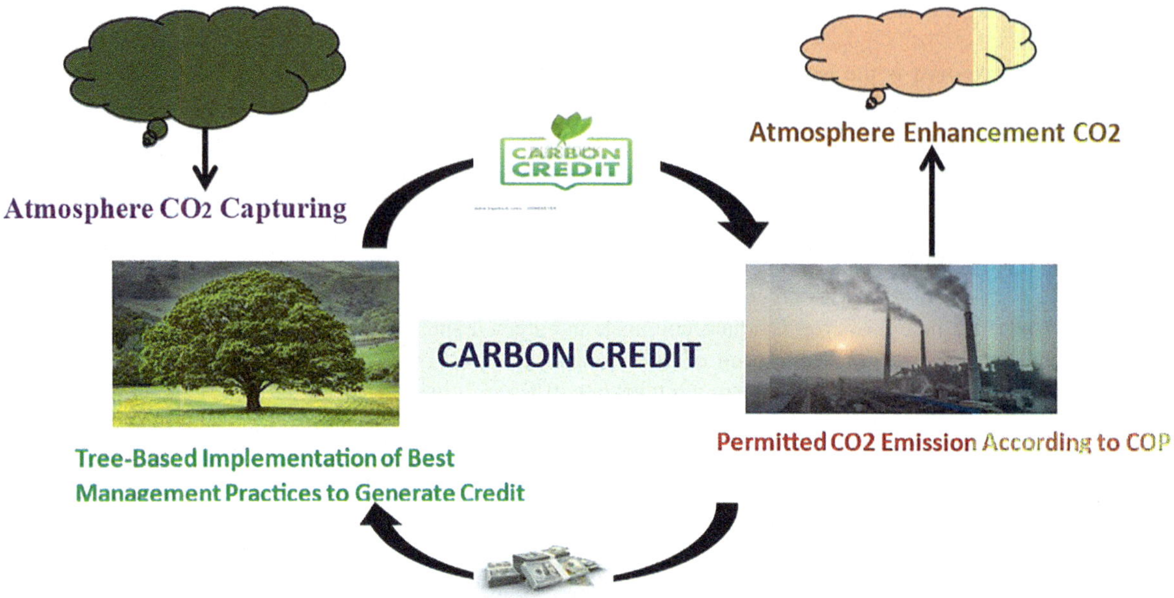

FIGURE 1.3 Carbon credit mechanism.

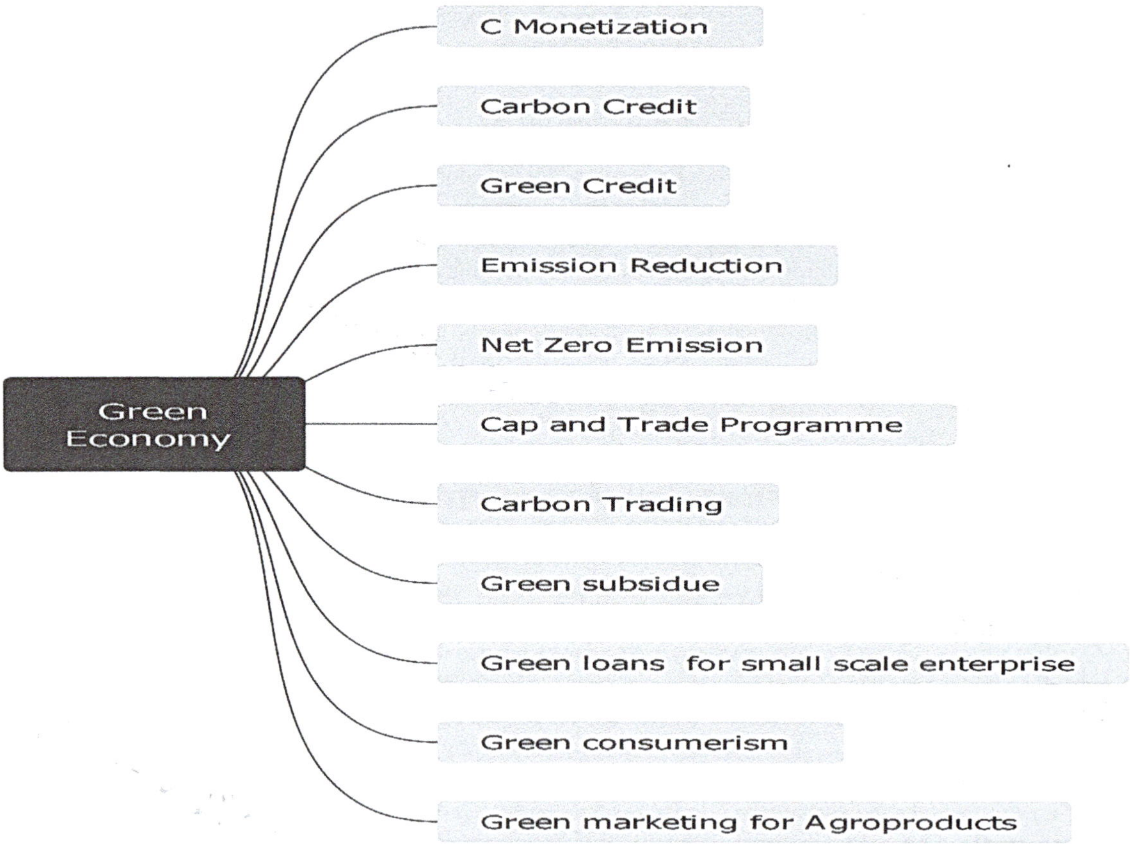

FIGURE 1.4 Green economy and its components.

At the national level, the use of renewable and eco-friendly energy resources would also help reduce GHG emissions across various sectors. In this connection, one example is Pakistan's approach to adopting mitigatory measures, which offer monetary benefits of up to 14 billion USD. Furthermore, under the Green Pakistan program, 100 million trees were planted to enhance forest cover, which could potentially reduce the atmospheric CO_2 load. This reduction can subsequently be converted into C credits, yielding additional economic benefits and carbon monetization (Syed et al., 2022).

1.6 Strategic decision-making for the economic benefit of the farming community through carbon credits

Monetization of C through farming practices is an important aspect that would address the problems of hunger, food crises, and food security, as well as serve as an economic incentive for the rural farming community. Such approaches would promote greater willingness among the farming community to reduce atmospheric CO_2 levels through agricultural interventions. All these things are an important part of climate change that will move toward C neutrality and net-zero emissions. Among the policy issues, strategies such as C_{seq} in agricultural soils are particularly important, as they would help reduce emissions and provide the farming community with additional economic benefits. Reducing GHG emissions through regenerative agriculture is urgently needed (Fig. 1.5). It will promote agricultural sustainability through net-C-negative emissions, as shown in Fig. 1.6

Global emission reduction is the need of the hour, as GHG emissions continue to rise across sectors. Policy formulation for combating climate change requires a reduction in CO_2 emissions, which is subsequently monetized in the name of CO_2 credits or certificates for emission reduction. Another breakthrough in policymaking requires the screening and adoption of green technologies, that is, a commitment to C emission reduction with proper subsidies at the government level (Abbass et al., 2022). Involvement in sectoral policy development is needed to promote such C initiatives among the farming community in developed and developing nations (Quandt et al., 2023).

Now there is a need to set up a proper C market for monetization of C credits, both in developed and developing nations. Indirect reduction of the C footprint can also be achieved through this C trading mechanism, which will be a

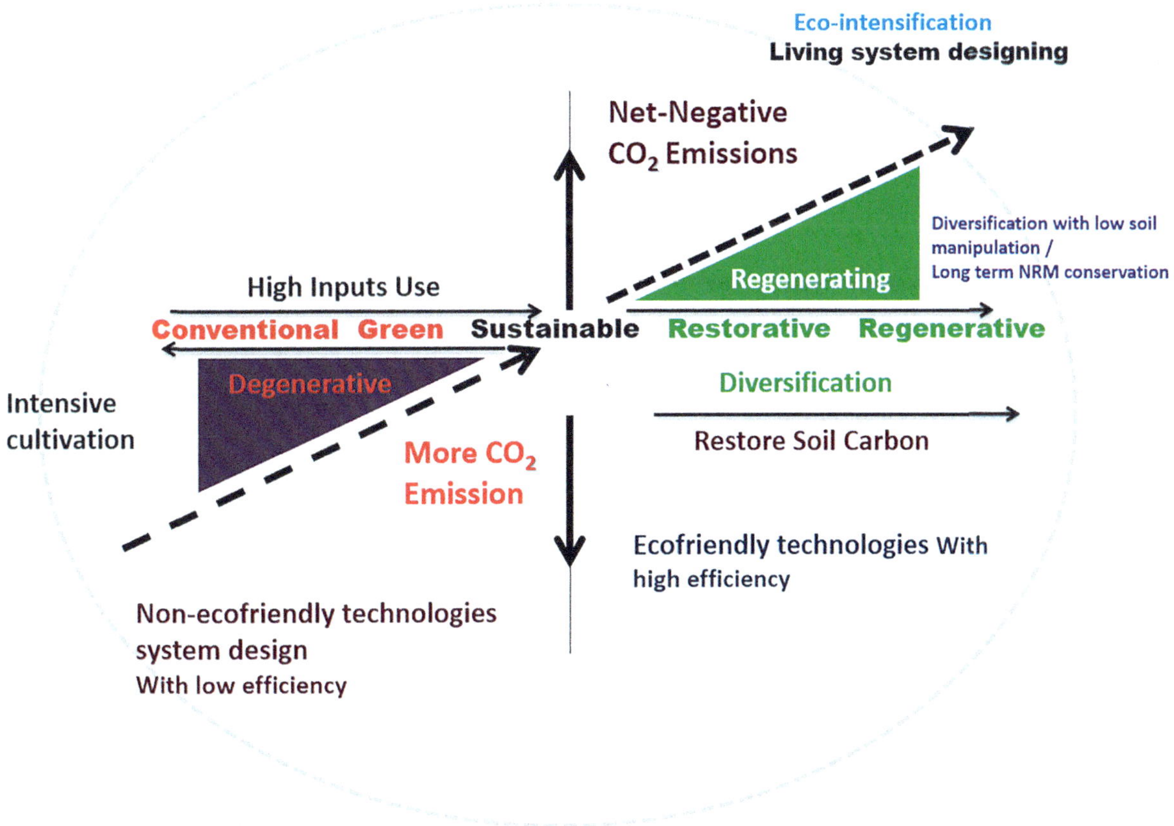

FIGURE 1.5 Greenhouse gas emissions reduction through regenerative.

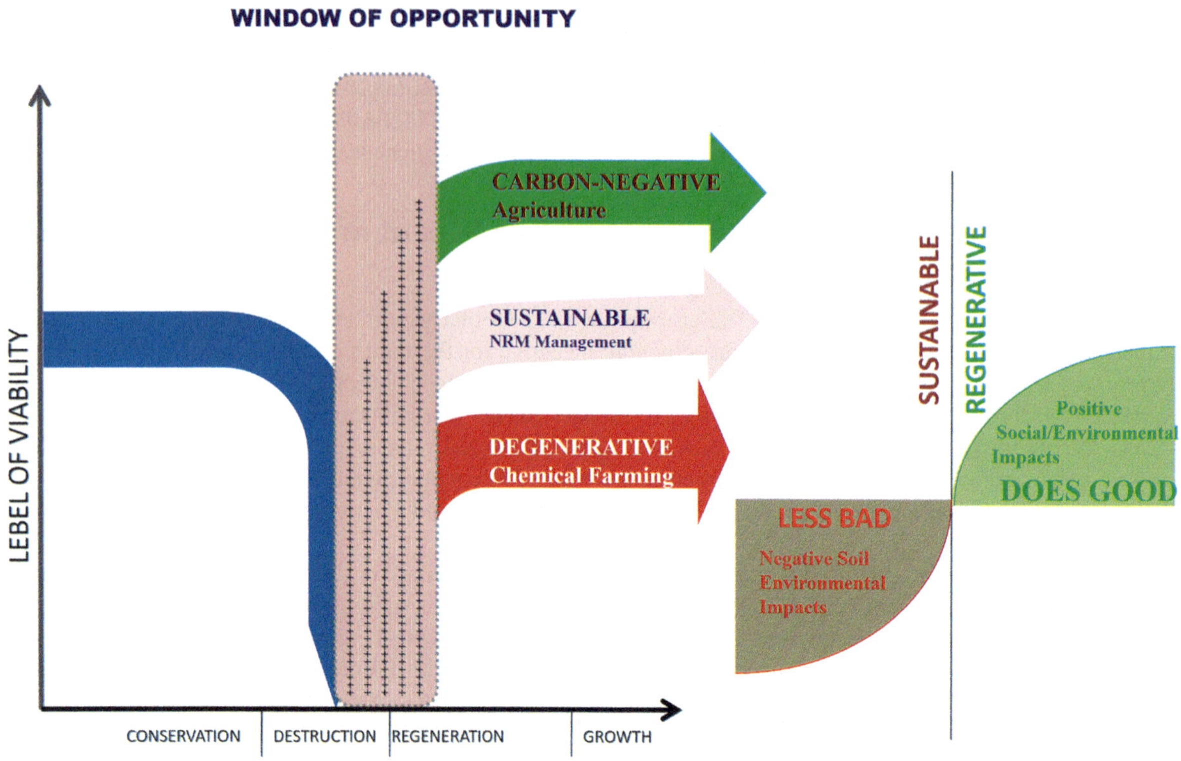

FIGURE 1.6 Regenerate the soil through net-carbon negative approaches.

win–win situation for achieving sustainability and fulfilling the SDGs (Jhariya et al., 2022). Nature-based solutions through C monetization would be a significant step toward combating climate change. Through such an approach, the primary focus would be to capture CO_2 into various segments of the agroecosystem and achieve the target of C neutrality or net-zero emissions. At both the global and national levels, multiple forms of economic incentive programs can be implemented to encourage the farming community to actively participate in the soil Cseq process through on-farm activities. Farmers showing a positive inclination toward and response to adopting soil Cseq should be provided with economic incentives in due time to bolster moral responsibilities and socio-economic well-being. Another major aspect is that farmers adopting on-farm C_{seq} practices should be provided with immediate financial incentives to gain their greater responsiveness toward the mechanism of C monetization and sustainable development (Barbato and Strong, 2023).

1.7 R&D and future research toward monetizing carbon credits through agricultural land use

In the European Union Framework, a framework for agri-environmental and climate resilience has been developed, under which various certification schemes have been developed to promote soil C_{seq} (Paul et al., 2023). The farming community may engage commercial certificate providers to help them assess the C credit and increase the SOC level for a specified time frame. Furthermore, after procuring the C credit valuation, the farming community can sell it to multinational companies and corporations as a C Credit-Neutral policy or a C offset mechanism. Various forms of certified emission reduction (CER) have already been implemented in the forestry sector under the CDM, Kyoto Protocol, and REDD+ (Reducing Emissions from Deforestation and Forest Degradation) within international treaties (Marion Suiseeya and Caplow, 2013).

Agriculture, since prehistoric times, has been an economic activity in which the primary focus has always been on crop yield and productivity, along with net financial gain for the farming community. Monetization of C emission reduction in the agricultural sector is an important aspect as well as an emerging area where various forms of green initiatives, such as nature-based farming, climate-based farming, and eco-friendly technology, need to be implemented. Priority should be given to monetizing C credits through transformation in agricultural technology (Kumar et al., 2023). All these efforts worldwide will help achieve the net-zero emission targets as outlined in the Paris Climate Convention. Further, it would provide additional benefits to the farming community in terms of financial gain through proper C monetization. Future R&D efforts should focus on the valuation methodology of C credits across various agricultural practices and agroecosystems. Furthermore, the effective implementation of best agricultural practices, accompanied by sustainable yields and significant monetary benefits, is necessary for the popularization and effective execution of C monetization in the global agricultural system (Jhariya et al., 2021a, 2021b). Such approaches should be repeated across various regions and sectors to capture spatial and temporal variation in the output through R&D activities. Further, in this assessment process, the total GHG emissions and their impact need to be evaluated under agro-practices comprising Cseq, as this would help evaluate the actual monetary benefits in terms of emission reduction. It would also enable assessment of the emissions that need to be reduced across various C farming practices. However, all these require proper assessment of other factors, such as transportation, emissions, fertilizer use, and mechanization of agricultural practices.

1.8 Policy framework for carbon monetization

Policy formulation regarding the development of C incentives in the agricultural system is the biggest challenge for the effective implementation of C monetization in an agroecosystem. Different policies were framed by countries at different times for the effective implementation of C monetization (Table 1.1).

TABLE 1.1 Carbon monetization policies across the globe (Compiled from Wuike et al. (2024)).

S.N.	Country	Policy	Program
1	India	Energy Conservation (Amendment) Act of 2022	Promotion of C markets in the agricultural practices.
		Nature-based solutions in agriculture	Combating climate change
		National Action Plan on Climate Change (NAPCC)	Developing a climate-resilient agroecosystem
		National Policy on Agriculture (NPA)	GHG emission reduction
		Intended Nationally Determined Contributions (INDC)	Promotion of C credits
		Paramparagat Krishi Vikas Yojana (PKVY)	Promotion of organic farming and soil C_{seq}
		Soil Health Card Scheme	Balance fertilizer uses and soil health
		Pradhan Mantri Krishi Sinchayee Yojana (PMKSY)	Irrigation management and water use.
		National Mission on Oilseeds and Oil Palm (NMOOP) and National Food Security Mission (NFSM)	Promote C_{seq} in soil through cultivation practices.
		National Mission on Sustainable Agriculture (NMSA): Promoting Climate-Smart Agriculture.	Climate-smart agriculture practices.
		Forest Conservation and Afforestation Efforts: Compensatory Afforestation Fund Management and Planning Authority (CAMPA)	Biodiversity conservation and C capture through afforestation and reforestation activities.
2	United States	California Cap-and-Trade	Promotion of C credit-generating agricultural practices.
		Conservation Programs	Conservation approach for soil C_{seq}
3	European Union (EU)	Common Agricultural Policy (CAP): Includes	Sustainable practices for emission reduction.
		Green Deal	Promotion of agroecological principles
4	Australia	Carbon Farming Initiative	Promotion of C offsets
		Emissions Reduction Fund	Monetization for emission reduction.
5	New Zealand	Emissions Trading Scheme	Monetization for emission reduction and C_{seq}
6	Brazil	Low Carbon Agriculture Program (ABC)	Agricultural practices for a low-C economy.
		Forest Code	Restoration of native vegetation.
7	Africa	Climate-Smart Agriculture	Develop climate resiliency.

Various policies have been framed across diverse nations at different times. However, every policy has its own limitations and constraints that hinder the effective implementation of C monetization in agricultural activities. Effectively, the measurement and assessment of C stock in the agricultural system is required for better output of the C monetization mechanism. Further, as more benefits accrue to farming communities through C credits, economic and cost viability increase, which will play a significant role in developing a sustainable C economy for agroecosystems. Both a competitive market and pricing of the C credits would be highly helpful for the effective implementation of C monetization through climate-resilient practices. Technological advancement and integration in pricing and the C economy would facilitate proper evaluation and benefit sharing, as well as the socioeconomic upliftment of the farming community. Transformation in policies and legal framework for practical functionality of the C monetization process would be highly fruitful across developed and developing nations. Commitment to adopting the C monetization process and its benefits by the farming community on a spatial-temporal scale can be achieved through education, awareness generation, creation of a knowledge base, and sensitization for the green economy. Furthermore, the sociocultural perspective and landholdings are key to the effective implementation of C monetization of agricultural production (Wuike et al., 2024; Roy et al., 2022).

1.9 Conclusion

The futuristic approach to C monetisation has emerged as a promising avenue for advancing agricultural sustainability. It would also help to develop green initiatives in agriculture and build climate resilience and smart farming practices. Considering the challenge of food crisis and security at the international and national levels, collaboration, financial subsidies, policy support systems, and new innovations would be the driving forces for the C monetization mechanism in the agriculture sector. Such approaches in India are helping to integrate climate-smart practices into mainstream agricultural production and economic well-being. In this process, capacity building for rural livelihoods, followed by technology transfer, is crucial to achieving the SDGs, given the climate change mega-event. From a future perspective, promoting Cseq would help achieve the target of emission reduction, accompanied by the assessment of C credits, which would be highly beneficial and necessary in the climate change era for human adaptability and future progress. Now, at present, humanity is moving toward a green economy and C neutrality; therefore, a proper understanding of C dynamics in agroecosystems is very important to create a market for C monetization that would bring welfare to the farming community. This would also help to fulfill the 2030 Agenda for achieving the SDGs.

References

Abagandura, G.O., Sekaran, U., Singh, S., Singh, J., Ibrahim, M.A., Subramanian, S., et al., 2020. Intercropping kura clover with prairie cordgrass mitigates soil greenhouse gas fluxes. Scientific Reports 10 (1). Available at https://doi.org/10.1038/s41598-020-64182-2.

Abbass, K., Qasim, M.Z., Song, H., Murshed, M., Mahmood, H., Younis, I., 2022. A review of the global climate change impacts, adaptation, and sustainable mitigation measures. Environmental Science and Pollution Research 29 (28), 42539–42559. Available at https://doi.org/10.1007/s11356-022-19718-6.

Babu, S., Singh, R., Avasthe, R., Kumar, S., Rathore, S.S., Singh, V.K., et al., 2023. Soil carbon dynamics under organic farming: Impact of tillage and cropping diversity. Ecological Indicators 147. Available at https://doi.org/10.1016/j.ecolind.2023.109940.

Baghel, Y., Singh, L., Jhariya, M.K., Singh, A.K., Singh, J.S., Chaturvedi, R.K., et al., 2025. Impact of anthropogenic disturbance on storage and flux of carbon in tropical dry deciduous forest of Chhattisgarh. Plant Ecology 226 (6), 673–687. Available at https://doi.org/10.1007/s11258-025-01522-0.

Banerjee, A., Meena, R.S., Jhariya, M.K., Yadav, D.K., 2021. Agroecological Footprints Management for Sustainable Food System. Springer, Singapore. Available at https://doi.org/10.1007/978-981-15-9496-0.

Banerjee, A., Jhariya, M.K., Bargali, S.S., Palit, D., 2023. Ecorestoration for Sustainability. Wiley-Scrivener, USA.

Banerjee, A., Jhariya, M.K., Raj, A., Mechergui, T., 2024. Ecosystem Management: Climate Change and Sustainability Ecosystem Management: Climate Change and Sustainability. wiley, India, pp. 1–669. Available at https://doi.org/10.1002/9781394231249.

Barbato, C.T., Strong, A.L., 2023. Farmer perspectives on carbon markets incentivizing agricultural soil carbon sequestration. npj Climate Action 2 (1). Available at https://doi.org/10.1038/s44168-023-00055-4.

Blaufelder, C., Levy, C., Mannion, P., Pinner, D., 2021. A blueprint for scaling voluntary carbon markets to meet the climate challenge. McKinsey & Company.

Chen, J., Heiling, M., Resch, C., Mbaye, M., Gruber, R., Dercon, G., 2018. Does maize and legume crop residue mulch matter in soil organic carbon sequestration? Agriculture, Ecosystems & Environment 265, 123–131. Available at https://doi.org/10.1016/j.agee.2018.06.005.

Crippa, M., Solazzo, E., Guizzardi, D., Monforti-Ferrario, F., Tubiello, F.N., Leip, A., 2021. Food systems are responsible for a third of global anthropogenic GHG emissions. Nature Food 2 (3), 198–209. Available at https://doi.org/10.1038/s43016-021-00225-9.

Derpsch, R., Friedrich, T., Kassam, A., Hongwen, L., 2010. Current status of adoption of no-till farming in the world and some of its main benefits. International Journal of Agricultural and Biological Engineering 3 (1), 1–25. Available at https://doi.org/10.3965/j.issn.1934-6344.2010.01.001-025.

European Commission, 2019. The European Green Deal. Available from:. https://eur-lex.europa.eu/resource.html?uri=cellar:b828d165-1c22-11ea-8c1f-01aa75ed71a1.0002.02/DOC_1&format=PDF. (accessed 03.08.22).

Food and Agriculture Organization of the United Nations (FAO), 2018. FAO Rice Market Monitor, vol. 21, pp. 1–38. Available from: https://www.fao.org/rice-trade-monitor/en/.

Frank, S., Havlík, P., Stehfest, E., van Meijl, H., Witzke, P., Pérez-Domínguez, I., et al., 2019. Agricultural non-CO_2 emission reduction potential in the context of the 1.5 °C target. Nature Climate Change 9 (1), 66–72. Available at https://doi.org/10.1038/s41558-018-0358-8.

Gopalakrishnan, T., Hasan, M., Haque, A., Jayasinghe, S., Kumar, L., 2019. Sustainability of coastal agriculture under climate change. Sustainability 11 (24), 7200. Available at https://doi.org/10.3390/su11247200.

Green, L., Webb, E., Johnson, E., Wynn, S., Bogen, C., 2023. Cost-effective approach to explore key impacts on the environment from agricultural tools to inform sustainability improvements: inversion tillage as a case study. Environmental Sciences Europe 35 (1). Available at https://doi.org/10.1186/s12302-023-00784-7.

Gross, A., Bromm, T., Glaser, B., 2021. Soil organic carbon sequestration after biochar application: a global meta-analysis. Agronomy 11 (12), 2474. Available at https://doi.org/10.3390/agronomy11122474.

Harindintwali, J.D., Zhou, J., Muhoza, B., Wang, F., Herzberger, A., Yu, X., 2021. Integrated eco-strategies towards sustainable carbon and nitrogen cycling in agriculture. Journal of Environmental Management 293, 112856. Available at https://doi.org/10.1016/j.jenvman.2021.112856.

IISD, 2020. International Institute for sustainable development: World population to reach 9.9 billion by 2050. Available from:. https://sdg.iisd.org/news/world-population-to-reach-9-9-billion-by-2050/.

Jhariya, M.K., Meena, R.S., Banerjee, A., 2021. Ecological Intensification of Natural Resources for Sustainable Agriculture. Springer, Singapore, pp. 1–655. Available at https://doi.org/10.1007/978-981-33-4203-3.

Jhariya, M.K., Singh, L., Toppo, S., 2024b. Wildfires and carbon budget of certain seasonally dry forests in India. Land Degradation and Development 35 (12), 3771–3789. Available at https://doi.org/10.1002/ldr.5166.

Jhariya, M.K., Banerjee, A., Yadav, D.K., Raj, A., 2018. Leguminous trees an innovative tool for soil sustainability. Legumes for Soil Health and Sustainable Management. Springer Nature, India, pp. 315–345. Available at https://doi.org/10.1007/978-981-13-0253-4_10.

Jhariya, M.K., Meena, R.S., Banerjee, A., Meena, S.N., 2022. Natural Resources Conservation and Advances for Sustainability. Elsevier. Available at https://doi.org/10.1016/C2019-0-03763-6.

Jhariya, M.K., Banerjee, A., Swaroop Meena, R., Kumar, S., Raj, A., 2021. Sustainable Intensification for Agroecosystem Services and Management. Springer, Singapore, pp. 1–870. Available at https://doi.org/10.1007/978-981-16-3207-5.

Jhariya, M.K., Meena, R.S., Banerjee, A., Kumar, S., Raj, A., 2024a. Agroforestry for Carbon and Ecosystem Management. Elsevier. Available at https://doi.org/10.1016/C2021-0-02916-3.

Kumar, R., Kumar, S., Yashavanth, B.S., Venu, N., Meena, P.C., Dhandapani, A., et al., 2023. Natural farming practices for chemical-free agriculture: implications for crop yield and profitability. Agriculture 13 (3), 647. Available at https://doi.org/10.3390/agriculture13030647.

Kumar, S., Meena, R.S., Jhariya, M.K., 2020. Resources Use Efficiency in Agriculture. Springer, Singapore. Available at https://doi.org/10.1007/978-981-15-6953-1.

Kumar, S., Meena, R.S., Sheoran, P., Jhariya, M.K., 2024. Regenerative Agriculture for Sustainable Food Systems. Springer, Singapore, pp. 1–493. Available at https://doi.org/10.1007/978-981-97-6691-8.

Lee, H., Calvin, K., Dasgupta, D., Krinner, G., Mukherji, A., Thorne, P., et al., 2023. IPCC, 2023: Summary for Policymakers. Climate Change 2023: Synthesis Report. Contribution of Working Groups I, II and III to the Sixth Assessment Report of the Intergovernmental Panel on Climate Change. IPCC, Geneva, pp. 1–35. Available at 10.59327/IPCC/AR6-9789291691647.001.

Leifeld, J., 2023. Carbon farming: climate change mitigation via non-permanent carbon sinks. Journal of Environmental Management 339, 117893. Available at https://doi.org/10.1016/j.jenvman.2023.117893.

Liu, D.L., Chan, K.Y., Conyers, M.K., 2009. Simulation of soil organic carbon under different tillage and stubble management practices using the Rothamsted carbon model. Soil and Tillage Research 104 (1), 65–73. Available at https://doi.org/10.1016/j.still.2008.12.011.

Marion Suiseeya, K.R., Caplow, S., 2013. In pursuit of procedural justice: lessons from an analysis of 56 forest carbon project designs. Global Environmental Change 23 (5), 968–979. Available at https://doi.org/10.1016/j.gloenvcha.2013.07.013.

Meena, R.S., Pradhan, G., 2023. Industrial garbage-derived biocompost enhances soil organic carbon fractions, CO_2 biosequestration, potential carbon credits and sustainability index in a rice-wheat ecosystem. Environmental Research 235, 116525. Available at https://doi.org/10.1016/j.envres.2023.116525.

Meena, R.S., Pradhan, G., Kumar, S., 2025. Energy flow, eco-efficiency, and economic circulation with recycled industrial waste compost application in wheat and subsequent rice farming. Science of the Total Environment 967. Available at https://doi.org/10.1016/j.scitotenv.2025.178779.

Meena, R.S., Pradhan, G., Kumar, S., Lal, R., 2023. Using industrial wastes for rice-wheat cropping and food-energy-carbon-water-economic nexus to the sustainable food system. Renewable and Sustainable Energy Reviews 187, 113756. Available at https://doi.org/10.1016/j.rser.2023.113756.

Meena, R.S., Yadav, A., Kumar, S., Jhariya, M.K., Jatav, S.S., 2022. Agriculture ecosystem models for CO_2 sequestration, improving soil physicochemical properties, and restoring degraded land. Ecological Engineering 176. Available at https://doi.org/10.1016/j.ecoleng.2022.106546.

Meena, R.S., Lal, R., Kumar, S., Pradhan, G., Srinivasarao, C., Singh, A.K., et al., 2024a. Potential of Indian agriculture for capturing atmospheric CO_2 and monetizing carbon credits to the farmers: an overview and policy framework. Advances in Agronomy 188, 101–206. Available at https://doi.org/10.1016/bs.agron.2024.06.001.

Meena, R.S., Pradhan, G., Singh, K., Kumar, S., Singh, A.K., Shashidhar, K.S., et al., 2024c. Agriculture models for restoring degraded land to enhance CO_2 biosequestration and carbon credits in the Vindhyan region of India. Science of the Total Environment 929. Available at https://doi.org/10.1016/j.scitotenv.2024.172661.

Meena, R.S., Pradhan, G., Nalani, Kumar, S., Singh, A.K., Verma, S.K., et al., 2024b. Diversified cropping modules designed for soil restoration, CO_2 sequestration, and generating carbon credits. Land Degradation and Development 35 (14), 4314–4329. Available at https://doi.org/10.1002/ldr.5224.

Morizet-Davis, J., Marting Vidaurre, N.A., Reinmuth, E., Rezaei-Chiyaneh, E., Schlecht, V., Schmidt, S., et al., 2023. Ecosystem services at the farm level—overview, synergies, trade-offs, and stakeholder analysis. Global Challenges 7 (7). Available at https://doi.org/10.1002/gch2.202200225.

Outhwaite, C.L., McCann, P., Newbold, T., 2022. Agriculture and climate change are reshaping insect biodiversity worldwide. Nature 605 (7908), 97–102. Available at https://doi.org/10.1038/s41586-022-04644-x.

Painkra, S., Singh, K.P., Singh, B., Jhariya, M.K., Banerjee, A., Sahu, K., et al., 2025. Energy budgeting and economic analysis of coriander crop production in central India. Environment, Development and Sustainability. Available at https://doi.org/10.1007/s10668-025-06217-0.

Paul, C., Bartkowski, B., Dönmez, C., Don, A., Mayer, S., Steffens, M., et al., 2023. Carbon farming: are soil carbon certificates a suitable tool for climate change mitigation? Journal of Environmental Management 330, 117142. Available at https://doi.org/10.1016/j.jenvman.2022.117142.

Poore, J., Nemecek, T., 2018. Reducing food's environmental impacts through producers and consumers. Science (New York, N.Y.) 360 (6392), 987–992. Available at https://doi.org/10.1126/science.aaq0216.

Pradhan, G., Meena, R.S., 2023b. Utilizing waste compost to improve the atmospheric CO_2 capturing in the rice-wheat cropping system and energy-cum-carbon credit auditing for a circular economy. Science of the Total Environment 892. Available at https://doi.org/10.1016/j.scitotenv.2023.164572.

Pradhan, G., Meena, R.S., 2023a. Interaction impact of biocompost on nutrient dynamics and relations with soil biota, carbon fractions index, societal value of CO_2 equivalent and ecosystem services in the wheat-rice farming. Chemosphere 339. Available at https://doi.org/10.1016/j.chemosphere.2023.139695.

Prasad, R., Jhariya, M.K., Banerje, A., 2021. Advances in Sustainable Development and Management of Environmental and Natural Resources: Economic Outlook and Opinions: Volume 1. **1**. Apple Academic Press, India, pp. 1–428. Available at https://doi.org/10.1201/9781003187455.

Quandt, A., Grafton, D., Gorman, K., Dawson, P.M., Ibarra, C., Mayes, E., et al., 2023. Mitigation and adaptation to climate change in San Diego County, California. Mitigation and Adaptation Strategies for Global Change 28 (1). Available at https://doi.org/10.1007/s11027-022-10041-6.

Raj, A., Jhariya, M.K., Meena, R.S., 2025. Agroforestry for Monetising Carbon Credits. Springer Nature, Switzerland.

Raj, A., Jhariya, M.K., Banerjee, A., Nema, S., Bargali, K., 2023. Land and Environmental Management Through Forestry. Wiley-Scrivener, USA.

Raj, A., Jhariya, M.K., Banerjee, A., Jha, R.K., Singh, K.P., 2024. Agroforestry. Wiley-Scrivener, USA.

Robinson, Sa, 2020. Climate change adaptation in SIDS: a systematic review of the literature pre and post the IPCC fifth assessment report. Wiley Interdisciplinary Reviews: Climate Change 11 (4). Available at https://doi.org/10.1002/wcc.653.

Rodias, E., Aivazidou, E., Achillas, C., Aidonis, D., Bochtis, D., 2021. Water-energy-nutrients synergies in the agrifood sector: a circular economy framework. Energies 14 (1), 159. Available at https://doi.org/10.3390/en14010159.

Rodríguez-Espinosa, T., Papamichael, I., Voukkali, I., Gimeno, A.P., Candel, M.B.A., Navarro-Pedreño, J., et al., 2023. Nitrogen management in farming systems under the use of agricultural wastes and circular economy. Science of the Total Environment 876. Available at https://doi.org/10.1016/j.scitotenv.2023.162666.

Roy, O., Meena, R.S., Kumar, S., Jhariya, M.K., Pradhan, G., 2022. Assessment of land use systems for CO_2 sequestration, carbon credit potential, and income security in Vindhyan region, India. Land Degradation and Development 33 (4), 670–682. Available at https://doi.org/10.1002/ldr.4181.

Sarkar, R., Corriher-Olson, V., Long, C., Somenahally, A., 2020. Challenges and Potentials for soil organic carbon sequestration in forage and grazing systems. Rangeland Ecology and Management 73 (6), 786–795. Available at https://doi.org/10.1016/j.rama.2020.04.002.

Shang, Z., Abdalla, M., Xia, L., Zhou, F., Sun, W., Smith, P., 2021. Can cropland management practices lower net greenhouse emissions without compromising yield? Global Change Biology 27 (19), 4657–4670. Available at https://doi.org/10.1111/gcb.15796.

Soil Enrichment Protocol, 2020. Climate Action Reserve. Available from:. https://www.climateactionreserve.org/wp-content/uploads/2020/10/Soil-Enrichment-Protocol-V1.0.pdf.

Syed, A., Raza, T., Bhatti, T.T., Eash, N.S., 2022. Climate Impacts on the agricultural sector of Pakistan: risks and solutions. Environmental Challenges 6. Available at https://doi.org/10.1016/j.envc.2021.100433.

The Circularity Gap Report, 2019. Circle economy. Available from https://circulareconomy.europa.eu/platform/sites/default/files/circularity_gap_report_2019.pdf.

United Nations, 2021. Food systems summit 2021—about the summit. Available from: https://www.un.org/en/food-systems-summit/about.

Wolf, A., 2022. Sustainable carbon cycles: a framework for the ramp-up of carbon capture? Intereconomics 57 (4), 260–266. Available at https://doi.org/10.1007/s10272-022-1060-7.

Wuike, V.G., Gourkhede, P.H., Singh, N.K., Teja, R.R., 2024. Carbon monetization in agriculture: initiatives, policies, and global perspectives. Just Agriculture 4 (12), 60–70.

Xu, L., Solangi, Y.A., Wang, R., 2023. Evaluating and prioritizing the carbon credit financing risks and strategies for sustainable carbon markets in China. Journal of Cleaner Production 414. Available at https://doi.org/10.1016/j.jclepro.2023.137677.

Yadav, V.S., Yadav, S.S., Gupta, S.R., Meena, R.S., Lal, R., Sheoran, N.S., et al., 2022. Carbon sequestration potential and CO_2 fluxes in a tropical forest ecosystem. Ecological Engineering 176. Available at https://doi.org/10.1016/j.ecoleng.2022.106541.

Zelikova, J., Chay, F., Freeman, J., Cullenward, D., 2021. A buyer's guide to soil carbon offsets. carbon plan. Carbonplan. Available at https://carbonplan.org/research/soil-protocols-explainer.

Section II

Agriculture and soil carbon

Chapter 2

Challenges and solutions in carbon credits generated through carbon farming

Miracle Uwa Livinus[1], Musa Ojeba Innocent[2], Mustapha Abdulsalam[2], Sunday Zeal Bala[3], Joshua Oluwatobi Akinola[4], Madinat Hassan[5], Katimu Yusuf[6], and Nafisa Dayyabu[7]

[1]*Department of Biochemistry, School of Sciences and Information Technology, Skyline University Nigeria, Kano, Nigeria,* [2]*Department of Microbiology, School of Sciences and Information Technology, Skyline University Nigeria, Kano, Nigeria,* [3]*Department of Education, University of Maryland Eastern Shore, Princess Anne, MD, United States,* [4]*Department of Natural Sciences, University of Maryland, College Park, MD, United States,* [5]*Foresight Institute of Research and Translation (FIRAT), Kigali, Rwanda,* [6]*Department of Microbiology, Federal University Gashua, Gashua, Yobe State, Nigeria,* [7]*Faculty of Earth and Environmental Sciences, Bayero University, Kano, Nigeria*

Highlights

- Carbon farming can be a way toward a climate mitigation strategy.
- Express the challenges in carbon credit implementation.
- Proposed solutions for enhancing carbon farming.
- Deals with the monitoring, evaluation, and verification processes.
- Showcase the future direction and research needs.

2.1 Introduction

Climate change remains one of the most pressing global challenges, which urgently requires a mitigation strategy to reduce greenhouse gas (GHG) emissions and increase carbon sequestration (Cseq) (Filonchyk et al., 2024; Meena et al., 2025b). Carbon (C) farming is one of the mitigation strategies and is being looked upon as a nature-based solution (NbS) that combines sustainable agricultural and land management practices to store atmospheric carbon dioxide (CO_2) in soils and vegetation (Sharma et al., 2024). Techniques involved in C farming include cover cropping, agroforestry (AF), reduced tillage, the use of biochar or other C-enhancing, and regenerative grazing (Fig. 2.1), which either store C in soils or biomass (Shoudho et al., 2024; Meena et al., 2024a). But among various other benefits, these practices also mitigate climate change while helping to maintain soil health, biodiversity, and water retention, and, as such, this turns out to be beneficial for the environment and agricultural productivity (Hannam, 2024).

C farming as an option is one of the key incentives, as it allows the farmer or landowner to earn C credits for verified Cseq activities (Sharma et al., 2024). Credits are tradable units of one metric ton of CO_2 equivalent (tCO_2e) removed from or prevented from going into the atmosphere. Most of these credits can be sold in compliance markets (subject to international frameworks such as the Kyoto Protocol and the Paris Agreement) or voluntary markets (where individual clients or even consumers purchase them to offset their emissions) (Kornek et al., 2021; Lokuge and Anders, 2022). As a result of the increasing corporate demand for C offsets, there has been an increased interest in C farming-based C credits to fulfill C neutrality and corporate social responsibility commitments (Boiral et al., 2024; Meena et al., 2024b).

Several obstacles stand in the way of the viability of C farming as a C-credit mechanism, even in its potential. Key technical barriers to the measurement and verification of soil Cseq are the fluctuations in soil C stocks owing to environmental factors, land-use history, and management practices (Pudasaini et al., 2025). They also face additional challenges in establishing reliable baselines and securing the permanence of C storage, as Cseq in soils is reversible, especially when land-use changes or extreme weather events occur (Zhang et al., 2024b). Additionally, prohibiting due to the high costs of monitoring, reporting, and verification (MRV) means that smallholder farmers often lack financial incentives to participate in C credit schemes (Batjes et al.,

Carbon Incentives, Plans, and Policies in Agriculture. DOI: https://doi.org/10.1016/B978-0-443-33557-0.00008-2

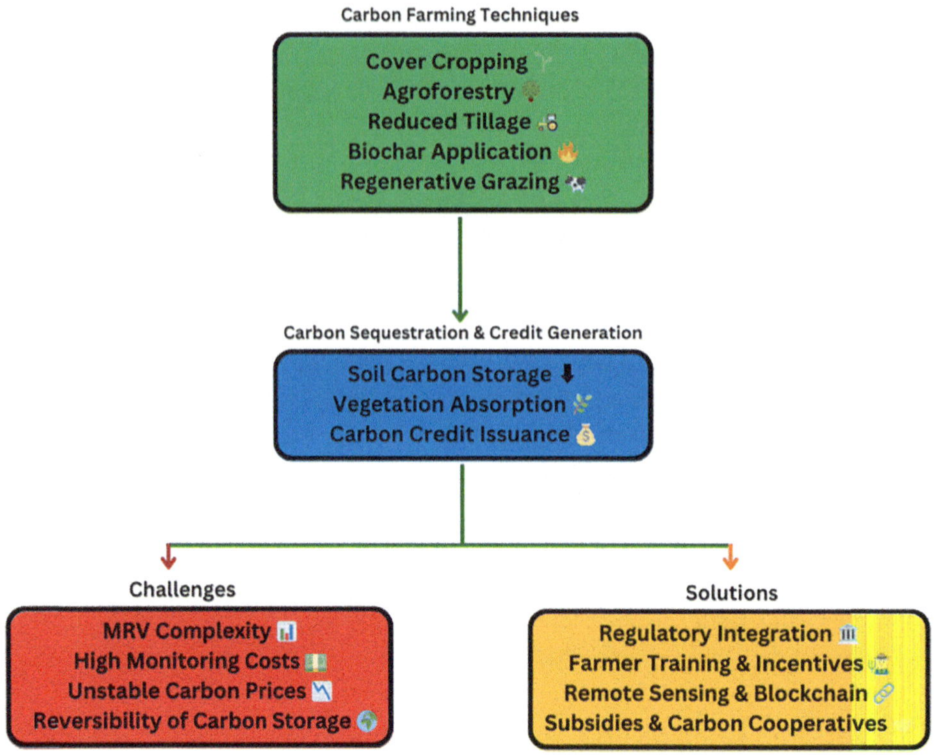

FIGURE 2.1 Carbon farming process: opportunities and challenges.

2024). C farming also operates under limitations of technical and economic barriers, as well as competing land-use and policy uncertainties. Whilst the concept of C farming is innovative and has the potential to contribute to climate change mitigation in many developing nations, the agricultural land in these countries is mainly used in food production, and therefore scaling up to C farming could come at the expense of food security (Raina et al., 2024; Meena et al., 2024c). Further, inconsistencies in C credit pricing and validation across various regions degenerate market confidence (Sharma et al., 2024; Raj et al., 2025). The high level of volatility of C markets compounds these challenges, as long-term investments in C farming are uncertain for both farmers and investors (Baumber et al., 2024).

In light of these challenges, several novel solutions to advance C credit accounting have been proposed, such as improvements made in remote sensing, machine learning, and blockchain technologies to increase accuracy and transparency (Baumber et al., 2024). Lower participation costs due to financial mechanisms such as subsidies, C farming cooperatives, or ecosystem service payments are being incentivized for smallholder farmers. Policy reforms that make C farming an integral part of national climate strategies and routinely undertake capacity-building programs among farmers can be exceptionally beneficial to the effectiveness of C credit initiatives. There are also the engagements of stakeholders, such as governments, the private sector, and local communities, for equitable benefits and the sustainable implementation of C farming practices (Günther et al., 2024; Meena and Pradhan, 2023).

This chapter takes a deep dive into the challenges involved in establishing C farming, C credit, and what practices can avoid the barriers. The chapter examines methodological aspects of assessing C credit, including policy framework, financial mechanism, and technological innovation, and proposes practical insights to enhance the scalability and efficiency of C farming by examining these four aspects. This also helps in the further direction of future research and emerging trends in C farming integration with biodiversity conservation and agroecology. These challenges need to be tackled, and innovative solutions are required to be applied for C farming to be a feasible strategy for global climate change mitigation, consistent with sustainable agricultural development.

2.2 Carbon farming in developed and developing nations

The fluttering characteristics of the C farming adoption and implementation across developed and developing nations are due to differences in financial resources, technological infrastructures, policies, and institutional capacity. Developed countries have more monitoring technologies, while in developing countries there are barriers to development, such as a lack of financial incentives, a lack of technical knowledge in numbers, and a lack of institutional support (Sharma et al., 2024). Although there are disparities, C farming remains a good tactic to advance global Cseq as well as to configure a climate change mitigation strategy for both contexts, which can be important to boost soil health and agricultural resilience (Shoudho et al., 2024; Livinus et al., 2024a).

2.2.1 Carbon farming in developed nations

C farming has advanced in its implementation in the United States, Canada, Australia, and European Union member states, largely due to a strong regulatory framework and a well-developed C credit mechanism (Pudasaini et al., 2025). In these countries, very strong institutional support, financial investments, and advanced technological solutions for the measurement and MRV of soil Cseq are available. One of the key enablers for some C farming to take place is the existence of well-structured C markets in developed countries. For example, there are compliance and voluntary C markets in the United States where farmers can obtain C credits for sustainable land management practices. The California Cap-and-Trade Program and voluntary initiatives such as the Climate Action Reserve (CAR) offer financial incentives to farmers to be a part of C farming projects (Sharma et al., 2024; Meena et al., 2023). Similar to other efforts to allow Cseq revenue-harvesting opportunities, the European Union Emissions Trading System stimulates such Cseq harvestable revenue opportunities by incentivizing farmers to use climate-smart agricultural practices (Kornek et al., 2021; Lokuge and Anders, 2022).

The deployment of sophisticated technologies in coordinated monitoring to achieve a better and faster assessment of C credits for C farming is being undertaken by developed nations. Remote sensing, satellite imagery, and blockchain-based platforms are necessary to track changes in soil C and ensure transparency in C credit transactions. The carbon farming initiative is implemented in Australia through the Emissions Reduction Fund (ERF), implemented through the use of Geographic Information System (GIS) technologies and soil sampling techniques (Baumber et al., 2024). Progress in C farming, nonetheless, is much slower than in the European nations. Despite having high implementation costs, complex regulatory costs, and land-use contracts, there are still large barriers (Batjes et al., 2024). It can be time-consuming and expensive for farmers to confer with the C-credit verification processes. In addition, there are competing land-use priorities, such as urban expansion or biofuel production, that complicate the large-scale adoption of C farming practices (Günther et al., 2024).

2.2.2 Carbon farming in developing nations

In contrast to developed countries, as seen in Table 2.1, C farming in developing countries across Africa, Latin America, and South Asia is limited by financial reasons, the lack of technical knowledge, and weak institutional frameworks. C farming presents exclusive opportunities for these regions to boost agricultural sustainability, soil fertility rehabilitation, and the generation of alternative sources of income for smallholder farmers (Sharma et al., 2024). In terms of large financial resource barriers, C farming is one of the biggest challenges for developing countries. The dominant C credit

TABLE 2.1 Comparison of carbon farming adoption in developed versus developing nations.

Factor	Developed nations	Developing nations
Financial resources	High government and private funding	Limited funding and investment
Monitoring technologies	Advanced (remote sensing, GIS, and blockchain)	Limited or underdeveloped
Policy and regulations	Strong regulatory frameworks	Weak or inconsistent policies
Institutional support	Well-developed institutions	Lacks technical and institutional capacity
Farmer participation	Higher adoption due to incentives	Lower adoption due to financial constraints
Carbon credit market	Established and structured	Underdeveloped or emerging

GIS, Geographic Information System.

markets haven't developed in these regions, so funding smallholder farmers to adopt sustainable land management practices is difficult. Nevertheless, many developing countries have limited government subsidies and private investment in C farming initiatives, similar to those in developed countries (Boiral et al., 2024). It will also require knowledge of soil management, C_{seq}, and MRV methodologies.

In many developing countries, for example, farmers are not trained or educated in the best ways to adopt climate-smart adaptive practices. Unlike C_{seq} under these technological infrastructures (e.g., soil testing laboratories and remote sensing systems), which can precisely evaluate the C_{seq} and produce verifiable C credits (Zhang et al., 2024b). Another challenge to C farming in developing nations involves the weaknesses of institutional and policy frameworks that govern and promote C credit markets in developing countries. Many governments lack clear guidelines or national climate policies for incorporating C farming (Raina et al., 2024). A lack of regulatory structure, which hinders the participation of farmers and investors, also hinders C-credit generation and trading (Lokuge and Anders, 2022).

2.2.3 Opportunities for carbon farming in developing nations

The C farming effort can address several challenges and could be beneficial to improving the sustainability of agriculture and boosting rural livelihoods in developing nations. Smallholder farmers need to sequester C, earn additional income from C credits, and build resilience to climate change, all of which are aware of the negative consequences of climate change on smallholder farmers (Jat et al., 2022). Different initiatives have been implemented at the grassroots level; C farming is feasible and demonstrates its viability for large-scale adoption. Among the successful integrations, the Kenya Agricultural Carbon Project (KACP) has applied Cseq techniques in the smallholder farming systems. KACP is supported by the World Bank and the Vi AF program contributions to allow farmers to practice sustainable land management, which will store soil C and increase agricultural productivity (Nyberg et al., 2020; Jhariya et al., 2023; Jhariya et al., 2022). However, this initiative also resulted in C credit generation, improved soil fertility, increased crop yield, and increased farmer resilience to climate variability (Batjes et al., 2024).

The World Bank is involved in a Cseq program through the BioCarbon Fund Initiative for Sustainable Forest Landscapes, which encourages AF and sustainable land use in countries such as Ethiopia and Zambia. It helps smallholder farmers and communities by restoring degraded landscapes, increasing tree cover, and adopting climate-resilient agricultural methods (Gizachew et al., 2017; Jat et al., 2022). The program has aligned financial incentives with environmental conservation, thereby documenting the widespread adoption of C farming and contributing to the inclusion of sustainable practices within local agricultural systems (Sharma et al., 2024; Abdulsalam et al., 2024; Meena et al., 2025a). Strategic investments in financial incentives, technical training, and legal reform are needed to scale up C farming in developing countries. Supportive regulatory frameworks that encourage sustainability in land use and enable smallholder farmers to participate in C markets need to be set at both the government and international organization levels. Additionally, mobile-based extension services can help fill the knowledge gap by providing real-time information on what is best for soil Cseq. C credit cooperatives can help make community-level C markets simpler to access, to ensure that small-scale farmers receive their fair share of C from C farming projects. In addition, public C farming could be financially accessible through public–private partnerships (PPPs), in which resources from government, nongovernmental organizations (NGOs), and private investors can be used (Pudasaini et al., 2025).

2.3 Methodological aspects of carbon credit assessment in carbon farming

The assessment of C credits in C farming is a complex process due to the various methodologies required to ensure accuracy, reliability, and compliance with international standards. C farming is any process that aims at either promoting Cseq or reducing GHGs produced by farming and agricultural practices (Raina et al., 2024; Pradhan and Meena, 2023a,b). C credits generated by Cseq practices are measured, verified, and certified using robust methods to determine the amount of C credits. The methodologies comprise Cseq measurement techniques, criteria for baseline and additionality, and verification and certification processes (Michaelowa et al., 2019; West et al., 2024). These methodological aspects are illustrated in Fig. 2.2

2.3.1 Carbon sequestration measurement

Accurate measurement of C_{seq} establishes the quantity of C credits issued to farmers as scientifically backed and verifiable sequestration claims. There are various techniques for assessing soil and biomass C storage, providing both direct and indirect measurements (Paul et al., 2023). Although soil core examinations for soil organic C need to be recorded at various depths and involve heavy worker contribution and multiple sessions of testing for monitoring C changes over time

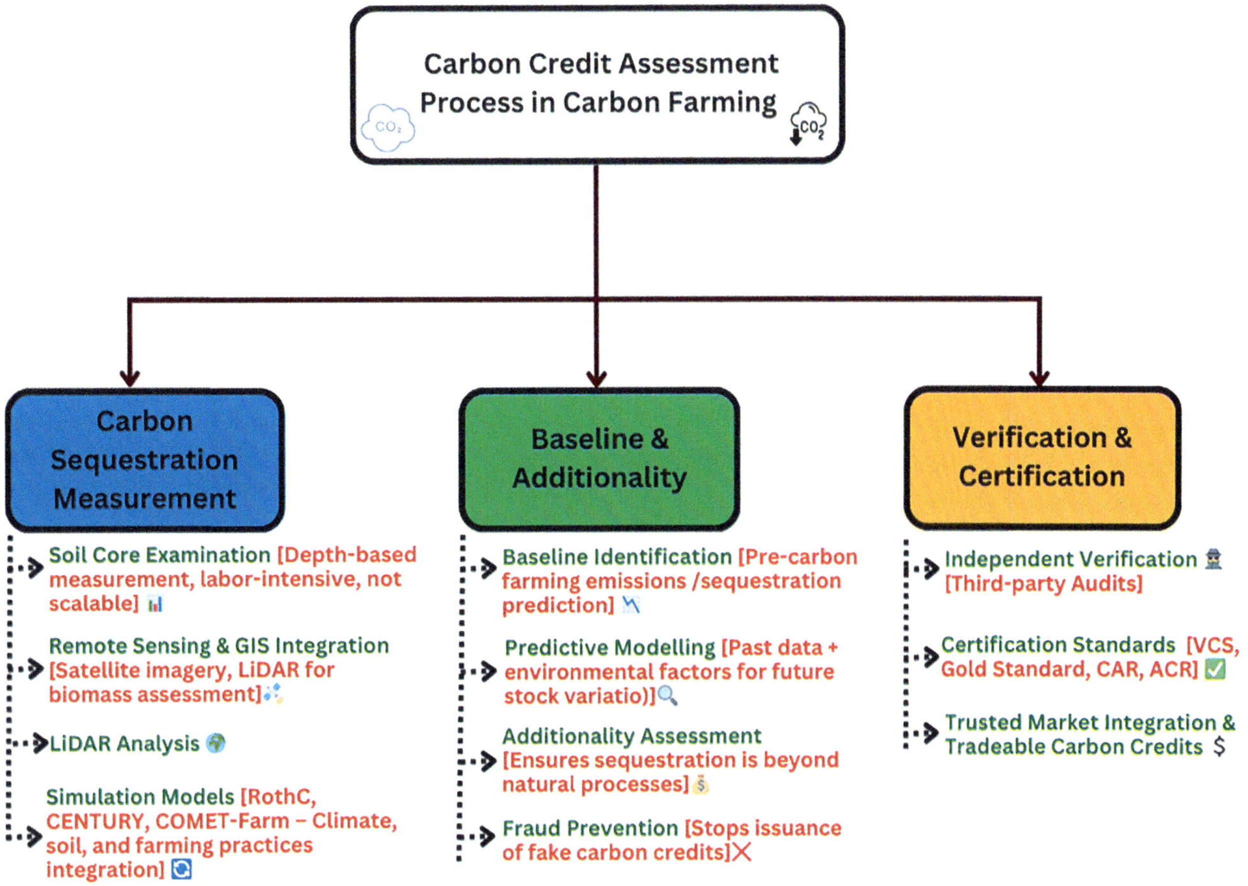

FIGURE 2.2 Methodological aspects of carbon credit assessment in carbon farming.

(Davis et al., 2018), the process is not feasible. The integration of remote sensing, GIS, and LiDAR (light detection and ranging) permits the satellite imagery and light-detecting and ranging data to evaluate the large extent of C_{seq} by using land cover analysis of biomass storage patterns (Schick et al., 2023). The combined use of the RothC, CENTURY, and COMET-Farm models represents a relatively cost-effective option for simulating soil C change by integrating climate patterns, soil characteristics, and agricultural practices. Different forms of assessment work together to yield a more precise, expandable, and time-efficient C_{seq} analysis that needs to assign C credits properly within C farming systems (Shoudho et al., 2024).

2.3.2 Baseline and additionality

Proper baseline identification and additional evaluation are necessary in the process of C credit assessment for C farming to demonstrate that the reduction is actually above natural development. Baselines are the expected C emission and Cseq numbers to be established before C farming activities start. Future C stock variation was predicted by combining past data analysis and environmental evaluations with predictive modeling as it relates to not using C farming strategies (Pudasaini et al., 2025). Having an established baseline is also essential in verifying that C gains from specific farming methods are being won. The procedure of additionality looks at whether Cseq projects would happen naturally by analyzing incentives from policies and monetary backing perspectives. Key to these prevention efforts is C credit assessments due to the large number of fake credits being issued without reducing emissions. Economic, technological, and regulatory constraints that would end C farming techniques are such evaluations that decide on the additionality (Michaelowa et al., 2019; Pradhan et al., 2023). Proper assessment of baseline scenarios and evaluation of additionality is critical for C credit programs to avoid losing credibility while enabling them to contribute effectively to climate change mitigation (West et al., 2024).

2.3.3 Verification and certification

C farming projects need to be independently verified and then certified before they can be considered authoritative (reputable) C credits, as these processes confirm that actual C_{seq} has occurred and that projects have provided adequate disclosure (Batjes et al., 2024). Third-party organizations verify that the C_{seq} data reported by the project adhere to established methodologies and baseline estimates through independent third-party evaluations. Independent auditing teams are put through examinations, scientific studies, and data audits to confirm the actual C reductions are being reported (West et al., 2024). The certification procedure verifies that C credits following international quality criteria are tradeable within compliance-based and voluntary programs. The issuance of C credits falls under the regulations of various certification standards. The Verra's Verified Carbon Standard (VCS) serves as the primary standard in voluntary C markets for establishing robust project methodologies and validating the environmental impacts of land-based C programs (Verra, 2022).

The Gold Standard exists to expand sustainable development benefits, in addition to C capture, through the certification of programs with positive societal and ecological impacts (Gold Standard, 2014, 2019). The CAR and American Carbon Registry conduct strict scientific tests on C accounting procedures and project authorization procedures in North America (Climate Action Reserve, 2021). Following the verification procedures and certification systems allows C farming projects to protect environmental integrity while building trust with market stakeholders and helping to achieve wider C credit program expansion. Farmers and landowners can obtain premium C-market prices through certification under these recognized frameworks. Verification procedures, along with certification, maintain transparency while preventing fake credit issuance to establish trust among investors in C farming operations (Pan et al., 2022; Batjes et al., 2024).

2.4 Challenges of carbon farming-based carbon credits

The general acceptance of C farming as a climate change solution faces multiple obstacles that limit its general acceptance and performance benefits. Various obstacles associated with C farming incorporate complex measurement procedures and baseline development problems, together with monetary restrictions and conflicts regarding land-use, policy inconsistencies, and environmental trade-offs, alongside volatile market conditions. C farming will achieve its maximum potential for GHG emission reduction after overcoming the various barriers that affect farmers' economic viability (Prasad et al., 2021).

2.4.1 Measurement and verification

Soil C_{seq} measurement proves to be the major hurdle when generating C credits through C farming systems. The soil C storage quantity shows wide variation due to differences in soil composition, climate, and agricultural methods (Sharma et al., 2024). The garden variability creates obstacles to designing a standard measurement platform that enables precise C_{seq} tracking in diverse scenes. Soil sampling techniques that use traditional methods deliver accurate results, though they prove expensive and labor-intensive, especially to smallholder farmers (Pudasaini et al., 2025). Digital modeling, together with remote sensing technologies, provides better measurement options, although complex technical requirements and high costs accompany them. Satellite imaging with high-resolution capabilities, LiDAR technology, and machine learning systems require significant costs to build infrastructure and support professional staff (Brummitt et al., 2024). Retailers find it challenging to accept C credits due to the lack of credibility caused by differences in data interpretation among different verification standards (Kreibich and Hermwille, 2021). Accurately measuring C_{seq} is a complex task that requires various methodologies, each with its advantages and limitations. Table 2.2 provides a comparison of the most commonly used methods for measuring soil C_{seq}.

2.4.2 Baseline establishment

The process of determining actual Cseq from C farming relies heavily on creating believable starting points. The definition of baseline emissions remains uncertain due to environmental factors combined with changing land-use patterns and their dynamic nature (Shoudho et al., 2024). It is difficult to set accurate reference levels due to historical data inconsistencies, differences in soil C dynamics, and regional differences in agricultural activities (West et al., 2024). Proving additionality is another key issue, as it requires that the Cseq counted be additional to what would occur in the absence of a policy. Difficulties in C farming projects demonstrating additionality have arisen from the fact that some farming practices would have been pursued regardless of C credit incentives. Given the legitimacy of the C credits, this challenge suggests an uncertain contribution to climate mitigation (Michaelowa et al., 2019; Sharma et al., 2024).

TABLE 2.2 Comparison of measurement methods for C_{seq}.

Measurement method	Description	Advantages	Limitations
Soil sampling	Directly collects and analyzes soil samples.	High accuracy, well-established method.	Expensive, labor-intensive, and limited scalability.
Remote sensing and satellite imaging	Uses satellite and drone images to assess soil C levels.	Large-scale monitoring, noninvasive.	Requires high-resolution technology, costly.
Machine learning models	AI-driven models predict soil C_{seq}.	Efficient, reduces manual sampling costs.	Needs extensive training data, computationally demanding.
Blockchain for verification	Secure, transparent record-keeping of C credits.	Reduces fraud, ensures traceability.	Complex setup requires internet access.

Cseq, carbon sequestration; *AI*, artificial intelligence.

2.4.3 Financial and economic barriers

High transaction costs involved in C markets discourage many small-scale farmers from adopting C farming. These costs include verification fees, administrative expenses, and the cost of compliance with certification requirements (Baumber et al., 2024). Furthermore, significant amounts of upfront financing that farmers lack mean that they cannot invest in C farming techniques like AF, cover cropping, and no-till agriculture (Raina et al., 2024). Finally, market price volatility affects C credit revenues, which are interdependent with C payments to the farmers. C credit price fluctuations are determined by regulatory, supply and demand, and corporate sentiment regarding becoming C neutral (Lovcha et al., 2022; Meena et al., 2025a). The unpredictability of where harvests will occur and the available conditions for that particular year creates an immoderate burden on farmers, and they cannot plan their long-term investments in sustainable agricultural practices (Raina et al., 2024).

2.4.4 Land use and competition

One primary concern in C farming is the competition for land between food production and C_{seq} initiatives. Table 2.3 presents the potential benefits and drawbacks of different land-use options. C farming could decrease food production and, therefore, competition between C farming and C_{seq} initiatives, as arable land is scarce. In developing countries, land C farming could be costly, and they may have to trade away land for essential African crops, exacerbating a problem already facing food security (Sharma et al., 2024). Some C farming projects also encourage monoculture plantations of fast-growing tree species for C_{seq}. These plantations can further increase C storage, but they usually entail biodiversity loss and ecosystem imbalance resulting from the displacement of native plant species and the loss of habitat for wildlife (Hannam, 2024).

2.4.5 Policy and regulatory frameworks

Regulatory frameworks across different jurisdictions are very inconsistent, which complicates C credit transactions in both voluntary and compliance markets. The policy on C pricing, verification standards, and the eligibility of credits varies by country (Boiral et al., 2024), making it difficult for farmers to maneuver through the system. Unreliable institutional capacity contributes to poor enforcement of C credit agreements in many developing nations, thereby reducing trust in the market (Batjes et al., 2024). Additionally, there are not enough government incentives to encourage widespread adoption of C farming. In comparison to developing nations, which often lack the financial means to implement sustainable farming practices, C_{seq} practices are not widely encouraged. Such policy reforms can find ways to integrate C farming into national climate strategies that may help to bridge the gap and pave the way for the required support for farmers (Sharma et al., 2024).

2.4.6 Environmental risks and trade-offs

C farming can mitigate climate change, but on the other hand, there are certain environmental risks to be taken care of. The reliance of afforestation projects on the replacement of native ecosystems with nonnative tree species poses a question

TABLE 2.3 Land use trade-offs in carbon farming.

Land use option	Benefits	Drawbacks
C farming (AF, cover cropping, no-till, etc.)	Increases soil C storage, improves soil fertility, and enhances biodiversity.	May reduce the land available for staple food crops and requires long-term commitment.
Traditional agriculture	Ensures food security, supports rural livelihoods.	May contribute to soil degradation and GHG emissions.
Monoculture C_{seq} (treelantations)	High C_{seq} potential, attracts C credit investments.	Biodiversity loss, displacement of native species, and competition for water resources.

AF, agroforestry; *Cseq*, carbon sequestration.

as to whether or not they are being appropriately implemented, if this occurs, as this can potentially disrupt local biodiversity (Kirschbaum et al., 2024). Such afforestation may create unintended consequences, such as a reduction in available water and soil nutrient depletion. In addition, intensive farming of C can harm the soil if not properly managed. The application of some sequestration techniques, such as biochar application and deep tillage, may also induce changes in soil properties and microbial communities and cause long-term soil fertility loss. C farming strategies must be ecologically sustainable to avoid unintended negative consequences (Shoudho et al., 2024; Zhang et al., 2024b).

2.4.7 Market volatility and risk management

Farming and investing in C credits are very unpredictable and risky. However, changes in global C policies, demand from corporations, and the volume of C credits in the voluntary market are all sources of price fluctuations (Cariappa and Krishna, 2025). Such volatility makes it hard for farmers to figure out how much revenue they might earn from C farming programs, so it discourages them from entering these types of programs in the long-term (Hannam, 2024). There is also the risk of oversupply in voluntary C markets, such that C credits will be devalued, resulting in a corresponding reduction in financial returns to farmers. Cseq has an economic incentive when prices are high, as it prevents too many C credits from entering the market, thereby maintaining prices (Jat et al., 2022). Therefore diversified revenue streams and long-term contractual agreements are effective risk management strategies for farmers to protect them against market uncertainty (Jat et al., 2022; Zhang et al., 2024b).

2.5 Solutions and best practices toward carbon farming-based carbon credits

C farming is an attractive mechanism for mitigating climate change via not only increasing soil C_{seq} but also "C farming" for generating C credits. Nevertheless, a solution that combines technological, financial, policy-driven, and community-based approaches is necessary to ensure its effectiveness. C farming, when solved for key barriers, for example, verification challenges, financial constraints, and stakeholder coordination, can be a scalable and sustainable option in global C markets.

2.5.1 Technological innovations

C farming practices are enhanced through technological advancements, which improve the accuracy and efficiency of the techniques used by them. Among these, innovation is the creation of artificial intelligence (AI)-driven tools to monitor soil C with the use of machine learning models predicting the level of soil C-based on environmental factors and historical data (Spilker and Nugent, 2022; Sharma et al., 2024). However, these tools facilitate reductions in the cost of performing manual soil sampling and also increase measurement accuracy. Furthermore, blockchain technology has made a huge difference in the transparency of C credit transactions. Blockchain has the benefits of eliminating fraud and double-counting issues in C trading markets. Moreover, satellite imaging and remote sensing advancements enable pure-time land-use changes and soil Cseq rates to be observed (Abdulraheem et al., 2023; Zhang et al., 2024a; Boumaiza and Maher, 2024).

2.5.2 Capacity building and education

Technical knowledge and skills are the main barriers to farmers' widespread adoption of C farming, particularly in developing regions. Agricultural communities are receiving the knowledge needed to adopt farming techniques that sequester C, including no-till farming, AF, and biochar use (Block et al., 2024a). Such programs should be developed in partnership with agricultural research institutions in a way that provides access to large-scale and smallholder farmers. Also, there is a need for awareness campaigns in the public that will encourage participation in C farming initiatives. These campaigns are to showcase success stories and explain the economic benefits of C credits, which should encourage more farmers, policymakers, and investors to support sustainable agricultural practices (Shoudho et al., 2024).

2.5.3 Financial mechanisms

C farming is heavily reliant on high initial costs, which keep both smallholder farmers and private investors away from engaging in it. Subsidies and grants can help farmers cover these costs and appropriately adopt C-friendly practices (Batjes et al., 2024). A good way for governments and international organizations to influence C_{seq} is to create incentive programs that reward farmers for measurable C_{seq} and motivate them to maintain these pledges. In addition, such an establishment can grant farmers collective power in bargaining over C, since they can collect resources and cover costs associated with carbonate credit confirmation and market entry. The most viable means to get C farming off the ground, to begin with, for small-scale farmers is reduced transaction costs and economies of scale brought on by cooperatives (Block et al., 2024a).

2.5.4 Policy reforms

C farming-based C credits require a robust policy framework to ensure both credibility and long-term sustainability of the credits. There is no consensus among regulatory bodies to design a C market framework that refers explicitly to the standardized regulatory framework, resulting in a lack of C credit valuation and trading criteria (Günther et al., 2024). The internationally recognized standards for Cseq verification and reporting development will promote cross-border C credit trading and increase investor confidence (Boiral et al., 2024). In addition, national climate action plans and policies may facilitate the integration of C farming into the overall sustainability goals of the Paris Agreement. Mandatory C offset policies should also be considered by governments that require parts of industries' C credits to be sourced from verified agricultural projects (Michaelowa et al., 2019; Cariappa and Krishna, 2025).

2.5.5 Stakeholder engagement

C farming initiatives succeed when supported by a collaboration of key stakeholders, such as governments, the private sector, research institutions, and farming communities. Specifically, PPPs are very successful ways of marshaling resources and expertise to reach C farming projects (Baumber et al., 2024). A mutually beneficial system is created as companies provide funding and technical support to farmers who must offset their emissions (Lovcha et al., 2022). NGOs and research institutions can also help in facilitating knowledge transfer, piloting projects, and ensuring the sustainability of C farming. There will be multistakeholder platforms where different actors can share best practices and collaborate to accelerate the global adoption of C farming (Kreibich and Hermwille, 2021).

2.5.6 Ecosystem-based approaches

C farming should involve not only Cseq but also other ecological benefits and is best done holistically. AF and regenerative agriculture are the two most effective routes to C storage in soil and biodiversity benefits (Pudasaini et al., 2025; Raj et al., 2024; Kumar et al., 2024). Farming with trees, shrubs, and perennial crops in the landscape improves soil fertility and promotes the capture and storage of C, as well as weather extremes (decreases in temperature, snowmelt, expanding and contracting stream flows, etc.). This is also promising because landscape-level C farming projects, such as wetland restoration and afforestation, can provide a significantly greater opportunity for Cseq while supporting vegetation cover services, such as water retention and pollination (Sharma et al., 2024; Banerjee et al., 2023, 2024).

2.5.7 Social and cultural considerations

To ensure the longevity of C farming initiatives, the projects must be inclusive and equitable. Barriers to accessing C markets are faced by many smallholder farmers, especially those in marginalized communities. Community-led C farming initiatives that combine traditional land management practices with these initiatives can increase cultural acceptability and the effectiveness of these projects (Hannam, 2024). The inclusion of indigenous knowledge, which for so long has prioritized the sustainable use of land, should be incorporated into C farming strategies. In addition, improving the distribution of C credits to address equity issues and avoid excluding small-scale farmers from C allocation solely in favor of large agribusinesses. Such programs need to be designed so that women and disadvantaged groups enjoy equal access to training, financial resources, and market opportunities (Jat et al., 2022).

2.6 Monitoring, evaluation, and assessment of carbon farming for carbon credits

It is important to maintain the credibility and sustainability of C farming initiatives through effective monitoring, evaluation, and assessment (MEA). Various complexities involved in soil C_{seq} necessitate having robust MEA frameworks to track the fulfillment of C credit generation and compliance with the regulatory standard, as well as transparency in the market. The rise of technology has contributed to advances in the way C farming outcomes are measured and scored.

2.6.1 Establishment of standardized monitoring frameworks

To achieve this, it is necessary to develop standardized monitoring frameworks for C credit assessments (Pudasaini et al., 2025). The cost-effective and scalable solutions of remote sensing technologies, including satellite imaging and drones, in monitoring land-use changes and soil organic C levels are more attractive. Additionally, by utilizing machine learning algorithms to predict Cseq trends with high precision, trends can be enhanced through analysis of historical data on climate and soils (Sharma et al., 2024). Integration of technologies is possible, allowing for more reliable measurements and reducing dependence on field-based assessments, which are very expensive. On the other hand, blockchain technology is also being leveraged as a powerful tool to increase transparency and traceability in credit transactions. The application of blockchain in this way means that Cseq activities are recorded as immutable to prevent fraudulent reporting and that C credits represent real, verifiable emission reductions. There needsto be unified standards in governments and international organizations for monitoring C that follow the prevailing framework of C in the global C market (Shahi et al., 2025; Sharma et al., 2024; West et al., 2024).

2.6.2 Periodic evaluation of carbon sequestration outcomes

Regular C_{seq} evaluation beyond monitoring is also needed to determine the long-term effectiveness of C farming practices. A role played by independent third-party verification agencies is to verify C credits and assure conformity to established protocols (Block et al., 2024a). In addition, the use of dynamic assessment models incorporating environmental, economic, and social factors would enable the evaluation of the overall C farming benefit. Policymakers should commit to periodic program policy reviews and systems of adaptive management that explore and accommodate changes in C credit programs in the light of the new scientific evidence (Fuller et al., 2025). Stakeholders can continue to build trust in C farming initiatives and improve their contribution to the global climate goals through careful improvement of MEA processes (Okoli and Birkenberg, 2024).

2.7 Solutions and future thrust areas for carbon credits and carbon farming

Innovative solutions are needed to overcome current limitations in their effectiveness, as well as to nurture sustainable and equitable participation in C farming and credits for the future. Given the growing global effort to mitigate climate change, the C credit system can be strengthened by dynamic C pricing models and broadening C farming in underrepresented regions. These approaches will not only add to farmers' fiscal stability but also contribute to broader environmental and socio-economic benefits.

2.7.1 Development of dynamic carbon pricing models to enhance financial stability

Price uncertainty is a major challenge in the C-credit market, making investment for farmers and investors uncertain. This suggests a potential solution for developing dynamic C pricing models that first take into account real-time market conditions, demand fluctuations, and C_{seq} performance (Piris-Cabezas et al., 2023). In contrast to the way fixed pricing determines emissions rates, dynamic models can adapt as C offset demand shifts, there are changes in the regulations, and technological developments in C measurement tools. Farmers would be compensated fairly for sequestration efforts, and their risk of price fluctuations would be limited at the same time (Block et al., 2024a,b).

The integration of AI and blockchain technology into the market also enables the enhancement of transparency and pricing efficiency. In theory, the future C credit values can be forecasted to allow farmers and investors to make informed decisions. It is proven that C markets become robust by designing blockchain and secure and traceable transactions that prevent fraud and verifying C credits (Zhang et al., 2024a; Boumaiza and Maher, 2024). Government policies and international cooperation will also be of help in stabilizing C-credit prices. Extreme swings and speculations can be controlled using C price floors and ceilings, resulting in a market that is more sustainable in the long-term. Additionally, linking to C markets from an international convention would allow regional markets to become more integrated and universal in relation to pricing, as well as the encouragement of Cseq on a global scale (Piris-Cabezas et al., 2023).

2.7.2 Expansion of carbon farming initiatives in underrepresented regions

C farming, although popularized in developed countries, lacks supportive infrastructure and means in many underrepresented regions, mainly in Africa, South Asia, and Latin America (Sharma et al., 2024). In these areas, several advantages exist in expanding C farming activities, ranging from better soil health to increased biodiversity to empowering local farmers economically. Capacity building and educational advancement are key steps toward expansion. In developing nations, many smallholder farmers are unaware of the opportunity for an economically beneficial form of C farming. Training of farmers in sustainable agricultural practices and C_{seq} methods, as well as market participation, have to be invested in by governments, NGOs, and research institutions. Within the proper digital platforms, digital information on C credit eligibility and verification processes can also be provided (Raina et al., 2024).

Financial support mechanisms, including low-interest loans and C farming subsidies, as well as microfinance programs, pave the way for the participation of small-scale farmers. Initiatives that reduce upfront investment costs can help more farmers adopt climate-smart practices, thereby increasing C_{seq} and enhancing rural development. It can also be integrated into national climate policies and lead to large-scale adoption (Bhatnagar et al., 2024). C farming should be considered as a viable method of climate mitigation by governments that provide policy support, such as tax incentives and direct payments for ecological services. Countries can contribute maximally to the global climate targets by aligning C farming with national C reduction goals (Mehmood et al., 2024).

2.8 Research and development in carbon farming and credits

Research and development (R&D) is important to the advancement of C farming practices and improvement of C credit systems. Biochar applications and microbial solutions offer a means to increase C storage while maintaining sustainable agricultural productivity to a much greater degree than innovations in this field conventionally achieve (Alli et al., 2024). However, with increasing investment in R&D, varying levels of breakthrough C capture are possible, and optimization of market participation in C credit schemes is achievable (Ayaz et al., 2021).

2.8.1 Advanced research into biochar applications for long-term soil carbon storage

Biochar, a C-rich material, has recently shown significant promise as a long-term C storage in the soil. The research has shown that biochar helps to improve soil fertility by increasing the activity of microbes and considerably extending the residence time of soil C (Shoudho et al., 2024). Biochar is different from traditional organic matter, which decomposes with time. Still, instead, biochar remains stable in the soil for hundreds of years, a process that is thought to make it useful on its own as a tool to promote Cseq in agricultural systems (Ayaz et al., 2021; Innocent et al., 2024).

The improvement in the retention of biochar has been a focus of recent studies through the optimization of biochar properties. The effectiveness of biochar is determined by the feedstock type, pyrolysis temperature, and soil conditions. For example, biomass used to produce biochar may be woody (C stable) or herbaceous (less C stable) (Yaashikaa et al., 2020). Additionally, research has demonstrated that incorporating biochar with organic fertilizers enhances nutrient availability while reducing GHG emissions from the soil (Schick et al., 2023). Another important area of research is to scale up biochar production. To realize large-scale production and end use amounting to the proven benefits of biochar, efficient production and distribution channels are required. Improving access to biochar by smallholder farmers involves studies on how to introduce mobile pyrolysis units and decentralized biochar production models (Tisserant and Cherubini, 2019; Patel and Panwar, 2023). Furthermore, the use of biochar may also attract additional financial incentives for incorporation into C credit markets and increase the application of biochar (Ayaz et al., 2021).

2.8.2 Exploration of microbial and biotechnological solutions for improving soil carbon sequestration

Microbial activity is responsible for the C stabilization, decomposition, and nutrient cycles; thus, these activities are driven by soil C dynamics. Recent research on soil microbiomes has suggested the use of microbial inoculants as well as genetically engineered microbes for Cseq. Improving root biomass and decreasing C loss with respiration through beneficial microbes, such as nitrogen-fixing bacteria and mycorrhizal fungi, may improve soil organic C (Rebello et al., 2021; Livinus et al., 2024b).

Active C capturing and storage in soil are being explored as a possible biotechnological enterprise through microbial strain advancement using synthetic biology and gene editing. For instance, scientists are working with engineered bacteria that produce extracellular polymeric substances that bind soil particles and lead to long-term C retention (Onyeaka and Ekwebelem, 2023). Also, microbial consortia that promote stimulation of plant root exudation can induce C deposition in deeper soil layers and long-term storage of C remains (Mason et al., 2023). Challenges presented by the need to increase food production, particularly in developing countries, include soil degradation and the increase in occurrences of nutrient-imbalanced food, desalination of drinking water, and limited water resources from precipitation, landslides, and flooding. Although cover cropping, reduced tillage, and compost application favor microbial communities, and these can sequester C, this has not been confirmed. The current approach to C farming strategies helps include microbial-based soil amendments to build holistic approaches that store maximum C and enhance soil health (Romdhane et al., 2019; Khangura et al., 2023; Rodrigues et al., 2023).

2.9 Future outlook and advancements

The future of C farming and C credits is being set for maximum growth and use as the world accelerates its efforts to reduce its C footprint. The effectiveness of C farming will be the driver of its success, which will include the integration of innovative technologies, expansion of NbS, and strengthening of international collaborations. The adoption of blockchain-based C trading, large-scale deployment of NbS, and the adoption of global C farming standards will enable more transparency, scalability, and inclusivity in C_{seq} efforts.

2.9.1 Potential integration of carbon farming into blockchain-based carbon trading platforms

The widespread adoption of blockchain technology can enhance transparency, traceability, and transaction efficiency in the C farming credit market. As has happened with traditional C credit markets, the lack of transparency, double-counting, and difficulty in verifying Cseq claims have all been leveled at it (Zhang et al., 2024a). These challenges can be addressed using blockchain, which generates decentralized, tamper-proof records, allowing the entire cycle of a C credit to be tracked from generation to retirement. Smart contracts produce another key advantage of blockchain in C trading: automatic transactions. Smart contracts enable the automatic execution of agreements based on predefined conditions, reducing administrative costs and minimizing fraud. Finally, tokenized C credits facilitate smallholder farmers entering the market as fractional owners and microtransaction by making C farming financially inclusive (Javaid et al., 2022; Boumaiza, 2024; Ballesteros-Rodríguez et al., 2024).

Within the space of blockchain-based C trading platforms, which include KlimaDAO, CarbonX, and Toucan Protocol, using decentralized finance (DeFi) to facilitate C credits in liquid markets through the purchasing of C credits has already started (Ballesteros-Rodríguez et al., 2024). Currently, blockchain can bring standardization to the entire process of verifying C credits in global C markets, eliminating intermediaries and making these markets as efficient as possible. CarbonX, KlimaDAO, and Toucan Protocol are already some of these blockchain C-trading platforms that are integrating C offsets into DeFi and developing a liquid C-credits market. If widely adopted, blockchain could standardize the entire process of C credit and validate certifications in global C markets, thereby eliminating intermediaries and increasing the efficiency of these markets (Ballesteros-Rodríguez et al., 2024).

2.9.2 Scaling up nature-based solutions for carbon sequestration

NbS are economically feasible and sustainable means of combatting climate change. Such solutions include AF and reforestation, wetland restoration, regenerative agriculture, and any other efforts that increase C_{seq}, biodiversity, and the resilience of ecosystems (Campbell-Verduyn, 2024). The future of C farming will be in scaling up these solutions, and these have to be attractive to adopt and adequately financed. Such expansion could be based on the main strategy of implementing AF systems; the forests are positioned alongside agricultural landscapes to improve C storage. It is known that AFs can sequester up to 5.5 gigatons of CO_2 annually, which ranks among the most effective NbS for C removal (Cassou et al., 2022). AF initiatives linked to C credit markets have already been done on a large scale in countries such as Brazil and Kenya, which provide evidence that this approach is feasible (Araya et al., 2023).

Soil C_{seq} through regenerative agriculture is also a critical NbS addition. These practices have been found to both increase soil organic C to levels favorable for C_{seq} as soil health and resilience to climate change in soil (Shoudho et al.,

2024). C-credit programs for regenerative agriculture have been developed by governments and private organizations that may promote the mainstreaming of sustainable farming practices. To scale NbS, policy and financial support, as well as technological innovation, need to be strong. Simplifying NbS is an emerging tool that includes satellite imaging, remote sensing, and the utilization of AI in soil analysis, which can improve how C based on NbS C_{seq} is monitored and, therefore, produce assured, verifiable C credit reporting (Acuña-Alonso et al., 2022).

2.9.3 Strengthening international collaborations to create global carbon farming standards

C farming faces one of the key challenges without harmonized global standards for MRV Cseq. The valuation of C credits depends on different countries, and C credit organizations have different methodologies, resulting in inconsistency (Batjes et al., 2024). International collaborations can help establish standardized frameworks for the existence of C farming credits, thereby strengthening them and making them more credible and market-acceptable (Batjes et al., 2024; Brummitt et al., 2024).

The United Nations Framework Convention on Climate Change, the International Carbon Reduction and Offset Alliance, and Verra are engaged in the development of uniform C credit certification standards at the moment. Bringing in globally recognized C farming protocols would broaden engagement by farmers, corporations, and governments in C markets and thus scale (Tavonvunchai et al., 2024). Bilateral and multilateral agreements between nations can increase cross-border trade of C credits. Perhaps Article 6 of the Paris Agreement allows countries to trade C credits and use this to incentivize investment in international C farming projects (Michaelowa et al., 2019). Meanwhile, establishing joint initiatives for knowledge sharing, capacity building, and technology transfer can also further accelerate the adoption of C farming around the world. C farming projects are also supported by low-interest loans and grants from international financial institutions, such as the World Bank and the Green Climate Fund. Financial cooperation between ministries, NGOs, and the private sector can expand the resources necessary for farmers to adapt to sustainable C_{seq} practices (Tavonvunchai et al., 2024).

2.10 Conclusion

C credits derived from C farming and its related nodes are a promising mechanism for mitigating climate change by promoting soil C_{seq} for sustainable agriculture. Despite this, implementation is hampered by several barriers, such as measurement, verification, and additionality determination, with few exceptions. Participation is complicated by financial constraints, market volatility, and policy inconsistencies, particularly during a period of food market redistribution and agricultural product production. To tackle these problems, there is a need for a multifaceted approach that combines technological zeniths such as AI-driven audience surveillance, remote sensing, and blockchain for increased transparency and efficiency. In addition, policy reforms and financial mechanisms such as subsidies and incentives can be important for simplifying access to C farming. Improved government, private, and local community collaboration will be needed to achieve the long-term success of C-credit initiatives. Emergency is nothing new to C farming; as global efforts to combat climate change continue, only research, regulatory standardization, and innovative solutions can secure its future. C farming would overcome the challenges of the past and enhance opportunities of the future, making a grand contribution to the C_{seq}, biodiversity conservation, and sustainable agricultural development.

References

Abdulraheem, M.I., Zhang, W., Li, S., Moshayedi, A.J., Farooque, A.A., Hu, J., 2023. Advancement of remote sensing for soil measurements and applications: a comprehensive review. Sustainability 15 (21), 15444. Available at https://doi.org/10.3390/su152115444.

Abdulsalam, M., Elelu, S.A., Innocent, M.O., Ibrahim, G.O., Livinus, M.U., Lateefat, S.O., et al., 2024. Women empowerment in environmental conservation. Prospects for Soil Regeneration and Its Impact on Environmental Protection. Springer Nature, Switzerland, pp. 351–370.

Acuña-Alonso, C., Novo, A., Rodríguez, J.L., Varandas, S., Álvarez, X., 2022. Modelling and evaluation of land use changes through satellite images in a multifunctional catchment: social, economic and environmental implications. Ecological Informatics 71, 101777. Available at https://doi.org/10.1016/j.ecoinf.2022.101777.

Alli, Y.A., Bamisaye, A., Bamidele, M.O., Etafo, N.O., Chkirida, S., Lawal, A., et al., 2024. Transforming waste to wealth: harnessing carbon dioxide for sustainable solutions. Results in Surfaces and Interfaces 17. Available at https://doi.org/10.1016/j.rsurfi.2024.100321.

Araya, Y.N., Emmott, A., Rawes, W., Zuza, E.J., 2023. Promoting climate-smart sustainable agroforestry to tackle social and environmental challenges: the case of macadamia agroforestry in Malawi. Journal of Agriculture and Food Research 14. Available at https://doi.org/10.1016/j.jafr.2023.100846.

Ayaz, M., Feizienė, D., Tilvikienė, V., Akhtar, K., Stulpinaitė, U., Iqbal, R., 2021. Biochar role in the sustainability of agriculture and environment. Sustainability 13 (3), 1330. Available at https://doi.org/10.3390/su13031330.

Ballesteros-Rodríguez, A., De-Lucio, J., Sicilia, M.-Á., 2024. Tokenized carbon credits in voluntary carbon markets: the case of KlimaDAO. Frontiers in Blockchain 7. Available at https://doi.org/10.3389/fbloc.2024.1474540.

Banerjee, A., Jhariya, M.K., Bargali, S.S., Palit, D., 2023. Ecorestoration for Sustainability. Wiley-Scrivener, USA.

Banerjee, A., Jhariya, M.K., Raj, A., Mechergui, T., 2024. Ecosystem management: Climate change and global sustainability—an introduction. Ecosystem Management: Climate Change and Sustainability. wiley, India, pp. 1–21. Available at https://doi.org/10.1002/9781394231249.ch1.

Batjes, N.H., Ceschia, E., Heuvelink, G.B.M., Demenois, J., le Maire, G., Cardinael, R., et al., 2024. Towards a modular, multi-ecosystem monitoring, reporting and verification (MRV) framework for soil organic carbon stock change assessment. Carbon Management 15 (1). Available at https://doi.org/10.1080/17583004.2024.2410812.

Baumber, A., Cross, R., Ampt, P., Waters, C., Ringbauer, J., Bowdler, I., et al., 2024. Soil-based carbon farming: opportunities for collaboration. Journal of Rural Studies 108, 103268. Available at https://doi.org/10.1016/j.jrurstud.2024.103268.

Bhatnagar, S., Chaudhary, R., Sharma, S., Janjhua, Y., Thakur, P., Sharma, P., et al., 2024. Exploring the dynamics of climate-smart agricultural practices for sustainable resilience in a changing climate. Environmental and Sustainability Indicators 24, 100535. Available at https://doi.org/10.1016/j.indic.2024.100535.

Block, J.B., Michels, M., Mußhoff, O., Hermann, D., 2024a. How to reduce the carbon footprint of the agricultural sector? Factors influencing farmers' decision to participate in carbon sequestration programs. Journal of Environmental Management 359, 121019. Available at https://doi.org/10.1016/j.jenvman.2024.121019.

Block, J.B., Danne, M., Mußhoff, O., 2024b. Farmers' willingness to participate in a carbon sequestration program – a discrete choice experiment. Environmental Management 74 (2), 332–349. Available at https://doi.org/10.1007/s00267-024-01963-9.

Boiral, O., Brotherton, M.-C., Talbot, D., 2024. Achieving corporate carbon neutrality: a multi-perspective framework. Journal of Cleaner Production 467, 143040. Available at https://doi.org/10.1016/j.jclepro.2024.143040.

Boumaiza, A., 2024. A blockchain-centric P2P trading framework incorporating carbon and energy trades. Energy Strategy Reviews 54, 101466. Available at https://doi.org/10.1016/j.esr.2024.101466.

Boumaiza, A., Maher, K., 2024. Leveraging blockchain technology to enhance transparency and efficiency in carbon trading markets. International Journal of Electrical Power & Energy Systems 162, 110225. Available at https://doi.org/10.1016/j.ijepes.2024.110225.

Brummitt, C.D., Mathers, C.A., Keating, R.A., O'Leary, K., Easter, M., Friedl, M.A., et al., 2024. Solutions and insights for agricultural monitoring, reporting, and verification (MRV) from three consecutive issuances of soil carbon credits. Journal of Environmental Management 369. Available at https://doi.org/10.1016/j.jenvman.2024.122284.

Campbell-Verduyn, M., 2024. Conjuring a cooler world? Imaginaries of improvement in blockchain climate finance experiments. Environment and Planning C: Politics and Space 42 (5), 782–799. Available at https://doi.org/10.1177/23996544231162858.

Cariappa, A.A.G., Krishna, V.V., 2025. Carbon farming in India: are the existing projects inclusive, additional, and permanent? Climate Policy 25 (5), 756–771. Available at https://doi.org/10.1080/14693062.2024.2416497.

Cassou, E., Shangbin, G., Quanhui, W., Keqiang, Z., Weifeng, Z., Yaping, L., et al., 2022. Deepening pollution prevention and control on China's Farms. A compendium of thematic Papers 23.

Climate Action Reserve, 2021. Reserve offset program manual. Available from: https://www.climateactionreserve.org/wp-content/uploads/2021/03/Reserve_Offset_Program_Manual_March_2021.pdf (accessed 01.03.25).

Davis, M., Alves, B., Karlen, D., Kline, K., Galdos, M., Abulebdeh, D., 2018. Review of soil organic carbon measurement protocols: a US and Brazil comparison and recommendation. Sustainability 10 (1), 53. Available at https://doi.org/10.3390/su10010053.

Filonchyk, M., Peterson, M.P., Zhang, L., Hurynovich, V., He, Y., 2024. Greenhouse gases emissions and global climate change: examining the influence of CO_2, CH_4, and N_2O. Science of the Total Environment 935, 173359. Available at https://doi.org/10.1016/j.scitotenv.2024.173359.

Fuller, M.R., Ganjam, M., Baker, J.S., Abt, R.C., 2025. Advancing forest carbon projections requires improved convergence between ecological and economic models. Carbon Balance and Management 20 (1). Available at https://doi.org/10.1186/s13021-024-00290-0.

Gizachew, B., Astrup, R., Vedeld, P., Zahabu, E.M., Duguma, L.A., 2017. REDD+ in Africa: contexts and challenges. Natural Resources Forum 41 (2), 92–104. Available at https://doi.org/10.1111/1477-8947.12119.

Gold Standard, 2014. The real value of robust climate action: impact investment far greater than previously understood. Gold Standard.

Gold Standard, 2019. Gold Standard market report 2018. Available from: https://www.goldstandard.org/sites/default/files/market_report_2018.pdf.

Günther, P., Garske, B., Heyl, K., Ekardt, F., 2024. Carbon farming, overestimated negative emissions and the limits to emissions trading in land-use governance: the EU carbon removal certification proposal. Environmental Sciences Europe 36 (1). Available at https://doi.org/10.1186/s12302-024-00892-y.

Hannam, I., 2024. Sustainable soil management and soil carbon sequestration. In: Ginzky, H. (Ed.), International Yearbook of Soil Law and Policy, vol 2022. Springer, Cham, pp. 3–33. Available at https://doi.org/10.1007/978-3-031-40609-6_1.

Innocent, M.O., Mustapha, A., Abdulsalam, M., Livinus, M.U., Samuel, J.O., Shehu-Alimi, E., et al., 2024. Soil microbes and soil contamination. In: Aransiola, S.A., Atta, H.I., Maddela, N.R. (Eds.), Soil Microbiome in Green Technology Sustainability. Springer, Cham. Available at https://doi.org/10.1007/978-3-031-71844-1_1.

Jat, M.L., Chakraborty, D., Ladha, J.K., Parihar, C.M., Datta, A., Mandal, B., et al., 2022. Carbon sequestration potential, challenges, and strategies towards climate action in smallholder agricultural systems of South Asia. Crop and Environment 1 (1), 86–101. Available at https://doi.org/10.1016/j.crope.2022.03.005.

Javaid, M., Haleem, A., Singh, R.P., Suman, R., Khan, S., 2022. A review of blockchain technology applications for financial services. BenchCouncil Transactions on Benchmarks, Standards and Evaluations 2 (3), 100073. Available at https://doi.org/10.1016/j.tbench.2022.100073.

Jhariya, M.K., Meena, R.S., Banerjee, A., Kumar, S., Raj, A., 2023. Agroforestry for Carbon and Ecosystem Management Agroforestry for Carbon and Ecosystem Management. Elsevier, India, pp. 1–434. Available at https://doi.org/10.1016/C2021-0-02916-3.

Jhariya, M.K., Meena, R.S., Banerjee, A., Meena, S.N., 2022. Natural Resources Conservation and Advances for Sustainability. Elsevier. Available at: https://doi.org/10.1016/C2019-0-03763-6.

Khangura, R., Ferris, D., Wagg, C., Bowyer, J., 2023. Regenerative agriculture—a literature review on the practices and mechanisms used to improve soil health. Sustainability 15 (3), 2338. Available at https://doi.org/10.3390/su15032338.

Kirschbaum, M.U.F., Cowie, A.L., Peñuelas, J., Smith, P., Conant, R.T., Sage, R.F., et al., 2024. Is tree planting an effective strategy for climate change mitigation? Science of the Total Environment 909. Available at https://doi.org/10.1016/j.scitotenv.2023.168479.

Kornek, U., Klenert, D., Edenhofer, O., Fleurbaey, M., 2021. The social cost of carbon and inequality: when local redistribution shapes global carbon prices. Journal of Environmental Economics and Management 107, 102450. Available at https://doi.org/10.1016/j.jeem.2021.102450.

Kreibich, N., Hermwille, L., 2021. Caught in between: credibility and feasibility of the voluntary carbon market post-2020. Climate Policy 21 (7), 939–957. Available at https://doi.org/10.1080/14693062.2021.1948384.

Kumar, S., Meena, R.S., Sheoran, P., Jhariya, M.K., 2024. Regenerative Agriculture for Sustainable Food Systems Regenerative Agriculture for Sustainable Food Systems. Springer Science+Business Media, India, pp. 1–493. Available at https://doi.org/10.1007/978-981-97-6691-8.

Livinus, M.U., Bala, S.Z., Abdulsalam, M., Innocent, M.O., Hassan, M., Kini, P., 2024a. Natural occurrences of soil dilapidation. Prospects for Soil Regeneration and Its Impact on Environmental Protection. Springer, Cham, pp. 205–223.

Livinus, M.U., Bala, S.Z., Abdulsalam, M., Musa, I.O., Hassan, M.A., Kini, P., 2024b. Marine genetic resources of blue economy. In: Aransiola, S.A., Bamisaye, A., Abioye, O.P., Maddela, N.R. (Eds.), Marine Bioprospecting for Sustainable Blue-Bioeconomy. Springer, Cham. Available at https://doi.org/10.1007/978-3-031-68911-6_9.

Lokuge, N., Anders, S., 2022. Carbon-credit systems in agriculture: a review of literature. The School of Public Policy Publications 15, 1–44. Available at https://doi.org/10.11575/sppp.v15i1.74591.

Lovcha, Y., Perez-Laborda, A., Sikora, I., 2022. The determinants of CO_2 prices in the EU emission trading system. Applied Energy 305, 117903. Available at https://doi.org/10.1016/j.apenergy.2021.117903.

Mason, A.R.G., Salomon, M.J., Lowe, A.J., Cavagnaro, T.R., 2023. Microbial solutions to soil carbon sequestration. Journal of Cleaner Production 417, 137993. Available at https://doi.org/10.1016/j.jclepro.2023.137993.

Meena, R.S., Pradhan, G., 2023. Industrial garbage-derived biocompost enhances soil organic carbon fractions, CO_2 biosequestration, potential carbon credits and sustainability index in a rice-wheat ecosystem. Environmental Research 235, 116525. Available at https://doi.org/10.1016/j.envres.2023.116525.

Meena, R.S., Pradhan, G., Kumar, S., Lal, R., 2023. Using industrial wastes for rice-wheat cropping and food-energy-carbon-water-economic nexus to the sustainable food system. Renewable and Sustainable Energy Reviews 187, 113756. Available at https://doi.org/10.1016/j.rser.2023.113756.

Meena, R.S., Pradhan, G., Nalani, Kumar, S., Singh, A.K., Verma, S.K., et al., 2024a. Diversified cropping modules designed for soil restoration, CO_2 sequestration, and generating carbon credits. Land Degradation and Development 35 (14), 4314–4329. Available at https://doi.org/10.1002/ldr.5224.

Meena, R.S., Pradhan, G., Singh, K., Kumar, S., Singh, A.K., Shashidhar, K.S., et al., 2024b. Agriculture models for restoring degraded land to enhance CO_2 biosequestration and carbon credits in the Vindhyan region of India. Science of the Total Environment 929. Available at https://doi.org/10.1016/j.scitotenv.2024.172661.

Meena, R.S., Lal, R., Kumar, S., Pradhan, G., Srinivasarao, C., Singh, A.K., et al., 2024c. Potential of Indian agriculture for capturing atmospheric CO_2 and monetizing carbon credits to the farmers: an overview and policy framework. Advances in Agronomy 188, 101–206. Available at https://doi.org/10.1016/bs.agron.2024.06.001.

Meena, R.S., Pradhan, G.S., Kumar, S., Jhariya, M.K., Rao, C.S., 2025a. Carbon-Negative Agriculture. Springer Nature, Singapore.

Meena, R.S., Pradhan, G., Kumar, S., 2025b. Energy flow, eco-efficiency, and economic circulation with recycled industrial waste compost application in wheat and subsequent rice farming. Science of the Total Environment 967. Available at https://doi.org/10.1016/j.scitotenv.2025.178779.

Mehmood, S., Zaman, K., Khan, S., Ali, Z., Khan, Hu.R., 2024. The role of green industrial transformation in mitigating carbon emissions: exploring the channels of technological innovation and environmental regulation. Energy and Built Environment 5 (3), 464–479. Available at https://doi.org/10.1016/j.enbenv.2023.03.001.

Michaelowa, A., Hermwille, L., Obergassel, W., Butzengeiger, S., 2019. Additionality revisited: guarding the integrity of market mechanisms under the paris agreement. Climate Policy 19 (10), 1211–1224. Available at https://doi.org/10.1080/14693062.2019.1628695.

Nyberg, Y., Musee, C., Wachiye, E., Jonsson, M., Wetterlind, J., Öborn, I., 2020. Effects of agroforestry and other sustainable practices in the Kenya agricultural carbon project (KACP). Land 9 (10), 389. Available at https://doi.org/10.3390/land9100389.

Okoli, A.O., Birkenberg, A., 2024. Monitoring soil carbon in smallholder carbon projects: insights from Kenya. Climatic Change 177 (9). Available at https://doi.org/10.1007/s10584-024-03796-1.

Onyeaka, H., Ekwebelem, O.C., 2023. A review of recent advances in engineering bacteria for enhanced CO_2 capture and utilization. International Journal of Environmental Science and Technology 20 (4), 4635–4648. Available at https://doi.org/10.1007/s13762-022-04303-8.

Pan, C., Shrestha, A., Innes, J.L., Zhou, G., Li, N., Li, J., et al., 2022. Key challenges and approaches to addressing barriers in forest carbon offset projects. Journal of Forestry Research 33 (4), 1109–1122. Available at https://doi.org/10.1007/s11676-022-01488-z.

Patel, M.R., Panwar, N.L., 2023. Biochar from agricultural crop residues: environmental, production, and life cycle assessment overview. Resources, Conservation and Recycling Advances 19. Available at https://doi.org/10.1016/j.rcradv.2023.200173.

Paul, C., Bartkowski, B., Dönmez, C., Don, A., Mayer, S., Steffens, M., et al., 2023. Carbon farming: are soil carbon certificates a suitable tool for climate change mitigation? Journal of Environmental Management 330, 117142. Available at https://doi.org/10.1016/j.jenvman.2022.117142.

Piris-Cabezas, P., Lubowski, R.N., Leslie, G., 2023. Estimating the potential of international carbon markets to increase global climate ambition. World Development 167, 106257. Available at https://doi.org/10.1016/j.worlddev.2023.106257.

Pradhan, G., Meena, R.S., 2023a. Interaction impact of biocompost on nutrient dynamics and relations with soil biota, carbon fractions index, societal value of CO_2 equivalent and ecosystem services in the wheat-rice farming. Chemosphere 339. Available at https://doi.org/10.1016/j.chemosphere.2023.139695.

Pradhan, G., Meena, R.S., 2023b. Utilizing waste compost to improve the atmospheric CO_2 capturing in the rice-wheat cropping system and energy-cum-carbon credit auditing for a circular economy. Science of the Total Environment 892. Available at https://doi.org/10.1016/j.scitotenv.2023.164572 Available from.

Pradhan, G., Meena, R.S., Kumar, S., Lal, R., 2023. Utilizing industrial wastes as compost in wheat-rice production to improve the above and below-ground ecosystem services. Agriculture, Ecosystems & Environment 358, 108704. Available at https://doi.org/10.1016/j.agee.2023.108704.

Prasad, R., Jhariya, M.K., Banerje, A., 2021. Advances in Sustainable Development and Management of Environmental and Natural Resources: Economic Outlook and Opinions: Volume 11. Apple Academic Press, India, pp. 1–428. Available at https://doi.org/10.1201/9781003187455.

Pudasaini, K., Bhattarai, T., Rolfe, J., 2025. Exploring the barriers to farmer participation in soil carbon projects under the Australian Carbon Credit Unit Scheme. Australasian Journal of Environmental Management 32 (2), 97–115. Available at https://doi.org/10.1080/14486563.2024.2393246.

Raina, N., Zavalloni, M., Viaggi, D., 2024. Incentive mechanisms of carbon farming contracts: a systematic mapping study. Journal of Environmental Management 352, 120126. Available at https://doi.org/10.1016/j.jenvman.2024.120126.

Raj, A., Jhariya, M.K., Banerjee, A., Jha, R.K., Singh, K.P., 2024. Agroforestry. Wiley-Scrivener, USA.

Raj, A., Jhariya, M.K., Meena, R.S., 2025. Agroforestry for Monetising Carbon Credits. Springer Nature, Switzerland.

Rebello, S., Nathan, V.K., Sindhu, R., Binod, P., Awasthi, M.K., Pandey, A., 2021. Bioengineered microbes for soil health restoration: present status and future. Bioengineered 12 (2), 12839–12853. Available at https://doi.org/10.1080/21655979.2021.2004645.

Rodrigues, C.I.D., Brito, L.M., Nunes, L.J.R., 2023. Soil carbon sequestration in the context of climate change mitigation: a review. Soil Systems 7 (3), 64. Available at https://doi.org/10.3390/soilsystems7030064.

Romdhane, S., Spor, A., Busset, H., Falchetto, L., Martin, J., Bizouard, F., et al., 2019. Cover crop management practices rather than composition of cover crop mixtures affect bacterial communities in no-till agroecosystems. Frontiers in Microbiology 10 (JULY). Available at https://doi.org/10.3389/fmicb.2019.01618.

Schick, M., Griffin, R., Cherrington, E., Sever, T., 2023. Utilizing LiDAR to quantify aboveground tree biomass within an urban university. Urban Forestry & Urban Greening 89, 128098. Available at https://doi.org/10.1016/j.ufug.2023.128098.

Shahi, T.B., Balasubramaniam, T., Sabir, K., Nayak, R., 2025. Pasture monitoring using remote sensing and machine learning: a review of methods and applications. Remote Sensing Applications: Society and Environment 37. Available at https://doi.org/10.1016/j.rsase.2025.101459.

Sharma, M., Kaushal, R., Kaushik, P., Ramakrishna, S., 2024. Correction: Sharma et al. Carbon farming: prospects and challenges. Sustainability 2021, 13, 11122. Sustainability 16 (21), 9523. Available at https://doi.org/10.3390/su16219523.

Shoudho, K.N., Khan, T.H., Ara, U.R., Khan, M.R., Shawon, Z.B.Z., Hoque, M.E., 2024. Biochar in global carbon cycle: towards sustainable development goals. Current Research in Green and Sustainable Chemistry 8. Available at https://doi.org/10.1016/j.crgsc.2024.100409.

Spilker, G., Nugent, N., 2022. Voluntary carbon market derivatives: growth, innovation & usage. Borsa Istanbul Review 22, S109. Available at https://doi.org/10.1016/j.bir.2022.11.008.

Tavonvunchai, N., Niner, H.J., McQuatters-Gollop, A., Attrill, M.J., Rees, S., 2024. Global Review of the Status and Trend of Verified Blue Carbon Projects. SSRN, United Kingdom SSRN. doi: 10.2139/ssrn.4875179.

Tisserant, A., Cherubini, F., 2019. Potentials, limitations, co-benefits, and trade-offs of biochar applications to soils for climate change mitigation. Land 8 (12), 179. Available at https://doi.org/10.3390/land8120179.

Verra, 2022. Methodologies. Verra [website]. Available from: https://verra.org/methodologies-main/#vcs-methodologies (accessed 21.02.25).

West, T.A.P., Bomfim, B., Haya, B.K., 2024. Methodological issues with deforestation baselines compromise the integrity of carbon offsets from REDD+. Global Environmental Change 87. Available at https://doi.org/10.1016/j.gloenvcha.2024.102863.

Yaashikaa, P.R., Kumar, P.S., Varjani, S., Saravanan, A., 2020. A critical review on the biochar production techniques, characterization, stability and applications for circular bioeconomy. Biotechnology Reports 28, e00570. Available at https://doi.org/10.1016/j.btre.2020.e00570.

Zhang, G., Chen, S.C.-I., Yue, X., 2024a. Blockchain technology in carbon trading markets: impacts, benefits, and challenges—a case study of the Shanghai environment and energy exchange. Energies 17 (13), 3296. Available at https://doi.org/10.3390/en17133296.

Zhang, K., Liu, Z., McCarl, B.A., Fei, C.J., 2024b. Enhancing agricultural soil carbon sequestration: a review with some research needs. Climate 12 (10). Available at https://doi.org/10.3390/cli12100151.

Chapter 3

Understanding greenhouse gas emissions from agriculture through land-use practices and policy perspectives

S. Sarvade, S.B. Agrawal, R. Bajpai, and Vishnu K. Solanki
Department of Forestry, Jawaharlal Nehru Agricultural University, Jabalpur, Madhya Pradesh, India

Highlights

- Globally, agriculture accounts for nearly 24% of GHG emissions, with India contributing 18% of its national emissions, primarily driven by livestock, rice paddies, and fertilizer.
- Developed countries utilize taxes and Emissions Trading Systems (ETS) to cut industrial emissions by up to 35%; India primarily depends on indirect approaches such as the Perform, Achieve, and Trade (PAT) scheme.
- India's FAME scheme has driven a 300% surge in electric vehicle sales over five years, significantly reducing transportation-related emissions.
- India's green R&D investment is lower, remaining at 1% compared to the global norm of 2.5%, there is a pressing need to bolster financial commitments to drive innovation.
- India's Renewable Energy Purchase Obligation (RPO) and Bharat Stage VI vehicle emission norms are aligning with global standards, but policy enforcement challenges remain.
- Strengthening international collaborations, enhancing policy enforcement, and integrating carbon pricing can accelerate India's transition to a sustainable, low-emission economy.

3.1 Introduction

Agriculture serves as a cornerstone of global food security but remains a significant driver of greenhouse gas (GHG) emissions. Key agricultural activities, including crop cultivation, livestock rearing, agroforestry (AF), and fisheries, release substantial amounts of carbon dioxide (CO_2), methane (CH_4), and nitrous oxide (N_2O), intensifying climate change. These emissions originate from various processes, including soil degradation, enteric fermentation in ruminants, synthetic fertilizer decomposition, and deforestation associated with agricultural expansion (Borrelli et al., 2020; Scott et al., 2023). With rising global food demands (Sarvade et al., 2014; Sarvade and Singh, 2014; Sarvade et al., 2019), comprehending the emission contributions of different agricultural sectors is crucial for formulating sustainable mitigation strategies (Clark et al., 2020; Jhariya et al., 2023; Jhariya et al., 2022; Meena et al., 2025b).

GHG emissions vary significantly among nations due to differences in industrial output, energy consumption, and population dynamics. Table 3.1 presents an overview of total GHG emissions for selected countries, expressed in kilotons (kt) of CO_2 equivalent, based on data from the Emissions Database for Global Atmospheric Research (Crippa et al., 2023).

While nations like China and India exhibit high total GHG emissions due to their vast populations and rapid industrialisation, per capita emissions often remain lower than those of smaller, highly developed economies. A country's emissions footprint is shaped not only by its population size but also by key factors such as its energy mix, ranging from coal and natural gas to renewables; its economic structure; and the effectiveness of its environmental regulations (Banerjee et al., 2024). The steady rise in atmospheric GHG concentrations continues to drive global warming and climate change. Table 3.2 provides a summarized average concentration of major GHGs over recent years, based on data from the World Meteorological Organization and the Copernicus Climate Change Service (2024).

Carbon Incentives, Plans, and Policies in Agriculture. DOI: https://doi.org/10.1016/B978-0-443-33557-0.00023-9

TABLE 3.1 Main GHG emitters (kt CO_2e).

Country	GHG emissions (kt CO_2e)
China	12,667,428
United States	4,853,780
India	2,693,034
Russia	1,909,039
Japan	1,082,645
Indonesia	692,236
Iran	686,415

Note: These figures represent the total GHG emissions, encompassing carbon dioxide (CO_2), methane (CH_4), nitrous oxide (N_2O), and fluorinated gases. GHG, greenhouse gas; O_2e, carbon dioxide equivalent.
Source: (Crippa et al., 2023).

TABLE 3.2 GHG concentrations in the atmosphere (World Meteorological Organization, 2024; Copernicus Climate Change Service, 2024).

Year	Carbon dioxide (CO_2)	Methane (CH_4)	Nitrous oxide (N_2O)
2019	410.5 ppm	1877 ppb	332 ppb
2020	412.5 ppm	1889 ppb	334 ppb
2021	414.7 ppm	1904 ppb	336 ppb
2022	416.1 ppm	1911 ppb	337 ppb
2023	419.3 ppm	1902 ppb	-

Note: *ppm*, parts per million; *ppb*, parts per billion; *GHG*, greenhouse gas.

India's GHG emissions have grown significantly from 1990 to 2020. In 1990, the total GHG emissions were approximately 1,237,963 kt of CO_2 equivalent, and by 2020, they had risen to about 3,200,821 kt CO_2e (The Global Economy, 2023) (Fig. 3.1).

Livestock farming stands as a major driver of GHG emissions, with CH_4 released from enteric fermentation in ruminants and manure decomposition being the primary culprits. Globally, livestock account for approximately 14.5% of total human-induced GHG emissions, with beef and dairy production leading the way (Gerber et al., 2013). Beyond direct emissions, the expansion of grazing land and feed-crop cultivation accelerates deforestation and soil degradation, thereby amplifying CO_2 emissions (IPCC, 2019). To curb livestock-related emissions, researchers are exploring strategies such as optimizing efficiency, implementing rotational grazing practices, and incorporating CH_4-reducing feed additives (Reuters, 2024; Meena et al., 2024a).

Crop production is a significant contributor to GHG emissions, largely due to the overuse of synthetic fertilizers, which release N_2O, a potent GHG with nearly 300 times the global warming potential of CO_2 (Reay et al., 2012). Additional factors, such as poor soil management, intensive monocropping, and crop residue burning, further increase emissions. However, sustainable approaches like precision agriculture, organic soil amendments, cover cropping, and conservation tillage offer viable solutions to minimize the environmental impact (FAO, 2013a, b; Kumar et al., 2024). Integrating AF, where trees are cultivated alongside crops, can further enhance carbon sequestration (Cseq), mitigate emissions, and improve ecosystem resilience (Smith et al., 2007; Borenstein, 2018b; Jhariya et al., 2022; Raj et al., 2025).

Fisheries and aquaculture areas, while often overlooked, also contribute to GHG emissions, particularly through fuel consumption in fishing vessels and CH_4 release from aquaculture ponds. Overfishing and habitat destruction, such as

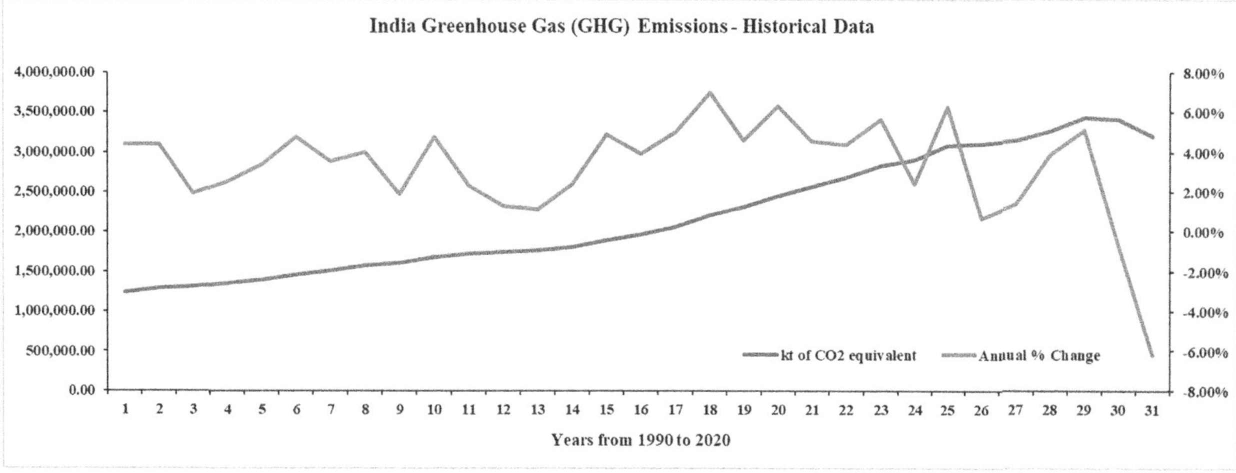

FIGURE 3.1 The last three decades' data on GHG emissions in India (*GHG*, greenhouse gas).

mangrove deforestation for shrimp farming, further exacerbate the problem (IPCC, 2019; IPCC, 2021). Transitioning to low-carbon (C) fishing technologies, sustainable aquaculture practices, and marine conservation efforts can help lower emissions from this sector (Tilman et al., 2011; Meena et al., 2024b).

Addressing GHG emissions from agriculture requires a holistic approach that integrates sector-specific mitigation strategies with broader climate adaptation measures. Policies promoting C farming, incentives for low-emission livestock production, and investment in climate-smart technologies are essential to reduce agriculture's environmental impact (European Commission, 2022; Meena et al., 2025a). By understanding the emission profiles of different agricultural sectors, policymakers and farmers can work toward a more sustainable food system that balances productivity with climate resilience.

3.2 Crop production and greenhouse gas emissions in developed and developing nations

While crop production remains fundamental to global agriculture, it is also a significant source of GHGs, with key sources including the use of synthetic fertilizers, land-use changes, and burning of crop residues. These processes release CO_2, CH_4, and N_2O, contributing to climate change (IPCC, 2019). However, emission patterns vary considerably between developed and developing nations, influenced by differences in farming techniques, technological innovations, and policy interventions. While developed countries focus on precision agriculture and emission mitigation, developing nations often face challenges related to resource constraints, rapid land conversion, and reliance on traditional farming practices (FAO, 2021).

In countries such as the United States, Canada, and those in the European Union, emissions from crop production are primarily driven by intensive mechanical monoculture farming and heavy reliance on chemical inputs (West and Marland, 2002). However, advancements in precision agriculture, controlled-release fertilizers, and conservation tillage have helped mitigate emissions to some extent (Lal, 2020; Meena et al., 2024c). Strict environmental regulations and C credit markets further drive emission reductions (Smith et al., 2008). Additionally, alternative fertilizers fertilize sequences, and digital farming tech minimizes climate impact (Paustian et al., 2016). Despite these efforts, large-scale agricultural operations in developed countries still contribute significantly to global emissions due to high fossil fuel use and intensive farming systems (Searchinger et al., 2018).

Developing nations such as India, Brazil, and several African countries exhibit a distinct emission profile shaped by small-scale farming, limited access to modern inputs, and traditional agricultural techniques (Ghosh et al. 2022). Scientists estimate that land-use changes, including slower transitions and lower land productivity, further exacerbate emissions, while practices such as open-field residue burning, deforestation for farmland expansion, and inadequate irrigation management significantly increase emissions (van der Werf et al., 2009). Moreover, rice cultivation, a dietary staple in many of these regions, is a key source of CH4 due to prolonged waterlogging and anaerobic decomposition in paddy fields (Yan et al., 2009). The challenge is compounded by financial limitations and weak regulatory enforcement, making the adoption of low-emission agricultural technologies difficult (FAO, 2021).

TABLE 3.3 GHG emissions from different crops in India.

Crop	GHG emissions (kg CO_2e/ha)	GHG emissions (kg CO_2e/kg yield)
Oryza sativa (Paddy/Rice)	7593.74	4.701
Saccharum officinarum (Sugarcane)	5083.94	0.061
Solanum tuberosum (Potato)	4210.67	0.185
Gossypium hirsutum (Cotton)	1656.79	1.123
Allium cepa (Onion)	1610.65	0.095
Corchorus olitorius (Jute)	1483.59	0.592
Sesamum indicum (Sesamum)	1416.62	2.127
Triticum aestivum (Wheat)	947.71	0.313
Zea mays (Maize)	797.50	0.315
Eleusine coracana (Ragi)	751.45	0.473
Pennisetum glaucum (Bajra)	636.50	0.377
Vigna radiata [Moong (Green gram)]	597.69	1.339
Hordeum vulgare (Barley)	561.18	0.180
Pisum sativum (Peas)	504.33	0.385
Sorghum bicolor (Jowar)	494.95	0.601
Brassica juncea (Mustard)	481.42	0.397
Cicer arietinum (Gram)	444.06	0.540
Cajanus cajan [Arhar (Red gram)]	422.91	0.513
Vigna mungo [Urad (Black gram)]	352.09	0.630
Arachis hypogaea (Groundnut)	344.04	0.306
Glycine max (Soyabean)	321.83	0.249
Cocos nucifera (Coconut)	363.34	0.00029
Lens culinaris [Masur (Lentil)]	260.25	0.331

Note: *CO_2e*, carbon dioxide equivalent; *ha*, hectare; *GHG*, greenhouse gas.
Source: (Pathak et al., 2016).

In India, GHG emissions from crop production fluctuate widely depending on cropping patterns, shaped by differences in farming methods, fertilizer, and irrigation efficiency. Table 3.3 highlights the emissions associated with major crops in India (Pathak et al., 2016).

Despite these hurdles, developing nations are increasingly adopting mitigation strategies to curb GHG emissions from agriculture. Integrated nutrient management (INM), along with the use of bio-fertilisers and organic amendments, enhances soil fertility while reducing N_2O emissions (Pathak et al., 2016). Additionally, sustainable intensification methods, such as the System of Rice Intensification (SRI) and alternate wetting and drying (AWD) in paddy cultivation, have shown promise in reducing CH_4 emissions (Richards and Sander, 2014). AF and conservation agriculture are also gaining traction as viable solutions for C_{seq} and emission control (Garrity et al., 2010). However, for widespread adoption, there is a need for greater investment in agricultural research, capacity building, and policy support (FAO, 2021).

Closing the technological and policy gap between developed and developing nations is essential for achieving meaningful global emission reductions. Knowledge transfer, international cooperation, and financial incentives for climate-smart agriculture can facilitate the transition of developing countries toward low-emission farming (Lipper et al., 2014). While developed nations must continue refining their precision agriculture techniques and Cseq efforts, developing nations require support in adopting efficient and sustainable farming practices (Paustian et al., 2016). A balanced approach that considers the socio-economic realities of both regions is essential for mitigating GHG emissions from crop production while ensuring food and nutritional security and agricultural resilience (Sarvade et al., 2020; Prasad and Jhariya, 2021; Jharia et al., 2021).

3.3 Animal husbandry practices and greenhouse gas emissions in developed and developing nations

The livestock sector is a major contributor to GHG emissions, with its impact varying between developed and developing nations. Annually, animal husbandry generates approximately 7.1 gigatons (Gt) of CO_2-equivalent emissions, accounting for roughly 14.5% of global human-induced emissions (Food and Agriculture Organization (FAO), 2013a). Key emission sources include enteric fermentation, manure handling, feed production, and land-use alterations (FAO, 2013a).

In developed nations, intensive livestock production systems enhance efficiency and productivity, thereby reducing GHG emissions per unit of output (FAO, 2013a). For example, dairy farming in Europe and North America emits approximately 1.6–1.9 kg CO_2-equivalent per kilogram of fat- and protein-corrected milk (FPCM) (FAO, 2022). Conversely, developing countries predominantly rely on extensive livestock systems, which generate higher per-unit emissions due to lower productivity and inefficient feed conversion (FAO, 2013a). In sub-Saharan Africa, for instance, dairy production emits an average of 6.5 kg CO_2-equivalent per kg FPCM, significantly exceeding emissions from developed regions (FAO, 2022).

GHG emissions from livestock vary across countries due to differences in livestock populations, management practices, and agricultural systems. Table 3.4 below summarizes emissions from livestock in India and other selected countries, based on published studies (Patra, 2012; Zhou et al., 2022; Nugrahaeningtyas et al., 2024).

3.4 Aquaculture practices and greenhouse gas emissions

Aquaculture, the cultivation of aquatic organisms such as mollusks, crustaceans, and aquatic plants, has also seen rapid global expansion, enhancing both food security and economic growth. However, its GHG emissions differ across nations due to variations in farming intensity, technological adoption, and production scale.

In 2017 aquaculture accounted for roughly 0.49% of global anthropogenic GHG emissions, comparable to emissions from sheep farming. This relatively low footprint is attributed to the absence of enteric CH_4 emissions, a major factor in terrestrial livestock, as well as the efficient feed conversion and high reproductive rates of finfish and shellfish (Gephart et al., 2020).

Developed countries often employ intensive aquaculture systems that depend on high-quality formulated feeds and advanced farming technologies. For instance, in China, the world's largest aquaculture producer, feed production is the

TABLE 3.4 Major livestock contributing to total GHG emissions (Compiled: Patra (2012); Zhou et al. (2022); Nugrahaeningtyas et al. (2024)).

Country	Total livestock GHG emissions (Mt CO_2-eq/year)	Major livestock contributors	Particulars
India	222.7 (2021)	Cattle	India has one of the world's highest livestock densities, with cattle predominantly raised for dairy production. However, the efficiency and productivity of Indian cattle are lower than those of other cattle globally.
	-	Cattle, Buffalo, and Goats	Enteric fermentation is responsible for 89.2% of total livestock-related GHG emissions, with contributions from various animal categories: dairy cattle (30.52%), nondairy cattle (24.0%), buffaloes (37.7%), goats (4.34%), sheep (2.09%), and other livestock (3.52%).
Global	-	Various livestock	Livestock contributes around 32% of global anthropogenic methane emissions, primarily originating from enteric fermentation (CH_4) and manure management (CH_4 and N_2O).

Notes: For India, the data indicate significant GHG emissions from livestock, priddairy cattle, and dairy products. Globally, livestock production contributes substantially to anthropogenic CH_4 emissions, highlighting the sector's impact on climate change. *GHG*, greenhouse gas; *CO_2e*, carbon dioxide equivalent.

most significant contributor to industry-related GHG emissions. A study indicated that intensive shrimp farming systems emit approximately 500 tons of CO_2 per hectare annually, with electricity consumption accounting for 82% and feed contributing 18% of total emissions (Xu et al., 2022). In contrast, developing countries often employ extensive or semi-intensive aquaculture systems, which are less reliant on external feed inputs and energy. For example, in Vietnam's Bac Lieu province, extensive shrimp farming practices have been associated with lower GHG emissions. Intervention strategies, including optimized techniques and enhanced water treatment systems, have proven effective in reducing emissions, achieving a 17% decline in extensive aquaculture systems and an 11% reduction in intensive systems within nine months (Vietnam Agriculture News, 2023).

Globally, aquaculture accounted for about 0.49% of anthropogenic GHG emissions, amounting to approximately 263 million tons of CO_2 equivalent (CO_2e) in 2017 (Gephart et al., 2020). The following table outlines GHG emissions from aquaculture across different categories, categorized by group.

Table 3.5 highlights the significant GHG contributions from regions such as East Asia and South Asia, where intensive rearing practices, particularly for cyprinids, shrimp, and prawns, are prevalent. Mitigation strategies such as optimization, energy-efficient aeration, and improved water quality management are crucial for reducing emissions across the sector.

3.5 Agrochemical use and global greenhouse gas emissions

Agrochemicals, particularly synthetic fertilizers and pesticides, are fundamental to modern agricultural systems, enhancing crop yields and protecting against pests and diseases. However, their production and application are major contributors to global GHG emissions, exacerbating climate change (Zhang et al., 2022; Reay et al., 2021). The synthetic nitrogen (fertilizer) supply chain alone generated approximately 1129.1 million tons of CO_2e emissions in 2018. These emissions originate from various stages, including manufacturing (438.5 Mt CO_2e), transportation (29.8 Mt CO_2e), direct N_2O emissions from soil (379.9 Mt CO_2e), CO_2 release from urea application (86.0 Mt CO_2e), and indirect losses (volatilization and leaching) (196.4 Mt CO_2e) (Zhang et al., 2022).

China led global fertilizer emissions with 316.1 Mt CO_2e, followed by India (165.5 Mt CO_2e), the United States (115.5 Mt CO_2e), and the European Union (102.4 Mt CO_2e). Regional disparities in C intensity were notable, with Latin America exhibiting nearly twice the global average, primarily due to fertilizer use and climate-related factors (Zhang et al., 2022).

The production of pesticides is also highly energy-intensive, with the manufacture of 1 kg of pesticide requiring, on average, approximately 10 times more energy than 1 kg of fertilizer (Pretty and Bharucha, 2018). This high energy demand results in substantial GHG emissions. Additionally, certain pesticides, such as soil fumigants, can significantly increase N_2O emissions from soils, further exacerbating their climate impact (Reay et al., 2021). Research indicates that chloropicrin, a commonly used soil fumigant, can increase soil N_2O emissions by up to seven times compared to untreated soils (Liu et al., 2020).

The continued rise in N_2O emissions, primarily from agricultural practices, poses a significant challenge to achieving global climate targets. The Global N_2O Assessment warns that, without decisive action, N_2O emissions could surge by 30% from 2020 levels by 2050, jeopardizing efforts to cap global warming at 1.5°C as per the Paris Agreement (Davidson and Kanter, 2021). In India, agrochemicals, especially synthetic fertilizers and pesticides, are vital for agricultural productivity but also represent a primary source of GHG fertilizer production and application release significant amounts of N_2O and CO_2, amplifying the sector's environmental footprint (Pathak, 2015; Meena and Pradhan, 2023). Table 3.6 details these emission contributions, highlighting the urgent need for sustainable nutrient management strategies.

3.5.1 Greenhouse gas emissions from pesticides

Though pesticide-related GHG emissions are lower than those from fertilizers, they still contribute significantly due to their energy-intensive production and application (Chandra and Pathak, 2018). Increased pesticide use in India, driven by climate change-induced pest infestations, has further contributed to agricultural emissions.

3.6 Altered land-use practices and global greenhouse gas emissions

3.6.1 Deforestation and carbon emissions

Deforestation is a primary driver of GHG emissions, as it releases huge quantities of CO_2 stored in perennials and soil. Historically, land-use changes, particularly deforestation, have contributed about one-third of CO_2 emissions since 1750 (Zhao and Jackson, 2020). Despite a decline in recent decades, emissions from deforestation remain a significant factor in global C budgets (Winrock International, 2017). According to the Global Carbon Project (2023), CO_2 emissions from land-use changes were approximately 3.2 Gt in 2022, underscoring the urgent need for mitigation strategies.

TABLE 3.5 GHG emissions from aquaculture by region and culture group (2017, × 1000 tons CO_2eq).

Region	Bivalves	Catfish	Cyprinids	Freshwater fish	Indian major carps	Marine fish	Salmonids	Shrimps/ Prawns	Tilapias	Total emissions
East Asia	16,775	13,018	70,264	13,468	0	20,695	0	43,782	15,319	1,93,319
South Asia	0	2788	3763	3144	12,743	0	0	5270	0	27,708
Sub-Saharan Africa	0	530	74	160	0	0	0	0	812	1576
West Asia & N. Africa	0	0	592	0	0	2203	263	0	3288	6346
Central & S. America	389	0	0	522	0	0	4215	2418	1017	8561
Oceania	126	0	0	0	0	215	133	0	0	474
Eastern Europe	0	0	176	0	0	0	49	0	0	225
Western Europe	639	0	0	0	0	0	4902	0	0	5542
North America	228	356	0	0	0	0	420	295	0	1299
Russian Federation	0	0	189	0	0	0	119	0	0	307
World Total	18,157	16,692	75,057	17,743	12,743	23,112	10,102	51,764	20436	2,45,357

Note: Values are in thousand tons of CO_2e for the year 2017. *GHG*, greenhouse gas; *CO_2e*, carbon dioxide equivalent.
Source: (Boyd et al., 2022).

TABLE 3.6 GHG emissions from agrochemical use in India (2014–2023).

Year	Total agricultural GHG emissions (Tg CO_2eFertilizer-related)	ted emissions (Tg CO_2eq)
2014	658	74.3
2016	670	78.1
2018	700	82.5
2020	720	86.0
2022	740	89.7
2023	750	91.2

Emissions of nitrous oxide (N_2O) from fertilizer use: Inorganic fertilizers are a foremost contributor to India's agricultural GHG emissions. In 2014 fertilizer use accounted for 73.29% of total N_2O emissions in the sector (Pathak, 2015). By 2023 N_2O emissions fertilizer application had increased due to rising fertilizer consumption, exacerbating India's total GHG emissions (Chachei, 2024; Meena et al., 2023). *GHG*, greenhouse gas; *CO_2e*, carbon dioxide equivalent.
Source: (Pathak, 2015; Chandra and Pathak, 2018; Chachei, 2024).

3.6.2 Urbanization and greenhouse gas emissions

Urban sprawl disrupts natural C sinks while amplifying emissions from construction activities and rising energy demands. A recent AI-driven analysis of 29 years of data confirmed urbanization is a key driver of escalating GHG emissions, emphasizing the urgency of sustainable urban development (Zhang et al., 2023). Similarly, the European Environment Agency (2023) highlighted that land-use changes driving urban expansion were associated with increased emissions across Europe, underscoring the widespread climate implications of unchecked urban growth.

Agricultural expansion and emissions expansion and emissions of forests and grasslands into farmland release CO_2 due to soil disruption, while CH4 and N_2O emissions arise from livestock activities and fertilizer application (Jackson, 2020; Pradhan and Meena, 2023a). In the European Union, the land use, land-use change, and forestry (LULUCF) sector functioned as both a C sink and a source in 2022, sequestering a net 236 million tonnes of CO2 equivalent while still contributing to emissions in specific areas (European Environment Agency, 2023).

3.6.3 Global trends in land-use emissions

From 1961 to 2017, emissions from land-use change increased by 24%, primarily due to deforestation, agricultural expansion, and infrastructure growth (Zhao and Jackson, 2020). This persistent rise highlights the challenge of balancing economic growth with ecological preservatio recognizinging this, the United Nations Framework Convention on Climate Change underscores the need for strategic land management policies that enhance land-use efficiency while aligning with global climate mitigation targets (United Nations Framework Convention on Climate Change UNFCCC, 2022).

3.7 Methodological assessment of greenhouse gas emissions in agroecosystems

Evaluating GHG emissions in agroecosystems is crucial for quantifying agriculture's climate impact and formulating practical mitigation approaches (Denmead, 2008). A range of measurement techniques has been developed, each tailored to specific applications, offering unique benefits while presenting certain limitations.

3.7.1 Chamber-based methods

Chamber-based techniques are commonly employed to quantify soil emissions of GHGs such as CO_2, CH_4, and N_2O (Minamikawa et al., 2015). These methods involve enclosing a soil sample in a chamber to trap gases for subsequent analysis.

3.7.1.1 Closed chambers

Available in static and dynamic designs, these chambers provide controlled flux measurements. Static chambers are an economical option but may disrupt natural gas gradients if left in place for extended periods (Denmead, 2008). In contrast, dynamic chambers maintain continuous airflow, thereby mitigating concentration buildup, but require more advanced instrumentation. Both must be carefully designed minimizeize soil disturbance and maintain measurement accuracy.

3.7.1.2 Open chambers

Unlike closed systems, open chambers facilitate uninterrupted gas exchange, preventing accumulation within the enclosure (Schlesinger and Bernhardt, 2013). While they offer continuous monitoring capabilities, their reliance on sophisticated equipment makes them more expensive and technically demanding to operate.

3.7.2 Micrometeorological techniques

Micrometeorological techniques enable the measurement of GHG fluxes over extensive areas without disturbing natural soil or vegetation conditions (Baldocchi et al., 1988). These methods are particularly valuable for capturing spatial and temporal variations in gas exchange at ecosystem scales.

3.7.2.1 Eddy covariance

This approach quantifies vertical turbulent fluxes in the atmospheric boundary layer, analyzing high-frequency wind and gas concentration data (Burba, 2013). It offers continuous, real-time monitoring across large landscapes but demands sophisticated instrumentation and complex data processing (Denmead, 2008).

3.7.2.2 Aerodynamic method

Gas fluxes are estimated by assessing wind-speed variations and gas-concentration gradients at different heights (Monteith and Unsworth, 2013). Since air density is influenced by heat and moisture fluxes, stability corrections are necessary to refine measurements.

3.7.2.3 Bowen ratio/energy balance method

This method calculates gas fluxes by determining the ratio between sensible and latent heat fluxes, relying on energy balance principles (Schlesinger and Bernhardt, 2013). However, fluctuating radiation levels—especially under cloudy conditions or at night—can compromise its accuracy.

3.7.3 Remote sensing

Remote sensing technologies, including satellite imagery and drone-mounted sensors, offer a nonintrusive approach to assessing GHG emissions over extensive landscapes (Del Grosso et al., 2001). These methods enable large-scale monitoring without direct field interventions, enhancing spatial coverage and temporal resolution. Satellite-based observations utilize spectral data and atmospheric gas measurements to estimate emissions at regional and global scales. Advanced sensors, such as those on the GOSAT and Sentinel satellites, track CO_2 and CH_4 concentrations with high precision. Meanwhile, drone-mounted sensors offer finer spatial resolution, enabling localizedized assessments of emission hotspots, agricultural fields, and wetlands. By integrating remote sensing with ground-based measurements, researchers can refine emission models, improve accuracy, and support climate mitigation strategies with real-time data insights. Satellites measure the intensity of sunlight reflection at specific wavelengths to detect near-surface concentrations of CO_2 and CH_4 (Monteith and Unsworth, 2013). Meanwhile, drones equipped with gas sensors capture high-resolution atmospheric data, facilitating targeted mitigation efforts.

3.7.4 Emerging techniques

Innovative approaches are being explored to refine GHG emission measurement. For instance, researchers employ walk-in chambers to monitor CH_4 emissions from livestock and identify forage types that reduce CH_4 production (Denmead, 2008). Such studies help mitigate agricultural CH_4 emissions, particularly in developing countries.

3.8 Research and development in agricultural sectors related to greenhouse gas emissions

Agriculture plays a pivotal role in global GHG emissions, accounting for approximately 24% of total emissions. Key contributors include enteric fermentation, fertilizer use, rice cultivation, and land-use alterations (Beach et al., 2008). As the sector strives to meet rising food demands, it simultaneously faces mounting pressure to mitigate its environmental impact.

This challenge has spurred significant investment in research and development (R&D) to create sustainable, low-emission farming practices. Comparing global and Indian R&D initiatives offers valuable insights into their efficacy, adaptability, and scalability. While global efforts focus on advanced precision agriculture, bioengineering, and CH_4-reducing livestock diets, Indian research emphasizes resource-efficient technologies like AWD in rice farming, INM, and AF-based C_{seq}. Evaluating these approaches can guide policy and innovation pathways for sustainable agricultural transitions worldwide.

3.8.1 Global versus Indian research and development initiatives in sustainable farming

Globally, sustainable farming practices have evolved through precision agriculture, conservation tillage, and advanced irrigation systems. In the United States, public R&D funding has facilitated the adoption of land-efficient farming techniques, reducing the need for deforestation-related expansion (Islam et al., 2020b). Similarly, in Asia, early-season drainage methods in rice fields have emerged as a crucial mitigation strategy, leading to significant reductions in CH_4 emissions (Islam et al., 2020a). In India, R&D efforts have increasingly focused on improving soil health and water management. Conservation agriculture techniques such as zero tillage and residue retention have demonstrated a 30%–50% reduction in emissions while enhancing long-term productivity (Jat et al., 2020). However, compared to developed nations, the large-scale adoption of precision agriculture remains limited due to cost constraints and fragmented landholding structures.

3.8.2 Livestock management—a comparative perspective

CH_4 emissions from livestock, primarily driven by enteric fermentation, present a significant challenge worldwide. In developed nations, innovative dietary strategies, such as incorporating seaweed-based feed additives, have demonstrated CH_4 reductions under integrated conditions (Beerling et al., 2020). Additionally, integrating AI-driven precision livestock farming, leveraging reduced monitoring and data analytics, has enhanced feed efficiency and reduced emissions (Wang et al., 2024).

In contrast, India, where livestock serves as an economic backbone for rural communities, has focused on more cost-effective mitigation strategies. Research on tannin-enriched fodder and probiotic-infused feed supplements has reported a reduction in CH_4 emissions of approximately 30% (Saleem et al., 2025). However, selective breeding initiatives targeting low-emission cattle breeds remain in theglobal advancesdersa disparity between India's progress and the advancements seen in global livestock management.

3.8.3 Methane reduction in rice cultivation—global and Indian strategies

Rice cultivation remains a major contributor to agricultural CH_4 emissions, primarily due to prolonged flooding in conventional paddy fields. On a global scale, innovative water management techniques such as AWD have emerged as effective mitigation strategies, achieving CH_4 reductions of nearly 48% (Gupta et al., 2021). Similarly, the SRI, which emphasizes optimized plant spacing and controlled irrigation, has been shown to lower CH_4 emissions by up to 35% while improving water efficiency (Mishra et al., 2021). In India, AWD adoption has been slower due to concerns over yield stability and water availability. However, localized trials indicate that with proper implementation, AWD can be equally effective in Indian agroecosystems. Research has also highlighted the role of biochar application in flooded fields in mitigating CH_4 emissions, although cost barriers remain a significant challenge (Beerling et al., 2020).

3.8.4 Fertilizer use and nitrous oxide emissions—a policy-driven approach

N-based fertilizers are a major driver of N_2O emissions globally, with their impact magnified by the gas's staggering global warming potential—approximately 300 times that of CO_2. Developed nations have prioritized precision agriculture solutions, including controlled-release fertilizers and nitrification inhibitors, which have been shown to reduce emissions by up to 40%. The European Union's stringent N regulations have further contributed to measurable declines in agricultural emissions. In contrast, India's heavy reliance on subsidized synthetic fertilizers presents ongoing challenges. Nevertheless, advancements in research and development have fostered the growing adoption of bio-fertilizers, organic soil amendments, and integrated nutrient management strategies, offering a promising pathway to curb emissions while sustaining soil health (Solangi and Magazzino, 2025; Pradhan et al., 2023).

3.8.5 Agroforestry and carbon sequestration—an opportunity for scale-up

AF has emerged as a vital approach for capturing atmospheric CO_2 while offering cobenefits such as biodiversity conservation and improved soil health. Research from Europe and North America indicates that tree-based agricultural systems can sequester up to 4.5 Mg CO_2 per hectare annually (Chlela and Selosse, 2025). In India, the promotion of AF has accelerated, supported by government policies and initiatives. Traditional models integrating fruit and timber trees with staple crops have shown significant Cseq potential (Singh et al., 2015; Chaturvedi et al., 2016; Sarvade et al., 2016a; Sarvade et al., 2016b; Jat et al., 2020). Despite these advancements, barriers such as restrictive land tenure policies and insufficient financial incentives remain key challenges to broader implementation.

3.8.6 Climate-resilient crops—addressing global and regional needs

The development of climate-resilient crop varieties is crucial for minimizing emissions linked to excessive irrigation and fertilizer use. In developed countries, genetic engineering and marker-assisted breeding have led to the commercialisation of drought-resistant and N-efficient crop strains (Pradhan and Meena, 2023b). In India, research institutions have focused on improving traditional staple crops such as rice and wheat. Recent studies show that enhanced rice varieties require 30% less water, reducing CH_4 emissions, while high-yield wheat varieties have led to low fertilizer requirements (Adjah et al., 2025). These efforts align with national food security goals while contributing to global emission reduction strategies.

3.8.7 Policy interventions—India's approach versus global frameworks

Globally, policies promoting C-neutral agriculture have gained traction. The United States and European nations have adopted C-pricing mechanisms that incentivise farmers to reduce emissions . Subsidies for sustainable agricultural practices and targeted research funding have played a crucial role in driving the success of climate-smart farming globally. India has made strides in this direction with initiatives such as the National Adaptation Fund for Climate Change, which supports climate-resilient farming techniques. Furthermore, the emerging C credit market offers a promising incentive for practices that reduce emissions (Bhatnagar et al., 2024). However, the widespread adoption of these practices faces significant hurdles, primarily due to financial constraints and infrastructural gaps that hinder large-scale implementation.

3.9 Policy perspectives for greenhouse gas reduction toward a green economy

Shifting toward a green economy requires robust policy frameworks designed to curb GHG emissions. Across the globe, including in India, governments are implementing a mix of regulatory measures, market-driven strategies, and technological advancements to move toward C neutrality. The Paris Agreement emphasizes the crucial role of policy interventions in limiting global temperature rise to below 2°C (Tanaka et al., 2020). While developed countries have capitalised on C pricing and advanced technology investments, India has prioritised large-scale renewable energy deployment, regulatory reforms, and targeted financial incentives to align sustainability with economic development.

3.9.1 Carbon pricing as a policy instrument

C pricing, which includes mechanisms such as C taxes and cap-and-trade systems, is widely regarded as an effective strategy for emission reduction. Studies suggest that a $10-per-ton increase in the C tax can lead to a 1.3% decline in per-capita CO_2 emissions in the short term, with reductions reaching up to 4.6% over the long term (Borenstein, 2018a). The European Union Emissions Trading System (EU ETS) has demonstrated significant success in limiting emissions within the industrial and energy sectors (Monbiot, 2024).

India's approach to C pricing is still developing. While no direct C tax exists, the Perform, Achieve, and Trade (PAT) scheme functions as an alternative market-based framework. Through PAT, energy-intensive industries must achieve mandated efficiency targets, with surplus achievers trading energy-saving certificates (Singh et al., 2023). Additionally, the coal cess introduced in 2010, later restructured as the GST compensation cess, acts as an implicit C pricing tool by discouraging coal consumption (Ghosh, 2022). Though India's C pricing remains indirect, its policies are progressively aligning with international market-driven approaches.

3.9.2 Regulatory approaches

Governments worldwide implement regulatory frameworks to curb GHG emissions, employing measures such as emission performance standards, renewable energy mandates, and vehicle efficiency norms. In the United States, the

Clean Power Plan has significantly curtailed emissions from coal-fired power plants (Niiler, 2024). Meanwhile, China's stringent industrial emission caps have contributed to a 12% reduction in CO_2 emissions across key sectors over the past decade (Monde, 2024).

India's regulatory landscape prioritizes the integration of renewable energy and the control of industrial emissions. The Renewable Energy Purchase Obligation requires power utilities to source a designated portion of their electricity from renewable sources, thereby accelerating the adoption of solar and wind energy (Sharma and Verma, 2023). In the transportation sector, the Bharat Stage VI emission norms, introduced in 2020, align with European standards and have played a pivotal role in curbing vehicular pollution (Gupta, 2023). Additionally, India's Energy Conservation Building Code mandates energy-efficiency measures for commercial buildings, reinforcing the nation's commitment to sustainable development .

3.9.3 Financial incentives and green subsidies

Financial incentives serve as powerful catalysts in steering economies toward low-C development by fostering investments in clean energy and sustainable technologies. The United States has taken a bold approach with the Inflation Reduction Act of 2022, which offers significant tax credits for electric vehicles and renewable energy projects, with estimates suggesting a potential 40% reduction in national emissions by 2030 (Bistline et al., 2023). Germany's Renewable Energy Act (EEG) has similarly played a transformative role in scaling up wind and solar power, cementing the nation's leadership in green energy (Zhang, 2023).

India has strategically implemented financial mechanisms to drive its clean energy transition. The Production-Linked Incentive (PLI) scheme incentivizes domestic manufacturing of high-efficiency solar modules, aiming to reduce reliance on imports while bolstering the renewable sector . To accelerate electric vehicle adoption, the Faster Adoption and Manufacturing of Hybrid and Electric Vehicles (FAME) initiative provides targeted subsidies, making EVs more accessible (Mukherjee, 2023). Additionally, the viability gap funding program bridges financial shortfalls in solar and wind energy projects, ensuring their economic viability and long-term sustainability (Solangi and Magazzino, 2025).

3.9.4 Role of green investments and research and development

Investing in green technology R&D is a cornerstone of sustainable progress, with nations dedicating at least 2% of their GDP to climate-focused R&D, witnessing significant declines in C intensity (Bistline et al., 2024). Public–private collaborations in carbon capture and storage have emerged as a viable strategy for curbing industrial emissions, underscoring the role of innovation in achieving climate goals (Tanaka et al., 2020).

India's green investment landscape is expanding rapidly, with the National Green Hydrogen Mission spearheading R&D efforts in hydrogen-based energy solutions (Ghosh, 2023). The Indian Renewable Energy Development Agency has played a crucial role in facilitating low-interest financing for renewable projects, thereby strengthening investments in solar, wind, and bioenergy . Additionally, the country's ambitious space-based solar power initiative is exploring cutting-edge approaches to sustainable energy generation from orbit (Kumbhar et al., 2026). While India's overall climate R&D expenditure still lags behind global frontrunners, recent policy-driven advancements reflect a growing commitment to pioneering green innovation.

3.9.5 Equitable and just transition policies

Effective climate policies must incorporate social equity to protect vulnerable populations from the unintended economic burdens of environmental transitions. Without careful structuring, C pricing mechanisms, such as C taxes, risk placing a heavier financial strain on low-income groups (Bistline et al., 2024). Globally, revenue redistribution strategies—where C tax proceeds are funnelled back to households—have proven effective in mitigating these disparities (Monde, 2024).

India's approach to equitable climate action encompasses targeted initiatives that strike a balance between sustainability and socio-economic welfare. The Unnat Jyoti by Affordable LEDs for All program has facilitated the distribution of over 360 million LED bulbs, significantly reducing energy costs and emissions, particularly benefiting lower-income households (Gupta, 2023). Meanwhile, the Mahatma Gandhi National Rural Employment Guarantee Act has integrated climate-resilient employment schemes, ensured financial security, and fostered ecological sustainability . Additionally, India's ethanol blending policy not only curbs dependence on fossil fuels but also provides economic opportunities for farmers by creating a steady market for biofuel crops (Sharma and Verma, 2023). These policies exemplify how climate strategies can be designed to drive both environmental and social progress.

3.10 Conclusion

The changeover to a green economy requires a balanced integration of policy mechanisms that address GHG emissions while fostering economic growth. A comparative perspective reveals that while developed nations rely heavily on C pricing, stringent regulations, and investments in advanced green technology, India has focused on expanding renewable energy, offering financial incentives, and implementing equitable policy frameworks. Regulatory measures such as the EU ETS and India's PAT scheme exemplify distinct approaches to market-driven emission reductions. Simultaneously, subsidies such as EEG and India's PLI scheme for solar manufacturing underscore the role of financial interventions in driving sustainable development. To bridge the gap, India must enhance its C pricing mechanisms, increase investments in climate-focused research and development, and strengthen policy enforcement across industries. The success of India's National Green Hydrogen Mission and the Faster Adoption and Manufacturing of Electric Vehicles scheme underscores the potential of targeted policies in achieving low-C transitions. However, a just transition must prioritize social equity, ensuring that vulnerable populations benefit from green economic policies. Strengthening international collaborations, aligning policy frameworks with global best practices, and fostering public–private partnerships will be critical in accelerating India's journey toward C neutrality. By integrating robust policy frameworks with sustainable technological innovations, India can emerge as a global leader in climate action while securing long-term economic resilience.

Acknowledgment

The authors express their sincere gratitude to the professor and head of the Department of Forestry, as well as the Dean of the College of Agriculture, Jawaharlal Nehru Krishi Vishwa Vidyalaya, Jabalpur, for their invaluable support and insightful guidance throughout this work.

References

Adjah, K.L., Semwal, V.K., Amoah, N.K.A., Tawiah, I., Zenna, N., Elnamaky, R., Futakuchi, K., Dossou-Yovo, E.R., Yadav, S., 2025. Designing Climate-Resilient Rice Production Systems: Leveraging Genomics for Low-Emission Rice Varieties. Rice Science 32, 813–830. Available at https://doi.org/10.1016/j.rsci.2025.08.003.

Baldocchi, D.D., Hicks, B.B., Meyers, T.P., 1988. Measuring biosphere-atmosphere exchanges of biologically related gases with micrometeorological methods. Ecology 69 (5), 1331–1340. Available at https://doi.org/10.2307/1941631.

Banerjee, A., Jhariya, M.K., Raj, A., Mechergui, T., 2024. Ecosystem management: climate change and global sustainability—an introduction. Ecosystem Management: Climate Change and Sustainability. wiley, India, pp. 1–21. Available at https://doi.org/10.1002/9781394231249.ch1.

Beach, R.H., DeAngelo, B.J., Rose, S., Li, C., Salas, W., DelGrosso, S.J., 2008. Mitigation potential and costs for global agricultural greenhouse gas emissions. Agricultural Economics 38 (2), 109–115. Available at https://doi.org/10.1111/j.1574-0862.2008.00286.x.

Beerling, D.J., Kantzas, E.P., Lomas, M.R., Wade, P., Eufrasio, R.M., Renforth, P., et al., 2020. Potential for large-scale CO_2 removal via enhanced rock weathering with croplands. Nature 583 (7815), 242–248. Available at https://doi.org/10.1038/s41586-020-2448-9.

Bhatnagar, S., Chaudhary, R., Sharma, S., Janjhua, Y., Thakur, P., Sharma, P., Keprate, A., 2024. Exploring the dynamics of climate-smart agricultural practices for sustainable resilience in a changing climate. Environmental and Sustainability Indicators 24, 100535. Available at https://doi.org/10.1016/j.indic.2024.100535.

Bistline, J., Blanford, G., Brown, M., Burtraw, D., Domeshek, M., Farbes, J., et al., 2023. Emissions and energy impacts of the inflation reduction act. Science (New York, N.Y.) 380 (6652), 1324–1327. Available at https://doi.org/10.1126/science.adg3781.

Bistline, J., Onda, C., Browning, M., Emmerling, J., Iyer, G., Mahajan, M., et al., 2024. Equity implications of net-zero emissions: a multi-model analysis of energy expenditures across income classes under economy-wide decarbonizationation policies. Energy and Climate Change 5. Available at https://doi.org/10.1016/j.egycc.2023.100118.

Borenstein, S., 2018a. Carbon pricing and market-based climate policies. Review of Environmental Economics and Policy 12 (3), 349–372.

Borenstein, S., 2018b. The cost of reducing greenhouse gas emissions. Journal of Economic Perspectives 32 (4), 53–72.

Borrelli, P., Robinson, D.A., Panagos, P., Lugato, E., Yang, J.E., Alewell, C., et al., 2020. Land use and climate change impacts on global soil erosion by water (2015-2070). Proceedings of the National Academy of Sciences of the United States of America 117 (36), 21994–22001. Available at https://doi.org/10.1073/pnas.2001403117.

Boyd, C.E., D'Abramo, L.R., Glencross, B.D., Hargreaves, J.A., Kaushik, S.J., 2022. Environmental Best Management Practices for Aquaculture. Wiley-Blackwell.

Burba, G., 2013. Eddy Covariance Method for Scientific, Industrial, Agricultural and Regulatory Applications: A Field Book on Measuring Ecosystem Gas Exchange and Areal Emission Rates. LI-COR Biosciences.

Chachei, K., 2024. Greenhouse gas emissions in the Indian agriculture sector and mitigation by best management practices and smart farming technologies—a review. Environmental Science and Pollution Research 31 (32), 44489–44510. Available at https://doi.org/10.1007/s11356-024-33975-7.

Chandra, N., Pathak, H., 2018. The carbon footprint of agricultural crop cultivation in India. Carbon Management 9 (3), 233–244. Available at https://doi.org/10.1080/17583004.2018.1457908.

Chaturvedi, O.P., Handa, A.K., Kaushal, R., Uthappa, A.R., Sarvade, S., Panwar, P., 2016. Biomass production and carbon sequestration through agroforestry. Range Management and Agroforestry 37 (2), 116–127.

Clark, M.A., Domingo, N.G.G., Colgan, K., Thakrar, S.K., Tilman, D., Lynch, J., et al., 2020. Global food system emissions could preclude achieving the 1.5° and 2°C climate change targets. Science (New York, N.Y.) 370 (6517), 705–708. Available at https://doi.org/10.1126/science.aba7357.

Chlela, S., Selosse, S., 2025. The co-benefits of integrating carbon dioxide removal in the energy system: A review from the prism of natural climate solutions. Science of The Total Environment 976, 179271. Available at https://doi.org/10.1016/j.scitotenv.2025.179271.

Copernicus Climate Change Service, 2024. European State of the Climate 2023. Available at https://climate.copernicus.eu/esotc/2023/greenhouse-gas-concentrations.

Crippa, M., Oreggioni, G., Guizzardi, D., Muntean, M., Schaaf, E., Lo Vullo, E., et al., 2023. EDGAR v7.0 Greenhouse Gas Emissions. European Commission, Joint Research Centre (JRC). Available at https://edgar.jrc.ec.europa.eu.

Davidson, E.A., Kanter, D., 2021. Nitrous oxide: a growing climate threat. Nature Climate Change 11 (8), 562–563. Available at https://doi.org/10.1038/s41558-021-01055-9.

Denmead, O.T., 2008. Approaches to measuring fluxes of methane and nitrous oxide between landscapes and the atmosphere. Plant and Soil 309 (1-2), 5–24. Available at https://doi.org/10.1007/s11104-008-9599-z.

European Commission, 2022. EU Emissions Trading System (EU ETS). Available at https://ec.europa.eu/clima/policies/ets_en (accessed 08.03.25).

European Environment Agency, 2023. Greenhouse gas emissions from land use, land use change, and forestry. Available at eea.europa.eu.

FAO, 2013a. Climate-Smart Agriculture: Sourcebook. Food and Agriculture Organization of the United Nations, Rome, Italy.

FAO, 2013b. Tackling Climate Change Through Livestock: A Global Assessment of Emissions and Mitigation Opportunities. Food and Agriculture Organization, Rome, Italy.

FAO, 2021. The State of Food and Agriculture 2021: Making Agri-food Systems More Resilient to Shocks and Stresses. Food and Agriculture Organization, Rome, Italy.

FAO, 2022. Climate-smart Agriculture Sourcebook. Food and Agriculture Organization, Rome, Italy.

Garrity, D.P., Akinnifesi, F.K., Ajayi, O.C., Weldesemayat, S.G., Mowo, J.G., Kalinganire, A., et al., 2010. Evergreen agriculture: a robust approach to sustainable food security in Africa. Food Security 2 (3), 197–214. Available at https://doi.org/10.1007/s12571-010-0070-7.

Gephart, J.A., Froehlich, H.E., Branch, T.A., 2020. Quantifying greenhouse gas emissions from global aquaculture. Scientific Reports 10 (1), 11679. Available at https://doi.org/10.1038/s41598-020-68231-8.

Gerber, P.J., Steinfeld, H., Henderson, B., Motter, A., Opio, C., Dijkman, J., et al., 2013. Tackling Climate Change Through Livestock: A Global Assessment of Emissions and Mitigation Opportunities. Food and Agriculture Organization, Rome, Italy, pp. 109.

Ghosh, A., 2022. India's carbon pricing strategies and coal cess framework. Economic & Political Weekly 57 (13), 92–105.

Ghosh, A., 2023. National green hydrogen mission: India's roadmap to sustainable energy. Renewable Energy Review 28 (2), 134–150.

Ghosh, S., Sarkar, S., Dutta, S., 2022. Agriculture and greenhouse gas emissions: a systematic review of mitigation strategies. Journal of Environmental Management 303, 114104.

Global Carbon Project, 2023. Annual CO_2 emissions from land-use change. Available at ourworldindata.org.

Del Grosso, S.J., Parton, W.J., Mosier, A.R., Hartman, M.D., Keough, C.A., Peterson, G.A., et al., 2001. Simulated effects of land use, soil texture, and precipitation on N gas emissions using DAYCENT. In: Follett, R.F., Hatfield, J.L. (Eds.), Nitrogen in the Environment: Sources, Problems, and Management Elsevier Science. Publishers, pp. 413–431.

Gupta, S., 2023. Energy efficiency and climate resilience in India. Sustainable Development Reports 31 (2), 178–194.

Gupta, K., Kumar, R., Baruah, K.K., Hazarika, S., Karmakar, S., Bordoloi, N., 2021. Greenhouse gas emission from rice fields: a review from Indian context. Environmental Science and Pollution Research 28, 30551–30572. Available at https://doi.org/10.1007/s11356-021-13935-1.

IPCC, 2019. Climate Change and Land: An IPCC Special Report on Climate Change, Desertification, Land Degradation, Sustainable Land Management, Food Security, and Greenhouse Gas Fluxes in Terrestrial Ecosystems. Intergovernmental Panel on Climate Change.

IPCC, 2021. Climate Change 2021: The Physical Science Basis. Contribution of Working Group I to the Sixth Assessment Report of the Intergovernmental Panel on Climate Change. Cambridge University Press.

Islam, S.Fu, de Neergaard, A., Sander, B.O., Jensen, L.S., Wassmann, R., van Groenigen, J.W., 2020a. Reducing greenhouse gas emissions and grain arsenic and lead levels without compromising yield in organically produced rice. Agriculture, Ecosystems and Environment 295. Available at https://doi.org/10.1016/j.agee.2020.106922.

slam, S.F.U., van Groenigen, J.W., Jensen, L.S., Sander, B.O., de Neergaard, A., 2020b. The effective mitigation of greenhouse gas emissions from rice paddies without compromising yield by early-season drainage. arXiv, Netherlands arXiv. Available at https://doi.org/10.48550/arxiv.2006.02917.

Jhariya, M.K., Banerjee, A., Meena, R.S., Kumar, S., Raj, A., 2021a. Sustainable Intensification for Agroecosystem Services and Management. Springer, Singapore, pp. 1–870. Available at https://doi.org/10.1007/978-981-16-3207-5.

Jhariya, M.K., Meena, R.S., Banerjee, A., 2021b. Ecological Intensification of Natural Resources for Sustainable Agriculture. Springer, Singapore, pp. 1–655. Available at https://doi.org/10.1007/978-981-33-4203-3.

Jhariya, M.K., Meena, R.S., Banerjee, A., Meena, S.N., 2022. Natural Resources Conservation and Advances for Sustainability. Elsevier. Available at https://doi.org/10.1016/C2019-0-03763-6.

Jat, M.L., Chakraborty, D., Ladha, J.K., Rana, D.S., Gathala, M.K., McDonald, A., Gerard, B., 2020. Conservation agriculture for sustainable intensification in South Asia. Nature Sustainability 3, 336–343. Available at https://doi.org/10.1038/s41893-020-0500-2.

Jhariya, M.K., Meena, R.S., Banerjee, A., Kumar, S., Raj, A., 2023. Agroforestry for Carbon and Ecosystem Management Agroforestry for Carbon and Ecosystem Management. Elsevier, India, pp. 1–434. Available at https://doi.org/10.1016/C2021-0-02916-3.

Kumar, S., Meena, R.S., Sheoran, P., Jhariya, M.K., 2024. Regenerative Agriculture for Sustainable Food Systems Regenerative Agriculture for Sustainable Food Systems. Springer Science+Business Media, India, pp. 1–493. Available at https://doi.org/10.1007/978-981-97-6691-8.

Kumbhar, Sanjay R., Satpute, Sanjay T., Patil, Yogesh S., 2026. Harnessing green power: A comprehensive analysis of India's renewable energy growth and future outlook. Next Energy 10, 100493. Available at https://doi.org/10.1016/j.nxener.2025.100493.

Lal, R., 2020. Regenerative agriculture for food and climate. Journal of Soil and Water Conservation 75 (5). Available at https://doi.org/10.2489/jswc.2020.0620a.

Lipper, L., McCarthy, N., Zilberman, D., Asfaw, S., Branca, G., 2014. Climate-smart agriculture. Natural Resource Management and Policy 52, 1–29.

Liu, S., Linquist, B.A., Brookes, P.C., 2020. Effects of fumigation on nitrous oxide emissions in agricultural soils: a meta-analysis. Soil Biology and Biochemistry 140, 107644. Available at https://doi.org/10.1016/j.soilbio.2020.107644.

Meena, R.S., Pradhan, G., Kumar, S., Lal, R., 2023. Using industrial wastes for rice-wheat cropping and food-energy-carbon-water-economic nexus to the sustainable food system. Renewable and Sustainable Energy Reviews 187, 113756. Available at https://doi.org/10.1016/j.rser.2023.113756.

Meena, R.S., Pradhan, G., 2023. Industrial garbage-derived biocompost enhances soil organic carbon fractions, CO_2 biosequestration, potential carbon credits and sustainability index in a rice-wheat ecosystem. Environmental Research 235, 116525. Available at https://doi.org/10.1016/j.envres.2023.116525.

Meena, R.S., Pradhan, G., Nalani, Kumar, S., Singh, A.K., Verma, S.K., et al., 2024a. Diversified cropping modules designed for soil restoration, CO_2 sequestration, and generating carbon credits. Land Degradation and Development 35 (14), 4314–4329. Available at https://doi.org/10.1002/ldr.5224.

Meena, R.S., Pradhan, G., Singh, K., Kumar, S., Singh, A.K., Shashidhar, K.S., et al., 2024b. Agriculture models for restoring degraded land to enhance CO_2 biosequestration and carbon credits in the Vindhyan region of India. Science of the Total Environment 929. Available at https://doi.org/10.1016/j.scitotenv.2024.172661.

Meena, R.S., Lal, R., Kumar, S., Pradhan, G., Srinivasarao, C., Singh, A.K., et al., 2024c. , Potential of Indian agriculture for capturing atmospheric CO_2monetizingizing carbon credits to the farmers: an overview and policy framework. Advances in Agronomy 188, 101–206. Available at https://doi.org/10.1016/bs.agron.2024.06.001.

Meena, R.S., Pradhan, G.S., Kumar, S., Jhariya, M.K., Rao, C.S., 2025a. Carbon-negative Agriculture. Springer Nature, Singapore.

Meena, R.S., Pradhan, G., Kumar, S., 2025b. Energy flow, eco-efficiency, and economic circulation with recycled industrial waste compost application in wheat and subsequent rice farming. Science of the Total Environment 967. Available at https://doi.org/10.1016/j.scitotenv.2025.178779.

Minamikawa, K., Yagi, K., Tokida, T., 2015. Field measurement of greenhouse gases: an introduction. In Measuring Emission of Agricultural Greenhouse Gases and Developing Mitigation Options using Nuclear and Related Techniques Springer. pp. 1-13.

Mishra, J.S., Poonia, S.P., Kumar, Rakesh, Dubey, Rachana, Kumar, Virender, Mondal, Surajit, Dwivedi, S.K., Rao, K.K., Kumar, Rahul, Tamta, Manisha, Verma, Mausam, Saurabh, Kirti, Kumar, Santosh, Bhatt, B.P., Malik, R.K., McDonald, Andrew, Bhaskar, S., 2021. An impact of agronomic practices of sustainable rice-wheat crop intensification on food security, economic adaptability, and environmental mitigation across eastern Indo-Gangetic Plains. Field Crops Research 267, 108164. Available at https://doi.org/10.1016/j.fcr.2021.108164.

Monbiot, G., 2024. Out of 1,500 Global Climate Policies, only 63 have Really Worked. The Guardian.

Monde, L., 2024. Industrial emission caps and climate policies in China. Climate Watch 36 (1), 12–23.

Monteith, J.L., Unsworth, M.H., 2013. Principles of Environmental Physics: Plants, Animals, and the Atmosphere, fourth ed. Academic Press.

Mukherjee, R., 2023. Policy impact of India's FAME scheme on EV adoption. Transportation Research Journal 51 (3), 307–322.

Niiler, E., 2024. The U.S. clean power plan and emission reductions. Science & Policy Today 17 (4), 211–228.

Nugrahaeningtyas, E., Lee, J.-S., Park, K.-H., 2024. Greenhouse gas emissions from livestock: sources, estimation, and mitigation. Journal of Animal Science and Technology 66 (6), 1083–1098. Available at https://doi.org/10.5187/jast.2024.e86.

Pathak, H., 2015. Greenhouse gas emission and mitigation in Indian agriculture: trends, drivers and mitigation strategies. Proceedings of the Indian National Science Academy 81 (5), 1133–1149. Available at https://doi.org/10.16943/ptinsa/2015/v81i5/48333.

Pathak, H., Jain, N., Bhatia, A., Patel, J., Aggarwal, P.K., 2016. Carbon footprints of Indian food items. Agriculture, Ecosystems & Environment 226, 132–141.

Patra, A.K., 2012. Estimation of methane and nitrous oxide emissions from Indian livestock. Journal of Environmental Monitoring 14 (10), 2673–2684. Available at https://doi.org/10.1039/c2em30396e.

Paustian, K., Lehmann, J., Ogle, S., Reay, D., Robertson, G.P., Smith, P., 2016. Climate-smart soils. Nature 532 (7597), 49–57. Available at https://doi.org/10.1038/nature17174.

Pradhan, G., Meena, R.S., 2023a. Interaction impact of biocompost on nutrient dynamics and relations with soil biota, carbon fractions index, societal value of CO_2 equivalent and ecosystem services in the wheat-rice farming. Chemosphere 339. Available at https://doi.org/10.1016/j.chemosphere.2023.139695.

Pradhan, G., Meena, R.S., 2023b. Utilizingizing waste compost to improve the atmospheric CO_2 capturing in the rice-wheat cropping system and energy-cum-carbon credit auditing for a circular economy. Science of the Total Environment 892. Available at https://doi.org/10.1016/j.scitotenv.2023.164572.

Prasad, R., Jhariya, M.K., Banerje, A., 2021. Advances in Sustainable Development and Management of Environmental and Natural Resources: Economic Outlook and Opinions: Volume 1. pp. 1–428. Available at https://doi.org/10.1201/9781003187455.

Pradhan, G., Meena, R.S., Kumar, S., Lal, R., 2023. Utilizingizing industrial wastes as compost in wheat-rice production to improve the above and below-ground ecosystem services. Agriculture, Ecosystems & Environment 358, 108704. Available at https://doi.org/10.1016/j.agee.2023.108704.

Pretty, J., Bharucha, Z.P., 2018. Sustainable intensification in agricultural systems. Annals of Botany 114 (3), 157–172. Available at https://doi.org/10.1093/aob/mcx138.

Raj, A., Jhariya, M.K., Meena, R.S., 2025. Agroforestry for Monetising Carbon Credits. Springer Nature, Switzerland.

Reay, D.S., Davidson, E.A., Smith, K.A., Smith, P., Melillo, J.M., Dentener, F., et al., 2012. Global agriculture and nitrous oxide emissions. Nature Climate Change 2 (6), 410–416. Available at https://doi.org/10.1038/nclimate1458.

Reay, D.S., Davidson, E.A., Smith, K.A., Smith, P., Melillo, J.M., Dentener, F., et al., 2021. Global agriculture and nitrous oxide emissions. Nature Climate Change 3 (5), 315–321. Available at https://doi.org/10.1038/nclimate1694.

Reuters, 2024. Brazilian farmers adopting regenerative agriculture to combat climate change. Available at https://www.reuters.com (accessed 08.03.25).

Richards, M., Sander, B.O., 2014. Alternate wetting and drying in irrigated rice: Implementation guidance for policymakers and investors. CGIAR Research Program on Climate Change, Agriculture and Food Security (CCAFS).

Saleem, A.S.A., Al-Marakby, K.M., Elaref, M.Y., Bassiony, S.M., Helal, A.A., Abdel-Monem, U.M., Abdelnour, S.A., 2025. Effects of quadric probiotic blends on rumen fermentation, nutrient degradability, and methane emission in sheep: an in vitro study. AMB Express 15. Available at https://doi.org/10.1186/s13568-025-01955-w.

Sarvade, S., Singh, R., 2014. Role of agroforestry in food security. Popular Kheti 2, 25–29.

Sarvade, S., Singh, R., Vikas, G., Kachawaya, D.S., Khachi, B., 2014. Agroforestry: an approach for food security. Indian Journal of Ecology 41 (1), 95–98.

Sarvade, S., Gupta, B., Singh, M., 2016a. Composition, diversity and distribution of tree species in response to changing soil properties with increasing distance from water source—a case study of Gobind Sagar Reservoir in India. Journal of Mountain Science 13 (3), 522–533. Available at https://doi.org/10.1007/s11629-015-3493-y.

Sarvade, S., Gupta, B., Singh, M., 2016b. Soil carbon storage potential of different land use systems in upstream catchment area of Gobind Sagar reservoir, Himachal Pradesh. Indian Journal of Soil Conservation 44 (2), 112–119.

Sarvade, S., Gautam, D.S., Upadhyay, V.B., Sahu, R.K., Shrivastava, A.K., Kaushal, R., et al., 2019. Agroforestry and soil health: an overview. In: Inder, D., Asha, R., Kumar, N., Singh, R., Kumar, D., Uthappa, A.R., Handa, A.K., Chaturvedi, O.P. (Eds.), Agroforestry for Climate Resilience and Rural Livelihood. Scientific Publishers, Jodhpur, Rajasthan, India, pp. 275–297.

Sarvade, S., Shrivastava, A.K., Rai, S.K., Bisen, S., Bisen, U., Bisen, N.K., et al., 2020. Socio-economic study of farming communities, their knowledge on climate change and agroforestry systems in the cluster of villages of Chhattisgarh Plain region, Madhya Pradesh. Journal of Pharmacognosy and Phytochemistry 9 (1), 2158–2166.

Schlesinger, W.H., Bernhardt, E.S., 2013. Biogeochemistry: An Analysis of Global Change, third ed. Academic Press.

Scott, D., Hall, C.M., Rushton, B., Gössling, S., 2023. A review of the IPCC Sixth Assessment and implications for tourism development and sectoral climate action. Journal of Sustainable Tourism 32 (9), 1725–1742. Available at https://doi.org/10.1080/09669582.2023.2195597.

Searchinger, T.D., Wirsenius, S., Beringer, T., Dumas, P., 2018. Assessing the efficiency of changes in land use for mitigating climate change. Nature 564 (7735), 249–253. Available at https://doi.org/10.1038/s41586-018-0757-z.

Sharma, A., Verma, K., 2023. Renewable energy policies and emission mitigation in India. Energy Studies Journal 39 (1), 67–89.

Singh, M., Gupta, B., Sarvade, S., Awasthe, R.K., 2015. Biomass and carbon sequestration potential in different agroforestry systems in Giri cathment of North Western Indian Himalaya. Indian Journal of Agroforestry 17 (2), 42–48.

Smith, P., Martino, D., Cai, Z., Gwary, D., Janzen, H., Kumar, P., et al., 2007. Agriculture. In: Metz, B., Davidson, O.R., Bosch, P.R., Dave, R., Meyer, L.A. (Eds.), Climate Change 2007: Mitigation of Climate Change. Contribution of Working Group III to the Fourth Assessment Report of the Intergovernmental Panel on Climate Change. Cambridge University Press, Cambridge, UK.

Smith, P., Martino, D., Cai, Z., Gwary, D., Janzen, H., Kumar, P., et al., 2008. Greenhouse gas mitigation in agriculture. Philosophical Transactions of the Royal Society B: Biological Sciences 363 (1492), 789–813. Available at https://doi.org/10.1098/rstb.2007.2184.

Solangi, Y.A., Magazzino, C., 2025. Evaluating financial implications of renewable energy for climate action and sustainable development goals. Renewable and Sustainable Energy Reviews 212, 115390. Available at https://doi.org/10.1016/j.rser.2025.115390.

Anaka, K., Boucher, O., Ciais, P., Johansson, D.J.A., Morfeldt, J., 2020. Cost-effective implementation of the Paris agreement using flexible greenhouse gas metrics. arXiv, France arXiv. Available at https://doi.org/10.48550/arxiv.2006.01901.

The Global Economy, 2023. India: Greenhouse gas emissions. Available at https://www.theglobaleconomy.com/India/greenhouse_gas_emissions.

Tilman, D., Balzer, C., Hill, J., Befort, B.L., 2011. Global food demand and the sustainable intensification of agriculture. Proceedings of the National Academy of Sciences 108 (50), 20260–20264. Available at https://doi.org/10.1073/pnas.1116437108.

United Nations Framework Convention on Climate Change (UNFCCC), 2022. Introduction to Land Use. Retrieved from unfccc.int.

Vietnam Agriculture News, 2023. Greenhouse gas emissions from extensive shrimp farming ponds decreased by 17%. Vietnam Agriculture News.

Wang, Z., Xiao, S., Wang, J., Parab, A., Patel, S., 2024. Intelligent agricultural management considering N_2O emission and climate variability with uncertainties. arXiv, United States arXiv. Available at https://doi.org/10.48550/arXiv.2402.08832.

van der Werf, G.R., Morton, D.C., DeFries, R.S., Olivier, J.G.J., Kasibhatla, P.S., Jackson, R.B., et al., 2009. CO_2 emissions from forest loss. Nature Geoscience 2 (11), 737–738. Available at https://doi.org/10.1038/ngeo671.

West, T.O., Marland, G., 2002. Net carbon flux from agricultural ecosystems: methodology for full carbon cycle analyses. Environmental Pollution 116 (3), 439–444. Available at https://doi.org/10.1016/s0269-7491(01)00221-4.

Winrock International, 2017. Greenhouse gas emissions from land use and land use change. Available at winrock.org.

World Meteorological Organization, 2024. WMO Greenhouse Gas Bulletin: The State of Greenhouse Gases in the Atmosphere Based on Global Observations through 2023. Available from https://public.wmo.int/en/resources/library/wmo-greenhouse-gas-bulletin.

Xu, C., Su, G., Zhao, K., Xu, X., Li, Z., Hu, Q., et al., 2022. Current status of greenhouse gas emissions from aquaculture in China. Water Biology and Security 1 (3), 100041. Available at https://doi.org/10.1016/j.watbs.2022.100041.

Yan, X., Akiyama, H., Yagi, K., Akimoto, H., 2009. Global estimations of the inventory and mitigation potential of methane emissions from rice cultivation conducted using the 2006 IPCC Guidelines. Global Biogeochemical Cycles 23 (2), GB2002.

Zhang, H., 2023. Leveraging policy instruments and financial incentives to reduce embodied carbon in energy retrofits. arXiv, Canada arXiv. Available at https://doi.org/10.48550/arXiv.2304.03403.

Zhang, X., Zhang, Y., Cui, Z., Chen, X., Ju, X., 2022. Nitrfertilizerlizer supply chain emissions and climate change mitigation potential. Environmental Research Letters 17 (4), 045005. Available at https://doi.org/10.1088/1748-9326/ac5b23.

Zhang, Y., Li, C., Wang, J., 2023. Global land use change and its impact on greenhouse gas emissions: evidence from 29 years of data and ai modeling. Environmental Research Letters 18 (4), 045003.

Zhao, K., Jackson, R.B., 2020. Global and regional drivers of land-use emissions in 1961–2017. Nature 580, 453–457.

Zhou, Y., Jain, N., Jha, G.K., Begho, T., 2022. Nitrogen pollution from cattle production in India: a review of the social, cultural and economic influences. The Journal of Agricultural Science 160 (1-2), 98–106. Available at https://doi.org/10.1017/s0021859622000120.

Chapter 4

Soil microbiomes in carbon farming: Pathways to enhance soil carbon sequestration

Debabrata Nath[1], Vandana Kumari[2], and Ram Swaroop Meena[3]

[1]*International Rice Research Institute (IRRI)-South Asia Regional Centre (ISARC), Varanasi, Uttar Pradesh, India,* [2]*Department of Soil Science, Advance Centre on Sericulture, Dr. Kalam Agricultural College, Bihar Agricultural University, Kishanganj, Bihar, India,* [3]*Department of Agronomy, Institute of Agricultural Sciences, Banaras Hindu University, Varanasi, Uttar Pradesh, India*

Highlights

- Soil microbiomes regulate carbon cycling and greenhouse gas (GHG) emissions.
- Temperature, moisture, pH, and organic matter control microbial activity and carbon turnover.
- Microbial enzymes and necromass drive soil organic matter formation and carbon sequestration.
- Microbial interactions influence carbon allocation and soil respiration dynamics.
- Microalgae biofertilizers enhance carbon sinks and stabilize microbial communities.
- Advancing soil microbiome research supports carbon farming and sustainable agriculture.

4.1 Introduction

Soil is the second-largest carbon (C) reservoir on Earth, holding approximately 2000 Pg of C as soil organic carbon (SOC) (Janzen, 2004). Due to their long-term stability, the SOC pools, both labile and nonlabile, are crucial parts of the global C cycle. SOC production and degradation can directly impact climate control and provide opportunities for climate change mitigation, as they are essential for either releasing or sequestering CO_2 (Lal, 2008). The SOC pools (labile and nonlabile) and their long-term stability make them essential components of the global C cycle. As SOC production and degradation play critical roles in either sequestering or releasing CO2, they can directly affect climate regulation and present opportunities to mitigate climate change (Lal, 2008; Meena et al., 2025a). In addition to its climate implications, SOC is crucial for supporting ecosystem health by providing the nutrients and energy needed to sustain plant growth and soil biota (Jansson and Hofmockel, 2020; Meena et al., 2024c). Therefore, SOC transformation mechanisms have become a major focus of research, particularly in light of climate change and global warming (Trumbore, 1997; Lal, 2004). SOC has traditionally been associated with plant residues; however, recent studies suggest that microbial biomass may also play a significant role, with microbial necromass accounting for 50%–80% of SOC (Liang et al., 2019). Approximately one quarter of Earth's biodiversity is represented by soil microbiome, which contain millions of bacteria, archaea, fungi, viruses, and micro-eukaryotic species, but only a fraction of them have been thoroughly characterized (Sokol et al., 2022). Only a small portion of the millions of bacteria, archaea, fungi, viruses, and micro-eukaryotic species that make up soil microbiomes, which account for about 25% of Earth's biodiversity, have been fully described (Sokol et al., 2022). Biological and abiotic elements coexist in a microbiome through the concept of C, which provides genetic and phenotypic links. In soil, microorganisms produce organic matter (OM) and release C through respiration and fermentation, processes that involve in vivo and ex vivo modification (Liang et al., 2017). To maintain C cycling, optimizing C storage is crucial, since C fixation saturation sets the stage for subsequent decomposition and assimilation. Abiotic and biotic factors shape soil microbial communities and their environments through successive transformations that lead to an equilibrium in soil C cycling. Researchers are currently focused on understanding the mechanisms by which soil microorganisms contribute to the formation of soil organic matter (SOM), how these processes are influenced by microbial metabolic capacities, and

Carbon Incentives, Plans, and Policies in Agriculture. DOI: https://doi.org/10.1016/B978-0-443-33557-0.00019-7

the impact of climate change (Sokol et al., 2022; Meena et al., 2024b). Despite this fragmented research on soil microbiomes, there is a need for a unified approach. This chapter summarizes how soil microbial metabolism, community dynamics, and environmental factors influence C cycling. It explores how soil microorganisms release and sequester C through ecological processes. This chapter provides a broader perspective on the function of soil microbiomes by addressing current gaps and potential future directions in C cycling, as determined through a bibliometric analysis.

4.2 Understanding carbon farming: prospects and challenges

Farm-to-fork forms a critical part of the EU's roadmap toward net-zero emissions by 2050. By promoting sustainable practices in food production and consumption, it aims to minimize the environmental and C impacts of the EU food system. This strategy includes specific targets, such as reducing pesticide use by 50%, reducing nutrient losses by 50%, and reducing fertilizer use by 20% by 2030. To achieve these objectives, agricultural practices must shift toward environmental sustainability at a systemic level (Meena et al., 2025b; Raj et al., 2025). Additionally, the strategy seeks to expand organic farming by 2030, which will boost biodiversity and reduce the C footprint of the agricultural sector. A management strategy known as "Carbon farming" uses methods to store atmospheric C in soils, as well as in plant roots, wood, and leaves. C farming is defined as a management approach in which techniques are employed to sequester atmospheric C into soils, as well as into plant roots, wood, and foliage. Reducing global warming is the primary objective, as soil absorbs CO_2 from the atmosphere and stores it for extended periods (Rodias et al., 2021; Jhariya et al., 2023; Meena et al., 2024a). The primary goal of capturing CO2 from the atmosphere and storing it in the soil for an extended period is to reduce global warming (Rodias et al., 2021).

Within the context of the global C cycle, the evolution of soil environments and ecological responses remains a largely unresolved scientific inquiry. Alterations in environmental conditions impact microbial development and evolution (Ruan et al., 2023). Global warming, driven by rising atmospheric CO2 levels, amplifies the greenhouse effect and increases microbial respiration, thereby releasing more CO_2 into the atmosphere (Dong et al., 2021; Dove et al., 2021). Soil respiration can therefore affect climate change by reducing or increasing atmospheric CO_2 levels (Lei et al., 2021; Meena and Pradhan, 2023). As a result of increasing atmospheric CO_2 concentrations, researchers assert that global warming is exacerbated, thereby enhancing microbial respiration (Dong et al., 2021). Thus fluctuations in soil respiration can either reduce or increase atmospheric CO_2 levels, thereby affecting climate change (Lei et al., 2021).

The soil ecosystem exhibits high complexity and is affected by a variety of landscape-scale disturbances that influence soil C retention or its release into the atmosphere (Jansson and Hofmockel, 2020; Jhariya et al., 2022). Soil organisms, notably soil microbes, determine the eventual fate of SOC in combination with plants. Soil microbiome is well known for its various roles in C cycling (Thompson et al., 2017). Still, the majority is uncultivated and has largely unknown functions (Goel et al., 2018) (Fig. 4.1). Only a few soil microbes have been documented to date. Still, ongoing discoveries continue to expand our understanding (Schulz et al., 2018). Microbial contributions to SOC cycling are uncertain due to this knowledge gap, complicating the development of global C flux predictions (Friedlingstein et al., 2006). Therefore constant

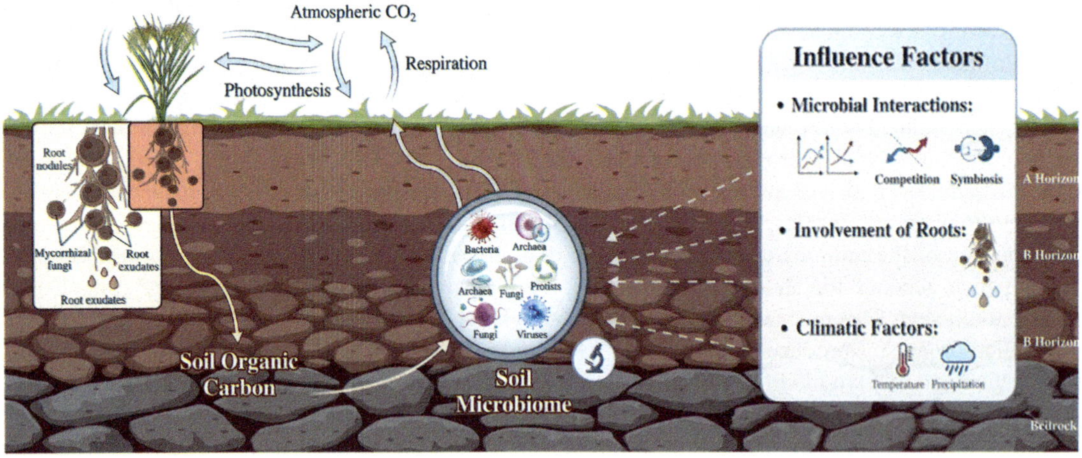

FIGURE 4.1 Soil microbiomes-driven carbon cycling by decomposing and transforming organic matter and stabilizing soil organic carbon.

refinement is needed for the soil microbiome's biochemical potential and SOC's metabolic fate (Naylor et al., 2020). It is particularly challenging to account for shifts in microbial activity induced by environmental changes because of limited information on microbial metabolic capacities. Plant C can stimulate microbial activity, accelerating SOC decomposition beyond predictions of C flux models (Kuzyakov et al., 2019; Meena et al., 2023). A microbial C pump model was developed to clarify the role of soil microbes in the transformation of soil OM (Liang et al., 2017). In the model, microbial C turnover occurs through two pathways: ex vivo enzyme-mediated modification and in vivo necromass accumulation. In catabolic processes, CO_2 is produced, while in anabolic processes, stable C compounds are produced. In the microbial C pump model, an understudied component of soil C flux is illuminated (Naylor et al., 2020; Liang et al., 2017). There is considerable uncertainty regarding the influence of plant-derived C on SOC storage mediated by microorganisms, as well as its role in accelerating SOC decomposition (Sulman et al., 2014). For instance, C from leaf litter and needle litter can facilitate microbial growth in forest soils. However, variations in litter chemistry and pH among different plant types, such as root litter compared to foliar litter (Harmon et al., 2009) or deciduous litter versus coniferous litter (Qualls, 2016), can significantly affect SOC levels by modulating decomposition rates. Furthermore, increased diversity within plant communities enhances rhizodeposition, stimulating microbial activity and promoting SOC storage (Naylor et al., 2020; Lange et al., 2015). Nonetheless, soil ultimately reaches a saturation point, beyond which additional C storage becomes unfeasible (Stewart et al., 2007). To improve our understanding of soil C flux, we must consider local environmental and biogeochemical factors.

Microorganisms also alternate between active and dormant states, which affects metabolic rates (Naylor et al., 2020; Joergensen and Wichern, 2018). The metabolism of SOC and biomass turnover are still carried out by some microorganisms during dormancy (Naylor et al., 2020; Joergensen and Wichern, 2018; Jones and Lennon, 2010), though at reduced rates. Active microbial members primarily drive biogeochemical transformations and have shifted the focus from taxonomic profiles to functional pathways and phenotypes within soil microbiomes (Jansson and Hofmockel, 2018; Naylor et al., 2020). It is essential to note, however, that current sequencing technologies measure both dormant and exogenous DNA (Carini et al., 2017), which may bias the results against members of the active community. By improving functional assessment approaches, model accuracy is expected to be improved. The chemical and physical structure of soils also influences SOC decomposition. Traditional explanations for slower C turnover rates include physical protection within microaggregates or mineral associations (Balesdent et al., 2000) and chemical recalcitrance (Marschner et al., 2008). A current understanding includes the sorption of biopolymers from microbial and plant necromass onto soil particles (Cotrufo et al., 2013) and deep SOM, primarily composed of microbial products (Erich et al., 2012). Dynamic and heterogeneous soil structures result in "hot spots" of microbial activity that are spatially and temporally variable (Bach and Hofmockel, 2016). C cycling is primarily limited to sufficiently hydrated areas or to desiccation-tolerant microbes that produce extracellular polymers to retain moisture, for instance (Vurukonda et al., 2016). Several factors influence SOC mineralization, including the distinction between anaerobic and aerobic microsites (where anaerobic C respiration is less energetically efficient), the availability of electron acceptors, and soil redox status (Keiluweit et al., 2017). A compound that resists anaerobic breakdown, such as lipids, is more likely to persist in anaerobic aggregates (Keiluweit et al., 2017). To better understand soil metabolism and C flux, comprehensive comparisons of SOC decomposition are needed.

Carbon farming initiatives require supportive agro-environmental policies to encourage farmers to adopt sustainable practices. However, farmer participation is often limited due to the complexity of program design, conflicting goals with policymakers, and personal motivations or farm-specific characteristics that affect adoption (Liu et al., 2018a; Salas Castelo, 2017). Barriers also include limited skills, political instability, and insufficient awareness of C farming options (Ingram et al., 2016; Toensmeier, 2016). High input costs and uncertainties about productivity impacts further hinder participation, along with the lack of standardized methods, administrative burdens, and certification challenges (Macintosh and Waugh, 2012). Additional obstacles involve capital requirements, potential impacts on financial assistance eligibility, fluctuating C prices, and uncertainties in C market benefits and monitoring (Narassimhan et al., 2018; Toensmeier, 2016; Kragt et al., 2017).

4.3 Understanding the role of soil microbes for carbon farming

The primary determinant of soil C fluxes is microbial metabolism; however, microbial interactions have been rarely explored. Changes in phenotypic patterns can affect C allocation and utilization efficiency, which in turn impact resource availability, community structure, and evolutionary pressures that feed back into the microbial metabolic cycle (Ramin and Allison, 2019). To address biodiversity loss and climate change, it is crucial to understand the composition and interactions of soil microorganisms. The thermodynamics of metabolism set fundamental limits on the fluxes of energy and material in microbial metabolism (Brown et al., 2004). These constraints are shaped by microbial interactions, which affect growth potential and responses to resource scarcity or chemical challenges (Ramin and Allison, 2019). OM flow and resource allocation are affected directly by key interaction mechanisms, including exploitative and interference competition, mutualism, and predation (García et al., 2023; Sokol et al., 2022).

High-efficiency absorbers are weaker competitors for resources, leading to exploitative competition among microbial populations. Microbes can grow rapidly even at low substrate concentrations (Schmidt and Konopka, 2009). Exploitative tactics are further supported by interference competition, which involves producing toxins to limit competitors (Wu et al., 2024; Ogawa et al., 2022). Using interference competition to limit competitors further supports exploitative strategies (Wu et al., 2024; Ogawa et al., 2022). There are often broader ecological niches and complex networks among antagonists in interference competition (Russel et al., 2017). The complementarity of niches can optimize resource use, but competition can impact ecosystem function (Yu et al., 2019). In soil microbes and wood-decaying fungi, mutual interference reduces carbon use efficiency (CUE) (Wu et al., 2024; Maynard et al., 2017). Alternatively, resource-based competition that leads to niche differentiation can enhance diversity and reduce soil respiration (Chen et al., 2019; Wu et al., 2024). Metabolic exchange can be improved by beneficial mutations in donor microbes or mutualistic partnerships (Giri et al., 2022). Several studies have shown that symbiotic relationships between bacteria (especially those that help in fermentation) and methane (CH_4)-producing archaea facilitate OM degradation and accelerate soil C turnover (McInerney et al., 2009). Additionally, interactions between Arbuscular mycorrhizal (AM) fungi and saprophytic bacteria promote litter mineralization (Cao et al., 2022b; Wu et al., 2024). By maintaining microbial partners in close proximity, biofilms enhance biomass production and nutrient turnover (Jansson and Wu, 2022). The activity of biofilms can significantly increase soil respiration (Wu et al., 2019) and affect SOC composition (Wang et al., 2022). Studies have shown that soil C dynamics are influenced by microbial diversity (Schmitz and Leroux, 2020). Incorporating microbial interactions into C cycling models could improve predictions despite their complexity. Unexplored microbial interactions, such as those involving radiation bacteria acting as parasites or symbionts, may also contribute significantly to nutrient cycling (Naud et al., 2022).

4.4 Involvement of soil microbiomes in carbon cycling: species, functions, and mechanisms

4.4.1 Different species of soil microbes in carbon cycling

C cycling in soil is a dynamic, nonequilibrium process that requires microbial C pumps for storage (Luo and Weng, 2011; Zhu et al., 2020). Soil C dynamics are influenced by bacteria, archaea, fungi, and microeukaryotes (protozoa) (Table 4.1). As microbial cell residues accumulate in soil over time, they contribute to biogeochemical processes and form one of the planet's largest OM reservoirs (Sokol et al., 2022). Soils emit about 75-100 Pg of CO_2 each year, and OC decomposition accounts for 60 Pg of that (Jiang et al., 2022). The release and respiration of soil C is influenced by various environmental factors. OM decomposition can be accelerated by increased temperatures, leading to greater CO_2 emissions; however, this effect depends on factors specific to each region and ecosystem. Additionally, changes in human activity affect soil respiration rates, which in turn influence C dynamics.

TABLE 4.1 Microbes perform important functions in the soil carbon cycle.

Species	Major and specific roles in the carbon cycle	References
Bacteria	Primary decomposers in soil ecosystems break down dead plants, animals, and fungi. C sources differ between Gram-negative and Gram-positive bacteria.	Wu et al. (2024), Bardgett and Van Der Putten (2014)
Archaea	Methanogenic archaea play an important role in climate change and the C cycle in anoxic environments.	Wu et al. (2024), Nazaries et al. (2013), Naitam and Kaushik (2021)
Fungi	Soil fungi contribute significantly to global soil C stores by digesting complex C-rich substrates.	Wu et al. (2024), Žifčáková et al. (2016), Bar-On et al. (2018); Žifčáková and Větrovský, 2016; undefined
Protozoa	By feeding selectively, protozoa have a substantial impact on microbial community dynamics and the C cycle in soils.	Wu et al. (2024), Xiong et al. (2018), Thakur and Geisen (2019), Sokol et al. (2022)

Source: Adapted from (Wu et al., 2024).

4.4.2 Functions of the soil microbiomes in carbon cycling

Plant productivity, nutrient cycling, and decomposition are positively correlated with microbial diversity. The majority of soil biomass and microbial diversity is composed of bacteria and fungi (Bardgett and Van Der Putten, 2014; Fierer, 2017; Wu et al., 2024). As a result, their necromass contributes significantly to SOM. Soil bacteria can decompose dead OM, including that of animals, plants, and fungi. Plant C is used by Gram-negative bacteria, whereas Gram-positive bacteria use SOM-derived C. Throughout the soil matrix, microbial distribution may affect substrate preferences (Kramer and Gleixner, 2008). A slow-growing, oligotrophic bacterium (Acidobacterium) with high metabolic flexibility can decompose complex C compounds in recalcitrant SOM pools despite low resource availability (Lladó et al., 2017). Additionally, predatory bacteria influence the structure of the soil food web and C cycling, with obligatory predators acquiring C more rapidly than nonpredatory bacteria (Hungate et al., 2021). In anoxic, organic-rich niches, methanogenic archaea play a crucial role in C cycling and climate regulation (Wu et al., 2024; Nazaries et al., 2013; Naitam and Kaushik, 2021). A dominant microbial community in soil is fungi that metabolize complex C-rich substrates for soil C storage (Bar-On et al., 2018). Protozoa contribute to soil C cycling by feeding on microbial communities (Wu et al., 2024; Xiong et al., 2018).

4.4.3 Mechanisms of carbon fixation by the soil microbiomes

Soil microbiomes use various C fixation pathways. They are categorised as light-driven, chemically driven, and photoelectron-driven microorganisms based on their sources of energy (Wu et al., 2024; Mistry et al., 2019; Jiang et al., 2022). Through various metabolic pathways, including anaerobic sulfide oxidation, photosynthesis, ureolysis, ammonification, denitrification, sulfate reduction, and CH_4 oxidation, soil microorganisms sequester CO_2. In these processes, pH increases and dissolved inorganic carbon (DIC) levels rise. Carbonate minerals, such as $CaCO_3$, are formed from DIC (Wu et al., 2024; Zhu and Dittrich, 2016; Jiang et al., 2022). In addition, soil microbes store CO_2 as SOC. Through the Calvin-Benson (CB) cycle, the reductive tricarboxylic acid (rTCA) cycle, the 3-hydroxypropionate (3-HP)/4-hydroxybutyrate cycle, the dicarboxylate/4-hydroxybutyrate cycle, the 3-HP cycle, and the Wood–Ljungdahl (WL) pathway, autotrophs sequester inorganic C (Wu et al., 2024; Liu et al., 2018b). The CB cycle is the primary C fixation pathway in soils and is utilized by plants, algae, cyanobacteria, and various proteobacteria (Berg, 2011). Several oxygen-sensitive enzymes are present in the rTCA pathway, the dicarboxylate/4-hydroxybutyrate pathway, and the WL pathway (Wu et al., 2024; Liu et al., 2018b). Furthermore, the composition of C species and the distribution of microbial populations can affect the activation of specific C fixation pathways (Berg, 2011). Assimilatory carboxylation and anaplerotic reactions within the tricarboxylic acid (TCA) cycle can also promote C fixation through heterotrophic processes, commonly referred to as "dark fixation" (Alonso-Sáez et al., 2010; Erb, 2011; Wu et al., 2024). CO_2 fixation occurs across various soil types, including those found in Mediterranean regions, semiarid deserts, Arctic tundra, temperate peatlands, paddy fields, temperate meadows, and croplands (Spohn et al., 2020). Through autotrophic and heterotrophic pathways, microorganisms fix C in multiple ways. Considering soil heterogeneity, other C-fixation mechanisms may also exist.

4.5 Current challenges to optimize soil microbes used for CO_2 fixation

The soil ecosystem is very complex and susceptible to landscape-scale perturbations that might influence C retention or release (Jansson and Hofmockel, 2020). Soil bacteria, including those associated with plants, play an important role in determining the fate of SOC. Although soil microbiomes perform several biogeochemical functions related to cycling (Fig. 4.2) (Thompson et al., 2017), most of them remain uncultivated and have largely unknown functions. Soil microbes and viruses have been discovered in a growing number of places, but a mere fraction of them has been cataloged thus far (Goel et al., 2018). Soil microbes contribute to SOC cycling in an uncertain manner, hindering the development of accurate models of global C flux under climate change (Friedlingstein et al., 2006; Pradhan and Meena, 2023a). The biochemical potential of the soil microbiome and the metabolic fate of SOC are constantly being refined. Despite the scarcity of knowledge on the soil microbiome's metabolic capability, it is especially challenging to account for alterations in microbial activity in response to environmental changes. Microbial activity, for example, can be triggered by plant-derived C inputs, resulting in higher rates of C decomposition than predicted (Kuzyakov et al., 2019). The balance between catabolic and anabolic activities impacts SOC retention rates. The catabolic pathway releases CO_2, while the anabolic pathway produces stable C compounds. Furthermore, there is considerable uncertainty regarding how plant-derived C influences microbial SOC storage or accelerates its decomposition (Naylor et al., 2020; Sulman et al., 2014). Forest soils have high C content due to leaf and needle litter accumulation; however, the chemical

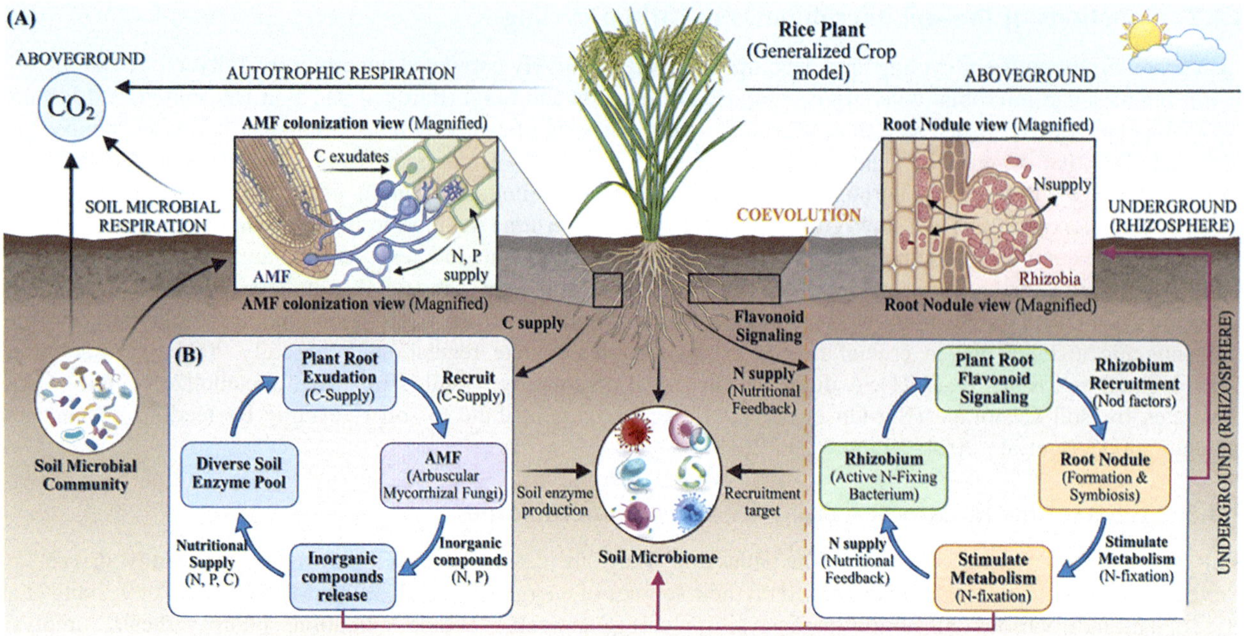

FIGURE 4.2 Influence of soil microbiomes on carbon cycling: (A) schematic representation of the interactions between the soil microbiomes and (B) carbon transformation and nutrient exchange. (Arbuscular mycorrhizal (AM) fungi enhance soil microbial decomposers by releasing labile carbon into their intermycelial soils. In exchange, these fungi acquire N and P from microbial sources, transporting these essential nutrients to the host plant).

composition of litter and its pH can vary significantly depending on vegetation type, such as the differences between root litter and foliar litter (Harmon et al., 2009; Naylor et al., 2020) or between deciduous and coniferous forest litter (Naylor et al., 2020; Qualls, 2016). These biochemical variations impact SOC levels by altering decomposition dynamics. Moreover, a diverse plant community can enhance rhizodeposition rates, stimulating microbial activity and promoting SOC storage (Naylor et al., 2020; Lange et al., 2015). It is also important to note that soil can reach a saturation point beyond which additional C storage becomes infeasible (Stewart et al., 2007; Pradhan and Meena, 2023b). Current gaps in understanding net C flux in soil may be addressed by incorporating local environmental and biogeochemical variables into research considerations.

4.6 Role of soil microbes in CO_2 fixation in soil and plants

Soil microbes contribute to the SIC and SOC pools of soil by fixing CO_2. SIC is formed by bacteria, fungi, algae, and protozoa (Gadd, 2010). During photosynthesis, cyanobacteria facilitate exchanges of bicarbonate (HCO_3^-) and hydroxide (OH^-) ions across their cell membranes, increasing pH within their microenvironment and promoting carbonate precipitation (Zhu and Dittrich, 2016). Extracellular polymeric substances (EPS) produced by microorganisms enhance carbonate precipitation by capturing divalent cations such as calcium (Ca^{2+}) and magnesium (Mg^{2+}) (Kremer et al., 2008). For instance, *Bacillus subtilis* absorbs substantial quantities of metal ions, promoting carbonate precipitation (Zhu and Dittrich, 2016). Bacteria facilitate carbonate formation by using protons (H^+) and CO_2 (Zhu and Dittrich, 2016). The surface soil microbial community includes microeukaryotes, prokaryotes, and soil algae.

Through photosynthesis, plants sequester atmospheric C to support their growth (Jassey et al., 2022). Although algae contribute minimally to soil microbiome biomass (Jassey et al., 2022), they may make up to 40% of eukaryotic and 4% of prokaryotic communities (Oliverio et al., 2020). C is primarily derived from soil algae within photoautotrophic biocrusts (Maier et al., 2018). Chemoautotrophic bacteria contribute to global climate change, CH_4 generation and oxidation, and nitrogen cycling, but their ability to sequester C is little understood (Naylor et al., 2020). In particular, habitat conditions, such as those that favor soil chemoautotrophs, can sequester C (Zhang et al., 2022; Yu et al., 2023).

In soil, CO_2 fixation is primarily driven by autotrophic microbes, which utilise both phototrophic and heterotrophic sources of energy. For chemoautotrophs and photoautotrophs, inorganic. At the same time, heterotrophs and polytrophs bind inorganic C into carbon, which is the primary C source. In contrast, heterotrophs and polytrophs incorporate inorganic C into their metabolic pathways through carboxylation reactions, a process known as dark CO_2 fixation (Erb, 2011). In addition to OC availability, dark CO_2 fixation is associated with heterotrophic microbes (Šantrůčková et al., 2018). Soil type determines the structure of the microbial community, leading to variation in the dominant species involved in dark CO_2 fixation. Gram-positive bacteria fix CO_2 primarily in temperate forests (Spohn et al., 2020), whereas Gram-negative bacteria fix CO_2 in arctic permafrost ecosystems (Šantrůčková et al., 2018). Dark CO_2 fixation is attributed more prominently to bacteria than to fungi in both studies. Heterotrophic CO_2 fixation may not result in net C accumulation in the soil (Alonso-Sáez et al., 2010), but dark CO_2 fixation by microorganisms significantly influences the C cycle (Akinyede et al., 2020).

4.7 The complex interactions between the soil microbiome and soil carbon

Soil C fluxes are driven by microbial metabolism, yet microbial interactions often receive less attention than microbial metabolism. These interactions can change C allocation patterns and utilization efficiency by triggering phenotypic responses. Resources are now limited, resulting in changes in community structure and resource availability that foster evolutionary adaptations, which are then reflected in microbial metabolism (Ramin and Allison, 2019). Understanding soil microbial composition, soil–microbe interactions, and C cycling is the first step to managing biodiversity loss and climate change. Microbial processes are governed by metabolic thermodynamics, which sets limits on energy and material fluxes (Brown et al., 2004; Wu et al., 2024). The strategies microorganisms employ to utilize these resources significantly affect ecological interactions, resulting in the production of essential molecules for growth and responses to resource depletion or chemical stressors (Ramin and Allison, 2019). In soil ecosystems, microbial communities engage in various interactions, including exploitative competition, interference competition, mutualism, and predation (Wu et al., 2024; Sokol et al., 2022). Whether beneficial or antagonistic, these microbial interactions affect the soil's chemical composition, OM flow, and C distribution. C cycling can be intensified by strong interspecies interactions, leading to enhanced feedback loops among populations (García et al., 2023).

A microbial population competes indirectly for the same resources when it engages in exploitative competition. When a species has superior resource-acquisition capabilities, it can grow more rapidly, thereby limiting weaker competitors' access to resources. Despite low substrate concentrations, some microbes have relatively high growth rates, allowing them to gain a competitive advantage over their competitors (Schmidt and Konopka, 2009). It has become common for bacteria to exploit resources disproportionately, concentrating their efforts on biomass growth rather than respiration, resulting in lower metabolism efficiency (Lipson, 2015). This exploitative advantage can be reinforced by interference competition, such as the production of toxins to suppress competitors (Ogawa et al., 2022). Researchers have discovered that antagonists that utilize interference competition have larger metabolic niches and ecological networks (Russel et al., 2017). Any competitive contact might impact ecosystem function despite niche complementarity reducing resource overlap (Yu et al., 2023). Microbial CUE is reduced by interactions between wood-decaying fungi and soil microbes, which overcome growth barriers (Maynard et al., 2017). Microbial competition increases soil respiration entropy, carbohydrate degradation, and metabolic activity (Chen et al., 2019). Mutations in donors or recipient organisms may favor the donor during community succession. Mutualism occurs when donor cells benefit from partner organisms' metabolic activities or secretions (such as vitamins and amino acids) (Giri et al., 2022). In soil, numerous symbiotic interactions boost OM production or accelerate substrate decomposition. By breaking down complex organic molecules, bacterial and archaeal mutualism accelerates soil C turnover (McInerney et al., 2009). AM fungi and saprotrophic bacteria promote litter mineralization and primary productivity (Cao et al., 2022a; Cao et al., 2022b; Nath and Meena, 2018; Nath et al., 2026). A biofilm hosts a wide range of metabolically active microbial communities, making it a hotspot for symbioses. In over 60% of the consortia, biomass levels were higher than those in single-species biofilms (Jansson and Wu, 2022). Increasing nutrient turnover is enabled by biofilms, which facilitate close interactions among partners. Based on the analysis of dynamic respiration, biofilm=induced 23 times higher substrate activity than free-moving bacteria (Wu et al., 2019). A biofilm might affect SOC composition and provide a substrate for CH_4 production and soil respiration (Wang et al., 2022). Microbial aggregates play a crucial role in soil C biogeochemistry, and they should be considered when refining models that predict global C fluxes. Recent studies link microbial diversity to soil C cycling (Trivedi et al., 2016; Schmitz and Leroux, 2020; Pradhan et al., 2023). Because microbial interactions are closely tied to ecosystem biogeochemistry, incorporating them into ecological models would improve predictions of global C cycling. Additionally, soils may harbour biological interactions that remain to be classified, significantly influencing the dynamics of microbial communities and nutrient recycling (Naud et al., 2022).

4.8 Microbial-mediated soil carbon cycling driven by soil aggregates

4.8.1 Soil aggregates and carbon cycle relationships

In SOC cycling, aggregates play a crucial role in storing and redistributing SOC. The interaction between OM and soil minerals in the organic–inorganic matrix leads to various chemical bonds and associations, including covalent bonds, hydrogen bonds, and adsorption phenomena from both biological and abiotic processes. Additionally, these interactions, along with Van der Waals forces and microbial hyphal binding, contribute to the formation of aggregates of varying particle sizes within the soil. As OC transforms and decomposes, soil aggregates and minerals physically and chemically protect it (Kopittke et al., 2022). To enhance soil fertility and sustain soil C stocks, different aggregate sizes must be distributed and stable. There are two classifications of soil aggregates: macroaggregates, which are larger than 0.25 mm, and microaggregates, which are smaller than 0.25 mm. Microaggregates contain a higher proportion of recalcitrant OC, which is less susceptible to degradation and can facilitate the growth of soil microorganisms (Tiemann et al., 2015; Lin et al., 2019). Microorganism communities are shaped by the nutrient content, pore size, and the capacity for water and gas exchange within soil aggregates. Fungi and bacteria produce polysaccharides in soil that contribute significantly to aggregate stability. In soil, bacteria and fungi are directly involved in particle aggregation through their hyphae. In pores (>10 m) or aggregate surfaces colonized by actinomycetes and fungi, hyphal networks promote macroaggregate formation (Liao et al., 2018). Biofilms and bacterial colonies are typically found on aggregate surfaces or in pores (Gupta and Germida, 2015). When together, these microbes play an essential role in the soil C cycle. Furthermore, soil aggregates influence microbial interactions, such as cooperation, competition, and predation. Fungal–bacterial interaction networks become less complex as aggregate particle size decreases (Liao et al., 2020), which impacts SOC cycling.

4.8.2 Effect of soil aggregate composition on carbon cycle

Soil aggregates contain OM and mineral particles. A majority of OM comes from microbial sources (e.g., bacterial extracellular polymers, fungal secretions, and AM fungi), humic substances (e.g., humic acid), and plant root exudates (Galloway et al., 2018; Deka et al., 2019). A primary role for microorganisms in aggregate formation is that the OM contributes substantially to this process (Fig. 4.3) (Li et al., 2022). Minerals serve as inorganic binders in soil aggregates. Biological polysaccharides act as cementing agents at the micron scale, forming microaggregates, while hyphal networks enclose these microaggregates, forming macroaggregates (Bronick and Lal, 2005). By encouraging polysaccharide production within the cell, kaolinite promotes bacterial growth, whereas other minerals, such as montmorillonite, remain only loosely bound (Cai et al., 2018). Minerals and microorganisms can interact, triggering quorum sensing and EPS secretion, enhancing aggregate formation and SOC stability (Ma et al., 2017; Sokol et al., 2022).

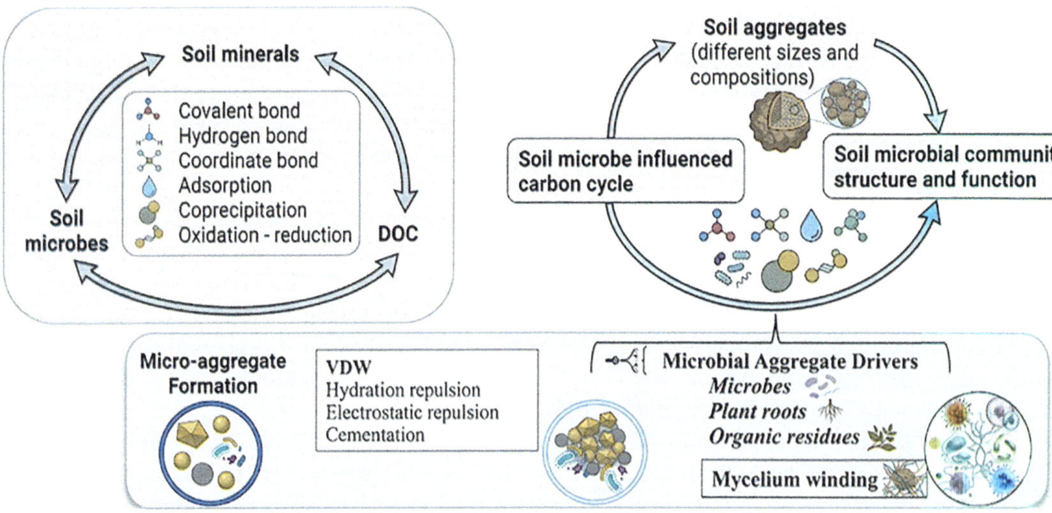

FIGURE 4.3 Effect of soil aggregates on the soil carbon cycle.

4.9 Microbially mediated soil emission reduction mechanisms through the application of microalgae biofertilizers

Microalgae inhabit a diverse range of ecosystems, including terrestrial, freshwater, and marine environments. Due to their high photosynthetic efficiency, microalgae are used for energy production, wastewater treatment, and the reduction of atmospheric CO_2 levels. Microalgal biofertilizers typically include eukaryotic green algae, known for their superior photosynthetic efficiency, and prokaryotic cyanobacteria, which are nitrogen-fixing. Biofertilizers are rich in vital trace elements and are highly efficient, environmentally sustainable, and C sequestering, all contributing to reduced greenhouse gas (GHG) emissions (Alvarez et al., 2021). Microalgae have a positive impact on soil and GHG mitigation, as shown in Fig. 4.4. By fixing atmospheric CO_2 while simultaneously releasing oxygen, microalgae increase soil oxygen levels by up to 10 times greater than terrestrial plants (de Siqueira et al., 2017). As a result of enzymatic processes, microalgae activate immobilized phosphorus and potassium, enhance the bioavailability of mineral elements, and promote the accumulation and transformation of photosynthetic products within soil environments. Through the capture of atmospheric CO_2 and N_2, microbial extracellular polysaccharides, together with those produced by microalgae, form a biofilm layer that contributes to the accumulation of soil C and nitrogen (Bharti et al., 2017). Agricultural soils can form photosynthetic biofilms faster when Chlorella culture suspensions are added (Marks et al., 2017).

C and nitrogen fixation are dual functions of cyanobacteria. The process of photosynthesis in green algae is similar to that of plants in fixing atmospheric CO_2. Among cyanobacteria, there are both vegetative and heterocyst cells, which produce a nitrogenase enzyme that converts nitrogen (N_2) into ammonia (NH_3). In vegetative cells, the enzymes nitrite reductase and nitrate reductase facilitate the reduction of nitrate and nitrite to ammonia (NH_3), thereby augmenting soil nitrogen reserves (Knoche et al., 2017). Cyanobacteria secrete nitrogen-rich compounds during their growth, including amino acids, sugars, and polysaccharides (Mallappa et al., 2016; Renuka et al., 2016). In a study spanning two consecutive rice-growing seasons, Ali et al. (2014) showed that the introduction of Azolla and cyanobacteria led to a 12% reduction in CH_4 emissions in rice soils in Bangladesh (Ali et al., 2014). Additionally, Anabaena biofilms, specifically *Anabaena-Trichoderma* and *Anabaena-Pseudomonas* aeruginosa, were introduced in rice fields in New Delhi, resulting in a 50%-80% reduction in CH_4 emissions (Prasanna et al., 2015). Although microalgal biofertilizers did not significantly increase wheat yield compared to urea, they reduced nitrogen oxide (N_2O) and nitrogen monoxide (NO) emissions in the soil (Shrestha et al., 2022). Furthermore, researchers have reported that combining microalgal biofertilizers with biochar or organic fertilizers enhances the capacity of microalgal biofertilizers to sequester C (Hu et al., 2019; Zhang et al., 2018). With biochar and OM added to microalgae, glucose becomes more readily available to soil microorganisms, thereby improving the soil's ability to sequester C (Nath et al., 2025). Microalgal biofertilizers not only fix nitrogen and sequester C, but their decaying cells also enrich SOM, increasing fertility and crop yield (Bhardwaj et al., 2014). Microalgae-based carbon sequestration (C_{seq}) can also be used to treat emissions from coal-fired power plants; microalgae absorb CO_2 through photosynthesis and take up NO_x and SO_x as they grow (Yen et al., 2015; Zeraatkar et al., 2016). Microalgal biofertilizers can be produced from industrial C-fixation biomass, thereby reducing GHG emissions both in industrial and agricultural settings. Microalgae present significant potential for future development, particularly as the world moves toward sustainable agriculture.

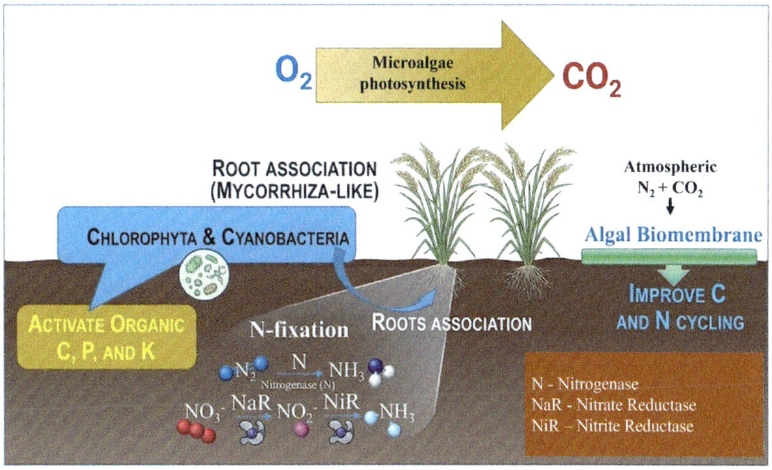

FIGURE 4.4 Positive impacts of microalgae on GHGs mitigation and C_{seq}. *GHG*, greenhouse gas; *C_{seq}*, carbon sequestration.

4.10 Future directions for integrating soil microbiome research into sustainable agricultural practices

Future advances in sustainable agriculture and climate change mitigation will depend on the deeper integration of soil microbiome research into both management practices and predictive frameworks. Microbial carbon (C) cycling models must be more explicitly incorporated into large-scale Earth system and climate models, particularly those that represent microbial metabolic pathways and trait-based functions. Such integration will enhance the accuracy of predictions related to soil carbon sequestration (Cseq) and greenhouse gas (GHG) fluxes under diverse land-use and climate change scenarios.

Targeted microbial management strategies offer promising pathways to reduce CO_2, CH_4, and N_2O emissions from agricultural soils. Practices such as the application of biochar, organic amendments, and microalgae-based biofertilizers can modify soil physicochemical conditions, promote beneficial microbial communities, and enhance long-term soil carbon storage. In parallel, improving microbial resilience to climatic stressors including temperature extremes, moisture variability, and nutrient imbalances could stabilize GHG emissions and sustain soil C sequestration capacity under changing environmental conditions.

Enhancing microbiome stability requires a deeper understanding of microbial diversity, functional redundancy, and adaptive capacity within soil ecosystems. Future research should emphasize resilience-based frameworks that recognize not only resistance to disturbance but also recovery potential following environmental perturbations. Accordingly, soil health assessment metrics must evolve to incorporate indicators of microbial functional resilience and ecosystem stability, particularly in the context of climate change.

Greater insight into interactions among bacteria, fungi, and plants is also critical for advancing soil carbon management. Microbe-plant interactions in the rhizosphere play a pivotal role in soil organic carbon (SOC) dynamics, as plant root exudates shape microbial community composition and regulate carbon stabilization pathways. Understanding these interactions across spatial and temporal scales will clarify the mechanisms governing soil C sinks and inform adaptive management strategies that optimize carbon farming outcomes. Ultimately, linking microbial ecology with agronomic practices, climate modeling, and ecosystem resilience concepts will provide a robust foundation for sustainable soil management. Such integrative approaches will be essential for enhancing agricultural productivity, strengthening soil carbon sinks, and supporting long-term climate change mitigation efforts.

4.11 Conclusion

Soil microbiomes play a fundamental role in regulating carbon (C) cycling, greenhouse gas (GHG) emissions, and overall soil health. Interactions between microbial communities and key environmental drivers govern the emissions of CO_2, CH_4, and N_2O, thereby exerting a direct influence on global warming potential from agricultural soils. Recent research has identified several promising mitigation strategies in soil microbial management, including targeted soil amendments and biologically informed remediation practices that can reduce GHG emissions while enhancing soil carbon storage. However, the intrinsic complexity of soil microbiomes marked by high taxonomic diversity, functional redundancy, and context-dependent responses poses significant challenges for modeling and predicting ecosystem-level outcomes, particularly under accelerating climate change. To fully harness the mitigation potential of soil microbiomes, future efforts must prioritize integrative and systems-based approaches that link microbial ecology with climate-smart soil management. Such strategies will be essential for enhancing the resilience of agroecosystems, advancing sustainable agricultural practices, and achieving long-term climate change mitigation goals.

Acknowledgments

We sincerely appreciate the support and guidance of our colleagues, whose insights have enriched this chapter. We acknowledge the contributions of researchers in this field, whose work has been cited in shaping the content of this chapter. Finally, the authors are deeply grateful to the anonymous reviewers whose valuable suggestions have significantly improved the chapter.

References

Akinyede, R., Taubert, M., Schrumpf, M., Trumbore, S., Küsel, K., 2020. Rates of dark CO_2 fixation are driven by microbial biomass in a temperate forest soil. Soil Biology and Biochemistry 150, 107950. Available at https://doi.org/10.1016/j.soilbio.2020.107950.

Ali, M.A., Sattar, M.A., Islam, M.N., Inubushi, K., 2014. Integrated effects of organic, inorganic and biological amendments on methane emission, soil quality and rice productivity in irrigated paddy ecosystem of Bangladesh: field study of two consecutive rice growing seasons. Plant and Soil 378 (1-2), 239–252. Available at https://doi.org/10.1007/s11104-014-2023-y.

Alonso-Sáez, L., Galand, P.E., Casamayor, E.O., Pedrós-Alió, C., Bertilsson, S., 2010. High bicarbonate assimilation in the dark by Arctic bacteria. ISME Journal 4 (12), 1581–1590. Available at https://doi.org/10.1038/ismej.2010.69.

Alvarez, A.L., Weyers, S.L., Goemann, H.M., Peyton, B.M., Gardner, R.D., 2021. Microalgae, soil and plants: a critical review of microalgae as renewable resources for agriculture. Algal Research 54. Available at https://doi.org/10.1016/j.algal.2021.102200.

Bach, E.M., Hofmockel, K.S., 2016. A time for every season: soil aggregate turnover stimulates decomposition and reduces carbon loss in grasslands managed for bioenergy. GCB Bioenergy 8 (3), 588–599. Available at https://doi.org/10.1111/gcbb.12267.

Balesdent, J., Chenu, C., Balabane, M., 2000. Relationship of soil organic matter dynamics to physical protection and tillage. Soil and Tillage Research 53 (3-4), 215–230. Available at https://doi.org/10.1016/s0167-1987(99)00107-5.

Bardgett, R.D., Van Der Putten, W.H., 2014. Belowground biodiversity and ecosystem functioning. Nature 515 (7528), 505–511. Available at https://doi.org/10.1038/nature13855.

Bar-On, Y.M., Phillips, R., Milo, R., 2018. The biomass distribution on Earth. Proceedings of the National Academy of Sciences of the United States of America 115 (25), 6506–6511. Available at https://doi.org/10.1073/pnas.1711842115.

Berg, I.A., 2011. Ecological aspects of the distribution of different autotrophic CO_2 fixation pathways. Applied and Environmental Microbiology 77 (6), 1925–1936. Available at https://doi.org/10.1128/AEM.02473-10.

Bhardwaj, D., Ansari, M.W., Sahoo, R.K., Tuteja, N., 2014. Biofertilizers function as key player in sustainable agriculture by improving soil fertility, plant tolerance and crop productivity. Microbial Cell Factories 13 (1). Available at https://doi.org/10.1186/1475-2859-13-66.

Bharti, A., Velmourougane, K., Prasanna, R., 2017. Phototrophic biofilms: diversity, ecology and applications. Journal of Applied Phycology 29 (6), 2729–2744. Available at https://doi.org/10.1007/s10811-017-1172-9.

Bronick, C.J., Lal, R., 2005. Soil structure and management: a review. Geoderma 124 (1-2), 3–22. Available at https://doi.org/10.1016/j.geoderma.2004.03.005.

Brown, J.H., Gillooly, J.F., Allen, A.P., Savage, V.M., West, G.B., 2004. Ecological society of America United States Toward a metabolic theory of ecology. Ecology 85 (7), 1771–1789. Available at https://doi.org/10.1890/03-900000129658.

Cai, P., Liu, X., Ji, D., Yang, S., Walker, S.L., Wu, Y., et al., 2018. Impact of soil clay minerals on growth, biofilm formation, and virulence gene expression of *Escherichia coli* O157:H7. Environmental Pollution 243, 953–960. Available at https://doi.org/10.1016/j.envpol.2018.09.032.

Cao, M.M., Liu, S.Y., Bi, L., Chen, S.J., Wu, H.Y., Ge, Y., et al., 2022a. Distribution characteristics of soil viruses under different precipitation gradients on the Qinghai-Tibet Plateau. Frontiers in Microbiology 13. Available at https://doi.org/10.3389/fmicb.2022.848305.

Cao, T., Fang, Y., Chen, Y., Kong, X., Yang, J., Alharbi, H., et al., 2022b. Synergy of saprotrophs with mycorrhiza for litter decomposition and hotspot formation depends on nutrient availability in the rhizosphere. Geoderma 410, 115662. Available at https://doi.org/10.1016/j.geoderma.2021.115662.

Carini, P., Marsden, P.J., Leff, J.W., Morgan, E.E., Strickland, M.S., Fierer, N., 2017. Relic DNA is abundant in soil and obscures estimates of soil microbial diversity. Nature Microbiology 2 (3). Available at https://doi.org/10.1038/nmicrobiol.2016.242.

Chen, L., Jiang, Y., Liang, C., Luo, Y., Xu, Q., Han, C., et al., 2019. Competitive interaction with keystone taxa induced negative priming under biochar amendments. Microbiome 7 (1). Available at https://doi.org/10.1186/s40168-019-0693-7.

Cotrufo, M.F., Wallenstein, M.D., Boot, C.M., Denef, K., Paul, E., 2013. The Microbial Efficiency-Matrix Stabilization (MEMS) framework integrates plant litter decomposition with soil organic matter stabilization: Do labile plant inputs form stable soil organic matter? Global Change Biology 19 (4), 988–995. Available at https://doi.org/10.1111/gcb.12113.

Deka, P., Goswami, G., Das, P., Gautom, T., Chowdhury, N., Boro, R.C., et al., 2019. Bacterial exopolysaccharide promotes acid tolerance in *Bacillus amyloliquefaciens* and improves soil aggregation. Molecular Biology Reports 46 (1), 1079–1091. Available at https://doi.org/10.1007/s11033-018-4566-0.

Dong, X., Liu, C., Ma, D., Wu, Y., Man, H., Wu, X., et al., 2021. Organic carbon mineralization and bacterial community of active layer soils response to short-term warming in the great hing'an mountains of northeast China. Frontiers in Microbiology 12. Available at https://doi.org/10.3389/fmicb.2021.802213.

Dove, N.C., Torn, M.S., Hart, S.C., Taş, N., 2021. Metabolic capabilities mute positive response to direct and indirect impacts of warming throughout the soil profile. Nature Communications 12 (1). Available at https://doi.org/10.1038/s41467-021-22408-5.

Erb, T.J., 2011. Carboxylases in natural and synthetic microbial pathways. Applied and Environmental Microbiology 77 (24), 8466–8477. Available at https://doi.org/10.1128/AEM.05702-11.

Erich, M.S., Plante, A.F., Fernández, J.M., Mallory, E.B., Ohno, T., 2012. Effects of profile depth and management on the composition of labile and total soil organic matter. Soil Science Society of America Journal 76 (2), 408–419. Available at https://doi.org/10.2136/sssaj2011.0273.

Fierer, N., 2017. Embracing the unknown: disentangling the complexities of the soil microbiome. Nature Reviews. Microbiology 15 (10), 579–590. Available at https://doi.org/10.1038/nrmicro.2017.87.

Friedlingstein, P., Cox, P., Betts, R., Bopp, L., von Bloh, W., Brovkin, V., et al., 2006. Climate–carbon cycle feedback analysis: results from the C4MIP model intercomparison. Journal of Climate 19 (14), 3337–3353. Available at https://doi.org/10.1175/jcli3800.1.

Gadd, G.M., 2010. Metals, minerals and microbes: geomicrobiology and bioremediation. Microbiology (Reading, England) 156 (3), 609–643. Available at https://doi.org/10.1099/mic.0.037143-0.

Galloway, A.F., Pedersen, M.J., Merry, B., Marcus, S.E., Blacker, J., Benning, L.G., et al., 2018. Xyloglucan is released by plants and promotes soil particle aggregation. New Phytologist 217 (3), 1128–1136. Available at https://doi.org/10.1111/nph.14897.

García, F.C., Clegg, T., O'Neill, D.B., Warfield, R., Pawar, S., Yvon-Durocher, G., 2023. The temperature dependence of microbial community respiration is amplified by changes in species interactions. Nature Microbiology 8 (2), 272–283. Available at https://doi.org/10.1038/s41564-022-01283-w.

Giri, S., Yousif, G., Shitut, S., Oña, L., Kost, C., 2022. Prevalent emergence of reciprocity among cross-feeding bacteria. ISME Communications 2 (1). Available at https://doi.org/10.1038/s43705-022-00155-y.

Goel, R., Kumar, V., Suyal, D.C., Narayan Soni, R., 2018. Toward the unculturable microbes for sustainable agricultural production. In: Meena, V.S. (Ed.), Role of Rhizospheric Microbes in Soil. Springer, Singapore, pp. 107–123. Available at https://doi.org/10.1007/978-981-10-8402-7_4.

Gupta, V.V.S.R., Germida, J.J., 2015. Soil aggregation: influence on microbial biomass and implications for biological processes. Soil Biology and Biochemistry 80, A3. Available at https://doi.org/10.1016/j.soilbio.2014.09.002.

Harmon, M.E., Silver, W.L., Fasth, B., Chen, H., Burke, I.C., Parton, W.J., et al., 2009. L. Blum, Long-term patterns of mass loss during the decomposition of leaf and fine root litter: an intersite comparison. Global Change Biology 15 (5), 1320–1338. Available at https://doi.org/10.1111/j.1365-2486.2008.01837.x.

Hu, J., Guo, H., Xue, Y., Gao, Mt, Zhang, S., Tsang, Y.F., et al., 2019. Using a mixture of microalgae, biochar, and organic manure to increase the capacity of soil to act as carbon sink. Journal of Soils and Sediments 19 (11), 3718–3727. Available at https://doi.org/10.1007/s11368-019-02337-z.

Hungate, B.A., Marks, J.C., Power, M.E., Schwartz, E., Groenigen, K.J.V., Blazewicz, S.J., et al., 2021. The functional significance of bacterial predators. mBio 12 (2). Available at https://doi.org/10.1128/mBio.00466-21.

Ingram, J., Mills, J., Dibari, C., Ferrise, R., Ghaley, B.B., Hansen, J.G., et al., 2016. Communicating soil carbon science to farmers: Incorporating credibility, salience and legitimacy. Journal of Rural Studies 48, 115–128. Available at https://doi.org/10.1016/j.jrurstud.2016.10.005.

Jansson, J.K., Hofmockel, K.S., 2018. The soil microbiome — from metagenomics to metaphenomics. Current Opinion in Microbiology 43, 162–168. Available at https://doi.org/10.1016/j.mib.2018.01.013.

Jansson, J.K., Hofmockel, K.S., 2020. Soil microbiomes and climate change. Nature Reviews. Microbiology 18 (1), 35–46. Available at https://doi.org/10.1038/s41579-019-0265-7.

Jansson, J.K., Wu, R., 2022. Soil viral diversity, ecology and climate change. Nature Reviews. Microbiology. Available at https://doi.org/10.1038/s41579-022-00811-z.

Janzen, H.H., 2004. Carbon cycling in earth systems—a soil science perspective. Agriculture, Ecosystems & Environment 104 (3), 399–417. Available at https://doi.org/10.1016/j.agee.2004.01.040.

Jassey, V.E.J., Walcker, R., Kardol, P., Geisen, S., Heger, T., Lamentowicz, M., et al., 2022. Contribution of soil algae to the global carbon cycle. New Phytologist 234 (1), 64–76. Available at https://doi.org/10.1111/nph.17950.

Jhariya, M.K., Meena, R.S., Banerjee, A., Meena, S.N., 2022. Natural Resources Conservation and Advances for Sustainability. Elsevier. Available at https://doi.org/10.1016/C2019-0-03763-6.

Jhariya, M.K., Meena, R.S., Banerjee, A., Kumar, S., Raj, A., 2023. Agroforestry for Carbon and Ecosystem Management. Academic Press, pp. 1–434. Available at 10.1016/C2021-0-02916-3.

Jiang, P., Xiao, L.Q., Wan, X., Yu, T., Liu, Y.F., Liu, M.X., 2022. Research progress on microbial carbon sequestration in soil: a review. Eurasian Soil Science 55 (10), 1395–1404. Available at https://doi.org/10.1134/s1064229322100064.

Joergensen, R.G., Wichern, F., 2018. Alive and kicking: why dormant soil microorganisms matter. Soil Biology and Biochemistry 116, 419–430. Available at https://doi.org/10.1016/j.soilbio.2017.10.022.

Jones, S.E., Lennon, J.T., 2010. Dormancy contributes to the maintenance of microbial diversity. Proceedings of the National Academy of Sciences of the United States of America 107 (13), 5881–5886. Available at https://doi.org/10.1073/pnas.0912765107.

Keiluweit, M., Wanzek, T., Kleber, M., Nico, P., Fendorf, S., 2017. Anaerobic microsites have an unaccounted role in soil carbon stabilization. Nature Communications 8 (1). Available at https://doi.org/10.1038/s41467-017-01406-6.

Knoche, K.L., Aoyama, E., Hasan, K., Minteer, S.D., 2017. Role of nitrogenase and ferredoxin in the mechanism of bioelectrocatalytic nitrogen fixation by the cyanobacteria anabaena variabilis SA-1 mutant immobilized on indium tin oxide (ITO) electrodes. Electrochimica Acta 232, 396–403. Available at https://doi.org/10.1016/j.electacta.2017.02.148.

Kopittke, P.M., Berhe, A.A., Carrillo, Y., Cavagnaro, T.R., Chen, D., Chen, Q.L., et al., 2022. Ensuring planetary survival: the centrality of organic carbon in balancing the multifunctional nature of soils. Critical Reviews in Environmental Science and Technology 52 (23), 4308–4324. Available at https://doi.org/10.1080/10643389.2021.2024484.

Kragt, M.E., Dumbrell, N.P., Blackmore, L., 2017. Motivations and barriers for Western Australian broad-acre farmers to adopt carbon farming. Environmental Science and Policy 73, 115–123. Available at https://doi.org/10.1016/j.envsci.2017.04.009.

Kramer, C., Gleixner, G., 2008. Soil organic matter in soil depth profiles: distinct carbon preferences of microbial groups during carbon transformation. Soil Biology and Biochemistry 40 (2), 425–433. Available at https://doi.org/10.1016/j.soilbio.2007.09.016.

Kremer, B., Kazmierczak, J., Stal, L.J., 2008. Calcium carbonate precipitation in cyanobacterial mats from sandy tidal flats of the North Sea. Geobiology 6 (1), 46–56. Available at https://doi.org/10.1111/j.1472-4669.2007.00128.x.

Kuzyakov, Y., Horwath, W.R., Dorodnikov, M., Blagodatskaya, E., 2019. Review and synthesis of the effects of elevated atmospheric CO_2 on soil processes: no changes in pools, but increased fluxes and accelerated cycles. Soil Biology and Biochemistry 128, 66–78. Available at https://doi.org/10.1016/j.soilbio.2018.10.005.

Lal, R., 2004. Soil carbon sequestration impacts on global climate change and food security. Science (New York, N.Y.) 304 (5677), 1623–1627. Available at https://doi.org/10.1126/science.1097396.

Lal, R., 2008. Carbon sequestration. Philosophical Transactions of the Royal Society B: Biological Sciences 363, 815–830. Available at https://doi.org/10.1098/rstb.2007.2185.

Lange, M., Eisenhauer, N., Sierra, C.A., Bessler, H., Engels, C., Griffiths, R.I., et al., 2015. Plant diversity increases soil microbial activity and soil carbon storage. Nature Communications 6. Available at https://doi.org/10.1038/ncomms7707.

Lei, J., Guo, X., Zeng, Y., Zhou, J., Gao, Q., Yang, Y., 2021. Temporal changes in global soil respiration since 1987. Nature Communications 12 (1). Available at https://doi.org/10.1038/s41467-020-20616-z.

Li, X., Zhang, X., Xu, M., Ye, Q., Gao, H., He, X., 2022. Improved tolerance of *Artemisia ordosica* to drought stress via dark septate endophyte (DSE) symbiosis. Journal of Fungi 8 (7), 730. Available at https://doi.org/10.3390/jof8070730.

Liang, C., Schimel, J.P., Jastrow, J.D., 2017. The importance of anabolism in microbial control over soil carbon storage. Nature Microbiology 2 (8). Available at https://doi.org/10.1038/nmicrobiol.2017.105.

Liang, C., Amelung, W., Lehmann, J., Kästner, M., 2019. Quantitative assessment of microbial necromass contribution to soil organic matter. Global Change Biology 25 (11), 3578–3590. Available at https://doi.org/10.1111/gcb.14781.

Liao, H., Zhang, Y., Zuo, Q., Du, B., Chen, W., Wei, D., et al., 2018. Contrasting responses of bacterial and fungal communities to aggregate-size fractions and long-term fertilizations in soils of northeastern China. Science of The Total Environment 635, 784–792. Available at https://doi.org/10.1016/j.scitotenv.2018.04.168.

Liao, H., Zhang, Y., Wang, K., Hao, X., Chen, W., Huang, Q., 2020. Complexity of bacterial and fungal network increases with soil aggregate size in an agricultural Inceptisol. Applied Soil Ecology 154, 103640. Available at https://doi.org/10.1016/j.apsoil.2020.103640.

Lin, Y., Ye, G., Kuzyakov, Y., Liu, D., Fan, J., Ding, W., 2019. Long-term manure application increases soil organic matter and aggregation, and alters microbial community structure and keystone taxa. Soil Biology and Biochemistry 134, 187–196. Available at https://doi.org/10.1016/j.soilbio.2019.03.030.

Lipson, D.A., 2015. The complex relationship between microbial growth rate and yield and its implications for ecosystem processes. Frontiers in Microbiology 6. Available at https://doi.org/10.3389/fmicb.2015.00615.

Liu, T., Bruins, R., Heberling, M., 2018a. Factors influencing farmers' adoption of best management practices: a review and synthesis. Sustainability 10 (2), 432. Available at https://doi.org/10.3390/su10020432.

Liu, Z., Sun, Y., Zhang, Y., Feng, W., Lai, Z., Fa, K., et al., 2018b. Metagenomic and 13 C tracing evidence for autotrophic atmospheric carbon absorption in a semiarid desert. Soil Biology and Biochemistry 125, 156–166. Available at https://doi.org/10.1016/j.soilbio.2018.07.012.

Lladó, S., López-Mondéjar, R., Baldrian, P., 2017. Forest soil bacteria: diversity, involvement in ecosystem processes, and response to global change. Microbiology and Molecular Biology Reviews: MMBR 81, e00063. Available at https://doi.org/10.1128/MMBR.00063-16.

Luo, Y., Weng, E., 2011. Dynamic disequilibrium of the terrestrial carbon cycle under global change. Trends in Ecology and Evolution 26 (2), 96–104. Available at https://doi.org/10.1016/j.tree.2010.11.003.

Ma, W., Peng, D., Walker, S.L., Cao, B., Gao, C.H., Huang, Q., et al., 2017. *Bacillus subtilis* biofilm development in the presence of soil clay minerals and iron oxides. npj Biofilms and Microbiomes 3 (1). Available at https://doi.org/10.1038/s41522-017-0013-6.

Macintosh, A., Waugh, L., 2012. An introduction to the carbon farming initiative: key principles and concepts. Environmental and Planning Law Journal 29 (6), 439–461.

Maier, S., Tamm, A., Wu, D., Caesar, J., Grube, M., Weber, B., 2018. Photoautotrophic organisms control microbial abundance, diversity, and physiology in different types of biological soil crusts. ISME Journal 12 (4), 1032–1046. Available at https://doi.org/10.1038/s41396-018-0062-8.

Mallappa, M., Amrita, K., Kunal, R., Siddarthan, V., Radha, P., Balasubramanian, R., et al., 2016. Beneficial cyanobacteria and eubacteria synergistically enhance bioavailability of soil nutrients and yield of okra. Heliyon 2, e00066.

Marks, E.A.N., Miñón, J., Pascual, A., Montero, O., Navas, L.M., Rad, C., 2017. Application of a microalgal slurry to soil stimulates heterotrophic activity and promotes bacterial growth. Science of the Total Environment 605-606, 610–617. Available at https://doi.org/10.1016/j.scitotenv.2017.06.169.

Marschner, B., Brodowski, S., Dreves, A., Gleixner, G., Gude, A., Grootes, P.M., et al., 2008. How relevant is recalcitrance for the stabilization of organic matter in soils? Journal of Plant Nutrition and Soil Science 171 (1), 91–110. Available at https://doi.org/10.1002/jpln.200700049.

Maynard, D.S., Crowther, T.W., Bradford, M.A., 2017. Fungal interactions reduce carbon use efficiency. Ecology Letters 20 (8), 1034–1042. Available at https://doi.org/10.1111/ele.12801.

McInerney, M.J., Sieber, J.R., Gunsalus, R.P., 2009. Syntrophy in anaerobic global carbon cycles. Current Opinion in Biotechnology 20 (6), 623–632. Available at https://doi.org/10.1016/j.copbio.2009.10.001.

Meena, R.S., Pradhan, G., 2023. Industrial garbage-derived biocompost enhances soil organic carbon fractions, CO_2 biosequestration, potential carbon credits and sustainability index in a rice-wheat ecosystem. Environmental Research 235, 116525. Available at https://doi.org/10.1016/j.envres.2023.116525.

Meena, R.S., Pradhan, G., Kumar, S., Lal, R., 2023. Using industrial wastes for rice-wheat cropping and food-energy-carbon-water-economic nexus to the sustainable food system. Renewable and Sustainable Energy Reviews 187, 113756. Available at https://doi.org/10.1016/j.rser.2023.113756.

Meena, R.S., Lal, R., Kumar, S., Pradhan, G., Srinivasarao, C., Singh, A.K., et al., 2024a. Potential of Indian agriculture for capturing atmospheric CO_2 and monetizing carbon credits to the farmers: an overview and policy framework. Advances in Agronomy 188, 101–206. Available at https://doi.org/10.1016/bs.agron.2024.06.001.

Meena, R.S., Pradhan, G., Singh, K., Kumar, S., Singh, A.K., Shashidhar, K.S., et al., 2024b. Agriculture models for restoring degraded land to enhance CO_2 biosequestration and carbon credits in the Vindhyan region of India. Science of the Total Environment 929. Available at https://doi.org/10.1016/j.scitotenv.2024.172661.

Meena, R.S., Pradhan, G., Nalani, Kumar, S., Singh, A.K., Verma, S.K., et al., 2024c. Diversified cropping modules designed for soil restoration, CO_2 sequestration, and generating carbon credits. Land Degradation and Development 35 (14), 4314–4329. Available at https://doi.org/10.1002/ldr.5224.

Meena, R.S., Pradhan, G., Kumar, S., 2025a. Energy flow, eco-efficiency, and economic circulation with recycled industrial waste compost application in wheat and subsequent rice farming. Science of the Total Environment 967. Available at https://doi.org/10.1016/j.scitotenv.2025.178779.

Meena, R.S., Pradhan, G.S., Kumar, S., Jhariya, M.K., Rao, C.S., 2025b. Carbon-Negative Agriculture. Springer Nature, Singapore.

Mistry, A.N., Ganta, U., Chakrabarty, J., Dutta, S., 2019. A review on biological systems for CO_2 sequestration: organisms and their pathways. Environmental Progress and Sustainable Energy 38 (1), 127–136. Available at https://doi.org/10.1002/ep.12946.

Naitam, M.G., Kaushik, R., 2021. Archaea: an agro-ecological perspective. Current Microbiology 78 (7), 2510–2521. Available at https://doi.org/10.1007/s00284-021-02537-2.

Narassimhan, E., Gallagher, K.S., Koester, S., Alejo, J.R., 2018. Carbon pricing in practice: a review of existing emissions trading systems. Climate Policy 18 (8), 967–991. Available at https://doi.org/10.1080/14693062.2018.1467827.

Nath, D., Meena, V.S., 2018. Mycorrhizae: a potential microorganism and its implication in agriculture. In: Meena, V. (Ed.), Role of Rhizospheric Microbes in Soil: Stress Management and Agricultural Sustainability. Springer, Singapore, pp. 251–276. Available at https://doi.org/10.1007/978-981-10-8402-7_10.

Nath, D., Kumari, V., Durga, M.L., 2026. Sustainability Sciences in Asia and Africa,Elicitors for Sustainable Crop Production. pp. 169–185. Available at 10.1007/978-981-95-3533-0_7.

Nath, D., Durga, M.L., Kumari, V., Singh, S., 2025. Biochar: A Precious Resource from Biological Waste. pp. 305–319. Available at 10.1007/978-981-95-0425-1_12.

Naud, S., Ibrahim, A., Valles, C., Maatouk, M., Bittar, F., Alou, M.T., et al., 2022. Candidate phyla radiation, an underappreciated division of the human microbiome, and its impact on health and disease. Clinical Microbiology Reviews 35 (3). Available at https://doi.org/10.1128/cmr.00140-21.

Naylor, D., Sadler, N., Bhattacharjee, A., Graham, E.B., Anderton, C.R., McClure, R., et al., 2020. Soil microbiomes under climate change and implications for carbon cycling. Annual Review of Environment and Resources 45, 29–59. Available at https://doi.org/10.1146/annurev-environ-012320-082720.

Nazaries, L., Murrell, J.C., Millard, P., Baggs, L., Singh, B.K., 2013. Methane, microbes and models: fundamental understanding of the soil methane cycle for future predictions. Environmental Microbiology 15 (9), 2395–2417. Available at https://doi.org/10.1111/1462-2920.12149.

Ogawa, A., Golé, C., Bermudez, M., Habarugira, O., Joslin, G., McCain, T., et al., 2022. Extracellular DNAses facilitate antagonism and coexistence in bacterial competitor-sensing interference competition. Applied and Environmental Microbiology 88 (23). Available at https://doi.org/10.1128/aem.01437-22.

Oliverio, A.M., Geisen, S., Delgado-Baquerizo, M., Maestre, F.T., Turner, B.L., Fierer, N., 2020. The global-scale distributions of soil protists and their contributions to belowground systems. Science Advances 6 (4). Available at https://doi.org/10.1126/sciadv.aax8787.

Pradhan, G., Meena, R.S., 2023a. Interaction impact of biocompost on nutrient dynamics and relations with soil biota, carbon fractions index, societal value of CO_2 equivalent and ecosystem services in the wheat-rice farming. Chemosphere 339. Available at https://doi.org/10.1016/j.chemosphere.2023.139695.

Pradhan, G., Meena, R.S., 2023b. Utilizing waste compost to improve the atmospheric CO_2 capturing in the rice-wheat cropping system and energy-cum-carbon credit auditing for a circular economy. Science of the Total Environment 892. Available at https://doi.org/10.1016/j.scitotenv.2023.164572.

Pradhan, G., Meena, R.S., Kumar, S., Lal, R., 2023. Utilizing industrial wastes as compost in wheat-rice production to improve the above and below-ground ecosystem services. Agriculture, Ecosystems & Environment 358, 108704. Available at https://doi.org/10.1016/j.agee.2023.108704.

Prasanna, R., Adak, A., Verma, S., Bidyarani, N., Babu, S., Pal, M., et al., 2015. Cyanobacterial inoculation in rice grown under flooded and SRI modes of cultivation elicits differential effects on plant growth and nutrient dynamics. Ecological Engineering 84, 532–541. Available at https://doi.org/10.1016/j.ecoleng.2015.09.033.

Qualls, R.G., 2016. Long-term (13 Years) decomposition rates of forest floor organic matter on paired coniferous and deciduouswatersheds with contrasting temperature regimes. Forests 7 (10). Available at https://doi.org/10.3390/f7100231.

Raj, A., Jhariya, M.K., Meena, R.S., 2025. Agroforestry for Monetising Carbon Credits. Springer Nature, Switzerland.

Ramin, K.I., Allison, S.D., 2019. Carbon cycle implications of soil microbial interactions. In: Hurst, C.J. (Ed.), Understanding Terrestrial Microbial Communities, Advances in Environmental Microbiology. Springer International Publishing, Cham, pp. 1–29. Available at https://doi.org/10.1007/978-3-030-10777-2_1.

Renuka, N., Prasanna, R., Sood, A., Ahluwalia, A.S., Bansal, R., Babu, S., et al., 2016. Exploring the efficacy of wastewater-grown microalgal biomass as a biofertilizer for wheat. Environmental Science and Pollution Research 23 (7), 6608–6620. Available at https://doi.org/10.1007/s11356-015-5884-6.

Rodias, E., Aivazidou, E., Achillas, C., Aidonis, D., Bochtis, D., 2021. Water-energy-nutrients synergies in the agrifood sector: a circular economy framework. Energies 14 (1), 159. Available at https://doi.org/10.3390/en14010159.

Ruan, Y., Kuzyakov, Y., Liu, X., Zhang, X., Xu, Q., Guo, J., et al., 2023. Elevated temperature and CO_2 strongly affect the growth strategies of soil bacteria. Nature Communications 14 (1). Available at https://doi.org/10.1038/s41467-023-36086-y.

Russel, J., Røder, H.L., Madsen, J.S., Burmølle, M., Sørensen, S.J., 2017. Antagonism correlates with metabolic similarity in diverse bacteria. Proceedings of the National Academy of Sciences of the United States of America 114 (40), 10684–10688. Available at https://doi.org/10.1073/pnas.1706016114.

Salas Castelo, E.M., 2017. The role of factors that influence the adoption of the Australian carbon farming initiative-emissions reduction fund: a mixed methods study. Ph.D. Thesis, James Cook University, Douglas, Australia. https://doi.org/10.4225/28/5afb73ab1fb44.

Šantrůčková, H., Kotas, P., Bárta, J., Urich, T., Čapek, P., Palmtag, J., et al., 2018. Significance of dark CO_2 fixation in arctic soils. Soil Biology and Biochemistry 119, 11–21. Available at https://doi.org/10.1016/j.soilbio.2017.12.021.

Schmitz, O.J., Leroux, S.J., 2020. Food webs and ecosystems: linking species interactions to the carbon cycle. Annual Review of Ecology, Evolution, and Systematics 51, 271–295. Available at https://doi.org/10.1146/annurev-ecolsys-011720-104730.

Schmidt, T.M., Konopka, A.E., 2009. Physiological and ecological adaptations of slow-growing, heterotrophic microbes and consequences for cultivation. In: Epstein, S.S. (Ed.), Uncultivated Microorganisms, Microbiology Monographs. Springer, Berlin Heidelberg, pp. 257–276. Available at https://doi.org/10.1007/978-3-540-85465-4_11.

Schulz, F., Alteio, L., Goudeau, D., Ryan, E.M., Yu, F.B., Malmstrom, R.R., et al., 2018. Hidden diversity of soil giant viruses. Nature Communications 9 (1). Available at https://doi.org/10.1038/s41467-018-07335-2.

Shrestha, R.C., Ghazaryan, L., Poodiack, B., Zorin, B., Gross, A., Gillor, O., et al., 2022. The effects of microalgae-based fertilization of wheat on yield, soil microbiome and nitrogen oxides emissions. Science of the Total Environment 806 (Pt 3), 151320. Available at https://doi.org/10.1016/j.scitotenv.2021.151320.

de Siqueira, C.J., Lucia, C.M., Peixoto, A.P., Roberto, C.P., Rodrigues, D.A.I., Jose, R.V., 2017. Microalgae biofilm in soil: greenhouse gas emissions, ammonia volatilization and plant growth. The Science of the Total Environment 574, 1640–1648.

Sokol, N.W., Slessarev, E., Marschmann, G.L., Nicolas, A., Blazewicz, S.J., Brodie, E.L., et al., 2022. Life and death in the soil microbiome: how ecological processes influence biogeochemistry. Nature Reviews Microbiology 20 (7), 415–430. Available at https://doi.org/10.1038/s41579-022-00695-z.

Spohn, M., Müller, K., Höschen, C., Mueller, C.W., Marhan, S., 2020. Dark microbial CO_2 fixation in temperate forest soils increases with CO_2 concentration. Global Change Biology 26 (3), 1926–1935. Available at https://doi.org/10.1111/gcb.14937.

Stewart, C.E., Paustian, K., Conant, R.T., Plante, A.F., Six, J., 2007. Soil carbon saturation: concept, evidence and evaluation. Biogeochemistry 86 (1), 19–31. Available at https://doi.org/10.1007/s10533-007-9140-0.

Sulman, B.N., Phillips, R.P., Oishi, A.C., Shevliakova, E., Pacala, S.W., 2014. Microbe-driven turnover offsets mineral-mediated storage of soil carbon under elevated CO_2. Nature Climate Change 4 (12), 1099–1102. Available at https://doi.org/10.1038/nclimate2436.

Thakur, M.P., Geisen, S., 2019. Trophic regulations of the soil microbiome. Trends in Microbiology 27 (9), 771–780. Available at https://doi.org/10.1016/j.tim.2019.04.008.

Thompson, L.R., Sanders, J.G., McDonald, D., Amir, A., Ladau, J., Locey, K.J., et al., 2017. A communal catalogue reveals Earth's multiscale microbial diversity. Nature 551 (7681), 457–463. Available at https://doi.org/10.1038/nature24621.

Tiemann, L.K., Grandy, A.S., Atkinson, E.E., Marin-Spiotta, E., McDaniel, M.D., Hooper, D., 2015. Crop rotational diversity enhances belowground communities and functions in an agroecosystem. Ecology Letters 18 (8), 761–771. Available at https://doi.org/10.1111/ele.12453.

Toensmeier, E., 2016. The Carbon Farming Solution: A Global Toolkit of Perennial Crops and Regenerative Agriculture Practices for Climate Change Mitigation and Food Security. Chelsea Green Publishing, London, UK, pp. 2016.

Trivedi, P., Delgado-Baquerizo, M., Trivedi, C., Hu, H., Anderson, I.C., Jeffries, T.C., et al., 2016. Microbial regulation of the soil carbon cycle: evidence from gene-enzyme relationships. ISME Journal 10 (11), 2593–2604. Available at https://doi.org/10.1038/ismej.2016.65.

Trumbore, S.E., 1997. Potential responses of soil organic carbon to global environmental change. Proceedings of the National Academy of Sciences of the United States of America 94 (16), 8284–8291. Available at https://doi.org/10.1073/pnas.94.16.8284.

Vurukonda, S.S.K.P., Vardharajula, S., Shrivastava, M., SkZ, A., 2016. Enhancement of drought stress tolerance in crops by plant growth promoting rhizobacteria. Microbiological Research 184, 13–24. Available at https://doi.org/10.1016/j.micres.2015.12.003.

Wang, S., Sun, P., Zhang, G., Gray, N., Dolfing, J., Esquivel-Elizondo, S., et al., 2022. Contribution of periphytic biofilm of paddy soils to carbon dioxide fixation and methane emissions. The Innovation 3 (1), 100192. Available at https://doi.org/10.1016/j.xinn.2021.100192.

Wu, Y., Cai, P., Jing, X., Niu, X., Ji, D., Ashry, N.M., et al., 2019. Soil biofilm formation enhances microbial community diversity and metabolic activity. Environment International 132, 105116. Available at https://doi.org/10.1016/j.envint.2019.105116.

Wu, H., Cui, H., Fu, C., Li, R., Qi, F., Liu, Z., et al., 2024. Unveiling the crucial role of soil microorganisms in carbon cycling: a review. Science of the Total Environment 909, 168627. Available at https://doi.org/10.1016/j.scitotenv.2023.168627.

Xiong, W., Jousset, A., Guo, S., Karlsson, I., Zhao, Q., Wu, H., et al., 2018. Soil protist communities form a dynamic hub in the soil microbiome. ISME Journal 12 (2), 634–638. Available at https://doi.org/10.1038/ismej.2017.171.

Yen, H.W., Ho, S.H., Chen, C.Y., Chang, J.S., 2015. CO_2, NOx and SOx removal from flue gas via microalgae cultivation: a critical review. Biotechnology Journal 10 (6), 829–839. Available at https://doi.org/10.1002/biot.201400707.

Yu, X., Polz, M.F., Alm, E.J., 2019. Interactions in self-assembled microbial communities saturate with diversity. ISME Journal 13 (6), 1602–1617. Available at https://doi.org/10.1038/s41396-019-0356-5.

Yu, X., Qian, L., Tu, Q., Peng, Y., Wang, C., Wu, D., et al., 2023. Chemoautotrophic sulphur oxidizers dominate microbial necromass carbon formation in coastal blue carbon ecosystems. Functional Ecology 37 (10), 2634–2651. Available at https://doi.org/10.1111/1365-2435.14417.

Zeraatkar, A.K., Ahmadzadeh, H., Talebi, A.F., Moheimani, N.R., McHenry, M.P., 2016. Potential use of algae for heavy metal bioremediation, a critical review. Journal of Environmental Management 181, 817–831. Available at https://doi.org/10.1016/j.jenvman.2016.06.059.

Zhang, S., Wang, L., Wei, W., Hu, J., Mei, S., Zhao, Q., et al., 2018. Enhanced roles of biochar and organic fertilizer in microalgae for soil carbon sink. Biodegradation 29 (4), 313–321. Available at https://doi.org/10.1007/s10532-017-9790-0.

Zhang, C., Liu, S., Hussain, S., Li, L., Baiome, B.A., Xiao, S., et al., 2022. Fe(II) addition drives soil bacterial co-ocurrence patterns and functions mediated by anaerobic and chemoautotrophic taxa. Microorganisms 10 (3). Available at https://doi.org/10.3390/microorganisms10030547.

Zhu, T., Dittrich, M., 2016. Carbonate precipitation through microbial activities in natural environment, and their potential in biotechnology: a review. Frontiers in Bioengineering and Biotechnology 4. Available at https://doi.org/10.3389/fbioe.2016.00004.

Zhu, X., Jackson, R.D., DeLucia, E.H., Tiedje, J.M., Liang, C., 2020. The soil microbial carbon pump: From conceptual insights to empirical assessments. Global Change Biology 26 (11), 6032–6039. Available at https://doi.org/10.1111/gcb.15319.

Žifčáková, L., Větrovský, T., Howe, A., Baldrian, P., 2016. Microbial activity in forest soil reflects the changes in ecosystem properties between summer and winter: seasonal dynamics of a soil microbial community. Environmental Microbiology 18, 288–301. Available at https://doi.org/10.1111/1462-2920.13026.

Chapter 5

Farm waste recycling for the green economy

Musa Ojeba Innocent[1], Job Oloruntoba Samuel[2], Adams Mustapha[2], Mustapha Abdulsalam[1], Miracle Uwa Livinus[3], Asmau M. Maude[2], Bata Timothy[4], Ijah U.J.J.[2], Abioye O. Peter[2], and Abdulhakeem Idris Abdulhakeem[5]

[1]*Department of Microbiology, School of Sciences and Information Technology, Skyline University Nigeria, Kano, Nigeria,* [2]*Department of Microbiology, Federal University of Technology Minna, Minna, Nigeria,* [3]*Department of Biochemistry, School of Sciences and Information Technology, Skyline University Nigeria, Kano, Nigeria,* [4]*Department of Geology, Skyline University Nigeria, Kano, Nigeria,* [5]*Department of Fisheries and Aquaculture, Federal University Dutsin-Ma, Dutsin-Ma, Katsina, Nigeria*

Highlights

- Farm waste recycling is an eco-friendly approach.
- Green credit schemes may favor farm waste recycling.
- Financial incentives may promote sustainable farming practices.
- Policy formulation and regulatory frameworks are key for effective implementation.

5.1 Introduction

Agriculture is a cornerstone of human civilization, providing food, raw materials, and livelihoods for billions of people worldwide (Sawicka, 2020). However, it also generates significant waste, contributing to environmental degradation if not managed properly. Farm waste recycling provides a sustainable solution by converting waste into valuable resources, including compost, biogas, and biofertilizers. Green credit plans offer financial incentives for environmentally friendly practices, thereby further enhancing sustainability (Meena et al., 2025; Raj et al., 2025). The integration of farm waste recycling into these plans allows farmers to adopt more sustainable methods to reduce their environmental footprint while also benefiting economically (Mo et al., 2016; Meena et al., 2025).

Farm waste recycling involves collecting, processing, and reusing agricultural waste materials such as crop residues, animal manure, and agro-industrial by-products. This practice helps to mitigate environmental pollution by reducing the waste that would otherwise contribute to soil and water contamination (De Corato, 2020). By transforming waste into valuable products such as compost, biogas, and biofertilizers, farm waste recycling enhances soil fertility, provides renewable energy sources, and reduces the need for synthetic fertilizers). This sustainable approach not only addresses waste management issues but also supports more efficient and eco-friendly farming practices (Yang et al., 2021).

Piñeiro et al. (2020) stated that green credit plans are financial incentives specifically designed to promote environmentally friendly agriculture practices. These plans offer lower interest rates, favorable loan terms, or subsidies to farmers who adopt sustainable practices, such as farm waste recycling. By providing financial support, green credit plans make it more feasible for farmers to invest in environmentally beneficial technologies and methods (Zhang et al., 2021). One of the primary goals of green credit plans is to reduce agriculture's environmental footprint (Banerjee et al., 2020). Traditional agricultural practices often rely heavily on chemical fertilizers, pesticides, and fossil fuels, which can lead to soil degradation, water contamination, and increased greenhouse gas (GHG) emissions. By incentivizing sustainable practices, green credit plans encourage farmers to adopt methods that are less environmentally harmful. For example, farm waste recycling reduces the need for chemical fertilizers by providing natural alternatives, such as compost and biofertilizers, which enhance soil health and reduce pollution (Su et al., 2022). Moreover, green credit plans promote long-term sustainability in agriculture. Sustainable practices not only

Carbon Incentives, Plans, and Policies in Agriculture. DOI: https://doi.org/10.1016/B978-0-443-33557-0.00014-8

protect the environment but also improve the economic resilience of farms. By reducing dependency on synthetic inputs and optimizing resource use, sustainable practices can lower production costs and increase farm profitability. This financial benefit is crucial for the long-term viability of farms, particularly in the face of challenges such as climate change and resource scarcity (Prasad et al., 2021; Zhang et al., 2021). Additionally, green credit plans support broader environmental and social goals. By promoting practices that enhance biodiversity, improve soil and water quality, and reduce carbon (C) emissions, these plans contribute to global efforts to combat climate change and protect natural ecosystems. Socially, they can improve the quality of life in rural communities by fostering sustainable livelihoods and creating jobs related to green technologies and practices (Solazzo et al., 2016). This section provides a comprehensive analysis of farm waste recycling practices, the role of green credit plans in agriculture, and the potential benefits of integrating these two approaches. It will also discuss best practices for enhancing sustainability in agriculture through these innovative approaches, highlighting how farm waste recycling can be beneficial in green credit plans for agriculture.

5.2 Farm waste recycling practices

Farm waste recycling practices refer to the various methods and techniques employed to manage and repurpose the diverse types of waste generated by agricultural activities (Yang et al., 2021; Meena et al., 2024b). These practices aim to reduce the environmental impact of farm waste while also deriving value from these materials. One prominent practice is composting, where organic waste, such as crop residues, animal manure, and spoiled produce, undergoes controlled decomposition, resulting in a nutrient-rich compost that can be used as a natural fertilizer. Anaerobic digestion is another method that breaks down organic waste in the absence of oxygen, producing biogas for energy generation and a digestate that can serve as a soil amendment (Adegbeye et al., 2020). Farm waste can also be utilized as mulch, with crop residues such as straw and husks spread on fields to conserve soil moisture and suppress weeds. Certain residues and by-products can be used as animal fodder, reducing waste while providing a valuable feed source for livestock (Diacono et al., 2019). Energy production is another avenue for recycling farm waste, where materials such as crop residues and manure can be converted into renewable energy through processes such as combustion, gasification, or pyrolysis. Additionally, programs exist to collect and recycle agricultural plastics, such as greenhouse covers and mulch films, thereby reducing the amount of plastic waste sent to landfills (Muise et al., 2016). On-site waste management practices, such as vermicomposting, biogas plants, and waste-to-energy systems, are also implemented by many farms to manage waste locally and reduce the need for off-site disposal. These practices not only mitigate environmental pollution but also contribute to sustainable agriculture by creating valuable products and promoting resource conservation (De Corato, 2020).

5.2.1 Methods of recycling farm waste

There are various methods available for recycling different types of farm waste, each with its own advantages and considerations. Here are some of the standard methods of recycling farm waste:

5.2.1.1 Composting

Composting is one of the most widely adopted methods for recycling organic farm waste, including crop residues, animal manure, and spoiled produce. The process involves controlled decomposition of these materials by microorganisms, resulting in the production of nutrient-rich compost that can be used as a soil amendment and organic fertilizer (Sayara et al., 2020). There are different composting techniques, including:

a. **Windrow composting:** The organic waste is piled into long rows (windrows) and periodically turned to ensure proper aeration and decomposition (Gavilanes-Terán et al., 2016).
b. **Aerated static pile composting:** The waste is piled and aerated using a network of perforated pipes or a forced air system, reducing the need for frequent turning (Michel et al., 2022).
c. **In-vessel composting:** The composting process takes place in an enclosed vessel or reactor, allowing for better control of temperature, moisture, and aeration (Ajmal et al., 2020).

The study by Ayilara et al. (2020) reported that composting not only recycles organic waste but also helps to reduce GHG emissions that would otherwise occur from the decomposition of these materials.

5.2.1.2 Anaerobic digestion

Anaerobic digestion is a biological process that involves the breakdown of organic matter, such as animal manure and crop residues, in the absence of oxygen (Manyi-Loh et al., 2019). This process produces biogas, a mixture of methane and carbon dioxide (CO_2), which can be utilized to generate electricity or heat. The remaining digested material, known as digestate, is rich in nutrients and can be used as a soil amendment or fertilizer. Anaerobic digesters can be designed for on-farm use or as centralized facilities that process waste from multiple farms. The biogas produced can be used on-site for energy needs or fed into the natural gas grid (Bharathiraja et al., 2018).

5.2.1.3 Mulching and land application

Certain farm waste materials, such as crop residues like straw and husks, can be directly applied to agricultural fields as mulch. Mulching helps retain soil moisture, suppresses weed growth, and eventually decomposes, providing nutrients to the soil (Sarkar et al., 2020). Similarly, animal manure can be directly applied to agricultural fields as a natural fertilizer, provided it is properly treated and managed to prevent environmental contamination and odor issues (Iqbal et al., 2020).

5.2.1.4 Animal feed

Some crop residues and by-products, such as wheat straw, corn stover, and vegetable tops, can be used as fodder for livestock, reducing waste and providing a valuable source of animal feed. However, it is essential to ensure that the materials are free from harmful contaminants and have appropriate nutritional value for the specific livestock (Akram and Fırıncıoğlu, 2019).

5.2.1.5 Agricultural plastics recycling

Agricultural activities often involve the use of plastic materials, such as mulch films, greenhouse covers, and irrigation pipes. Many regions have established programs for collecting and recycling these plastics, reducing the amount of waste sent to landfills or incinerated. The collected plastics are typically sent to specialized recycling facilities, where they are sorted, cleaned, and processed into new plastic products or used as feedstock for energy recovery processes (El-Beltagi et al., 2022; Salama and Geyer, 2023; Musa et al., 2023).

5.2.1.6 Bioenergy production

Farm waste, including crop residues, animal manure, and woody waste from pruning and trimming, can be used as feedstock for bioenergy production through various processes, as stated by Duca and Toscano (2022), which include: (1) Combustion: Direct burning of the waste materials to generate heat or electricity. (2) Gasification: Conversion of the waste into a combustible gas (syngas) that can be used for electricity generation or as a fuel source. (3) Pyrolysis: Thermal decomposition of the waste in the absence of oxygen, producing bio-oil, syngas, and biochar (a solid C-rich material).

Bioenergy production not only provides a renewable energy source but also helps to reduce the volume of waste and associated disposal costs.

5.2.1.7 On-site waste management systems

Many farms implement on-site waste management systems to handle their waste locally and reduce the need for off-site disposal. These systems can include:

a. Vermicomposting: Utilizing earthworms to decompose organic waste, producing nutrient-rich vermicompost. This process enhances soil fertility and promotes robust plant growth, offering an efficient and eco-friendly solution for organic waste management on farms (Bellitürk, 2018).
b. Biogas plants: Small-scale anaerobic digesters break down organic matter to produce biogas, a renewable energy source. This biogas can be used for heating, cooking, or generating electricity on the farm, thereby reducing reliance on fossil fuels and promoting sustainability (Obileke et al., 2021).
c. Waste-to-energy systems: Facilities convert farm waste into heat or electricity through processes like combustion or gasification. These systems provide a sustainable energy source, reduce waste volume, and efficiently manage agricultural by-products, contributing to the farm's energy self-sufficiency and environmental sustainability (Foster et al., 2021).

Effective farm waste recycling requires a combination of these methods, tailored to the specific types and quantities of waste generated on each farm. Additionally, proper handling, treatment, and storage of waste materials are crucial for preventing environmental contamination and ensuring the safety of recycled products (Koul et al., 2022). It is also important to consider local regulations, infrastructure availability, and economic feasibility when selecting the most appropriate recycling methods for farm waste. Collaboration among farmers, waste management companies, and government agencies can help promote and facilitate the adoption of sustainable waste recycling practices in agriculture (Sharma et al., 2020).

5.2.2 Types of farm waste suitable for recycling

Farm waste recycling is an essential aspect of sustainable agriculture, as it not only reduces the environmental impact of agricultural activities but also creates valuable resources and promotes resource conservation (Alvarenga et al., 2017). There are various types of farm waste suitable for recycling, each with its own unique properties and recycling methods. Here are some of the major categories of farm waste and their potential for recycling:

5.2.2.1 Crop residues

Crop residues, such as stalks, leaves, and husks, are generated in abundance after harvesting crops like rice, wheat, corn, and cotton. These residues are rich in organic matter and can be effectively recycled through several methods, as reported by Hsu (2021) and De Corato (2020). They are as follows:

Composting: Crop residues can be composted along with other organic materials, such as animal manure and food waste, to produce a nutrient-rich compost that can be used as a soil amendment and fertilizer.

Mulching: Crop residues can be spread as mulch on fields, where they help retain soil moisture, suppress weed growth, and gradually decompose, providing nutrients to the soil.

Animal bedding: Certain crop residues, such as straw, can be used as bedding for livestock and later composted or used as a soil amendment.

Biofuel production: Crop residues can be utilized as feedstock for biofuel production through processes such as combustion, gasification, or pyrolysis, thereby providing a renewable energy source.

5.2.2.2 Animal manure

Malomo et al. (2018) found that livestock farming generates substantial amounts of animal manure, which can be a valuable resource if managed appropriately. Recycling methods for animal manure include:

Anaerobic digestion: Animal manure can undergo anaerobic digestion, a process where microorganisms break down the organic matter in the absence of oxygen, producing biogas (a mixture of methane and CO_2) and a nutrient-rich digestate that can be used as a soil amendment (Bharathiraja et al., 2018).

Composting: Animal manure can be composted, either alone or in combination with other organic wastes, to produce compost that can be used as a soil conditioner and fertilizer (Hsu, 2021).

Direct land application: In some cases, animal manure can be directly applied to agricultural fields as a natural fertilizer, provided it is appropriately treated and managed to prevent environmental contamination (Srivastav et al., 2024).

5.2.2.3 Agricultural plastics

Agricultural activities often involve the use of various plastic materials, such as mulch films, greenhouse covers, and irrigation pipes (Shah and Wu, 2020). These plastics can be recycled through dedicated collection and recycling programs:

Collection programs: Many regions have established programs to collect and recycle used agricultural plastics, ensuring these materials are properly managed and diverted from landfills or open burning (Shah and Wu, 2020).

Recycling facilities: Specialized recycling facilities process collected agricultural plastics, converting them into new plastic products or using them as feedstock for energy recovery processes (Shah and Wu, 2020).

5.2.2.4 Fruit and vegetable waste

Spoiled or unsold fruits and vegetables can contribute to significant food waste in the agricultural sector (Bilska et al., 2020). These organic wastes can be recycled through various methods:

Composting: Fruit and vegetable waste can be composted, either on-site or at centralized facilities, to produce nutrient-rich compost for agricultural use (Hsu, 2021).

Anaerobic digestion: Like animal manure, fruit and vegetable waste can be processed through anaerobic digestion to produce biogas and digestate (Bharathiraja et al., 2018).

Animal feed: In some cases, certain types of fruit and vegetable waste can be used as animal feed, particularly for livestock like pigs and cattle, reducing waste and providing a valuable nutrient source (Valdez-Arjona and Ramírez-Mella, 2019).

5.2.2.5 Pruning and trimming waste

Viswanath and Lubina (2018) reported that agricultural activities like pruning fruit trees, trimming hedges, and maintaining landscapes generate significant amounts of woody waste. This waste can be recycled through:

Chipping and mulching: Pruning and trimming waste can be chipped or shredded and used as mulch in agricultural fields or landscaping, helping to retain soil moisture and suppress weed growth (Prem et al., 2020).

Composting: Woody waste can be composted, although the process may take longer due to the higher lignin content in woody materials (Hsu, 2021).

Bioenergy production: Pruning and trimming waste can be used as feedstock for bioenergy production through processes like combustion or gasification, providing a renewable energy source (Martinez-Valencia et al., 2021).

Effective farm waste recycling practices not only reduce the environmental impact of agricultural activities but also create valuable resources, such as compost, biofuels, and renewable energy. By implementing these practices, farmers can contribute to sustainable agriculture, reduce waste disposal costs, and potentially generate additional revenue streams from the sale of recycled products or energy (Sharma et al., 2019; Koul et al., 2022).

According to Chia et al. (2020), the specific recycling methods employed may vary depending on factors such as the type and quantity of waste, local regulations, and available infrastructure. Additionally, proper handling and treatment of farm waste are crucial for preventing environmental contamination and ensuring the safety of recycled products.

5.3 Green economy plans in agriculture

Green economy plans for financial initiatives and products specifically designed to promote and support environmentally sustainable practices in the agriculture sector (Zhang et al., 2021). These plans aim to incentivize farmers and agricultural businesses to adopt eco-friendly and resource-efficient practices by providing them with favorable financial terms and conditions (Zhang et al., 2021).

Tian et al. (2022) stated that the primary objective of green credit plans is to encourage the adoption of sustainable agricultural practices that minimize negative environmental impacts while ensuring long-term viability and productivity. These plans typically offer benefits such as:

Reduced interest rates: Farmers who implement sustainable practices like organic farming, water conservation, or renewable energy initiatives may qualify for lower interest rates on agricultural loans or credit facilities.

Grants and subsidies: Green credit plans may offer grants, subsidies, or cost-sharing programs to offset the upfront costs of transitioning to sustainable practices, such as installing renewable energy systems, implementing precision agriculture technologies, or upgrading to energy-efficient equipment.

Preferential loan terms: Agricultural businesses that prioritize environmental sustainability may be eligible for preferential loan terms, such as repayment periods or more flexible collateral requirements, which facilitate their access to financing for green initiatives.

Technical assistance: Some green credit plans offer technical assistance and advisory services to farmers, helping them understand and implement sustainable practices effectively and navigate the application processes for green financing. Green credit plans are often offered by government agencies, development banks, or private financial institutions in collaboration with environmental organizations and agricultural authorities. These plans aim to create a financial incentive structure that encourages farmers and farming businesses to adopt sustainable practices, thereby contributing to environmental conservation, climate change mitigation, and long-term agricultural productivity (Su et al., 2022).

Gu et al. (2021) stated that providing access to favorable financing and support, green credit plans help to address the financial barriers that may hinder the adoption of sustainable agriculture practices. They also serve as a catalyst for the transition toward a more environmentally responsible and resilient agricultural sector.

Current trends in green credit schemes are shaping the future of sustainable finance and environmental conservation (Mohd and Kaushal, 2018). Green credit schemes are moving beyond carbon sequestration (C_{seq}) and emission reduction to encompass a broader range of environmental obligations. Activities like sustainable agriculture, water conservation, waste management, and biodiversity protection are now eligible for green credits. This expansion allows for more comprehensive environmental benefits and incentivizes a broader range of stakeholders to participate (Lee, 2020). Governments are increasingly adopting market-based approaches to incentivize voluntary ecological actions. Green credit programs leverage competitive forces by creating tradable green credits that can be bought and sold on domestic or international platforms. This creates a financial incentive for individuals, companies, and local bodies to invest in green projects and earn credits (Lyon and Maxwell, 2019). There is a growing emphasis on restoring degraded ecosystems rather than just planting trees. Green credit schemes prioritize the use of native species and promote activities that enhance biodiversity and ecosystem services. This shift recognizes that simply increasing tree cover does not automatically lead to a healthy and functional ecosystem (Chistyakova et al., 2019). Some green credit schemes allow the generated credits to be used for C trading in addition to compliance with environmental regulations. This creates a dual incentive for green projects by allowing them to earn both green credits and C credits. However, concerns persist about the equivalence between green credits and C credits, as well as the potential for greenwashing (Anukwonke and Abazu, 2022). Green credit schemes offer an opportunity to transform existing agricultural subsidy regimes that have become detrimental to natural resources. By linking economic incentives to the adoption of sustainable farming practices, green credits can encourage farmers to transition to more eco-friendly methods while still receiving financial support. This can help to address issues such as groundwater depletion and air pollution caused by unsustainable agricultural practices (Sun and Zeng, 2023). As green credit schemes continue to evolve, it will be crucial to establish robust methodologies, standards, and monitoring mechanisms to ensure their effectiveness and prevent greenwashing. Engaging stakeholders through public consultations and fostering sufficient demand for green credits will also be the key to the success of these innovative financing mechanisms (Sun and Zeng, 2023).

5.4 Linking and relating farm waste recycling to the green economy

Farm waste recycling encourages the adoption of sustainable agricultural practices by promoting the reuse and repurposing of waste materials, such as crop residues and manure (Alan and Köker, 2023). By recycling farm waste, farmers can improve soil health, reduce their reliance on chemical fertilizers, and enhance overall farm productivity, aligning with the goal of sustainable agriculture under green credit schemes. Recycling farm waste also helps reduce pollution, conserve natural resources, and protect public health through proper waste management practices. Green credit plans often incentivize activities that contribute to waste reduction and environmental protection, making farm waste recycling a valuable component in achieving these objectives (Kumar et al., 2024). Certain farm waste recycling practices, such as converting organic waste into biogas or bioenergy, contribute to renewable energy production and reduce dependence on fossil fuels. Green credit schemes that promote renewable energy initiatives can support and incentivize farmers to engage in waste-to-energy projects, thereby promoting the goals of sustainable energy production. To ensure effective alignment of farm waste recycling with green credit objectives, addressing challenges such as a lack of infrastructure, financial constraints, and behavioral change is crucial. Investing in on-farm recycling facilities, providing economic incentives through green credit plans, and offering educational programs to raise awareness about the benefits of waste recycling can help to overcome these challenges (Alan and Köker, 2023). The primary goal of farm waste recycling is to minimize the environmental impact of waste and generate renewable energy, making the green credit scheme applicable to farm waste recycling (Kumar et al., 2024). Here's how farm waste recycling supports green credit objectives, as per Kumar et al. (2024), are as follows:

5.4.1 Promoting sustainable agriculture

Farm waste recycling aligns closely with the objectives of green credit schemes in promoting sustainable agricultural practices. By recycling agricultural waste materials, such as crop residues and manure, farmers can improve soil health, reduce their reliance on chemical fertilizers, and enhance overall farm productivity. This directly supports the goal of sustainable agriculture under green credit plans, which incentivize farmers to adopt eco-friendly farming methods. The recycling of farm waste helps maintain nutrient balances, increase soil organic matter, and reduce reliance on energy-intensive synthetic inputs. In this way, farm waste recycling contributes to building a more sustainable and resilient agricultural system, a key objective of green credit schemes.

5.4.2 Reducing environmental impact

Recycling farm waste is a crucial component in reducing the environmental impact of agriculture and aligning with the objectives of green credit schemes. By properly managing agricultural waste through recycling practices, farmers can help mitigate pollution, conserve natural resources, and protect public health. Green credit plans often incentivize activities that contribute to waste reduction and environmental stewardship. When farmers adopt farm waste recycling methods, they not only improve their own farming practices but also contribute to broader environmental goals, making their operations more sustainable and eligible for green credit incentives. This alignment between farm waste recycling and green credit objectives is crucial for promoting a more eco-friendly agricultural sector.

5.4.3 Generating renewable energy

Certain farm waste recycling practices, such as converting organic waste into biogas or bioenergy, contribute to renewable energy production and reduce dependence on fossil fuels. Green credit schemes that promote renewable energy initiatives can support and incentivize farmers to engage in waste-to-energy projects, promoting the goals of sustainable energy production. It is imperative to address issues such as inadequate infrastructure, budgetary limitations, and behavioral change in order to guarantee the efficient alignment of farm waste recycling with green credit goals. To properly manage garbage, one must invest in recycling facilities on the farm. Waste-to-energy systems, biogas plants, and composting units can all be part of these facilities. Farmers can now more easily adopt recycling practices due to financial incentives offered by green credit plans, which help offset the high initial costs of installing these systems (Sharma et al., 2021).

Educational programs are essential for spreading awareness about the benefits of recycling. Farmers can acquire the information and skills from these programs to successfully establish and operate recycling systems. By demonstrating long-term economic and environmental benefits, these programs can persuade more farmers to participate in waste recycling (Pietzsch et al., 2017). Furthermore, cooperative efforts of governments, financial institutions, and agricultural associations can contribute to the development of a favourable legislative environment for waste recycling. Further encouraging the use of sustainable waste management techniques are financial incentives, market development support, and simplified regulatory procedures (Pietzsch et al., 2017).

5.5 Benefits of integrating farm waste recycling

5.5.1 Environmental benefit

Integrating farm waste recycling into green credit plans offers significant environmental benefits. Firstly, recycling agricultural waste reduces pollution by minimizing the need for chemical fertilizers and pesticides, which often contaminate soil and water bodies (Zhang et al., 2021). This helps to preserve local ecosystems and biodiversity. Secondly, farm waste recycling practices, such as composting and anaerobic digestion, reduce GHG emissions by capturing methane and converting it into biogas. This process not only mitigates climate change but also promotes cleaner energy use (Galanakis et al., 2022). Thirdly, recycled waste products like compost and biochar enhance soil health by improving soil structure, fertility, and water retention capacity. This leads to more sustainable farming practices, which in turn reduce erosion and agricultural land degradation. Overall, integrating farm waste recycling into green credit plans supports a healthier, more resilient environment and alligns agricultural practices with broader environmental conservation goals (Zhang et al., 2021).

5.5.2 Economic advantages for farmers

Farm waste recycling provides numerous economic advantages for farmers when integrated into green credit plans. By converting agricultural waste into valuable products such as compost, biofertilizers, and biogas, farmers can significantly reduce their reliance on expensive synthetic fertilizers and energy sources. This reduction in input costs leads to improved profitability and financial sustainability. Additionally, the sale of surplus recycled products, like compost or biogas, can generate new revenue streams, further enhancing farm income. Green credit plans can also offer lower interest rates and favorable loan to farmers for investing in recycling infrastructure, reducing the financial burden of initial setup costs. Enhancing soil health and fertility through organic amendments can lead to higher crop yields and higher-quality produce, thereby improving market competitiveness. Overall, farm waste recycling enhances economic resilience and profitability, making sustainable agriculture a viable and attractive option for farmers (Musa et al., 2024).

5.5.3 Social impacts on rural communities

Integrating farm waste recycling into green credit plans can have profound social impacts on rural communities. By promoting sustainable waste management practices, these initiatives create new job opportunities in areas such as compost production, biogas plant operation, and maintenance of waste-to-energy systems. This job creation stimulates local economies and reduces rural unemployment, contributing to improved livelihoods and community well-being. Additionally, reducing the use of chemical fertilizers and pesticides leads to healthier environmental conditions, decreasing health risks for farmers and residents. Improved environmental quality and the adoption of sustainable farming practices enhance food security and nutrition, ultimately benefiting the broader community. Educational programs and capacity-building initiatives associated with green credit plans also empower farmers with the knowledge and skills necessary to foster a culture of sustainability and innovation. Overall, the integration of farm waste recycling into green credit schemes strengthens rural communities, promoting social equity and long-term resilience.

5.6 Policy recommendations and best practices

5.6.1 Policy interventions to promote farm waste recycling

From a policy intervention perspective, effective promotion of farm waste recycling requires several policy interventions (Cattaneo et al., 2021). Governments should introduce subsidies and tax incentives for farmers who invest in recycling infrastructure such as composting units, biogas plants, and waste-to-energy systems. These financial incentives can significantly reduce the initial costs and encourage widespread adoption. Additionally, implementing regulatory frameworks that mandate waste management practices and provide clear guidelines can help standardize and streamline recycling efforts. These regulations can ensure that all agricultural operations adhere to best practices in waste management, thereby promoting uniformity and accountability across the sector (Bhuvaneshwari et al., 2019).

Establishing public–private partnerships can also enhance resource pooling and expertise sharing, making recycling initiatives more viable and sustainable. Such collaborations can leverage private-sector efficiency and innovation while ensuring public-sector oversight and support. Furthermore, government-funded education and training programs can raise awareness about the benefits of farm waste recycling and provide farmers with the necessary skills and knowledge. These programs can include workshops, seminars, and hands-on training sessions that empower farmers to implement effective recycling practices on their farms (Storey et al., 2015; Yeboah-Assiamah et al., 2017). Moreover, developing markets for recycled products through government procurement policies and certifications can create demand, making waste recycling more economically attractive. For example, governments can prioritize the purchase of compost and biogas generated from farm waste for public landscaping and energy projects. Certification schemes can assure buyers of the quality and environmental benefits of recycled products, boosting market confidence and demand (Xiong et al., 2020).

5.6.2 Strategies for incentivizing for the green economy

To incentivize green credit schemes, financial institutions should offer preferential loan terms, such as lower interest rates and extended repayment periods, for farmers adopting sustainable practices (Jin et al., 2022). These favourable terms can significantly reduce the financial burden on farmers, making it easier for them to invest in eco-friendly technologies and practices (Jhariya et al., 2022, 2023). Additionally, implementing eco-score systems to assess and reward environmentally friendly practices can motivate farmers to improve their sustainability efforts. These systems can provide a clear, measurable framework for evaluating the environmental impact of agricultural activities and offer tangible incentives to those who achieve high eco-scores (Jin et al., 2022). Oyewole et al. (2024) reported that governments can support these schemes by providing guarantees and risk-sharing mechanisms to reduce the financial risk for lenders. By mitigating potential losses, these measures encourage financial institutions to extend credit to farmers who might otherwise be deemed too risky. Additionally, integrating green credit with digital platforms can simplify the application process, making it more accessible to smallholder farmers. Digital tools can streamline documentation, reduce processing times, and enhance transparency, thereby increasing the appeal and uptake of green credit schemes.

Collaborations between banks, agribusinesses, and technology providers can also facilitate the dissemination of information and resources. These partnerships can leverage the strengths of each sector, providing farmers with comprehensive support, including financial products, technological solutions, and market access. Promoting awareness campaigns to highlight the benefits of green credit and showcasing successful case studies can further encourage participation. These campaigns can demonstrate the practical benefits of sustainable practices, building trust and interest among farmers (Akomea-Frimpong et al., 2022).

5.6.3 Best practices for sustainable agriculture and waste management

Adopting best practices in sustainable agriculture and waste management involves several key strategies. Firstly, implementing integrated pest management reduces the reliance on chemical pesticides, promoting ecological balance. Utilizing crop rotation and cover cropping enhances soil fertility and structure, reducing erosion and nutrient depletion. Incorporating organic farming techniques, such as compost and biofertilizers, improves soil health and productivity. Additionally, efficient water management practices, such as drip irrigation and rainwater harvesting, conserve water resources and increase efficiency. Establishing waste-recycling systems, such as vermicomposting and biogas production, turns farm waste into valuable resources, thereby minimizing environmental impact. Ultimately, engaging in continuous education and training ensures that farmers remain informed about the latest sustainable practices and technologies, thereby fostering a culture of innovation and sustainability in agriculture (Velten et al., 2015; Zeweld et al., 2017).

5.7 Conclusion

Integrating farm waste recycling into green credit plans represents a pivotal strategy for promoting sustainable agriculture. This approach aligns environmental, economic, and social benefits, making it a holistic solution to the challenges faced by the agricultural sector. By converting agricultural waste into valuable resources, farmers can reduce their reliance on synthetic inputs, cut costs, and enhance soil health. Green credit plans provide the necessary financial incentives and support to make these practices economically viable. Additionally, policy interventions and best practices play a crucial role in facilitating the adoption of farm waste recycling, ensuring that farmers have access to the necessary infrastructure, knowledge, and markets to succeed. The combined efforts of governments, financial institutions, and agricultural communities can drive significant progress toward a more sustainable, resilient, and prosperous agricultural future. As stakeholders continue to invest in and support these initiatives, the positive impacts on environmental conservation, rural development, and food security will become increasingly evident, reinforcing the value of integrating farm waste recycling into green credit plans.

References

Adegbeye, M.J., Salem, A.Z.M., Reddy, P.R.K., Elghandour, M.M.M., Oyebamiji, K.J., 2020. Waste Recycling for the Eco-friendly Input Use Efficiency in Agriculture and Livestock Feeding. In: Kumar, S., Meena, R.S., Jhariya, M.K. (Eds.), Resources Use Efficiency in Agriculture. Springer, Singapore, pp. 1–45. Available at https://doi.org/10.1007/978-981-15-6953-1_1.

Ajmal, M., Aiping, S., Uddin, S., Awais, M., Faheem, M., Ye, L., et al., 2020. A review on mathematical modeling of in-vessel composting process and energy balance. Biomass Conversion and Biorefinery 12, 4201–4213. Available at https://doi.org/10.1007/s13399-020-00883-y.

Akomea-Frimpong, I., Adeabah, D., Ofosu, D., Tenakwah, E.J., 2022. A review of studies on green finance of banks, research gaps and future directions. Journal of Sustainable Finance and Investment 12 (4), 1241–1264. Available at https://doi.org/10.1080/20430795.2020.1870202.

Akram, M.Z., Fırıncıoğlu, S.Y., 2019. The use of agricultural crop residues as alternatives to conventional feedstuffs for ruminants: a review. Eurasian Journal of Agricultural Research 3 (2), 58–66.

Alan, H., Köker, A.R., 2023. Analyzing and mapping agricultural waste recycling research: an integrative review for conceptual framework and future directions. Resources Policy 85. Available at https://doi.org/10.1016/j.resourpol.2023.103987.

Alvarenga, P., Palma, P., Mourinha, C., Farto, M., Dôres, J., Patanita, M., et al., 2017. Recycling organic wastes to agricultural land as a way to improve its quality: a field study to evaluate benefits and risks. Waste Management 61, 582–592. Available at https://doi.org/10.1016/j.wasman.2017.01.004.

Anukwonke, C.C., Abazu, C.I., 2022. Green business through carbon credits. In: Dervash, M.A., Wani, A.A. (Eds.), Climate Change Alleviation for Sustainable Progression: Floristic Prospects and Arboreal Avenues as a Viable Sequestration Tool. CRC Press, Boca Raton, pp. 315–332. Available at 10.1201/9781003106982-16.

Ayilara, M., Olanrewaju, O., Babalola, O., Odeyemi, O., 2020. Waste management through composting: challenges and potentials. Sustainability 12 (11), 4456. Available at https://doi.org/10.3390/su12114456.

Banerjee, A., Meena, R.S., Jhariya, M.K., Yadav, D.K., 2020. Agroecological Footprints Management for Sustainable Food System. Springer, Singapore, pp. 1–514. Available at https://doi.org/10.1007/978-981-15-9496-0.

Bellitürk, K., 2018. Vermicomposting in Turkey: challenges and opportunities in future. Eurasian Journal of Forest Science 6 (4), 32–41. Available at https://doi.org/10.31195/ejejfs.476504.

Bharathiraja, B., Sudharsana, T., Jayamuthunagai, J., Praveenkumar, R., Chozhavendhan, S., Iyyappan, J., 2018. Biogas production – a review on composition, fuel properties, feed stock and principles of anaerobic digestion. Renewable and Sustainable Energy Reviews 90, 570–582. Available at https://doi.org/10.1016/j.rser.2018.03.093.

Bhuvaneshwari, S., Hettiarachchi, H., Meegoda, J.N., 2019. Crop residue burning in India: policy challenges and potential solutions. International Journal of Environmental Research and Public Health 16 (5). Available at https://doi.org/10.3390/ijerph16050832.

Bilska, B., Tomaszewska, M., Kołożyn-Krajewska, D., Szczepański, K., Łaba, R., Łaba, S., 2020. Environmental aspects of food wastage in trade – a case study. Environmental Protection and Natural Resources 31 (2), 24–34. Available at https://doi.org/10.2478/oszn-2020-0009.

Cattaneo, A., Sánchez, M.V., Torero, M., Vos, R., 2021. Reducing food loss and waste: five challenges for policy and research. Food Policy 98. Available at https://doi.org/10.1016/j.foodpol.2020.101974.

Chia, W.Y., Chew, K.W., Le, C.F., Lam, S.S., Chee, C.S.C., Ooi, M.S.L., et al., 2020. Sustainable utilization of biowaste compost for renewable energy and soil amendments. Environmental Pollution 267. Available at https://doi.org/10.1016/j.envpol.2020.115662.

Chistyakova, G., Rolgayzer, A., Bondareva, E., Schlee, I., Tyulenev, M., Zhironkin, S., et al., 2019. International practice of environmental challenges regulation. E3S Web of Conferences 105, 2024. Available at https://doi.org/10.1051/e3sconf/201910502024.

De Corato, U., 2020. Agricultural waste recycling in horticultural intensive farming systems by on-farm composting and compost-based tea application improves soil quality and plant health: a review under the perspective of a circular economy. Science of the Total Environment 738, 139840. Available at https://doi.org/10.1016/j.scitotenv.2020.139840.

Diacono, M., Persiani, A., Testani, E., Montemurro, F., Ciaccia, C., 2019. Recycling agricultural wastes and by-products in organic farming: biofertilizer production, yield performance and carbon footprint analysis. Sustainability 11 (14), 3824. Available at https://doi.org/10.3390/su11143824.

Duca, D., Toscano, G., 2022. Biomass energy resources: feedstock quality and bioenergy sustainability. Resources 11 (6), 57. Available at https://doi.org/10.3390/resources11060057.

El-Beltagi, H.S., Basit, A., Mohamed, H.I., Ali, I., Ullah, S., Kamel, E.A.R., et al., 2022. Mulching as a sustainable water and soil saving practice in agriculture: a review. Agronomy 12 (8). Available at https://doi.org/10.3390/agronomy12081881.

Foster, W., Azimov, U., Gauthier-Maradei, P., Molano, L.C., Combrinck, M., Munoz, J., et al., 2021. Waste-to-energy conversion technologies in the UK: processes and barriers – a review. Renewable and Sustainable Energy Reviews 135. Available at https://doi.org/10.1016/j.rser.2020.110226.

Galanakis, C.M., Brunori, G., Chiaramonti, D., Matthews, R., Panoutsou, C., Fritsche, U.R., 2022. Bioeconomy and green recovery in a post-COVID-19 era. Science of the Total Environment 808, 18791026. Available at https://doi.org/10.1016/j.scitotenv.2021.152180.

Gavilanes-Terán, I., Jara-Samaniego, J., Idrovo-Novillo, J., Bustamante, M.A., Moral, R., Paredes, C., 2016. Windrow composting as horticultural waste management strategy - a case study in Ecuador. Waste Management 48, 127–134. Available at https://doi.org/10.1016/j.wasman.2015.11.026.

Gu, B., van Grinsven, H.J.M., Lam, S.K., Oenema, O., Sutton, M.A., Mosier, A., et al., 2021. A credit system to solve agricultural nitrogen pollution. Innovation 2 (1). Available at https://doi.org/10.1016/j.xinn.2021.100079.

Hsu, E., 2021. Cost-benefit analysis for recycling of agricultural wastes in Taiwan. Waste Management 120, 424–432. Available at https://doi.org/10.1016/j.wasman.2020.09.051.

Iqbal, R., Raza, M.A.S., Valipour, M., Saleem, M.F., Zaheer, M.S., Ahmad, S., et al., 2020. Potential agricultural and environmental benefits of mulches—a review. Bulletin of the National Research Centre 44 (1). Available at https://doi.org/10.1186/s42269-020-00290-3.

Jhariya, M.K., Meena, R.S., Banerjee, A., Meena, S.N., 2022. Natural Resources Conservation and Advances for Sustainability. Elsevier. Available at https://doi.org/10.1016/C2019-0-03763-6.

Jhariya, M.K., Meena, R.S., Banerjee, A., Kumar, S., Raj, A., 2023. Agroforestry for Carbon and Ecosystem Management. Academic Press, pp. 1–434. Available at 10.1016/C2021-0-02916-3.

Jin, W., Ding, W., Yang, J., 2022. Impact of financial incentives on green manufacturing: loan guarantee vs. interest subsidy. European Journal of Operational Research 300 (3), 1067–1080. Available at https://doi.org/10.1016/j.ejor.2021.09.011.

Koul, B., Yakoob, M., Shah, M.P., 2022. Agricultural waste management strategies for environmental sustainability. Environmental Research 206. Available at https://doi.org/10.1016/j.envres.2021.112285.

Kumar, P., Raj, A., Kumar, V.A., 2024. Approach to reduce agricultural waste via sustainable agricultural practices. In: Srivastav, A.L., Bhardwaj, A.K., Kumar, M. (Eds.), Valorization of Biomass Wastes for Environmental Sustainability: Green Practices for the Rural Circular Economy. Springer, Cham, pp. 21–50. Available at https://doi.org/10.1007/978-3-031-52485-1_2.

Lee, J.W., 2020. Green finance and sustainable development goals: the case of China. Journal of Asian Finance Economics and Business 7 (7), 577–586.

Lyon, T.P., Maxwell, J.W., 2019. "Voluntary" approaches to environmental regulation. Economic Institutions and Environmental Policy. Routledge, London, pp. 75–120.

Malomo, G.A., Madugu, A.S., Bolu, S.A., 2018. Sustainable animal manure management strategies and practices. Agricultural Waste and Residues 119, 145–151.

Manyi-Loh, C.E., Mamphweli, S.N., Meyer, E.L., Okoh, A.I., 2019. Microbial anaerobic digestion: process dynamics and implications from the renewable energy, environmental and agronomy perspectives. International Journal of Environmental Science and Technology 16 (7), 3913–3934. Available at https://doi.org/10.1007/s13762-019-02380-w.

Martinez-Valencia, L., Camenzind, D., Wigmosta, M., Garcia-Perez, M., Wolcott, M., 2021. Biomass supply chain equipment for renewable fuels production: a review. Biomass and Bioenergy 148, 106054. Available at https://doi.org/10.1016/j.biombioe.2021.106054.

Meena, R.S., Pradhan, G.S., Kumar, S., Jhariya, M.K., Rao, C.S., 2025. Carbon-Negative Agriculture. Springer, Singapore, pp. 1–500. Available at https://doi.org/10.1007/978-981-95-1483-0.

Michel, F., O'Neill, T., Rynk, R., Gilbert, J., Smith, M., Aber, J., et al., 2022. Chapter 6 - Forced aeration composting, aerated static pile, and similar methods. In: Rynk, R. (Ed.), The Composting Handbook: A how-to and why manual for farm, municipal, institutional and commercial composters. Academic Press, pp. 197–269. Available at https://doi.org/10.1016/B978-0-323-85602-7.00007-8.

Mo, K.H., Alengaram, U.J., Jumaat, M.Z., Yap, S.P., Lee, S.C., 2016. Green concrete partially comprised of farming waste residues: a review. Journal of Cleaner Production 117, 122–138. Available at https://doi.org/10.1016/j.jclepro.2016.01.022.

Mohd, S., Kaushal, V.K., 2018. Green finance: a step towards sustainable development. MUDRA: Journal of Finance and Accounting 5 (1). Available at https://doi.org/10.17492/mudra.v5i01.13036.

Muise, I., Adams, M., Côté, R., Price, G.W., 2016. Attitudes to the recovery and recycling of agricultural plastics waste: a case study of Nova Scotia, Canada. Resources, Conservation and Recycling 109, 137–145. Available at https://doi.org/10.1016/j.resconrec.2016.02.011.

Musa, O.I., Auta, H.S., Ilyasu, U.S., Aransiola, S.A., Makun, H.A., Adabara, N.U., et al., 2023. Micro- and nanoplastics in environment: degradation, detection, and ecological impact. International Journal of Environmental Research 18, 1. Available at https://doi.org/10.1007/s41742-023-00551-9.

Musa, O.I., Samuel, J.O., Adams, M., Abdulsalam, M., Nathaniel, V., Maude, A.M., et al., 2024. Soil erosion, mineral depletion and regeneration. Prospect for Soil Regeneration and Its Impact on Environmental Protection. Springer, pp. 159–172. Available at https://doi.org/10.1007/978-3-031-53270-2_7.

Obileke, K.C., Nwokolo, N., Makaka, G., Mukumba, P., Onyeaka, H., 2021. Anaerobic digestion: technology for biogas production as a source of renewable energy—a review. Energy & Environment 32 (2), 191–225. Available at https://doi.org/10.1177/0958305x20923117.

Oyewole, A.T., Adeoye, O.B., Addy, W.A., Okoye, C.C., Ofodile, O.C., 2024. Enhancing global competitiveness of U.S. SMEs through sustainable finance: a review and future directions. International Journal of Management & Entrepreneurship Research 6 (3), 634–647. Available at https://doi.org/10.51594/ijmer.v6i3.876.

Pietzsch, N., Ribeiro, J.L.D., de Medeiros, J.F., 2017. Benefits, challenges and critical factors of success for Zero Waste: a systematic literature review. Waste Management 67, 324–353. Available at https://doi.org/10.1016/j.wasman.2017.05.004.

Piñeiro, V., Arias, J., Dürr, J., Elverdin, P., Ibáñez, A.M., Kinengyere, A., et al., 2020. A scoping review on incentives for adoption of sustainable agricultural practices and their outcomes. Nature Sustainability 3 (10), 809–820. Available at https://doi.org/10.1038/s41893-020-00617-y.

Prasad, R., Jhariya, M.K., Banerje, A., 2021. Advances in Sustainable Development and Management of Environmental and Natural Resources: Economic Outlook and Opinions, 2-volume set. Apple Academic Press, New York, pp. 1–918. Available at https://doi.org/10.1201/9781003187455.

Prem, M., Ranjan, P., Seth, N., Patle, G.T., 2020. Mulching techniques to conserve the soil water and advance the crop production - a review. Current World Environment Special Issue (1). Available at https://doi.org/10.12944/cwe.15.special-issue1.02.

Raj, A., Jhariya, M.K., Meena, R.S., 2025. Agroforestry for Monetising Carbon Credits. Springer, Cham, pp. 1–609. Available at https://doi.org/10.1007/978-3-031-97151-8.

Salama, K., Geyer, M., 2023. Plastic mulch films in agriculture: their use, environmental problems, recycling and alternatives. Environments 10 (10), 179. Available at https://doi.org/10.3390/environments10100179.

Sarkar, S., Skalicky, M., Hossain, A., Brestic, M., Saha, S., Garai, S., et al., 2020. Management of crop residues for improving input use efficiency and agricultural sustainability. Sustainability 12 (23), 9808. Available at https://doi.org/10.3390/su12239808.

Sawicka, B., Leal Filho, W., Azul, A., Brandli, L., Özuyar, P., Wall, T., 2020. Resilient agricultural practices. In: Leal Filho, W., Azul, A., Brandli, L., Özuyar, P., Wall, T. (Eds.), Zero Hunger. Encyclopedia of the UN Sustainable Development Goals. Springer, Cham, pp. 1–13. Available at https://doi.org/10.1007/978-3-319-69626-3_42-1.

Sayara, T., Basheer-Salimia, R., Hawamde, F., Sánchez, A., 2020. Recycling of organic wastes through composting: process performance and compost application in agriculture. Agronomy 10 (11), 1838. Available at https://doi.org/10.3390/agronomy10111838.

Shah, F., Wu, W., 2020. Use of plastic mulch in agriculture and strategies to mitigate the associated environmental concerns. Advances in Agronomy 164, 231–287. Available at https://doi.org/10.1016/bs.agron.2020.06.005.

Sharma, B., Vaish, B., Monika, Singh, U.K., Singh, P., Singh, R.P., 2019. Recycling of organic wastes in agriculture: an environmental perspective. International Journal of Environmental Research 13 (2), 409–429. Available at https://doi.org/10.1007/s41742-019-00175-y.

Sharma, H.B., Vanapalli, K.R., Cheela, V.S., Ranjan, V.P., Jaglan, A.K., Dubey, B., et al., 2020. Challenges, opportunities, and innovations for effective solid waste management during and post COVID-19 pandemic. Resources, Conservation and Recycling 162. Available at https://doi.org/10.1016/j.resconrec.2020.105052.

Sharma, H.B., Vanapalli, K.R., Samal, B., Cheela, V.R.S., Dubey, B.K., Bhattacharya, J., 2021. Circular economy approach in solid waste management system to achieve UN-SDGs: solutions for post-COVID recovery. Science of the Total Environment 800. Available at https://doi.org/10.1016/j.scitotenv.2021.149605.

Solazzo, R., Donati, M., Tomasi, L., Arfini, F., 2016. How effective is greening policy in reducing GHG emissions from agriculture? Evidence from Italy. Science of the Total Environment 573, 1115–1124. Available at https://doi.org/10.1016/j.scitotenv.2016.08.066.

Srivastav, A.L., Patel, N., Rani, L., Kumar, P., Dutt, I., Maddodi, B.S., et al., 2024. Sustainable options for fertilizer management in agriculture to prevent water contamination: a review. Environment, Development and Sustainability 26 (4), 8303–8327. Available at https://doi.org/10.1007/s10668-023-03117-z.

Storey, D., Santucci, L., Fraser, R., Aleluia, J., Chomchuen, L., 2015. Designing effective partnerships for waste-to-resource initiatives: lessons learned from developing countries. Waste Management and Research 33 (12), 1066–1075. Available at https://doi.org/10.1177/0734242X15602964.

Su, C.W., Li, W., Umar, M., Lobonţ, O.R., 2022. Can green credit reduce the emissions of pollutants? Economic Analysis and Policy 74, 205–219. Available at https://doi.org/10.1016/j.eap.2022.01.016.

Sun, C., Zeng, Y., 2023. Does the green credit policy affect the carbon emissions of heavily polluting enterprises? Energy Policy 180, 113679. Available at https://doi.org/10.1016/j.enpol.2023.113679.

Tian, C., Li, X., Xiao, L., Zhu, B., 2022. Exploring the impact of green credit policy on green transformation of heavy polluting industries. Journal of Cleaner Production 335, 130257. Available at https://doi.org/10.1016/j.jclepro.2021.130257.

Valdez-Arjona, L.P., Ramírez-Mella, M., 2019. Pumpkin waste as livestock feed: impact on nutrition and animal health and on quality of meat, milk, and egg. Animals: An Open Access Journal from MDPI 9 (10). Available at https://doi.org/10.3390/ani9100769.

Velten, S., Leventon, J., Jager, N., Newig, J., 2015. What is sustainable agriculture? A systematic review. Sustainability 7 (6), 7833–7865. Available at https://doi.org/10.3390/su7067833.

Viswanath, S., Lubina, P.A., 2018. Traditional agroforestry systems. Agroforestry: Anecdotal to Modern Science. Springer, Singapore, pp. 91–119. Available at 10.1007/978-981-10-7650-3_3.

Xiong, W., Chen, B., Wang, H., Zhu, D., 2020. Public–private partnerships as a governance response to sustainable urbanization: lessons from China. Habitat International 95, 102095. Available at https://doi.org/10.1016/j.habitatint.2019.102095.

Yang, L., Xiao, X., Gu, K., 2021. Agricultural waste recycling optimization of family farms based on environmental management accounting in Rural China. Sustainability 13 (10), 5515. Available at https://doi.org/10.3390/su13105515.

Yeboah-Assiamah, E., Asamoah, K., Kyeremeh, T.A., 2017. Decades of public-private partnership in solid waste management: a literature analysis of key lessons drawn from Ghana and India. Management of Environmental Quality: An International Journal 28 (1), 78–93. Available at https://doi.org/10.1108/MEQ-05-2015-0098.

Zeweld, W., Van Huylenbroeck, G., Tesfay, G., Speelman, S., 2017. Smallholder farmers' behavioural intentions towards sustainable agricultural practices. Journal of Environmental Management 187, 71–81. Available at https://doi.org/10.1016/j.jenvman.2016.11.014.

Zhang, S., Wu, Z., Wang, Y., Hao, Y., 2021. Fostering green development with green finance: an empirical study on the environmental effect of green credit policy in China. Journal of Environmental Management 296, 113159. Available at https://doi.org/10.1016/j.jenvman.2021.113159.

Chapter 6

Advocating carbon-based agroforestry systems

Manoj Kumar Jhariya[1], Arnab Banerjee[2], Ram Swaroop Meena[3], Abhishek Raj[4], Annpurna Devi[1], Pankaj[1], Bhimappa Honnappa Kittur[5], Suchismita Satapathy[3], and Sanjeev Kumar[3]

[1]*Department of Farm Forestry, Sant Gahira Guru Vishwavidyalaya, Ambikapur, Chhattisgarh, India,* [2]*School of Studies in Environmental Science, Pt. Ravishankar Shukla University, Raipur, Chhattisgarh, India,* [3]*Department of Agronomy, Institute of Agricultural Sciences, Banaras Hindu University, Varanasi, Uttar Pradesh, India,* [4]*Pt. Deendayal Upadhyay College of Horticulture and Forestry, Dr. Rajendra Prasad Central Agricultural University, Samastipur, Bihar, India,* [5]*Faculty of Technical Forestry, Indian Institute of Forest Management, Bhopal, Madhya Pradesh, India*

Highlights

- Agroforestry as a carbon incentive practice for agricultural sustainability.
- Agroforestry has reflected higher carbon sequestration potential under diverse conditions.
- Agroforestry serves as a tool for emission reduction, a builder of the soil carbon pool.
- Agroforestry integrates tree–crop–animal systems for sustainable production and a green economy.
- Monetization of AFS leads to a greener and cleaner economy.

6.1 Introduction

At the present time, the world is moving toward reducing carbon (C) emissions and has targets to achieve a low-C economy across various sectors. Therefore AF practice would be a nature-based solution (NBS) to achieve a sustainable future for the upcoming times (Lehmann et al., 2025). Hence, agroforestry (AF) becomes a mechanism or tool that contributes significantly to the development of climate-resilient agroecosystems, along with the management and protection of natural ecosystems (Meena et al., 2025; Jhariya et al., 2022). The major benefit of AF practices, apart from mitigating and combating climate change, is increase yields and productivity, as well as promote biodiversity conservation, generate livelihoods for rural stakeholders, and provide other ecosystem services (Jhariya et al., 2023).

Presently, agricultural activity acts as the primary reason behind deforestation globally, as in most cases, forest land has been rapidly converted into agricultural land to improve productivity and yield. Under such circumstances, AF acts to reverse the forest loss to a considerable extent (Raj et al., 2024). One typical example, such as the degraded eastern part of the Amazon forest, has been supplemented by AF plantations, which have shown significant improvement in the reforestation process. Deforestation is associated with soil degradation, and AF also aims to restore the soil quality and its ecosystem services for the welfare and benefits of society (Steinfeld et al., 2024).

Under traditional practice, primary emphasis was given to promoting native plant species in building the structure of the forest ecosystem (Maezumi et al., 2018). Therefore AF has a significant role to play in degraded land restoration (Wolz et al., 2018). The process of soil degradation is severe in the tropics, due to gross mismanagement of soil, and hence AF could be a suitable solution for restoring soil quality under degraded conditions (Lal, 2015; Muchane et al., 2020). AF is a diverse system that integrates crops and woody species on the same piece of land with diverse applications based on the geographical range. Such an integrated approach helps to improve biomass, nutrient flow, and soil carbon pool (SCP) through litter deposition on the forest floor (Fonte and Six, 2010). AF is considered as NBS; however, the proper potential to mitigate climate change is yet to be adequately explored. Screening of the best practices that would promote C mitigation, along with uncertainty in the carbon sequestration (C_{seq}) process, often creates problems regarding the assessment of AF potential.

Carbon Incentives, Plans, and Policies in Agriculture. DOI: https://doi.org/10.1016/B978-0-443-33557-0.00022-7

AF incorporates traditional practices with technological advancement and has been practiced since prehistoric times. Cultivation of olive trees in the Amazonian basin by Roman farmers, along with cacao and coffee crops, is a typical example of a tree-crop integration system of AF (Maezumi et al., 2018). AF, therefore, promotes ecosystem services along with gains in yield and production, followed by direct benefits to the farming community. Hence, future thrust should be developing proper research and development (R&D) along with funding to promote AF at the grassroots level for climate resilience and sustainable food systems globally. Hence, the present chapter aims to evaluate the potential role of AF in the C economy of agroecosystems, along with advocating for sustainable agriculture through climate change mitigation and adaptation via AF practices.

6.2 Agroforestry scenarios in developed and developing nations

Widely AF can be considered as an integrated land use management system in which tree, agricultural crop, and animal husbandry practices are done in the same piece of land, leading to multifaceted benefits. The main principles behind AF applications include diversity and integration, followed by various forms of ecological services in the form of C capture, storage, and sequestration (Meena et al., 2022, 2024). This, in turn, helps in developing climate resilience in the AFS, followed by an increase in the production potential of the concerned land area. Such an integrated approach helps to create a complex form of agroecosystem that provides sustainable yield and productivity (Banerjee et al., 2023).

Through AF applications, it was observed that effective management takes place through optimum resource utilization, followed by minimizing the negative consequences of climate change. The tree–crop interaction in the AFS significantly influences the soil ecosystem by preventing erosion and improving soil quality, health, and conservation of water. Biomass in the form of litter and other vegetal components significantly influences the cycling of nutrients and the fertility of soil. Thus AF ensures sustainable use of natural resources and its effective output through multifaceted benefits (Singh et al., 2024).

The scenario of AF in developed and developing nations seems to be different, as the purpose of implementing AF varies depending on the nature of output that is expected from a particular practice. For developed countries, AF acts as a supplementary practice within the agricultural system and mainly focuses on biodiversity and C_{seq}. On the other hand, in developing nations, it is an integral component of the farming process and has significant potential for livelihood generation and also income diversification for small land-holding farmers (Kumar et al., 2020, 2024).

In developed countries, AF is explored on a larger scale with modern techniques to increase the economic output. On the other hand, in developing countries, it is implemented on smaller plots through traditional approaches. Further, the environmental benefits of AF are usually focused on in developed nations, and on the other hand, the economic and associated benefits, such as livelihood generation, are being focused on by developing countries (Prasad et al., 2021). As developed countries are financially and technologically sound, they have developed significant policy formulation for the effective implementation of AF with R&D approaches. The scenario is different for developing nations, as they require proper funding and technological development to implement AF practices.

Globally, a lack of technical education, along with awareness among farming communities, often hinders the application of AF. Simultaneously, the lack of proper market and value chain processes for AF products is another bigger challenge for AF implementation. Further, in many places, there are issues with proper land allotment and tenure, and the conflicting nature of policies related to land use (Jhariya et al., 2024). The limited scope and application of developing financial mechanisms or credit systems hinder the promotion of AF. There is a significant level of variation in motivation and implementation strategies from a socio-economic perspective across regions, which is also a critical issue for the effective implementation of AF.

6.3 Carbon dynamics in an agroforestry system

C_{seq} is an important process that is being explored in various sectors as a mitigative strategy for climate change. It simply implies storage and capture of CO_2 from the ambient air and incorporation into the trees of agricultural systems (Meena et al., 2022, 2024). Thus AF becomes an integrated system of tree–crop–livestock nexus that provides multifaceted benefits from economic, social, and ecological perspectives. Within the system, the trees serve as a potential C sink through CO_2 absorption, which is converted into biomass and the C pool of the plant and soil. C_{seq} is a process that helps to mitigate climate change through AF practices. Hence, it involves the mechanism of C capture and storage from the atmosphere through tree–crop interactions (Kittur et al., 2023a; Meena et al., 2024). AF seems to provide multi-dimensional benefits from a socio-economic to an environmental perspective, leading to the fulfillment of Sustainable Development Goals (SDGs) (Kittur et al., 2023b). The tree components act as a potential C sink through photosynthesis

TABLE 6.1 Various agroforestry models promote carbon stock across the globe.

S.No.	Agroforestry model	Details	Location	References
1	28-years-old AFS	– AFS allocated more carbon (54.2%–60.6%) to the passive pool than to the cropland (53.3%). – AFS increased C stocks by 23.7%–35.6% in relation to cropland in the top 1-m soil. – Posses higher SOC stock (65.3–71.6 Mg/ha) than cropland (52.8 Mg/ha).	Tripura, India	Yadav et al. (2021)
2	31-years-old tree-based intercropping (TBI) system	– Red oak TBI systems had the highest SOC, followed by black walnut, Norway spruce, and white cedar, with values of 93.2, 83.9, 78.1, and 72.2 t C ha^{-1}, respectively. – Red oak TBI systems also had the highest mean tree C content, followed by black walnut, Norway spruce, and white cedar, with values of 299.7, 285.8, 255.4, and 70.1 kg C $tree^{-1}$, respectively.	Southern Ontario, Canada	Ijzerman et al. (2022)
3	Various global AFS	The C stock remains below 30 cm soil depth, which can be as high as 50%–60%. About 44% of the total C pool is present in 1 m of soil depth.	Central and Eastern European soils	Soussana and Lemaire (2014) and Orgill et al. (2014)
4	Tree-based (Mango, Litchi, Populus, Eucalyptus, Dalbergia, Prunus) ecosystem	The C stock of different components varies from 4.51 to 43.39 Mg ha^{-1}.	Central Himalayan Trai Region, India	Kanime et al. (2013)
5	Eucalyptus and Mahogany woodlot-based AFS	Higher SOC (1.60%) was found in Eucalyptus woodlot-based AFSand the least in Mahogany woodlot-based AFS (0.29%).	Bangladesh	Ali et al. (2022)
6	10–25 years tropical AFS	The soil carbon sequestration values range from 5 to 10 kg C ha^{-1} in approximately 25 years in extensive tree-intercropping systems of arid and semiarid lands to 100–250 kg C ha^{-1} in approximately 10 years in species-intensive, multistrata, shaded perennial systems and homegardens of the humid tropics.	Tropical AFS	Nair et al. (2009)

AFS, agroforestry systems.

mechanism and help store C in biomass and soil. Several processes were implemented for C_{seq} by the AF practice. Within the system, the different vegetal parts of trees act as a potential source of a C sink, which helps in the reduction of atmospheric CO_2 load and storage of C in the wood material. Further, the tree roots that get inserted deep in the soil liberate organic matter (OM) and enrich soil C in the soil profile. Hence, the soil quality and C stock improve significantly (Table 6.1). Litter and mulching are important components of forest and agroecosystems, which add a significant amount of OM to the soil and help in SOC build-up. The nexus of tree-crop-livestock promotes beneficial interactions by increasing productivity, reducing fertilizer use, and resulting in a low C footprint in the agroecosystem (Meena and Pradhan, 2023; Banerjee et al., 2020).

6.4 Potential and prospects of agroforestry systems as a carbon sink

The proper selection of species aiding in C_{seq} is essential to explore the potential of AFS (agroforestry systems). For example, the growth rate of trees is an important criterion, as species with a higher growth rate can sequester more C per unit area in comparison with species with a slow growth rate. Further, this would be regulated by the lifespan of the tree species, depending upon the storage of C on a long-term basis (Devi et al., 2024).

AF, apart from providing production and economic output, also helps to develop climate resiliency for small landholding farmers of the developing world against climate change (Lin, 2014). It is actually a NBS toward sustainable use of natural resources and provides multifaceted benefits through tree integration within the agricultural system (Torquebiau, 2017). Further, AF contributes to the SCP through C storage in the biomass of the different vegetal layers (IPCC, 2019). The integrated AF system that integrates crop and animal husbandry practices has shown significant promise in terms of enhanced C_{seq} and storage in vegetal components. It is a well-known fact that various forms of anthropogenic activities, such as fossil fuel combustion and deforestation, act as triggering agents of global climate change (Ripple et al., 2020). It also affects livelihood generation and hampers the well-being of human beings. The prevalence of natural disasters significantly hampers various sectors of human development, and thus, combating such mega events is a big challenge (Tucker et al., 2015). Various climate-resilient agricultural practices tend to improve the SCP of agricultural land and minimize the C losses (Meena et al., 2023). Such a mechanism can be significantly enhanced through AFS (Griscom et al., 2017). Literature studies at global AFS have shown efficiency in removing $1.1–2.2 \times 10^{15}$ g C from the ambient air in the upcoming times, as revealed from C stock analysis globally (Albrecht and Kandji, 2003). Under various geoclimatic conditions, the C stock value of AFS ranged between 9 and 63 Mg C per hectare (Montagnini and Nair, 2004). AF has such potential that it can reduce the pressure on natural forests by acting as a potential C sink. Besides their huge potential, very little emphasis was given to the integration of AFS into the agroecosystem or any other natural resource-based ecosystem (Sileshi and Mafongoya, 2006). However, through the application of ecological economics, promotion and adoption of AFS were studied in South Africa (Ajayi et al., 2007). The active role of AF involves the reduction of atmospheric CO_2 load in the pathway for combating climate change (Labata et al., 2012). Another advantage of such CO_2 fixation by AFS is the long-term storage of C. According to one estimate, AF aims to restore a significant amount of C both under forested and agricultural land in tropical conditions (Palm et al., 1999). AFS has a higher capacity of 2–3 times more C storage than the farmland on an annual basis (Peichl et al., 2006; Labata et al., 2012).

6.5 Methodological assessment of carbon-based agroforestry systems

Proper methodology for assessing C_{seq} and monitoring C stock is the need of the hour, which would help formulate policies for climate change mitigation (Nair, 2011). Mathematical applications in the form of allometric equation are widely practiced for the estimation of C stock and biomass in the above ground part of the trees (Nair, 2011). Various parameters such as height, wood density, and diameter are considered for assessing C stock value. Further, in the soil component, assessing soil C stock is an important part of the assessment methodology. For this, soil samples were collected from different soil depths, and soil C estimation was done following standard protocols (Gómez et al., 2022; Jhariya et al., 2024; Pradhan and Meena, 2023a). Recent advancements include the application of remote sensing and geographic information systems systems based on toposheets, aerial photography, and satellite imagery, which help in the assessment of tree cover under AFS, along with evaluating C stock.

Estimation of C stock in AFS across the globe varied significantly depending on the methodological approach. Also, the scientific database in this regard has huge lacunae, which is reflected in the literature review (Nair, 2011). Usually, C is present in soil, and the dry weight of the living beings is considered as sequestered C. Both above ground and below ground C stocks reflect significant variation among forestry, agriculture, and AFS. Hence, there are some inherent problems in the methodological assessment of C dynamics in AFS because of the nexus among forestry and agriculture

(Gómez et al., 2022). Further, a huge lacuna exists in understanding the ecological services of AFS, which would have an impact on the C financing mechanism under AFS (Nair, 2011). Further, the methodological component of assessment includes identification of C pools in different soil–plant components, various models of AFS under varied agroclimatic conditions, fulfill the specific objectives of C management, etc. A collection of proper data is an essential component of such analysis. However, such analysis is dependent upon some factors such as the nature of sampling, plot area, proper tree measurements, sampling of biomass and soil, etc., followed by their analysis. Regarding interpretation and data analysis, the use of the allometric equations, the calculation of C_{seq} and stock values, and their statistical interpretation for current and future predictions are the most important parts of the methodological assessment of C under AFS (Gómez et al., 2022). Some key considerations for the methodological assessment of C under AFS are shown in Fig. 6.1.

6.6 Carbon-based agroforestry systems toward a cleaner and greener economy

The ever-increasing human population has put significant pressure on the agricultural production unit, which has impacted the agrobiodiversity across the globe. It was observed that the ecological conversion of diverse habitats in terrestrial ecosystems is the triggering agent for biodiversity loss (Wilting et al., 2017). Implementation of AF practices would help in mixed culture practice development that would provide better production, ecosystem health, and an enhanced level of net gain. AF is itself a driver of eco-friendly technologies or NBS that contribute significantly to the green economy (Fig. 6.2). As the process is eco-friendly, it causes the least environmental damage, promotes cleanliness, and supports ecologically knowledge-based management of natural resources, thereby promoting the concept of cleaner production and environmental sustainability (Burton et al., 2002). As a component of sustainability, AFS addresses various dimensions of sustainable development along with keeping harmony with nature for a cleaner and greener future (Devi et al., 2024). One can consider the economy of AFS to be a C-based economy that aims toward C neutralization and offsets for achieving the SDGs. Throughout the studies, it was observed that when we discussed AFS, the major emphasis was given to the potential role of AFS toward C_{seq}, which at present times can be monetized in the form of C credits or green credits. Further, these C credits will act as the building blocks of the green economy, and there will be no externalities issues in terms of negative consequences of AFS in the overall growth and development of mankind. AF also performs the tasks of disaster/risk reduction, adaptation, and vulnerability reduction toward climatic hazards, followed by natural resource management at the local level and social well-being for people at risk from the vagaries of climate change (Tompkins and Neil Adger, 2005).

Hence, proper stress should be given to screening climate-resilient agricultural practices, followed by adaptability to altered climatic conditions (Lott et al., 2009). The crisis in chemical fertilizer production, as well as environmental degradation of land ecosystems, has forced a diversion of the policy of agricultural output from food security to sustainable practices that intensify the farm production in an eco-friendly way (Kiptot and Franzel, 2012; Krausmann et al., 2013; Jhariya et al., 2021b, 2021a). Various tree-based AFS can be effectively implemented that would promote emission reduction, atmospheric load of GHGs, development of biomass, and SCP (Zomer et al., 2009; Smith and

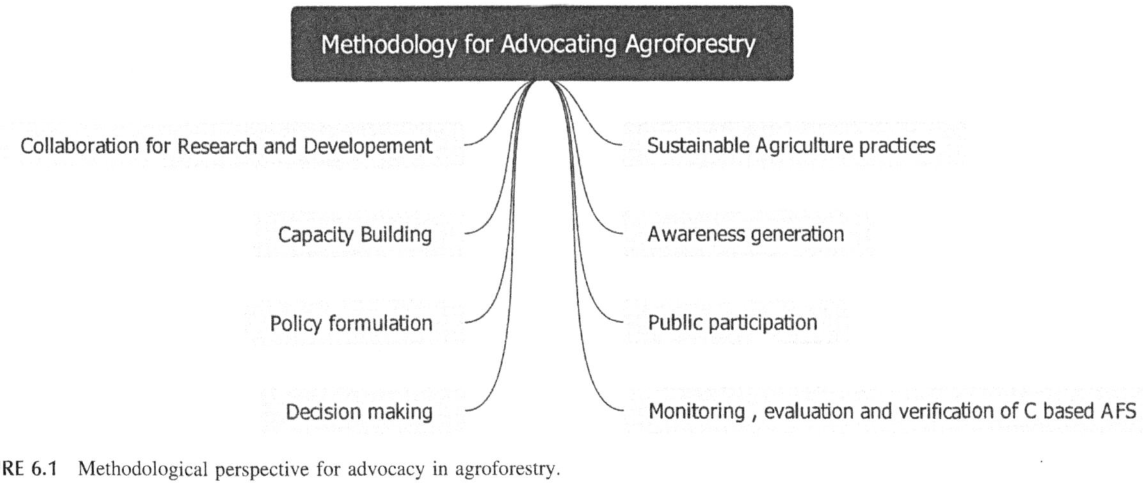

FIGURE 6.1 Methodological perspective for advocacy in agroforestry.

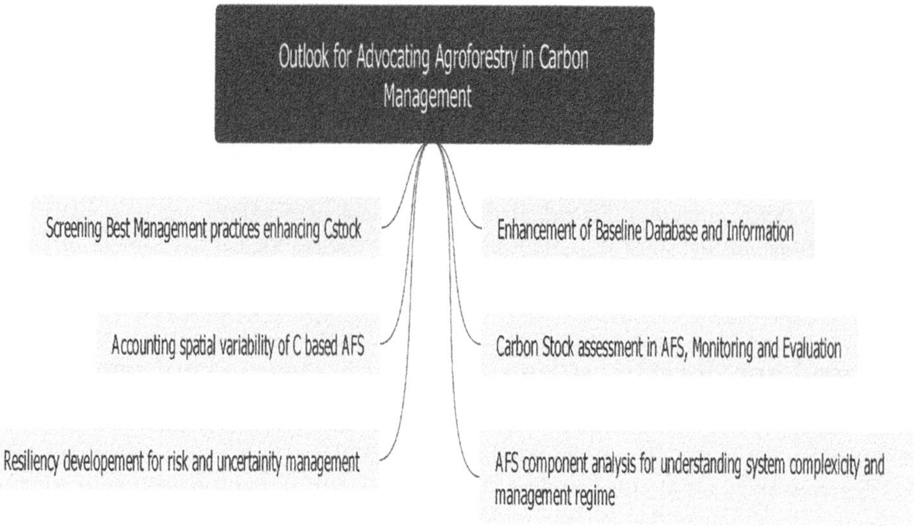

FIGURE 6.2 Carbon management through agroforestry: an empirical outlook.

Wollenberg, 2011). AFS acts as a critical nexus between the soil and atmosphere by acting as a potential C sink and maintaining a balance between various biophysical processes that regulate the C balance in the Earth's ecosystem (Li et al., 2015; Chavan et al., 2016; Pradhan and Meena, 2023b).

6.7 Advocacy of carbon-based agroforestry systems toward sustainable agriculture

From the present perspective, the availability of land for cultivation practices is becoming scarce and hence requires prior conservation. Due to the promotion of agricultural production, it was observed that various constraints have emerged that require prior attention in terms of species conservation. At present, agriculture is viewed as a triggering agent of deforestation, loss of habitat, and associated alteration in biodiversity (Hoang et al., 2023). As per a report, 1/4th of the area of forest degradation across the globe may be attributed to agricultural activity. Therefore traditional practices and indigenous knowledge are highly useful for sustainable land management that supports forest and biodiversity conservation and the maintenance of ecosystem services (Estrada et al., 2022). The cultural diversity across the globe is quite diverse in terms of language and cultural belief systems (World Bank, 2022). But still, there is a lack of proper recognition of traditional knowledge, and very little of these aspects are properly documented. Indigenous people, through their traditional knowledge, are capable of managing natural resources very efficiently using traditional, conventional practices (World Bank, 2022). All the welfare and benefits for mankind are related to the existing natural resources along with their proper management (Nath et al., 2022). Under traditional practices, AF is an age-old practice since prehistoric times due to its multiple forms of benefits under different environmental conditions.

AF and sustainable agricultural practices are intricately related to each other, and hence advocacy toward C-based AFS would promote sustainable agriculture. Advocating C-based AF would help to develop climate-resilient agricultural practices, provide an option for sustainable land management, and address issues of livestock management, food security, and nutritional improvement, followed by providing diverse forms of ecosystem services (Kittur et al., 2023b). Further, the process of advocating C-based AFS requires awareness generation, public participation, proper decision-making, and capacity building of local community stakeholders to implement C-based AF practices at the grassroots level, along with proper policy formulation that would support the purpose of implementing C-based AFS (Raj et al., 2021). Further, collaboration at local, regional, national, and international levels would help to promote C-based AFS as well as in the policy formulation process.

6.8 Monetization of carbon-based agroforestry systems

Monetizing the amount of emission reduction of GHGs is a most acceptable process for income generation through C credits and thus can be promoted for more CO_2 sequestration from the ambient air through proper plantation and AF practices (Raj et al., 2025; Pradhan et al., 2023). This would be a dual benefit for the farming community that would help the soil ecosystem as well as bring economic benefits to the farmer, and above all, it would be a big step toward mitigating climate change (Banerjee et al., 2024).

On a global basis, many organizations in the corporate sector are incapable of achieving the C neutrality goals, as their policy involves lacunae in the C offset mechanism to mitigate climate change. In this connection, C credits earned from AFS would be a promising tool to enhance the C offset mechanism in the corporate sector and provide organizations with an opportunity to achieve their target of C neutrality and move toward sustainable development (Meena et al., 2025).

C credit seems to show a growing nature among the sectors of forestry, agriculture, and other land uses, which have shown promise to earn an amount of 6283 million dollars. Further, it may increase to 97,100 million dollars by 2033. Hence, there is urgency to implement innovative practices and technological developments to achieve the SDGs (Meena et al., 2025a). In the energy sector, C credits are earned through energy conservation or forest protection. C credits in the AF sector focus on tree growth within the farming systems. Hence, it is more advantageous than the energy sector or forest protection projects, as it provides dual benefits: emission reduction and an increase in the productivity of agricultural systems. Apart from economic, ecological, and environmental benefits, the AF C credits have also long-term benefits in terms of developing a climate-resilient system, habitat preservation, and financial gain for the rural community (Fig. 6.3). Such systems thus become highly helpful for the corporate sectors that want to achieve the target of C neutrality and enhance their corporate social responsibilities toward the environment (Gómez et al., 2022).

6.9 Research and development toward carbon-based agroforestry systems

C-based AFS is a new area of research in which the proper exploration of C_{seq} potential can be assessed across different ecosystems with special reference to agroecosystems. Another important area is to explore the incorporation of AF practices into policy and planning for climate resilience and for the development of sustainable agroecosystems. New areas and innovations need to be addressed to screen flood-tolerant tree systems, multistrata cultivation systems, and integration of fruit trees, followed by community participation and environmental consequences (Roy et al., 2022). AF is associated with numerous problems related to pests, diseases, and natural resources; hence, these are the major challenges and research issues that need to be addressed for its effective implementation. Furthermore, a major emphasis on R&D activities should be placed on proper technology design, capacity building, and the formulation of suitable policies for effective land management. AF also highlights the formulation of strategies for SD and the development of adaptability in climate-vulnerable areas as a climate-resilient strategy or application. Further, AF is an NBS approach that strives toward finding new concepts and applications for developing sustainable agricultural practices (Lehmann et al., 2025).

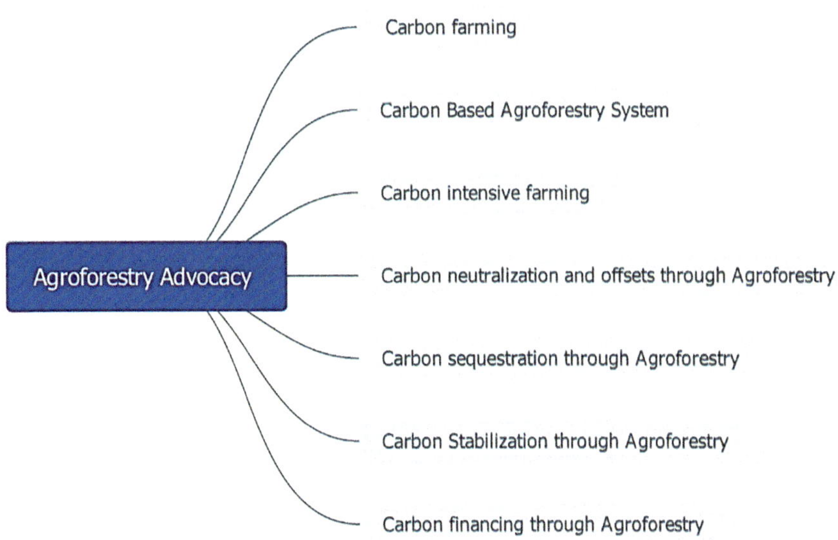

FIGURE 6.3 Various approaches for advocating agroforestry.

R&D activities for promoting C-based AFS have some key components that needs to be adequately addressed for effective implementation of C-credits in AFS. The first and foremost is the methodological assessment and monitoring of C_{seq} potential of different AF species under varied agroclimatic conditions. Following C_{seq} assessment of biomass, C stock is the secondary parameter that needs to be considered for identifying and screening potential AFS playing an important role in the C economy. Soil is an important component of AF practices, and hence a proper understanding of soil C dynamics is a new area in the R&D sector of AF, which is yet to be explored properly (Singh et al., 2024). In terms of decision-making through R&D in AFS, it requires proper species selection along with their management attributes for smooth and effective functioning. Intercropping practices between tree–crop systems under AFS from the perspective of biodiversity and C dynamics, need prior attention in terms of R&D activities (Moreira et al., 2024). If we consider AF on a component basis, specific strategies need to be formulated through R&D activities for the effective functioning of the system. For example, R&D activities in soil health and fertility maintenance, sustainable water management, initiatives for climate-resilient agricultural practices, and establishment and functionality of C markets, followed by livelihood generation and net economic return, are the most significant part that, for future perspectives, needs to be taken care of for the betterment of the C-based AFS. The R&D policies should also focus on critical issues such as nutritional security and food quality, specific models under different climatic conditions, implementation of practices for practicing AF in unmanaged agricultural lands, biocarbon projects, and, above all, accounting of the C in all diverse AFS is the need of the hour. Integration of C accounting in sustainable corporate management to fulfill the target of emission reduction and combating climate change is another most important R&D sector for C-based AFS (Sitompul et al., 2023).

6.10 Policy perspective for carbon-based agroforestry systems

Policy formulations and implementation for C-based AFS are the biggest challenges that need to be addressed properly in both the agriculture and forestry sectors. India has established a stepping stone with the formation of the National Agroforestry Policy in 2014 on a global basis, which was followed by neighboring countries like Nepal in 2019. On a regional basis, a meeting was held by the ASEAN Ministry of Agriculture and Forestry in 2015, which formulated a vision plan for 2016–2025. Such a policy identifies some critical issues, which include developing climate resilience and adaptation to natural disasters and climatic extremes, as well as screening suitable climate-resilient AFS that would be economically and ecologically suitable (Rizvi et al., 2024). Furthermore, at the 20th meeting of ASEAN Senior Officials of Forestry, policy guidelines for implementing AF were developed, which were subsequently adopted at the 39th meeting of ASEAN Ministers of Agriculture and Forestry in 2017. The entire policy reported 75 guidelines under the 14 principles, which address different dimensions of AF applications. Special emphasis was given to the other sustainable dimensions of AF in the ASEAN subcontinents through R&D activities. Various ASEAN countries were explored to identify the economic and ecological services of AF services, as well as identify the perception of the farming community for AF management (Park et al., 2022).

Policy initiatives promoting C-based AFS have key implications, including climate change mitigation and adaptation, sustainable use of land resources, and the maintenance of livelihood and nutritional security, among others. Some key policy instruments to promote C-based AFS include the generation of economic gain and profit through C-credit revenue generation, the establishment of voluntary C-markets, suitable land-use planning, the promotion of extension services, capacity building, skill development, public participation, ensuring the rights over land, followed by ready access for markets, proper evaluation of socio-environmental impacts, etc. (Meena et al., 2025; Raj et al., 2025). Various forms of C financing projects that would firmly establish the C market are the most potential area for policy making for effective implementation of C-based AFS. All such policies can help achieve the SDGs and lead to a sustainable future.

6.11 Conclusion

AF practices as an integrated approach have revealed huge potential for C_{seq} up to a level that it can be regarded as a C-based AFS. Furthermore, the tree–crop interaction of AFS, combined with animal husbandry practices, may enhance C storage, followed by sustainable land management practices. AF has also shown significant promise in above-ground, below-ground, and soil components and thus helps in CO_2 emission reduction in the atmosphere, followed by major C storage in the natural C sink of forests and agroecosystems on a long-term basis. For the mitigation of a changing climate, AF can be a significant ecosystem from a future perspective. Additionally, the higher potential of Cseq by the AFS is accompanied by various forms of ecosystem services, including improved soil health, sustainable yields and production, and the conservation of agrobiodiversity. Hence, considering all the potentialities of AF tends to make it an NBS and help humanity to move toward a low-C economy.

References

Ajayi, O.C., Akinnifesi, F.K., Sileshi, G., Chakeredza, S., 2007. Adoption of renewable soil fertility replenishment technologies in the southern African region: lessons learnt and the way forward. Natural Resources Forum 31 (4), 306–317. Available at https://doi.org/10.1111/j.1477-8947.2007.00163.x.

Albrecht, A., Kandji, S.T., 2003. Carbon sequestration in tropical agroforestry systems. Agriculture, Ecosystems and Environment 99 (1-3), 15–27. Available at https://doi.org/10.1016/S0167-8809(03)00138-5.

Ali, A.J., Rahman, S., Raihan, A., 2022. Soil carbon sequestration in agroforestry systems as a mitigation strategy of climate change: a case study from Dinajpur, Bangladesh. LIDSEN Publishing Inc, Bangladesh Advances in Environmental and Engineering Research 3 (4). Available at https://doi.org/10.21926/aeer.2204056.

Banerjee, A., Jhariya, M.K., Bargali, S.S., Palit, D., 2023. Ecorestoration for Sustainability. Wiley-Scrivener, USA, pp. 1–441.

Banerjee, A., Jhariya, M.K., Raj, A., Mechergui, T., 2024. Ecosystem management: climate change and global sustainability—an introduction. In: Banerjee, A., Jhariya, M.K., Raj, A, Mechergui, T. (Eds.), Ecosystem Management: Climate Change and Sustainability. Wiley, India, pp. 1–21. Available at 10.1002/9781394231249.ch1.

Banerjee, A., Meena, R.S., Jhariya, M.K., Yadav, D.K., 2020. Agroecological Footprints Management for Sustainable Food System. Springer, Singapore, pp. 1–514. Available at https://doi.org/10.1007/978-981-15-9496-0.

Burton, I., Huq, S., Lim, B., Pilifosova, O., Schipper, E.L., 2002. From impacts assessment to adaptation priorities: the shaping of adaptation policy. Climate Policy 2 (2-3), 145–159. Available at https://doi.org/10.1016/s1469-3062(02)00038-4.

Chavan, S.B., Uthappa, A.R., Sridhar, K.B., Keerthika, A., Handa, A.K., Newaj, R., et al., 2016. Trees for life: creating sustainable livelihood in Bundelkhand region of central India. Current Science 111 (6), 994. Available at https://doi.org/10.18520/cs/v111/i6/994-1002.

Devi, A., Jhariya, M.K., Raj, A., Banerjee, A., Singh, K.P., Singh, B., A. Raj, M.K. Jhariya, A. Banerjee, S. Nema and K. Bargali), 2024. Acacia nilotica: A Promising Species for Soil Sustainability. In: Raj, A., Jhariya, M.K., Banerjee, A., Nema, S., Bargali, K. (Eds.), Land and Environmental Management through Forestry. Wiley, India, pp. 339–353. Available at https://doi.org/10.1002/9781119910527.ch14.

Estrada, A., Garber, P.A., Gouveia, S., Fernandez-Llamazares, A., Ascensao, F., Fuentes, A., et al., 2022. Global importance of Indigenous Peoples, their lands, and knowledge systems for saving the world's primates from extinction. American Association for the Advancement of Science, Mexico. Science Advances 8 (32). Available at https://doi.org/10.1126/sciadv.abn2927.

Fonte, S.J., Six, J., 2010. Ecosystem services, earthworms, and litter inputs in the Quesungual agroforestry system. Ecological Applications 20, 1061–1073.

Griscom, B.W., Adams, J., Ellis, P.W., Houghton, R.A., Lomax, G., Miteva, D.A., et al., 2017. Natural climate solutions. Proceedings of the National Academy of Sciences of the United States of America 114 (44), 11645–11650. Available at https://doi.org/10.1073/pnas.1710465114.

Gómez, M.U., Bueno, A.L., León, A.C., Uribe Bernal, J.I., Aguirre, S.A.H., 2022. Traditional agroforestry systems: a methodological proposal for its analysis, intervention, and development. Agroforestry Systems 96 (3), 491–503. Available at https://doi.org/10.1007/s10457-021-00692-w.

Hoang, N.T., Taherzadeh, O., Ohashi, H., Yonekura, Y., Nishijima, S., Yamabe, M., et al., 2023. Mapping potential conflicts between global agriculture and terrestrial conservation. Proceedings of the National Academy of Sciences of the United States of America 120 (23). Available at https://doi.org/10.1073/pnas.2208376120.

Ijzerman, M.M., Bazrgar, A.B., Gordon, A.M., Thevathasan, N.V., 2022. Quantification of the carbon sequestration potential of a 31-year-old tree-based intercropping system in Southern Ontario, Canada. Advances in Environmental and Engineering Research 3 (4). Available at https://doi.org/10.21926/aeer.2204043.

IPCC, 2019. Technical summary. In: Climate change and Land: an IPCC special report on climate change, desertification, land degradation, sustainable land management, food security, and greenhouse gas fluxes in terrestrial ecosystems.

Jhariya, M.K., Banerjee, A., Swaroop Meena, R., Kumar, S., Raj, A., 2021a. Sustainable intensification for agroecosystem services and management. Sustainable Intensification for Agroecosystem Services and Management. Springer, Singapore, pp. 1–870. Available at https://doi.org/10.1007/978-981-16-3207-5.

Jhariya, M.K., Meena, R.S., Banerjee, A., Kumar, S., Raj, A., 2023. Agroforestry for Carbon and Ecosystem Management. Academic Press, pp. 1–434. Available at 10.1016/C2021-0-02916-3.

Jhariya, M.K., Meena, R.S., Banerjee, A., 2021. Ecological Intensification of Natural Resources for Sustainable Agriculture. Springer, Singapore, pp. 1–655. Available at https://doi.org/10.1007/978-981-33-4203-3.

Jhariya, M.K., Meena, R.S., Banerjee, A., Meena, S.N., 2022. Natural Resources Conservation and Advances for Sustainability. Elsevier. Available at https://doi.org/10.1016/C2019-0-03763-6.

Jhariya, M.K., Raj, A., Banerjee, A., Bargali, S.S., Bargali, K., 2024. Plantation Forestry for Ecorestoration: A Working Premises. In: Pandey, V.C. (Ed.), Biodiversity and Ecosystem Services on Post-Industrial Land. Wiley, pp. 171–187. Available at https://doi.org/10.1002/9781394187416.ch6.

Jhariya, M.K., Singh, L., Toppo, S., 2024. Wildfires and carbon budget of certain seasonally dry forests in India. Land Degradation and Development 35 (12), 3771–3789. Available at https://doi.org/10.1002/ldr.5166.

Kanime, N., Kaushal, R., Tewari, S.K., Raverkar, K.P., Chaturvedi, S., Chaturvedi, O.P., 2013. Biomass production and carbon sequestration in different tree-based systems of Central Himalayan Tarai region. Forests, Trees and Livelihoods 22 (1), 38–50. Available at https://doi.org/10.1080/14728028.2013.764073.

Kiptot, E., Franzel, S., 2012. Gender and agroforestry in Africa: a review of women's participation. Agroforestry Systems 84 (1), 35–58. Available at https://doi.org/10.1007/s10457-011-9419-y.

Kittur, B.H., Upadhyay, A.P., Jhariya, M.K., Raj, A., Banerjee, A., 2023. Chapter 2 - Agroforestry for resource diversification and sustainable development. In: Jhariya, M.K., Meena, R.S., Banerjee, A., Kumar, S., Raj, A. (Eds.), Agroforestry for Carbon and Ecosystem Management. Academic Press, pp. 19–32. Available at https://doi.org/10.1016/B978-0-323-95393-1.00028-2.

Kittur, B.H., Upadhyay, A.P., Jhariya, M.K., Raj, A., Banerjee, A., 2023. Chapter 26 - Process-based models for tree–crop interaction. In: Jhariya, M.K., Meena, R.S., Banerjee, A., Kumar, S., Raj, A. (Eds.), Agroforestry for Carbon and Ecosystem Management. Academic Press, pp. 365–374. Available at https://doi.org/10.1016/B978-0-323-95393-1.00002-6.

Krausmann, F., Erb, K.H., Gingrich, S., Haberl, H., Bondeau, A., Gaube, V., et al., 2013. Global human appropriation of net primary production doubled in the 20th century. Proceedings of the National Academy of Sciences of the United States of America 110 (25), 10324–10329. Available at https://doi.org/10.1073/pnas.1211349110.

Kumar, S., Meena, R.S., Jhariya, M.K., 2020. Resources Use Efficiency in Agriculture. Springer, Singapore, pp. 1–760. Available at https://doi.org/10.1007/978-981-15-6953-1.

Kumar, S., Meena, R.S., Sheoran, P., Jhariya, M.K., 2024. Regenerative Agriculture for Sustainable Food Systems. Springer, Singapore, pp. 1–493. Available at https://doi.org/10.1007/978-981-97-6691-8.

Labata, M., Aranico, E., Tabaranza, A., Patricio, J.H.P., Amparado Jr., R.F., 2012. Carbon stock assessment of three selected agroforestry systems in Bukidnon, Philippines. AES Bioflux 4 (1), 5–11.

Lal, R., 2015. Restoring soil quality to mitigate soil degradation. Sustainability 7 (5), 5875–5895. Available at https://doi.org/10.3390/su7055875.

Lehmann, I., Grosinger, J., Bauer, S., Francisco, J.C.R.de, Negacz, K., Hein, J., 2025. Time in and for nature-based solutions. No Quick Fix Solutions for Complex Ecological and Social Processes. Nature-Based Solutions 7, 100219. Available at https://doi.org/10.1016/j.nbsj.2025.100219.

Lin, B., 2014. Agroforestry adaptation and mitigation options for smallholder farmers vulnerable to climate change. In: Benkeblia, N., Lesueur-Jannoyer, M., Ozier-Lafontaine, H. (Eds.), Agroecology, Ecosystems, and Sustainability. Taylor & Francis, pp. 221–238.

Li, P., Zhou, G., Du, H., Lu, D., Mo, L., Xu, X., et al., 2015. Current and potential carbon stocks in Moso bamboo forests in China. Journal of Environmental Management 156, 89–96. Available at https://doi.org/10.1016/j.jenvman.2015.03.030.

Lott, J.E., Ong, C.K., Black, C.R., 2009. Understorey microclimate and crop performance in a Grevillea robusta-based agroforestry system in semi-arid Kenya. Agricultural and Forest Meteorology 149 (6-7), 1140–1151. Available at https://doi.org/10.1016/j.agrformet.2009.02.002.

Maezumi, S.Y., Alves, D., Robinson, M., de Souza, J.G., Levis, C., Barnett, R.L., et al., 2018. The legacy of 4,500 years of polyculture agroforestry in the eastern Amazon. Nature Plants 4 (8), 540–547. Available at https://doi.org/10.1038/s41477-018-0205-y.

Meena, R.S., Lal, R., Kumar, S., Pradhan, G., Srinivasarao, C., Singh, A.K., et al., 2024. Potential of Indian agriculture for capturing atmospheric CO2 and monetizing carbon credits to the farmers: an overview and policy framework. Advances in Agronomy 188, 101–206. Available at https://doi.org/10.1016/bs.agron.2024.06.001.

Meena, R.S., Pradhan, G.S., Kumar, S., Jhariya, M.K., Rao, C.S., 2025. Carbon-Negative Agriculture. Springer, Singapore, pp. 1–500. Available at https://doi.org/10.1007/978-981-95-1483-0.

Meena, R.S., Pradhan, G., Kumar, S., Lal, R., 2023. Using industrial wastes for rice-wheat cropping and food-energy-carbon-water-economic nexus to the sustainable food system. Renewable and Sustainable Energy Reviews 187, 113756. Available at https://doi.org/10.1016/j.rser.2023.113756.

Meena, R.S., Pradhan, G., 2023. Industrial garbage-derived biocompost enhances soil organic carbon fractions, CO2 biosequestration, potential carbon credits and sustainability index in a rice-wheat ecosystem. Environmental Research 235, 116525. Available at https://doi.org/10.1016/j.envres.2023.116525.

Meena, R.S., Pradhan, G., Kumar, S., 2025. Energy flow, eco-efficiency, and economic circulation with recycled industrial waste compost application in wheat and subsequent rice farming. Science of the Total Environment 967. Available at https://doi.org/10.1016/j.scitotenv.2025.178779.

Meena, R.S., Pradhan, G., Nalani, Kumar, S., Singh, A.K., Verma, S.K., et al., 2024. Diversified cropping modules designed for soil restoration, CO2 sequestration, and generating carbon credits. Land Degradation and Development 35 (14), 4314–4329. Available at https://doi.org/10.1002/ldr.5224.

Meena, R.S., Pradhan, G., Singh, K., Kumar, S., Singh, A.K., Shashidhar, K.S., et al., 2024. Agriculture models for restoring degraded land to enhance CO2 biosequestration and carbon credits in the Vindhyan region of India. Science of the Total Environment 929. Available at https://doi.org/10.1016/j.scitotenv.2024.172661.

Meena, R.S., Yadav, A., Kumar, S., Jhariya, M.K., Jatav, S.S., 2022. Agriculture ecosystem models for CO2 sequestration, improving soil physicochemical properties, and restoring degraded land. Ecological Engineering 176. Available at https://doi.org/10.1016/j.ecoleng.2022.106546.

Montagnini, F., Nair, P.K.R., 2004. Carbon sequestration: An underexploited environmental benefit of agroforestry systems. Agroforestry Systems 61, 281–295. Available at https://doi.org/10.1023/B:AGFO.0000029005.92691.79.

Moreira, B., Gonçalves, A., Pinto, L., Prieto, M.A., Carocho, M., Caleja, C., et al., 2024. Intercropping systems: an opportunity for environment conservation within nut production. Agriculture 14 (7), 1149. Available at https://doi.org/10.3390/agriculture14071149.

Muchane, M.N., Sileshi, G.W., Gripenberg, S., Jonsson, M., Pumariño, L., Barrios, E., 2020. Agroforestry boosts soil health in the humid and sub-humid tropics: a meta-analysis. Agriculture, Ecosystems and Environment 295. Available at https://doi.org/10.1016/j.agee.2020.106899.

Nair, P.K.R., 2011. Methodological challenges in estimating carbon sequestration potential of agroforestry systems. In: Kumar, B., Nair, P. (Eds.), Carbon Sequestration Potential of Agroforestry Systems. Advances in Agroforestry. Springer, Dordrecht. Available at https://doi.org/10.1007/978-94-007-1630-8_1 vol 8.

Nair, P.K.R., Nair, V.D., Kumar, B.M., Haile, S.G., 2009. Soil carbon sequestration in tropical agroforestry systems: a feasibility appraisal. Environmental Science and Policy 12 (8), 1099–1111. Available at https://doi.org/10.1016/j.envsci.2009.01.010.

Nath, A.J., Reang, D., Sileshi, G.W., 2022. The shifting cultivation juggernaut: an attribution problem. Global Challenges 6 (8). Available at https://doi.org/10.1002/gch2.202200051.

Orgill, S.E., Condon, J.R., Conyers, M.K., Greene, R.S.B., Morris, S.G., Murphy, B.W., 2014. Sensitivity of soil carbon to management and environmental factors within Australian perennial pasture systems. Geoderma 214-215, 70–79. Available at https://doi.org/10.1016/j.geoderma.2013.10.001.

Palm, C.A., Woomer, P.L., Alegre, J., Arevalo, L., Castilla, C., Cordeiro, D.G., et al., 1999. Carbon Sequestration and Trace Gas Emissions in Slash-and-Burn and Alternative Land Uses in the Humid Tropics. ASB Climate Change Working Group, Nairobi, Kenya. Available at https://www.cifor-icraf.org/publications/downloads/Publications/PDFS/RP99031.pdf.

Park, M.S., Baral, H., Shin, S., 2022. Systematic approach to agroforestry policies and practices in Asia. Forests 13 (5). Available at https://doi.org/10.3390/f13050635.

Peichl, M., Thevathasan, N.V., Gordon, A.M., Huss, J., Abohassan, R.A., 2006. Carbon sequestration potentials in temperate tree-based intercropping systems, southern Ontario, Canada. Agroforestry Systems 66 (3), 243–257. Available at https://doi.org/10.1007/s10457-005-0361-8.

Pradhan, G., Meena, R.S., 2023a. Interaction impact of biocompost on nutrient dynamics and relations with soil biota, carbon fractions index, societal value of CO2 equivalent and ecosystem services in the wheat-rice farming. Chemosphere 339. Available at https://doi.org/10.1016/j.chemosphere.2023.139695.

Pradhan, G., Meena, R.S., 2023b. Utilizing waste compost to improve the atmospheric CO2 capturing in the rice-wheat cropping system and energy-cum-carbon credit auditing for a circular economy. Science of the Total Environment 892. Available at https://doi.org/10.1016/j.scitotenv.2023.164572.

Pradhan, G., Meena, R.S., Kumar, S., Lal, R., 2023. Utilizing industrial wastes as compost in wheat-rice production to improve the above and below-ground ecosystem services. Agriculture, Ecosystems & Environment 358, 108704. Available at https://doi.org/10.1016/j.agee.2023.108704.

Prasad, R., Jhariya, M.K., Banerje, A., 2021. Advances in Sustainable Development and Management of Environmental and Natural Resources: Economic Outlook and Opinions, 2-volume set. Apple Academic Press, New York, pp. 1–428. Available at https://doi.org/10.1201/9781003187455 vol. 1.

Raj, A., Jhariya, M.K., Banerjee, A., Jha, R.K., Singh, K.P., 2024. Agroforestry. Wiley-Scrivener, USA.

Raj, A., Jhariya, M.K., Banerjee, A., Meena, R.S., Nema, S., Khan, N., et al., 2021. Chapter 13 - Agroforestry a model for ecological sustainability. In: Jhariya, M.K., Meena, R.S., Banerjee, A., Meena, S.N. (Eds.), Natural Resources Conservation and Advances for Sustainability. Elsevier, India, pp. 289–307. Available at https://doi.org/10.1016/B978-0-12-822976-7.00002-8.

Raj, A., Jhariya, M.K., Meena, R.S., 2025. Agroforestry for Monetising Carbon Credits. Springer Nature, Switzerland.

Ripple, W.J., Wolf, C., Newsome, T.M., Barnard, P., Moomaw, W.R., 2020. World scientists' warning of a climate emergency. BioScience 70 (1), 8–12. Available at https://doi.org/10.1093/biosci/biz088.

Rizvi, J., Dhyani, S.K., Arunachalam, A., Rouge, J.C., Singh, A., 2024. The Success of a National Agroforestry Policy in India has Become a Model for Other Countries. CIFOR-ICRAF, Bogor, Indonesia and Nairobi, Kenya, pp. 1–8. Available at https://doi.org/10.17528/cifor-icraf/009240.

Roy, O., Meena, R.S., Kumar, S., Jhariya, M.K., Pradhan, G., 2022. Assessment of land use systems for CO2 sequestration, carbon credit potential, and income security in Vindhyan region, India. Land Degradation and Development 33 (4), 670–682. Available at https://doi.org/10.1002/ldr.4181.

Sileshi, G., Mafongoya, P.L., 2006. Long-term effects of improved legume fallows on soil invertebrate macrofauna and maize yield in eastern Zambia. Agriculture, Ecosystems & Environment 115 (1-4), 69–78. Available at https://doi.org/10.1016/j.agee.2005.12.010.

Singh, K.P., Singh, B., Jhariya, M.K., Khemraj, Raj, A., Banerjee, A., et al., 2024. Soil properties and carbon dynamics under coffee-based agroforestry system in Bastar region of Chhattisgarh, India. Environment, Development and Sustainability 26 (5), 13411–13428. Available at https://doi.org/10.1007/s10668-023-04230-9.

Sitompul, M., Suroso, A.I., Sumarwan, U., Zulbainarni, N., 2023. Revisiting the impact of corporate carbon management strategies on corporate financial performance: a systematic literature review. Economies 11 (6). Available at https://doi.org/10.3390/economies11060171.

Smith, P., Wollenberg, E., 2011. Achieving mitigation through synergies with adaptation. In: Wollenberg, E., Nihart, A., Grieg-Gran, M., Tapio-Bistrom, M.-L. (Eds.), Climate Change Mitigation and Agriculture. Taylor and Francis, pp. 50–57.

Soussana, J.F., Lemaire, G., 2014. Coupling carbon and nitrogen cycles for environmentally sustainable intensification of grasslands and crop-livestock systems. Agriculture, Ecosystems and Environment 190, 9–17. Available at https://doi.org/10.1016/j.agee.2013.10.012.

Steinfeld, J.P., Miatton, M., Creamer, R.E., Ehbrecht, M., Valencia, V., Ballester, M.V.R., et al., 2024. Identifying agroforestry characteristics for enhanced nutrient cycling potential in Brazil. Agriculture, Ecosystems and Environment 362. Available at https://doi.org/10.1016/j.agee.2023.108828.

Tompkins, E.L., Adger, W.N., 2005. Defining response capacity to enhance climate change policy. Environmental Science and Policy 8 (6), 562–571. Available at https://doi.org/10.1016/j.envsci.2005.06.012.

Torquebiau, E., 2017. Multifunctional agriculture – achieving sustainable development in Africa. Forests, Trees and Livelihoods 26 (4), 286–287. Available at https://doi.org/10.1080/14728028.2017.1376000.

Tucker, J., Daoud, M., Oates, N., Few, R., Conway, D., Mtisi, S., et al., 2015. Social vulnerability in three high-poverty climate change hot spots: what does the climate change literature tell us? Regional Environmental Change 15 (5), 783–800. Available at https://doi.org/10.1007/s10113-014-0741-6.

Wilting, H.C., Schipper, A.M., Bakkenes, M., Meijer, J.R., Huijbregts, M.A.J., 2017. Quantifying biodiversity losses due to human consumption: a global-scale footprint analysis. Environmental Science and Technology 51 (6), 3298–3306. Available at https://doi.org/10.1021/acs.est.6b05296.

Wolz, K.J., Lovell, S.T., Branham, B.E., Eddy, W.C., Keeley, K., Revord, R.S., et al., 2018. Frontiers in alley cropping: transformative solutions for temperate agriculture. Global Change Biology 24 (3), 883–894. Available at https://doi.org/10.1111/gcb.13986.

World Bank, 2022. Indigenous peoples. https://www.worldbank.org/en/topic/indigenouspeoples#1/.

Yadav, G.S., Kandpal, B.K., Das, A., Babu, S., Mohapatra, K.P., Devi, A.L., et al., 2021. Impact of 28-year-old agroforestry systems on soil carbon dynamics in Eastern Himalayas. Journal of Environmental Management 283, 111978.

Zomer, R.J., Trabucco, A., Coe, R., Place, F., 2009. Trees on farm: analysis of global extent and geographical patterns of agroforestry. ICRAF Working Paper No. 89. World Agroforestry Centre, Nairobi.

Chapter 7

Carbon credits drive sustainable agriculture

Janus Ruszkowski
Institute of Political Science and Security Studies, University of Szczecin, Szczecin, Poland

Highlights

- Carbon financing or credit is an integral part of sustainable agriculture.
- Presently, carbon credits also act as an income addition for the farming community.
- Carbon credits have become a potential tool for combating the changing climate.
- Climate financing through carbon credits leads to carbon management and emission reduction.

7.1 Introduction

From a policy perspective, integrating carbon (C) credits, sustainable agriculture, and climate currency involves creating a robust framework that promotes environmental sustainability, economic viability, and social equity. The concept of C credits (or C offset certificates) emerged as part of a broader strategy to address climate change by creating financial incentives for reducing greenhouse gas (GHG) emissions (Cariappa et al., 2024; Gordon-Smith, 2024). It is a tool for addressing a changing climate through supportive, climate-friendly actions. Many activities produce unwanted, harmful GHGs, but some undertakings can absorb these emissions (Meena et al., 2025). This absorption can be measured in tons, and each ton is equivalent to one C credit or offset. Individuals and organizations can freely move into the C financing mechanism with the aim of emission reduction by adopting sustainable development (SD) projects that otherwise would not have happened.

The origins of the credit system in the C economy have a long history, initiating globally from environmental agreements and evolving economic theories aimed at mitigating the undesirable effects of climate change (UNFCCC, 1992; Kyoto Protocol, 1998; Paris Agreement, United Nation Organization, 2015). Article 6 the Paris Agreement included legal provisions to facilitate international C markets and cooperative approaches, emphasizing SD and environmental integrity. One of the many qualitative components obtained from the Paris Agreement in relation to C offset projects as an additional category often creates a question mark about the effectiveness of such a financial mechanism in relation to emission reduction in the form of C credits (Raj et al., 2025; Meena et al., 2025). It would be a flexible mechanism that can participate in the market economy through open trade and end-use. The additional benefit for SD projects ensures the purchase of C credits by companies that fulfill SDGs.

The analysis aims to evaluate the role of C credit in promoting sustainable agriculture and to assess the importance of C credit as a specific supranational climate currency for climate protection in the context of the emergence of the C credit market.

The analysis covers several main issues:

1. C credits as a tool for influencing the process of agricultural sustainability (analysis based on the theory of impact), implementation practices, promotion of sustainable agriculture, offset emissions, and perspectives.
2. Monetization of C credits toward a supranational climate currency and the importance of financial climate incentives and research and development (R&D) for C credits and sustainable agriculture under the pressures of a changing climate toward C footprint reduction and climate protection (analysis based on the theory of change). Climate currency

Carbon Incentives, Plans, and Policies in Agriculture. DOI: https://doi.org/10.1016/B978-0-443-33557-0.00007-0

is an emerging concept in which the value of a currency is tied to climate-related metrics, such as C-reduction achievements. In practice, C credits are a climate engineering tool (e.g., as a tool for trade and exchange between industry and agriculture).

3. C credits as a source of cofinancing for sustainable agriculture (market for trading CC as a climate currency); therefore, C credits are a supranational climate currency and a market instrument for reducing the C footprint and climate protection (analysis based on the theory of change).

Thanks to the participatory approach, which demonstrates stakeholder engagement in the policy-making process, including farmers, researchers, businesses, and civil society. This ensures policies are well-informed and widely accepted. Creating effective policies for C credits, sustainable agriculture, and climate currency requires a comprehensive approach that combines regulatory measures, financial incentives, and educational initiatives. By fostering innovation and promoting best practices through a transparent and accountable system, proper decision-making and policy formulation would be possible toward sustainability.

7.2 Carbon credits and sustainable agriculture

C credits (also known as emission units, emission certificates, or compensation certificates) are transaction documents that confirm the reduction of 1 ton of CO_2 emissions as part of a reduction project. The Paris Agreement, adopted in 2015, contains regulations (Article 6) facilitating cooperative approaches in international C markets, linking them to SD and environmental integrity (Paris Agreement, United Nations Organization, 2015). According to this position, C credits, as a key instrument of governance in international C markets, are a tool for managing sustainable agriculture and a tool for influencing sustainable agriculture (Gordon-Smith, 2024; Chen et al., 2024; Meena et al., 2024a).

C credit has wide variability based on the types of projects in which it was assessed:

1. Renewable energy credits: It is an economic valuation of the benefit obtained in the energy sector from generating 1 megawatt-hour (MWh) from nonconventional sources in comparison to conventional sources.
2. Carbon sequestration (C_{seq}) credits: It is a form of credit that is generated from the projects that physically attain emission reduction from the atmosphere and look toward safe storage. For example, various forms of afforestation schemes, reforestation approaches, social forestry, farm forestry, and the cap-and-trade program may be worth mentioning.
3. Methane capture credits: Generated by projects that capture methane emissions (a potent GHG) from sources like landfills, coal mines, or livestock operations, preventing their release into the atmosphere.
4. Avoided emissions credits: These come from projects that prevent the release of GHGs that would have otherwise occurred, such as through the development of renewable energy projects that replace electricity generation based on fossil fuels.

C credits can be an excellent strategy for offsetting the costs associated with sustainable agriculture practices. Generated through the Emissions Trading Scheme and certification process, these credits enable monitoring and reporting of standards of practices and their impact on farms and the environment. C credits are a tool to encourage farmers to implement sustainable land management practices, increasing the amount of C stored underground (Jhariya et al., 2022, 2023). By generating C credits, farmers are motivated to adopt practices such as no-till farming, maintaining soil cover through the use of cover crops, and reducing fertilizer use (Meena et al., 2024b). These actions have a significant impact on the effectiveness of the fight against climate change. Proper certification through standard organizations such as Gold Standard (GS) or Verra and the CDM (Clean Development Mechanism) makes the credit system more flexible in the C market economy, enabling it to be purchased or sold at the industry and individual levels to achieve the target of emission reduction.

The proper definition of a C credit involves a form of certificate or permit that ensures the emission of 1 ton equivalent of CO_2 or other GHGs by the credit holder. These credits can be earned through activities that reduce, avoid, or sequester GHG emissions (Li, 2000; Smith et al., 2008; Erickson and Lazarus, 2013). The genesis of C credits is rooted in the growing awareness that climate change is a major global challenge and urgently requires the development of an international framework to address it (Parnphumeesup and Kerr, 2015). From the early scientific awareness and policy responses to the establishment of the Kyoto Protocol and the expansion of C markets, C credits have evolved into a crucial tool for driving emission reductions and supporting SD. As the world continues to strive for ambitious climate targets, the role of C credits and markets will likely expand, supported by technological advancements and increased corporate and public engagement.

C credits work by assigning a specific amount of CO_2 emissions to a specific emission source, allowing companies and organizations to offset these emissions by investing in projects that reduce emissions (e.g., modernizing drinking water processes, protecting biodiversity, or generating energy from renewable sources) and thus have a positive impact on local communities. Generating C credits on a given farm is based on the calculation of reduced CO_2 emissions and increased levels of CO_2 absorption into the soil. An external verifier checks the data. C credits promote the SD of agriculture by financing the developmental projects in the energy and forestry sectors. They are thus becoming a tool for moving toward a sustainable economy. Using C credits, companies and individuals can offset their environmental impact by investing in sustainable projects worldwide. This is how C offsetting occurs: by purchasing C credits, you can offset your C footprint, contributing to global efforts to combat climate change (Sahoo et al., 2023; CAST, 2004). One of the applications of C credits aims to reduce the C footprint at the organizational and individual levels. It is highly applicable for the rural stakeholders and the farming community to achieve the target of emission reduction through climate-resilient agricultural practices. They are available on various marketplaces and platforms, which allow them to be traded and support environmental protection efforts. As agriculture becomes more advanced in terms of sustainable practices, C credits have the potential to play a role in a cleaner future for the agriculture industry.

Thanks to the theory of impact, C credits are not only a tool for influencing the process of agricultural sustainability implementation practices, promotion of sustainable agriculture, offset emissions, and perspectives, but also serve as a vital mechanism in promoting sustainable agriculture by acting as a form of climate currency (Meena et al., 2024a). C credits affect sustainable agriculture through the creation and implementation of good practices.

The first is by incentivizing sustainable practices, including revenue generation and cost offset. On the one hand, revenue generation, farmers and agricultural businesses can earn money C credits by adopting practices that reduce GHG emissions or sequester C. These credits can then be sold in C markets, providing an additional revenue stream. On the other hand, cost offset is related to income from C credits, which can offset the costs of implementing sustainable practices, such as transitioning to organic farming, adopting no-till agriculture, or integrating agroforestry (AF) (Brown et al., 2017; Meena and Pradhan, 2023; Van Brunt et al., 2012).

The second is by encouraging $\mathbf{C_{seq}}$, which improves soil health. Eco-friendly practices such as the use of cover crops, minimum tillage, and organic farming to supplement soil organic matter, which sequesters C. Healthier soils also retain more moisture and nutrients, boosting productivity. In turn, integrating trees and shrubs into farmland not only sequesters C but also improves biodiversity, reduces erosion, and enhances microclimates, as observed in AF (Broekhoff and Zyla, 2008; Updegraff et al., 2010; Hijbeek et al., 2019). The Sustainable Development Goals (SDGs) focus on preventing, managing, and resolving the multiple conflicts—as well as many interconnected challenges, notably poverty, inequality, and climate change—that force millions of people to leave their homes (Ansari et al., 2024).

The third is reducing GHG emissions (e.g., methane reduction). Implementing better manure management practices and altering livestock diets can significantly reduce methane emissions from ruminants. Efficient fertilizer use is also important here because techniques such as precision farming aim to apply fertilizers at the optimal levels, leading to a reduction in GHG emissions.

The fourth is by driving technological innovations, including sustainable technologies as well as data and monitoring tools, which need to be integrated into C financing. The demand for C credits encourages the development and adoption of technologies such as biochar production, methane digesters, and renewable energy installations on farms. Advances in remote sensing, satellite imagery, and soil C measurement techniques help accurately quantify $\mathbf{C_{seq}}$ and emissions reductions (Meena et al., 2023; Chen et al., 2024).

The fifth is by encouraging holistic farm management. The integrated farming systems can be used for such management. C credits promote holistic approaches, such as regenerative agriculture, which emphasizes soil health, biodiversity, and ecosystem services, ultimately leading to resilient and sustainable farming systems (Kumar et al., 2024). Moreover, biodiversity and ecosystem services serve new practices that enhance biodiversity and ecosystem services can earn C credits, thereby promoting the preservation of natural habitats and pollinators.

The sixth is supporting policy and market development, which includes regulatory support from governments and international bodies in order to create policies that support C credit markets, ensuring transparency, accountability, and fair pricing. Robust C markets provide the infrastructure needed for trading C credits, which would enable and stimulate farming communities to actively participate in the C economy and gain additional financial benefits.

The seventh aspect is community and social benefits, which include C credit projects that often involve local communities, providing education and employment opportunities, and enhancing social cohesion. It is a form of community engagement and a necessary practice of sustainable agriculture, as well as sustainable livelihoods through the diversification of income sources and promoting sustainable practices. In this way, C credits contribute to the long-term viability of rural communities (Allgood et al., 2024; Thompson, 2004).

The eight is global climate goals. The 17 global objectives, known as the SDGs, were adopted by the General Assembly of the UNO in 2015. SDGs are aimed at global economic, social, and environmental policies available for 15 years (Olsen and Arens, 2021; United Nations, 2015). SDGs are focused on preventing, managing, and resolving the multiple conflicts—as well as many interconnected challenges, notably poverty, inequality, and climate change—that force millions of people to leave their homes. In this context, we should indicate the Nationally Determined Contributions (NDCs). Sustainable agricultural practices (SAP) supported by C credits can contribute toward NDCs as per the Paris Agreement, helping meet global climate targets.

Two representative examples of successful implementations can be indicated: (1) C farming initiatives (programs like the Australian Carbon Farming Initiative encourage farmers to adopt practices that earn C credits), (2) Soil C projects (organizations like Indigo Ag promote soil $\mathbf{C_{seq}}$ projects, allowing farmers to earn C credits for increasing soil organic matter). By transforming SAP into economically viable activities, C credits serve as a powerful tool to combat climate change, enhance farm resilience, and improve food security.

7.3 Monetization of carbon credits toward climate currency

The C credit mechanism emphasizes creating a market-based mechanism (market-based flexible approach) to incentivize and finance the reduction, avoidance, and sequestration of GHGs. Under the mechanism, one metric ton of GHG emissions is permissible at the organizational or individual level. By assigning a monetary value to the reduction of emissions, C credits create financial incentives for businesses, governments, and individuals to invest in projects that mitigate climate change and facilitate the trading of emission reductions, and encourage investments in sustainable practices. Therefore C credits are a type of tradable commodity that can be sold in C markets. Entities that exceed their emission limits can purchase C credits from those with a surplus due to lower-emission activities. While the system offers significant environmental, economic, and social benefits, addressing challenges related to verification, market stability, and equity is essential for its long-term success and integrity.

Monetization of C credits toward a climate currency involves transforming C credits (a tangible asset) into a financial asset (a means of payment) that can be traded, invested in, and used to drive and support climate-friendly activities. This process may play an active role in climate change mitigation by generating additional economic benefits through GHG emission reductions (Banerjee et al., 2024; Pradhan and Meena, 2023a). C credits functioning as a supranational climate currency refer to their role as a measurable, tradeable commodity that incentivizes and finances climate-friendly activities. C credits can be purchased by both individuals and institutions (Van Brunt et al., 2012; Parnphumeesup and Kerr, 2015). It is also possible to transfer specific amounts to offset GHG emissions resulting from travel, for example, when buying a plane ticket. The revenue from selling C credits can fund sustainable practices, such as the use of renewable energy, and promote sustainable agriculture. C credits support scalable solutions by encouraging large-scale projects and aggregating smaller projects to make them economically viable. By providing a revenue stream, C credits reduce the financial risks associated with investing in new technologies and practices aimed at reducing C footprints (Gupta, 2014; Sabbatini, 2022; Sirilakshmi et al., 2024; Pradhan and Meena, 2023b).

Climate currency has two main financial instruments derived from C credits that can be used as a medium of exchange, an investment, or a store of value, similar to traditional currencies but with a focus on incentivizing climate-positive activities. The first are green bonds issued to fund projects that have positive environmental impacts, which can be linked with C credits. The second are C funds that buy and sell C credits can offer returns based on the appreciation of these credits.

C credits allow GHG emissions at a certain level in exchange for compensation. These credits are part of international efforts to limit climate change and are often traded on C markets. Early C markets began to emerge in the 2000s. The European Union Emissions Trading System (EU ETS) was launched in 2005. The EU ETS became the world's largest C market, setting a cap on emissions from major industrial sectors and allowing trading of emission allowances (Japan Organization for Metals and Energy Security (JOGMEC), 2023). This process may be highly effective in mitigating the changing climate by developing a suitable financial mechanism that would lead to emission reductions. Monetization of C credits (functioning as climate currency) is facilitated by the creation and functioning of the C credit market and can be done through several comprehensive mechanisms, such as valuation of C reductions, C credit markets, tokenization, or flexible mechanisms. Valuation of C reduction is possible thanks to market prices and financial incentives. C credits have a market value (market price) that can be evaluated through a demand–supply function in C markets. This price puts a monetary value on reducing or avoiding GHG emissions. This valuation is a financial approach for businesses and individuals for moving toward C reduction projects.

We can identify two main types of C credits markets: compliance carbon markets and voluntary carbon markets (Tjon, 2023; Wettenberg et al., 2024; Biermann et al., 2022). Regulation of C markets at various levels through emission reduction systems, such as the EU ETS, consists of regulated markets and cap-and-trade systems. Companies must comply with emission limits and use C credits to offset excess emissions. Regulated markets are governed by mandatory regulations that require entities to comply with legally binding emission reduction targets. Examples include the EU ETS and the California cap-and-trade program. Cap-and-trade systems rely on setting a limit on total emissions, and allowances are distributed or auctioned. Entities must have proper subsidies in the form of alliances to invest in emission reduction, and can trade allowances to meet their needs.

Alongside compliance markets, the second type voluntary C markets (especially voluntary offset markets) began to emerge and develop, where businesses, organizations, and individuals could purchase C credits to offset or voluntarily reduce their C footprint driven by corporate social responsibility goals and consumer demand for sustainability (Blaufelder et al., 2021; Howard and Greiner, 2021; Tjon, 2023; Sirilakshmi et al., 2024; Yin and Ghosh, 2023; Kreibich and Obergassel, 2019). This market operates outside of regulatory requirements and is driven by corporate social responsibility and consumer demand. Therefore voluntary markets based on internal standards and certification and registries help in market stability. Voluntary markets rely on standards like the Verified Carbon Standard (VCS) and the GS to ensure the quality and integrity of the credits. Those markets promote business activity at both individual and institutional levels through the purchase of C credits as a voluntary service that helps in emission reduction. The C markets operate on principles of supply and demand. Higher demand for credits incentivizes more C-reduction projects, expanding markets and increasing the overall impact on climate mitigation. A big threat to market stability is price volatility and regulatory uncertainty. On the one hand, C credit prices can be volatile, influenced by regulatory changes, market demand, and economic conditions impacting their stability as a currency; on the other hand, changes in policies and regulations can impact market stability and investor confidence. Varying regulations across different jurisdictions can complicate the standardization of C credits. A great challenge for C market stability are verification and fraud. Ensuring the authenticity and validity of C credits is crucial to maintaining trust in the system.

In the case of tokenizing, C credits on blockchain can provide transparency, traceability, and liquidity, making it easier to trade and convert C credits into a digital climate currency (Widholm and Stolyarova, 2023; Bettín-Díaz et al., 2018). Bit smart contracts can automate the trading process and ensure compliance with emission reduction goals. Entities that need to offset their emissions can purchase these credits. Sellers are typically project developers or entities with excess credits due to lower-than-expected emissions. C credits provide a cost-effective way for companies to achieve their emission reduction targets by investing in projects that reduce emissions at a lower cost than internal (in-house) measures. Selling C credits generates revenue for projects that reduce or sequester C, making these projects economically viable—the emergence of climate-focused investment funds and financial products aimed at supporting C reduction projects and C credits. While developing the structure of C credit, the technological intervention can be effectively used in diverse ways. Oracles, blockchain protocols, smart contracts, and token standards are all key basic tools to consider while issuing a C credit token (Widholm and Stolyarova, 2023). In practice, C credit tokenization involves converting traditional C credits into digital tokens that exist on a blockchain (Bettín-Díaz et al., 2018).

Flexibility in C financing helps in cost reduction through market-based mechanisms in the form of:

- Emissions Trading: Countries with surplus emission reduction are allowed to channelize their additional C credits to the needy countries and help them to achieve the target of emission reduction.
- Clean Development Mechanism: Under CDM, developed countries should promote emission reduction development projects in developing countries and may earn additional benefits in the form of CERs.
- Joint Implementation: This is another form of earning C credits in which one developed nation can invest in another developed country in the sector of emission reduction to earn Emission Reduction Units.

One of the main economic benefits of C credits as a supranational climate currency is the cost-effective reduction of GHG emissions by enabling entities to invest in the most economical projects available, regardless of location. Moreover, companies have the flexibility to choose how to meet their emission reduction targets, balancing internal reductions with the purchase of C credits (Allgood et al., 2024; Kragt et al., 2024).

Another benefit is closely tied to global participation and equity, as C credits facilitate global participation in climate action, enabling projects in developing countries to generate credits that can be purchased by entities in developed countries. Equity and development mechanism promotes SD in less industrialized regions, supporting local economies and contributing to social and environmental goals. Finally, we should pay attention to the benefits associated with technological and environmental innovation. The potential for revenue from C credits drives innovation in technologies and practices that reduce or sequester C, such as carbon capture and storage and regenerative agriculture. Projects that generate C credits often provide additional environmental benefits, such as the conservation of biodiversity and ecosystem services.

There are also some challenges to consider with C credits as a supranational currency. It should aim for a potential and permanent reduction of GHGs, which requires rigorous measurement, reporting, and verification (MRV) processes. It is also crucial to prevent greenwashing, where companies may claim C neutrality through questionable credits that do not represent real emission reductions. Other challenges are related to C credit pricing. C markets can experience price volatility, which can affect the predictability of revenue from C credits. Ensuring that C credits are reasonably priced to reflect the actual cost of C reduction and the social cost of C emissions is vital for the integrity of the market.

7.4 Climate incentives and subsidies for the management of agroecosystems

Climate incentives and subsidies for the management of agroecosystems are essential tools for promoting climate-resilient agricultural practice. By supporting farmers in adopting environmentally friendly practices, these financial mechanisms can lead to significant environmental, economic, and social benefits. However, successful implementation requires careful design, effective monitoring, and alignment of policies across different levels and sectors.

In practice, climate incentives and subsidies are financial support mechanisms provided by governments, nongovernmental organizations, and international bodies to encourage the adoption of SAP. These incentives can be in the form of direct payments, tax breaks, grants, or price supports. Incentives and subsidies for the management of agroecosystems are crucial in promoting SAP, enhancing biodiversity, and mitigating climate change (El Bakali et al., 2023; Piñeiro et al., 2020). These financial mechanisms can support farmers and land managers in adopting practices that help to build soil quality, emission reduction, and develop climate resiliency in agroecosystems. Agroecosystem management involves integrating ecological principles into agricultural practices to create sustainable farming systems. It encompasses various practices, such as organic farming, conservation agriculture, zero-tillage, AF, sustainable pest management, and crop diversification.

The Food and Agriculture Organisation (FAO) also provides directives for Incentives for Ecosystem Services: Supporting the transition to Sustainable Food System, Food and Agriculture Organization of the United Nations, s. 1, (n.d.) that would help the farming community to go for nature-based solutions in the cultivation practices. Holistically, ecosystem services provide all the essentials for the living planet. They provide many goods and services, such as raw materials, fresh water, healthy soils, pollination, and pest and disease regulation (Liu et al., 2022).

Consequently, the Incentives for Agro-Ecosystem Services Approach (IAESA) is based on improving and changing production practices to make them more beneficial to biodiversity and developing policy and regulatory frameworks that remove inappropriate subsidies and provide incentives for biodiversity-friendly and productive use of resources that do not degrade biodiversity (Kazemi et al., 2018). Incentives are often used in isolation, but when combined with the IAESA approach, they would ensure policy implementation and stimulate more investment in the sustainability of agroecosystems. Economic incentives from various sectors can be effectively used to achieve the larger target of SDGs. An example of an incentive package for agro-ecosystem services could be a tax incentive package including the rehabilitation of degraded areas, protection of soil and water resources, and marketing of agro-environmental products (Zabala et al., 2021).

Several types of incentives and subsidies warrant mention. The first type is direct payments and grants. Farmers receive payments for implementing specific environmental practices such as cover cropping, reduced tillage, or maintaining buffer strips (Bonne et al., 2019). Thanks to grant funding provided for research, development, and implementation of sustainable farming techniques. The second type is tax incentives, because we can see that tax reductions for farmers who invest in sustainable practices or equipment. Moreover, tax credits are important for activities that sequester C or reduce emissions. The third type is market-based incentives. The farming community would have an economic revenue in the C financing mechanism, which would help them in $\mathbf{C_{seq}}$ and GHG emission reduction, which can be sold on C markets. Moreover, thanks to premium prices for products certified as sustainably produced, it is creating eco-labeling and certification mechanisms.

Benefits of climate incentives and subsidies for the management of agroecosystems are environmental (reduction in GHG emissions, improved soil health, and enhanced biodiversity), economic (financial support for farmers, increased market opportunities, and long-term sustainability of agricultural production), and social nature improving rural livelihoods, food security, and community resilience. On the other hand, some challenges related to incentives and subsidies for the management of agroecosystems arise from ensuring proper implementation and monitoring of sustainable practices, which can be resource-intensive, balancing the economic needs of farmers with environmental goals, and aligning various policies acorss diverse sectors.

7.5 Accounting of carbon credits in sustainable agriculture

Accounting of C credits in sustainable agriculture is a critical process for quantifying the environmental benefits of sustainable practices and creating financial incentives for farmers. This process involves rigorous methodologies for the

MRV as well as certification and standard (CAS) to ensure transparency, credibility, and effecti veness of C credit systems as well as for trading C offsets, which are generated by agricultural practices that reduce or sequester GHG emissions. These C credits are part of a market mechanism designed to incentivize sustainable practices and contribute to climate change mitigation (Smith et al., 2008; Freeman and Zerriffi, 2014).

To generate tradable C credits, agricultural operations need to follow rigorous methodological protocols to MRV the emission reductions or $\mathbf{C_{seq}}$ (Ellis and Larsen, 2008; Niederberger and Kimble, 2011). These protocols include (1). baseline assessment that consists of establishing a baseline for emissions or C stocks before sustainable practices are implemented, (2) monitoring, that is, continuously tracking changes in GHG emissions or C storage over time, often using scientific methods such as soil sampling, remote sensing, or modeling, (3) reporting on all GHGs, including CO_2, nonstandardized factors of emission, mechanism of finance, source, and location-specific market, and (4) verification through independant assessment and auditing, and subsequently ensuring proper estimation of emission reduction and $\mathbf{C_{seq}}$, additional (i.e., wouldn't have occurred without the practice), and permanent.

Various CAS bodies govern the creation of C credits in sustainable agriculture. Within the framework of the CAS, we should point out a technocratic tool of a utilitarian nature and methodology standards, such as (1) VCS is used for a widely recognized standard for GHG reductions, including projects related to agriculture and forestry (Parnphumeesup and Kerr, 2015). Moreover, VCS provides specific methodologies for agricultural projects, such as improved cropland management, avoided grassland conversion, and AF. (2) GS, which focuses on projects and C credits that deliver social and environmental benefits, including sustainable agricultural practices, and serve C reductions along with SD cobenefits (Netter et al., 2022). GS has methodologies for sustainable land use and agriculture, (3) Climate Action Reserve, a U.S.-based standard that includes agriculture-related C-reduction projects, and (4) American Carbon Registry (ACR), which provides methodologies for agricultural $\mathbf{C_{seq}}$ and emission reductions. The methodology requires several steps from project inception to C credit issuance, following a systematic process of MRV as well as CAS.

The first step is project design and baseline assessment to define a baseline, which reflects the emissions or C stock level before sustainable practices are implemented. The baseline serves as the point of comparison to measure additional $\mathbf{C_{seq}}$ or reduced emissions (Updegraff et al., 2010; Ansari et al., 2024). Thanks to the project activities, the SAP to be adopted are specified, such as no-till farming, AF, improved manure management, or cover cropping. Finally, the project boundary allows identification of the geographical and operational scope of the project (e.g., specific fields, farms, or land parcels).

The second step involves quantifying GHG emission reductions or $\mathbf{C_{seq}}$. In this step, they are used standardized methodologies to calculate the expected or actual reductions in GHG emissions or the amount of C_{seq} in soil or biomass (Smith et al., 2008). This is achieved through direct measurement, utilizing sensors, soil sampling, or gas flux measurements to track changes in soil C or methane/nitrous oxide emissions. Equally important are modeling approaches, in this case using models such as COMET-Farm (it is widely used in U.S.-based agricultural C credit projects for soil $\mathbf{C_{seq}}$ and GHG accounting tool) developed by the United States Department of Agriculture to estimate $\mathbf{C_{seq}}$ and GHG reductions of farm practices (Miller, 2013) or DayCent (for nitrous oxide emissions) to estimate GHG impacts. It is necessary to remember that applying region-specific emission factors is used to estimate the effects of agricultural activities on GHG emissions, particularly for methane and nitrous oxide.

The third step is monitoring and data collection. Data on soil C levels, biomass, fertilizer usage, livestock management, and other parameters must be regularly collected. Monitoring ensures that the project continues to sequester C or reduce emissions as planned. Monitoring involves both field data collection and the broader use of remote sensing technologies (e.g., satellite imagery for land use change) to track C stocks and emissions over time.

The fourth step is reporting. All measured or modeled emissions reductions and $\mathbf{C_{seq}}$ must be reported transparently. Reporting typically includes documentation on baseline conditions, methodologies used for C calculations, monitoring processes, emission reductions, or $\mathbf{C_{seq}}$ achieved.

The fifth step is Third-Party Verification (Abhinandan, 2024). Independent third-party auditors verify the project's results, ensuring the claimed C credits reflect actual reductions or $\mathbf{C_{seq}}$. The verifier checks all aspects of the project, including adherence to the chosen methodology and accuracy of measurement and monitoring data, as well as compliance with additionality, permanence, and leakage standards.

The sixth step is certification and registration. Once the project is verified, the C credits are certified by the chosen C standard (e.g., VCS, GS) and registered on a C registry. The credits are then issued and can be sold or traded in C markets because each credit is assigned a unique serial number to prevent double-counting.

The seventh step is Post-Issuance Monitoring and Re-verification. Projects must continue to monitor emissions reductions and $\mathbf{C_{seq}}$ beyond the initial crediting period to maintain compliance. Periodic re-verification may be required depending on the standard and the project's duration (Malamas et al., 2024).

The framework for C credits accounting in sustainable agriculture was built on several key principles (Croker, 2024). These principles ensure that the C offsets generated are real and credible, verifiable, and marketable (Kaplan et al., 2023). These principles, similar to the methods, focus on transparency, accuracy, and the legitimacy of the C reductions or $\mathbf{C_{seq}}$. The principle of additionality is based on the assumption that C credits must reflect emissions reductions or $\mathbf{C_{seq}}$ that wouldn't have occurred without the implementation of specific sustainable practices. This means the agricultural activities should be "additional" to business-as-usual operations (Campbell et al., 2018). This is determined by comparing the project outcomes against a defined baseline scenario (i.e., what would have happened without the project).

The principle of permanence means that, for $\mathbf{C_{seq}}$ projects, credits are only valid if the C remains stored for an extended period (often measured in decades). The permanence principle addresses the risk of "reversal," where stored C could be released due to practices such as deforestation or tillage. Some standards require buffer credits or insurance mechanisms to address potential reversals. The principle of measurability indicates that C emissions reductions or $\mathbf{C_{seq}}$ must be measurable using scientifically accepted methods. This includes quantifying the actual reduction or $\mathbf{C_{seq}}$ in metric tons of CO_2 equivalent (tCO_2e). This requires robust data collection, modeling, and scientific methodologies to ensure accuracy.

The principle of verifiability emphasizes that it is essential for a transparent and neutral verification process for the C credits to ensure the real scenario of C transactions that have been accurately measured. This verification also helps in the integration of C credit. Projects must undergo regular audits to maintain certification and ensure the issuance of credits. The principle of transparency allows us to notice that all accounting, reporting, and documentation related to the C credit project must be transparent and made available to relevant stakeholders. This enhances trust in the C market. Project documentation should detail the methodology, baseline data, monitoring processes, and verification results.

The principle of leakage prevention refers to a situation that may lead to an increase in emissions in one area through emission reduction in another area. For example, the removal of forest cover in an area may lead to deforestation in another. The accounting methodology must address potential leakage and ensure that the overall climate benefit is maintained. The principle of consistency with C accounting standards explains that projects should comply with internationally recognized C accounting standards, such as the GHG protocol or standards set by specific C registries (e.g., VCS, GS, ACR). Consistent methodologies ensure credibility and facilitate the trading of C credits in both voluntary and compliance markets. The accounting of C credits in sustainable agriculture follows a structured process built on the seven principles and a dedicated multistep methodology. The multistep methodology involves designing projects with clear baselines, quantifying C reductions or $\mathbf{C_{seq}}$, continuous monitoring, and independent verification. These processes ensure that C credits generated from agricultural practices are credible and can be traded on C markets to incentivize further sustainable practices.

7.6 Financial incentives for climate adaptation of agroecosystems

The concept of agriculture combining soil protection with productive activity can be of great importance for understanding the climate adaptation of agroecosystems. But this concept is not new. Over the past two decades, financial incentives for implementing, for example, agri-environmental practices have been regularly developed in the regulatory environment of agriculture. What is new in the currently popularized C programs is the market element and the fact that regenerative agriculture schemes are financed by industry, governments, and local authorities, as well as the broadly understood business environment of SD. This creates a massive opportunity for the agricultural sector, which is moving from being the recipient of environmental claims to being a provider of environmental benefits and an active player in the global market for voluntary emission reductions. This is a complete game-changer, and it is important for farmers to learn how to operate in this market and what the financial instruments are. To earn money from C credits, farmers need not only information and a legal framework, but also financial incentives.

Financial incentives to support climate change adaptation in agroecosystems are crucial for promoting SAP and ensuring food security in the face of climate change (El Bakali et al., 2023; Pradhan et al., 2023; Siedenburg et al., 2012; Piñeiro et al., 2020; Financing the Green Transition, n.d.). These incentives can come from various sources, including government policies, international organizations, private sector investments, and community initiatives. We can build a catalog of multilevel financial incentives designed to encourage climate adaptation in agroecosystems: international incentives, national incentives, regional or local incentives, interlevel incentives (international/national and national/regional or local).

Some important and significant Global Climate Funds include the Global Environment Facility, which offers financial support for climate resilient practices and development projects in developing countries. It is equally important that development aid is provided by international aid organizations and development banks, which often fund climate adaptation initiatives in agriculture (Morel and Delbosc, 2012).

At the national level, financial incentives are focused on grants, subsidies, and taxes. Many governments offer grants and subsidies to farmers for sustainable inputs, encouraging them to adopt climate-resilient practices such as crop diversification, soil conservation, water management, and AF. Subsidies can reduce the cost of inputs like drought-resistant seeds, organic fertilizers, and efficient irrigation systems. When it comes to taxes, it's essential to consider tax credits and deductions. Governments can offer tax credits or deductions for spending on climate-resilient practices and technologies. Of course, tax exemptions can also be offered. Certain sustainable agricultural products or practices can be exempted from taxes, encouraging farmers to adopt them.

Another facilitation at the national level is C-credits programs and payment for ecosystem services (PES). C credit programs are designed for farmers who adopt practices that sequester C (e.g., no-till farming, cover cropping) and can earn C credits, which they can sell on C markets. PES programs compensate farmers for maintaining or improving ecosystem services such as water purification, biodiversity conservation, and $\mathbf{C_{seq}}$ (Fripp, 2014).

At the regional or local level, it should be noted that community-based approaches include community savings and loan groups. These groups can provide financial support for community-based climate adaptation projects. To this must be added local investment funds because communities can pool resources to fund local adaptation initiatives. Interlevel (international/national) financial incentives are insurance, loans, and credit. Climate risk insurance can protect farmers from losses caused by extreme weather events. Governments or international institutions can subsidize these programs to make them affordable. Equally important is index-based insurance. This includes a payout mechanism and additional benefits based on a predetermined index rather than actual losses, which can be more efficient and easier to manage. Interlevel loans and credit facilities based on low-interest loans. Financial institutions, often supported by government or international funds, may offer low-interest loans for investments in climate-resilient infrastructure and technologies (Morel and Delbosc, 2012). Small-scale loans can be crucial for smallholder farmers to invest in climate adaptation practices.

At the interlevel (national/regional or local), we should point out two main financial incentives. The first public–private partnerships (PPPs) can mobilize private-sector investment in climate-resilient agricultural technologies and infrastructure (collaborative investments). These partnerships can reduce risks and increase the scale of adaptation efforts (Smyth et al., 2021). In addition, companies can invest in climate adaptation projects as part of their corporate social responsibility initiatives. The second are market-based instruments. Certification schemes (e.g., organic, fair trade) can allow farmers to receive higher prices for sustainably produced goods (i.e., in practice, premium prices for sustainable products). Markets for services such as water regulation and biodiversity (ecosystem service markets) can provide financial incentives for farmers to adopt sustainable practices (Table 7.1).

Financial incentives are key to encouraging farmers and other stakeholders to invest in climate change adaptation measures in agroecosystems. A multilevel combination of subsidies, insurance, loans, market-based instruments, and community initiatives can create a supportive environment for SAP that enhance resilience to climate change. Implementing these incentives effectively requires coordination among governments, private sector actors, international organizations, and local communities.

7.7 Research and development toward carbon credits and sustainable agriculture under a changing climate

R&D on the development of C credits and sustainable agriculture in the face of a changing climate are key areas that contribute to climate change mitigation and agroecosystem adaptation. This research encompasses a range of technological, ecological, economic, and policy-oriented approaches. For example, community and participatory methods, such as community-based monitoring that is based on engaging local communities in monitoring environmental changes and the effectiveness of adaptation strategies, help in fine-tuning practices and policies (Trouwloon et al., 2023). Also, farmer field schools with programs educate farmers on sustainable practices and climate adaptation strategies. Participatory research ensures that innovations are tailored to local contexts. Most of this research is interdisciplinary and collaborative in nature. Both PPP and international cooperation use triangulation of approaches (Smyth et al., 2021). On the one hand, collaborations between research institutions, governments, and private companies can accelerate the development and implementation of sustainable technologies and practices. On the other side, global initiatives and networks (e.g., Consultative Group for International Agricultural Research or FAO) facilitate the sharing of knowledge, technologies, and best practices across regions and countries.

Thanks to this research, technological innovations for sustainable agriculture have emerged. At this point, it is worth mentioning precision agriculture, that is, utilizing GPS, IoT sensors, and data analytics to optimize input use (water, fertilizers, and pesticides), reduceing waste, enhancing efficiency, and minimizing environmental impacts. It can't be

TABLE 7.1 Multilevel financial incentives for climate adaptation of agroecosystems.

Multilevel financial incentives					
International level	**Green climate fund (GCF)**	**Global environment facility (GEF)**			
National level	Grants	Subsidies	Tax	C credits programs	Payment for ecosystem services (PES)
Regional or local level	Community savings and loans groups	Climate adaptation projects	Local investment funds		
Interlevel	Insurance	Loans	Credit	Public–private partnerships (PPPs)	Market-based instruments

Source: own elaboration.

missed that climate-resilient crops, consisting of breeding and genetic modification efforts, are directed toward developing crops that can withstand extreme weather, pests, and diseases. Traits like drought tolerance, heat resistance, and flood tolerance are priorities. In practice, R&D in integrated pest management (IPM) aims to combine biological, physical, and chemical tools to control pests in an environmentally and economically sustainable manner. R&D is not possible without economic and political instruments. This toolbox consists of market development for C credits, subsidies and incentives, as well as regulatory frameworks (Sternberg, 2017).

Establishing transparent and robust C markets is crucial for market development and for C credits. This includes setting clear methodologies for measuring, reporting, and verifying $\mathbf{C_{seq}}$ in agriculture. Policy research focuses on designing effective subsidy programs (e.g., C credits programs) and financial incentives that encourage farmers to adopt sustainable practices. These programs are focused on soil $\mathbf{C_{seq}}$, AF, and methane, as well as nitrous oxide reduction. Research focuses on practices that increase soil organic C, such as no-till, cover crops, and biochar applications. These practices not only sequester C but also improve soil health and productivity. Integrating trees and shrubs into agricultural landscapes can significantly sequester C. R&D efforts aim to identify optimal species and management practices that maximize C storage and cobenefits. Methane and nitrous oxide reduction is connected with innovations in livestock management (e.g., feed additives to reduce enteric fermentation), and optimized fertilizer use can lower methane and nitrous oxide emissions, respectively. These practices can be monetized through C credits (Mirzaee et al., 2022).

Finally, developing multilevel regulations that support SAP and the integration of C credits into national, international and global climate policies is creating regulatory frameworks for R&D. Within the framework of the SAP we can find conservation agriculture, that is, practices such as reduced tillage, crop rotation, and intercropping are being refined and promoted to maintain soil structure, enhance biodiversity, and reduce erosion (Korsching and Malia, 1991). It is also worth noting that water management and organic farming, and irrigation innovations (e.g., drip irrigation, rainwater harvesting), as well as water-efficient crop varieties, are essential to cope with variable rainfall patterns. Research is exploring ways to improve the productivity and sustainability of organic farming systems, including soil fertility management and natural pest control methods. In the areas of R&D, future-oriented and innovative solutions and directions cannot be ignored. Here, it is necessary to mention digital agriculture, including leveraging AI, machine learning, and big data analytics to predict climate impacts, optimize resource use, and support decision-making in real-time. Similar to Climate-Smart Agriculture (CSA), which integrate climate change mitigation, adaptation, and resilience into farming systems. R&D in CSA focuses on practices and technologies that provide triple wins. Innovative solutions like Ecosystem-based Adaptation are based on research and promot the role of biodiversity and ecosystem services to enhance the resilience of agricultural landscapes (Munang et al., 2013; Nalau et al., 2018).

Investments in R&D for C credits and sustainable agriculture are vital for addressing the dual challenges of food security and climate change. By advancing technologies, practices, policies, and programs we can create resilient agroecosystems that contribute to global C mitigation efforts and adapt to a changing climate. Collaboration across disciplines and sectors will be key to achieving these goals.

7.8 Policy perspectives for carbon credits, sustainable agriculture, and climate currency

Developing effective policies for C credits, sustainable agriculture, and climate currency requires a multifaceted approach that integrates environmental, economic, and social objectives. A comprehensive policy perspective and trends on these critical areas encompass assumed policy goals and the planned use of instruments.

The main goals in the area of policy perspective for C credit are: (1) promoting the use of agricultural practices that sequester C and reduce GHG emissions, (2) ensuring transparency, integrity, and efficiency in C credit markets, and (3) facilitating farmers' participation in C markets, especially smallholders and marginalized communities. The catalog of policy instruments intended for the area of C credits includes regulatory frameworks, incentives and subsidies, capacity building, and market development. Regulatory frameworks based on establishing clear guidelines and standards for MRV $\mathbf{C_{seq}}$ and emissions reductions in agriculture. This includes defining eligible practices, baseline methodologies, and certification processes.

Incentives and subsidies involve providing financial incentives in the form of subsidies or tax breaks for farmers who implement $\mathbf{C_{seq}}$ practices. Subsidies can cover the cost of transitioning to sustainable practices and adopting new technologies. Capacity building that is in practice, investing in training and extension services to educate farmers about C farming, C markets, and sustainable practices. This includes technical support for MRV and participation in C credit programs.

Market development, in fact, supports the development of robust and transparent C markets. This involves creating platforms for trading C credits and ensuring fair prices that reflect the true value of $\mathbf{C_{seq}}$ (Carney et al., 2020; Michaelowa et al., 2019; Blaufelder et al., 2021). The main policy goals in the area of sustainable agriculture are: (1) enhance agricultural productivity while minimizing environmental impacts, (2) promote resilience and adaptability of agricultural systems to climate change, and (3) support the livelihoods of farmers and rural communities. Sustainable practices incentives based on implementing policies that incentivize SAP, such as conservation tillage, crop rotation, organic farming, and IPM. Financial incentives can include grants, low-interest loans, and subsidies.

The catalog of policy instruments intended for the area of sustainable agriculture includes sustainable practices incentives, R&D, extension services, market access, and support for risk management.

Next instrument is R&D. Here, the most important is on the one side investment in R&D to develop climate-resilient crop varieties, sustainable farming technologies, and efficient resource management practices, and on the other hand, encourage PPP to leverage innovation. In the case of extension services, this instrument is used to strengthen these agricultural services and to disseminate knowledge and best practices. This includes training programs, demonstration projects, and farmer field schools. Market access and support are instruments intended to develop infrastructure and policies to improve market access for sustainably produced agricultural products. This can include certification schemes, premium pricing for organic and fair-trade products, and support for local food systems (Michaelowa et al., 2019).

Finally establish insurance schemes and risk management tools is needed to protect farmers from climate-related risks (Howden et al., 2007; Reidsma and Ewert, 2008). This includes index-based insurance, weather derivatives, and disaster relief funds (Disaster, 2024) (may direct, coordinate, manage funds eligible finance qualified response and recovery activities related to major natural disasters and national emergencies that exceed state resources) (Disaster, 2024).

Three policy goals set the perspective in climate currency. The first is facilitating investments in climate adaptation and mitigation through innovative financial instruments. The second is based on mobilizing private sector capital for climate-related projects. The third is to enhance the economic viability of sustainable practices and technologies. The climate currency toolbox includes, first of all, financial instruments (green finance) such as green bonds and climate bonds, C pricing mechanisms, or C credit futures. Issue bonds are specifically designed to fund climate adaptation and mitigation projects. Governments can offer tax incentives or guarantees to attract investors. Similarly, implementing C pricing mechanisms, such as C taxes or cap-and-trade systems, aims to capture the cost of C emissions and incentivize action to reduce them (Bhutta et al., 2022). In this context, it may be important to digitalization of climate currency by exploringe the development of digital currencies or tokens that represent C credits or climate-positive actions (Oyewale and Bassey, 2024). This can facilitate trade, transparency, and tracking of climate investments. Moreover, establishing national and regional climate funds and PPP is important to pool $\mathbf{C_{seq}}$ initiatives. Therefore it is also necessary to encourage partnerships between governments, financial institutions, and private companies to finance climate-friendly projects. This includes cofinancing arrangements and risk-sharing mechanisms. Finally, C credit futures (via futures contracts) allow the purchaser to offset emissions through C offset projects without having to directly invest in those projects at the time (Demiralay, 2022).

Finally, C credit futures contracts (via futures contracts) allow the purchaser to offset emissions through C offset projects (Table 7.2).

TABLE 7.2 Goals and instruments of policy perspectives for carbon credits, sustainable agriculture, and climate currency.

	Matrix of policy perspectives		
	Carbon credits	**Sustainable agriculture**	**Climate currency**
Goals	1. Promoting the use of agricultural practices that sequester C and reduce GHG emissions 2. Ensuring transparency, integrity, and efficiency 3. Facilitating farmers' participation	1. Enhancing agricultural productivity 2. Promote resilience and adaptability of agricultural systems 3. Support the livelihoods of farmers and rural. Communities	1. Facilitating investments in climate adaptation and mitigation 2. Mobilizing private sector capital for climate-related projects 3. Increasing the economic viability of sustainable practices and technologies
Instruments	1. Regulatory frameworks 2. Incentives and subsidies 3. Capacity building 4. Market development	1. Sustainable practices incentives 2. Research and development 3. Extension services 4. Market access 5. Support risk management	1. Green bonds and climate bonds 2. C pricing mechanisms 3. C credits futures

GHG, greenhouse gases.
Source: Author's own work.

Effective policies on C credits, sustainable agriculture, and climate currency are essential for addressing climate change and promoting SD. These policies should be designed to create synergies between environmental sustainability, economic growth, and social equity. By leveraging regulatory frameworks, financial incentives, market mechanisms, and innovative financial instruments, governments can foster an environment that promotes sustainable practices and climate resilience. Collaboration among stakeholders, including governments, the private sector, civil society, and international organizations, is crucial for the successful implementation of these policies and for their prospects.

7.9 Conclusion

The term "C credits" became widely used with the establishment and implementation of a market for GHG reduction under the Kyoto Protocol, as well as frameworks aimed at controlling and mitigating climate impacts. It describes tradable certificates that represent a significant reduction of GHGs. The concept and terminology have been rapidly adopted by businesses, policymakers, and environmental markets, becoming an integral part of combating measures for changing the climate and achieving SDGs.

C credits are an excellent strategy to offset the costs associated with sustainable farming practices. Generated through the emissions trading scheme and certification process, these credits enable monitoring, reporting on standards of practice, and assessing their impact on farms and the environment. With agriculture responsible for a significant share of global emissions, C credits offer a way to monetize environmental stewardship, effectively aligning economic and ecological goals. By earning C credits, farmers can access a new revenue stream, reducing financial barriers to sustainable transitions (thanks to climate and financial incentives as well as R&D investments). For corporations and countries, these credits provide a verified means to offset emissions and meet regulatory targets. The process of monetization of C credits (functioning as climate currency) is facilitated by the creation and functioning of the C credit market and can be done through several comprehensive mechanisms, such as valuation of C reductions, C credit markets, tokenization, or flexible mechanisms. In the pursuit of a more sustainable and low-C future, purchasing C certificates has become a key way to support this transition. While pricing C credits can seem complicated, especially when determining the value of a single credit, it is important to understand that not all C credits are created equal. The prices of credits generated from different projects can vary, even though they all represent a reduction of one ton of CO_2 in the atmosphere. C credits are becoming a transformative tool in sustainable agriculture by assigning economic value to practices that reduce GHG emissions. This "climate currency" incentivizes farmers to adopt sustainable methods, such as $\mathbf{C_{seq}}$ through cover cropping, no-till farming, and AF, which not only ensure soil quality and biodiversity but could also be an effective measure for changing the climate. Despite challenges, with robust regulatory frameworks and market oversight, C credits can significantly contribute to achieving the climate goals globally, along with sustainability.

By implementing, developing, and setting perspectives for comprehensive policies on C credits, sustainable agriculture, and climate currency, governments may provide suitable conditions for adopting sustainable agriculture and effective C credit systems, while also promoting broader participation in climate currency initiatives. This holistic approach can help mitigate climate change, enhance food security, and foster economic growth.

References

Abhinandan, 2024. What is third-party verification (TPV? All you need to know, September 13, 2024). Available from: outhbridge.com (accessed 12.10.24).

Allgood, B., Waugh, J., Talnage, C.A., Musikanski, L., Weeraman, D., 2024. How will we know when local communities benefit from carbon offset schemes? (commentary), Mongabay, 5 January 2024. Available from: https://news.mongabay.com/2024/01/ (accessed 23.09.24).

Ansari, M.A., Ravisankar, N., Shamim, M., Rani, M., Prusty, A.K., Singh, R., et al., 2024. Accounting of carbon sequestration and tradeoff under various climatic scenarios in alternative agricultural system: a comprehensive framework toward carbon neutrality. Frontiers in Sustainable Food Systems 8. Available at https://doi.org/10.3389/fsufs.2024.1371255.

Banerjee, A., Jhariya, M.K., Raj, A., Mechergui, T., A. Banerjee, M.K. Jhariya, A. Raj and T. Mechergui), 2024. Ecosystem Management: Climate Change and Global Sustainability—An Introduction. In: Banerjee, A., Jhariya, M.K., Raj, A., Mechergui, T. (Eds.), Ecosystem Management: Climate Change and Sustainability. Wiley, India, pp. 1–21. Available at https://doi.org/10.1002/9781394231249.ch1.

Bettín-Díaz, R., Rojas, A.E., Mejía-Moncayo, C., 2018. Methodological approach to the definition of a blockchain system for the food industry supply chain traceability. In: Gervasi, O., Murgante, B. (Eds.), Computational Science and Its Applications – ICCSA 2018. ICCSA 2018. Lecture Notes in Computer Science, vol 10961. Springer, Cham, pp. 19–33. Available at https://doi.org/10.1007/978-3-319-95165-2_2.

Bhutta, U.S., Tariq, A., Farrukh, M., Raza, A., 2022. Muhammad Khalid Iqbal, green bonds for sustainable development, Review of literature on development and impact of green bonds. Technological Forecasting and Social Change 175.

Biermann, F., Gupta, A., Mason, M., 2022. Carbon Market Challenge: Preventing Abuse Through Effective Governance. Cambridge University Press.

Blaufelder, C., Levy, C., Mannion, P., Pinner, D., 2021. Sustainability and Risk Practices: A Blueprint for Scaling Voluntary Carbon Markets to Meet the Climate Challenge. McKinsey & Company, Zurich.

Bonne, L., Roldan-Ruiz, I., Linden, V.V., Muylle, H., Dewulf, J., 2019. Environmental sustainability of conventional and organic farming: accounting for ecosystem services in life cycle assessment. Science of Total Environment 695, 133841. Available at https://doi.org/10.1016/j.scitotenv.2019.133841.

Broekhoff, D., Zyla, K., 2008. Outside the Cap: opportunities and Limitations of Greenhouse Gas Offsets, Climate and Energy Policy Series. World Resources Institute, Washington, DC.

Brown, T., Lee, C.M., Kruger, C.E., Reganold, J.P., Huggins, D.R., 2017. Comparison of green-house gas offset quantification protocols for nitrogen management in dryland wheat cropping systems of the pacific northwest. Frontiers in Environmental Science 5, 72. Available at 10.3389/fenvs.2017.00072.

Campbell, J., Herremans, I.M., Kleffner, A., 2018. Barriers to achieving additionality in carbon offsets: a regulatory risk perspective. Journal of Environmental Planning and Management 61 (14), 2570–2589. Available at https://doi.org/10.1080/09640568.2017.1406340.

Cariappa, A.A.G., Konath, N.C., Sapkota, T.B., Krishna, V.V., 2024. Evaluating the potential and eligibility of conservation agriculture practices for carbon credits. Scientific Reports 14, 9193. Available at https://doi.org/10.1038/s41598-024-59262-6.

Carney, M., Adams, T., Winters, B., Nazareth, A., 2020. Taskforce on scaling voluntary carbon markets, Consultation Document. Available from: https://custom.cvent.com/8644FD66069649369747A352DBAB07C3/files/ (accessed 01.10.24).

CAST, 2004. Climate Change and Greenhouse Gas Mitigation: Challenges and Opportunities for Agriculture. Task Force Report No. 141. Council for Agricultural Science and Technology, Ames, IA.

Chen, W., Xie, Q., Guo, L., 2024. Innovative approaches in agricultural sustainability and environmental impact management: challenges and opportunities. Agriculture 14 (12), 2316. Available at https://doi.org/10.3390/agriculture14122316.

Croker, T., 2024. 5 Principles to consider for Calculation Carbon Emission, Climatiq, January 31, 2024. Available from: climatiq.io (accessed 01.12.24).

Demiralay, S., Gencer, H.G., Bayraci, S., 2022. Hatice Carbon credit futures as an emerging asset: hedging diversification and downside risks. Energy Economics 113, 106196. Available at https://doi.org/10.1016/j.eneco.2022.106196.

Disaster Relief Fund, Monthly Reports, FEMA, 1 October 2024. Available from: fema.gov (accessed 02.12.24).

El Bakali, I., El Mekki, A.A., Maatala, N., Harbouze, R., 2023. A systemic review on the impact of incentives on the adoption of conservation agriculture, new guideline for policymakers and researchers. International Journal of Agricultural Sustainability 21 (1), 2290415. Available at https://doi.org/10.1080/14735903.2023.2290415.

Ellis, J., Larsen, K., 2008. Measurement, Reporting and Verification of Mitigation Actions and Commitments, OECD, Environment Directorate, International Energy Agency, November 2008.

Erickson, P.A., Lazarus, M., 2013. Implications of international GHG offsets on global climate change mitigation. Climate Policy 13 (4), 433–450. Available at https://doi.org/10.1080/14693062.2013.777632.

Financing the Green Transition, n.d. Mobilising investments to accelerate the transition to carbon-neutral and climate-resilient economy, White paper for A Green Transition, State of Green, COnnet. Inspire. Sgare. Think Denmark, 29. November 2023.

Food and Agriculture Organization of the United Nations, 2024. Incentives for Ecosystem Services: Supporting the transition to Sustainable Food System, Food and Agriculture Organization of the United Nations, s. 1, n.d. www.fao.org/in-action/incentives-for-ecosystem-services/en (accessed 04.09.24).

Freeman, O.E., Zerriffi, H., 2014. How you count carbon matters: implications of differing cookstove carbon credit methodologies for climate and development cobenefits. Environmental Science and Technology 48 (24), 14112–14120. Available at https://doi.org/10.1021/es503941u.

Fripp, E., 2014. Payments for Ecosystem Services (PES). A practical guide to assessing the feasibility of PES projects. Center for International Forestry Research (CIFOR).

Gordon-Smith, H., 2024. The Carbon Credit Dilemma in Agriculture: Progress, Pitfalls, and the Future, 25 September 2024. Available from: https://www.linkedin.com/pulse/carbon-credit-dilemma-agriculture-progress-pitfalls-gordon-smith (accessed 02.09.24).

Gupta, R., 2014. Aggregator Business Model for Carbon Credits Monetization, Indore Smart City Development Limited, CarbonCreditsAggregatorModel_INDORE.pdf (18.12.2014), (accessed 02.10.24).

Hijbeek, R. et al., 2019. Fertiliser use and soil carbon sequestration: trade-offs and opportunities. Working Paper No. 264. CGIAR Research Program on Climate.

Howard, A., Greiner, S., 2021. Accounting Approaches for the Voluntary Carbon Market. VCM Global Dialogue, Amsterdam.

Howden, S.M., Soussana, J.F., Tubiello, F.N., Chhetri, N., Dunlop, M., Meinke, H., 2007. Adapting agriculture to climate change. Proceedings of the National Academy of Sciences of the United States of America 104 (50), 19691–19696. Available at https://doi.org/10.1073/pnas.0701890104.

Japan Organization for Metals and Energy Security (JOGMEC), 2023. Handbook for CCS Carbon Credits: Workshop report of "Global Carbon Market and CCS: Towards ASEAN Decarbonization" Japan Organization for Metals and Energy Security (JOGMEC), Minato-ku, Tokyo. Available at https://www.jogmec.go.jp/content/300382578.pdf.

Jhariya, M.K., Meena, R.S., Banerjee, A., Meena, S.N., 2022. Natural Resources Conservation and Advances for Sustainability. Elsevier. Available at https://doi.org/10.1016/C2019-0-03763-6.

Jhariya, M.K., Meena, R.S., Banerjee, A., Kumar, S., Raj, A., 2023. Agroforestry for Carbon and Ecosystem Management. Academic Press, pp. 1–434. Available at 10.1016/C2021-0-02916-3.

Kaplan, R.S., Ramanna, K., Roston, M., 2023. Accounting for Carbon Credits, Balancing Bold, Fast and, Responsible, AI Deployment. July-August 2023. Available from: hbr.org. (accessed 01.10.24).

Kazemi, H., Klug, H., Kamkar, B., 2018. New services and roles of biodiversity in modern agroecosystems: a review. Ecological Indicators 93, 1126–1135. Available at https://doi.org/10.1016/j.ecolind.2018.06.018.

Korsching, P.F., Malia, J.E., 1991. Institutional support for practicing sustainable agriculture. American Journal of Alternative Agriculture 6 (1), 17–22. Available at https://doi.org/10.1017/S0889189300003751.

Kragt, M.E., Gibson, F.L., Maseyk, F., Wilson, K.A., 2024. The Core Carbon Principles, Integrity Council. Available from: icvcm.org (accessed 12.09.24).

Kreibich, N., Obergassel, W., 2019. The Voluntary Carbon Market: What may be Its Future Role and Potential Contributions to Ambition Raising? Available from: https://epub.wupperinst.org/frontdoor/deliver/index/docId/7396/file/7396_Carbon_Market.pdf (accessed 10.11.24).

Kumar, S., Meena, R.S., Sheoran, P., Jhariya, M.K., 2024. Regenerative Agriculture for Sustainable Food Systems. Springer, Singapore, pp. 1–493. Available at 10.1007/978-981-97-6691-8.

Kyoto Protocol to the United Nations Framework. Convention on Climate Change, United Nations 1998.

Li, C., 2000. Modeling trace gas emissions from agricultural ecosystems. Nutrition Cycling Agroecosystem 58, 259–276. Available at https://doi.org/10.1023/A:1009859006242.

Liu, Q., Sun, X., Wu, W., Liu, Z., Fang, G., Yang, P., 2022. Agroecosystem services: a review of concepts, indicators, assessment methods and future research perspectives. Ecological Indicators 142. Available at https://doi.org/10.1016/j.ecolind.2022.109218.

Malamas, V., Dasaklis, T.K., Arakelian, V., Chondrokoukis, G., 2024. A blockchain framework for digitizing securities issuance: the case of green bonds. Journal of Sustainable Finance and Investment 14 (3), 569–595. Available at https://doi.org/10.1080/20430795.2023.2275212.

Meena, R.S., Pradhan, G., 2023. Industrial garbage-derived biocompost enhances soil organic carbon fractions, CO_2 biosequestration, potential carbon credits and sustainability index in a rice-wheat ecosystem. Environmental Research 235, 116525. Available at https://doi.org/10.1016/j.envres.2023.116525.

Meena, R.S., Pradhan, G., Kumar, S., Lal, R., 2023. Using industrial wastes for rice-wheat cropping and food-energy-carbon-water-economic nexus to the sustainable food system. Renewable and Sustainable Energy Reviews 187, 113756. Available at https://doi.org/10.1016/j.rser.2023.113756.

Meena, R.S., Lal, R., Kumar, S., Pradhan, G., Srinivasarao, C., Singh, A.K., et al., 2024a. Potential of Indian agriculture for capturing atmospheric CO_2 and monetizing carbon credits to the farmers: an overview and policy framework. Advances in Agronomy 188, 101–206. Available at https://doi.org/10.1016/bs.agron.2024.06.001.

Meena, R.S., Pradhan, G., Singh, K., Kumar, S., Singh, A.K., Shashidhar, K.S., et al., 2024b. Agriculture models for restoring degraded land to enhance CO_2 biosequestration and carbon credits in the Vindhyan region of India. Science of the Total Environment 929. Available at https://doi.org/10.1016/j.scitotenv.2024.172661.

Meena, R.S., Pradhan, G., Kumar, S., 2025. Energy flow, eco-efficiency, and economic circulation with recycled industrial waste compost application in wheat and subsequent rice farming. Science of the Total Environment 967. Available at https://doi.org/10.1016/j.scitotenv.2025.178779.

Michaelowa, A., Shishlov, I., Brescia, D., 2019. Evolution of international carbon markets: lessons for the Paris agreement. Wiley Interdisciplinary Reviews: Climate Change 10 (6), e613. Available at https://doi.org/10.1002/wcc.613.

Miller, S.,2013. COMET-FARMtm: Conservation, Calculation, U.S. Department of Agriculture, August 21, 2013. Available from: usda.org (accessed 02.09.24).

Mirzaee, H., Samarghandi, H., Willoughby, K., 2022. A three-player game theory model for carbon cap-and-trade mechanism with stochastic parameters. Computers and Industrial Engineering 169. Available at https://doi.org/10.1016/j.cie.2022.108285.

Morel, R., Delbosc, A., 2012. Assessing the financial efficiency of the green climate fund: leverages ratios from theory to practice. Climate Brief: Focus on the Economics of Climate Change 19.

Munang, R., Thiaw, I., Alverson, K., Mumba, M., Liu, J., Rivington, M., 2013. Climate change and ecosystem based adaptation: a new pragmatic approach to buffering climate change impacts. Current Opinion in Environmental Sustainability 5 (1), 67–71. Available at https://doi.org/10.1016/j.cosust.2012.12.001.

Nalau, J., Becken, S., Mackey, B., 2018. Ecosystem-based adaptation: a review of the constraints. Environmental Science & Policy 89, 357–364. Available at https://doi.org/10.1016/j.envsci.2018.08.014.

Netter, L., Luedeling, E., Whitney, C., 2022. Agroforestry and reforestation with the gold standard-decision analysis of a voluntary carbon offset label. Mitigation and Adaptation Strategies for Global Change 27, 17. Available at https://doi.org/10.1007/s11027-021-09992-z.

Niederberger, A.A., Kimble, M., 2011. MRV under the UN climate regime: paper tiger or catalyst for continual improvement. Greenhouse Gas Measurement and Management 1 (1), 47–54. Available at https://doi.org/10.3763/ghgmm.2010.0009.

Olsen, K.H., Arens, C., 2021. Promoting Sustainable Development in Article 6 pilot activities. Available at https://orbit.dtu.dk/en/publications/promoting-sustainable-development-in-article-6-pilot-activities/.

Oyewale, K., Bassey, K.E., 2024. Climate action and social equity: mitigation strategies and carbon credits. World Journal of Advanced Research and Reviews 24 (2), 187–201. Available at https://doi.org/10.30574/wjarr.2024.24.2.3325.

Paris Agreement, United Nation Organization, 2015. Available from: unfccc.int (accessed 01.09.24).

Parnphumeesup, P., Kerr, S.A., 2015. Willingness to pay for gold standard carbon credits. Energy Sources, Part B: Economics, Planning, and Policy 10 (4), 412–417. Available at https://doi.org/10.1080/15567249.2010.551251.

Piñeiro, V., Arias, J., Dürr, J., Elverdin, P., Ibáñez, A.M., Kinengyere, A., et al., 2020. A scoping review on incentives for adoption of sustainable agricultural practices and their outcomes. Nature Sustainability 3 (10), 809–820. Available at https://doi.org/10.1038/s41893-020-00617-y.

Pradhan, G., Meena, R.S., 2023a. Interaction impact of biocompost on nutrient dynamics and relations with soil biota, carbon fractions index, societal value of CO_2 equivalent and ecosystem services in the wheat-rice farming. Chemosphere 339. Available at https://doi.org/10.1016/j.chemosphere.2023.139695.

Pradhan, G., Meena, R.S., 2023b. Utilizing waste compost to improve the atmospheric CO_2 capturing in the rice-wheat cropping system and energy-cum-carbon credit auditing for a circular economy. Science of the Total Environment 892, 164572. Available at https://doi.org/10.1016/j.scitotenv.2023.164572.

Pradhan, G., Meena, R.S., Kumar, S., Lal, R., 2023. Utilizing industrial wastes as compost in wheat-rice production to improve the above and below-ground ecosystem services. Agriculture, Ecosystems & Environment 358, 108704. Available at https://doi.org/10.1016/j.agee.2023.108704.

Raj, A., Jhariya, M.K., Meena, R.S., 2025. Agroforestry for Monetising Carbon Credits. Springer, Cham, pp. 1–609. Available at https://doi.org/10.1007/978-3-031-97151-8.

Reidsma, P., Ewert, F., 2008. Regional farm diversity can reduce vulnerability of food production to climate change. Ecology and Society 13 (1), 38. Available at https://doi.org/10.5751/ES-02476-130138.

Sabbatini, M., 2022. What is 'monetizing carbon credits? And how would it work? Juneau Empire December. 19, 2022.

Sahoo, G., Majid Wani, A., Prusty, M., Ray, M., 2023. Effect of globalization and climate change on forest – a review. Materials Today: Proceedings 80, 2060–2063. Available at https://doi.org/10.1016/j.matpr.2021.06.113.

Siedenburg, J., Martin, A., McGuire, S., 2012. The power of "farmer friendly" financial incentives to deliver climate smart agriculture a critical data gap. Journal of Integrative Environmental Sciences 9 (4), 201–217. Available at https://doi.org/10.1080/1943815X.2012.748304.

Sirilakshmi, Y., Gogoi, B.P., Ashwini, T., Dkhar, D., 2024. A review on carbon foot printing in sustainable agriculture as a cutting-edge approach to environmental responsibility. International Journal of Environment and Climate Change 14 (10), 261–269. Available at https://doi.org/10.9734/ijecc/2024/v14i104485.

Smith, P., Martino, D., Cai, Z., Gwary, D., Janzen, H., Kumar, P., et al., 2008. Greenhouse gas mitigation in agriculture. Philosophical Transactions of the Royal Society B: Biological Sciences 363 (1492), 789–813. Available at https://doi.org/10.1098/rstb.2007.2184.

Smyth, S.J., Webb, S.R., Phillips, P.W.B., 2021. The role of public-private partnerships in improving global food security. Global Food Security 31, 100588. Available at https://doi.org/10.1016/j.gfs.2021.100588.

Sternberg, J.A., 2017. A conceptual framework for integrated pest management. Trends in Plant Science 22 (9), 759–769. Available at https://doi.org/10.1016/j.tplants.2017.06.010.

Thompson, J., 2004. The social cost of carbon credits, Hakai Magazine Audio Edition, July 16. Available from: https://hakaimagazine.com/features/the-social-cost-of-carbon-credits/, (accessed 18.09.24).

Tjon Akon, M., 2023. The role of market operators in scaling up voluntary carbon markets. Capital Markets Law Journal 18 (2), 259–275. Available at https://doi.org/10.1093/cmlj/kmad001.

Trouwloon, D., Streck, C., Chagas, T., Martinus, G., 2023. Understanding the use of carbon credits by companies: a review of the defining elements of corporate climate claims. Global Challenges 7 (4). Available at https://doi.org/10.1002/gch2.202200158.

United Nations Framework Convention on Climate Change (UNFCCC), United Nations 1992.

United Nations, 2015. Transforming our world: The 2030 Agenda for Sustainable Development, A/RES/70/1, United Nations. Available from: sustainabledevelopment.un.org (accessed 01.10.24).

Updegraff, K., Zimmerman, P.R., Kozak, P., Chen, D.-G., Price, M., 2010. Estimating the uncertainty of modelled carbon sequestration: the greencert™ system. Environ. Modelling Software 25 (12), 1565–1572. Available at https://doi.org/10.1016/j.envsoft.2010.05.009.

Van Brunt, M., Kantor, B., Johnson, N., 2012. Generating and selling carbon offset credits. Proceedings of the 20th Annual North American Waste-to-Energy Conference. 20th Annual North American Waste-to-Energy Conference. Portland, Maine, USA. April 23–25, 2012. ASME, pp. 159–165. Available at https://doi.org/10.1115/NAWTEC20-7048.

Wettenberg, K., Ellis, J., Schneider, L., 2024. The interplay between voluntary and compliance markets: Implications for environmental integrity, OECD Environment Working Papers, 244.

Widholm, B., Stolyarova, E., 2023. Tokenization is the process of utilizing blockchain technology throughout an asset's lifecycle, state street, July 2023. Available from: statestree.com (accessed 11.11.24).

Yin, I., Ghosh, A., 2023. Voluntary carbon markets faces headwind as article 6 mechanism gains traction. S&P Global, 12 May 2023. Available from: https://www.spglobal.com/commodity-insights/en/news-research/latest-news/energy-transition (accessed 07.09.24).

Zabala, J.A., Martínez-Paz, J.M., Alcon, F., 2021. A comprehensive approach for agroecosystem services and disservices valuation. Science of the Total Environment 768. Available at https://doi.org/10.1016/j.scitotenv.2020.144859.

Section III

Management practices in agriculture for carbon incentives

Chapter 8

Agronomic and technological interventions toward carbon farming

Shashi S. Udumann, D.M.N.S. Dissanayaka, Tharindu D. Nuwarapaksha, and Anjana J. Atapattu
Agronomy Division, Coconut Research Institute, Lunuwila, Sri Lanka

Highlights

- Carbon farming has a significant impact on agroecosystem sustainability.
- Agrotechnology can help to lower the negative impact of climate change and enhance resilience.
- Improving the carbon sink through various approaches can lead toward a green economy.

8.1 Introduction

Carbon (C) farming has become an essential method to mitigate climate change and maintain agricultural productivity. C farming practices show a strong capability to capture significant amounts of carbon dioxide (CO_2) when practiced at scale (Jansson et al., 2021; Meena et al., 2025). This includes agricultural strategies that involve specific management to facilitate carbon sequestration (C_{seq}) in some soil components, including plant structures, as well as to maintain sustainable agricultural production and ecosystem services (Bayata and Mulatu, 2024; Raj et al., 2025; Meena et al., 2025).

Applied on agricultural land for C_{seq}, practices of C farming methods are essential elements. Under these agronomic practices, conservation tillage, cover cropping, and improved crop rotation improve soil organic carbon (SOC) content (Kumar et al., 2020, 2024). It is also mentioned in recent studies that when these practices come into effect, soil C reserves increase, and they improve soil structure quality and increase water retention ability (Atapattu et al., 2025; Chivenge et al., 2007). Perennial crops integration with agroforestry systems (AFs) has great potential to store C for the long term (Atapattu and Udumann, 2024; Lorenz and Lal, 2014; Dissanayaka et al., 2023a; Jhariya et al., 2022, 2023).

The implementation of effective monitoring for the practice of C farming has changed, and the practice has been transformed by agrotechnology. By using advanced sensing technologies, machine learning (ML) algorithms, and precision agriculture (PA) tools, farmers can optimize C_{seq} at an unparalleled level of precision (Nuwarapaksha et al., 2026). The use of artificial intelligence (AI) with remote sensing techniques reduces assessment and monitoring costs, and this approach aligns with research showing that combining approaches that utilize remote sensing, tends to lower costs compared to using these approaches alone (Atapattu et al., 2024; Mor et al., 2021). Through smart farming, which involves the implementation of Internet of Things (IoT) sensors and an automated management system, producers can achieve optimal crop sequence and yield performance (Polymeni et al., 2024). IoT sensors and automated management systems in smart farming let producers get the most out of their C_{seq} potential and yield potential.

8.2 Agronomic practices and carbon farming

Increased SOC storage can be achieved without adverse effects from climate change and improved soil health through agronomic practices (Fig. 8.1). C farming has become a major sustainable option to reduce atmospheric CO_2 that involves practicing agriculture to store C either in soils or vegetation (Meena et al., 2025). The unique combination of agronomic practices contributes to C_{seq} through improved soil structure, enhanced organic matter (OM) input, and minimized C losses.

Carbon Incentives, Plans, and Policies in Agriculture. DOI: https://doi.org/10.1016/B978-0-443-33557-0.00018-5

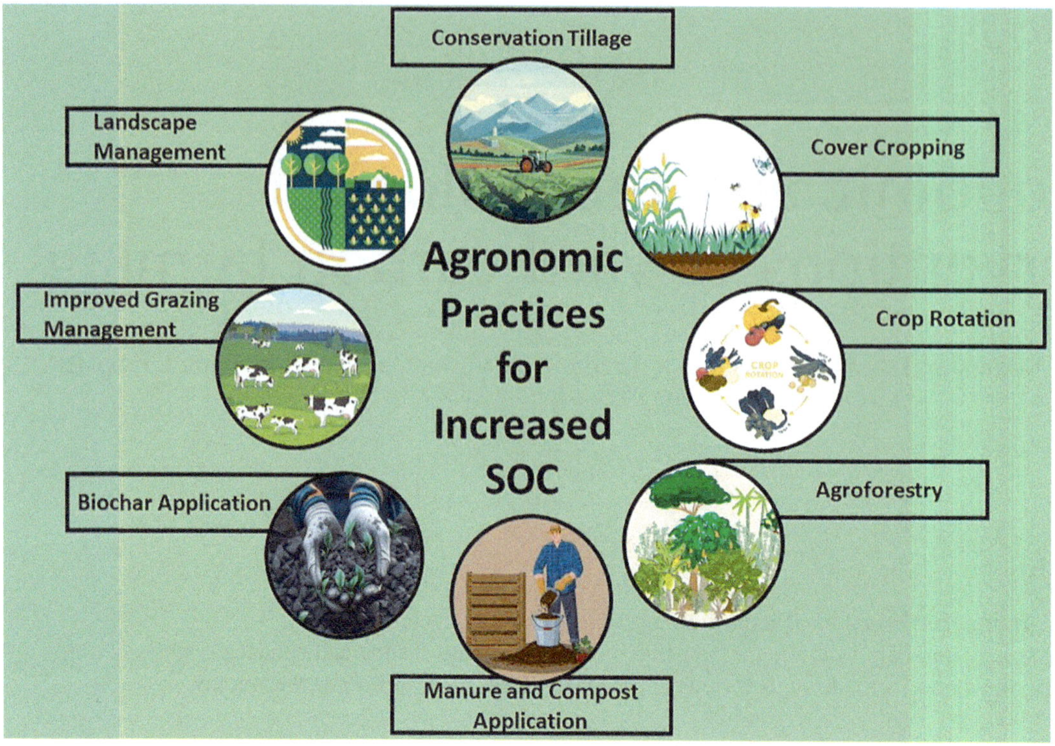

FIGURE 8.1 Agronomic practices for enhanced soil organic carbon and carbon farming.

8.2.1 Conservation tillage

The adoption of conservation tillage methods modifies agricultural management and produces substantial benefits for enhancing C_{seq} in soil systems. These practices minimize soil disturbance, which helps to maintain soil structure, along with preserving OM from damaging decomposition and oxidation processes (Hussain et al., 2021). Integrating conservation tillage practices leads to increased C stabilization in agricultural soil due to both reduced erosion risks and enhanced structural integrity (Lal, 2015). Studies demonstrate that applying no-till methods helps develop 0.1–0.5 Mg C ha^{-1} yr^{-1} SOC stocks, which are most prominent in surface soil layers (West and Post, 2002).

Soil C_{seq} processes in conservation tillage systems work through two mechanisms. They have limited oxygen access, which slows OM decomposition by microorganisms and the physical protection of C as a result of enhanced soil aggregates. The practice of reduced tillage allows more OM residue to remain on soil surfaces as continuous C inputs to the soil system. The reduction of tillage operations enables fungal soil food webs to dominate, which stabilizes soil C accumulation (Chen et al., 2020).

The success rate of conservation tillage methods depends on different soil types, regional climates, and agricultural cropping systems (Six et al., 2004). A global metaanalysis found that conservation tillage maintains optimal C storage capacity mainly in humid soil regions having fine-textured soils. The C_{seq} potential decreased significantly in arid regions, which contain coarse-textured soils (Thapa et al., 2023a).

8.2.2 Cover cropping

The practice of cover cropping provides agricultural systems with multiple benefits, together with C_{seq} enhancements in the soil through increased biomass accumulation. The nonharvested crops planted between main crop cycles supply substantial biomass to agricultural soils, which enhances soil C_{seq} while accelerating microbial processes (Schmidt et al., 2018; Dissanayaka et al., 2022).

The ability of cover crops to sequester C depends heavily on the specific plant species chosen. Leguminous cover crops contribute nitrogen to the system while building soil C (Dissanayaka et al., 2024b), whereas nonleguminous species enhance SOC by adding OM through root exudates and biomass decomposition (Poeplau and Don, 2015). Cover crops play a vital role in C farming as they reduce soil erosion and enhance water conservation, in addition to building soil C (Blanco-Canqui et al., 2015).

8.2.3 Crop rotation

Crop rotation means rotating out various plant types from one plot of land to improve soil health and reduce pest populations. Good crop rotations are a key foundation for sustainable agriculture because they exert multiple impacts on soil C storage. Good crop rotation systems involve growing several crops that have unique root systems, each requiring a different nutrient composition and each with a different type of growth pattern, which together will help by increasing soil structure, building microbial diversity, and improving OM content.

Multiple crop rotations do encourage more SOC by creating more valuable OM that returns to the soil (McDaniel et al., 2014). For example, legumes added to rotational farming systems help increase SOC content through their rich nitrogen content, which decomposes quickly (Drinkwater et al., 1998; Jhariya et al., 2018). Long-term studies show that crop rotations enhance SOC levels by 10%–20% more than farming with a single crop (Smith et al., 2008).

Crop rotations enhance agricultural resilience by increasing pest control and nutrient recycling while minimizing the use of synthetic materials (Atapattu et al., 2026; Dissanayaka et al., 2023c). These advantages boost C capture capabilities (Dissanayaka et al., 2023a, 2023b).

8.2.4 Agroforestry

Agroforestry (AF) combines trees and shrubs with agricultural operations to produce favorable ecological interactions and economic value. The C_{seq} capabilities of AF depend heavily on system design and management approaches (Nuwarapaksha et al., 2025). Silvopastoral systems were proven to be highly effective in C_{seq} when livestock grazing combines with tree integration (Atapattu et al., 2024c, 2025). Working together, aboveground trees and underground tree structures carry out the process of C_{seq}, along with fallen leaves and roots, as they store soil C (Ramachandran Nair et al., 2009). Tree selection and management approaches determine the C_{seq} capability of AFs ranging from 1.5 to 3.5 Mg C ha^{-1} yr^{-1} (Montagnini and Nair, 2004). Furthermore, AF increases both biological diversity and provides valuable ecosystem services, such as shade and windbreaks (Atapattu and Udumann, 2024; Dissanayaka et al., 2023c).

8.2.5 Manure and compost application

One straightforward way to increase soil C content and recycle nutrients simultaneously is through the use of organic amendments, such as animal manure and compost. The soil C is stabilized by these OM into long-duration stable C forms (Atapattu et al., 2025; Diacono and Montemurro, 2011). Different material processing methods, as well as variations in chemical composition, result in significant differences in organic amendment C_{seq} rates. Through humification processes, raw manures are better stabilized as soil C pools compared to unprocessed materials in composting, thereby increasing soil C stability (Udumann et al., 2023). When manure is applied over many years, scientists have revealed that SOC stocks increase by 20%–40% when compared to untreated soil (Maillard and Angers, 2014). The stabilization process of OM through composting helps reduce greenhouse gas (GHG) emissions before agricultural use (Bernal et al., 2009).

8.2.6 Biochar application

The C_{seq} technique of biochar production uses organic waste pyrolysis to produce a persistent, C-dense product that remains in soil for extended timescales of hundreds to thousands of years (Dissanayaka et al., 2023c; Maitra et al., 2025). Because biochar has an aromatic C structure, it stands out from typical organic amendments by being resistant to microbial decomposition, which leads to prolonged C storage (Shamila et al., 2024; Dissanayaka et al., 2023c). Biochar application increases SOC levels by 20%–50% and improves soil fertility through better nutrient retention and water storage capacity (Sohi et al., 2010). Biochar delivers multiple soil benefits beyond C trapping because it enhances water retention, improves nutrient availability, and creates suitable habitats for microbes (Dissanayaka et al., 2025). The multiple benefits of biochar applications promote increased plant productivity and soil health, creating continuous positive feedback processes that increase C_{seq} (Dissanayaka et al., 2023c). However, biochar effectiveness varies based on both feedstock type and pyrolysis conditions, as well as soil properties (Spokas et al., 2012).

8.2.7 Improved grazing management

Approximately 70% of the world's total agricultural land is grazing land that can capture considerable amounts of C by applying improved management methods (Abdalla et al., 2018). The implementation of adaptive grazing systems, together with the strategic management of grazing pressure distribution, enables better $C_{seq,}$ resulting in improved forage production and ecosystem health. Soil C accumulation occurs through proper grazing methods that stimulate root development together with the accumulation of plant litter (Conant et al., 2001). SOC levels in rotational grazing systems increase by 0.5–1.0 Mg C ha^{-1} yr^{-1} higher than those under continuous grazing (Teague et al., 2011). Additionally, the implementation of proper grazing management techniques reduces soil compaction while simultaneously decreasing erosion and supporting an increased C_{seq} (Sanderman et al., 2017).

Implementing improved grazing management requires specific adjustments to local environmental constraints and operational settings to achieve its substantial C_{seq} potential (Nuwarapaksha et al., 2023). The effects of grazing management on C_{seq} differ substantially across climate types based on a global metaanalysis and exhibit maximum impact in temperate and semiarid grasslands (Jones and Donnelly, 2004). Modern precision grazing technologies, such as virtual fencing systems, together with remote monitoring tools, now make adaptive grazing practices more practical across many different landscapes (Monteiro et al., 2021; Atapattu et al., 2024a).

8.2.8 Landscape management

C farming practices across different land units at the landscape scale combine various strategies to optimize C storage while maintaining ecosystem functions and agricultural productivity levels (Nuwarapaksha et al., 2024b). These strategies acknowledge that C processes depend on factors extending beyond individual field edges, including landscape position, hydrology, and ecological interactions (Evans et al., 2015). Landscape management includes several operations such as land restoration of degraded sites, buffer zone development, and implementation of agroecological practices (Maitra et al., 2025; Lal, 2004). Riparian buffers, along with wetlands, function as C sinks while providing multiple ecosystem services, including water purification and habitat preservation (Moreno-Mateos et al., 2012). Landscape-level programs serve as key strategies for achieving climate change mitigation targets and boosting the resilience of ecosystems (Scherr et al., 2012; Banerjee et al., 2023, 2024).

To provide a clear overview of how different agronomic practices contribute to C farming, Table 8.1 summarizes their mechanisms of C_{seq}, potential C storage rates, and additional benefits for soil health and ecosystem resilience.

8.3 Agrotechnology and carbon farming

The agricultural fields undergo a transformation through agrotechnology by implementing innovative tools that boost productivity, along with sustainability and environmental protection measures (Nuwarapaksha et al., 2025). The implementation of agrotechnology optimizes C_{seq} practices as part of C farming, thus mitigating climate change and promoting soil health (Meena et al., 2025). This section evaluates three important technical advancements in C farming, which are PA, along with remote sensing and Geographic Information Systems (GIS), and AI with ML.

8.3.1 Precision agriculture

Using technology-enabled PA allows stakeholders to optimize agricultural practices, thus achieving higher productivity through efficient resource management and sustainable environmental outcomes. PA implements global positioning systems together with sensors and variable-rate technology to precisely distribute fertilizers, water, and pesticides to specific outdoor locations at specific time periods, thus promoting waste reduction and environmental sustainability (Zhang and Kovacs, 2012). Soil C_{seq} within C farming depends heavily on PAs ability to optimize farming practices. Turbo-charged fertilizer distribution practices through precision nutrient management optimize soil nutrient balance and drive better plant development along with enhanced root growth, resulting in higher organic C accumulation in soil (Gebbers and Adamchuk, 2010). The implementation of conservation tillage and cover cropping becomes simpler through PA since the technology provides live information on soil status and crop progression (Getahun et al., 2024). Multiple research studies indicate that PA raises SOC by 10%–15% in intensive agricultural fields (Robertson et al., 2000).

8.3.2 Remote sensing and geographic information systems

Remote sensing and GIS offer effective tools for tracking and managing C-farming operations across multiple geographic scales and time periods. High-resolution remote sensing technologies, which consist of satellite imagery and drone imagery, enable scientists to collect data on vegetation health and soil moisture, as well as track land-use changes, thus determining C stock and flux dynamics (Mulla, 2013). GIS systems link spatial data into comprehensive soil C distribution maps, which enable farmers to locate their land areas most suitable for C storage (Paustian et al., 2016). Remote sensing technologies track the evolution of aboveground biomass to assess the effectiveness of AFs C_{seq} (Thapa et al., 2023a). GIS-based prediction models help determine the impacts of land management practices on SOC dynamics, enabling

TABLE 8.1 Contribution of different agronomic practices to carbon farming.

Agronomic practice	Mechanism of carbon sequestration	Additional benefits
Conservation tillage	• Reduces soil disturbance • Preserves OM • Enhances soil aggregates • Limits decomposition	• Reduces erosion • Improves soil structure • Stabilizes C accumulation
Cover cropping	• Adds biomass to soil through nonharvested crops • Enhances microbial processes • Reduces erosion.	• Improves water conservation • Reduces erosion • Enhances soil health
Crop rotation	• Diversifies OM input • Improves soil structure • Enhances microbial diversity.	• Enhances pest control • Nutrient recycling • Reduces synthetic input use
Agroforestry	• Combines trees and shrubs with crops/livestock • C_{seq} in aboveground and belowground biomass	• Increases biodiversity • Provides ecosystem services • Improves soil fertility
Manure and compost application	• Stabilizes carbon through organic amendments • Enhances humification processes	• Recycles nutrients • Improves soil fertility • Reduces greenhouse gas emissions
Biochar application	• Converts organic waste into stable, aromatic C resistant to decomposition	• Enhances water retention • Nutrient availability • Microbial habitats
Improved grazing management	• Stimulates root development • Increases plant litter • Reduces soil compaction and erosion	• Improves forage production and ecosystem health • Reduces erosion
Landscape management	• Integrates C farming practices across land units, including restoration and buffer zones	• Enhances ecosystem resilience • Water purification • Habitat preservation

OM, organic matter.

decisions on sustainable C farming approaches (Smith et al., 2020). Remote sensing, together with GIS tools, establishes a complete system to enhance C farming approaches and achieve climate change reduction objectives.

8.3.3 Artificial intelligence and machine learning

The technology of AI and ML is revolutionizing C farming through predictive soil C management models that effectively analyze complex datasets. AI algorithms function by analyzing vast sensor data, satellite information, and field-based observations to find patterns that standard methods cannot detect (Kamilaris and Prenafeta-Boldú, 2018). Soil C level predictions have become possible through ML models that analyze soil characteristics together with climate measurements and agricultural management methods, provide farmers with useful information on increasing C capture levels (Padarian et al., 2020). AI decision support tools optimize both crop rotation schedules and irrigation maintenance alongside fertilizer programs to keep farming practices in line with C-based practices (Liakos et al., 2018; Nuwarapaksha et al., 2024a). AI tools enable continuous C credit monitoring along with project verification services, which strengthen transparency throughout C markets (Foley et al., 2011). The combination of AI and ML technology enables stakeholders to achieve better precision, efficiency, and scalability when they implement C farming approaches for climate change mitigation.

8.4 Agrotechnology for enhancing carbon credits

Advanced agrotechnologies enable revolutionary changes in C credit generation while improving monitoring and trading processes to achieve greater transparency, enhanced efficiency, and scalability in C farming. This part examines two revolutionary technologies that transform C credit systems through blockchain and IoT-enabled sensor networks.

8.4.1 Blockchain technology

Blockchain technology serves as an emerging solution that transforms C credit markets through enhanced transparency and traceability, as well as increased trust features. Because blockchain technology provides an unchangeable decentralized ledger, transactions for C credits become safe, easily proven, and completely resistant to tampering (Saberi et al., 2019). Blockchain technology enables the verification of C_{seq} processes through its ability to trace and authenticate activities such as storing C in soils, planting new forests, and practicing sustainable land management (Kshetri, 2021). Smart contracts on blockchain platforms allow automated C credit handling through verified remote sensing, IoT sensor, and field measurement data (Kouhizadeh et al., 2021). The implementation of blockchain technology decreases fraudulent activities and duplicate reporting throughout C market operations, as these issues commonly arise when operating without blockchain. Smallholder farmers can now join C credit programs through blockchain technology because it provides an open system for documenting and obtaining financial benefits from their C_{seq} activities (Tian, 2016). Blockchain technology strengthens C credit systems; therefore, it will become essential for the massive global expansion of C farming projects.

8.4.2 Internet of Things and sensor networks

Soil C dynamics and crop performance data collection, along with environmental condition assessment, remain possible through IoT technology and sensor networks during C farming operations (Fig. 8.2). The continuous monitoring capabilities of IoT devices, including soil moisture sensors, weather stations, and drones, generate data streams useful for C_{seq} practice improvement (Nuwarapaksha et al., 2024c; Ray, 2017). Soil sensors strategically positioned in the field permit organic C measurement as well as temperature and moisture tracking to enable farmers to utilize PA practices, which boost SOC storage (Boursianis et al., 2022). IoT networks also facilitate the integration of data from multiple sources, such as satellites. The integration becomes possible through IoT networks because they combine satellite imagery data and field sensor readings to construct complete C flux models (Hasan et al., 2024). The use of data-driven methods advances the precision of C credit assessment; thus, C farming operations maintain scientific validity. IoT technology platforms generate actionable insights for farmers, who can utilize this information to implement sustainable practices that optimize C_{seq} efforts (Wolfert et al., 2017). The combination of IoT and sensor networks enables C farming to operate more efficiently, becoming scalable enough to achieve greater success in mitigating climate change.

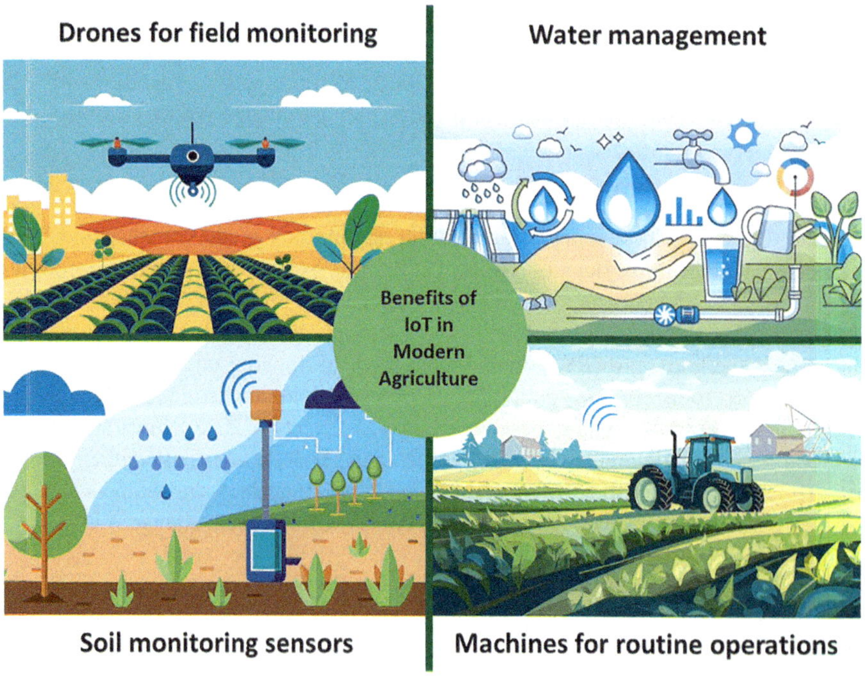

FIGURE 8.2 Uses of IoT in modern agriculture *IOT*, Internet of Things.

8.5 Issues and concerns of carbon farming

The implementation of C farming faces numerous challenges that prevent it from reaching its full effectiveness and universal adoption. A deep analysis shows important matters, which include difficulties with accessible data and interoperable programs, the technological gap between communities, verification obstacles, and regulatory frameworks that need enhancement to support successful implementation.

8.5.1 Data accessibility and interoperability

C farming faces its main challenge from difficulties related to data accessibility across disparate platforms while ensuring they function together seamlessly. Decision-making effectiveness suffers because inadequate standardized high-quality data blocks proper measurement and verification of C_{seq} (Lobell et al., 2015). Data type incompatibility produces problems when remote sensing data joins IoT device information with field measurement records because these platforms keep their data in different incompatible formats (Wolfert et al., 2017). Some developing regions face an obstacle because their smallholder farmers lack the digital infrastructure necessary to participate in C farming programs (Dhillon and Moncur, 2023). Open data standards and interoperable platforms, alongside capacity-building programs, need to be developed to enable stakeholders across the board to effectively utilize C farming data (Hasan et al., 2024).

8.5.2 Technology adoption and digital divide

Sophisticated C farming technology adoption faces resistance from the digital divide, which mostly impacts smallholder farmers, together with inhabitants of rural settings. Most farmers do not possess sufficient financial capabilities, together with the needed technology expertise and proper digital infrastructure to deploy PA, IoT, or blockchain systems (Bronson and Knezevic, 2016). The digital divide restricts the expansion of C farming projects while weakening their ability to include all farming groups, especially those in developing nations (Carolan, 2017). The recent fast rate of technological advancement presents obstacles for farmers because they find it challenging to keep pace with new technological systems (Eastwood et al., 2019). The digital divide requires investment in rural infrastructure, together with training programs and affordable solutions that meet farming requirements (Choruma et al., 2024).

8.5.3 Verification and accounting challenges

The accuracy in C_{seq} verification and accounting determines the credibility, and success of C-farming programs. Measuring soil C stocks and fluxes remains complex because of multiple variables in soil properties and climate factors, and different management methods affect the results (Paustian et al., 2016). The current methods for verification, which include soil sampling and remote sensing, result in high costs and incomplete data accuracy, which diminishes confidence in C credit systems (Smith et al., 2020). Standardized C accounting procedures must be established to eliminate diverse methods and reporting approaches in measuring C credits (Ogle et al., 2014). The challenges in C farming require careful implementation of better measurement methods, together with unified accounting protocols and independent verification processes (Sanderman et al., 2017).

8.5.4 Policy and regulatory framework

C farming requires specific supportive policies and regulatory frameworks that encourage sustainable farming methods while ensuring fairness in the distribution of benefits. Although several countries have struggled to establish transparent policies regarding C credits and land management alongside soil C practices, this creates complex conditions for agricultural producers and those investing in C credits (Amundson and Biardeau, 2018). Complimentary regulations between different regions act as barriers that prevent global C market expansion, which constrains the growth of C farming (Hsu et al., 2017). A proper policy structure must include provisions that guarantee all farmers, including small-scale holders and marginalized groups, equal access to available C credit systems (Chhatre et al., 2012; Meena et al., 2025). The development of strong policy frameworks requires collaboration between government, research, and stakeholders to establish C_{seq} incentives alongside clear C credit trading rules and equal opportunities for C farming participation (Scherr et al., 2012).

8.6 Technological interventions and sustainable approaches toward carbon farming

C farming progress requires both technological advancements and sustainable methods to manage climate change and food security issues. The section analyzes how traditional agricultural methods join modern agrotechnologies and how active farmer involvement through extension services and public–private partnerships (PPPs) enhances C farming initiatives. Farmers who adopt precision technologies, such as remote sensing and IoT, alongside conservation agriculture, improve resource efficiency and enhance soil C storage potential.

8.6.1 Integrating agronomic practices and agrotechnology

Agriculture requires a combination of agronomic practices and advanced agrotechnology systems to realize its full potential for C farming while implementing sustainable agricultural methods. Traditional agricultural approaches, such as conservation tillage, cover cropping, and crop rotation, can work in harmony with modern technological solutions, including PA, IoT systems, and remote sensing techniques, to facilitate simultaneous soil C_{seq}, increased resource efficiency, and enhanced productivity (Zhang and Kovacs, 2012). Precision nutrient management systems achieve two objectives by optimizing fertilizer use, which decreases environmental emissions and boosts soil–C storage (Gebbers and Adamchuk, 2010; Thapa and Ghimire, 2023b). Real-time soil C measurement through IoT-enabled sensors enables farmers to adjust their management practices based on live data (Boursianis et al., 2022). Through the combined use of agronomic practices and agrotechnology, farmers achieve the peak potential of C_{seq} while maintaining sustainable farming systems for the future (Lal, 2015).

8.6.2 Participatory research and extension

Success in the adoption of C farming requires active involvement between farmers and researchers through participatory research and extension approaches, especially for smallholder farmers. Complex practices unite local farmers together with researchers alongside extension agents to create specific strategies for C farming that match regional characteristics and cultural traditions (Hildebrand, 1991). Participatory research methods enable better innovation because they blend local wisdom with specific solutions to farmer obstacles (Vanlauwe et al., 2014). Extension services function crucially as knowledge distribution services, together with training provision and resource and technology facilitation (Anderson, 2004). Farmer field schools combined with demonstration plots serve as effective methods to promote soil C management and conservation agriculture best practices (Bruges and Smith, 2007). The adoption and impact of C farming initiatives improve when farmers receive participatory support.

8.6.3 Public–private partnerships and collaborations

Substantiated growth of C farming requires the collaborative involvement of government organizations and private enterprise partners to resolve technical, financial, and institutional barriers. PPP unites governmental agencies with private firms, along with research facilities and civil organizations, to develop and implement innovative methods for C farming (Wang and Ma, 2021). The private sector uses its investments in agrotechnology to create platforms with advanced C-monitoring tools for farmers to use (Bruce and Costa, 2019). Research institutions that combine forces with agribusinesses foster the development of innovative technologies to increase soil C storage levels (Scherr et al., 2012). Through PPPs, farmers gain access to sustainable-practice incentives through the creation of C credit markets (Wang and Ma, 2021). Partnership programs between sectors help speed up the implementation of climate-smart agriculture (CSA), coupled with C farming initiatives.

Table 8.2 highlights the collaborative efforts required to scale C farming effectively, emphasizing the roles and outcomes of each stakeholder in the PPP framework.

8.7 Future of agrotechnology-based carbon farming and green economy

Modern agriculture needs the blending of agrotechnology with C farming and a green economy to provide a powerful reply regarding climate change while securing the food supply and protecting environmental resources. The worldwide adoption of sustainable development requires agrotechnology-based C farming to function as a core strategy, which will enable climate resilience and boost ecosystem services alongside economic development (Prasad et al., 2021). This section examines C farming relationships with circular and bioeconomy frameworks together with innovative climate agriculture functions for resilience development, while discussing both food security enhancement through sustainable intensification (SI) methods and ecological advantages from C farming practices (Udumann et al., 2023). Agrotechnology unifies agricultural practices with green economy targets to create a fundamental change in food systems that prioritize sustainability and environmental regeneration.

TABLE 8.2 Role of public–private partnerships (PPPs) in scaling carbon farming.

Stakeholder	Role	Outcome
Government	Develop supportive policies and regulatory frameworks.	Creates an enabling environment for C farming.
	Provide funding and incentives for C farming initiatives.	Ensures equitable access to C credit systems.
Private sector	Invest in agrotechnology development	Drives innovation in C farming technologies.
	Develop C credit markets and trading platforms.	Enhances scalability and transparency in C markets.
Research institutions	Conduct research on C_{seq} methods.	Provides a scientific basis for C farming practices.
	Develop standardized measurement and verification protocols.	Improves accuracy in C accounting.
Farmers	Implement C farming practices (e.g., conservation tillage, agroforestry).	Increases adoption of sustainable practices.
	Participate in C credit programs.	Improves livelihoods through C credit earnings.
NGOs and civil society	Facilitate capacity-building and training programs for farmers.	Bridges the gap between technology and smallholder farmers.
	Advocate for inclusive and equitable C farming policies.	Ensures marginalized groups benefit from C farming.
International organizations	Provide technical and financial support for global C farming initiatives.	Promotes global adoption of C farming.
	Foster cross-border collaborations.	Supports climate change mitigation targets.

C_{seq}, carbon sequestration.

8.7.1 Integration with circular economy and bioeconomy

A complete strategy for resource efficiency and sustainability emerges through combining C farming with circular economy and bioeconomy principles. The circular economy system promotes resource regeneration through recycling and reuse, alongside resource rebirth, which minimizes waste and environmental harm (Geissdoerfer et al., 2017). Transforming agricultural waste materials like crop residues into bioenergy, compost, and biochar creates cycles of nutrient conservation together with soil C capture (Venkata Mohan et al., 2016). The bioeconomy works hand in hand with C farming by developing sustainability-focused biological resource practices that produce bio-based products while decreasing C emissions (Bugge et al., 2016). Integrating food production with biomass generation in AF ecosystems creates systems that lower C while supporting bioeconomic growth activities (Ramachandran Nair et al., 2009). C farming, when connected to circular and bioeconomy principles, results in agricultural systems that possess higher levels of resource efficiency and environmental sustainability while becoming resilient.

8.7.2 Climate-smart agriculture and resilience

CSA is a necessary framework to increase the climate resilience of farm systems against climate change and to sequester C at the same time. The three areas that CSA brings together are increasing agricultural productivity, adaptation to climate change, and GHG mitigation (Dissanayaka et al., 2025). These objectives can be attained through agrotechnology, which allows for precision management of resources and optimization of crop yields, as well as emission reductions (Campbell et al., 2014). IoT-enabled weather forecasting and soil monitoring systems can help farmers adapt to changing climate conditions, and conservation tillage and cover cropping can increase soil C storage (Branca et al., 2013). By practicing CSA in their farmer households, these farmers can be resilient to climate variability and better their livelihoods, as well as assist in realizing global climate goals (Nuwarapaksha et al., 2024b).

8.7.3 Sustainable intensification and food security

Since SI is a means for increasing agricultural productivity without compromising environmental integrity, it will be critical to achieve food security in a changing climate (Jhariya et al., 2021a,b). However, SI focuses on using inputs in the most efficient way possible, for example, water, fertilizers, energy, and so forth, while staying in tune with ecosystem services and soil health (Pretty et al., 2018). SI practices utilize C farming practices, including AF, integration of crop and livestock, and the use of organic amendments to improve soil fertility, water retention, and C storage (Garnett et al., 2013). The application of agrotechnology also supports SI by sustaining SI through the way of PA that allows for the optimal use of inputs and reduces environmental impact (Zhang and Kovacs, 2012). For instance, drone-based crop monitoring and variable rate fertilization systems can improve yields while reducing resource use and emissions (Basso and Antle, 2020). SI, if coupled with C farming, allows us to continue to feed the planet while protecting natural resources and addressing climate change.

8.7.4 Ecosystem services and environmental co-benefits

C farming offers a broad set of ecosystem services and environmental cobenefits that go beyond C_{seq}. Improved soil health, increased biodiversity, water regulation, and diminished erosion are included in them (Lal, 2016). For example, AFs sequester C while also providing a habitat for pollinators, improving water infiltration, and reducing nutrient runoff (Dissanayaka et al., 2024a). Just as conservation agriculture practices like no-till farming and cover cropping improve soil structure, storage water capacity, and decrease GHG emissions (Atapattu et al., 2024b). However, through agrotechnology, one can translate these benefits by means of precise monitoring and management of ecosystem services. For instance, remote and GIS tools can monitor shifts in soil moisture, vegetation cover, and biodiversity, enabling farmers to harness multiple benefits from the land (Díaz et al., 2018). C farming serves the green economy and supports the achievement of sustainable development goals, as it enhances ecosystem services.

8.8 Conclusion

Innovative agronomic practices and advanced agrotechnology associated with C farming represent a novel tool to address and mitigate climate change, while increasing agricultural productivity and sustainability. Using examples in agriculture and nutrient recovery, C farming integrates conservation tillage, cover cropping, crop rotation, AF, and organic amendments with PA, remote sensing, IoT, and AI to maximize soil C_{seq} along with increasing efficiency of resources and resilience. In addition to reducing GHG emissions, these practices help build soil health, biodiversity, and ecosystem services, which help create a green economy as well. Yet achieving the success of C farming hinges on overcoming technical and policy blanks, data accessibility, and technology adoption, as well as verification complexities. To scale up C farming initiatives globally, bridging the digital divide, fostering participatory research, and strengthening PPPs are also essential for ensuring inclusivity. Secondly, C farming can be integrated into a circular economy and bioeconomy principles as well as CSA to support SI and food security through the achievement of global climate targets. When we think to the future, the merging of agrotechnology, C farming, and sustainable development will potentially lead to resilient and regenerative agricultural systems. Through technological innovations and by fostering collaboration among different sectors, C farming can become a game changer for a sustainable, low-C future that people and the planet require.

References

Abdalla, M., Hastings, A., Chadwick, D.R., Jones, D.L., Evans, C.D., Jones, M.B., et al., 2018. Critical review of the impacts of grazing intensity on soil organic carbon storage and other soil quality indicators in extensively managed grasslands. Agriculture, Ecosystems & Environment 253, 62–81. Available at https://doi.org/10.1016/j.agee.2017.10.023.

Amundson, R., Biardeau, L., 2018. Soil carbon sequestration is an elusive climate mitigation tool. Proceedings of the National Academy of Sciences of the United States of America 115 (46), 11652–11656. Available at https://doi.org/10.1073/pnas.1815901115.

Anderson, J.R., 2004. Agricultural extension: good intentions and hard realities. The World Bank Research Observer 19 (1), 41–60. Available at https://doi.org/10.1093/wbro/lkh013.

Atapattu, A.J., Udumann, S.S., 2024. Leveraging agroforestry principles for nature-based climate-smart solutions for coconut cultivation. Handbook of Nature-Based Solutions to Mitigation and Adaptation to Climate Change. Springer International Publishing, pp. 1–28. Available at https://doi.org/10.1007/978-3-030-98067-2_166-1.

Atapattu, A.J., Perera, L.K., Nuwarapaksha, T.D., Udumann, S.S., Dissanayaka, N.S., 2024. Artificial Intelligence Techniques in Smart Agriculture. Springer Nature, Singapore, pp. 7–34. Available at https://doi.org/10.1007/978-981-97-5878-4_2.

Atapattu, A.J., Dissanayaka, N.S., Udumann, S.S., Nuwarapaksha, T.D., 2025. Agroforestry for Monetising Carbon Credits. Springer Nature, Switzerland, pp. 147–179. Available at https://doi.org/10.1007/978-3-031-97151-8_7.

Atapattu, A.J., Dissanayaka, N.S., Udumann, S.S., & Nuwarapaksha, T.D., 2026. Reducing nitrogen fertilizer to mitigate climate change. In Point Source Nitrogen Pollution (pp. 453–475). Elsevier. https://doi.org/10.1016/B978-0-443-27352-0.00025-2.

Atapattu, A.J., Perera, L.K., Nuwarapaksha, T.D., Udumann, S.S., Dissanayaka, N.S., 2024a. Challenges in achieving artificial intelligence in agriculture. In: Chouhan, S.S., Saxena, A., Singh, U.P., Jain, S. (Eds.), Artificial Intelligence Techniques in Smart Agriculture. Springer Nature, Singapore, pp. 7–34. Available at https://doi.org/10.1007/978-981-97-5878-4_2.

Atapattu, A.J., Ranasinghe, C.S., Nuwarapaksha, T.D., Udumann, S.S., Dissanayaka, N.S., 2024b. Sustainable agriculture and Sustainable Development Goals (SDGs). In: Garwi, J., Masengu, R., Chiwaridzo, O. (Eds.), Emerging Technologies and Marketing Strategies for Sustainable Agriculture. IGI Global, Hershey PA, pp. 1–27. Available at https://doi.org/10.4018/979-8-3693-4864-2.ch001.

Atapattu, A.J., Nuwarapaksha, T.D., Udumann, S.S., Dissanayaka, N.S., 2024c. Integrated farming systems: a holistic approach to sustainable agriculture. In: Babu, S., Singh, R., Rathore, S.S., Das., A., Singh, V.K. (Eds.), Agricultural Diversification for Sustainable Food Production. Sustainability Sciences in Asia and Africa. Springer, Singapore, pp. 89–127. Available at https://doi.org/10.1007/978-981-97-7517-0_4.

Atapattu, A.J., Nuwarapaksha, T.D., Udumann, S.S., Dissanayaka, N.S., 2025. Integrating organic fertilizers in coconut farming: best practices and application techniques. Crops 5 (2), 17. Available at https://doi.org/10.3390/crops5020017.

Banerjee, A., Jhariya, M.K., Bargali, S.S., Palit, D., 2023. Ecorestoration for Sustainability. Wiley-Scrivener, USA, pp. 1–441.

Banerjee, A., Jhariya, M.K., Raj, A., Mechergui, T., 2024. Ecosystem management: climate change and global sustainability—an introduction. In: Banerjee, A., Jhariya, M.K., Raj, A, Mechergui, T. (Eds.), Ecosystem Management: Climate Change and Sustainability. Wiley, India, pp. 1–21. Available at https://doi.org/10.1002/9781394231249.ch1.

Basso, B., Antle, J., 2020. Digital agriculture to design sustainable agricultural systems. Nature Sustainability 3 (4), 254–256. Available at https://doi.org/10.1038/s41893-020-0510-0.

Bayata, A., Mulatu, G., 2024. Carbon farming, climate smart agriculture practice and current climate change mitigation strategy- in the case of Ethiopia. International Journal of Environmental Monitoring and Analysis 12 (6), 149–156. Available at https://doi.org/10.11648/j.ijema.20241206.12.

Bernal, M.P., Alburquerque, J.A., Moral, R., 2009. Composting of animal manures and chemical criteria for compost maturity assessment. A review. Bioresource Technology 100 (22), 5444–5453. Available at https://doi.org/10.1016/j.biortech.2008.11.027.

Blanco-Canqui, H., Shaver, T.M., Lindquist, J.L., Shapiro, C.A., Elmore, R.W., Francis, C.A., et al., 2015. Cover crops and ecosystem services: insights from studies in temperate soils. Agronomy Journal 107 (6), 2449–2474. Available at https://doi.org/10.2134/agronj15.0086.

Boursianis, A.D., Papadopoulou, M.S., Diamantoulakis, P., Liopa-Tsakalidi, A., Barouchas, P., Salahas, G., et al., 2022. Internet of Things (IoT) and agricultural Unmanned Aerial Vehicles (UAVs) in smart farming: a comprehensive review. Internet of Things 18, 100187. Available at https://doi.org/10.1016/j.iot.2020.100187.

Branca, G., Lipper, L., McCarthy, N., Jolejole, M.C., 2013. Food security, climate change, and sustainable land management. A review. Agronomy for Sustainable Development 33 (4), 635–650. Available at https://doi.org/10.1007/s13593-013-0133-1.

Bronson, K., Knezevic, I., 2016. Big Data in food and agriculture. Big Data and Society 3 (1). Available at https://doi.org/10.1177/2053951716648174.

Bruce, K., Costa, H., 2019. Enabling environment for PPPs in agricultural extension projects: policy imperatives for impact. Journal of Rural Studies 70, 87–95. Available at https://doi.org/10.1016/j.jrurstud.2019.07.005.

Bruges, M., Smith, W., 2007. Participatory approaches for sustainable agriculture: a contradiction in terms? Agriculture and Human Values 25 (1), 13–23. Available at https://doi.org/10.1007/s10460-007-9058-0.

Bugge, M., Hansen, T., Klitkou, A., 2016. What is the bioeconomy? A review of the literature. Sustainability 8 (7), 691. Available at https://doi.org/10.3390/su8070691.

Campbell, B.M., Thornton, P., Zougmoré, R., van Asten, P., Lipper, L., 2014. Sustainable intensification: what is its role in climate smart agriculture? Current Opinion in Environmental Sustainability 8, 39–43. Available at https://doi.org/10.1016/j.cosust.2014.07.002.

Carolan, M., 2017. Publicising food: big data, precision agriculture, and co-experimental techniques of addition. Sociologia Ruralis 57 (2), 135–154. Available at https://doi.org/10.1111/soru.12120.

Chen, H., Dai, Z., Veach, A.M., Zheng, J., Xu, J., Schadt, C.W., 2020. Global meta-analyses show that conservation tillage practices promote soil fungal and bacterial biomass. Agriculture, Ecosystems and Environment 293. Available at https://doi.org/10.1016/j.agee.2020.106841.

Chhatre, A., Lakhanpal, S., Larson, A.M., Nelson, F., Ojha, H., Rao, J., 2012. Social safeguards and co-benefits in REDD+: a review of the adjacent possible. Current Opinion in Environmental Sustainability 4 (6), 654–660. Available at https://doi.org/10.1016/j.cosust.2012.08.006.

Chivenge, P.P., Murwira, H.K., Giller, K.E., Mapfumo, P., Six, J., 2007. Long-term impact of reduced tillage and residue management on soil carbon stabilization: implications for conservation agriculture on contrasting soils. Soil and Tillage Research 94 (2), 328–337. Available at https://doi.org/10.1016/j.still.2006.08.006.

Choruma, D.J., Dirwai, T.L., Mutenje, M.J., Mustafa, M., Chimonyo, V.G.P., Jacobs-Mata, I., et al., 2024. Digitalisation in agriculture: a scoping review of technologies in practice, challenges, and opportunities for smallholder farmers in sub-Saharan Africa. Journal of Agriculture and Food Research 18. Available at https://doi.org/10.1016/j.jafr.2024.101286, https://www.sciencedirect.com/science/journal/26661543.

Conant, R.T., Paustian, K., Elliott, E.T., 2001. Grassland management and conversion into grassland: effects on soil carbon. Ecological Applications 11 (2), 343–355. Available at https://doi.org/10.1890/1051-0761(2001)011[0343:GMACIG]2.0.CO;2, www.esajournals.org.

Dhillon, R., Moncur, Q., 2023. Small-scale farming: a review of challenges and potential opportunities offered by technological advancements. Sustainability 15 (21), 15478. Available at https://doi.org/10.3390/su152115478.

Diacono, M., Montemurro, F., Lichtfouse, E., Hamelin, M., Navarrete, M., Debaeke, P., 2011. Long-term effects of organic amendments on soil fertility. In: Lichtfouse, E., Hamelin, M., Navarrete, M., Debaeke, P. (Eds.), Sustainable Agriculture Volume 2. Springer, Dordrecht, pp. 761–786. Available at https://doi.org/10.1007/978-94-007-0394-0_34.

Díaz, S., Pascual, U., Stenseke, M., Martín-López, B., Watson, R.T., Molnár, Z., et al., 2018. Assessing nature's contributions to people. Science (New York, N.Y.) 359 (6373), 270–272. Available at https://doi.org/10.1126/science.aap8826.

Dissanayaka, D.M.N.S., Nuwarapaksha, T.D., Udumann, S.S., Dissanayake, D.K.R.P.L., Anjana, J., Atapattu, 2022. A sustainable way of increasing productivity of coconut cultivation using cover crops: a review. Circular Agricultural Systems 2 (1), 1–9. Available at https://doi.org/10.48130/cas-2022-0007.

Dissanayaka, N.S., Udumann, S.S., Nuwarapaksha, T.D., Atapattu, A.J., 2023a. Agroforestry: an avenue for resilient and productive farming through integrated crops and livestock production. In: Kiba, D.I. (Ed.), Transitioning to Zero Hunger. MDPI Books, pp. 115–136. Available at https://doi.org/10.3390/books978-3-03897-863-3-5.

Dissanayaka, D.M.N.S., Dissanayake, D.K.R.P.L., Udumann, S.S., Nuwarapaksha, T.D., Atapattu, A.J., 2023b. Agroforestry—a key tool in the climate-smart agriculture context: a review on coconut cultivation in Sri Lanka. Frontiers in Agronomy 5. Available at https://doi.org/10.3389/fagro.2023.1162750.

Dissanayaka, D.M.N.S., Udumann, S.S., Nuwarapaksha, T.D., Atapattu, A.J., 2023c. Effects of pyrolysis temperature on chemical composition of coconut-husk biochar for agricultural applications: a characterization study. Technology in Agronomy 3. Available at https://doi.org/10.48130/tia-2023-0013.

Dissanayaka, D.M.N.S., Udumann, S.S., Atapattu, A.J., 2024. Synergies between tree crops and ecosystems in tropical agroforestry. Agroforestry. Wiley, pp. 49–87. Available at https://doi.org/10.1002/9781394231164.ch3.

Dissanayaka, D.M.N.S., Udumann, S.S., Nuwarapaksha, T.D., Atapattu, A.J., 2024b. Harnessing the potential of Mucuna cover cropping: a comprehensive review of its agronomic and environmental benefits. Circular Agricultural Systems. 4. Available at https://doi.org/10.48130/cas-0024-0001.

Dissanayaka, N.S., Rajaratnam, K., Udumann, S., Nuwarapaksha, T.D., Shamila, S.K., Jayamali, A., Gayadari, M.P., Gunarathna, N.K., Atapattu, A.J., 2025. Evaluation of the nutritional composition of king coconut husk waste biochar and ash soil conditioners: a comprehensive analysis. Technology in Agronomy 5. Available at https://doi.org/10.48130/tia-0024-0034.

Dissanayaka, N.S., Nimanthika, N., Nuwarapaksha, T.D., Udumann, S.S., Atapattu, A.J., 2025. Agroforestry and climate-smart agriculture. Agroforestry for a Climate-Smart Future. IGI Global, pp. 173–199. Available at https://doi.org/10.4018/979-8-3693-8282-0.ch006.

Drinkwater, L.E., Wagoner, P., Sarrantonio, M., 1998. Legume-based cropping systems have reduced carbon and nitrogen losses. Nature 396 (6708), 262–265. Available at https://doi.org/10.1038/24376.

Eastwood, C., Klerkx, L., Ayre, M., Dela Rue, B., 2019. Managing socio-ethical challenges in the development of smart farming: from a fragmented to a comprehensive approach for responsible research and innovation. Journal of Agricultural and Environmental Ethics 32 (5-6), 741–768. Available at https://doi.org/10.1007/s10806-017-9704-5.

Evans, M.C., Carwardine, J., Fensham, R.J., Butler, D.W., Wilson, K.A., Possingham, H.P., et al., 2015. Carbon farming via assisted natural regeneration as a cost-effective mechanism for restoring biodiversity in agricultural landscapes. Environmental Science and Policy 50, 114–129. Available at https://doi.org/10.1016/j.envsci.2015.02.003.

Foley, J.A., Ramankutty, N., Brauman, K.A., Cassidy, E.S., Gerber, J.S., Johnston, M., et al., 2011. Solutions for a cultivated planet. Nature 478 (7369), 337–342. Available at https://doi.org/10.1038/nature10452.

Garnett, T., Appleby, M.C., Balmford, A., Bateman, I.J., Benton, T.G., Bloomer, P., et al., 2013. Sustainable intensification in agriculture: premises and policies. Science (New York, N.Y.) 341 (6141), 33–34. Available at https://doi.org/10.1126/science.1234485.

Gebbers, R., Adamchuk, V.I., 2010. Precision agriculture and food security. Science (New York, N.Y.) 327 (5967), 828–831. Available at https://doi.org/10.1126/science.1183899.

Geissdoerfer, M., Savaget, P., Bocken, N.M.P., Hultink, E.J., 2017. The circular economy – a new sustainability paradigm? Journal of Cleaner Production 143, 757–768. Available at https://doi.org/10.1016/j.jclepro.2016.12.048.

Getahun, S., Kefale, H., Gelaye, Y., Vasanthavigar, M., 2024. Application of precision agriculture technologies for sustainable crop production and environmental sustainability: a systematic review. The Scientific World Journal 2024 (1). Available at https://doi.org/10.1155/2024/2126734.

Hasan, H.R., Musamih, A., Salah, K., Jayaraman, R., Omar, M., Arshad, J., et al., 2024. Smart agriculture assurance: IoT and blockchain for trusted sustainable produce. Computers and Electronics in Agriculture 224, 109184. Available at https://doi.org/10.1016/j.compag.2024.109184.

Hildebrand, P., 1991. Farmer first: farmer innovation and agricultural research. 1989. Edited by Robert Chambers, Arnold Pacey, and Lori Ann Thrupp. Intermediate Technology Publications, London, and The Bootstrap Press, New York, New York, xx + 218 pp. $23.50, paper. American Journal of Alternative Agriculture 6 (3), 150. Available at https://doi.org/10.1017/s0889189300004136.

Hsu, A., Weinfurter, A.J., Xu, K., 2017. Aligning subnational climate actions for the new post-Paris climate regime. Climatic Change 142 (3-4), 419–432. Available at https://doi.org/10.1007/s10584-017-1957-5.

Hussain, S., Hussain, S., Guo, R., Sarwar, M., Ren, X., Krstic, D., et al., 2021. Carbon sequestration to avoid soil degradation: a review on the role of conservation Tillage. Plants 10 (10), 2001. Available at https://doi.org/10.3390/plants10102001.

Jansson, C., Faiola, C., Wingler, A., Zhu, X.G., Kravchenko, A., de Graaff, M.A., et al., 2021. Crops for carbon farming. Frontiers in Plant Science 12. Available at https://doi.org/10.3389/fpls.2021.636709.

Jhariya, M.K., Banerjee, A., Yadav, D.K., Raj, A., 2018. Leguminous trees an innovative tool for soil sustainability. Legumes for Soil Health and Sustainable Management. Springer Nature, India, pp. 315–345. Available at https://doi.org/10.1007/978-981-13-0253-4_10.

Jhariya, M.K., Banerjee, A., Swaroop Meena, R., Kumar, S., Raj, A., 2021a. Sustainable Intensification for Agroecosystem Services and Management. Springer, Singapore, pp. 1–870. Available at https://doi.org/10.1007/978-981-16-3207-5.

Jhariya, M.K., Meena, R.S., Banerjee, A., 2021b. Ecological Intensification of Natural Resources for Sustainable Agriculture. Springer, Singapore, pp. 1–655. Available at https://doi.org/10.1007/978-981-33-4203-3.

Jhariya, M.K., Meena, R.S., Banerjee, A., Meena, S.N., 2022. Natural Resources Conservation and Advances for Sustainability. Elsevier. Available at https://doi.org/10.1016/C2019-0-03763-6.

Jhariya, M.K., Meena, R.S., Banerjee, A., Kumar, S., Raj, A., 2023. Agroforestry for Carbon and Ecosystem Management. Elsevier, India, pp. 1–434. Available at https://doi.org/10.1016/C2021-0-02916-3.

Jones, M.B., Donnelly, A., 2004. Carbon sequestration in temperate grassland ecosystems and the influence of management, climate and elevated CO2. New Phytologist 164 (3), 423–439. Available at https://doi.org/10.1111/j.1469-8137.2004.01201.x.

Kamilaris, A., Prenafeta-Boldú, F.X., 2018. Deep learning in agriculture: a survey. Computers and Electronics in Agriculture 147, 70–90. Available at https://doi.org/10.1016/j.compag.2018.02.016.

Kouhizadeh, M., Saberi, S., Sarkis, J., 2021. Blockchain technology and the sustainable supply chain: theoretically exploring adoption barriers. International Journal of Production Economics 231, 107831. Available at https://doi.org/10.1016/j.ijpe.2020.107831.

Kshetri, N., 2021. Blockchain and sustainable supply chain management in developing countries. International Journal of Information Management 60, 102376. Available at https://doi.org/10.1016/j.ijinfomgt.2021.102376.

Kumar, S., Meena, R.S., Jhariya, M.K., 2020. Resources Use Efficiency in Agriculture. Springer Nature, Singapore, pp. 760. Available at https://doi.org/10.1007/978-981-15-6953-1.

Kumar, S., Meena, R.S., Sheoran, P., Jhariya, M.K., 2024. Regenerative Agriculture for Sustainable Food System. Springer.

Lal, R., 2004. Soil carbon sequestration impacts on global climate change and food security. Science (New York, N.Y.) 304 (5677), 1623–1627. Available at https://doi.org/10.1126/science.1097396.

Lal, R., 2015. Sequestering carbon and increasing productivity by conservation agriculture. Journal of Soil and Water Conservation 70 (3), 55A–62A. Available at https://doi.org/10.2489/jswc.70.3.55A.

Lal, R., 2016. Soil health and carbon management. Food and Energy Security 5 (4), 212–222. Available at https://doi.org/10.1002/fes3.96.

Liakos, K., Busato, P., Moshou, D., Pearson, S., Bochtis, D., 2018. Machine learning in agriculture: a review. Sensors 18 (8), 2674. Available at https://doi.org/10.3390/s18082674.

Lobell, D.B., Thau, D., Seifert, C., Engle, E., Little, B., 2015. A scalable satellite-based crop yield mapper. Remote Sensing of Environment 164, 324–333. Available at https://doi.org/10.1016/j.rse.2015.04.021.

Lorenz, K., Lal, R., 2014. Soil organic carbon sequestration in agroforestry systems. A review. Agronomy for Sustainable Development 34 (2), 443–454. Available at https://doi.org/10.1007/s13593-014-0212-y.

Maillard, E., Angers, D.A., 2014. Animal manure application and soil organic carbon stocks: a meta-analysis. Global Change Biology 20 (2), 666–679. Available at https://doi.org/10.1111/gcb.12438.

Maitra, S., Ray, S., Sagar, L., Sairam, M., Pramanick, B., Gitari, H., Santosh, D.T., Jagannath, G.D., Atapattu, A., Chappa, L.R., 2025. Sustainability Sciences in Asia and Africa,Climate-Smart Agricultural Technologies. Springer, Singapore, pp. 327–356. Available at https://doi.org/10.1007/978-981-96-7699-6_15.

McDaniel, M.D., Tiemann, L.K., Grandy, A.S., 2014. Does agricultural crop diversity enhance soil microbial biomass and organic matter dynamics? A meta-analysis. Ecological Applications 24 (3), 560–570. Available at https://doi.org/10.1890/13-0616.1.

Meena, R.S., Pradhan, G.S., Kumar, S., Jhariya, M.K., Rao, C.S., 2025. Carbon-Negative Agriculture. Springer, pp. 1–500.

Montagnini, F., Nair, P.K.R., 2004. Carbon sequestration: An underexploited environmental benefit of agroforestry systems. In: Nair, P.K.R., Rao, M.R., Buck, L.E. (Eds.), New Vistas in Agroforestry. Advances in Agroforestry, vol 1. Springer, Dordrecht, pp. 281–295. Available at https://doi.org/10.1007/978-94-017-2424-1_20.

Monteiro, A., Santos, S., Gonçalves, P., 2021. Precision agriculture for crop and livestock farming—brief review. Animals 11 (8), 2345. Available at https://doi.org/10.3390/ani11082345.

Mor, S., Madan, S., Prasad, K.D., 2021. Artificial intelligence and carbon footprints: roadmap for Indian agriculture. Strategic Change 30 (3), 269–280. Available at https://doi.org/10.1002/jsc.2409.

Moreno-Mateos, D., Power, M.E., Comín, F.A., Yockteng, R., 2012. Structural and functional loss in restored wetland ecosystems. PLoS Biology 10 (1). Available at https://doi.org/10.1371/journal.pbio.1001247.

Mulla, D.J., 2013. Twenty five years of remote sensing in precision agriculture: key advances and remaining knowledge gaps. Biosystems Engineering 114 (4), 358–371. Available at https://doi.org/10.1016/j.biosystemseng.2012.08.009.

Nuwarapaksha, T.D., Rajapaksha, U.N., Ekanayake, J., Weerasooriya, S.A., Atapattu, A.J., 2023. Exploring the economic viability of integrating Jamnapari goat into underutilized pastures under coconut cultivations in Coconut Research Institute, Sri Lanka. Biology and Life Sciences Forum 27 (1), 27. Available at https://doi.org/10.3390/IECAG2023-14990.

Nuwarapaksha, T.D., Udumann, S.S., Dissanayaka, N.S., Dilshan, R.M.N., Atapattu, A.J., 2024a. AI-driven solutions for sustainable irrigation: exploring smart technologies to enhance conservation and efficiency. In: Garwi, J., Dzingirai, M., Masengu, R. (Eds.), Integrating Agriculture, Green Marketing Strategies, and Artificial Intelligence. IGI Global Scientific Publishing, Hershey, PA, pp. 1–32. Available at https://doi.org/10.4018/979-8-3693-6468-0.ch001.

Nuwarapaksha, T.D., Udumann, S.S., Atapattu, A.J., 2024b. Fostering food and nutritional security through agroforestry practices. In: Raj, A., Jhariya, M.K., Banerjee, A., Jha, R.K., Singh, K.P. (Eds.), Agroforestry. Wiley, Sri Lanka, pp. 285–318. Available at https://doi.org/10.1002/9781394231164.ch10.

Nuwarapaksha, T.D., Udumann, S.S., Dissanayaka, N.S., Dilshan, R.M.N., Atapattu, A.J., 2024c. Revolutionizing agriculture by advanced water and irrigation management technologies. Emerging Trends and Technologies in Water Management and Conservation. IGI Global, pp. 285–318. Available at https://doi.org/10.4018/979-8-3693-6920-3.ch009.

Nuwarapaksha, T.D., Udumann, S.S., Dissanayaka, N.S., Atapattu, A.J., 2026. Climate change impacts and adaptation strategies in coconut plantations: integrating remote sensing and real-time monitoring. Frontiers in Climate 8. Available at https://doi.org/10.3389/fclim.2026.1762364.

Nuwarapaksha, T.D., Dilhan, R.M.C.P., Udumann, S.S., Ranasinghe, C.S., Egodawatta, W.C.P., Dissanayaka, N.S., Raveendra, T., Atapattu, A.J., 2025. Estimation of the total carbon sequestration potential of coconut-gliricidia mixed cropping systems in Sri Lanka. Carbon Footprints 4. Available at https://doi.org/10.20517/cf.2024.31.

Ogle, S.M., Olander, L., Wollenberg, L., Rosenstock, T., Tubiello, F., Paustian, K., et al., 2014. Reducing greenhouse gas emissions and adapting agricultural management for climate change in developing countries: providing the basis for action. Global Change Biology 20 (1), 1–6. Available at https://doi.org/10.1111/gcb.12361.

Padarian, J., Minasny, B., McBratney, A.B., 2020. Machine learning and soil sciences: a review aided by machine learning tools. SOIL 6 (1), 35–52. Available at https://doi.org/10.5194/soil-6-35-2020.

Paustian, K., Lehmann, J., Ogle, S., Reay, D., Robertson, G.P., Smith, P., 2016. Climate-smart soils. Nature 532 (7597), 49–57. Available at https://doi.org/10.1038/nature17174.

Poeplau, C., Don, A., 2015. Carbon sequestration in agricultural soils via cultivation of cover crops - a meta-analysis. Agriculture, Ecosystems and Environment 200, 33–41. Available at https://doi.org/10.1016/j.agee.2014.10.024.

Polymeni, S., Skoutas, D.N., Sarigiannidis, P., Kormentzas, G., Skianis, C., 2024. Smart agriculture and greenhouse gas emission mitigation: a 6G-IoT perspective. Electronics 13 (8), 1480. Available at https://doi.org/10.3390/electronics13081480.

Prasad, R., Jhariya, M.K., Banerje, A., 2021. Advances in Sustainable Development and Management of Environmental and Natural Resources: Economic Outlook and Opinions, 2-volume set. Apple Academic Press, New York, pp. 1–428. Available at https://doi.org/10.1201/9781003187455.

Pretty, J., Benton, T.G., Bharucha, Z.P., Dicks, L.V., Flora, C.B., Godfray, H.C.J., et al., 2018. Global assessment of agricultural system redesign for sustainable intensification. Nature Sustainability 1 (8), 441–446. Available at https://doi.org/10.1038/s41893-018-0114-0.

Raj, A., Jhariya, M.K., Meena, R.S., 2025. Agroforestry for Monetising Carbon Credits. Springer Nature, Switzerland.

Ramachandran Nair, P.K., Mohan Kumar, B., Nair, V.D., 2009. Agroforestry as a strategy for carbon sequestration. Journal of Plant Nutrition and Soil Science 172 (1), 10–23. Available at https://doi.org/10.1002/jpln.200800030.

Ray, P.P., 2017. Internet of things for smart agriculture: technologies, practices and future direction. Journal of Ambient Intelligence Smart Environments 9 (4), 395–420. Available at https://doi.org/10.3233/AIS-170440.

Robertson, G.P., Paul, E.A., Harwood, R.R., 2000. Greenhouse gases in intensive agriculture: contributions of individual gases to the radiative forcing of the atmosphere. Science (New York, N.Y.) 289 (5486), 1922–1925. Available at https://doi.org/10.1126/science.289.5486.1922.

Saberi, S., Kouhizadeh, M., Sarkis, J., Shen, L., 2019. Blockchain technology and its relationships to sustainable supply chain management. International Journal of Production Research 57 (7), 2117–2135. Available at https://doi.org/10.1080/00207543.2018.1533261.

Sanderman, J., Hengl, T., Fiske, G.J., 2017. Soil carbon debt of 12,000 years of human land use. Proceedings of the National Academy of Sciences 114 (36), 9575–9580. Available at https://doi.org/10.1073/pnas.1706103114.

Scherr, S.J., Shames, S., Friedman, R., 2012. From climate-smart agriculture to climate-smart landscapes. Agriculture & Food Security 1, 12. Available at https://doi.org/10.1186/2048-7010-1-12.

Schmidt, R., Gravuer, K., Bossange, A.V., Mitchell, J., Scow, K., Grosch, R., 2018. Long-term use of cover crops and no-till shift soil microbial community life strategies in agricultural soil. PLoS One 13 (2), e0192953. Available at https://doi.org/10.1371/journal.pone.0192953.

Shamila, S.K., Udumann, S.S., Dissanayaka, N.S., Rajaratnam, K., Atapattu, A.J., 2024. Assessing the impact of king coconut husk ash and biochar, combined with chemical fertilizer application, on enhancing soil fertility in coconut plantations. Crops 4 (2), 227–241. Available at https://doi.org/10.3390/crops4020017.

Six, J., Ogle, S.M., Jay breidt, F., Conant, R.T., Mosier, A.R., Paustian, K., 2004. The potential to mitigate global warming with no-tillage management is only realized when practised in the long term. Global Change Biology 10 (2), 155–160. Available at https://doi.org/10.1111/j.1529-8817.2003.00730.x.

Smith, P., Martino, D., Cai, Z., Gwary, D., Janzen, H., Kumar, P., et al., 2008. Greenhouse gas mitigation in agriculture. Philosophical Transactions of the Royal Society B: Biological Sciences 363 (1492), 789–813. Available at https://doi.org/10.1098/rstb.2007.2184.

Smith, P., Soussana, J.F., Angers, D., Schipper, L., Chenu, C., Rasse, D.P., et al., 2020. How to measure, report and verify soil carbon change to realize the potential of soil carbon sequestration for atmospheric greenhouse gas removal. Global Change Biology 26 (1), 219–241. Available at https://doi.org/10.1111/gcb.14815.

Sohi, S.P., Krull, E., Lopez-Capel, E., Bol, R., 2010. A review of biochar and its use and function in soil. Advances in Agronomy 105, 47–82. Available at https://doi.org/10.1016/S0065-2113(10)05002-9.

Spokas, K.A., Cantrell, K.B., Novak, J.M., Archer, D.W., Ippolito, J.A., Collins, H.P., et al., 2012. Biochar: a synthesis of its agronomic impact beyond carbon sequestration. Journal of Environmental Quality 41 (4), 973–989. Available at https://doi.org/10.2134/jeq2011.0069.

Teague, W.R., Dowhower, S.L., Baker, S.A., Haile, N., DeLaune, P.B., Conover, D.M., 2011. Grazing management impacts on vegetation, soil biota and soil chemical, physical and hydrological properties in tall grass prairie. Agriculture, Ecosystems & Environment 141 (3-4), 310–322. Available at https://doi.org/10.1016/j.agee.2011.03.009.

Thapa, B., Lovell, S., Wilson, J., 2023a. Remote sensing and machine learning applications for aboveground biomass estimation in agroforestry systems: a review. Agroforestry Systems 97 (6), 1097–1111. Available at https://doi.org/10.1007/s10457-023-00850-2.

Thapa, V.R., Ghimire, R., Adhikari, K.P., Lamichhane, S., 2023b. Soil organic carbon sequestration potential of conservation agriculture in arid and semi-arid regions: A review. Journal of Arid Environments 217, 105028. Available at https://doi.org/10.1016/j.jaridenv.2023.105028.

Tian, F., 2016. An agri-food supply chain traceability system for China based on RFID & blockchain technology. 2016 13th International Conference on Service Systems and Service Management (ICSSSM), Kunming. IEEE, pp. 1–6. Available at https://doi.org/10.1109/ICSSSM.2016.7538424.

Udumann, S.S., Dissanayaka, D.M.N.S., Nuwarapaksha, T.D., Dissanayake, D.K.R.P.L., Atapattu, A.J., 2023. Megathyrsus maximus as a raw material for organic fertilizer production: A feasibility study. Technology in Horticulture 3. Available at https://doi.org/10.48130/tih-2023-0009.

Vanlauwe, B., Coyne, D., Gockowski, J., Hauser, S., Huising, J., Masso, C., et al., 2014. Sustainable intensification and the African smallholder farmer. Current Opinion in Environmental Sustainability 8, 15–22. Available at https://doi.org/10.1016/j.cosust.2014.06.001.

Venkata Mohan, S., Nikhil, G.N., Chiranjeevi, P., Nagendranatha Reddy, C., Rohit, M.V., Kumar, A.N., et al., 2016. Waste biorefinery models towards sustainable circular bioeconomy: critical review and future perspectives. Bioresource Technology 215, 2–12. Available at https://doi.org/10.1016/j.biortech.2016.03.130.

Wang, N., Ma, M., 2021. Public–private partnership as a tool for sustainable development – what literatures say? Sustainable Development 29 (1), 243–258. Available at https://doi.org/10.1002/sd.2127.

West, T.O., Post, W.M., 2002. Soil organic carbon sequestration rates by tillage and crop rotation. Soil Science Society of America Journal 66 (6), 1930–1946. Available at https://doi.org/10.2136/sssaj2002.1930.

Wolfert, S., Ge, L., Verdouw, C., Bogaardt, M.J., 2017. Big data in smart farming – a review. Agricultural Systems 153, 69–80. Available at https://doi.org/10.1016/j.agsy.2017.01.023.

Zhang, C., Kovacs, J.M., 2012. The application of small unmanned aerial systems for precision agriculture: a review. Precision Agriculture 13 (6), 693–712. Available at https://doi.org/10.1007/s11119-012-9274-5.

Chapter 9

Reducing CO_2 emissions with best management practices

Zied Haj-Amor[1], Amna Bouri[2], and Salem Bouri[1]

[1]*Water, Energy, and Environment Laboratory, National Engineering School of Sfax, Sfax, Tunisia,* [2]*Higher Institute of Water Sciences and Techniques, University of Gabès, Gabès, Tunisia*

Highlights

- CO_2 emission impacts were analyzed in both developed and developing countries.
- Adaptation practices were identified.
- Collaborative efforts to reduce CO_2 emissions are strongly needed.

9.1 Introduction

The progressive increase in carbon dioxide (CO_2) emissions over the last few decades was considered one of the most serious environmental threats facing humanity because of its huge contribution to the global warming issue (Nunes, 2023). CO_2, a primary greenhouse gas (GHG), is emitted into the atmosphere through various human interventions, including the burning of fossil fuels, farming (e.g., livestock production), deforestation, and industrial processes (Meena et al., 2025). Among all human interventions, burning fossil fuels for energy generation is the primary contributor to global CO_2 pollution (Eldesouki et al., 2023). On a global basis, the combustion of fossil fuels and industrial processes has increased total CO_2 emissions from about 24 gigatons (Gt) in 2000 to around 36 Gt in 2021 (IEA, 2021; Eldesouki et al., 2023), leading to a warmer atmosphere and extreme climate. Hugh et al. (2018) investigated the relationship between CO_2 emissions and changes in climate variables, finding that higher CO_2 concentrations in the Northern Hemisphere have intensified extreme summertime temperatures, heatwaves, and tropical rainfall events. Moreover, increasing CO_2 concentrations to 550 ppm relative to preindustrial levels could raise the global average temperature in 2050 by up to 4.5°C (IPCC, 2018). The ongoing rise in global temperature indicates that we are not far from a critical global warming scenario by 2050 (Liu et al., 2022). Furthermore, the slow effort in implementing the available protective actions shows that we are still off course from achieving net-zero carbon (C) emissions projected for the year 2050. This highlights the pressing need for more stringent climate policies to curb CO_2 emissions in the years to come (Meena et al., 2025a; Raj et al., 2025).

The implications of increasing CO_2 emissions are critical and profound, affecting not only climatic conditions but also biodiversity and ecosystem services (e.g., Locatelli, 2016; Meena et al., 2024a), agricultural productivity and global food security (e.g., Haj-Amor et al., 2023), human health (e.g., Dong et al., 2021), and economic growth (e.g., Çobanoğulları, 2024). Therefore to cope with these effects, adaptation strategies must be developed and implemented (Jhariya et al., 2022, 2023). To design flexible and effective adaptation measures, historical and current CO2 emissions, as well as regional inequality in CO2 emissions, such as between developed and developing countries, should be assessed (Wang and Feng, 2022). This is essential to ensure the fair allocation of CO_2-emission reduction tasks. Developed countries, with their huge historical CO_2 emissions, face pressure to lead in reduction tasks, while developing countries face difficulty in reducing CO_2 emissions because of many reasons, such as slow economic growth, backward technology, insufficient funding support, and moderate international assistance and cooperation (Wang et al., 2024; Meena et al., 2024b). Overall, several adaptation measures, such as the productive use of renewable energy, CO2 capture and storage, sustainable agricultural practices, and afforestation, can help facilitate CO2 emission reduction (Fawzy et al., 2020). All these strategies require close collaboration between policymakers, businesses, and civil society to be performant.

Carbon Incentives, Plans, and Policies in Agriculture. DOI: https://doi.org/10.1016/B978-0-443-33557-0.00016-1

Based on the above discussion, the objective of this chapter is to present a thorough analysis of CO_2 emissions and their impacts on ecological systems, as well as the best management practices needed to reduce them effectively. By analyzing current emission scenarios, examining the global impacts of CO_2 emissions, and outlining mitigation strategies, this chapter aims to contribute to better climate action and sustainability.

9.2 CO_2 emissions in developed and developing countries

9.2.1 Emissions in developed countries

Since the middle of the 20th century, developed countries have contributed the most to global pollution, particularly CO_2. This is primarily due to their significant economic growth and excessive use of natural resources to increase industrial production (Mitić et al., 2023). Overall, high- and upper-middle-income nations are accountable for about 80% of the world's CO_2 emissions (Ritchie, 2023). The United States of America (USA) and the European Union (EU) are the highest CO_2 emitters (Ritchie, 2023). The USA is the second largest CO_2 emitter. The country emitted around 5.2 billion metric tons of GHGs in 2021, 79% of which were CO2 (Ritchie and Rosado, 2025). In 2018 the EU, together with the United Kingdom (UK), emitted more than 5 billion metric tons of carbon dioxide equivalent (CO_{2e}) (EEA, 2020), placing the region as the third-largest contributor to the issue.

However, between 1990 and 2020, the EU achieved a 31% reduction in emissions (EEA, 2020), mainly during the COVID-19 pandemic (Haj-Amor et al., 2023). Initiatives like the Clean Air Act have significantly contributed to this progress (Underwood et al., 2024), supporting Europe's goal of becoming the first C-neutral bloc by 2050.

Developed countries can achieve net-zero emissions by 2050 through measures such as decarbonization, energy efficiency, and renewable energy. Skilled individuals will be key to this transition (Haj-Amor et al., 2023).

9.2.2 Emissions in developing countries

Between 1995 and 2015, developing countries accounted for 43% of global emissions (Meng et al., 2023a). However, it is essential to note that there is a significant disparity in the contributions each makes to the issue. Some countries, such as China and India, are the largest emitters, whereas less industrialized countries, such as African countries, have relatively low CO_2 emissions (Haj-Amor et al., 2023). In other words, China, as a highly energy-dependent developing country, is the world's top emitter of CO_2, with approximately 9.7 billion metric tons in 2019 (Huang et al., 2024). This represents 26.8% of total emissions (Meng et al., 2023a). Moreover, CO2 emissions are expected to exceed 10 billion metric tons in 2028 (Yang et al., 2017). In 2017 Africa, with its 54 countries, contributed only 4% of total pollution (Ayompe et al., 2020).

In addition to their own emissions, trade between the South (developing countries) and the North (developed countries) has exacerbated pollution issues in many developing countries. Indeed, over the last few decades, rising South–North trade led to increased global emissions, as production shifted to China, where goods were cheaper but more polluted. China and India were the top CO_2 emitters (Meng et al., 2023b). In both countries, reliance on fossil fuels continues to be a major barrier to reducing CO_2 emissions.

9.3 CO_2 emission impacts on ecological systems

Increased CO_2 concentrations in the atmosphere can have various impacts on ecosystems (Meena et al., 2024). The most consistent impacts include ocean acidification, biodiversity loss, and changes in plant growth and development. A proper understanding of these impacts is essential for determining how ecosystems respond to changes in climate variables.

9.3.1 Ocean acidification

Overall, the world's oceans uptake more than 25% of the atmospheric CO_2 (Secretariat of the Convention on Biological Diversity, 2014), and as concentrations of atmospheric CO_2 increase, so do the concentrations in the seawater. Since carbonate chemistry is one of the most important factors affecting air–sea equilibration, increased atmospheric CO2 concentrations have also entered the ocean, making it more acidic and reducing the abundance of carbonate ions (CO32−) (Schulz and Maher, 2023). In other words, atmospheric CO_2 dissolves in seawater, forming carbonic acid (H_2CO_3), which releases H^+ ions and lowers pH (Fig. 9.1). This is known as ocean acidification (Pelejero et al., 2010).

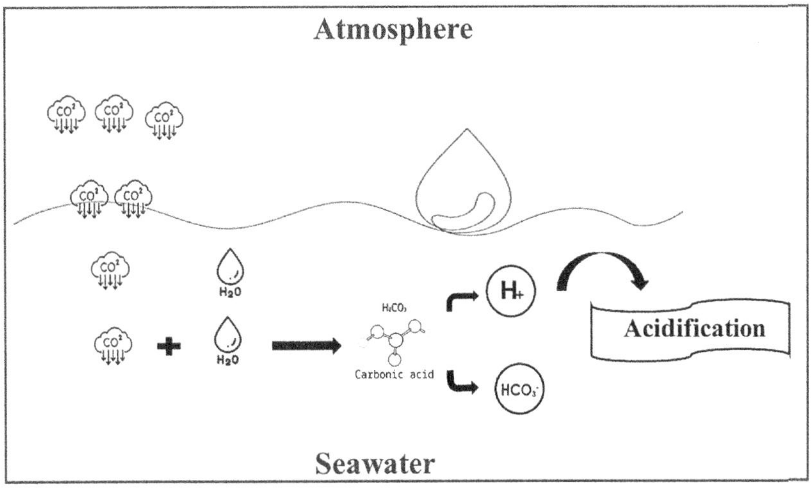

FIGURE 9.1 Ocean acidification process.

Since the early 19th century, ocean acidification has increased by about 30% (Schulz and Maher, 2023). Moreover, it is projected that by 2100, the average ocean pH will decrease by 0.1–0.4 units, depending on CO_2 emission scenarios (Garcia-Soto et al., 2021), potentially worsening the acidification issue.

As CO_3^{2-} ions are essential building structures such as seashells and coral skeletons, ocean acidification process often has critical implications for many marine species (Schulz and Maher, 2023). Specifically, reduced CO_3^{2-} ions make it harder for shell-building species to keep their shells. Furthermore, it can have indirect impacts, such as altering species interactions and disrupting key ecological processes, resulting in significant changes in marine ecosystems (Zunino et al., 2019).

9.3.2 Biodiversity loss

Global warming, resulting from increased CO_2 levels in the atmosphere, has effects on terrestrial ecosystems, destabilizing natural habitats and leading to concerning biodiversity loss (Wudu et al., 2023). For instance, it can alter precipitation patterns, with negative effects on water availability and plant distribution. Areas that were previously suitable for certain plant species may become less favorable, pushing plants to spread and animals to migrate to areas with better climate conditions, causing a change in the geographical distribution of species. This shift in geographical species distribution increases the risk of extinction for species unable to adapt or relocate to new habitats (Wudu et al., 2023). Without significant reductions in CO_2 emissions over the coming few decades, the Intergovernmental Panel on Climate Change estimates that, by 2100, millions of species could be at high risk of extinction.

Over the period 1995–2019, the total CO_2 emissions increased by about 55% (Dong et al., 2021), threatening the hydrological system, disease control and prevention, pollination, water purification, and C sequestration (Table 9.1), all of which are crucial for human survival and the ecological environment (Dong et al., 2021).

9.3.3 Changes in plant growth and development

Studies (e.g., Silva et al., 2020) showed both positive and negative impacts of increased CO_2 emissions into the atmosphere on plant growth and development (Fig. 9.2) were revealed:

- On the one hand, for several crops, especially C3 crops (e.g., wheat, rice, and soybeans), increased CO2 emission is considered a fertiliser as it causes an increased rate of photosynthesis. This is known as the fertilization effect. Specifically, in addition to water molecules (H2O) absorbed from the surrounding soil, increased CO2 emissions enable plants to obtain sufficient CO2 molecules from the atmosphere to form glucose (C6H12O6) through photosynthesis. This, in turn, leads to better plant growth as plants use glucose as a main source of metabolic energy. Many studies worldwide have correlated rising CO2 levels with higher crop yields. Overall, Taub (2010) reported that increasing CO_2 from 475 to 600 ppm can increase leaf photosynthetic rates and crop yields by approximately 40% and 17%, respectively. Moreover, studies (Liu et al., 2019; Fan et al., 2020) have shown that increased emission of CO_2 can decrease and even avoid the impacts of water stress on some crop such as tomato and pepper, given that it promotes osmotic adjustment, providing more C to plant root expansion, and thus allowing roots to absorb more water from the soil.

TABLE 9.1 Key effects of increased CO_2 emissions on ecosystem services.

Ecosystem service	Key effects of CO_2 emissions	Case study
Hydrological regulation	Altered precipitation patterns Changed evapotranspiration, change in water storage Change in surface runoff Change in soil moisture	Stagl et al. (2014)
Disease control	High disease risks Transmission of viruses Distribution of zoonotic disease	Filho et al. (2022)
Pollination	Changed plant–pollinator interactions Higher pollen production in plants Altered flowering patterns	Crowley et al. (2021)
Water purification	Altered freshwater chemistry Altered balance between saline and freshwater	Brown et al. (2020)
Carbon sequestration	Change in the amount of carbon sequestered	Rodrigues et al. (2023)

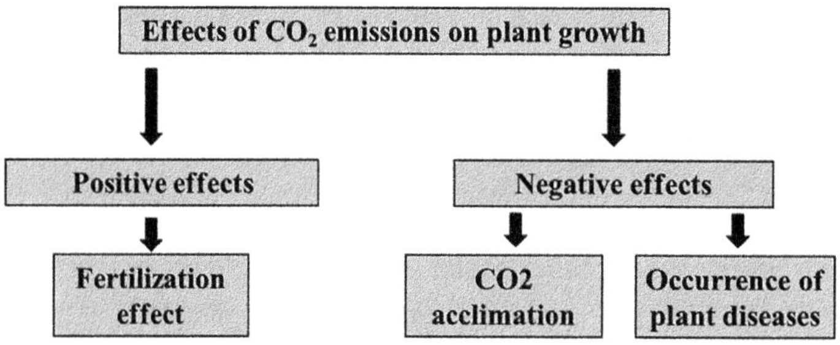

FIGURE 9.2 Effects of increased CO_2 levels on plant growth and development.

- On the other hand, many issues can limit the benefits of the CO_2 fertilization effect. For instance, long-term exposure to elevated CO2 concentrations may increase plants' ability to accumulate carbohydrates and reduce their accumulation of protein and minerals (e.g., iron, zinc, and magnesium), thereby worsening the nutritional quality of food crops. This phenomenon is termed "CO_2 acclimation" (e.g., Loladze, 2014). Also, higher atmospheric CO2 levels may lead to plant diseases (Gullino et al., 2018). In other words, increased CO_2 concentrations may result in denser canopies with higher humidity, which favors a variety of plant diseases, including foliar pathogens, powdery mildews, and downy mildews (Gullino et al., 2018).

9.4 Adaptation strategies for CO_2 emission reduction

Typically, CO_2 emissions can be reduced through the following key strategies: transitioning to renewable energy, improving energy efficiency, promoting sustainable transportation (ST), adopting sustainable agricultural practices, and fostering community engagement. Overall, these strategies target the main sources of CO_2 emissions (Afifa et al., 2024).

9.4.1 Transition to renewable energy

This adaptation option can significantly decrease CO_2 emissions. For instance, Yao et al. (2019) found that a 10% increase in renewable energy can lead to a 1.6% decrease in CO_2 emissions. In recent years, renewable energy technologies have made significant progress, becoming more efficient and cost-effective. Specifically, the share of renewable energy in total consumption increased from 6.6% in 1990 to 10.5% in 2017 (Li et al., 2023). However, it is essential to note that the transition is a complex process, as it is driven by two main factors: the growing demand for finite fossil fuels and the imperative to protect the environment (Neacsa et al., 2022). This requires innovation and policy development, especially for developing countries, to achieve significant CO_2 emission reductions (Cholewa et al., 2022). Tax incentives, financial support, energy mandates, C pricing, and worker training programs are key strategies for that transition. By opting for renewable energy sources over fossil fuels, several economic benefits can result.

9.4.2 Energy efficiency improvement

This strategy can contribute significantly to a reduction in CO_2 emissions, as it accounts for approximately 25% of total global energy consumption (Fawkes et al., 2016). By investing in improving industrial energy efficiency, industries can achieve substantial savings in energy consumption, enhanced economic growth, and better environmental protection (Fawkes et al., 2016; Meena and Pradhan, 2023). According to the International Energy Agency, energy efficiency improvements could reduce emissions by 40% by 2040. However, regulatory barriers in many countries, especially developing ones, hinder industries from fully benefiting (Sarkar and Singh, 2010). Many policies and programs, such as the adoption of energy-efficient practices, C pricing, tax incentives, low-interest loans for small enterprises, and the promotion of technological innovation, can be adopted to help overcome these obstacles (Fawkes et al., 2016; Pradhan et al., 2023).

9.4.3 Sustainable transportation

ST is also needed to reduce CO_2 emissions, given that this sector accounts for around 64% of global oil consumption and 27% of global energy use (Mead, 2021). In this sector, both energy consumption and CO_2 emissions are rising faster than in any other sector (Singh et al., 2021). In recent years, many countries, especially those in the EU, have integrated SD into their transportation policy. Overall, ST should maintain clean streets, protect the environment, and foster a dynamic economy (Abdel Wahed Ahmed and Abd El Monem, 2020). Achieving green and ST requires integrating urban transportation with broader urban life aspects and addressing key factors such as mobility, accessibility, affordability, social equity, and environmental protection. Sustainable public transport options such as subways and metro lines offer safe, reliable, and environmentally friendly solutions, while nonmotorized options, such as walking and cycling, promote health and reduce emissions (Brand et al., 2021). Cities that have invested in bike-sharing programs and pedestrian-friendly infrastructure have seen a significant decrease in vehicular emissions.

9.4.4 Sustainable agriculture

Shifting from conventional to sustainable agriculture is crucial for reducing GHG emissions (CH_4, N_2O), adapting to climate change, and meeting rising food demands (Meena et al., 2023; Haj-Amor et al., 2023). With the global population expected to reach at least 9.4 billion by 2050, sustainable farming is essential (van Dijk et al., 2021). Practices such as precision agriculture, improved irrigation, agroforestry, and efficient nitrogen use can enhance productivity while reducing emissions (Jhariya et al., 2021a, 2021b; Kumar et al., 2024). These measures could achieve net-zero agricultural emissions by 2050, especially in the EU, the USA, and Canada. Strong collaboration among policymakers, researchers, and farmers is vital for success (Haj-Amor et al., 2023).

9.4.5 Community engagement

In addition to technical measures, reducing CO_2 emissions depends on active community involvement, effective policies, and widespread public education (Afifa et al., 2024). Raising public awareness and fostering community participation are essential for successful emission reduction. Educational campaigns that highlight sustainable practices can inspire individuals and communities to take responsibility, helping to cultivate a culture of sustainability. Local schools and organizations can spearhead initiatives to promote recycling, energy efficiency, and sustainable consumption habits, ultimately driving long-term environmental change.

9.5 Management perspectives and initiatives for CO_2 emission reduction

Effective management practices are necessary for reducing CO_2 emissions across various sectors. These practices encompass a range of strategies, including government policies, corporate responsibility, and community engagement.

9.5.1 Government policies

Governments have an important role in implementing policies that promote a sharp reduction in CO_2 emissions. This includes developing regulations and implementing suitable adaptation strategies (e.g., public transportation and sustainable agriculture). In other words, governments should enforce strict environmental regulations, improve energy efficiency, support green innovation, and raise awareness about sustainable development. By combining regulation, innovation, and education, they can strike a balance between economic growth and environmental protection (Bakhsh et al., 2024).

It is essential to indicate that the relationship between government intervention and CO_2 emissions is multifaceted, with advantages and disadvantages (Zhang et al., 2023). For instance, in some regions, enhanced environmental performance appraisals have encouraged measures to reduce CO_2 emissions, demonstrating the positive effect of government intervention and the potential of effective government action. On the other hand, the adverse effect could be evident in short-term decision-making by local governments, such as supporting energy-intensive economic activities, which have further exacerbated the pollution issue (Zhang et al., 2023). Overall, effective intervention from governments requires an effort to understand the diverse consequences of CO_2 emissions, as well as the costs and benefits of potential mitigation and adaptation measures (Prasad et al., 2021).

9.5.2 Corporate responsibility

Businesses can significantly reduce their C footprints via sustainable practices and transparency in reporting their CO_2 emissions. In this context, several companies are currently adopting corporate social responsibility (CSR) strategies that incorporate sustainability into their core operations. CSR has great potential to enhance the reputation of companies (Ahmad et al., 2023). Rules limiting CO_2 emissions from companies and shaping their social responsibilities can guide policymakers in developing balanced regulations that align CSR efforts with environmental goals (Chen et al., 2019). Subsidies can incentivize CSR activities that support emission reduction, while optimal emission restrictions can prevent counterproductive overproduction driven by excessive social concern. Flexible regulatory frameworks further enable the dual achievement of CO_2 emissions reductions and improved social welfare without imposing undue burdens on firms, ensuring that economic and social benefits are maintained alongside environmental targets (Chen et al., 2019).

9.5.3 Collaboration and partnerships

Cooperation among governments, businesses, and nongovernmental organizations (NGOs) is highly required for reducing CO_2 emissions and ensuring the successful implementation of various intervention options. Specifically, the flow of information, materials, and energy between governments, businesses, and NGOs should be optimized in order to achieve the positive effects of each intervention aimed at reducing CO_2 emissions (Harangozó and Zilahy, 2015). Overall, cooperative strategies constitute an important and effective component toward a sustainable environment (Harangozó and Zilahy, 2015).

9.6 Transition to a green economy

This adaptation measure is needed to achieve sustainable development and reduce CO_2 emissions. A green economy emphasizes low-C, resource-efficient, and socially inclusive practices, aiming to improve human well-being and protect the environment.

9.6.1 The role of research and development

Investing in research and development (R&D) is essential for identifying new technologies and designing innovative measures that can support a smooth transition to a green economy. In other words, the benefits of R&D, such as knowledge advancement, low-cost innovation, and development of optimal scientific solutions, play a critical role in driving a green economy (Sarpong et al., 2023; Meena et al., 2023). R&D efforts not only improve knowledge but also prepare firms to capitalize on future innovation opportunities. R&D initiatives foster cooperation among governments, businesses, and NGOs, driving the development of innovative scientific concepts, particularly in eco-innovation (Sarpong et al., 2023). This cooperation is fundamental to implementing new scientific ideas and contributing to long-term socio-economic benefits (Banerjee et al., 2023; Banerjee et al., 2024).

Global R&D investment is becoming increasingly competitive, with the USA and China leading the way in research and technology. Increased R&D effort has enabled both countries to dominate many research areas, encouraging other countries to invest more in scientific research (Sarpong et al., 2023). Worldwide, governments and the private sector are focusing on sustainable innovations to tackle CO_2 emissions, such as the green hydrogen strategy (Velazquez Abad and Dodds, 2020). However, to achieve lasting benefits, R&D investments need to be aligned with the development of skilled talent and institutions (Sarpong et al., 2023). In addition to conventional R&D, technologies are key for driving sustainable innovation (Hermann and Hermann, 2022; Meena et al., 2023; Pradhan and Meena, 2023a, 2023b).

9.6.2 Policy frameworks

Creating supportive policy frameworks that promote sustainable practices is necessary for facilitating the transition to a green economy (Söderholm, 2020). Policies should focus on promoting investments in clean technologies, enhancing workforce skills, and encouraging sustainable consumption patterns. Comprehensive strategies that align economic growth with environmental sustainability can foster resilience against climate change.

9.6.3 Social inclusion

Ensuring social equity and inclusion in the transition to a green economy is essential (Aslam and Ghouse, 2023). Policymakers must address the needs of marginalized communities disproportionately impacted by pollution. By prioritizing equity in environmental policies and ensuring access to green jobs and opportunities, a smooth transition to a green economy can be achieved.

9.7 Cap-and-trade schemes

This scheme is one form of pricing GHG emissions. More specifically, it is a market-based approach to reducing GHG emissions by setting a fixed limit, or cap, on total emissions while allowing entities to trade emission allowances. Each allowance represents 1 ton of emissions, and companies that emit less than their allocated share can sell excess allowances to those exceeding their limits. This leads to a financial incentive for emission reductions, rewarding firms that adopt cleaner technologies. The trading mechanism ensures that emissions are reduced at the lowest possible cost, as businesses with cheaper reduction options sell allowances to those facing higher compliance costs. Unlike a C tax, the scheme guarantees a specific decrease in emissions while allowing market forces to determine allowance prices, promoting cost-effective compliance and flexibility for industries (Narassimhan et al., 2018). Studies (e.g., Chang, 2015) reported that cap-and-trade is more effective and cost-efficient than a C tax or regulatory standards. While a C tax sets a fixed price on emissions, providing cost predictability and stability, cap-and-trade ensures a defined emissions reduction by leveraging market dynamics to minimize compliance costs.

9.8 Future of CO_2 emission management and the green economy

The green economy prioritizes sustainability through low-C, eco-friendly, and resource-efficient development. It emphasizes efficient resource use and C emission control, where waste is minimized and resources are recycled. Technological advancements such as C capture will aid in achieving net-zero emissions. As climate concerns intensify, green sectors (such as energy, electricity, and agriculture) will drive economic growth. With rising global commitment, policies, innovations, and investments in clean energy will shape a future where economic progress aligns with environmental responsibility. If these efforts continue to accelerate, CO_2 emissions could be reduced rapidly, leading to a long-term decline through enhanced energy efficiency, widespread adoption of clean technologies, and stricter emission regulations (Peng et al., 2023). However, achieving deep decarbonization will require sustained collaboration, economic incentives, and large-scale infrastructure shifts to ensure a lasting reduction in global carbon footprints.

9.9 Conclusion

Addressing CO_2 emissions is urgently needed to mitigate climate change and promote a sustainable environment. By implementing best management practices, fostering collaboration, and transitioning to a green economy, it is possible to achieve substantial reductions in CO_2 emissions and enhanced environmental protection. In this context, governments, businesses, and individuals should ensure close collaboration, recognizing that every action counts in addressing climate change. As the challenges ahead are faced, the adoption of innovative technologies and sustainable practices will be paramount in ensuring a sustainable environment. Finally, it is important to note that searching for innovative alternatives to better protect the environment and avoid critical future environmental disasters due to rising CO2 emissions is always necessary.

References

Abdel Wahed Ahmed, M.M., Abd El Monem, N., 2020. Sustainable and green transportation for better quality of life case study greater Cairo–Egypt. HBRC Journal 16 (1), 17–37. Available at https://doi.org/10.1080/16874048.2020.1719340.

Afifa, Arshad, K., Hussain, N., Ashraf, M.H., Saleem, M.Z., 2024. Air pollution and climate change as grand challenges to sustainability. The Science of the Total Environment 928, 172370.

Ahmad, K., Irshad Younas, Z., Manzoor, W., Safdar, N., 2023. Greenhouse gas emissions and corporate social responsibility in USA: a comprehensive study using dynamic panel model. Heliyon 9 (3), e13979. Available at https://doi.org/10.1016/j.heliyon.2023.e13979.

Aslam, A., Ghouse, G., 2023. Targeting the new sustainable inclusive green growth: a review. Cleaner and Responsible Consumption 11, 100140. Available at https://doi.org/10.1016/j.clrc.2023.100140.

Ayompe, L.M., Davis, S.J., Egoh, B.N., 2020. Trends and drivers of African fossil fuel CO2 emissions 1990-2017. Environmental Research Letters 15 (12). Available at https://doi.org/10.1088/1748-9326/abc64f.

Bakhsh, S., Zhang, W., Ali, K., Oláh, J., 2024. Strategy towards sustainable energy transition: the effect of environmental governance, economic complexity and geopolitics. Energy Strategy Reviews 52, 101330. Available at https://doi.org/10.1016/j.esr.2024.101330.

Banerjee, A., Jhariya, M.K., Bargali, S.S., Palit, D., 2023. Ecorestoration for Sustainability. Wiley-Scrivener, USA, pp. 1–441. ISBN: 9781119879718.

Banerjee, A., Jhariya, M.K., Raj, A., Mechergui, T., 2024. Ecosystem management: climate change and global sustainability—an introduction. Ecosystem Management: Climate Change and Sustainability. wiley, India, pp. 1–21. Available at http://doi.org/10.1002/9781394231249.ch1.

Brand, C., Dons, E., Anaya-Boig, E., Avila-Palencia, I., Clark, A., de Nazelle, A., et al., 2021. The climate change mitigation effects of daily active travel in cities. Transportation Research Part D: Transport and Environment 93. Available at https://doi.org/10.1016/j.trd.2021.102764.

Brown, T.R.W., Lajeunesse, M.J., Scott, K.M., 2020. Strong effects of elevated CO2 on freshwater microalgae and ecosystem chemistry. Limnology and Oceanography 65 (2), 304–313. Available at https://doi.org/10.1002/lno.11298.

Chang, A., 2015. The politics and future of carbon cap-and-trade: lessons from the European Union. Claremont-UC Undergraduate Research Conference on the European Union 2014 (1), 75–90. Available at https://doi.org/10.5642/urceu.201401.07.

Chen, Y.H., Nie, P.Y., Wang, C., Meng, Y., 2019. Effects of corporate social responsibility considering emission restrictions. Energy Strategy Reviews 24, 121–131. Available at https://doi.org/10.1016/j.esr.2019.02.002.

Cholewa, M., Mammadov, F., Nowaczek, A., 2022. The obstacles and challenges of transition towards a renewable and sustainable energy system in Azerbaijan and Poland. Mineral Economics 35 (1), 155–169. Available at https://doi.org/10.1007/s13563-021-00288-x.

Çobanoğulları, G., 2024. Exploring the link between CO2 emissions, health expenditure, and economic growth in Türkiye: evidence from the ARDL model. Environment, Development and Sustainability 26 (11), 29605–29619. Available at https://doi.org/10.1007/s10668-024-04835-8.

Crowley, L.M., Sadler, J.P., Pritchard, J., Hayward, S.A.L., 2021. Elevated co2 impacts on plant–pollinator interactions: a systematic review and free air carbon enrichment field study. Insects 12 (6). Available at https://doi.org/10.3390/insects12060512.

Dong, H., Xue, M., Xiao, Y., Liu, Y., 2021. Do carbon emissions impact the health of residents? Considering China's industrialization and urbanization. Science of the Total Environment 758, 143688. Available at https://doi.org/10.1016/j.scitotenv.2020.143688.

EEA, 2020. Overall greenhouse gas emission trends and key statistics. EEA Report No 03/2020, European Environment Agency.

Eldesouki, M.H., Rashed, A.E., El-Moneim, A.A., 2023. A comprehensive overview of carbon dioxide, including emission sources, capture technologies, and the conversion into value-added products. Clean Technologies and Environmental Policy 25 (10), 3131–3148. Available at https://doi.org/10.1007/s10098-023-02599-9.

Fan, X., Cao, X., Zhou, H., Hao, L., Dong, W., He, C., et al., 2020. Carbon dioxide fertilization effect on plant growth under soil water stress associates with changes in stomatal traits, leaf photosynthesis, and foliar nitrogen of bell pepper (*Capsicum annuum* L). Environmental and Experimental Botany 179, 104203. Available at https://doi.org/10.1016/j.envexpbot.2020.104203.

Fawkes, S., Oung, K., Thorpe, D., Zhu, X., De La Rue Du Can, S., Farrell, T.C.,2016. Best Practices and Case Studies for Industrial Energy Efficiency Improvement: An Introduction for Policy Makers. UNEP DTU Partnership.

Fawzy, S., Osman, A.I., Doran, J., Rooney, D.W., 2020. Strategies for mitigation of climate change: a review. Environmental Chemistry Letters 18 (6), 2069–2094. Available at https://doi.org/10.1007/s10311-020-01059-w.

Filho, W.L., Ternova, L., Parasnis, S.A., Kovaleva, M., Nagy, G.J., 2022. Climate change and zoonoses: a review of concepts, definitions, and bibliometrics. International Journal of Environmental Research and Public Health 19 (2). Available at https://doi.org/10.3390/ijerph19020893.

Garcia-Soto, C., Cheng, L., Caesar, L., Schmidtko, S., Jewett, E.B., Cheripka, A., et al., 2021. An overview of ocean climate change indicators: sea surface temperature, ocean heat content, ocean pH, dissolved oxygen concentration, arctic sea ice extent, thickness and volume, sea level and strength of the AMOC (Atlantic Meridional Overturning Circulation). Frontiers in Marine Science 8. Available at https://doi.org/10.3389/fmars.2021.642372.

Gullino, M.L., Pugliese, M., Gilardi, G., Garibaldi, A., 2018. Effect of increased CO2 and temperature on plant diseases: a critical appraisal of results obtained in studies carried out under controlled environment facilities. Journal of Plant Pathology 100 (3), 371–389. Available at https://doi.org/10.1007/s42161-018-0125-8.

Haj-Amor, Z., Kim, D.G., Bouri, S., 2023. Sustainable Agriculture: Adaptation Strategies to Address Climate Change by 2050. CRC Press, Tunisia, pp. 1–229. Available at http://doi.org/10.1201/9781003404194.

Harangozó, G., Zilahy, G., 2015. Cooperation between business and non-governmental organizations to promote sustainable development. Journal of Cleaner Production 89, 18–31. Available at https://doi.org/10.1016/j.jclepro.2014.10.092.

Hermann, E., Hermann, G., 2022. Artificial intelligence in research and development for sustainability: the centrality of explicability and research data management. AI and Ethics 2 (1), 29–33. Available at https://doi.org/10.1007/s43681-021-00114-8.

Huang, C., Gan, X., Wan, Y., Jin, L., Teng, J., Li, Z., 2024. China contributed to low-carbon development: carbon emission increased but carbon intensity decreased. Frontiers in Ecology and Evolution 12. Available at https://doi.org/10.3389/fevo.2024.1338742.

Hugh, S.B., Richard, J.M., David, J.K., Urs, B., Benoit, P.G., Dann, M., et al., 2018. Higher CO2 concentrations increase extreme event risk in a 1.5◦ C world. Nature Climate Change 8, 604–608.

IEA, 2021. Global energy review: CO2 emissions in 2021 analysis.

IPCC, 2018. Global Warming of 1.5°C. An IPCC special report on the impacts of global warming of 1.5°C above pre-industrial levels and related global greenhouse gas emission pathways, in the context of strengthening the global response to the threat of climate change.

Jhariya, M.K., Banerjee, A., Swaroop Meena, R., Kumar, S., Raj, A., 2021. Sustainable Intensification for Agroecosystem Services and Management. Springer, Singapore, pp. 1–870. Available at https://doi.org/10.1007/978-981-16-3207-5.

Jhariya, M.K., Meena, R.S., Banerjee, A., 2021b. Ecological Intensification of Natural Resources for Sustainable Agriculture. Springer, Singapore, pp. 1–655. Available at http://doi.org/10.1007/978-981-33-4203-3.

Jhariya, M.K., Meena, R.S., Banerjee, A., Kumar, S., Raj, A., 2023. Agroforestry for Carbon and Ecosystem Management. Elsevier, India, pp. 1–434. Available at http://doi.org/10.1016/C2021-0-02916-3.

Jhariya, M.K., Meena, R.S., Banerjee, A., Meena, S.N., 2022. Natural Resources Conservation and Advances for Sustainability. Elsevier. Available at https://doi.org/10.1016/C2019-0-03763-6.

Kumar, S., Meena, R.S., Sheoran, P., Jhariya, M.K., 2024. Regenerative Agriculture for Sustainable Food Systems. Springer, Singapore, pp. 1–493. Available at https://doi.org/10.1007/978-981-97-6691-8.

Li, R., Wang, Q., lejia, L., 2023. Does renewable energy reduce per capita carbon emissions and per capita ecological footprint? New evidence from 130 countries. Energy Strategy Reviews 49, 101121. Available at https://doi.org/10.1016/j.esr.2023.101121.

Liu, H., Alharthi, M., Atil, A., Zafar, M.W., Khan, I., 2022. A non-linear analysis of the impacts of natural resources and education on environmental quality: green energy and its role in the future. Resources Policy 79, 102940. Available at https://doi.org/10.1016/j.resourpol.2022.102940.

Liu, J., Hu, T., Fang, L., Peng, X., Liu, F., 2019. CO2 elevation modulates the response of leaf gas exchange to progressive soil drying in tomato plants. Agricultural and Forest Meteorology 268, 181–188. Available at https://doi.org/10.1016/j.agrformet.2019.01.026.

Locatelli, B., 2016. Ecosystem services and climate change. In: Potschin, M., Haines-Young, R., Fish, R., Turner, R.K. (Eds.), Routledge Handbook of Ecosystem Services. Routledge, London and New York, pp. 481–490.

Loladze, I. (2014) Hidden shift of the ionome of plants exposed to elevated CO2 depletes minerals at the base of human nutrition. eLife 3:e02245. Available at https://doi.org/10.7554/eLife.02245.

Mead, L., 2021. The Road to Sustainable Transport, Brief #19. International Institute for Sustainable Development.

Meena, R.S., Pradhan, G., 2023. Industrial garbage-derived biocompost enhances soil organic carbon fractions, CO2 biosequestration, potential carbon credits and sustainability index in a rice-wheat ecosystem. Environmental Research 235, 116525. Available at https://doi.org/10.1016/j.envres.2023.116525.

Meena, R.S., Pradhan, G., Kumar, S., Lal, R., 2023. Using industrial wastes for rice-wheat cropping and food-energy-carbon-water-economic nexus to the sustainable food system. Renewable and Sustainable Energy Reviews 187, 113756. Available at https://doi.org/10.1016/j.rser.2023.113756.

Meena, R.S., Lal, R., Kumar, S., Pradhan, G., Srinivasarao, C., Singh, A.K., et al., 2024. Potential of Indian agriculture for capturing atmospheric CO2 and monetizing carbon credits to the farmers: an overview and policy framework. Advances in Agronomy 188, 101–206. Available at https://doi.org/10.1016/bs.agron.2024.06.001.

Meena, R.S., Pradhan, G., Kumar, S., 2025. Energy flow, eco-efficiency, and economic circulation with recycled industrial waste compost application in wheat and subsequent rice farming. Science of the Total Environment 967. Available at https://doi.org/10.1016/j.scitotenv.2025.178779.

Meena, R.S., Pradhan, G., Singh, K., Kumar, S., Singh, A.K., Shashidhar, K.S., et al., 2024a. Agriculture models for restoring degraded land to enhance CO2 biosequestration and carbon credits in the Vindhyan region of India. Science of the Total Environment 929. Available at https://doi.org/10.1016/j.scitotenv.2024.172661.

Meena, R.S., Pradhan, G.S., Kumar, S., Jhariya, M.K., Rao, C.S., 2025a. Carbon-Negative Agriculture. Springer, pp. 1–500.

Meena, R.S., Pradhan, G., Nalani, Kumar, S., Singh, A.K., Verma, S.K., et al., 2024b. Diversified cropping modules designed for soil restoration, CO2 sequestration, and generating carbon credits. Land Degradation and Development 35 (14), 4314–4329. Available at https://doi.org/10.1002/ldr.5224.

Meng, B., Liu, Y., Gao, Y., Li, M., Wang, Z., Xue, J., et al., 2023a. Developing countries' responsibilities for CO2 emissions in value chains are larger and growing faster than those of developed countries. One Earth 6 (2), 167–181. Available at https://doi.org/10.1016/j.oneear.2023.01.006.

Meng, J., Huo, J., Zhang, Z., Liu, Y., Mi, Z., Guan, D., et al., 2023b. The narrowing gap in developed and developing country emission intensities reduces global trade's carbon leakage. Nature Communications 14, 3775.

Mitić, P., Fedajev, A., Radulescu, M., Rehman, A., 2023. The relationship between CO2 emissions, economic growth, available energy, and employment in SEE countries. Environmental Science and Pollution Research 30 (6), 16140–16155. Available at https://doi.org/10.1007/s11356-022-23356-3.

Narassimhan, E., Gallagher, K.S., Koester, S., Alejo, J.R., 2018. Carbon pricing in practice: a review of existing emissions trading systems. Climate Policy 18 (8), 967–991. Available at https://doi.org/10.1080/14693062.2018.1467827.

Neacsa, A., Rehman Khan, S.A., Panait, M., Apostu, S.A., 2022. The transition to renewable energy—a sustainability issue? In: Khan, S.A.R., Panait, M., Puime Guillen, F., Raimi, L. (Eds.), Energy Transition. Industrial Ecology. Springer, Singapore, pp. 29–72. Available at https://doi.org/10.1007/978-981-19-3540-4_2.

Nunes, L.J.R., 2023. The rising threat of atmospheric CO2: a review on the causes, impacts, and mitigation strategies. Environments 10 (4), 66. Available at https://doi.org/10.3390/environments10040066.

Pelejero, C., Calvo, E., Hoegh-Guldberg, O., 2010. Paleo-perspectives on ocean acidification. Trends in Ecology & Evolution (Personal Edition) 25 (6), 332–344.

Peng, J., Hu, X., Fan, X., Wang, K., Gong, H., 2023. The impact of the green economy on carbon emission intensity: comparisons, challenges, and mitigating strategies. Sustainability 15 (14), 10965. Available at https://doi.org/10.3390/su151410965.

Pradhan, G., Meena, R.S., 2023a. Interaction impact of biocompost on nutrient dynamics and relations with soil biota, carbon fractions index, societal value of CO2 equivalent and ecosystem services in the wheat-rice farming. Chemosphere 339. Available at https://doi.org/10.1016/j.chemosphere.2023.139695.

Pradhan, G., Meena, R.S., 2023b. Utilizing waste compost to improve the atmospheric CO2 capturing in the rice-wheat cropping system and energy-cum-carbon credit auditing for a circular economy. Science of the Total Environment 892. Available at https://doi.org/10.1016/j.scitotenv.2023.164572.

Pradhan, G., Meena, R.S., Kumar, S., Lal, R., 2023. Utilizing industrial wastes as compost in wheat-rice production to improve the above and below-ground ecosystem services. Agriculture, Ecosystems & Environment 358, 108704. Available at https://doi.org/10.1016/j.agee.2023.108704.

Prasad, R., Jhariya, M.K., Banerje, A., 2021. Advances in Sustainable Development and Management of Environmental and Natural Resources: Economic Outlook and Opinions, 2-volume set. Apple Academic Press, New York, pp. 1–428. Available at https://doi.org/10.1201/9781003187455.

Raj, A., Jhariya, M.K., Meena, R.S., 2025. Agroforestry for Monetising Carbon Credits. Springer, Cham. Available at https://doi.org/10.1007/978-3-031-97151-8.

Ritchie, H. (2023). Global inequalities in CO2 emissions. Published online at OurWorldinData.org. Retrieved from: https://archive.ourworldindata.org/20260417-112857/inequality-co2.html [Online Resource].

Ritchie, H., Rosado, P. (2025). CO_2 and Greenhouse Gas Emissions. Published online at OurWorldinData.org. Retrieved from: https://ourworldindata.org/profile/co2/united-states [Online Resource].

Rodrigues, C.I.D., Brito, L.M., Nunes, L.J.R., 2023. Soil carbon sequestration in the context of climate change mitigation: a review. Soil Systems 7 (3). Available at https://doi.org/10.3390/soilsystems7030064.

Sarkar, A., Singh, J., 2010. Financing energy efficiency in developing countries-lessons learned and remaining challenges. Energy Policy 38 (10), 5560–5571. Available at https://doi.org/10.1016/j.enpol.2010.05.001.

Sarpong, D., Boakye, D., Ofosu, G., Botchie, D., 2023. The three pointers of research and development (R&D) for growth-boosting sustainable innovation system. Technovation 122, 102581. Available at https://doi.org/10.1016/j.technovation.2022.102581.

Schulz, K.G., Maher, D.T., 2023. Atmospheric carbon dioxide and changing ocean chemistry. In: Reichelt-Brushett, A. (Ed.), Marine Pollution – Monitoring, Management and Mitigation. Springer Textbooks in Earth Sciences, Geography and Environment. Springer, Cham. Available at https://doi.org/10.1007/978-3-031-10127-4_11.

Secretariat of the Convention on Biological Diversity, 2014. An updated synthesis of the impacts of ocean acidification on marine biodiversity. In: Hennige, S., Roberts, J.M., Williamson, P. (Eds.), Montreal, Technical Series No. 75, pp. 99.

Silva, R.G.da, Alves, R.de C., Zingaretti, S.M., 2020. Increased [CO2] causes changes in physiological and genetic responses in C4 crops: a brief review. Plants 9 (11), 1567. Available at https://doi.org/10.3390/plants9111567.

Singh, A., Gurtu, A., Singh, R.K., 2021. Selection of sustainable transport system: a case study. Management of Environmental Quality: An International Journal 32 (1), 100–113. Available at https://doi.org/10.1108/MEQ-03-2020-0059.

Stagl, J., Mayr, E., Koch, H., Hattermann, F.F., Huang, S., 2014. Effects of climate change on the hydrological cycle in central and Eastern Europe. In: In: Rannow, S., Neubert, M. (Eds.), Managing Protected Areas in Central and Eastern Europe Under Climate Change. Advances in Global Change Research 58 Springer, Dordrecht. Available at https://doi.org/10.1007/978-94-007-7960-0_3.

Söderholm, P., 2020. The green economy transition: the challenges of technological change for sustainability. Sustainable Earth 3 (1). Available at https://doi.org/10.1186/s42055-020-00029-y.

Taub, D., 2010. Effects of rising atmospheric concentrations of carbon dioxide on plants. Nature Education Knowledge 3 (10), 21.

Underwood, A., Marcantonio, R., Wood, D., Crippa, P., 2024. Interstate air pollution governance in the United States: exploring clean air act section 126. Environmental Management 74 (3), 401–413. Available at https://doi.org/10.1007/s00267-024-02002-3.

van Dijk, M., Morley, T., Rau, M.L., Saghai, Y., 2021. A meta-analysis of projected global food demand and population at risk of hunger for the period 2010–2050. Nature Food 2 (7), 494–501. Available at https://doi.org/10.1038/s43016-021-00322-9.

Velazquez Abad, A., Dodds, P.E., 2020. Green hydrogen characterisation initiatives: definitions, standards, guarantees of origin, and challenges. Energy Policy 138, 111300. Available at https://doi.org/10.1016/j.enpol.2020.111300.

Wang, M., Feng, C., 2022. Tracking the inequalities of global per capita carbon emissions from perspectives of technological and economic gaps. Journal of Environmental Management 315, 115144. Available at https://doi.org/10.1016/j.jenvman.2022.115144.

Wang, Z., Yan, H., Gao, X., Liang, Q., Mi, Z., Liu, L., 2024. Have consumption-based CO2 emissions in developed countries peaked? Energy Policy 184, 113894. Available at https://doi.org/10.1016/j.enpol.2023.113894.

Wudu, K., Abegaz, A., Ayele, L., Ybabe, M., 2023. The impacts of climate change on biodiversity loss and its remedial measures using nature based conservation approach: a global perspective. Biodiversity and Conservation 32 (12), 3681–3701. Available at https://doi.org/10.1007/s10531-023-02656-1.

Yang, T., Pan, Y., Yang, Y., Lin, M., Qin, B., Xu, P., et al., 2017. CO2 emissions in China's building sector through 2050: a scenario analysis based on a bottom-up model. Energy 128, 208–223. Available at https://doi.org/10.1016/j.energy.2017.03.098.

Yao, S., Zhang, S., Zhang, X., 2019. Renewable energy, carbon emission and economic growth: a revised environmental Kuznets Curve perspective. Journal of Cleaner Production 235, 1338–1352. Available at https://doi.org/10.1016/j.jclepro.2019.07.069.

Zhang, S., Li, J., Jiang, B., Guo, T., 2023. Government intervention, structural transformation, and carbon emissions: evidence from China. International Journal of Environmental Research and Public Health 20 (2), 1343. Available at https://doi.org/10.3390/ijerph20021343.

Zunino, S., Canu, D.M., Zupo, V., Solidoro, C., 2019. Direct and indirect impacts of marine acidification on the ecosystem services provided by coralligenous reefs and seagrass systems. Global Ecology and Conservation 18, e00625. Available at https://doi.org/10.1016/j.gecco.2019.e00625.

Chapter 10

Field-level issues and carbon management strategies to meet carbon credit program requirements

Mustapha Abdulsalam[1], Miracle Uwa Livinus[2], Musa Ojeba Innocent[1], Shehu-Alimi Elelu[3], Abdulsalam Khadija[4], Abdulhakeem Idris Abdulhakeem[5], Muhammed Lawal Subair[6], Muhammed Abdulazeez[1], Tijani Abiola Tajudeen[7], Husna Farouq Ibrahim[8], and Ganiyat Omotayo Ibrahim[9]

[1]*Department of Microbiology, School of Sciences and Information Technology, Skyline University Nigeria, Kano, Nigeria,* [2]*Department of Biochemistry, School of Sciences and Information Technology, Skyline University Nigeria, Kano, Nigeria,* [3]*Department of Chemistry, Howard University, Washington, DC, United States,* [4]*Department of Agricultural Sciences, University of Ilorin, Ilorin, Kwara, Nigeria,* [5]*Department of Fisheries and Aquaculture, Federal University Dutsin-Ma, Dutsin-Ma, Katsina, Nigeria,* [6]*Department of Accounting and Finance, Kwara State University, Malete, Kwara, Nigeria,* [7]*Department of Soil Science, Federal University of Technology, Minna, Nigeria,* [8]*Department of Microbiology, Bayero University, Kano, Nigeria,* [9]*Department of Chemistry, Nottingham Trent University, Nottingham, United Kingdom*

Highlights

- The effectiveness of the C credits program depends on C management strategies.
- C management at the field level is a major challenge.
- There is variation in C measurement in soil and agricultural practices.
- Nature-based practices would be fruitful for the green economy.

10.1 Introduction

10.1.1 Overview of carbon management in agriculture

Agriculture has become an important sector for both the challenge and the opportunity of managing carbon (C), as the global community works tirelessly to mitigate climate change. In the global C cycle, the agricultural sector encompasses crop cultivation, livestock activities, and land-use management, all of which play a major role. The greenhouse gases (GHGs) emitted from such activities include the biological processes, such as soil tillage, fertilizer application, and livestock digestion. Despite the potential for agriculture to help mitigate climate change through carbon sequestration (Cseq) and emission-reduction practices (Powlson et al., 2016), it has also provided significant amounts of manure nutrients that reach consumers through animal products (Meena et al., 2025).

Agricultural C management employs strategies that reduce GHG emissions while enhancing both soil C accumulation and plant-based C reserves (Meena et al., 2025a; Raj et al., 2025). Agriculture generates more C sinks than it consumes, thereby contributing to climate balance. Conservation practices, agroforestry systems (AFs), and cover crops improve soil C storage and promote ecosystem stability (Jhariya et al., 2023; Meena et al., 2024). These climate change mitigation (CCM) strategies offer additional benefits, including enhanced soil health, biodiversity, and water retention capacity (McGuire et al., 2022). Transitioning to sustainable agriculture requires achieving two primary objectives: reducing environmental emissions and increasing soil C content. Financial incentives, provided through innovative C credit programs, encourage the adoption of sustainable agricultural practices. Farmers and land managers can earn C credits by participating in these programs through C_{seq} or emission reductions. Engaging in C market trades enables sellers to utilize acquired C credits as financial compensation for implementing eco-friendly practices (Jhariya et al., 2022). The

Carbon Incentives, Plans, and Policies in Agriculture. **DOI:** https://doi.org/10.1016/B978-0-443-33557-0.00017-3

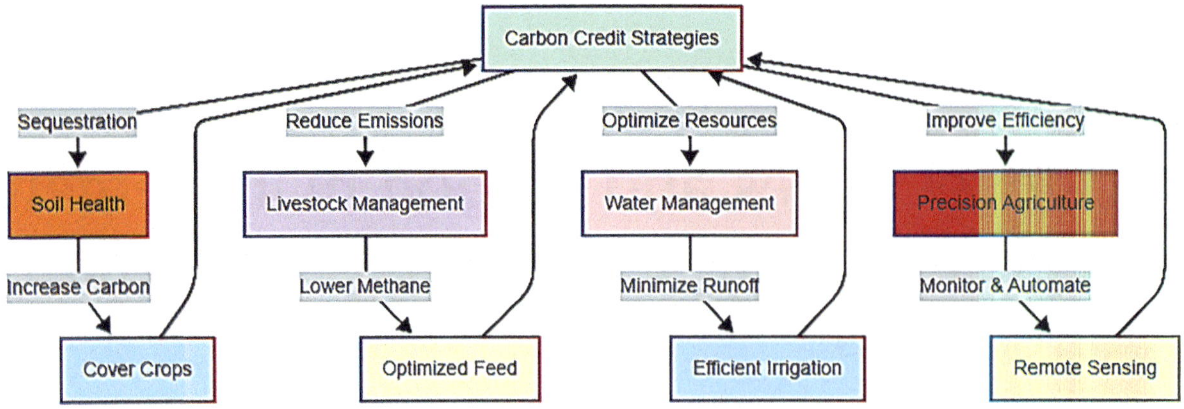

FIGURE 10.1 Carbon credit strategies and field-level implementation.

effectiveness of C credit programs depends on rigorous monitoring of C_{seq}, verification of emission reductions, and proper field-level C management (Paul et al., 2023).

10.1.2 Importance of effective field-level strategies for carbon credit programs

Field-level strategies are necessary for an effective C credit program to be effective. C management practices exhibit high variability in impacts due to considerable variation in local factors, including soil type, topography, climate, and management practices. Thus the C credit programs must be geared to respond to particular field conditions.

Soil health is of the utmost importance for C management. Healthy soils have a greater capacity to store greater quantities of organic matter (OM) and to maintain a better microenvironment for soil structure; therefore, they have a higher C_{seq} capacity (Meena et al., 2024a). For instance, conservation tillage reduces soil disturbance, conserves soil OM content, and increases C storage. In line with cover cropping, the organic material introduced by cover cropping can also increase soil C levels, and soil fertility in general, and is applied similarly to cover cropping (Qin et al., 2023; Meena et al., 2024b).

On this note, there are also crop and livestock management practices of C management. The same crop and livestock systems do not affect soil C levels and GHG emissions in the same way. Sending alternate crop types into the soil each season will improve the soil while also increasing C_{seq}. Livestock management practices such as moderating feed to reduce methane emissions and improving manure management (Kabange et al., 2023) also contribute to reducing GHGs and enhancing C stocks.

Field-level C management also involves controlling water management. Water usage and practices such as controlled drainage can influence GHG emissions. For instance, improperly maintained irrigation systems can emit nitrous oxide, a very potent GHG. Farmers can reduce these emissions and enhance C management strategies as a whole if they adopt water-saving technologies and optimize irrigation practices (Mehmood et al., 2023).

To implement field-level strategies successfully, one should understand local conditions and address the specific needs of different agricultural systems. Precision agriculture technologies, including GPS-guided machinery, remote sensing, and others, provide useful tools for field-level C monitoring and management. Farmers can utilize these technologies to optimize resource use, make data-driven decisions, and adopt C management practices tailored to their needs, with assistance in aligning with the objectives of these C credit programs (Karthika, 2024; Meena and Pradhan, 2023). Nevertheless, field-level issues and effective C management strategies need to be addressed for the success of C credit programs in the field. This chapter will examine these issues in depth, including the challenges and remedies in C management at the field level. This chapter aims to deepen our understanding of how to optimize C credit initiatives during the transition to a sustainable and green economy by providing insights into field-specific challenges and strategies. Field-level strategies enhance C credit programs by optimizing soil health, crop rotation, livestock management, and water use. Precision agriculture aids data-driven decisions, ensuring effective C_{seq} and sustainability (Fig. 10.1).

10.2 Understanding field-level issues

10.2.1 Soil degradation and carbon loss

Soil health and C dynamics are critical concerns in agriculture, particularly due to soil degradation. Factors such as erosion, compaction, salinization, and loss of OM contribute to this degradation. The decline in soil C_{seq} and storage capacity due to soil deterioration inevitably results in C loss and reduced potential for GHG mitigation (Kane, 2015).

Both wind and water erosion degrade topsoil—the surface layer rich in OM and C. As topsoil erodes, the C it contains is released into the atmosphere as CO_2. The use of heavy machinery often compacts the soil, reducing its porosity, disrupting soil structure, and diminishing microbial activity, which in turn lowers the rate of C_{seq}. Additionally, these factors contribute to salt accumulation in the soil, affecting fertility and C storage (Okur and Örçen, 2020).

Combating soil degradation and enhancing artificial C_{seq} rely not only on reduced tillage, cover cropping, and organic amendments but also on selecting appropriate crop types. Effective soil management strategies for reducing GHG emissions include maintaining soil structure, increasing OM content, and stabilizing soil organic carbon (SOC) levels to enhance C storage (Hussain et al., 2021; Meena et al., 2023).

10.2.2 Impact of agricultural practices on carbon sequestration

Agricultural practices influence soil C storage both positively and negatively. Conservation tillage enhances C_{seq} through three key mechanisms: preserving soil organic matter (SOM), stimulating microbial activity, and forming stable SOC compounds. Conversely, frequent plowing and soil disturbance release stored C into the atmosphere (Bhattacharyya et al., 2022; Pradhan and Meena, 2023a).

Cover crops, particularly vegetable crops, are widely recognized for their role in increasing C_{seq}. Farmers plant cover crops, which may have no direct economic value, during periods when the primary crop is not growing to enhance soil C storage. These crops improve soil structure, reduce erosion, and increase SOM levels (Zhu et al., 2024).

Crop rotation is another effective strategy for capturing soil C. Growing multiple crop species strengthens soil health by introducing diverse root structures, which help prevent pest and disease buildup while improving soil composition. Similarly, agroforestry (AF), which integrates trees into agricultural landscapes, aids in C_{seq} by storing C in tree biomass and enhancing SOM (Jhariya et al., 2021a; Singh et al., 2020; Jhariya et al., 2021b; Meena, Lal, et al., 2024).

Livestock management is also essential in controlling C dynamics. Optimizing feed utilization and manure management reduces GHG emissions while enhancing C storage. Research by Osman et al. (2022) identified two key strategies for reducing methane emissions in cattle: enhancing feed efficiency and effectively managing manure.

10.2.3 Challenges in quantifying and verifying carbon emissions and sequestration at the field level

The accurate quantification and verification of C emissions and C_{seq} at the field level present significant challenges. Measuring C stocks and fluxes in soils requires sophisticated techniques and equipment, and results can be affected by spatial and temporal variability. Soil C measurements are often subject to sampling errors, and the effects of management practices on soil C can take years to become apparent (Stockmann et al., 2013; Meena et al., 2023; Pradhan and Meena, 2023a, 2023b).

Verification of C_{seq} and emission reductions is also complex. C credit programs require rigorous monitoring and reporting to ensure that claimed reductions are real, additional, and permanent. This involves using standardized methods for measurement, accounting for potential leakage (i.e., the displacement of emissions from one area to another), and ensuring that C_{seq} benefits are sustained over time (Haya et al., 2023). To address these challenges, advances in technology and methodologies are being developed. Remote sensing technologies, such as satellite imagery and drones, provide new methods for monitoring soil conditions and tracking changes in C stocks over large areas. More so, improved modeling techniques and data integration can enhance the accuracy of C quantification and verification (Xiong et al., 2023). Accurately measuring and verifying C_{seq} is challenging due to variability and long-term effects. C credit programs require strict monitoring, while advances in remote sensing and modeling improve accuracy (Fig. 10.2).

10.2.4 Carbon management and credit programs

C management and credit programs are designed to incentivize and reward agricultural practices that enhance C_{seq} and reduce GHG emissions. These programs operate by providing financial rewards for verified reductions in emissions or increases in C storage, creating a market for C credits that can be traded or sold. For C credit programs to be effective, they must address field-level issues and ensure that C management practices are implemented correctly and monitored accurately. This involves developing clear guidelines for implementing the practice, establishing robust verification procedures, and providing support and training to farmers and land managers (Oldfield et al., 2022).

Successful C management and credit programs also require collaboration between stakeholders, including farmers, policymakers, and researchers. This collaboration can help to align program goals with field-level realities, ensure that practices are feasible and beneficial, and facilitate the exchange of knowledge and best practices (Shirrell et al., 2023). However, understanding field-level issues and implementing effective C management strategies are crucial for the success

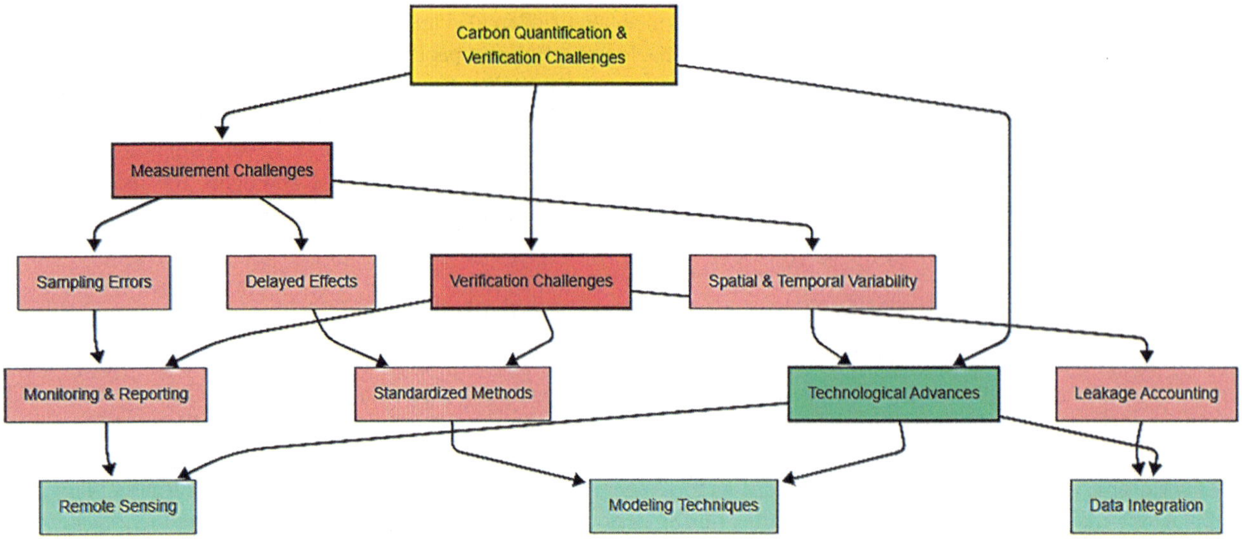

FIGURE 10.2 Carbon quantification and verification challenges.

of C credit programs in agriculture. By addressing soil degradation, optimizing agricultural practices, and overcoming challenges in quantification and verification, C credit programs can contribute to mitigating climate change and supporting the transition toward a green economy.

10.3 Carbon management strategies at the field level

10.3.1 Soil health improvement

10.3.1.1 Adoption of conservation tillage practices

Conservation tillage practices are so named because they minimize soil disturbance, thereby preserving or enhancing soil structure. Traditional tillage involves multiple layers of plowing and mixing the soil using various equipment. In contrast, conservation tillage includes methods such as no-till or minimum tillage, where the soil remains undisturbed mainly except for essential planting activities (Triplett, 2018).

The primary reason conservation tillage contributes to C_{seq} is its ability to preserve SOM by minimizing soil disturbance. This practice prevents the oxidation of SOC, allowing it to accumulate in the soil. Studies show that conservation tillage can increase SOC stocks by up to 25% compared with conventional tillage (Littrell et al., 2021). Moreover, it enhances soil structure, improves water infiltration, and minimizes erosion, all of which support soil health and increased C storage (Pradhan et al., 2023).

10.3.1.2 Organic matter addition through crop residue management

Crop residue management involves retaining or incorporating plant material left after harvest, a practice crucial for improving soil health and enhancing C_{seq}. Leaving crop residues on the field or incorporating them into the soil increases OM, thereby boosting soil fertility and C storage (Wang et al., 2020). Integrating crop residues into the soil supports long-term C_{seq} by fostering the development of stable SOM (Kumar et al., 2020). This approach also improves soil structure, stimulates microbial activity, and enhances nutrient cycling. Research suggests that leaving crop residues on the soil surface can increase SOC by 20% compared to incorporating them below the surface (Chen et al., 2016).

10.3.1.3 Cover cropping and green manure practices

Green manure involves growing specific crops that are later plowed into the soil, while cover cropping refers to cultivating plants that are not harvested but serve to protect and improve the soil. These practices significantly benefit soil health and C_{seq} (Scavo et al., 2022). Cover crops, including legumes, grasses, and brassicas, aid in OM accumulation and support C_{seq}. They also improve soil structure, minimize erosion, and enhance nutrient (Kumar et al., 2024). Likewise, green manure crops contribute significant OM and nutrients when incorporated into the soil. Research indicates that cover cropping and green manure practices can increase soil C stocks by up to 30% compared to bare fields (McClelland et al., 2021).

10.3.2 Precision agriculture techniques

10.3.2.1 Variable rate application of inputs

Variable-rate application (VRA) utilizes advanced technology to distribute inputs, such as fertilizers, pesticides, and water, at different rates across a field, adjusting to specific soil and crop needs. This precision approach optimizes input usage, reducing excess application and minimizing environmental impact (He, 2023).

By precisely targeting input applications, VRA helps to mitigate GHG emissions associated with excessive input use. For example, precision fertilization reduces nitrous oxide emissions resulting from excess nitrogen application. VRA also enhances soil health and crop productivity, indirectly supporting improved C management practices (Gorai et al., 2021).

10.3.2.2 Adoption of precision irrigation methods

Advanced irrigation methods, such as drip and sprinkler systems with automated control technologies, ensure precise water delivery according to crop needs. These systems improve water-use efficiency, enhance crop yields, and contribute to better C management (Zhang et al., 2021).

Efficient irrigation minimizes the risk of waterlogging, which can lead to increased emissions of methane and nitrous oxide. Precision irrigation supports healthy plant growth and enhances soil C_{seq}. Research indicates that precision irrigation can reduce GHG emissions by up to 20% compared to conventional irrigation methods (Jamali et al., 2021).

10.3.2.3 Remote sensing technologies for monitoring soil carbon levels

Remote sensing technologies, including satellites, drones, and ground-based sensors, enable the accurate monitoring of soil C levels and related parameters. These tools provide real-time data on soil conditions, crop health, and land use changes (Mandal et al., 2022).

By utilizing remote sensing, soil C stocks can be precisely estimated over time, allowing for improved C management and verification. Integrating remote sensing data with models and algorithms helps farmers and researchers in a better understanding of soil C dynamics and optimizes C_{seq} strategies (Li et al., 2023).

10.3.3 Agroforestry and silvopasture systems

10.3.3.1 Integration of trees and shrubs into agricultural landscapes

AF involves integrating tree-based systems into agricultural landscapes, offering both ecological and economic benefits, including enhanced C_{seq} (Abdulsalam et al., 2024).

Trees and shrubs capture atmospheric CO_2 in their biomass and root systems, contributing to the overall C_{seq}. AFs also improve soil structure, reduce erosion, and support biodiversity. Research shows that AF practices can sequester up to 50% more C than conventional agricultural systems (De Stefano and Jacobson, 2018).

10.3.3.2 Benefits of agroforestry in carbon sequestration

AFs offer significant benefits for C_{seq}. The presence of trees and shrubs in agricultural landscapes increases the total C pool by capturing and storing C in both aboveground biomass and soil OM. Moreover, AF practices can enhance soil fertility and structure, further supporting C_{seq} (Lorenz and Lal, 2014). Trees in AFs also provide shade and windbreaks, which can reduce crop stress and improve productivity. By integrating trees into agricultural systems, farmers can achieve both environmental and economic benefits while contributing to CCM (Lasco et al., 2014).

10.3.3.3 Carbon credit opportunities for agroforestry practices

AF practices present valuable opportunities for C credit programs. By implementing AFs, land managers can generate C credits through verified increases in C_{seq}. The credits from VCS are sold or traded in C-markets as a financial incentive to adopt sustainable land management practices (Narassimhan et al., 2018). To be part of the C credit programs, AF practices must be accurately monitored and verified. It entails evaluating the C_{seq} potential of pre-planted trees and shrubs of the AFs and the overall contribution of such systems to the soil C stock and GHG emissions (Nath et al., 2021). Field-level C management enhances C_{seq} through conservation tillage, crop residue management, and cover cropping. Precision agriculture optimizes inputs, while AF boosts C storage and sustainability (Table 10.1).

10.3.4 Livestock management practices

10.3.4.1 Implementing rotational grazing systems

Rotational grazing involves dividing livestock among multiple pasture areas, allowing grasslands to recover and regenerate. This method supports both the maintenance of healthy grasslands and the preservation of soil structure while promoting C_{seq} (Meena et al., 2019). By implementing rotational grazing, soil conditions improve as erosion decreases and OM increases, effectively managing animal grazing on pastures. This approach not only enhances pasture health but also strengthens grassland C storage capacity. Research indicates that soil C reserves increase by 15% under rotational grazing compared to continuous grazing (Alves et al., 2020).

10.3.4.2 Methane reduction strategies in livestock farming

Livestock, primarily through enteric fermentation during digestion, is a major source of methane emissions, contributing significantly to GHGs. Mitigation strategies include optimizing feed composition, incorporating feed additives, and improving manure management (Khoshnevisan et al., 2021). Additives like tannins and seaweed can help reduce methane production during digestion. Enhancing feed efficiency and refining livestock management practices further minimizes emissions. Moreover, manure management techniques such as anaerobic digestion and composting can lower methane release while promoting C_{seq} in manure (Osman et al., 2022).

10.3.4.3 Utilization of manure management techniques for carbon sequestration

Effective manure management is essential for C management. Techniques such as composting and anaerobic digestion convert manure into valuable organic fertilizers. These methods not only mitigate GHG emissions but also promote C_{seq}

TABLE 10.1 The key strategies and benefits of agroforestry and carbon sequestration.

Agroforestry strategy	Description	Carbon sequestration impact
Integration of trees and shrubs	Combining trees with crops to enhance biodiversity and soil health.	Sequesters up to 50% more C than conventional agriculture.
Agroforestry benefits	Improves soil fertility, provides windbreaks, and reduces crop stress.	Increases C storage in biomass and soil.
C credit opportunities	Enables participation in C markets through verified sequestration.	Generates financial incentives for sustainable practices.

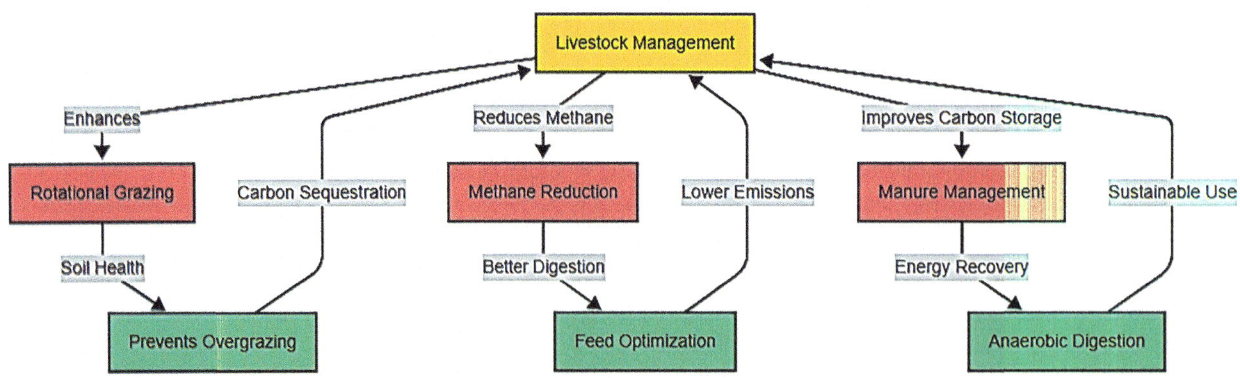

FIGURE 10.3 Livestock management and carbon sequestration strategies.

by stabilizing OM and enriching soil C levels (Navarro-Pedreño et al., 2021). Composting manure improves its nutrient content and reduces its potential for GHG emissions. Anaerobic digestion captures methane produced during the decomposition of manure, which can be utilized as a renewable energy source. Both techniques contribute to better C management and support sustainable agricultural practices (Obileke et al., 2021). Effective manure management enhances C_{seq} by converting waste into organic fertilizers through composting or anaerobic digestion. Composting enriches soil C, while anaerobic digestion captures methane for renewable energy, reducing GHG emissions and promoting sustainable agriculture (Fig. 10.3).

10.4 Challenges and opportunities

10.4.1 Economic viability of implementing carbon management practices

The financial feasibility of adopting C management practices poses a major challenge for many farmers and land managers. The upfront costs of implementing new technologies or practices can be high, while the economic returns from C credits may be delayed or uncertain.

- **Initial costs and investment**: Implementing practices such as conservation tillage, precision agriculture, or AF often requires upfront investment in equipment, technology, or infrastructure. For instance, adopting precision irrigation systems or installing remote sensing technologies can involve significant capital expenditure (Musa et al., 2024). Moreover, converting to conservation tillage or AFs may require adjustments to existing infrastructure or management practices.
- **Return on investment**: The financial returns from C credit programs may take time to materialize. The process of monitoring, reporting, and verifying C_{seq} or emission reductions is complex and can incur ongoing costs. Moreover, the C market can be volatile, with fluctuating prices for C credits that may affect the profitability of participating in these programs (Demiralay et al., 2022).
- **Economic risks and uncertainties**: Farmers may face risks stemming from variability in C credit prices and potential policy changes. Uncertainty in C markets and fluctuating prices can make it challenging to predict financial outcomes and may deter some farmers from investing in C management practices (Paul et al., 2023).

10.4.2 Technical feasibility and scalability issues

While the technical feasibility of many C management practices is well established, scaling these practices to larger areas or diverse agricultural systems presents several challenges.

- **Adaptability to different conditions:** Many C management practices need to be tailored to specific local conditions, including soil type, climate, and crop types. This can make it challenging to implement a one-size-fits-all solution across different regions or agricultural systems. For example, conservation tillage may be effective in some soil types but not in others due to differences in soil texture or structure (Zhang et al., 2021).
- **Technological limitations:** The adoption of precision agriculture technologies, such as remote sensing or variable rate application systems, may be limited by technological constraints or a lack of infrastructure. In some regions, access to advanced technologies and technical support may be restricted, hindering the widespread implementation of these practices (Greenhalgh et al., 2017).
- **Integration with existing practices:** Integrating new C management practices with existing agricultural systems can be challenging. Farmers may need to adjust their management practices, adapt to new technologies, or reconfigure their operations to accommodate C management strategies. This integration can require additional time, training, and resources (Stringer et al., 2020).

10.4.3 Policy and regulatory support for incentivizing field-level carbon management

Effective policy and regulatory support are crucial for incentivizing and promoting field-level C management practices. Policymakers play a key role in shaping the incentives, regulations, and frameworks that influence the adoption of C management practices.

- **C pricing and financial incentives:** Policies that incorporate financial mechanisms, such as C taxes and cap-and-trade systems, place a monetary value on C emissions, encouraging emission reductions. These frameworks create economic incentives for farmers to adopt practices that reduce emissions or improve C_{seq} (Green, 2021).
- **Subsidies and grants**: Government subsidies or grants can help offset the initial costs of implementing C management practices. Financial support can make it more feasible for farmers to implement new technologies or practices and

participate in C credit programs. For example, subsidies for precision agriculture equipment or grants for AF projects can encourage adoption and facilitate the transition to more sustainable practices (Livinus et al., 2024).
- **Technical assistance and training**: Providing technical assistance and training to farmers is essential for the successful implementation of C management practices. Extension services, workshops, and technical support can help farmers understand and adopt new practices, improve their management strategies, and navigate the complexities of C credit programs (Stringer et al., 2020).
- **Regulatory frameworks and standards**: Transparent and consistent regulatory frameworks are essential for maintaining the credibility and effectiveness of C credit programs. Establishing standardized methods for measuring, reporting, and verifying C_{seq} and emissions reductions can strengthen the integrity of C markets and foster trust among stakeholders (Ahonen et al., 2022).
- **Supporting research and innovation**: Investing in research and innovation can help address technical challenges and improve the scalability of C management practices. Research on new technologies, practices, and methodologies can lead to more effective and cost-efficient solutions for C management in agriculture (Mills et al., 2020).

10.4.4 Opportunities

Despite the challenges, there are significant opportunities for advancing C management in agriculture:

- **Expanded market access:** The rising demand for sustainable and climate-conscious products creates opportunities for farmers to tap into new markets and secure premium prices for their goods. Additionally, C-credit programs can provide supplementary revenue streams, thereby reinforcing the financial sustainability of eco-friendly agricultural practices (Paul et al., 2023).
- **Technological innovations:** Cutting-edge advancements, including precision agriculture tools, remote sensing, and data analytics, enhance C management strategies and improve their efficiency. These technologies offer critical insights into soil health, crop productivity, and C dynamics, facilitating more informed decision-making (Stringer et al., 2020).
- **Collaborative networks:** Partnerships among farmers, researchers, policymakers, and industry leaders encourage innovation, knowledge exchange, and the development of supportive policies for C management. Such collaborations can also help scale up successful practices and address shared challenges (Amin et al., 2020).
- **Public engagement and advocacy:** Growing public awareness and support for climate-friendly agriculture drive the adoption of C management practices. Consumer demand for environmentally responsible farming encourages farmers to adopt sustainable practices and participate in Ccredit programs (Zhu and Chen, 2024).

10.5 Case studies and success stories

Examples of successful C credit programs in agriculture

- **The Verified Carbon Standard program**

The Verified Carbon Standard (VCS) program is a widely recognized C credit certification program that includes a range of agricultural projects. One notable example is the Africa C exchange in Kenya, which focuses on soil C_{seq} projects. By implementing conservation tillage and improved land management practices, the project has successfully generated C credits and provided financial incentives for farmers (Bryan et al., 2010). The project has demonstrated significant soil C_{seq}, increased soil fertility, and enhanced farm productivity.

- **The American Carbon Registry program**

The American Carbon Registry (ACR) program has supported several successful agricultural C projects, including the California Almond Orchard Project. This project, based in California, focuses on soil health improvement through practices such as cover cropping and reduced tillage. By adopting these practices, the project has achieved substantial Cseq and earned C credits, which are traded in the California C market. The success of this project highlights the potential for integrating C management practices into commercial agriculture while generating economic benefits (Kaarakka et al., 2023).

- **The Climate Action Reserve program**

The Climate Action Reserve (CAR) program has played a crucial role in advancing the Sustainable Agriculture Initiative in Mexico. This initiative focuses on AF and silvopasture systems, incorporating trees and shrubs into agricultural landscapes to enhance Cseq. The project has effectively generated C credits while also enhancing soil health, improving water retention, and promoting biodiversity. It stands as a model for demonstrating how AF practices can support C management and environmental sustainability (Carlson, 2012).

10.6 Lessons learned from implementing field-level carbon management strategies

10.6.1 Importance of local adaptation

The successful implementation of C management strategies requires adaptation to local conditions. Projects that have considered local soil types, climate, and agricultural practices have generally been more successful. For instance, the effectiveness of conservation tillage or cover cropping can vary based on soil characteristics and climate conditions (Büchi et al., 2018). Tailoring strategies to specific local contexts enhance their effectiveness and adoption rates.

10.6.2 Need for comprehensive monitoring and verification

Accurate monitoring and verification are critical for the success of C credit programs. Successful projects have invested in robust monitoring systems and verification processes to ensure that C_{seq} claims are accurate and credible. For example, the VCS program emphasizes rigorous verification protocols to maintain the integrity of C credits and build trust among stakeholders (Battocletti et al., 2024).

10.6.3 Engagement and training of farmers

Effective implementation of C management practices often requires significant engagement and training of farmers. Projects that have provided training, support, and technical assistance to farmers have generally seen higher adoption rates and more successful outcomes. For instance, the ACR program's California Almond Orchard Project included farmer training as a key component of its success (Ory and Iles, 2021).

10.7 Research and development toward carbon management and carbon credits in agriculture

Research and development (R&D) plays a crucial role in advancing C management and C credit programs in agriculture. Key areas of focus include:

10.7.1 Development of new technologies

Ongoing R&D efforts are focused on developing new technologies for C measurement, monitoring, and verification. Innovations in remote sensing, soil sensors, and data analytics are improving the accuracy and efficiency of C assessments. For example, advancements in satellite technology and drones are providing new ways to monitor soil C levels and track changes over time (Inoue, 2020).

10.7.2 Improvement of carbon sequestration practices

Research is also focused on improving C_{seq} practices. Studies are exploring new crop varieties, soil management techniques, and AFs that enhance C storage. For instance, research on cover crop species and their impact on soil C_{seq} is helping to identify the most effective practices for different agricultural systems (Mandal et al., 2022).

10.7.3 Assessment of economic and environmental impacts

R&D efforts are assessing the economic and environmental impacts of C management practices. This includes evaluating the cost-effectiveness of different practices, their impact on farm productivity, and their contribution to overall GHG reduction goals. Understanding these impacts helps to refine C credit programs and ensure their effectiveness (Cornish and Miao, 2024).

10.8 Policy perspectives for carbon management and carbon credit in agriculture

Effective policy frameworks are essential for supporting C management and C credit programs in agriculture. Key policy considerations include:

10.8.1 Carbon pricing mechanisms

C pricing mechanisms, such as C taxes and cap-and-trade systems, provide financial incentives to encourage C management practices. These policies create a credit market that enables farmers to invest in strategies that either reduce emissions or enhance C storage, offering economic benefits in the process (Mandal et al., 2022).

10.8.2 Support for research and innovation

Policies that support research and innovation in C management are crucial for advancing the field. Government funding for research projects, the development of new technologies, and pilot programs can drive progress and improve the effectiveness of C management practices (Weiss et al., 2021).

10.8.3 Implementation of clear standards and regulations

Establishing clear standards and regulations for C credits is crucial to ensuring the credibility and effectiveness of C credit programs. Policies should define uniform methodologies for measuring, reporting, and verifying C_{seq} and emissions reductions. Consistent regulations enhance stakeholder confidence and uphold the integrity of C markets (Battocletti et al., 2024).

10.9 Future of carbon credit programs in agriculture

The future of C credit programs in agriculture holds significant promise:

10.9.1 Expansion of carbon markets

As awareness of climate change and sustainability grows, C markets are expected to expand, providing more opportunities for farmers to participate in C credit programs. This expansion will likely include new market mechanisms and trading platforms, increasing accessibility and participation (Paul et al., 2023).

10.9.2 Integration with broader sustainability goals

C credit programs will increasingly be integrated with broader sustainability goals, including biodiversity conservation, water management, and soil health. By aligning C management practices with other environmental objectives, programs can provide more comprehensive benefits and enhance overall sustainability (Lawal et al., 2023).

10.9.3 Technological advancements

Advancements in technology will continue to drive innovation in C management and credit programs. Emerging technologies, such as artificial intelligence and blockchain, have the potential to improve the monitoring, verification, and trading of C credits, making programs more efficient and transparent (Stringer et al., 2020).

TABLE 10.2 The future of carbon credit programs in agriculture.

Future aspect	Description	Impact on carbon credit programs
Expansion of C markets	Growth in market mechanisms and trading platforms.	Increased participation and accessibility for farmers.
Integration with sustainability goals	Aligning C credits with biodiversity, water, and soil health.	More comprehensive environmental benefits.
Technological advancements	Use of AI, blockchain, and digital tools for monitoring.	Improved efficiency, transparency, and verification.
Increased policy support	Stronger government incentives and regulatory frameworks.	Enhanced program effectiveness and financial support.

10.9.4 Increased policy support

Ongoing policy support and development will play a crucial role in shaping the future of C credit programs. Governments and international bodies are expected to continue developing supportive policies, providing financial incentives, and establishing regulations that support the growth and effectiveness of C credit programs (Zhu and Chen, 2024). The future of C credit programs in agriculture is promising, with expanding C markets, integration with broader sustainability goals, and technological advancements, such as AI and blockchain, enhancing efficiency. Increased policy support will further drive participation and effectiveness (Table 10.2).

10.10 Conclusion

This chapter provides a detailed analysis of C management strategies at the field level to ensure the effective implementation of agricultural C credit programs. Field studies indicate that adopting conservation tillage, cover cropping, and precision agriculture can enhance soil C storage while reducing greenhouse gas emissions. These techniques offer dual benefits by contributing to CCM while simultaneously improving soil quality and agricultural productivity. However, several challenges persist. The high initial costs of launching C credit initiatives, coupled with the unpredictability of C credit revenue, pose significant economic barriers. More so, technical challenges arise in adapting practices to regional conditions and integrating new technologies into existing systems. Programs such as the VCS, the ACR, and the CAR demonstrate that C management strategies can deliver environmental benefits when successfully implemented. Future directions for enhancing field-level C management involve expanding and innovating within C markets to increase accessibility and efficiency. Integrating C management practices with broader sustainability goals, such as biodiversity and water management, will provide more comprehensive benefits. Advancements in technology, including precision agriculture tools and remote sensing, will further improve the accuracy and effectiveness of C management. Supportive policies and regulatory frameworks will be crucial in driving adoption, providing financial incentives, and establishing clear standards for C credits. Moreso, investing in capacity building and engaging farmers through training and support will be essential for successful implementation. In summary, advancing field-level C management strategies offers significant opportunities for mitigating climate change, enhancing soil health, and promoting agricultural sustainability. By addressing the associated challenges and leveraging emerging opportunities, stakeholders can drive meaningful progress toward a greener and more resilient agricultural sector.

References

Abdulsalam, M., Elelu, S.A., Innocent, M.O., Ibrahim, G.O., Livinus, M.U., Lateefat, S.O., et al., Aransiola, S.A., Babaniyi, B.R., Aransiola, A.B., Maddela, N.R., 2024. Women empowerment in environmental conservation. In: Aransiola, S.A., Babaniyi, B.R., Aransiola, A.B., Maddela, N.R. (Eds.), Prospects for Soil Regeneration and Its Impact on Environmental Protection. Earth and Environmental Sciences Library. Springer, Cham, pp. 351–370. Available at https://doi.org/10.1007/978-3-031-53270-2_16.

Ahonen, H.M., Kessler, J., Michaelowa, A., Espelage, A., Hoch, S., 2022. Governance of fragmented compliance and voluntary carbon markets under the paris agreement. Politics and Governance 10 (1), 235–245. Available at https://doi.org/10.17645/pag.v10i1.4759.

Alves, L.A., Denardin, L.Gd.O., Martins, A.P., Bayer, C., Veloso, M.G., Bremm, C., et al., 2020. The effect of crop rotation and sheep grazing management on plant production and soil C and N stocks in a long-term integrated crop-livestock system in Southern Brazil. Soil and Tillage Research 203. Available at https://doi.org/10.1016/j.still.2020.104678.

Amin, M.N., Hossain, M.S., Lobry de Bruyn, L., Wilson, B., 2020. A systematic review of soil carbon management in Australia and the need for a social-ecological systems framework. Science of the Total Environment 719. Available at https://doi.org/10.1016/j.scitotenv.2019.135182.

Battocletti, V., Enriques, L., Romano, A., 2024. The voluntary carbon market: market failures and policy implications. University of Colorado Law Review 95, 519.

Bhattacharyya, S.S., Leite, F.F.G.D., France, C.L., Adekoya, A.O., Ros, G.H., de Vries, W., et al., 2022. Soil carbon sequestration, greenhouse gas emissions, and water pollution under different tillage practices. Science of the Total Environment 826. Available at https://doi.org/10.1016/j.scitotenv.2022.154161.

Bryan, E., Akpalu, W., Yesuf, M., Ringler, C., 2010. Global carbon markets: opportunities for sub-Saharan Africa in agriculture and forestry. Climate and Development 2 (4), 309–331. Available at https://doi.org/10.3763/cdev.2010.0057.

Büchi, L., Wendling, M., Amossé, C., Necpalova, M., Charles, R., 2018. Covering crops is important in alleviating the negative effects of reduced soil tillage and promoting soil fertility in a winter wheat cropping system. Agriculture, Ecosystems & Environment 256, 92–104.

Carlson, A.E., 2012. Regulatory capacity and state environmental leadership: california's climate policy. Fordham Environmental Law Review. 24, 63.

Chen, S., Xu, C., Yan, J., Zhang, X., Zhang, X., Wang, D., 2016. The influence of the type of crop residue on soil organic carbon fractions: an 11-year field study of rice-based cropping systems in southeast China. Agriculture, Ecosystems and Environment 223, 261–269. Available at https://doi.org/10.1016/j.agee.2016.03.009.

Cornish, B., Miao, R., 2024. Mitigating Climate Change with the Conservation Reserve Program (CRP): The Role of Carbon Credits and CRP Redesign. AgEcon Search.

Demiralay, S., Gencer, H.G., Bayraci, S., 2022. Carbon credit futures as an emerging asset: hedging, diversification, and downside risks. Energy Economics 113, 106196.

De Stefano, A., Jacobson, M.G., 2018. Soil carbon sequestration in agroforestry systems: a meta-analysis. Agroforestry Systems 92 (2), 285–299. Available at https://doi.org/10.1007/s10457-017-0147-9.

Gorai, T., Kumar Yadav, P., Choudhary, G.L., Kumar, A., 2021. Site-specific crop nutrient management for precision agriculture–a review. Current Journal of Applied Science and Technology 37–52. Available at https://doi.org/10.9734/cjast/2021/v40i1031357.

Green, J.F., 2021. Does carbon pricing reduce emissions? A review of ex-post analyses. Environmental Research Letters 16 (4), 043004. Available at https://doi.org/10.1088/1748-9326/abdae9.

Greenhalgh, T., Wherton, J., Papoutsi, C., Lynch, J., Hughes, G., A'Court, C., et al., 2017. Beyond adoption: a new framework for theorizing and evaluating nonadoption, abandonment, and challenges to the scale-up, spread, and sustainability of health and care technologies. Journal of Medical Internet Research 19 (11), e367. Available at https://doi.org/10.2196/jmir.8775.

Haya, B.K., Evans, S., Brown, L., Bukoski, J., Butsic, V., Cabiyo, B., et al., 2023. Comprehensive review of carbon quantification by improved forest management offset protocols. Frontiers in Forests and Global Change 6. Available at https://doi.org/10.3389/ffgc.2023.958879.

He, L., 2023. Variable rate technologies for precision agriculture. In: Zhang, Q. (Ed.), Encyclopedia of Digital Agricultural Technologies. Springer, Cham, pp. 1533–1542. Available at https://doi.org/10.1007/978-3-031-24861-0_34.

Hussain, S., Hussain, S., Guo, R., Sarwar, M., Ren, X., Krstic, D., et al., 2021. Carbon sequestration to avoid soil degradation: a review on the role of conservation tillage. Plants 10 (10), 2001. Available at https://doi.org/10.3390/plants10102001.

Inoue, Y., 2020. Satellite- and drone-based remote sensing of crops and soils for smart farming – a review. Soil Science and Plant Nutrition 66 (6), 798–810. Available at https://doi.org/10.1080/00380768.2020.1738899.

Jamali, M., Soufizadeh, S., Yeganeh, B., Emam, Y., 2021. A comparative study of irrigation techniques for energy flow and greenhouse gas (GHG) emissions in wheat agroecosystems under contrasting environments in south of Iran. Renewable and Sustainable Energy Reviews 139, 110704. Available at https://doi.org/10.1016/j.rser.2021.110704.

Jhariya, M.K., Banerjee, A., Swaroop Meena, R., Kumar, S., Raj, A., 2021. Sustainable Intensification for Agroecosystem Services and Management. Springer, Singapore, pp. 1–870. Available at https://doi.org/10.1007/978-981-16-3207-5.

Jhariya, M.K., Meena, R.S., Banerjee, A., 2021. Ecological Intensification of Natural Resources for Sustainable Agriculture. Springer, Singapore, pp. 1–655. Available at 10.1007/978-981-33-4203-3.

Jhariya, M.K., Meena, R.S., Banerjee, A., Kumar, S., Raj, A., 2023. Agroforestry for Carbon and Ecosystem Management. Academic Press, pp. 1–434. Available at 10.1016/C2021-0-02916-3.

Jhariya, M.K., Meena, R.S., Banerjee, A., Meena, S.N., 2022. Natural Resources Conservation and Advances for Sustainability. Elsevier. Available at https://doi.org/10.1016/C2019-0-03763-6.

Kaarakka, L., Rothey, J., Dee, L.E., Ashraf, M.I., 2023. Managing forests for carbon–status of the forest carbon offset markets in the United States. PLOS Climate 2 (7), e0000158. Available at https://doi.org/10.1371/journal.pclm.0000158.

Kabange, N.R., Kwon, Y., Lee, S.M., Kang, J.W., Cha, J.K., Park, H., et al., 2023. Mitigating greenhouse gas emissions from crop production and management practices, and livestock: a review. Sustainability (Switzerland) 15 (22), 15889. Available at https://doi.org/10.3390/su152215889.

Kane, D., 2015. Carbon sequestration potential on agricultural lands: a review of current science and available practices. National Sustainable Agriculture Coalition Breakthrough Strategies and Solutions, LLC, pp. 1–35.

Karthika, R., 2024. Chapter 8 - Role of precision agriculture in soil fertility and its application to farmers. In: Lamine, S., Srivastava, P.K., Kayad, A., Muñoz-Arriola, F., Pandey, P.C. (Eds.), Remote Sensing in Precision Agriculture: Transforming Scientific Advancement Into Innovation. Academic Press, pp. 137–147. Available at https://doi.org/10.1016/B978-0-323-91068-2.00017-5.

Khoshnevisan, B., Duan, N., Tsapekos, P., Awasthi, M.K., Liu, Z., Mohammadi, A., et al., 2021. A critical review on livestock manure biorefinery technologies: sustainability, challenges, and future perspectives. Renewable and Sustainable Energy Reviews 135, 110033.

Kumar, S., Meena, R.S., Jhariya, M.K., 2020. Resources use Efficiency in Agriculture. Springer, Singapore, pp. 1–760. Available at https://doi.org/10.1007/978-981-15-6953-1.

Kumar, S., Meena, R.S., Sheoran, P., Jhariya, M.K., 2024. Regenerative Agriculture for Sustainable Food Systems. Springer, Singapore, pp. 1–493. Available at https://doi.org/10.1007/978-981-97-6691-8.

Lasco, R.D., Delfino, R.J.P., Catacutan, D.C., Simelton, E.S., Wilson, D.M., 2014. Climate risk adaptation by smallholder farmers: the roles of trees and agroforestry. Current Opinion in Environmental Sustainability 6 (1), 83–88. Available at https://doi.org/10.1016/j.cosust.2013.11.013.

Lawal, T.O., Abdulsalam, M., Mohammed, A., Sundararajan, S., 2023. Economic and environmental implications of sustainable agricultural practices in arid regions: a cross-disciplinary analysis of plant science, management, and economics. International Journal of Membrane Science and Technology 10 (3), 3100–3114. Available at https://doi.org/10.15379/ijmst.v10i3.3027.

Li, T., Xia, A., McLaren, T.I., Pandey, R., Xu, Z., Liu, H., et al., 2023. Preliminary results in innovative solutions for soil carbon estimation: integrating remote sensing, machine learning, and proximal sensing spectroscopy. Remote Sensing 15 (23), 5571. Available at https://doi.org/10.3390/rs15235571.

Littrell, J., Xu, S., Omondi, E., Saha, D., Lee, J., Jagadamma, S., 2021. Long-term organic management combined with conservation tillage enhanced soil organic carbon accumulation and aggregation. Soil Science Society of America Journal 85 (5), 1741–1754. Available at https://doi.org/10.1002/saj2.20259.

Livinus, M.U., Bala, S.Z., Abdulsalam, M., Innocent, M.O., Hassan, M., Kini, P., Aransiola, S.A., Babaniyi, B.R., Aransiola, A.B., Maddela, N.R., 2024. Natural occurrences of soil dilapidation. In: Aransiola, S.A., Babaniyi, B.R., Aransiola, A.B., Maddela, N.R. (Eds.), Prospects for Soil Regeneration and Its Impact on Environmental Protection. Earth and Environmental Sciences Library. Springer, Cham, pp. 205–223. Available at https://doi.org/10.1007/978-3-031-53270-2_9.

Lorenz, K., Lal, R., 2014. Soil organic carbon sequestration in agroforestry systems. A review. Agronomy for Sustainable Development 34, 443–454.

Mandal, A., Majumder, A., Dhaliwal, S.S., Toor, A.S., Mani, P.K., Naresh, R.K., et al., 2022. Impact of agricultural management practices on soil carbon sequestration and its monitoring through simulation models and remote sensing techniques: a review. Critical Reviews in Environmental Science and Technology 52 (1), 1–49. Available at https://doi.org/10.1080/10643389.2020.1811590.

McClelland, S.C., Paustian, K., Schipanski, M.E., 2021. Management of cover crops in temperate climates influences soil organic carbon stocks: a meta-analysis. Ecological Applications 31 (3). Available at https://doi.org/10.1002/eap.2278.

McGuire, R., Williams, P.N., Smith, P., McGrath, S.P., Curry, D., Donnison, I., et al., 2022. Potential co-benefits and trade-offs between improved soil management, climate change mitigation and agri-food productivity. Food and Energy Security 11 (2). Available at https://doi.org/10.1002/fes3.352.

Meena, R.S., Pradhan, G., 2023. Industrial garbage-derived biocompost enhances soil organic carbon fractions, CO_2 biosequestration, potential carbon credits and sustainability index in a rice-wheat ecosystem. Environmental Research 235, 116525. Available at https://doi.org/10.1016/j.envres.2023.116525.

Meena, R.S., Pradhan, G., Kumar, S., Lal, R., 2023. Using industrial wastes for rice-wheat cropping and food-energy-carbon-water-economic nexus to the sustainable food system. Renewable and Sustainable Energy Reviews 187, 113756. Available at https://doi.org/10.1016/j.rser.2023.113756.

Meena, R.S., Kumar, S., Yadav, G.S., 2019. Soil carbon sequestration in crop production. In: Meena, R. (Ed.), Nutrient Dynamics for Sustainable Crop Production. Springer, Singapore, pp. 1–39. Available at https://doi.org/10.1007/978-981-13-8660-2_1.

Meena, R.S., Lal, R., Kumar, S., Pradhan, G., Srinivasarao, C., Singh, A.K., et al., 2024. Potential of Indian agriculture for capturing atmospheric CO_2 and monetizing carbon credits to the farmers: an overview and policy framework. Advances in Agronomy 188, 101–206. Available at https://doi.org/10.1016/bs.agron.2024.06.001.

Meena, R.S., Pradhan, G., Kumar, S., 2025. Energy flow, eco-efficiency, and economic circulation with recycled industrial waste compost application in wheat and subsequent rice farming. Science of the Total Environment 967. Available at https://doi.org/10.1016/j.scitotenv.2025.178779.

Meena, R.S., Pradhan, G., Singh, K., Kumar, S., Singh, A.K., Shashidhar, K.S., et al., 2024a. Agriculture models for restoring degraded land to enhance CO_2 biosequestration and carbon credits in the Vindhyan region of India. Science of the Total Environment 929. Available at https://doi.org/10.1016/j.scitotenv.2024.172661.

Meena, R.S., Pradhan, G., Nalani, Kumar, S., Singh, A.K., Verma, S.K., et al., 2024b. Diversified cropping modules designed for soil restoration, CO_2 sequestration, and generating carbon credits. Land Degradation and Development 35 (14), 4314–4329. Available at https://doi.org/10.1002/ldr.5224.

Meena, R.S., Pradhan, G.S., Kumar, S., Jhariya, M.K., Rao, C.S., 2025. Carbon-Negative Agriculture: Sustainability Sciences in Asia and Africa. Springer, Singapore, pp. 1–500. Available at https://doi.org/10.1007/978-981-95-1483-0.

Mehmood, F., Wang, G., Abubakar, S.A., Zain, M., Rahman, S.U., Gao, Y., et al., 2023. Optimizing irrigation management sustained grain yield, crop water productivity, and mitigated greenhouse gas emissions from the winter wheat field in North China Plain. Agricultural Water Management 290. Available at https://doi.org/10.1016/j.agwat.2023.108599.

Mills, J., Ingram, J., Dibari, C., Merante, P., Karaczun, Z., Molnar, A., et al., 2020. Barriers to and opportunities for the uptake of soil carbon management practices in European sustainable agricultural production. Agroecology and Sustainable Food Systems 44 (9), 1185–1211. Available at https://doi.org/10.1080/21683565.2019.1680476.

Musa, I.O., Samuel, J.O., Adams, M., Abdulsalam, M., Nathaniel, V., Maude, A.M., et al., 2024. Soil erosion, mineral depletion and regeneration. In: Aransiola, S.A., Babaniyi, B.R., Aransiola, A.B., Maddela, N.R. (Eds.), Prospects for Soil Regeneration and Its Impact on Environmental Protection. Earth and Environmental Sciences Library. Springer, Cham, pp. 159–172. Available at https://doi.org/10.1007/978-3-031-53270-2_7.

Narassimhan, E., Gallagher, K.S., Koester, S., Alejo, J.R., 2018. Carbon pricing in practice: a review of existing emissions trading systems. Climate Policy 8 (8), 967–991. Available at https://doi.org/10.1080/14693062.2018.1467827.

Nath, A.J., Sileshi, G.W., Laskar, S.Y., Pathak, K., Reang, D., Nath, A., et al., 2021. Quantifying carbon stocks and sequestration potential in agroforestry systems under divergent management scenarios relevant to India's Nationally determined contribution. Journal of Cleaner Production 281. Available at https://doi.org/10.1016/j.jclepro.2020.124831.

Navarro-Pedreño, J., Almendro-Candel, M.B., Zorpas, A.A., 2021. The increase of soil organic matter reduces global warming, myth or reality? Science 3 (1), 18. Available at https://doi.org/10.3390/sci3010018.

Obileke, K., Nwokolo, N., Makaka, G., Mukumba, P., Onyeaka, H., 2021. Anaerobic digestion: technology for biogas production as a renewable energy source—a review. Energy & Environment 32 (2), 191–225.

Okur, B., Örçen, N., 2020. Soil salinization and climate change. In: Climate Change and Soil Interactions. Elsevier, pp. 331–350.

Oldfield, E.E., Lavallee, J.M., Kyker-Snowman, E., Sanderman, J., 2022. The need for knowledge transfer and communication among stakeholders in the voluntary carbon market. Biogeochemistry 161 (1), 41–46. Available at https://doi.org/10.1007/s10533-022-00950-8.

Ory, J., Iles, A., 2021. Fostering healthy soils in California: farmer motivations and barriers. In: Gemmill-Herren, B., Baker, L.E., Daniels, P.A. (Eds.), True Cost Accounting for Food: Balancing the Scale. Routledge, London, pp. 137–146.

Osman, A.I., Fawzy, S., Farghali, M., El-Azazy, M., Elgarahy, A.M., Fahim, R.A., et al., 2022. Biochar for agronomy, animal farming, anaerobic digestion, composting, water treatment, soil remediation, construction, energy storage, and carbon sequestration: a review. Environmental Chemistry Letters 20 (4), 2385–2485. Available at https://doi.org/10.1007/s10311-022-01424-x.

Paul, C., Bartkowski, B., Dönmez, C., Don, A., Mayer, S., Steffens, M., et al., 2023. Carbon farming: are soil carbon certificates a suitable tool for climate change mitigation? Journal of Environmental Management 330, 117142. Available at https://doi.org/10.1016/j.jenvman.2022.117142.

Powlson, D.S., Stirling, C.M., Thierfelder, C., White, R.P., Jat, M.L., 2016. Does conservation agriculture deliver climate change mitigation through soil carbon sequestration in tropical agro-ecosystems? Agriculture, Ecosystems and Environment 220, 164–174. Available at https://doi.org/10.1016/j.agee.2016.01.005.

Pradhan, G., Meena, R.S., 2023a. Interaction impact of biocompost on nutrient dynamics and relations with soil biota, carbon fractions index, societal value of CO_2 equivalent and ecosystem services in the wheat-rice farming. Chemosphere 339. Available at https://doi.org/10.1016/j.chemosphere.2023.139695.

Pradhan, G., Meena, R.S., 2023b. Utilizing waste compost to improve the atmospheric CO_2 capturing in the rice-wheat cropping system and energy-cum-carbon credit auditing for a circular economy. Science of the Total Environment 892. Available at https://doi.org/10.1016/j.scitotenv.2023.164572.

Pradhan, G., Meena, R.S., Kumar, S., Lal, R., 2023. Utilizing industrial wastes as compost in wheat-rice production to improve the above and below-ground ecosystem services. Agriculture, Ecosystems & Environment 358, 108704. Available at https://doi.org/10.1016/j.agee.2023.108704.

Qin, Z., Guan, K., Zhou, W., Peng, B., Tang, J., Jin, Z., et al., 2023. Assessing long-term impacts of cover crops on soil organic carbon in the central US Midwestern agroecosystems. Global Change Biology 29 (9), 2572–2590. Available at https://doi.org/10.1111/gcb.16632.

Raj, A., Jhariya, M.K., Meena, R.S., 2025. Agroforestry for Monetising Carbon Credits. Springer, Cham. Available at https://doi.org/10.1007/978-3-031-97151-8.

Scavo, A., Fontanazza, S., Restuccia, A., Pesce, G.R., Abbate, C., Mauromicale, G., 2022. The role of cover crops in improving soil fertility and plant nutritional status in temperate climates. A review. Agronomy for Sustainable Development 42 (5). Available at https://doi.org/10.1007/s13593-022-00825-0.

Shirrell, M., Glazer, J.L., Duff, M., Freed, D., 2023. The winds of changes: how research alliances respond to and manage shifting field-level logics. American Educational Research Journal 60 (6), 1221–1257. Available at https://doi.org/10.3102/00028312231193401.

Singh, N.R., Kumar, D., Rao, K.K., Bhatt, B.P., 2020. Agroforestry: soil organic carbon and its carbon sequestration potential. In: Raj, A., Jhariya, M.K., Yadav, D.K., Banerjee, A. (Eds.), Climate Change and Agroforestry Systems: Adaptation and Mitigation Strategies. Apple Academic Press, Boca Raton, pp. 119–142. Available at 10.1201/9780429286759-5.

Stockmann, U., Adams, M.A., Crawford, J.W., Field, D.J., Henakaarchchi, N., Jenkins, M., et al., 2013. The knowns, known unknowns and unknowns of sequestration of soil organic carbon. Agriculture, Ecosystems and Environment 164, 80–99. Available at https://doi.org/10.1016/j.agee.2012.10.001.

Stringer, L.C., Fraser, E.D.G., Harris, D., Lyon, C., Pereira, L., Ward, C.F.M., et al., 2020. Adaptation and development pathways for different types of farmers. Environmental Science & Policy 104, 174–189. Available at https://doi.org/10.1016/j.envsci.2019.10.007.

Triplett, G.B., 2018. Tillage and crop productivity. In: Rechcigl, M. (Ed.), Handbook of Agricultural Productivity: Volume I: Plant Productivity. CRC Press, pp. 251–262.

Wang, H., Wang, S., Yu, Q., Zhang, Y., Wang, R., Li, J., et al., 2020. No-tillage increases soil organic carbon storage and decreases carbon dioxide emission in the crop residue-returned farming system. Journal of Environmental Management 261, 110261.

Weiss, G., Hansen, E., Ludvig, A., Nybakk, E., Toppinen, A., 2021. Innovation governance in the forest sector: reviewing concepts, trends and gaps. Forest Policy and Economics 130, 102506. Available at https://doi.org/10.1016/j.forpol.2021.102506.

Xiong, T., Liu, Y., Yang, C., Cheng, Q., Lin, S., 2023. Research overview of urban carbon emission measurement and future prospect for GHG monitoring network. Energy Reports 9, 231–242. Available at https://doi.org/10.1016/j.egyr.2023.04.047.

Zhang, J., Guan, K., Peng, B., Jiang, C., Zhou, W., Yang, Y., et al., 2021. Challenges and opportunities in precision irrigation decision-support systems for center pivots. Environmental Research Letters 16 (5), 053003. Available at https://doi.org/10.1088/1748-9326/abe436.

Zhang, Y., Tan, C., Wang, R., Li, J., Wang, X., 2021. Conservation tillage rotation enhanced soil structure and soil nutrients in long-term dryland agriculture. European Journal of Agronomy 131, 126379. Available at https://doi.org/10.1016/j.eja.2021.126379.

Zhu, S., Sainju, U.M., Zhang, S., Tan, G., Wen, M., Dou, Y., et al., 2024. Cover cropping promotes soil carbon sequestration by enhancing micro aggregate-protected and mineral-associated carbon. Science of the Total Environment 908, 168330.

Zhu, Y., Chen, J., 2024. How do we motivate farmers to adopt low-carbon production? Analysis of extrinsic incentives' internalization. Journal of Environmental Psychology 94, 102186. Available at https://doi.org/10.1016/j.jenvp.2023.102186.

Chapter 11

Green manuring practices for carbon-negative emissions

Anjana J. Atapattu, D.M.N.S. Dissanayaka, Tharindu D. Nuwarapaksha, and Shashi S. Udumann
Agronomy Division, Coconut Research Institute, Lunuwila, Sri Lanka

Highlights

- Green manuring presents a promising approach for achieving carbon-negative emissions in agriculture by sequestering atmospheric carbon dioxide and reducing reliance on synthetic fertilizers.
- Incorporating nutrient-rich plant biomass into the soil through green manuring enriches soil fertility and structure, consequently leading to enhanced crop yields.
- Breeding and selecting green manure crops, along with tailoring agronomic practices to diverse agroecological contexts, holds promise for optimizing and scaling the benefits of green manuring.

11.1 Introduction

Global warming, which is majorly influenced by human-induced greenhouse gas (GHG) emissions, has become one of the greatest challenges of the world in the 21st century (Yoro and Daramola, 2020). As the concentration of carbon dioxide (CO_2) and other GHGs rises in the atmosphere, the global average temperature increases, resulting in tremendous impacts on various ecosystems, economies, and societies around the globe (Kabir et al., 2023; Meena et al., 2025; Nuwarapaksha et al., 2026).

Over the preindustrial age, CO_2 in the atmosphere rose from about 280 ppm to about 410 ppm in recent years (Hari and Tyagi, 2022). It is widely associated with human activities such as the use of fossil fuels, deforestation, and industrial processes (Shakoor et al., 2020). Consequently, the Intergovernmental Panel on Climate Change has observed that projected global temperatures may deliver extreme climate consequences such as intensified weather events, sea level rise, species decline, and food and water insecurity if emissions of GHGs are not drastically reduced (Alley et al., 2007; Kumar et al., 2021; Atapattu et al., 2026). Carbon (C) emissions, especially CO_2, are instrumental in this climate change (Zheng et al., 2019). As C is a key component of the Earth's biogeochemical cycles, the fast flux of C from fossil fuel pools accumulated over millions of years in geologic time over short time horizons has outstripped the capabilities of the C sinks (Banerjee et al., 2024). Forests, oceans, and soils, which traditionally acted as C reservoirs, are struggling to keep pace with the rate of anthropogenic emissions (Abdullahi et al., 2018).

To this, the global community has waded in through agreements like the Paris Agreement, which sets an aspirational target of limiting climate change to below 2°C above preindustrial levels while pursuing efforts to limit the temperature increase to 1.5°C (Rogelj et al., 2016). However, current national commitments and actions are insufficient to achieve these targets. This has resulted in positive attitudes toward negative emission technologies and practices, which are the ability to remove CO_2 from the atmosphere and store it for a long-term period (Atapattu et al., 2025).

In this regard, the term "net zero" has emerged as a widely discussed topic more recently (Karlsson et al., 2020; Meena et al., 2025). This entails involving the emission of a similar quantity of GHGs to the removal of corresponding quantities of gases from the environment. Indeed, some sectors and countries have their eyes set on so-called "negative emission," whereby more C is captured from the atmosphere than is released.

Carbon Incentives, Plans, and Policies in Agriculture. DOI: https://doi.org/10.1016/B978-0-443-33557-0.00012-4

11.1.1 The role of agriculture in carbon emissions

Due to climate change concerns, many sectors have been involved in researching and developing ways to reduce emissions and increase C sink capacity. Considering the impacts of climate change and CO_2 emissions, it is necessary to examine the role of agriculture in this issue. The agricultural sector is one of the major sources of GHG emissions and, at the same time, the sector is the most sensitive to the effects of climatic change (Malhi et al., 2021). Agriculture contributes roughly between 10% and 12% of the global anthropogenic emissions of GHGs; hence, it is a priority for the reduction (Frank et al., 2017).

The C footprint of agriculture stems from various sources and practices (Fig. 11.1). Livestock production, particularly ruminant animals, is a major source of methane (CH_4) emissions through enteric fermentation (Min et al., 2022). Rice cultivation in flooded fields also contributes significantly to CH_4 release (Zhang et al., 2016). The widespread use of synthetic fertilizers leads to nitrous oxide (N_2O) emissions, a potent GHG with a global warming potential (GWP) far exceeding that of CO_2. Additionally, land-use changes associated with agriculture, such as deforestation for cropland expansion, result in substantial C releases from soil and biomass.

However, agriculture's relationship with C emissions is complex and multifaceted. While the sector is a significant emitter, it also has unique potential to act as a C sink. Agricultural soils, when appropriately managed, can sequester large amounts of C (Poeplau and Don, 2015). Practices such as conservation tillage, cover cropping, and agroforestry (AF) can enhance this sequestration potential (Dinakaran et al., 2022; Dissanayaka, et al., 2023a; Nuwarapaksha et al., 2025a). Moreover, the biomass produced in agricultural systems temporarily stores C, offering opportunities for C capture if managed strategically (Kumar et al., 2020a; Kumar et al., 2024).

The multifaceted nature of agriculture as both a source and a potential sink of C makes it rather complex when analyzed within the context of climate change mitigation (CCM). This duality highlights the need to establish and implement long-term, environmentally friendly strategies to increase the amount of C in the atmosphere within agricultural systems (Raj et al., 2023, 2024). In this context, green manuring stands out as a method that may be able to address many aspects of the C profile of agriculture, as well as provide additional advantages. The main objective of this chapter is to explore and analyze the potential of green manuring as a strategy for achieving negative C emissions in agriculture while providing multiple environmental and agronomic benefits.

11.1.2 Scenarios of green manure production and utilization in agroecosystems

The global landscape of green manure production and utilization varies significantly across different agricultural regions and farming systems. In developed countries, green manuring has been increasingly adopted as part of sustainable

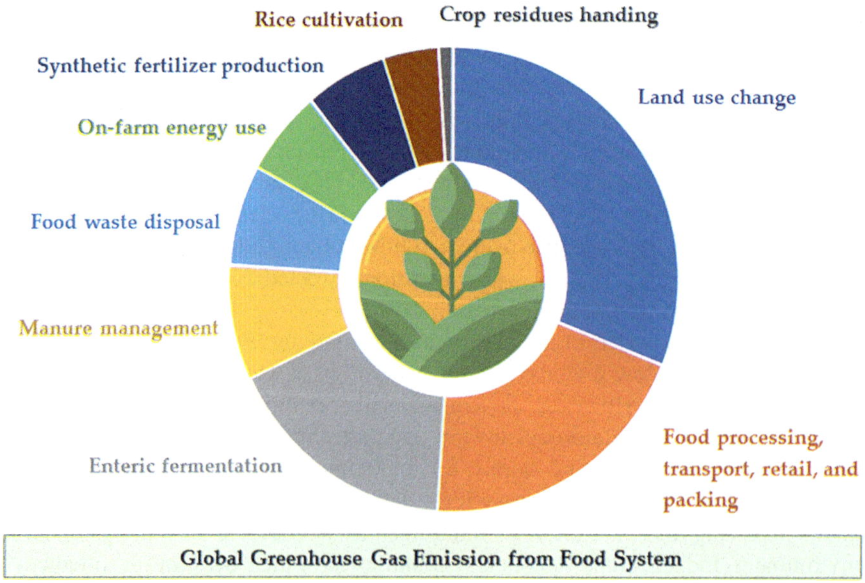

FIGURE 11.1 Shares of GHG emissions related to food systems in 2020 (Rosa and Gabrielli (2023). *GHG*, greenhouse gas.
Modified from Rosa, L., Gabrielli, P., 2023. Achieving net-zero emissions in agriculture: a review. Environmental Research Letters. 18 (6), 063002. https://doi.org/10.1088/1748-932.

agriculture and CCM strategies (Atapattu 2025; Meena et al., 2018). Developed nations like the United States, European Union countries, and Australia reflected a growing trend of integrating green manure crops into crop rotation systems, particularly in organic and conservation agriculture (Gill et al., 2020).

In contrast, developing countries, particularly in tropical and subtropical regions, have a long-standing tradition of green manuring, often driven by resource constraints and local agricultural knowledge. Countries in South Asia, such as India and Sri Lanka, have historically utilized leguminous green manure crops, including *Crotalaria juncea* L., *Pueraria phaseoloides*, and *Gliricidia sepium*, to enhance soil fertility in smallholder farming systems (Atapattu et al., 2025; Dissanayaka et al., 2022; Nuwarapaksha et al., 2023a; Udumann et al., 2023a, b).

The current global scenario of green manure production reveals several key trends: (1) increasing adoption in conservation agriculture; (2) growing interest from organic farming communities; (3) integration with precision agriculture technologies; (4) rising awareness of CCM potential; and (5) development of specialized green manure crop varieties. Meantime, challenges in the widespread adoption of green manure production include (1) limited farmer awareness; (2) initial investment costs; (3) lack of technical knowledge; (4) competition with cash crops for land use; and (5) regional variations in suitable green manure crop species.

11.2 Green manuring for carbon-negative emissions

The theory of green manure, an age-old method of crop rotation that has recently been revived in view of climate change mitigation, could create pathways for attaining C-negative emissions in agriculture (Forte et al., 2017). This practice entails planting crops such as legumes or other fast-growing crops and then turning them into the soil while they are still green and have high nutrient content (Kumar et al., 2020a; Singh et al., 2020). Green manuring has been an integral part of agricultural systems for thousands of years, with roots tracing back to ancient civilizations. These are age-old practices that have been synthesized with modern knowledge and innovations by farmers to sustain soil fertility and crop yields in many parts of the world (Meena et al., 2018).

11.2.1 Examples of green manure crops

These crops can be used individually or in mixtures; the choice is based on the desired outcomes of the green manuring practice, the region's climate, soil requirements, and the stage within the crop rotation. Mainly, it represents (1) legumes: clover, alfalfa, vetch, and cowpeas, because of their ability to fix nitrogen (N) (Rose et al., 2019); (2) growing grasses: rye, *Brachiaria* spp. due to their high rate of biomass accumulation (Snapp et al., 2022); and (3) indigenous plants: local species adapted to specific regional conditions (Table 11.1).

11.2.2 Selection of green manure crops

Overall, green manuring as a technique of C_{seq} and the realization of the negative C-emission rate depend on several factors. It is important to know these variables to enhance the practice further and to get the best results for their contribution to the solution of climate change problems. Green manure crop selection and their performance depend on several factors.

11.2.2.1 Type of green manure crop

As noted earlier, multiple plant types exhibit distinct potential for biomass accumulation and C storage (Kaul et al., 2010). All crops have various uses, and preferences may be based on primary goals, which include N-fixation, biomass production, or improvements in soil structure, as well as the particular regional needs of agriculture. Farmers can select the most suitable green manure crop by prioritizing the main needs. For instance, leguminous crops not only contribute to the addition of organic matter to soils but also play an important role in N fixation, making the use of artificial fertilizers almost irrelevant (Dinakaran et al., 2022). Inexhaustible plants like *Gliricidia*, which have extensive root systems, are likely to store C deeper, thus enhancing long-term storage (Nuwarapaksha et al., 2023a). Some crops, such as Mucuna or wild sunflower, can build biomass quickly and thus contribute more C inputs over the short term (Dissanayaka et al., 2024; Nuwarapaksha et al., 2023b).

11.2.2.2 Local climate conditions

Climate plays a significant role in determining the growth rate and biomass production of green manure crops, as well as other cash crops (Nuwarapaksha et al., 2022). The most suitable green manure crop can be selected after considering

TABLE 11.1 Common green manure crops.

Plant family	Name of the plant	Life span	Growth habit
Fabaceae	*Crotalaria juncea*	Annual	Herb
	Glycine max		Herb
	Lathyrus sativus L.		Herb
	Lathyrus tingitanus L.		Herb
	Lens culinaris Medikus		Herb
	Medicago spp.		Herb
	Pisum sativum		Herb
	Pisum sativum L.		Herb
	Trifolium alexandrinum L.		Herb
	Vicia faba L.		Herb
	Vicia villosa Roth ssp. *Varia*		Herb
	Vigna radiata		Creeping vine
	Melilotus officinalis L.	Biennial	Herb
	Vicia villosa Roth ssp. *Villosa*		Herb
	Baptisia australis	Perennial	Herb
	Calopogonium mucunoides		Creeping vine
	Centrosema pubescens		Creeping vine
	Gliricidia maculata		Tree
	Gliricidia sepium		Tree
	Leucaena leucocephala		Tree
	Macroptilium atropurpureum		Creeping vine
	Medicago sativa L.		Herb
	Mucuna spp.		Creeping vine
	Pongamia glabra		Tree
	Pueraria phaseoloides		Creeping vine
	Sesbania grandiflora		Tree
	Sesbania rostrata		Tree
	Trifolium pratense L.		Herb
Poaceae	*Zea mays*	Annual	Grass
	Brachiaria brizantha	Perennial	Grass
	Brachiaria ruziziensis		Grass
	Brachiaria milliformis		Grass
Sapotaceae	*Madhuca longifolia*	Perennial	Tree
Meliaceae	*Azadirachta indica*	Perennial	Tree
Asteraceae	*Tithonia diversifolia*	Perennial	Tree

Source: Dissanayaka et al., 2022; Singh et al., 2020; Thiessen Martens and Entz, 2011.

climatic factors such as rainfall, solar availability, relative humidity, and temperature (Dinakaran et al., 2022). In tropical and subtropical regions, higher temperatures and longer growing seasons can lead to increased biomass production and potentially higher C_{seq} rates (Oelbermann et al., 2004). For example, the biological potential of annual C_{seq} in tropical and temperate forests was noted as 55% and 31%, respectively (Dinakaran et al., 2022).

11.2.2.3 Soil type

Given that each plant has unique requirements for optimal performance, soil qualities should also be carefully considered before selecting the green manure crop. Furthermore, the physical and chemical characteristics of the soil to a large extent determine the growth and establishment of green manure crops (Scholberg et al., 2010). As shown by previous literature, in clay soils, C storage potential is often higher owing to the formation of strong organo-mineral associations (Bai et al., 2019). Sandy soils, although they will provide good root growth, may have a lower capacity to fix C. The level of C_{seq} also depends on the initial concentration of C in the soil, since depleted soils demonstrate the greatest extra C_{seq} when green manuring is used.

11.2.2.4 Management practices

Various management practices conserve and enhance the effectiveness of green manuring, leading to improved C_{seq} in the soil and overall soil health (Jhariya et al., 2023). Firstly, the timing of incorporation is critical; incorporating the green manure crop at the optimal stage of growth ensures the most effective nutrient release and C_{seq} (Dissanayaka et al., 2023c; Nuwarapaksha et al., 2024b; Udumann et al., 2023b). Tillage practices also significantly impact the success of green manuring, with reduced-tillage or no-till systems generally promoting better C retention in the soil compared to conventional tillage methods (Baker et al., 2007). Furthermore, integrating green manures into diverse crop rotations can enhance the overall resilience of the agricultural system and boost its C_{seq} potential (Qaswar et al., 2019). Lastly, the use of mixed species in green manuring can be particularly beneficial, as it combines the unique advantages of different plants, potentially leading to more comprehensive soil improvement and enhanced C_{seq} (Dissanayaka et al., 2022).

11.2.2.5 Duration and frequency

Long-term, consistent use of green manuring practices tends to result in higher cumulative C_{seq} (Hou et al., 2022). The frequency of green manure crops in the rotation can also impact the rate of C accumulation.

11.2.2.6 Interaction with other practices

The efficiency of green manuring can be enhanced when combined with other sustainable practices, such as composting, mulching, or integrated nutrient management (Dissanayaka et al., 2023b).

11.2.3 Common planting strategies for green manure crops

11.2.3.1 As a catch crop

Catch crops, a strategic component of green manuring practices, play a crucial role in sustainable agriculture by maximizing land use efficiency and enhancing soil health during periods that would otherwise be fallow (Piotrowska and Wilczewski, 2012). These fast-growing crops are deliberately planted between main crop seasons, serving as temporary yet highly beneficial soil cover (Vos and van der Putten, 2004). The primary function of catch crops is soil protection (Fageria et al., 2005). For example, planting *C. juncea* L. in a fallow period of the rice growing season is shown (Guimarães et al., 2018; Udumann et al., 2023b).

11.2.3.2 As an intercrop

Intercropping green manure crops is another efficient agricultural technique that involves planting these useful plants together with the main crop in the same plot (Li et al., 2023; Miyazawa et al., 2010). This approach is both unique and logical in the use of agricultural space, presenting numerous advantages that can help support increased productivity and the sustainability of agricultural systems. The example of this system can be illustrated by the planting of creeping legumes *Mucuna bracteata*, or grasses such as *Megathyrsus maximus*, and trees such as *Tithonia diversifolia* between rows of perennial crops (Fig. 11.2) (Dissanayaka et al., 2024; Nuwarapaksha et al., 2023b; Udumann et al., 2023a).

FIGURE 11.2 Different green manure cropping systems under coconut: (a) *Mucuna bracteata* with coconut cultivation; (b) *Gliricidia sepium* with coconut cultivation; and (c) *Tithonia diversifolia* with coconut cultivation.

11.2.3.3 In rotation

This involves incorporating green manure crops into a planned crop rotation system. As an example, a four-year rotation might include year 1—corn, year 2—soybeans, year 3—winter wheat, followed by a summer green manure crop, and year 4—another main crop. In this practice, green manure crops will not be present in the field every time.

Many successful farming systems employ a combination of these strategies, adapting their approach according to seasonal conditions and long-term sustainability goals.

11.3 Carbon offsetting of agroecosystems through green manuring practices

11.3.1 Mechanisms of carbon sequestration

Green manuring is an effective strategy for C_{seq} due to its multifaceted impact on soil and plant systems (Fig. 11.3) (Dissanayaka et al., 2024).

11.3.1.1 Photosynthetic carbon fixation

Photosynthesis is the fundamental process by which green plants convert sunlight, water, and CO_2 into glucose (a sugar) and oxygen (O_2) (Sekar and Ramasamy, 2015). The glucose produced serves as the primary source of energy and C for the plant, allowing it to grow and produce biomass. The C fixed during photosynthesis of the green manure crop is initially stored in the plant's biomass, including roots, stems, and leaves (Zhang et al., 2019). It is reported that green manure crops accumulate 938–1082 kg C/ha each year. While this C is temporarily removed from the atmosphere, its long-term impact depends on what happens to the plant material after it is returned to the soil. When green manure crops are cut and incorporated into the soil, the C contained within their biomass becomes an input into the soil C pool, initiating further C_{seq} processes (Zhang et al., 2019).

11.3.1.2 Soil organic matter accumulation

Soil organic matter (SOM) consists of decomposed plant and animal residues, cells and tissues of soil organisms, and substances synthesized by soil organisms. SOM is critical for soil health and plays a vital role in C_{seq} (Sarfraz et al., 2019). When green manure crops are incorporated into the soil, soil organisms (such as bacteria, fungi, and earthworms) break down the plant material (Ali et al., 2019). This decomposition process converts plant biomass into organic matter, increasing the soil's C content. A portion of the C from decomposed plant material forms stable organic compounds (like humus) that persist in the soil for decades or even centuries (Dissanayaka et al., 2025). This stable SOM is essential for long-term C storage, as it is less susceptible to being released back into the atmosphere as CO_2.

11.3.1.3 Root exudation

Root exudates are organic compounds released by plant roots into the soil. These compounds include sugars, amino acids, organic acids, and other C-rich molecules that serve various functions (Ma et al., 2022). Root exudates serve as a primary food source for soil microorganisms, stimulating their growth and activity. Active soil microbes decompose organic matter and can contribute to the formation of stable SOM by binding smaller organic molecules into larger, more stable

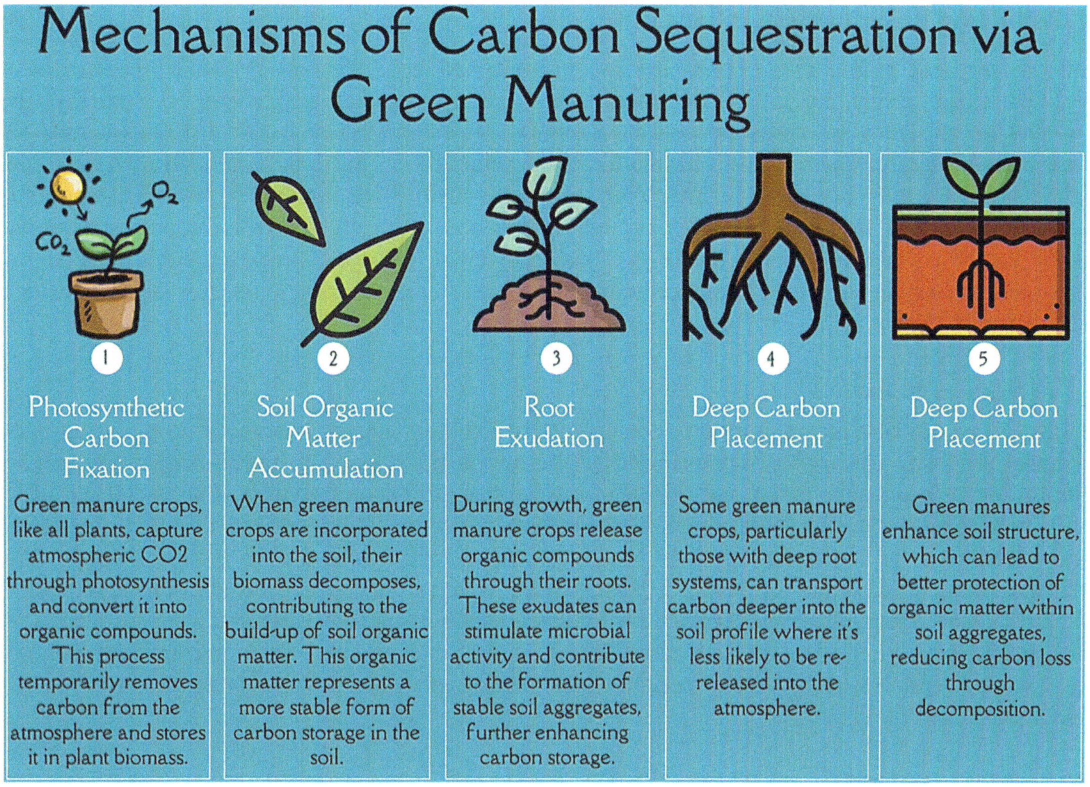

FIGURE 11.3 Mechanisms of carbon sequestration via green manuring.

forms. The interaction between root exudates and soil particles, aided by microbial activity, promotes the formation of soil aggregates (Costa et al., 2018). These aggregates are clusters of soil particles that protect organic matter from decomposition, thus enhancing C_{seq}. The C in root exudates can become part of the stable SOM, especially when it is incorporated into soil aggregates (Panchal et al., 2022). This process helps sequester C in the soil over the long term.

11.3.1.4 Deep carbon placement

Certain cover crops or green manure crops with deeper and more branching root systems, such as alfalfa, can move C down into the profile (Button et al., 2022). Hence, they can deliver C into the subsoil layers through root mass and root secretions. C deposited deeper in the soil is less susceptible to turning over (for instance, by tillage) or microbial breakdown and is more likely to be retained for longer periods (Jackson et al., 2017). C, which is locked in a deeper soil profile, is also shielded from direct interactions with physical factors that are characteristic of the ecosystem, such as temperature and O_2, which, when slow, the rate of decomposition facilitates long-term C storage (Luo et al., 2020).

11.3.1.5 Soil structure improvement

Green manure crops play a significant role in improving soil structure, which is crucial for enhancing C_{seq}. As mentioned earlier, the incorporation of green manure into the soil and the subsequent activity of soil organisms can lead to the formation of soil aggregates (Garcia-Franco et al., 2015). These aggregates enhance soil structural stability by strengthening organic matter (Karami et al., 2012). Soil aggregates provide physical protection for organic matter by limiting microbial access to it, thereby decreasing the rate of C mineralization. This protection is especially significant for the long-term preservation of the C stock in soils. Improved soil structure enhances porosity, allowing better air and water movement within the soil. This is beneficial for root development and microbial functions, thus enhancing the process of C_{seq} (Prasad et al., 2021). The sum of these changes further enhances the reliability of the agricultural production system for consistently high yields, even under adverse environmental conditions (Islam et al., 2019). Similarly, the benefits associated with green manuring practices, which aim to increase soil fertility, can be long-term and include increased land value and sustained production, thereby improving the economic security of farms and farmers (Meena et al., 2018).

11.3.2 Reduction of synthetic fertilizer use

Most synthetic fertilizers, particularly N-based ones, are manufactured through energy-intensive processes that emit large amounts of GHGs (Kayalıoğlu, 2022). NH_3 for N fertilizers is produced through the Haber–Bosch process, which uses a significant quantity of natural gas and emits considerable CO_2 (Nuwarapaksha et al., 2025; Bertilsson and Kirchmann, 2021). Thus by reducing the requirements for these fertilizers, emissions of this type from industries can be minimized. This practice provides a natural method for improving soil fertility and allows limited use of chemicals for crop growth and development (Fig. 11.4) (Meena et al., 2018). By reducing the need for synthetic fertilizers and pesticides, farmers can lower their input costs, making farming operations more economically viable. Green manuring also aligns with the growing market demand for sustainably produced crops, potentially opening up new revenue streams for farmers who adopt these practices (Kharel et al., 2023).

11.3.2.1 Biological N fixation

Biological N fixation is a key process in which certain plants, particularly legumes, convert atmospheric nitrogen gas (N_2) into a form that plants can use, namely ammonia (NH_3) (Olivares et al., 2013). This process occurs through a symbiotic relationship between leguminous plants and rhizobia bacteria that inhabit root nodules (Etesami, 2022). The rhizobia bacteria live in nodules on the roots of leguminous plants (such as clover, alfalfa, or beans). In exchange for C from the plant, these bacteria convert atmospheric N_2 into NH_3, which the plant can then use for growth. This naturally fixed N provides a significant portion of the N needed for plant growth, reducing the dependence on synthetic N fertilizers (Abd-Alla et al., 2023). Since the production of synthetic N fertilizers is energy-intensive and results in significant C emissions, reducing their use directly reduces GHG emissions. The N fixed by leguminous green manures not only benefits the current crop but also enriches the soil for subsequent crops, leading to a more sustainable agricultural system (Ohyama, 2017; Jhariya et al., 2021a, b).

11.3.2.2 Nutrient cycling

Nutrient cycling is a process whereby plants utilize nutrients from the soil and redeposit them back into the soil in forms suitable for future plant generations. Such crops include green manure, which has an important function in this cycle. Green manure crops fix nutrients in the soil during their vegetative growth stage (Cobo et al., 2002). When these crops are tilled into the soil or the residues left on the surface of the soil, the nutrients they contain are returned to the soil. It also makes important nutrients, such as N, phosphorus, and potassium, more available for the next crop (Dissanayaka et al., 2023d). Because of the ability to enhance the soil structure and microbial activity, green manures are capable of releasing and making nutrients that may not be readily available in the soil available for plant uptake. This minimizes the need for additional fertilization and improves the rate at which nutrients are used up in the crops. Slow solubilization of nutrients from decomposing green manure reduces nutrient leaching and runoff, which are undesirable effects observed with synthetic fertilizers that cause nutrients to be washed away, leading to and poor nutrient use efficiency (Cobo et al., 2002).

11.3.2.3 Soil organic matter contribution

While synthetic fertilizers readily supply nutrients that may be rapidly absorbed and metabolized, green manures release nutrients slowly as they go through the process of decomposition (Lei et al., 2022). As a result of this slow-release mechanism, uniform nutrient provision is achieved, which eliminates the need for constant application and usage of fertilizer (Cobo et al., 2002). The gradual decomposition of organic matter contributes to the availability of plant nutrients, as these nutrients are not easily leached or volatilized. This reduces the total amount of fertilizer needed to feed the crops and meet their nutrient requirements. Green manures also enhance the soil fertility in the long run by increasing the SOM content, which boosts the nutrient reserve and supply.

11.3.2.4 Soil health improvement

Green manuring contributes to another nontarget advantage of conserving and improving soil health, which helps reduce the use of synthetic fertilizers. Green manures and their crop residues help to enhance the rate of decomposition of organic matter and promote the growth of soil microorganisms that play a role in nutrient cycling and the development of soil structure (Ball et al., 2005). Organic matter and biologically active soils are more effective in retenting these nutrients and their availability to the plants, thus decreasing dependence on external supplies of fertilizers. Organic residues from green manures enhance soil texture, soil water retention capacity, and the cation exchange capacity (Nin et al., 2016). Such enhancements enable the soil to retain nutrients and make them more accessible to plants, thereby reducing the need for

FIGURE 11.4 Mechanisms of reduction of synthetic fertilizer usage via green manuring.

chemical fertilizers. Green manures can also have a positive effect on diseases in the soil as well as on improved root development, which can lead to increased nutrient uptake (Silva and Canellas, 2022). Healthier plants with stronger root systems are less dependent on synthetic fertilizers for optimal growth.

11.3.3 Mitigation of other greenhouse gass emissions

Green manuring practices can help mitigate emissions of other potent GHGs, particularly N_2O and CH_4.

11.3.3.1 *Reduced N_2O emissions*

N_2O is a GHG with a GWP of about 298 times that of CO_2 over a period of 100 years (Wang et al., 2022). Industrial N fertilizer is a major contributor to N_2O emissions, as it loses nutrients through processes such as nitrification and denitrification in the process of the N cycle (Bhatia et al., 2013). Green manuring, therefore, indirectly minimizes the usage of nitrogenous fertilizers through biological N fixation and hence reduces the emission of N_2O produced during fertilizer application. Besides, green manuring improves the physical properties of the soil, such as structure and organic matter, hence leading to better aeration and water retention. Enhanced aeration of the soil decreases the extent of the anaerobic state, which promotes the denitrification process and consequently reduces low rates of N_2O emission (Bhaduri et al., 2017). Also, higher levels of organic matter can directly reduce the amount of N through immobilization, which inhibits the conversion of N to N_2O.

11.3.3.2 *Methane emission reduction*

CH_4 is another well-known GHG, and its GWP is approximately 25 times that of CO_2 equivalent over 100 years (Bhatia et al., 2013). CH_4 emissions are presumed to be most dominant in anaerobic ecosystems, including flooded rice paddy fields (Bhaduri et al., 2017). In rice fields with standing water, green manure management practices, such as incorporating organic residues before flooding, may affect the redox potential of the rice root zone and provide different electron acceptors to methanogens. This can cut down the level of CH_4 emission during the rice-growing period. Through green manuring, soil structure is improved and hence reduces waterlogging problems (Manik et al., 2019). This improvement decreases the anaerobic status of the soil, which is required for CH_4 production, and subsequently decreases the emission of CH_4 from agricultural fields.

11.3.3.3 *Enhanced soil microbial diversity*

Green manuring has a positive impact on soil microbial diversity, which plays a crucial role in GHG mitigation. It enhances a better and enhanced population and growth of microbial populations in soil (Dissanayaka et al., 2022). This involves methanotrophs, which are bacteria that feed on methane, and this compound is toxic and is used by these bacteria as C to produce less toxic substances (Zhou et al., 2020). These methane-oxidizing bacteria can be enhanced by using green manures that absorb methane from the soil, thereby converting it into a sink. The application of organic matter from green manures provides a balanced microbial community that may be able to moderate processes that enhance the emission of GHGs, such as N_2O and CH_4, thereby lowering their emissions.

11.3.3.4 *Reduced fuel consumption*

Since green manures provide nutrients naturally, they minimize the rates and amounts of synthetic fertilizers and, therefore, decrease the fuel consumed by farming machinery in spreading them. Green manures can improve soil structure and SOM content, making soils more resilient and easier to work with. This can lead to reduced tillage needs, further lowering fuel consumption and associated CO_2 emissions from farm machinery.

11.3.3.5 *CO_2 emission reduction*

The production, transportation, and application of synthetic fertilizers are associated with significant CO_2 emissions. Green manuring helps reduce these emissions through several pathways. Green manures provide a natural source of nutrients, and they reduce the need for synthetic fertilizers. This reduction has a direct impact on the reduction of CO_2 emissions that are related to the energy-demanding processes of the production of these fertilizers. The extent of use of synthetic fertilizers is reduced, consequently resulting in a reduction in the transport of these products, which in turn leads to less CO_2 emission and a reduced need for vehicles. In addition, since green manuring helps to reduce the frequency and quantity of fertilizer applied, it also helps to reduce CO_2 emissions in all the machinery used in the application processes (Alluvione et al., 2010).

11.4 Research and development toward green manuring approaches for carbon-negative emissions

The potential of green manuring to contribute to C-negative emissions is an area of growing interest in agricultural research and development. Breeding and selecting appropriate green manure crops is crucial for enhancing their ability to C_{seq} and reduce GHG emissions (Glenn et al., 2017). Key areas for this include (1) the selection of high-biomass crops that produce a large amount of biomass above and below ground because they contribute more SOM (Ansari et al., 2022); (2) the development of green manure crops with deep and extensive root systems, as they transport C deeper into the soil profile where it is less likely to be released back into the atmosphere (Das et al., 2020); (3) enhancement in N fixation efficiency for leguminous green manure crops (Kebede, 2021); (4) developing green manure crops that are resilient to environmental stresses such as drought, salinity, and temperature extremes (Das et al., 2020); and (5) balancing decomposition rates with C_{seq} goals to ensure that the green manure can meet the nutrient needs of subsequent crops while still contributing to soil C levels.

Agronomic practices play a critical role in maximizing the C_{seq} potential of green manure crops. Research and development in this area focus on identifying and refining practices that optimize C capture and storage in agricultural soils: (1) exploring the best times to plant and incorporate green manures based on crop type, soil conditions, and climate (Nuwarapaksha et al., 2023b); (2) the potential of green manure crop cultivation and incorporation to soil under no-till and reduced tillage practices; and (3) research on planting density, irrigation, and crop rotation sequences and combinations to enhance C storage (Nuwarapaksha et al., 2025b). Other than that, accurately monitoring and measuring C_{seq} are essential for assessing the effectiveness of green manuring practices. Advances in soil C measurement technologies and methodologies are being developed to provide reliable data on the C_{seq} potential of different green manuring approaches (Chen et al., 2022; Smith et al., 2020).

11.5 Future of green manuring-based farming toward enhancing the green economy

The practice of green manuring is rising as one of the most important strategies in maintaining soil health and C_{seq} in sustainable agricultural systems, as well as the green economy value chain (Kumar et al., 2014). The future of green manuring-based farming lies in its integration with advanced agricultural concepts, its potential to contribute to C neutrality and negative emissions, and its alignment with global sustainable development goals (SDGs).

11.5.1 Integration with circular and regenerative agriculture

Circular and regenerative agriculture are two forward-looking approaches that aim to create sustainable and resilient agricultural systems by closing nutrient loops and restoring ecosystems. Green manuring fits seamlessly into these frameworks.

Circular agriculture is a relatively new concept that seeks to make the farm system cyclic in its use of resources, thereby adopting environmentally friendly practices (Jun and Xiang, 2011). In this system, there is no waste produced, but each output serves as an input to the next process, thus eliminating the need to import from the outside world and pollute the environment (Fan et al., 2018). This practice is particularly important for the recycling of nutrients by cycling valuable organic matter and nutrients to the field after crop harvest. While synthetic fertilizers require high energy for manufacture and can harm to the soil, green manures naturally build soil fertility. The nutrients infiltrated by the green manure crops are returned back to the soil through the decomposition of such crops, and this offers a slow-release form of nutrition for subsequent planting. Organic matter and nutrients on the farm are recycled through green manuring; therefore, it reduces importation (Kumar et al., 2020a). Because it helps to close nutrient loops and conserve soil, green manures are used to sustain the productivity of agricultural systems in the long run (Fan et al., 2018). This sustainable management of resources implies that the soils are made productive and available to future generations as the principles of circular agriculture are put into practice (Jhariya et al., 2022).

Since regenerative agriculture is more than sustainable agriculture, it actively aims at rebuilding the health of agricultural ecosystems (Muhie, 2022). This approach is centered on the reconstruction of soil structure, the creation of ecosystem diversification, and the re-establishment of the natural rhythms that support farm systems (Khangura et al., 2023). In the context of regenerative agriculture, green manuring is one of the most important practices, as it enhances soil structure and the load of organic matter and stimulates microbial activity. Implementing green manures gives the ground a better water-holding capacity, nutrients, and C, hence replenishing soils. By and large, the rebuilding of the soil health following the processes of its degradation may eventually turn the agricultural lands productive again. Since regenerative agriculture stresses the need for the use of a diverse population of species that consists of plants, animals, microbes, fungi, and other insects that act as decomposers, nutrient recycling agents, and suppressors of diseases and hazards above and below ground (Caban et al., 2018). However, using these crops on a rotation basis makes the plant and insect diversity in the farm higher, making the farm support more pollination, beneficial insects, and other wildlife (Dissanayaka et al., 2022). As green manure crops decompose, they add organic matter to the soil, creating a rich, fertile environment that supports plant growth (Das et al., 2020). Green manuring is part of a package of regenerative practices, including but not limited to cover cropping, AF, and minimal tillage. Thus since green manuring leads to better soil health and an increase biological diversity, there is an improvement in the ecosystem services farms provide, including water regulation, C storage, and habitat provision (Garcia et al., 2018). The following services are crucial to building an enduring agricultural landscape that can bear and adapt to the pressures of climate change for sustainable production.

11.5.2 Potential for carbon neutrality and negative emissions

It can be concluded that further development of green manuring-based farming has great potential for C neutrality and for attaining below-zero emissions (Cai et al., 2018; Khan et al., 2022). C neutrality involves balancing C emissions with C removal, and green manuring can be a key strategy in achieving this balance. Besides, C is stored in the soil, and the creation of bio-fertilizers through green manuring reduces the overall C footprint of the agriculture business. However, when integrated with other practices that reduce the C footprint on farms, such as the use of renewable energyor efficient water management, green manuring makes it possible for farms to go green and balance the C issue.

Negative emissions build on the C-neutrality concept but aim to have a higher rate of CO_2 removal from the atmosphere than emissions (Carton et al., 2020). Green manuring could also be used to reduce the amount of CO_2 in the atmosphere through the process of C_{seq} through the long-term storing of C in the soil through practices such as AF, deep-rooted perennials, and biochar application (Dissanayake et al., 2023; Ekanayaka et al., 2023; Nuwarapaksha et al., 2024a). These practices not only reduce emissions from agriculture but can also sequester atmospheric CO_2, making green manure indispensable to any plan to mitigate climate change.

The possibility of implementing green manuring practices across the different levels, from smallholder farmers to large-scale farmers, increases the prospects of achieving negative emissions across the world. Further research and investment in green manuring can increase adoption by a larger population, thus increasing the contribution toward the world's C cycles. With this, green manuring-based farming aligns closely with several of the United Nations SDGs, making it a valuable practice for advancing global sustainability efforts (Table 11.2) (Atapattu et al., 2024; Chang et al., 2022).

11.6 Policy perspectives for green manuring approachs for carbon-negative emissions

Green manuring is increasingly recognized as a crucial strategy for achieving negative C emissions within the global policy landscape. International climate agreements, including the Paris Agreement and subsequent climate conferences, have emphasized the importance of transforming the agricultural sector to meet global climate goals (Rogelj et al., 2016). Financial incentives represent a key policy recommendation for promoting green manuring practices. These include (1) establishing C credit mechanisms specifically for green manuring; (2) providing subsidies to farmers who adopt these practices; and (3) offering tax benefits for sustainable agricultural approaches. Such economic instruments can help overcome initial adoption barriers and incentivize long-term implementation (Raj et al., 2025). Research and development support is equally vital. Increased funding for green manuring research, development of region-specific green manure crop varieties, and technology transfer programs can drive innovation and improve effectiveness. These investments can lead to more efficient C_{seq} techniques tailored to local agricultural conditions.

Robust regulatory frameworks are also essential for widespread adoption. This includes (1) mandatory C accounting for agricultural sectors; (2) integration of green manuring in national climate adaptation plans; and (3) certification schemes for C-neutral agriculture. Such frameworks provide structure and accountability while creating market recognition for sustainable practices. Despite these recommendations, several challenges exist in policy implementation. These include (1) the complexity of measuring C_{seq} accurately, (2) the diversity of agricultural systems across regions, (3) economic constraints faced by developing countries, (4) limited technical capacity in some areas, and (5) the need for long-term policy commitments that transcend political cycles.

TABLE 11.2 Contributions of green manuring to the sustainable development goals (SDGs).

SDG	Goal	Contribution from green manuring
SDG 2	Zero Hunger	Improves soil fertility and crop yields Promotes sustainable agriculture Enhances crop resilience to climate variability
SDG 13	Climate Action	Sequesters of C in soil Reduces GHG emissions Enhances climate change adaptation
SDG 15	Life on Land	Promotes soil biodiversity Combats land degradation Supports sustainable land management
SDG 6	Clean Water and Sanitation	Improves water quality by reducing nutrient runoff Enhances water efficiency through improved soil structure
SDG 12	Responsible Consumption and Production	Promotes resource efficiency in agriculture, supports sustainable production systems Can contribute to waste reduction
SDG 1	No Poverty	Supports sustainable Reduces input costs for agriculture
SDG 3	Good Health and Well-being	Reduces exposure to agricultural chemicals May improve the nutritional quality of crops
SDG 9	Industry, Innovation and Infrastructure	Drives agricultural innovation Contributes to sustainable industrialization in agriculture
SDG 17	Partnerships for the Goals	Promotes knowledge sharing and partnerships Aligns with global climate change mitigation efforts
SDG 14	Life Below Water	Indirectly reduces ocean pollution by decreasing nutrient runoff
SDG 8	Decent Work and Economic Growth	Supports sustainable economic growth in agriculture Can create new jobs in sustainable farming practices

11.7 Conclusion

Green manuring is a unique and complex approach to meeting some of the climate change and sustainable agriculture issues. As has been demonstrated in this review, this practice that has great potential for negative C emissions and provides numerous agronomic and environmental benefits. As indicated above, there are various closely related ways by which green manuring enhances the C pool in the soil. Through photosynthetic C fixation, improvement of humus content, root exudation, and deep burial of C, green manure crops sequester C from the atmosphere and store it in the soil environment, thereby enhancing soil structure. This process not only reduces climate change but also improves soil characteristics and quality, thereby creating a cycle of both climate and soil improvement. However, a key factor to which this approach contributes is the reduction of the agricultural sector's dependence on synthetic fertilizers. Through the stimulation of biological N-fixation, increased nutrient turnover, and generally improving soil quality, green manuring procedures drastically reduce the need for chemical inputs. Besides, reducing the actual application rates of synthetic fertilizer not only decreases direct GHG emissions but also diminishes the amount of C expended on fertilizer production and distribution. Moreover, the practice can be viewed as part of circular or regenerative agriculture, represents a promising direction for developing more sustainable and restorative agricultural systems.

In this context, the potential of green manure to contribute to C neutrality and even negative emissions in agriculture in the future should be highlighted. Since green manuring is still in the experimental stage, and researchers' innovations improve each year with more efficient and effective crop varieties, the practice is likely to yield higher results in the future. The possibility of linking green manuring with other organic farming practices and other new technologies can go a long way in changing the face of emitting GHGs to that of a C sink in agriculture today. In addition, the fact that green manuring fits into more than one of the SDGs means that it is not only important in terms of ecological objectives but also for societal goals such as feeding the population, eradicating poverty, and achieving sustainable economic development.

References

Abd-Alla, M.H., Al-Amri, S.M., El-Enany, A.W.E., 2023. Enhancing rhizobium–legume symbiosis and reducing nitrogen fertilizer use are potential options for mitigating climate change. Agriculture (Switzerland) 13 (11), 2902. Available at https://doi.org/10.3390/agriculture13112092.

Abdullahi, A.C., Siwar, C., Shaharudin, M.I., Anizan, I., 2018. Carbon sequestration in soils: the opportunities and challenges. Carbon Capture, Utilization and Sequestration. InTech. Available at https://doi.org/10.5772/intechopen.79347.

Ali, Q., Ashraf, S., Kamran, M., Ijaz, M., Rehman, A., Tahir, M., et al., 2019. Organic Manuring for Agronomic Crops Agronomic Crops: Volume 2: Management Practices. Springer, Singapore, pp. 163–193. Available at 10.1007/978-981-32-9783-8.

Alley, R., Berntsen, T., Bindoff, N.L., Chen, Z., Chidthaisong, A., Friedlingstein, P., et al., 2007. Intergovernmental Panel On Climate Change Climate Change 2007: The Physical Science Basis Summary for Policymakers Contribution of Working Group I to the Fourth Assessment Report of the Intergovernmental Panel on Climate Change Summary for Policymakers IPCC WGI Fourth Assessment Report.

Alluvione, F., Bertora, C., Zavattaro, L., Grignani, C., 2010. Nitrous oxide and carbon dioxide emissions following green manure and compost fertilization in corn. Soil Science Society of America Journal 74 (2), 384–395. Available at https://doi.org/10.2136/sssaj2009.0092.

Ansari, M.A., Choudhury, B.U., Layek, J., Das, A., Lal, R., Mishra, V.K., 2022. Green manuring and crop residue management: effect on soil organic carbon stock, aggregation, and system productivity in the foothills of Eastern Himalaya (India). Soil and Tillage Research 218. Available at https://doi.org/10.1016/j.still.2022.105318.

Atapattu, A.J., Udumann, S.S., 2025. Handbook of Nature-Based Solutions to Mitigation and Adaptation to Climate Change. Springer Nature, Switzerland, pp. 1931–1958. Available at 10.1007/978-3-031-34967-6_166.

Atapattu, A.J., Ranasinghe, C.S., Nuwarapaksha, T.D., Udumann, S.S., Dissanayaka, N.S., 2024. Sustainable agriculture and sustainable development goals (SDGs). In: Garwi, J., Masengu, R., Chiwaridzo, O. (Eds.), Emerging Technologies and Marketing Strategies for Sustainable Agriculture. IGI Global, Hershey, PA, pp. 1–27. Available at 10.4018/979-8-3693-4864-2.ch001.

Atapattu, A.J., Udumann, S.S., Dissanayaka, N.S., Nuwarapaksha, T.D., Dissanayake, A.S.B., Tennakoon, A., Rasika, D.M.D., Dassanayake, D.H.B. R., Ekanayake, J., Karalliyadda, S.M.C.B., Bandara, N.P.S.N., Sankalpa, J.K.S., Vinujan, S., Arachchige, A.C.G., Kumarathunge, D.P., 2025. Transitioning to Climate Action. MDPI. Available at 10.3390/books978-3-03897-875-6-4.

Atapattu, A.J., Dissanayaka, N.S., Udumann, S.S., Nuwarapaksha, T.D., 2026. Point Source Nitrogen Pollution. Elsevier, pp. 453–475. Available at 10.1016/b978-0-443-27352-0.00025-2.

Bai, X., Huang, Y., Ren, W., Coyne, M., Jacinthe, P.A., Tao, B., et al., 2019. Responses of soil carbon sequestration to climate-smart agriculture practices: a meta-analysis. States Global Change Biology 25 (8), 2591–2606. Available at https://doi.org/10.1111/gcb.14658.

Baker, J.M., Ochsner, T.E., Venterea, R.T., Griffis, T.J., 2007. Tillage and soil carbon sequestration-what do we really know? Agriculture, Ecosystems and Environment 118 (1-4), 1–5. Available at https://doi.org/10.1016/j.agee.2006.05.014.

Ball, B.C., Bingham, I., Rees, R.M., Watson, C.A., Litterick, A., 2005. The role of crop rotations in determining soil structure and crop growth conditions. Canadian Journal of Soil Science 85 (5), 557–577. Available at https://doi.org/10.4141/s04-078.

Banerjee, A., Jhariya, M.K., Raj, A., Mechergui, T., 2024. Ecosystem management: climate change and global sustainability—an introduction. In: Banerjee, A., Jhariya, M.K., Raj, A., Mechergui, T. (Eds.), Ecosystem Management: Climate Change and Sustainability. Wiley, India, pp. 1–21. Available at 10.1002/9781394231249.ch1.

Bertilsson, G.O.B., Kirchmann, H., 2021. Sustainable N fertilizer production based on a loop: straw - biogas – 'Haber-Bosch' process. Agricultural Systems 190, 103100. Available at https://doi.org/10.1016/j.agsy.2021.103100.

Bhaduri, D., Mandal, A., Chakraborty, K., Chatterjee, D., Dey, R., 2017. Interlinked chemical-biological processes in anoxic waterlogged soil – a review. The Indian Journal of Agricultural Sciences 87 (12). Available at https://doi.org/10.56093/ijas.v87i12.76483.

Bhatia, A., Jain, N., Pathak, H., 2013. Methane and nitrous oxide emissions from Indian rice paddies, agricultural soils and crop residue burning. Greenhouse Gases: Science and Technology 3 (3), 196–211. Available at https://doi.org/10.1002/ghg.1339.

Button, E.S., Pett-Ridge, J., Murphy, D.V., Kuzyakov, Y., Chadwick, D.R., Jones, D.L., 2022. Deep-C storage: biological, chemical and physical strategies to enhance carbon stocks in agricultural subsoils. Soil Biology and Biochemistry 170. Available at https://doi.org/10.1016/j.soilbio.2022.108697.

Caban, J.R., Kuppusamy, S., Kim, J.H., Yoon, Y.E., Kim, S.Y., Lee, Y.B., 2018. Green manure amendment enhances microbial activity and diversity in antibiotic-contaminated soil. Applied Soil Ecology 129, 72–76. Available at https://doi.org/10.1016/j.apsoil.2018.04.013.

Cai, S., Pittelkow, C.M., Zhao, X., Wang, S., 2018. Winter legume-rice rotations can reduce nitrogen pollution and carbon footprint while maintaining net ecosystem economic benefits. Journal of Cleaner Production 195, 289–300. Available at https://doi.org/10.1016/j.jclepro.2018.05.115.

Carton, W., Asiyanbi, A., Beck, S., Buck, H.J., Lund, J.F., 2020. Negative emissions and the long history of carbon removal. WIREs Climate Change 11 (6). Available at https://doi.org/10.1002/wcc.671.

Chang, S.H.E., Yi, X.Y., Sauer, J., Yin, C.B., Li, F.D., 2022. Explaining farmers' reluctance to adopt green manure cover crops planting for sustainable agriculture in Northwest China. Journal of Integrative Agriculture 21 (11), 3382–3394. Available at https://doi.org/10.1016/j.jia.2022.09.005.

Chen, S., Arrouays, D., Leatitia Mulder, V., Poggio, L., Minasny, B., Roudier, P., et al., 2022. Digital mapping of GlobalSoilMap soil properties at a broad scale: a review. Geoderma 409, 115567. Available at https://doi.org/10.1016/j.geoderma.2021.115567.

Cobo, J.G., Barrios, E., Kass, D.C.L., Thomas, R.J., 2002. Decomposition and nutrient release by green manures in a tropical hillside agroecosystem. Plant and Soil 240 (2), 331–342. Available at https://doi.org/10.1023/a:1015720324392.

Costa, O.Y.A., Raaijmakers, J.M., Kuramae, E.E., 2018. Microbial extracellular polymeric substances: ecological function and impact on soil aggregation. Frontiers in Microbiology 9. Available at https://doi.org/10.3389/fmicb.2018.01636.

Das, K., Biswakarma, N., Zhiipao, R., Kumar, A., Ghasal, P.C., Pooniya, V., 2020. Significance and management of green manures. In: Giri, B., Varma, A. (Eds.), Soil Health. Soil Biology, vol 59. Springer, Cham, pp. 197–217. Available at https://doi.org/10.1007/978-3-030-44364-1_12.

Dinakaran, J., Abbas, N.S., Bhardwaj, S., Kaula, B.C., Vadakan, A.B., 2022. Soil carbon sequestration potential of terrestrial ecosystems: trends and soil priming effects. Current World Environment 17, 161–170. Available at https://doi.org/10.12944/CWE.17.1.14.

Dissanayaka, D.M.N.S., Nuwarapaksha, T.D., Udumann, S.S., Dissanayake, D.K.R.P.L., Atapattu, A.J., 2022. A sustainable way of increasing productivity of coconut cultivation using cover crops: a review. Maximum Circular Agricultural Systems 2. Available at https://doi.org/10.48130/CAS-2022-0007.

Dissanayaka, D.M.N.S., Dissanayake, D.K.R.P.L., Udumann, S.S., Nuwarapaksha, T.D., Atapattu, A.J., 2023a. Agroforestry—a key tool in the climate-smart agriculture context: a review on coconut cultivation in Sri Lanka. Frontiers in Agronomy 5. Available at https://doi.org/10.3389/fagro.2023.1162750.

Dissanayaka, D.M.N.S., Udumann, S.S., Dissanayake, D.K.R.P.L., Nuwarapaksha, T.D., Atapattu, A.J., 2023b. Review on aquatic weeds as potential source for compost production to meet sustainable plant nutrient management needs. Waste 1 (1), 264–280. Available at https://doi.org/10.3390/waste1010017.

Dissanayaka, D.M.N.S., Udumann, S.S., Nuwarapaksha, T.D., Atapattu, A.J., 2023c. Effects of pyrolysis temperature on chemical composition of coconut-husk biochar for agricultural applications: a characterization study. Technology in Agronomy 3. Available at https://doi.org/10.48130/TIA-2023-0013.

Dissanayaka, N.S., Udumann, S.S., Nuwarapaksha, T.D., Atapattu, A.J., 2023d. Agroforestry: an avenue for resilient and productive farming through integrated crops and livestock production. In: Kiba, D.I. (Ed.), Transitioning to Zero Hunger. MDPI Books, pp. 115–136. Available at https://doi.org/10.3390/books978-3-03897-863-3-5.

Dissanayake, D.K.R.P.L., Udumann, S.S., Dissanayaka, D.M.N.S., Nuwarapaksha, T.D., Atapattu, A.J., 2023. Effect of biochar application rate on macronutrient retention and leaching in two coconut growing soils. Technology in Agronomy 3. Available at https://doi.org/10.48130/TIA-2023-0005.

Dissanayaka, D.M.N.S., Udumann, S.S., Nuwarapaksha, T.D., Atapattu, A.J., 2024. Harnessing the potential of Mucuna cover cropping: a comprehensive review of its agronomic and environmental benefits. Circular Agricultural Systems 4. Available at https://doi.org/10.48130/cas-0024-0001.

Dissanayaka, N.S., Udumann, S.S., Nuwarapaksha, T.D., Atapattu, A.J., 2025. Microbial partnerships in agriculture: boosting crop health and productivity. Circular Agricultural Systems 5. Available at 10.48130/cas-0025-0011.

Ekanayaka, E.M.G.N., Dissanayake, D.K.R.P.L., Udumann, S.S., Dissanayaka, D.M.N.S., Nuwarapaksha, T.D., Herath, H.M.S.K., et al., 2023. Sustainable utilization of king coconut husk as a feedstock in biochar production with the highest conversion efficiency and desirable properties. IOP Conference Series: Earth and Environmental Science 1235 (1), 012009. Available at https://doi.org/10.1088/1755-1315/1235/1/012009.

Etesami, H., 2022. Root nodules of legumes: a suitable ecological niche for isolating non-rhizobial bacteria with biotechnological potential in agriculture. Current Research in Biotechnology 4, 78–86. Available at https://doi.org/10.1016/j.crbiot.2022.01.003.

Fageria, N.K., Baligar, V.C., Bailey, B.A., 2005. Role of cover crops in improving soil and row crop productivity. Communications in Soil Science and Plant Analysis 36 (19-20), 2733–2757. Available at https://doi.org/10.1080/00103620500303939.

Fan, W., Zhang, P., Xu, Z., Wei, H., Lu, N., Wang, X., et al., 2018. Life cycle environmental impact assessment of circular agriculture: a case study in fuqing, China. Sustainability 10 (6), 1810. Available at https://doi.org/10.3390/su10061810.

Forte, A., Fagnano, M., Fierro, A., 2017. Potential role of compost and green manure amendment to mitigate soil GHGs emissions in Mediterranean drip irrigated maize production systems. Journal of Environmental Management 192, 68–78. Available at https://doi.org/10.1016/j.jenvman.2017.01.037.

Frank, S., Havlík, P., Soussana, J.F., Levesque, A., Valin, H., Wollenberg, E., et al., 2017. Reducing greenhouse gas emissions in agriculture without compromising food security? Environmental Research Letters 12 (10). Available at https://doi.org/10.1088/1748-9326/aa8c83.

Garcia, L., Celette, F., Gary, C., Ripoche, A., Valdés-Gómez, H., Metay, A., 2018. Management of service crops for the provision of ecosystem services in vineyards: a review. Agriculture, Ecosystems and Environment 251, 158–170. Available at https://doi.org/10.1016/j.agee.2017.09.030.

Garcia-Franco, N., Albaladejo, J., Almagro, M., Martínez-Mena, M., 2015. Beneficial effects of reduced tillage and green manure on soil aggregation and stabilization of organic carbon in a Mediterranean agroecosystem. Soil and Tillage Research 153, 66–75. Available at https://doi.org/10.1016/j.still.2015.05.010.

Gill, K., Sandhu, S., Mor, M., Kalmodiya, T., Singh, M., 2020. Role of green manuring in sustainable agriculture: a review. European Journal of Molecular & Clinical Medicine 7, 2361–2366.

Glenn, K.C., Alsop, B., Bell, E., Goley, M., Jenkinson, J., Liu, B., et al., 2017. Bringing new plant varieties to market: plant breeding and selection practices advance beneficial characteristics while minimizing unintended changes. Crop Science 57 (6), 2906–2921. Available at https://doi.org/10.2135/cropsci2017.03.0199.

Guimarães, G.L., Buzetti, S., Da Silva, E.C., Arf, O., De Sá, M.E., Souza Bastos, A.V., 2018. Agronomic performance of cover crops and upland rice cultivars grown in sucession with different nitrogen rates in no-tillage system. Cultura Agronômica: Revista de Ciências Agronômicas 27 (2), 173–189. Available at https://doi.org/10.32929/2446-8355.2018v27n2p173-189.

Hari, M., Tyagi, B., 2022. Terrestrial carbon cycle: tipping edge of climate change between the atmosphere and biosphere ecosystems. Environmental Science: Atmospheres 2 (5), 867–890. Available at https://doi.org/10.1039/d1ea00102g.

Hou, P., Xue, L., Wang, J., Petropoulos, E., Xue, L., Yang, L., 2022. Green manure amendment in paddies improves soil carbon sequestration but cannot substitute the critical role of N fertilizer in rice production. Agronomy 12 (7), 1548. Available at https://doi.org/10.3390/agronomy12071548.

Islam, M.M., Urmi, T.A., Rana, M.S., Alam, M.S., Haque, M.M., 2019. Green manuring effects on crop morpho-physiological characters, rice yield and soil properties. Physiology and Molecular Biology of Plants 25 (1), 303–312. Available at https://doi.org/10.1007/s12298-018-0624-2.

Jackson, R.B., Lajtha, K., Crow, S.E., Hugelius, G., Kramer, M.G., Piñeiro, G., 2017. The ecology of soil carbon: pools, vulnerabilities, and biotic and abiotic controls. Annual Review of Ecology, Evolution, and Systematics 48, 419–445. Available at https://doi.org/10.1146/annurev-ecolsys-112414-054234.

Jhariya, M.K., Banerjee, A., Swaroop Meena, R., Kumar, S., Raj, A., 2021. Sustainable Intensification for Agroecosystem Services and Management. Springer, Singapore, pp. 1–870. Available at https://doi.org/10.1007/978-981-16-3207-5.

Jhariya, M.K., Meena, R.S., Banerjee, A., 2021. Ecological Intensification of Natural Resources for Sustainable Agriculture. Springer, Singapore, pp. 1–655. Available at 10.1007/978-981-33-4203-3.

Jhariya, M.K., Meena, R.S., Banerjee, A., Meena, S.N., 2022. Natural Resources Conservation and Advances For Sustainability. Elsevier. Available at https://doi.org/10.1016/C2019-0-03763-6.

Jhariya, M.K., Meena, R.S., Banerjee, A., Kumar, S., Raj, A., 2023. Agroforestry for Carbon and Ecosystem Management. Academic Press, pp. 1–434. Available at 10.1016/C2021-0-02916-3.

Jun, H., Xiang, H., 2011. Development of circular economy is a fundamental way to achieve agriculture sustainable development in China. Energy Procedia 5, 1530–1534. Available at https://doi.org/10.1016/j.egypro.2011.03.262.

Kabir, M., Habiba, U.E., Khan, W., Shah, A., Rahim, S., Rios-Escalante, P.R.D.L., et al., 2023. Climate change due to increasing concentration of carbon dioxide and its impacts on environment in 21st century; a mini review. Journal of King Saud University - Science 35 (5). Available at https://doi.org/10.1016/j.jksus.2023.102693.

Karami, A., Homaee, M., Afzalinia, S., Ruhipour, H., Basirat, S., 2012. Organic resource management: impacts on soil aggregate stability and other soil physico-chemical properties. Agriculture, Ecosystems and Environment 148, 22–28. Available at https://doi.org/10.1016/j.agee.2011.10.021.

Karlsson, I., Rootzén, J., Johnsson, F., 2020. Reaching net-zero carbon emissions in construction supply chains – analysis of a Swedish road construction project. Renewable and Sustainable Energy Reviews 120, 109651. Available at https://doi.org/10.1016/j.rser.2019.109651.

Kaul, M., Mohren, G.M.J., Dadhwal, V.K., 2010. Carbon storage and sequestration potential of selected tree species in India. Mitigation and Adaptation Strategies for Global Change 15 (5), 489–510. Available at https://doi.org/10.1007/s11027-010-9230-5.

Kayalıoğlu, A., 2022. Green approach to the fertiliser industry: low-carbon fertilisers. Kemija u industriji 71 (5-6), 347–357. Available at https://doi.org/10.15255/kui.2021.066.

Kebede, E., 2021. Contribution, utilization, and improvement of legumes-driven biological nitrogen fixation in agricultural systems. Frontiers in Sustainable Food Systems 5. Available at https://doi.org/10.3389/fsufs.2021.767998.

Khan, Z.A., Koondhar, M.A., Tiantong, M., Khan, A., Nurgazina, Z., Tianjun, L., et al., 2022. Do chemical fertilizers, area under greenhouses, and renewable energies drive agricultural economic growth owing the targets of carbon neutrality in China? Energy Economics 115, 106397. Available at https://doi.org/10.1016/j.eneco.2022.106397.

Khangura, R., Ferris, D., Wagg, C., Bowyer, J., 2023. Regenerative agriculture—a literature review on the practices and mechanisms used to improve soil health. Sustainability 15 (3), 2338. Available at https://doi.org/10.3390/su15032338.

Kharel, M., Raut, N., Dahal, B.M., 2023. An assessment of good agriculture practices for safe and sustainable vegetable production in mid-hills of Nepal. Journal of Agriculture and Food Research 11. Available at https://doi.org/10.1016/j.jafr.2023.100518.

Kumar, R., Mahajan, G., Srivastava, S., Sinha, A., 2014. Green manuring: a boon for sustainable agriculture and pest management – a review. Agricultural Reviews 35 (3), 196. Available at https://doi.org/10.5958/0976-0741.2014.00906.4.

Kumar, S., Meena, R.S., Jhariya, M.K., 2020a. Resources use Efficiency in Agriculture. Springer Nature, Singapore, pp. 760. Available at https://doi.org/10.1007/978-981-15-6953-1.

Kumar, A., Nagar, S., Anand, S., 2021. 1 - Climate change and existential threats. In: Singh, S., Singh, P., Rangabhashiyam, S., Srivastava, K.K. (Eds.), Global Climate Change. Elsevier, India, pp. 1–31. Available at https://doi.org/10.1016/B978-0-12-822928-6.00005-8.

Kumar, S., Meena, R.S., Sheoran, P., Jhariya, M.K., 2024. Regenerative Agriculture for Sustainable Food Systems. Springer, Singapore, pp. 1–493. Available at 10.1007/978-981-97-6691-8.

Lei, B., Wang, J., Yao, H., 2022. Ecological and environmental benefits of planting green manure in paddy fields. Agriculture 12 (2), 223. Available at https://doi.org/10.3390/agriculture12020223.

Li, H., Fan, Z., Wang, Q., Wang, G., Yin, W., Zhao, C., et al., 2023. Green manure and maize intercropping with reduced chemical N enhances productivity and carbon mitigation of farmland in arid areas. European Journal of Agronomy 145, 126788. Available at https://doi.org/10.1016/j.eja.2023.126788.

Luo, Z., Luo, Y., Wang, G., Xia, J., Peng, C., 2020. Warming-induced global soil carbon loss attenuated by downward carbon movement. Global Change Biology 26 (12), 7242–7254. Available at https://doi.org/10.1111/gcb.15370.

Ma, W., Tang, S., Dengzeng, Z., Zhang, D., Zhang, T., Ma, X., 2022. Root exudates contribute to belowground ecosystem hotspots: a review. Frontiers in Microbiology 13. Available at https://doi.org/10.3389/fmicb.2022.937940.

Malhi, G., Kaur, M., Kaushik, P., 2021. Impact of climate change on agriculture and its mitigation strategies: a review. Sustainability 13 (3), 1318. Available at https://doi.org/10.3390/su13031318.

Manik, S.M.N., Pengilley, G., Dean, G., Field, B., Shabala, S., Zhou, M., 2019. Soil and crop management practices to minimize the impact of waterlogging on crop productivity. Frontiers in Plant Science 10. Available at https://doi.org/10.3389/fpls.2019.00140.

Meena, B.L., Fagodiya, R.K., Prajapat, K., Dotaniya, M.L., Kaledhonkar, M.J., Sharma, P.C., et al., 2018. Legume green manuring: an option for soil sustainability. In: Meena, R.S., Das, A., Yadav, G.S., Lal, R. (Eds.), Legumes for Soil Health and Sustainable Management. Springer Singapore, Singapore, pp. 387–408. Available at https://doi.org/10.1007/978-981-13-0253-4_12.

Meena, R.S., Pradhan, G.S., Kumar, S., Jhariya, M.K., Rao, C.S., 2025. Carbon-Negative Agriculture. Springer, pp. 1–500.

Min, B.R., Lee, S., Jung, H., Miller, D.N., Chen, R., 2022. Enteric methane emissions and animal performance in dairy and beef cattle production: strategies, opportunities, and impact of reducing emissions. Animals 12 (8). Available at https://doi.org/10.3390/ani12080948.

Miyazawa, K., Murakami, T., Takeda, M., Murayama, T., 2010. Intercropping green manure crops-effects on rooting patterns. Plant and Soil 331 (1), 231–239. Available at https://doi.org/10.1007/s11104-009-0248-y.

Muhie, S.H., 2022. Novel approaches and practices to sustainable agriculture. Journal of Agriculture and Food Research 10, 100446. Available at https://doi.org/10.1016/j.jafr.2022.100446.

Nin, Y., Diao, P., Wang, Q., Zhang, Q., Zhao, Z., Li, Z., 2016. On-farm-produced organic amendments on maintaining and enhancing soil fertility and nitrogen availability in organic or low input agriculture. In: Larramendy, M., Soloneski, S. (Eds.), Organic Fertilizers - From Basic Concepts to Applied Outcomes. InTech. Available at https://doi.org/10.5772/62338.

Nuwarapaksha, T.D., Udumann, S.S., Dissanayaka, D.M.N.S., Dissanayake, D.K.R.P.L., Atapattu, A.J., 2022. Coconut based multiple cropping systems: an analytical review in Sri Lankan coconut cultivations. Circular Agricultural Systems 2. Available at https://doi.org/10.48130/CAS-2022-0008.

Nuwarapaksha, T.D., Dissanayaka, D.M.N.S., Udumann, S.S., Atapattu, A.J., 2023a. Gliricidia as a beneficial crop in resource-limiting agroforestry systems in Sri Lanka. Ind. Soc. Agroforestry 25 (1), 12–18.

Nuwarapaksha, T.D., Dissanayake, W.K., Gunathilaka, W.S., Udumann, S.S., Dissanayaka, N.S. and Atapattu, A.Js, 2023b. Assessing the Optimum Harvesting Stage of Tithonia diversifolia as Climate Smart Soil Amendment for Coconut Plantations. In: The 2nd International Online Conference on Agriculture. MDPI, Switzerland, p. 1. https://doi.org/10.3390/IOCAG2023-15967.

Nuwarapaksha, T.D., Dissanayaka, D.M.N.S., Udumann, S.S., Vinujan, S., Atapattu, A.J., 2024a. Exploring the impact of pyrolysis temperature on nutrient composition of *Gliricidia sepium* biochar: a comprehensive study. Technology in Agronomy 4. Available at https://doi.org/10.48130/tia-0024-0014.

Nuwarapaksha, T.D., Udumann, S.S., Dissanayaka, N.S., Atapattu, A.J., 2025. Nanotechnology in the Life Sciences,Green Nanomaterials. Springer Nature, Switzerland, pp. 55–77. Available at 10.1007/978-3-032-04492-1_3.

Nuwarapaksha, T.D., Udumann, S.S., Dissanayaka, N.S., Atapattu, A.J., 2024. Coconut-based livestock farming: a sustainable approach to enhancing food security in Sri Lanka. In: Kiba, D.I. (Ed.), Transitioning to Zero Hunger. MDPI Books, pp. 197–213. Available at https://doi.org/10.3390/books978-3-03897-863-3-8.

Nuwarapaksha, T.D., Dilhan, R.M.C.P., Udumann, S.S., Ranasinghe, C.S., Egodawatta, W.C.P., Dissanayaka, N.S., et al., 2025a. Estimation of the total carbon sequestration potential of coconut-gliricidia mixed cropping systems in Sri Lanka. Carbon Footprints 4 (1). Available at https://doi.org/10.20517/cf.2024.31.

Nuwarapaksha, T.D., Udumann, S.S., Dissanayaka, N.S., Dilshan, R.M.N., Atapattu, A.J., 2025b. AI-driven solutions for sustainable irrigation: exploring smart technologies to enhance conservation and efficiency. In: Garwi, J., Dzingirai, M., Masengu, R. (Eds.), Integrating Agriculture, Green Marketing Strategies, and Artificial Intelligence. IGI Global Scientific Publishing, pp. 1–32. Available at https://doi.org/10.4018/979-8-3693-6468-0.ch001.

Nuwarapaksha, T.D., Udumann, S.S., Dissanayaka, N.S., Atapattu, A.J., 2026. Climate change impacts and adaptation strategies in coconut plantations: integrating remote sensing and real-time monitoring. Frontiers in Climate 8. Available at 10.3389/fclim.2026.1762364.

Oelbermann, M., Paul Voroney, R., Gordon, A.M., 2004. Carbon sequestration in tropical and temperate agroforestry systems: a review with examples from Costa Rica and southern Canada. Agriculture, Ecosystems & Environment 104 (3), 359–377. Available at https://doi.org/10.1016/j.agee.2004.04.001.

Ohyama, T., 2017. The role of legume-rhizobium symbiosis in sustainable agriculture. In: Sulieman, S., Phan Tran, L.-S. (Eds.), Legume Nitrogen Fixation in Soils with Low Phosphorus Availability. Springer International Publishing, Cham, pp. 1–20. Available at https://doi.org/10.1007/978-3-319-55729-8_1.

Olivares, J., Bedmar, E.J., Sanjuán, J., 2013. Biological nitrogen fixation in the context of global change. Molecular Plant-Microbe Interactions 26 (5), 486–494. Available at 10.1094/MPMI-12-12-0293-CR.

Panchal, P., Preece, C., Peñuelas, J., Giri, J., 2022. Soil carbon sequestration by root exudates. Trends in Plant Science 27 (8), 749–757. Available at https://doi.org/10.1016/j.tplants.2022.04.009.

Piotrowska, A., Wilczewski, E., 2012. Effects of catch crops cultivated for green manure and mineral nitrogen fertilization on soil enzyme activities and chemical properties. Geoderma 189-190, 72–80. Available at https://doi.org/10.1016/j.geoderma.2012.04.018.

Poeplau, C., Don, A., 2015. Carbon sequestration in agricultural soils via cultivation of cover crops - A meta-analysis. Agriculture, Ecosystems and Environment 200, 33–41. Available at https://doi.org/10.1016/j.agee.2014.10.024.

Prasad, S., Malav, L.C., Choudhary, J., Kannojiya, S., Kundu, M., Kumar, S., et al., 2021. Soil microbiomes for healthy nutrient recycling. In: Yadav, A.N., Singh, J., Singh, C., Yadav, N. (Eds.), Current Trends in Microbial Biotechnology for Sustainable Agriculture. Springer, Singapore, pp. 1–21. Available at https://doi.org/10.1007/978-981-15-6949-4_1.

Qaswar, M., Huang, J., Ahmed, W., Li, D., Liu, S., Ali, S., et al., 2019. Long-term green manure rotations improve soil biochemical properties, yield sustainability and nutrient balances in acidic paddy soil under a rice-based cropping system. Agronomy 9 (12), 780. Available at https://doi.org/10.3390/agronomy9120780.

Raj, A., Jhariya, M.K., Banerjee, A., Nema, S., Bargali, K., 2023. Land and Environmental Management Through Forestry. Wiley-Scrivener, USA.

Raj, A., Jhariya, M.K., Banerjee, A., Jha, R.K., Singh, K.P., 2024. Agroforestry. Wiley-Scrivener, USA.

Raj, A., Jhariya, M.K., Meena, R.S., 2025. Agroforestry for Monetising Carbon Credits. Springer Nature, Switzerland.

Rogelj, J., Den Elzen, M., Höhne, N., Fransen, T., Fekete, H., Winkler, H., et al., 2016. Paris agreement climate proposals need a boost to keep warming well below 2 °c. Nature 534 (7609), 631–639. Available at https://doi.org/10.1038/nature18307.

Rosa, L., Gabrielli, P., 2023. Achieving net-zero emissions in agriculture: a review. Environmental Research Letters 18 (6), 063002. Available at https://doi.org/10.1088/1748-9326/acd5e8.

Rose, T.J., Kearney, L.J., Erler, D.V., van Zwieten, L., 2019. Integration and potential nitrogen contributions of green manure inter-row legumes in coppiced tree cropping systems. European Journal of Agronomy 103, 47–53. Available at https://doi.org/10.1016/j.eja.2018.11.008.

Sarfraz, R., Hussain, A., Sabir, A., Ben Fekih, I., Ditta, A., Xing, S., 2019. Role of biochar and plant growth promoting rhizobacteria to enhance soil carbon sequestration—a review. Environmental Monitoring and Assessment 191 (4). Available at https://doi.org/10.1007/s10661-019-7400-9.

Scholberg, J.M.S., Dogliotti, S., Leoni, C., Cherr, C.M., Zotarelli, L., Rossing, W.A.H., 2010. Cover Crops for Sustainable Agrosystems in the Americas. In: Lichtfouse, E. (Ed.), Genetic Engineering, Biofertilisation, Soil Quality and Organic Farming. Sustainable Agriculture Reviews, vol 4. Springer, Dordrecht, pp. 23–58. Available at https://doi.org/10.1007/978-90-481-8741-6_2.

Sekar, N., Ramasamy, R.P., 2015. Recent advances in photosynthetic energy conversion. Journal of Photochemistry and Photobiology C: Photochemistry Reviews 22, 19–33. Available at https://doi.org/10.1016/j.jphotochemrev.2014.09.004.

Shakoor, A., Ashraf, F., Shakoor, S., Mustafa, A., Rehman, A., Altaf, M.M., 2020. Biogeochemical transformation of greenhouse gas emissions from terrestrial to atmospheric environment and potential feedback to climate forcing. Environmental Science and Pollution Research 27 (31), 38513–38536. Available at https://doi.org/10.1007/s11356-020-10151-1.

Silva, R.M., Canellas, L.P., 2022. Organic matter in the pest and plant disease control: a meta-analysis. Chemical and Biological Technologies in Agriculture 9, 70. Available at https://doi.org/10.1186/s40538-022-00332-0.

Singh, T.B., Ali, A., Prasad, M., Yadav, A., Shrivastav, P., Goyal, D., et al., 2020. Role of organic fertilizers in improving soil fertility. In: Naeem, M., Ansari, A.A., Gill, S.S. (Eds.), Contaminants in Agriculture. Springer International Publishing, Cham, pp. 61–77. Available at https://doi.org/10.1007/978-3-030-41552-5_3.

Smith, P., Soussana, J.F., Angers, D., Schipper, L., Chenu, C., Rasse, D.P., et al., 2020. How to measure, report and verify soil carbon change to realize the potential of soil carbon sequestration for atmospheric greenhouse gas removal. Global Change Biology 26 (1), 219–241. Available at https://doi.org/10.1111/gcb.14815.

Snapp, S.S., Ugarte, C.M., Hunter, D.W., Wander, M.M., 2022. Cover crops for soil health. Improving Soil Health. Burleigh Dodds Science Publishing Limited, Cambridge, pp. 147–188. Available at 10.19103/AS.2021.0094.11.

Thiessen Martens, J., Entz, M., 2011. Integrating green manure and grazing systems: a review. Canadian Journal of Plant Science 91 (5), 811–824. Available at https://doi.org/10.4141/cjps10177.

Udumann, S.S., Dissanayaka, D.M.N.S., Nuwarapaksha, T.D., Dissanayake, D.K.R.P.L., Atapattu, A.J., 2023a. Megathyrsus maximus as a raw material for organic fertilizer production: a feasibility study. Technology in Horticulture 3. Available at https://doi.org/10.48130/TIH-2023-0009.

Udumann, S.S., Dissanayaka, N.S., Nuwarapaksha, T.D., Thelwadana, E.P., Atapattu, A.J., 2023b. Assessing the growth potential of Sunn hemp (*Crotalaria juncea* L.) as a cover crop for major coconut-growing soils. Trends in Horticulture 6 (2), 3579. Available at https://doi.org/10.24294/th.v6i2.3579.

Vos, J., van der Putten, P.E.L., 2004. Nutrient cycling in a cropping system with potato, spring wheat, sugar beet, oats and nitrogen catch crops. II. Effect of catch crops on nitrate leaching in autumn and winter. Nutrient Cycling in Agroecosystems 70 (1), 23–31. Available at https://doi.org/10.1023/B:FRES.0000049358.24431.0d.

Wang, J., Hussain, S., Sun, X., Chen, X., Ma, Z., Zhang, Q., et al., 2022. Nitrogen application at a lower rate reduce net field global warming potential and greenhouse gas intensity in winter wheat grown in semi-arid region of the Loess Plateau. Field Crops Research 280, 108475. Available at https://doi.org/10.1016/j.fcr.2022.108475.

Yoro, K.O., Daramola, M.O., 2020. Chapter 1 - CO_2 emission sources, greenhouse gases, and the global warming effect. In: Rahimpour, M.R., Farsi, M., Makarem, M.A. (Eds.), Advances in Carbon Capture: Methods, Technologies and Applications. Woodhead Publishing, pp. 3–28. Available at https://doi.org/10.1016/B978-0-12-819657-1.00001-3.

Zhang, B., Tian, H., Ren, W., Tao, B., Lu, C., Yang, J., et al., 2016. Methane emissions from global rice fields: magnitude, spatiotemporal patterns, and environmental controls. Global Biogeochemical Cycles 30 (9), 1246–1263. Available at https://doi.org/10.1002/2016GB005381.

Zhang, D., Yao, P., Zhao, N., Cao, W., Zhang, S., Li, Y., et al., 2019. Building up the soil carbon pool via the cultivation of green manure crops in the Loess Plateau of China. Geoderma 337, 425–433. Available at https://doi.org/10.1016/j.geoderma.2018.09.053.

Zheng, J., Mi, Z., Coffman, D.M., Shan, Y., Guan, D., Wang, S., 2019. The slowdown in China's carbon emissions growth in the new phase of economic development. One Earth 1 (2), 240–253. Available at https://doi.org/10.1016/j.oneear.2019.10.007.

Zhou, G., Gao, S., Xu, C., Dou, F., Shimizu, K.-yoshi, Cao, W., 2020. Rational utilization of leguminous green manure to mitigate methane emissions by influencing methanogenic and methanotrophic communities. Geoderma 361, 114071. Available at https://doi.org/10.1016/j.geoderma.2019.114071.

Chapter 12

Enhancing energy and water use efficiency to reduce carbon footprint

Inna Koblianska[1], Jan Buleca[2], Maryna Tatar[2], Qiang Li[3], and Lin Zhang[3]

[1]*Department of International Economic Relations, Sumy State University, Sumy, Ukraine,* [2]*Faculty of Economics, Technical University of Kosice, Kosice, Slovak Republic,* [3]*School of Energy and Environment, City University of Hong Kong, Hong Kong, P.R. China*

Highlights

- Integrating concepts of energy and water efficiency into agricultural decision-making should be prioritized, guiding technology and practice choices to enhance sustainability.
- Agriculture is a primary contributor to carbon and water footprints, both of which are increasing.
- Agrotechnologies offer considerable potential for improving energy and water efficiency. However, effective implementation requires stakeholder training and policy reforms that establish clear incentives and constraints.
- Current research emphasizes the development of alternative energy technologies (especially solar) and decision-support systems for irrigation management, both of which contribute to conserving energy and water.
- Policy decisions on energy and water resources in agriculture should adopt life-cycle thinking and material-flow analysis to ensure sustainable resource management.
- As global efforts to improve well-being, particularly for lower-income populations, lead to higher energy use and greenhouse gas emissions, the adoption of green energy sources becomes essential to reduce the environmental burden and achieve sustainable development goals.

12.1 Introduction

The rising energy demand, fueled by population growth and technological progress, represents a significant challenge for humanity. Modern society requires ever-greater energy resources to sustain its activities. Since fossil fuels remain the primary energy source, rising energy consumption intensifies climate change (Banerjee et al., 2024). Consequently, it is essential to find ways to increase energy efficiency, which would allow society to ensure a high standard of well-being for both current and future generations without compromising the climate system. Agriculture plays a vital role in this context, as an increased need for food accompanies population growth and, consequently, a greater demand for agricultural resources (Kumar et al., 2020; Meena et al., 2025). Energy consumed directly and indirectly in agri-food chains accounts for approximately one-third of global energy consumption. The consumption of energy resources is a significant contributor to climate change and has other environmental consequences, including air and water pollution, as well as land allocation for energy production. Thus improving energy efficiency, particularly in agriculture, is essential not only for reducing climate impact but also for ensuring the sustained availability of key resources necessary for food production and maintaining a high quality of life (Roy et al., 2022; Meena et al., 2026).

Another critical issue today is water scarcity. Water is a fundamental human need; however, 2.4 billion people currently reside in regions experiencing water stress, and projections are bleak: an estimated 1.8 billion people will experience absolute water scarcity, with up to two-thirds of the global population residing in water-deficient areas (UNEP, 2024). This situation will directly impact agriculture, where water is a vital input for food production. Agriculture is not only heavily dependent on water resources but is also the largest consumer and polluter of these resources. Pollution of water resources degrades aquatic ecosystems and disrupts their reproductive functions. Climate change, which contributes to decreased precipitation, coupled with impaired water ecosystems, exacerbates water scarcity. Therefore, the necessity for

Carbon Incentives, Plans, and Policies in Agriculture. DOI: https://doi.org/10.1016/B978-0-443-33557-0.00006-9

more efficient water utilization, especially in meeting current food demands, remains one of the most pressing challenges for the agricultural sector (Banerjee et al., 2023).

This chapter synthesizes and advances scientific knowledge on the energy and water efficiency concepts in agriculture, as well as the ecological footprint associated with these resources. It emphasizes the role of agricultural technologies in implementing these concepts, highlighting best practices, emerging scientific trends, and policy recommendations to improve water and energy resource management in agriculture, thereby attaining sustainable development goals (SDGs).

The first section explores the concept of energy efficiency with a focus on agricultural practices. The second section explores scientific perspectives on water efficiency and examines the impact of agricultural activities on the environment. The third section presents key indicators of the industry's environmental impact, as measured by the carbon (C) and water footprints, followed by an examination of the interconnectedness between water and energy use and environmental outcomes (Section 12.4). The fifth section highlights core agrotechnological approaches, accompanied by a review of practices that reduce environmental footprints and improve resource efficiency (Section 12.6). Section 12.7 analyzes research trends in energy and water use in agriculture based on patent data, and Section 12.8 provides an overview of tools for assessing and managing the industry's environmental footprint. Finally, Section 12.9 assesses the relationship between community well-being and C emissions, underscoring the necessity of immediate measures to improve water and energy use in agriculture for ensuring sustainable human development. The conclusions summarize the chapter's key findings.

12.2 Concept of energy use efficiency in agriculture

Energy efficiency has been a critical issue for humanity since the first signs of the energy crisis in the 60–70s of the 20th century. The rise in energy consumption driven by population growth and technological advancements has only intensified this challenge. Moreover, the environmental impact of energy dependence—particularly greenhouse gas (GHG) emissions tied to fossil fuel use—has escalated the climate change crisis, one of the most significant existential threats to humanity. Despite calls for energy rationalization and a shift toward renewable sources, fossil fuels remain the primary energy source for the global community (Fig. 12.1). In 2021 oil held the largest share of global energy supply at 31.7%, followed by coal at 26.5% and natural gas at 23.3%. Global carbon dioxide (CO_2) emissions from energy use grew by 2.2% in 2022, doubling the annual growth rate of 2010–19, even amid a general economic slowdown. Notable increases were observed in coal use (+8.8% in China), crude oil (+16% in Saudi Arabia), and oil products (+10% in India) (Enerdata, 2023).

Without ensuring proper measures for energy conservation and increasing the efficiency of energy use, humanity will not be able to maintain the achieved development rates. Searching for ways to enhance energy use efficiency is the most urgent task today. A special role here is assigned to agriculture, which heavily relies on energy consumption.

Energy efficiency, as evidenced by the approaches to the interpretation of this concept by various international organizations, is traditionally understood as the ability to achieve the same result with less energy consumption or to increase the amount of useful results with the same level of energy use. Energy efficiency involves reducing energy consumption and eliminating energy losses and waste (EEA, 2024; EESI, 2024). The most universal view of the concept of energy efficiency is its expression as the ratio of energy produced to energy consumed (Benedek et al., 2023). Experts from the International Energy Agency refer to energy efficiency as the "first fuel" in the transition to clean energy for its potential to reduce carbon dioxide (CO_2) emissions effectively and guarantee energy security (IEA, 2024).

The food sector accounts for approximately 30% of the total global energy consumption (FAO, 2011). The need for energy resources in agriculture encompasses all types of energy used in various sectors of agricultural production, forest management, and fishing, as well as indirectly in the food production and processing sector (Benedek et al., 2023; Meena et al., 2024a). In agriculture, a distinction is made between the direct and indirect consumption of energy resources. Direct consumption is energy used for agricultural production requirements at both the farm and processing plant levels, including irrigation, soil preparation, and harvesting. Indirect consumption covers the use of energy resources in the production of goods and services used in agriculture (machinery, fertilizers, and crop protection measures). The main share of energy consumption (about 80%) falls on the sectors of processing and distribution (43%), as well as trade, preparation, and cooking (36%). Moreover, this distribution is unequal for countries with high and low gross domestic product (GDP) levels (FAO, 2011).

Energy consumption is higher at the supply stage in developed countries, whereas in developing countries, it is more prominent at the agricultural stage. This is due to the larger proportion of the agricultural stage in the overall food production chain in these countries (underdeveloped processing). Scholars indicate that even with the growth of energy consumption, improvements in the energy efficiency of agriculture are observed (resulting from a reduction in the energy intensity of manufactured products), especially in developed countries (Wysokiński et al., 2020; Bajan et al., 2020). The input resource supply sector, for both developed and developing countries, is a source of high energy intensity, which

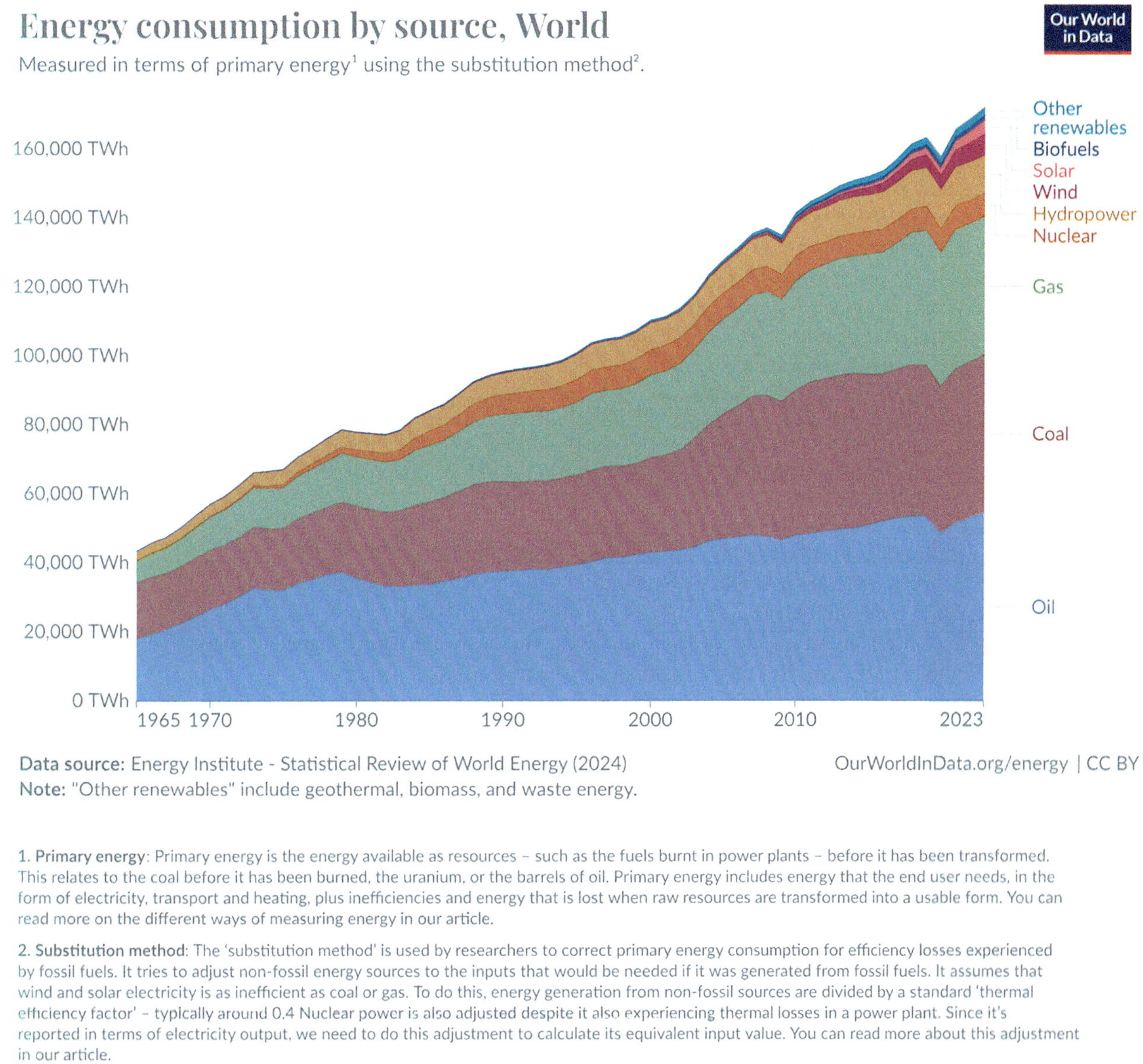

FIGURE 12.1 World energy consumption by source.
Ritchie et al. (2024).

indicates the potential for enhancing agricultural energy efficiency through the efficient management of input resources and the transformation of agricultural supply industries (Bajan et al., 2020; Meena et al., 2024b). The chemical production industry's contribution to agriculture's overall energy consumption is considerable. For example, in wheat production, the portion of total energy used due to chemical inputs can reach 47%; for rice, 43%; and for wheat, 45%. Moreover, about 60% of this volume is caused only by the use of nitrogen fertilizers (Chandra et al., 2020).

Energy efficiency can be understood as the effective utilization of energy as a resource (Wysokiński et al., 2020). In agriculture, energy also serves as a comprehensive metric of efficiency, as all inputs and outputs can be measured in energy units, enabling more holistic evaluations of productivity. This methodology, often referred to as energy analysis, is essential in studying agroecosystems to minimize energy use and enhance the environmental sustainability of production processes (Platis et al., 2019; Meena et al., 2024c).

Agriculture's energy use is unique in that energy serves a dual role: it is an essential input at each stage of the food and materials value chain, while the sector also generates energy, including biofuels, waste-derived energy, and energy from renewable sources (e.g., solar, wind, and small hydropower). The proportion of these renewable sources in total agricultural energy consumption is consistently rising (Fig. 12.1). Energy produced within the agricultural sector can be utilized to meet its own needs or provide an additional income stream for producers (FAO, 2011).

The total energy consumption in agriculture is influenced by several factors, including soil quality, crop types, farm size, climatic conditions, the type of technology and equipment used, fuel types, mechanization levels, and irrigation systems (Eskandari, 2023). Energy intensity also varies by farm size and structure. According to the FAO, subsistence farms typically exhibit low energy intensity and minimal dependence on fossil fuels, while small farms display a low to medium level of dependence on fossil fuels, with varying energy intensities. Nonlivestock small enterprises tend to be more energy-intensive and rely heavily on fossil fuels. In contrast, large corporations, often without livestock, are also highly energy-intensive and dependent on fossil fuels. This high dependency on fossil fuels in industrialized agriculture contributes significantly to GHG emissions. Consequently, FAO experts caution against agricultural industrialization with high fossil fuel usage as a sustainable development (SD) path for the sector. Instead, reducing energy inputs is essential to enhancing energy efficiency (Eskandari, 2023). The transition to more efficient systems and renewable energy sources is projected as the most viable direction for the coming decades (FAO, 2011).

Recent research by Benedek et al. (2023) identifies several key themes that shape the understanding of energy efficiency in agriculture. These themes, ranked by prevalence, include:

1. Agriculture as a source of alternative energy: Emphasizing energy efficiency in terms of using agricultural biomass and the effectiveness of these processes.
2. Energy efficiency as a component of overall farm efficiency: Evaluating energy efficiency within the broader context of farm performance, particularly in relation to climate change and SDGs.
3. Technological aspects of efficient use of energy: Exploring the role of alternative energy and other technologies in enhancing agricultural energy efficiency.
4. Economic aspects of agricultural energy supply: Assessing the cost efficiency of energy supply in agriculture and evaluating the economic effectiveness of energy-intensive technologies.
5. Environmental impact of energy consumption in agriculture: Focusing on the emissions of GHGs and the environmental impact of agricultural energy management.
6. Energy efficiency of the food supply system as a whole: Assessing the energy efficiency across the entire food supply chain.

The ongoing digital transformation, particularly in energy systems, is also reshaping perspectives on energy efficiency in agriculture (Benedek et al., 2023). Between 2010 and 2013, research emphasized biofuels and carbon sequestration (C_{seq}); from 2015 to 2017, the focus shifted to input resources, energy balances, system approaches, and tillage methods. Since 2020, energy efficiency has been increasingly examined from the perspective of advanced technologies, the Internet of Things, C emissions, and future-oriented approaches.

To evaluate energy efficiency in agriculture, a variety of indicators are utilized (Eskandari, 2023). The primary metric is the **Energy Ratio**, measuring the output energy (energy produced by the crop) relative to the input energy (energy consumed in all crop production stages). This ratio indicates the energy output per unit of input energy, with a higher ratio reflecting greater energy efficiency. The energy ratio is applicable for comparing different technologies and individual production systems (Eq. 12.1).

$$Energy\ ratio = \frac{Total\ energy\ output\ (\mathrm{kWh})}{Total\ energy\ input\ (\mathrm{kWh})} \tag{12.1}$$

Net energy output represents the difference between the energy produced through the harvest and the energy expended in its production. A positive and increasing net energy output over time is desirable, signifying an efficient system. Conversely, a negative value signals inefficiency within the system.

Energy productivity measures the quantity of products generated per unit of energy consumed. This metric is particularly useful for comparing production systems with similar characteristics (e.g., crop type). A higher value of energy productivity denotes a more efficient agronomic system.

The energy productivity of a specific agricultural product is determined by the formula given in Mushtaq et al. (2009) (Eq. 12.2), which facilitates efficiency comparisons across different product types.

$$Energy\ productivity = \frac{Grain\ yield\ (\mathrm{kg})}{Total\ energy\ input\ (\mathrm{kWh})} \tag{12.2}$$

Specific energy also referred to as energy intensity, Eq. (12.3) represents the energy required for each unit of output and is determined by dividing the total energy consumed by the quantity of products produced. A lower specific energy value indicates a more energy-efficient agronomic system (Eskandari, 2023; Mushtaq et al., 2009). This metric allows for direct assessment of energy efficiency, with smaller values signaling more sustainable production practices (Bajan et al., 2020).

$$\textit{Specific energy} = \frac{\textit{Total energy input}\ (\text{kWh})}{\textit{Output of grain yield}\ (\text{kg})} \tag{12.3}$$

Energy intensity represents the quantity of primary energy used for each unit of GDP, measured in kilowatt-hours per dollar. GDP is adjusted for inflation and variations in the cost of living between different countries.

12.3 Water use efficiency in agriculture

Water is a critical resource for food and energy production and is essential for sustainable agricultural development (Platis et al., 2019). Globally, agriculture accounts for 70% of freshwater consumption to irrigate 25% of its croplands (Bwambale et al., 2022). According to FAO experts, the strain on land and water ecosystems has reached a critical point, threatening both agricultural productivity and the well-being of those who rely on the sector. Groundwater, one of the primary sources of fresh water, supplies over 30% of the water used for irrigation, and this rate has been increasing by an average of 2.2% annually. About 70% of groundwater extraction is used for irrigating food, fiber, and industrial crops and for livestock production (FAO, 2021).

The intensive use and pollution of water resources, particularly from agricultural chemicals, contribute to increasing water scarcity, further aggravated by climate change. Worsening droughts are diminishing rainfed agricultural areas, thus increasing the demand for irrigation infrastructure (FAO, 2021). Currently, irrigated agriculture accounts for around 40% of global agricultural output. Over the past six decades, the irrigated area has expanded from 139 million hectares in 1961 to more than 328 million hectares in 2018, with roughly 33% of this irrigation supplied by groundwater (FAO, 2021).

This situation highlights the pressing need for effective water management in agriculture. Effective water use in this sector is crucial for maintaining food security, supporting livelihoods, and preserving environmental resources. With the global population rising and climate change intensifying water shortages, improving the efficiency of water usage in agriculture is becoming more crucial than ever (Yu et al., 2020; Meena and Pradhan, 2023). Addressing the challenge of producing more food for a growing population without escalating water usage highlights the need to enhance water efficiency in agriculture (Kilemo, 2022).

Water use efficiency (WUE) in agriculture refers to the relationship between crop yield and the volume of water consumed (An et al., 2022; Bwambale et al., 2022). Achieving high WUE means producing more crops per unit of water, which is essential for several reasons. Firstly, it helps conserve water resources, which are becoming increasingly scarce due to overextraction, pollution, and climate change. Secondly, efficient water use can lower the cost of irrigation, making farming more economically sustainable, particularly for smallholder farmers. Thirdly, it can mitigate the ecological effects of farming, including soil degradation, waterlogging, and salinization, which often result from inefficient irrigation practices (Du et al., 2010; Yu et al., 2020). Fourthly, improving WUE is crucial for mitigating the impacts of climate change-induced water stress, as it ensures that limited water resources are utilized more effectively, reducing wastage and preserving supplies for critical needs (An et al., 2022).

Enhancing the WUE in agriculture is crucial for achieving sustainable food production, conserving water resources, and protecting the environment (Yu et al., 2020; Meena et al., 2023; Bwambale et al., 2022). However, several challenges impede the attainment of high WUE in agriculture. One of the primary challenges is the variability in water availability due to climatic conditions (Konapala et al., 2020). In many regions, rainfall is unpredictable and unevenly distributed, making irrigation planning and management more challenging. Additionally, conventional irrigation techniques, such as flood irrigation, are often inefficient, resulting in significant water losses through evaporation, runoff, and deep percolation.

Another challenge is the lack of access to modern irrigation technologies and practices (Bwambale et al., 2022). Many farmers, particularly in developing nations, lack the finances or technical knowledge to implement efficient irrigation systems, such as drip or sprinkler irrigation. Furthermore, inadequate infrastructure, such as poorly maintained canals and distribution networks, can result in water losses and reduced efficiency.

Sustainable water management, a critical strategy for addressing water scarcity, is defined as utilizing water resources to meet human needs without compromising their availability for future generations (Kilemo, 2022; Pradhan and Meena, 2023a). This approach has reshaped perspectives on water efficiency, leading to the development of specific indicators to support effective water management and conservation. Key indicators include WUE, which measures the ratio or percentage of water absorbed by a crop relative to the total water applied, and water productivity, which indicates the quantity of food produced per unit of water used. Notably, these indicators serve distinct purposes and are not interchangeable. While higher WUE reflects optimal water absorption, it does not necessarily correlate with higher water productivity, as productivity levels are crop-specific and influenced by various factors (Kilemo, 2022).

More than 25% of the water used in buildings, construction sites, or industrial facilities is wasted. Such inefficiencies exacerbate water scarcity, leading to the need for more energy-intensive water supply methods that produce higher GHG emissions. Examples include large canals, extensive delivery pipelines, and desalination plants—each of which contributes to increased emissions and exacerbates climate change. The term "water intake" indicates the volume of water drawn from the source, whereas "water discharge" denotes the quantity returned. The difference between these two measures represents the amount of water consumed (Eq. 12.4).

$$\textit{Water intake} - \textit{Water discharge} = \textit{Consumption} \tag{12.4}$$

The total volume of water utilized is referred to as gross water use. The amount of recirculated water reflects the disparity between total water use and water intake (Eq. 12.5). This volume that is recirculated is commonly referred to as the recycling rate and measures the efficiency of water use.

$$\textit{Gross water use}-\textit{Water intake} = \textit{Amount recirculated (or recycling rate)} \tag{12.5}$$

Water productivity is a key metric for evaluating WUE. It is calculated as the marginal crop yield per unit of water consumed, reflecting the average crop output per volume of water consumed. Water productivity can be quantified by calculating the ratio of yield (in kilograms) to the volume of irrigation water applied (m^3), (Eq. 12.6).

$$\textit{Water productivity} = \frac{\textit{Grain yield}\ (\mathrm{kg})}{\textit{Volume of applied irregation water}\ (\mathrm{m}^3)} \tag{12.6}$$

Water productivity is measured as the amount of crop yield (in kilograms per hectare) relative to the amount of water applied, including both irrigation and precipitation. When no irrigation is used, precipitation during the crop's growing season can be considered instead. Water productivity can be evaluated across different spatial and temporal dimensions, encompassing levels from an individual site, plant, or ecosystem to broader scales such as biomes or even the global level (Kilemo, 2022). According to Pimentel (2022), modern water resource management in agriculture should prioritize optimizing WUE. This approach emphasizes using fewer water resources while simultaneously producing cost-effective plant-based protein, thereby enhancing resource conservation and affordability.

12.4 Carbon footprint of an agroecosystem

The economic efficiency of agroecosystems depends on how effectively resources such as water, energy, and land are used. By improving resource-use efficiency, farmers and producers can reduce costs while maintaining or even increasing productivity. For example, adopting more water-efficient irrigation practices or renewable energy technologies can lower operational costs, making agricultural production more cost-effective.

Water-intensive crops or high-energy farming methods (e.g., irrigation, mechanization) contribute significantly to the environmental footprint of agroecosystems. The costs associated with these inputs directly affect the cost structure of agricultural production. Efficiently managing these resources can reduce production costs, thereby improving the economic viability of farming.

Agricultural production also leads to GHG emissions, primarily from fertilizer use and livestock production. These emissions contribute to the C footprint of agroecosystems and often carry financial implications, such as increased taxation or regulatory costs related to C emissions. Implementing practices that reduce emissions (e.g., precision fertilization, agroforestry) can help to mitigate these costs (Kumar et al., 2024; Pradhan and Meena, 2023b).

The concepts of C and water footprints are essential for resource management and impact mitigation in agriculture (Jhariya et al., 2022, 2023). By assessing agricultural impacts in these terms, resource use can be better managed, supporting efforts to reduce GHG emissions within the industry (Platis et al., 2019). Evaluating these specific aspects, such as the C and water footprint, underscores the need for society to prioritize them in response to climate change.

The energy footprint quantifies the energy consumption associated with a particular product, organization, or region within a specified spatial and temporal framework (Banerjee et al., 2020). This assessment is typically conducted from a life cycle perspective. Energy sources can be categorized based on their nature and renewability into nonrenewable and renewable sources.

Agri-food systems are both influenced by climate change and play a major role in GHG emissions. These emissions occur throughout the entire agri-food supply chain, spanning from industrial processing and farming to household consumption, and contribute significantly to global warming (Meena et al., 2022).

The term "C footprint" describes the GHG emissions generated for each unit of a product or service, as well as emissions by pollutant types per unit produced, measured in mass units like kilograms or tons. The ISO 14067 standard provides guidelines for assessing the C footprint of products or services, ensuring consistency and comparability in measurement practices.

Major sources of emissions in agriculture stem from the application of chemicals, electricity consumption, and the burning of fossil fuels, all of which are driven by the increasing demands of a growing global population. In 2021 agrifood systems generated a total of 16.2 billion metric tons of CO_2 equivalent (Gt CO_2 eq.), marking a 10% increase—equivalent to an additional 1.5 Gt CO_2 eq.—compared to emissions in 2000. This growth is mainly attributed to higher emissions from both pre and postproduction processes, as well as on-farm activities, which have outweighed the decline in emissions resulting from land-use changes. Specifically, emissions from on-farm operations reached 7.8 Gt CO_2 eq., accounting for 48% of total emissions in 2021. Pre and postproduction activities contributed 5.3 Gt CO_2 eq., or 33%, while emissions from land-use changes made up 3.1 Gt CO_2 eq., or 19%.

Fig. 12.2 provides a breakdown of global GHG emissions associated with agricultural land.

Fig. 12.3 displays the distribution of GHG emissions from agricultural land by region. In 2020 Asia was the largest contributor to emissions, releasing 6.8 Gt CO_2 eq., which accounted for 42% of the global total. The Americas ranked second, emitting 4.3 Gt CO_2 eq. (26%), followed by Africa at 2.9 Gt CO_2 eq. (18%). Europe accounted for 1.9 Gt CO_2 eq. (12%), while Oceania had the lowest emissions, contributing 0.3 Gt CO_2 eq. (2%).

Global GHG emissions from agricultural land, categorized by type of GHG, are shown in Fig. 12.4.

CO_2 was the dominant GHG, making up half of the total emissions at 8.2 Gt CO_2 eq. Methane (CH_4) was the second most prevalent, contributing 5.3 Gt CO_2 eq., or 32%. Nitrous oxide (N_2O) followed with 2.3 Gt CO_2 eq., or 14%, and fluorinated gases (F-gases) accounted for 0.5 Gt CO_2 eq., or 3%.

Emissions from agricultural activities within the farm gate, which involve crop and livestock production, increased by 14% from 2000 to 2021, reaching 7.8 Gt CO_2 eq. (Fig. 12.5). Livestock-related activities were responsible for approximately 53% of these emissions, with enteric fermentation in ruminant animals alone contributing 37%, or 2.9 Gt

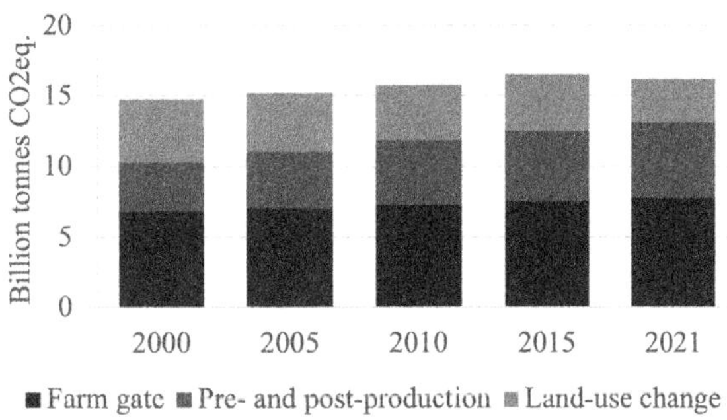

FIGURE 12.2 Global greenhouse gas emissions from agricultural land categorized by contributing factors. Own processing based on FAO UN (2024).

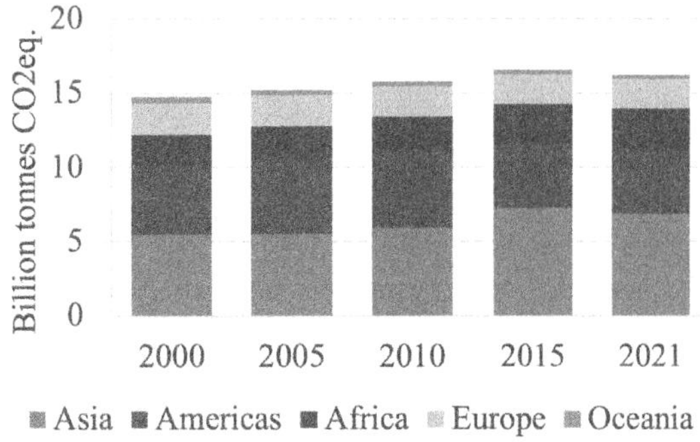

FIGURE 12.3 Global GHG emissions from agricultural land by geographical region. *GHG*, greenhouse gas. Own processing based on FAO UN (2024).

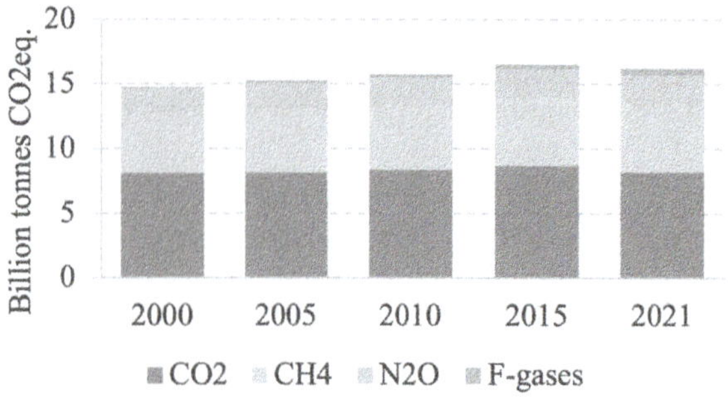

FIGURE 12.4 Global GHG emissions from agricultural land are categorized by type of GHG. *GHG*, greenhouse gas. Own processing based on FAO UN (2024).

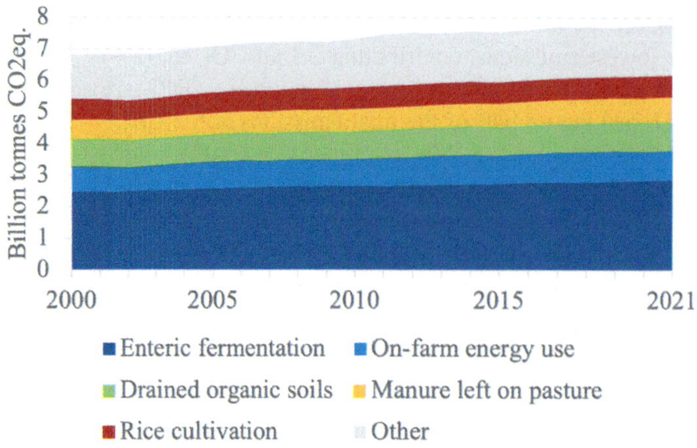

FIGURE 12.5 Global farm-gate GHG emissions categorized by agricultural activity. *GHG*, greenhouse gas. Own processing based on FAO UN (2024).

CO_2 eq. The application of fertilizers, including both synthetic and animal-based sources, was responsible for approximately 20% of agricultural emissions. Energy consumption on farms and emissions from drained organic soils each accounted for 12%–13% of farm-gate emissions between 2000 and 2021. Meanwhile, methane emissions generated from rice cultivation accounted for 9%–10% during the same period.

Assessing emission intensities enables the evaluation of GHG efficiency across various commodities, offering a clear metric for comparing production efficiency by country and over time. As shown in Fig. 12.6, in 2021, cattle meat and sheep meat had the highest C footprints, with emission intensities of 28 kg CO_2 eq./kg and 24 kg CO_2 eq./kg, respectively. In contrast, pig meat and chicken meat exhibited much lower emissions, at 1.7 kg CO_2 eq./kg and 0.5 kg CO_2 eq./kg, respectively.

For dairy products, cow milk had a global emissions intensity of 1 kg CO_2 eq./kg, while sheep milk was notably higher at 5.9 kg CO_2 eq./kg. Rice production had an emissions intensity of 1 kg CO_2 eq./kg, which is nearly six times greater than that of other cereals, which stood at 0.2 kg CO_2 eq./kg. Emission intensities also vary significantly by region, reflecting disparities in production efficiency. For instance, in Africa, the emissions intensity for cattle meat is over twice the global average at 57 kg CO_2 eq./kg, whereas in Europe, it is nearly half the global average at 17 kg CO_2 eq./kg.

In addition to absolute emissions data, indicator values are available for all countries, highlighting GHG emissions per capita, the emissions intensity of agricultural production, and the emissions intensity of land use. Table 12.1 provides an overview of the top and bottom 10 countries for each indicator, along with median values, focusing on nations with agricultural output valued at a minimum of USD 1 billion in 2021.

The C footprint impact of food and agriculture is presented in Fig. 12.7.

So, the conducted analysis of statistical data shows the growth of the world's C footprint from 2000 to 2021. The world farm-gate C footprint increased from all types of farming activities.

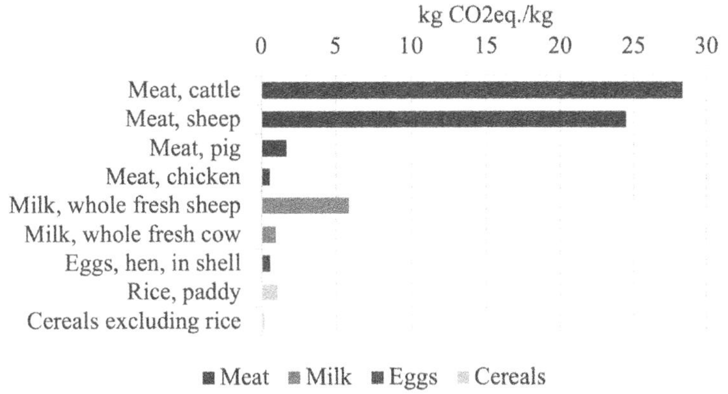

FIGURE 12.6 Global GHG emissions intensity of agricultural commodities in 2021. *GHG*, greenhouse gas. Own processing based on FAO UN (2024).

TABLE 12.1 Summary statistics of emissions indicators.

Indicator	Ten lowest values (range)	Ten highest values (range)	Median
GHG emissions from agri-food systems per person ($tCO_2eq./cap$)	0.5–0.8	8–18	2
Emissions from agricultural land relative to the value of agricultural production (kg CO_2 eq./USD)	0.7–1	13–51	3
Emissions from agricultural land per hectare of agricultural land ($tCO_2eq./ha$)	0.1–0.5	11–25	3

GHG, greenhouse gas.
Source: Own processing based on FAO UN (2024).

The "water footprint" refers to the overall amount of freshwater utilized in the production of a product or service, encompassing both direct and indirect water consumption throughout the entire production cycle (supply chain). This is measured in units of volume, such as liters or cubic meters. The water footprint encompasses all water used in the production process, from growing raw materials to manufacturing and distribution, as well as wastewater treatment.

There are three components of the water footprint:

1. Green water footprint: This refers to rainwater that is stored in the soil and used by plants for growth. It includes the water that evaporates or is transpired by plants.
2. Blue water footprint: This is the amount of freshwater from different sources (surface or ground) used in production. It can include water that is consumed and does not return to its original source, or is returned at a different time or location.
3. Gray water footprint: This represents the quantity of clean water needed to dilute contaminants to satisfy water quality standards. It is linked to the water required for cleaning or removing contaminants during the production process.

The overall water footprint associated with a product or service is the sum of these three components. In agriculture, it provides insight into the amount of water required to produce food and other agricultural products. The ISO 14046 standard outlines the methodology for conducting a water footprint assessment, ensuring consistency and accuracy in evaluating water use across industries and production processes.

The green water footprint represents the rainwater and soil moisture utilized by plants during their growth. In contrast, the blue water footprint encompasses water from surface water and groundwater that is either evaporated or absorbed by crops during cultivation.

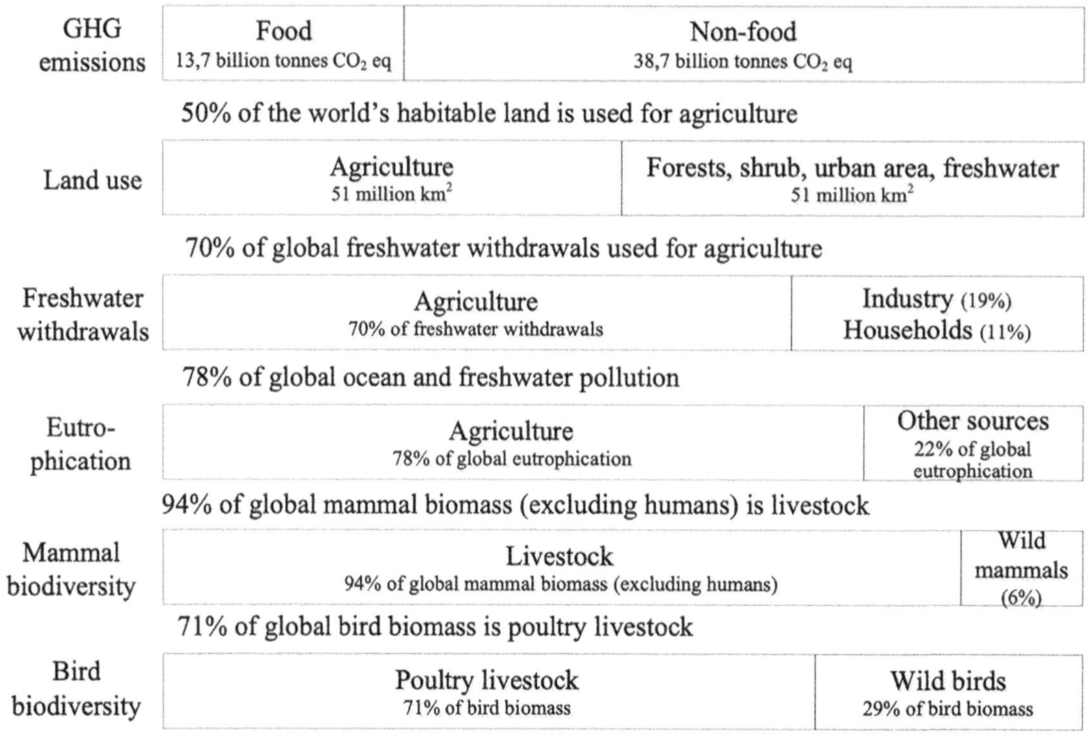

FIGURE 12.7 The footprints'impact of food and agriculture (Poore and Nemecek, 2018).
Own processing based on Bar-On et al. (2018).

Crop water demand encompasses both green and blue water components. Green water refers to rainfall and soil moisture that crops directly absorb, whereas "blue water" pertains to surface and groundwater utilized for irrigation (Fan et al., 2020a,b).

The theoretical green water requirement (CWR_{green}, $m^3\ ha^{-1}\ yr^{-1}$) and blue water requirement (CWR_{blue}, $m^3\ ha^{-1}\ yr^{-1}$) for crops are determined using the following calculations (Eqs. 12.7 and 12.8):

$$CRW_{green} = 10 \times \sum_{n}^{d=1} ET_{green\,(d)} \tag{12.7}$$

$$CRW_{blue} = 10 \times \sum_{n}^{d=1} ET_{blue\,(d)} \tag{12.8}$$

In this context, ET_{green} (mm d^{-1}) and ET_{blue} (mm d^{-1}) represent the crop evapotranspiration derived from precipitation and irrigation water, respectively. The variable "n" refers to the duration of the crop's growing season within a year. A constant value of 10 is applied to convert the units (i.e., from mm to $m^3\ ha^{-1}$). The values for ET_{green} and ET_{blue} are determined using the following formulas (Eqs. 12.9 and 12.10):

$$ET_{green} = \min(ET_c, P_e) \tag{12.9}$$

$$ET_{blue} = \max(0, ET_c - P_e) \tag{12.10}$$

In this context, ET_c (mm d^{-1}) represents the crop evapotranspiration for the specific month in question, while P_e refers to the effective precipitation accessible to each crop type (mm d^{-1}), excluding water lost through runoff and intercepted by plants. The parameter ET_c is determined using the following calculation (Eq. 12.11):

$$ET_c = K_c \times ET_0 \tag{12.11}$$

In this case, ET_0 (mm d^{-1}) represents the reference crop evapotranspiration for the given month, which is calculated using the Penman–Monteith equation as recommended by the FAO. K_c is a coefficient that indicates the variation in evapotranspiration between the cropped area and a reference grass surface. The value of P_e is determined using Eqs. (12.12) and (12.13):

$$P_e = P \times \frac{4,17 - 0,2 \times P}{4,17}, \ for\ P < 8,3 \text{ mm d}^{-1} \tag{12.12}$$

$$P_e = 4,17 + 0,1 \times P, \ for\ P \geq 8,3 \text{ mm d}^{-1} \tag{12.13}$$

Here, P represents the daily precipitation (mm d^{-1}), which is calculated by dividing the total monthly precipitation by the number of days in that particular month.

Agriculture accounts for 70% of the world's total freshwater usage. Additionally, agricultural activities contribute to 78% of oceanic and freshwater eutrophication, which involves the contamination of waterways with nutrient-rich runoff. In terms of mammal biomass, livestock constitute 94% of the total nonhuman mammal biomass, making them 15 times more prevalent than wild mammals. This proportion increases to 97% when only land-based mammals are considered. Likewise, poultry accounts for 71% of bird biomass, indicating that they outnumber wild birds by over 3-to-1.

The *land footprint* refers to the total area of land required to produce a product or service, including both direct land use (e.g., land used for cultivation or livestock) and indirect land use (e.g., land needed to produce inputs like feed, fertilizers, or other materials in the production process). It is measured in units of area, typically square meters or hectares.

The land footprint reflects the extent of natural land resources used by human activities in the production of goods and services. Similar to C and water footprints, it serves as a key measure for evaluating the sustainability of production methods. A larger land footprint typically indicates greater demand for land resources, which can result in environmental harm, habitat loss, and other detrimental effects on ecosystems.

In addition to calculating the land footprint for individual products or services, it can also be assessed at broader scales, such as for countries, cities, or organizations, or even on a per capita basis to understand environmental impact at a population level. This enables informed decisions about land use, agricultural practices, urban development, and resource management. Current agriculture utilizes half of the world's habitable land, which is defined as land that is not covered by ice or desert, illustrating the significance of the agricultural land footprint alongside with the C and water ones.

12.5 Interrelationship between energy–water use efficiency and Carbon footprint

Agriculture relies heavily on both energy and water for various activities such as irrigation, machinery operation, and processing. The food industry consumes a substantial amount of global energy, distributed across several areas:

1. agricultural production (this sector accounts for 6.6% of global energy use; energy is required for operations such as running tractors, maintaining greenhouses, primary irrigation, crop production, fertilizer manufacturing, livestock housing, crop storage, and fishing boats);
2. retail, preparation, and cooking (this category represents 11.6% of global energy consumption);
3. processing and distribution (these activities use 14.1% of global energy).

The water use effectiveness in agriculture is frequently associated with energy-demanding irrigation systems, and the energy footprint is closely related to the water required for the production and processing of agricultural products.

Agricultural practices are energy-intensive, involving significant energy for planting, harvesting, processing, and transportation. Improving energy efficiency in agriculture can help lower its C footprint, reduce expenses, and promote more sustainable farming practices. The agriculture–energy–water–C nexus is presented in Fig. 12.8.

The energy required for water production is significant due to the large-scale industrial processes involved (Dycian, 2022). Water is extracted from sources such as underground aquifers or reservoirs, and desalination processes, which are even more energy-intensive, may be used. Following extraction, water undergoes filtration and chemical treatment and is then distributed to consumers, sometimes over long distances. This entire process consumes substantial amounts of energy. Additionally, once water is used, it is directed to sewage treatment facilities, which also require significant energy. These sewage treatment processes contribute to the release of substantial GHGs.

These emissions are some of the most powerful GHGs, playing a major role in global warming. The overall GHG emissions linked to water usage can differ depending on factors such as the water's source and the distance it needs to travel to reach consumers. On average, every cubic meter of water used generates 23 pounds (10.6 kg) of C emissions. Fig. 12.9 shows the GHG emissions associated with water consumption.

Water usage in agriculture also involves considerable energy consumption. The amount of energy needed for irrigation depends on the volume of water used and the total pressure head, which is affected by factors such as the irrigation method

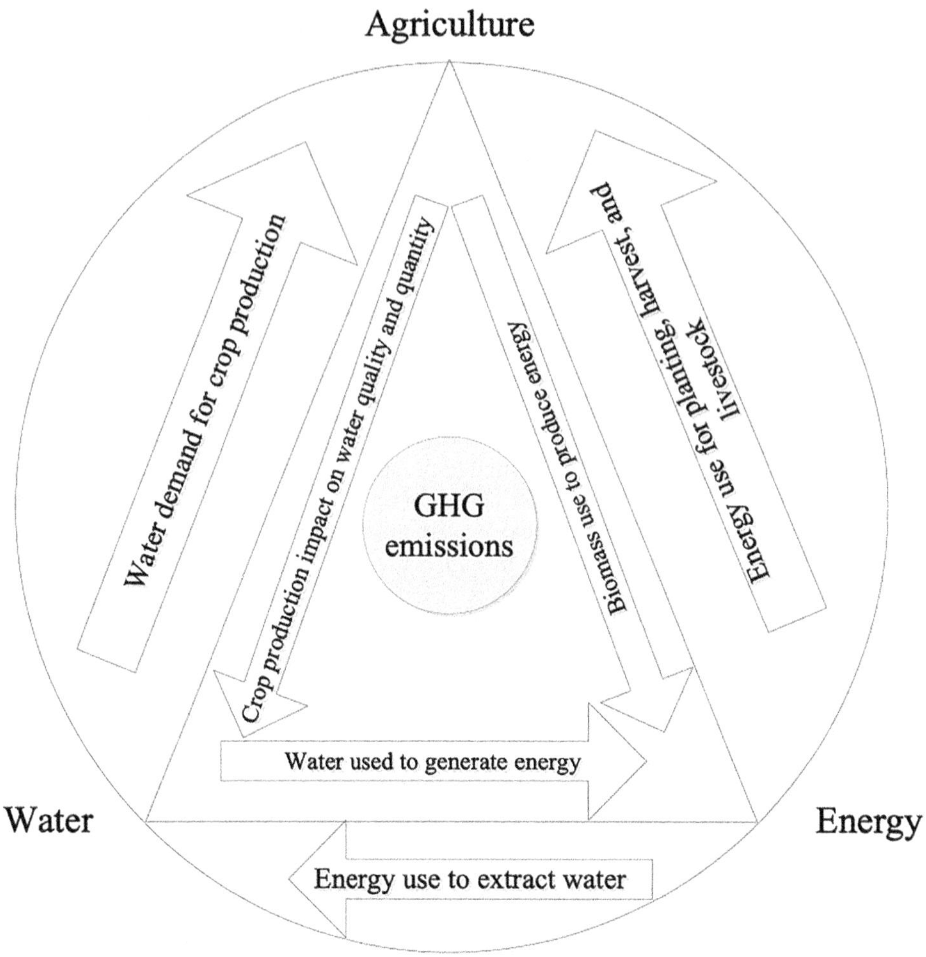

FIGURE 12.8 Agriculture–energy–water–carbon nexus.
Own processing.

(drip, sprinkler, or surface), the type of pumping systems employed (diesel, mechanical, or electric pumps), and the water source (surface water or groundwater).

For different irrigation methods, typical operating pressures are 0.41 bar for surface irrigation (with surface water sources set to 0), 1 bar for drip irrigation, and 3 bar for sprinkler systems. The friction losses in the piped distribution system, which are 0.69 bar, apply to all irrigation types, as noted by Brown (2002). The specific energy needs for these systems can be calculated using the methods described by Qin et al. (2024)–Eqs. (12.14) and (12.15).

$$EQ = \frac{V \times TH}{367 \times \eta} \tag{12.14}$$

$$TH = Lift + D + H + f_{losses} \tag{12.15}$$

The amount of energy required for irrigation, denoted as EQ (kWh), depends on several factors, including the volume of water used (V, m^3) and the total pressure head (TH, m). The total pressure head is determined by adding multiple components: the elevation difference from the groundwater table to the surface, the drawdown depth (D) caused by the cone of depression and reduced well efficiency, the operating pressure (H), and pipe friction losses (f_{losses}). The efficiency of the pumping system and its power source (η) also influence energy consumption. The prime mover's efficiency varies depending on whether the energy source is diesel or electricity. The groundwater table depth is a critical parameter in calculating the required lift for water extraction. The CO_2 emissions related to irrigation energy use can be determined by

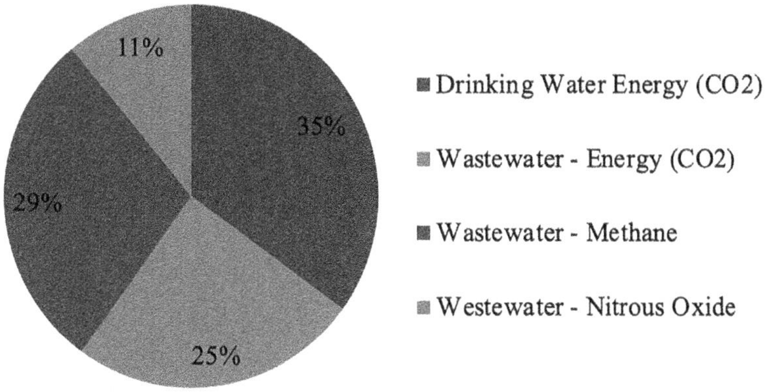

FIGURE 12.9 Greenhouse gas emissions from water use.
Water Intelligence. Dycian (2022).

multiplying the irrigation energy consumption by the relevant C emission factor. The equation for calculating CO_2 emissions from irrigation energy use can be calculated using Eq. (12.16).

$$C_e = E_c \times C_{ef} \tag{12.16}$$

where C_e (g CO_2) is CO_2 emissions; E_c (kWh) is energy consumption, and C_{ef} (g CO_2/kWh) is the CO_2 emission factor of energy consumption.

Energy consumption in irrigation is evaluated using both bottom-up and top-down methodologies. The bottom-up approach estimates direct energy consumption for activities such as land preparation, cultivation, and irrigation. Conversely, the top-down approach utilizes Leontief's input–output model to assess indirect energy consumption. Within irrigation, energy is primarily used for water extraction (pumping) and distribution. The direct energy use for irrigation ($DE_{irrigation}$, MJ ha^{-1} yr^{-1}) can be determined using the Eq. (12.17):

$$DE_{irrigation} = GIR \times R_{GW} \times EC_{irrigation} \times 3,6 \tag{12.17}$$

where GIR (m^3 ha^{-1} yr^{-1}) is the gross irrigation requirement for crops; R_{GW} is the proportion of irrigation sourced from groundwater; $EC_{irrigation}$ (kWh) is the energy required to lift 1 m^3 of water (with density of 1000 kg m^{-3}) by 1 m; and 3,6 MJ kWh^{-1} is the average low calorific value of electricity.

To better distinguish the role of irrigation in overall GHG emissions, the CO_2 emissions resulting from direct energy consumption (E_{direct_energy}) can be categorized into two segments: emissions specifically linked to irrigation and those arising from other agricultural operations excluding irrigation. The CO_2 emissions attributed to irrigation ($E_{irrigation}$, kg CO_2 eq.) can be determined using the following equation (Fan et al., 2020a,b)–(12.18):

$$E_{irrigation} = GIR \times R_{GW} \times \sum_{q}^{r=1} \left(EC_{irrigation} \times R_{energy(r)} \times EF_{energy\_GHG(r)} \right) \tag{12.18}$$

where r (r = 1, 2, …, q) is the energy type used in water pumping; $EC_{irrigation}$ (kWh) is the energy required to lift 1 m^3 of water (with density of 1000 kg m^{-3}) by 1 m; $R_{energy}(r)$ is the proportion of energy type r in the total energy consumption; and EF_{energy_GHG} (kg CO_2 eq. kWh^{-1}) is the GHG emission factor for energy consumption.

Similarly, CO_2 emissions from other on-farm energy use can be estimated by multiplying total energy consumption (TJ ha^{-1}), C content (t C TJ^{-1}), oxygenation efficiency, and a conversion factor for CO_2–C to CO_2. A summary of energy consumption and C emissions associated with water supply from various sources is provided in Table 12.2.

The process of water supply requires energy at multiple stages, including extraction, treatment, and distribution, with energy demands varying depending on the source of the water. For example, drawing groundwater from wells consumes 27% more energy than obtaining water from surface sources. Conversely, surface water requires additional treatment and chemical processing, resulting in a 31% increase in indirect energy use. Among all methods, seawater desalination is the most energy-intensive, demanding 14 times more energy than treating surface water to make it suitable for consumption.

TABLE 12.2 Energy consumption and carbon emissions to deliver water from different sources.

Water source	Energy consumption (kWh), per cubic meter	CO_2 emissions (kg) per cubic meter of water
Desalinated	11.6	8.2
Recycled	4.5	3.2
Surface	0.1	70 gr
Imported	4.7	3.4
Average	5.2	3.7

Source: Dycian (2022).

12.5.1 CO_2 emissions from groundwater degassing

Groundwater typically contains higher concentrations of dissolved CO_2 than the atmosphere above it. This difference in CO_2 levels causes degassing when groundwater is pumped to the surface, leading to CO_2 emissions. Groundwater degassing has been recognized as a contributor to CO_2 emissions from irrigation, but its impact has not been fully quantified. The CO_2 emissions resulting from groundwater degassing during irrigation can be estimated using Eq. (12.19) (Qin et al., 2024):

$$C_{GD} = V_{GD} \times R_{IWW} \times C_{GW} \tag{12.19}$$

In this context, C_{GD} represents the CO_2 emissions resulting from the release of dissolved CO_2 during groundwater extraction for irrigation. V_{GD} refers to the volume of groundwater utilized specifically for irrigation purposes (measured in cubic meters). R_{IWW} denotes the proportion of groundwater withdrawn for irrigation relative to the total groundwater extraction. C_{GW} signifies the concentration of CO_2 present in groundwater.

Ultimately, the total GHG emissions associated with agricultural production can be determined using the following equation, Eq. (12.20):

$$E_{agric} = E_{crops} + E_{fert} + E_{direct\_energy} + E_{indirect\_energy} \tag{12.20}$$

E_{crops}, E_{fert}, E_{direct_energy}, $E_{indirect\_energy}$ (g CO_2 eq.) represent the emissions of CH_4 from cultivated fields, N_2O from nitrogen fertilizer application on both irrigated and dryland farms, CO_2 from direct energy consumption, and CO_2 emitted from indirect energy usage.

Optimizing the use of water and energy not only leads to direct cost savings for businesses and industries but also plays a pivotal role in mitigating broader environmental impacts such as C emissions, water scarcity, and ecosystem degradation. By calculating and analyzing key economic indicators, businesses and policymakers can make informed decisions that balance environmental goals with economic performance. Furthermore, a focus on improving efficiency in both water and energy use contributes to long-term sustainability, enhances competitiveness, and minimizes external costs, making it a cornerstone of sustainable economic strategies. Economic implications of energy-water efficiency are presented in Table 12.3.

Also, several economic indicators can be derived when assessing the interrelationship between energy and WUE and the environmental footprints they generate. These indicators help in understanding both the costs and benefits associated with optimizing resource use and managing the environmental impacts (Table 12.4).

To reduce water and energy use, we advise implementing market- and policy-driven incentives. Governments may provide incentives, subsidies, or tax breaks for companies that reduce their water and energy consumption or C emissions. These financial incentives can enhance the profitability of adopting more sustainable practices and help offset the initial costs of implementing energy- and water-saving technologies. Conversely, industries with high footprints may face regulatory penalties or taxes (such as C taxes), making it economically advantageous to adopt efficient practices.

Adopting sustainable agricultural practices may require significant upfront investment in technologies and infrastructure, such as water-efficient irrigation systems, renewable energy sources, or precision farming equipment. While these investments can be costly initially, they may lead to long-term cost savings and increased resilience to environmental risks such as water shortages or energy price volatility.

TABLE 12.3 Economic implications of energy–water efficiency.

Economic indicator	Economic implications
Operational costs	Energy and water are essential inputs in many industrial processes. Improving the efficiency of both energy and water use can lead to substantial cost savings. For instance, a factory that optimizes its water use while simultaneously reducing energy consumption (through energy-efficient equipment or processes) can significantly lower its operational expenses.
Investment in efficiency technologies	The adoption of energy and water-efficient technologies (e.g., water recycling systems, low-energy irrigation techniques, or energy-efficient water treatment plants) often requires upfront investments. However, these investments can result in long-term economic benefits, such as lower operational costs, increased productivity, and compliance with environmental regulations, leading to a better return on investment (ROI).
Market demand and competitiveness	In markets where energy and water efficiency are increasingly valued, businesses that can demonstrate reduced environmental footprints will likely have a competitive edge. This can lead to increased market demand for environmentally responsible products or services, potentially leading to higher revenues. Consumers and businesses are increasingly aware of sustainability issues, so those who lead in energy–water efficiency are better positioned for growth.

Source: Own processing.

12.6 Agrotechnology for reducing the Carbon footprint of water and energy use in agroecosystems

The water productivity of plants is influenced by several factors, which can be categorized as specific to crops, climate-related, and induced by management practices. Plant-specific factors, such as genotype, play a crucial role in water productivity. Climatic factors include variables such as precipitation, atmospheric parameters, and temperature. While the influence of climatic factors is somewhat constrained, effective management practices can enhance water productivity by adapting to prevailing climatic conditions. These practices include the removal of excess water, irrigation, and crop selection, among others.

Management factors encompass tillage practices, water and land management, disease and weed control, soil management, and irrigation techniques. The latter aims to provide crops with the necessary amount of water while minimizing water resource loss. This goal is achieved through the proper scheduling of irrigation and the adoption of efficient technologies (Kilemo, 2022).

AF systems offer several advantages, including reducing dependence on nonrenewable energy sources, lowering CO_2 emissions, and promoting more efficient water use in agricultural production processes (Platis et al., 2019; Pradhan et al., 2023). It is well-established that soils with plant cover lose less water to the atmosphere, thereby contributing to water conservation. This underscores the importance of adopting technologies such as no-tillage systems, AF, and the use of green manure (Pimentel, 2022).

WUE can be assessed from physiological, agronomic, and engineering perspectives. Several key strategies for improving WUE can be identified (Pimentel, 2022; Kilemo, 2022):

1. Enhancing water absorption by plants: This can be achieved by selecting crops with well-developed root systems. Studies have shown that crops with higher starch content in their roots exhibit better water absorption and higher water-use efficiency, even in rainfed conditions.
2. Reducing water loss by plants: Selecting plants with high internal water-use efficiency can help minimize water loss.
3. Recommending drought-resistant crops: Traditional crops in arid regions tend to be more sustainable than newly adapted ones. Agricultural zoning in marginal regions, identifying crops best suited to specific conditions, and implementing appropriate production practices can help conserve water and improve WUE.
4. Monitoring physiological parameters for plant selection: Modern digital technologies enable real-time monitoring of physiological parameters in plants and the environment, which can optimize irrigation management and assist in selecting drought-resistant crops.

TABLE 12.4 Economic indicators that can be calculated in the context of energy–water use efficiency and footprint.

Economic indicator	Characteristic
Energy cost per unit of output	This indicator measures how much energy is consumed for each unit of product or service produced. It is a critical metric for businesses to assess their energy efficiency.
Water cost per unit of output	Similarly, this indicator tracks the amount of water used per unit of output, helping companies identify areas where water savings could reduce operational costs.
Energy–water efficiency ratio	This ratio calculates the relationship between the amount of energy consumed per unit of water used and vice versa. In sectors like agriculture or manufacturing, optimizing this ratio can lead to significant economic savings and environmental benefits. For example, in water-intensive industries (e.g., agriculture), increasing water efficiency while also reducing energy consumption can lower overall operational costs.
Water footprint	This refers to the total volume of water used to produce goods or services, including direct water use and virtual water embedded in supply chains. A smaller water footprint indicates that a company is using water more efficiently.
Energy footprint	This measures the total energy consumption associated with producing goods or services. An efficient energy use model minimizes the energy footprint, thus reducing the economic costs associated with energy use and improving sustainability.
C footprint	A company's or industry's C footprint quantifies the total GHG emissions associated with energy consumption. This metric is essential for understanding the indirect economic costs, such as potential C taxes or penalties.
ROI for efficiency projects	For businesses looking to invest in water-energy-saving technologies, calculating the ROI can help assess the financial feasibility of adopting these technologies. ROI considers the upfront investment, operational cost savings, and any potential revenue increases due to enhanced sustainability. A higher ROI indicates that the investment in efficiency improvements is financially beneficial.
Payback period	This is the time it takes for the cost savings from improved water–energy efficiency to pay back the initial investment. Shorter payback periods increase the attractiveness of these investments from an economic perspective.
Environmental externalities	One key aspect of assessing the economic relationship between energy-water use efficiency and the footprint is considering the external costs of environmental impacts, such as water scarcity, C emissions, and air pollution. For instance, the economic costs associated with the C emissions from energy use in water supply or irrigation systems could be internalized through C pricing mechanisms like taxes or cap-and-trade systems. By reducing resource consumption and environmental impacts, industries can minimize their exposure to such external costs and improve their economic performance.
Sustainability index	A sustainability index that considers factors like energy, water, and C footprints can provide a holistic view of the economic performance of a sector or company in terms of resource efficiency. A higher sustainability index suggests a lower resource usage and footprint, potentially leading to lower operating costs, better market positioning, and improved public perception.

5. Soil management: Proper soil management practices, which include providing essential nutrients that support biomass formation, are key to influencing water productivity. Effective soil and land management practices, like crop rotation, cover cropping, and selecting appropriate cultivation technologies, help to reduce soil erosion and minimize water loss.
6. Agronomic management: Such practices aimed at improving WUE include several well-established practices, such as no-tillage technologies, AF, interrow cropping, and high plant density. Modern agricultural technologies also incorporate plant growth-promoting microbes to promote the growth of plants and improve water efficiency, as well as embryo improvements to optimize crop development. Key agronomic techniques for increasing WUE include practices such as leaving straw on the field to conserve soil moisture. Research indicates that increasing plant density can lead to higher yields and improved WUE. AF, particularly the planting of tree legumes on fields, also contributes to better water management and overall productivity.

7. Irrigation management: The application of various techniques tailored to specific conditions, with the goal of enhancing WUE. By adjusting irrigation strategies based on environmental factors and crop needs, water use can be optimized, thereby improving overall resource management.

Finally, sustainable water resource management in agroecosystems focuses on balancing plant water requirements with the available water resources of appropriate quality and quantity, ensuring adequate water supply in both time and space, given the prevailing environmental conditions. A key aim of this strategy is to maintain hydrological resources for all users while promoting social well-being, sustainability, and harmony. Using indicators of efficiency and productivity of water provides a solid foundation for making informed decisions at various levels—from farmers to policymakers. These indicators help to ensure that water consumption remains at levels that do not deplete future water resources (Kilemo, 2022). For sustainable water management to be effective, appropriate practices must be implemented across all production systems, regardless of the level of water availability.

In developing nations, agriculture is often the primary economic activity, and water scarcity poses a significant concern. This scarcity is exacerbated by inadequate water management practices, which are frequently due to a lack of investment, knowledge, and skills. Many agricultural activities are organized on small-scale farms, which further contribute to inefficient use of water and water scarcity (Kilemo, 2022).

12.7 Best practices for enhancing water–energy use efficiency in agroecosystems

Energy efficiency in agriculture offers many advantages, such as lower costs, minimized ecological impact, increased overall performance, and, most importantly, a decrease in energy consumption. To achieve the goals of agriculture's high energy efficiency, it is crucial to optimize energy use through the following strategies (Benedek et al., 2023):

- Utilizing more effective water use methods like precision irrigation and drip systems, which lower water and energy waste.
- Adopting energy-efficient equipment, which helps in lowering energy consumption while maintaining productivity.
- Leveraging alternative energy sources like solar, wind, or biogas to decrease dependence on fossil fuels and minimize environmental impacts.
- Ensuring effective lighting, heating, and ventilation in agricultural buildings to optimize energy use and maintain optimal growing conditions.
- Using organic waste for energy production thus reduces waste and generates renewable energy for agricultural operations.
- Implementing precision farming technologies that enhance resource utilization, diminish energy waste, and improve crop yields.
- Optimizing fertilizer and pesticide usage to avoid overapplication, reduce energy costs, and decrease environmental harm.
- Auditing energy use to reveal inefficiencies and improvement opportunities, providing a basis for implementing more energy-efficient practices.
- Developing partnerships and collaborations with stakeholders, including technology providers and research institutions, to share knowledge and resources for enhancing energy efficiency.
- Training employees in energy-efficient practices, ensuring that all workers are aware of and skilled in methods that reduce energy consumption and costs.

In modern conditions, the role of digital technologies in achieving energy efficiency goals is becoming increasingly important. A promising approach in this context is the technology of energy-smart food systems, focusing on the efficient utilization of low-C energy systems (FAO, 2011).

The integration of smart agriculture technologies, combining 5G and blockchain platforms, has the potential to significantly improve energy efficiency in agriculture. A key challenge in rural areas is the lack of reliable communication infrastructure, which can be addressed through the use of drones providing 5G communication services. This approach can solve the communication problem in rural areas, facilitate the introduction of modern digital solutions, and ultimately enhance energy efficiency in agriculture (Kourtis et al., 2023).

Energy efficiency can be improved at all stages of the crop production process through effective management of input resources. High-quality seeds, minimum tillage, proper irrigation systems, and agronomic approaches to nutrition all contribute to increased energy efficiency. Eskandari (2023) discusses various strategies to enhance energy efficiency at different stages of plant growing, emphasizing the importance of these factors.

At the seed preparation stage, both the type of crop and the equipment used play crucial roles in determining energy efficiency. For example, in Japan, the use of manual labor in rice cultivation was found to be more energy-efficient than mechanized rice production in the United States, where the high energy consumption of equipment and the use of diesel fuel were major factors. Similarly, for wheat seed production, using less tillage can be more energy-efficient, considering the volume of the crop. Reducing mechanization is one possible pathway to higher energy efficiency (Eskandari, 2023).

During the sowing stage, the energy equivalent of the seed is an important consideration when selecting methods. For seeds with a high energy equivalent, using a larger quantity of seed results in higher energy consumption, thus reducing energy efficiency. Irrigation is another significant point of energy consumption for many crops, so making balanced decisions at this stage is crucial for overall energy efficiency. Effective irrigation management at irrigation sources, such as reservoirs, is one of the most energy-efficient practices (Eskandari, 2023).

Pesticide use should also be assessed from an energy efficiency perspective. While reducing chemical pesticide use may not always lead to increased energy efficiency, the latter can still be improved with more efficient nutrient provision, whether through organic or mineral fertilizers. Some studies show that, despite higher economic costs, using manure can reduce energy consumption by 36%–52% compared to chemical fertilizers (Eskandari, 2023). Although increasing pesticide use can contribute to greater energy efficiency by increasing yields, its impact is limited. Pesticide use must be strictly regulated to prevent negative environmental consequences and food contamination (Liu et al., 2021).

At the harvesting stage, where mechanized equipment is not always feasible, manual labor becomes the primary source of energy consumption, and improving energy efficiency in such operations is nearly impossible. Additionally, the use of lubricants for machinery maintenance can reduce energy consumption and contribute to higher energy efficiency (Eskandari, 2023).

To improve WUE, consider adopting the following strategies:

- reduce conveyance losses by lining channels or using closed conduits;
- avoid irrigating during midday to minimize evaporation;
- prevent runoff and percolation losses by avoiding excessive irrigation;
- implement effective irrigation management practices;
- utilize rainwater and header tanks;
- reassess the areas requiring water and evaluate the effectiveness of the current system;
- ensure drinking water points are efficient and reliable;
- upgrade reticulation systems.

Combining multiple water-saving and optimization techniques simultaneously is the best strategy for improving WUE.

To enhance WUE in agriculture, several strategies can be employed, encompassing technological and agronomic measures.

Technological advancements are crucial in improving WUE. One of the most effective technologies is drip irrigation (Fan et al., 2020a,b), which delivers water straight to the plant roots via a network of pipes and emitters, reducing water losses and ensuring efficient use. Similarly, sprinkler irrigation systems, which distribute water through overhead sprinklers, can also improve WUE by reducing evaporation and runoff. By adopting efficient irrigation technologies and practices, agriculture can maintain productivity even under fluctuating water availability, thereby enhancing resilience to climate variability.

Another promising technology involves utilizing soil moisture sensors and automated irrigation systems (Bwambale et al., 2022). These systems track soil moisture in real time and adjust irrigation schedules as needed, ensuring that crops receive the optimal amount of water. Additionally, the use of geographic information systems and remote sensing can help in mapping and monitoring water use patterns, enabling better management and planning (Cai et al., 2021).

Agronomic practices such as crop selection and soil management can also significantly impact WUE (Chen et al., 2018; Farooq et al., 2019). Selecting drought-resistant crop varieties that need minimal water can aid in minimizing water consumption. Additionally, adopting crop rotation and intercropping practices can enhance soil vitality and water retention, thereby enhancing WUE. Soil management techniques, including mulching and conservation tillage, can also contribute to higher WUE. Mulching involves applying organic or inorganic materials to the soil, which helps decrease evaporation and retain moisture. Conservation tillage reduces soil disturbance and enhances soil structure and water infiltration, leading to more efficient water use.

Agronomic practices offer significant potential for improving WUE, with various techniques that help optimize moisture retention and reduce water consumption. These practices include the careful selection of crops, soil management, intercropping, and various agronomic approaches aimed at maximizing water efficiency at every stage of the agricultural process (Alharbi et al., 2024).

- Crop selection and sowing timing: Choosing the right crops and sowing them at optimal times, in accordance with the evaporative demand of the environment, is essential for reducing reliance on groundwater. For instance, planting wheat at the correct time helps minimize water usage, improving the WUE.
- Moisture storage techniques: Implementing moisture storage techniques such as compartmental bunds, ridges, and furrows has been shown to increase yields, particularly in crops like sorghum. These practices help to capture and retain water in the soil, ensuring that crops have access to sufficient moisture throughout their growth cycle.
- Zero-tillage techniques: Zero-tillage, or no-tillage farming, plays an essential role in improving soil structure and retaining moisture. By leaving the soil undisturbed, these techniques enhance the organic matter content and reduce soil erosion, ultimately enhancing WUE in agriculture.
- Intercropping: In regions with limited rainfall, intercropping, the practice of growing different crops together, can significantly increase WUE. For example, combining corn with potatoes or corn with mung beans—rather than growing a single crop—improves overall water productivity and leads to higher yields per unit of resource.
- Spatial planning of sowing: proper spatial planning, including optimizing row spacing, enhances water efficiency. Studies show that chickpeas produce higher yields when planted with wider row spacing (45 cm instead of 30 cm), as this allows more space for moisture retention in the soil and enhances WUE.
- Weed control: Effective weed management is critical to maximizing water efficiency. In chickpea production, the use of weed-free treatments has been shown to enhance moisture use, while manual weeding in wheat fields leads to decreased weed density, allowing the remaining crops to utilize water more effectively.
- Maximizing green water use: Green water denotes precipitation that infiltrates the root zone of plants. The amount of green water available depends on soil structure and quality. Improving the soil's physical and chemical qualities by adding organic matter can reduce unproductive water losses and increase WUE. Besides straw, agricultural waste can also be utilized to improve the SOM.
- Cover crops and green manure: The use of cover crops or green manure helps improve soil characteristics and water dynamics in the soil. These crops tend to increase SOM, which aids in moisture retention and improves the overall soil's capacity of water retention.
- Canopy management: Effective canopy management techniques, such as optimizing leaf orientation and row management, can significantly influence water efficiency. Studies have shown that photosynthesis and transpiration are closely linked to canopy structure, particularly in crops with a high leaf area index (LAI). Shading between leaves can reduce water loss, and controlling the LAI to approximately four can help slow down soil surface evaporation, leading to better water efficiency. In contrast, leaves within the canopy that are overly shaded tend to have lower water efficiency.

Irrigation technology improvement and agronomic practices are not always adequate for securing agricultural water conservation. It can inadvertently lead to water rebound effects. This phenomenon occurs when savings from increased efficiency result in higher overall water consumption (Fan et al., 2020a,b). For instance, farmers who adopt efficient irrigation technologies may increase their irrigated areas or switch to more water-intensive crops, thereby offsetting the water savings achieved. Additionally, the perceived abundance of water due to efficiency gains can reduce the incentive to conserve, leading to less stringent water management practices. Consequently, while WUE improvements are crucial, they must be accompanied by policies and measures that prevent rebound effects, ensuring that the overall goal of water conservation is met. Specifically, it's important to take into account the variations in natural geography across regions when developing water conservation policies at a subregional level.

Modern technologies have the potential to significantly enhance the effectiveness of water and energy utilization in agriculture (Roy et al., 2023), necessitating a more comprehensive investigation into current research and development trends in this area. Additionally, researchers highlight the importance of institutional factors, such as farmer participation in regional irrigation schemes, the implementation of pricing mechanisms and legislative incentives to reduce water consumption, the provision of training and learning opportunities, and the acquisition and dissemination of knowledge about new technologies (Wani et al., 2020). Therefore, examining the political dimensions of promoting WUE and energy efficiency is a crucial aspect of this discourse.

12.8 Research and development toward energy–water use efficiency and carbon footprint management

To analyze current trends in research and development concerning energy and water efficiency in agriculture, we examined patent data from the Lens.org database (Lens, 2024). Two distinct search queries were conducted: Title: (energy AND agriculture) and Title: (water AND agriculture), with patents published between January 1, 2019, and January 1, 2024. The first query resulted in 847 patents, while the second yielded 3839.

An analysis of the yearly patent creation trends (Fig. 12.10) reveals similar patterns in both the energy (Fig. 10.10a) and water (Fig. 10.10b) sectors. The highest number of patents was published in 2021, with a noticeable decline in patent activity following this peak. Specifically, the number of energy-related patents in 2023 was nearly a third lower than in

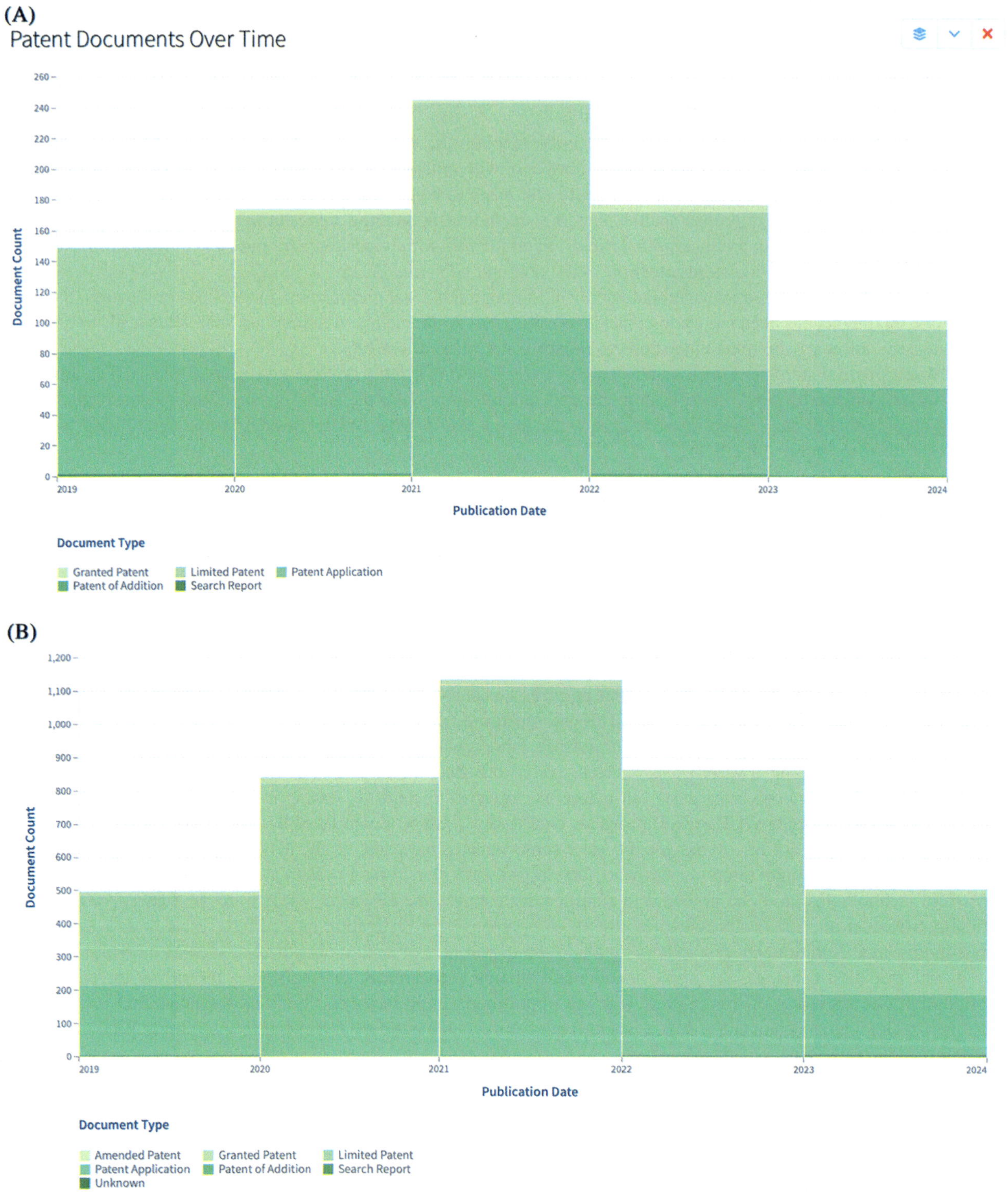

FIGURE 12.10 Dynamics of creating patents in the field of water and energy use in agriculture. (A) Number of patents found on request "energy AND agriculture" (B) Number of patents found on the request "water AND agriculture."
LENS (Lens, 2024).

2019, while water-related patents remained at the same level as in 2019, though more than half the number observed in 2021.

The energy sector consistently exhibited a larger share of patent applications, averaging 44%, compared to 30% in the water sector. In 2023 the share of patent applications in the energy sector peaked at 57%, indicating that research in this field is becoming increasingly applicable.

The leading organizations by the number of energy-related patents (top 5) are as follows: Rem Tec SRL (11 patents), Univ China Agricultural (10), Insolight SA (5), Zhu Jianhua (4), State Grid Corporation of China (4), Huadian Electric Power Research Institute Co. Ltd (4), and Hu Zefeng (4). By country, China stands out as the absolute leader, with 780 published patents, accounting for 92% of the total patents in this field. In comparison, the United States has only 6 patents, the EU has 10, and India did not produce any patents during this period.

For patents related to water use, the leading organizations are Evoqua Water Tech LLC (21 patents), China Institute of Water Resources and Hydropower (18), Univ Gansu Agricultural (10), Taian Jucheng Electronic Tech (8), and Guangdong Research Institute of Water Resources (8). Similar to the energy sector, China dominates the field of water-use patents, holding a total of 3560 patents, which account for 91% of the total. The United States has 44 patents, the EU has 25, and India has no patents in this area.

The primary research directions in these fields can be inferred from patent grouping according to the Cooperative Patent Classification system (Figs. 12.11 and 12.12).

The data presented in Fig. 12.11 indicate that the majority of patents in the energy sector are classified under the Y02 "Technologies or Applications for Mitigation or Adaptation Against Climate Change" group, accounting for 70% of the total number. Key technologies in this area include:

- GHG cooling systems (193 patents)
- Renewable energy sources, such as solar water pumping systems (142 patents)
- Energy storage measures, for example, those used in greenhouses (93 patents)
- Photovoltaic (PV) energy technologies (60 patents)
- Rainwater harvesting technologies (45 patents).

The A01 "Agriculture, Forestry, Animal Husbandry, Hunting, Trapping, and Fishing" group is relatively under-represented, comprising only 11% of the overall count of energy-related patents.

In terms of water-related technologies in agriculture (Fig. 12.11), patents from the Y02 group remain the leaders in terms of number. However, their share in the overall number of water-related patents is not substantial, making up only 26% of the total. The A01 "Agriculture, Forestry, Animal Husbandry, Hunting, Trapping, and Fishing" group, in contrast to the energy-related research, is more prominently represented in the water sector, comprising 16% of the total. Additionally, other groups, such as G06, also contribute to the patent landscape in this field.

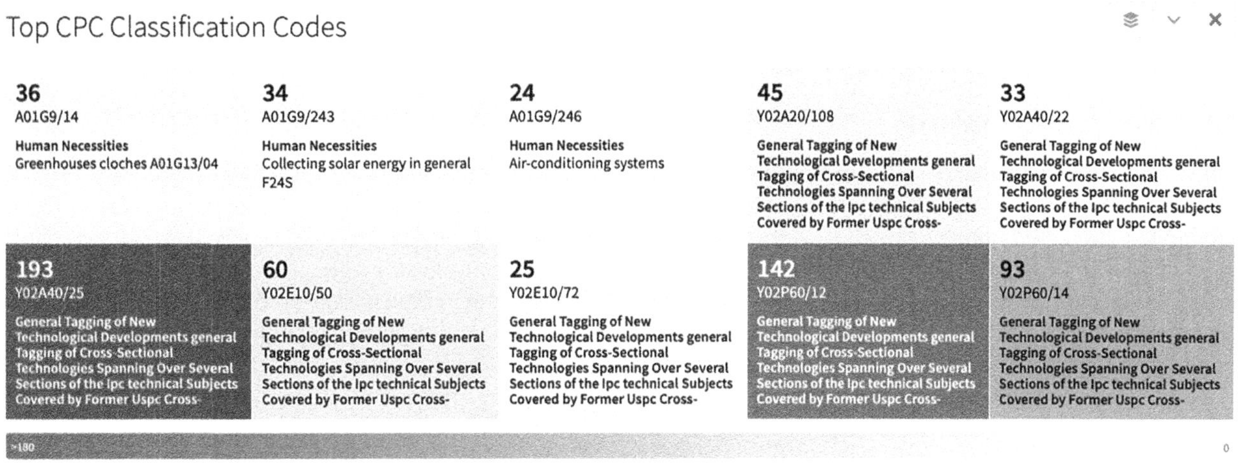

FIGURE 12.11 Grouping of patents related to energy in agriculture by CRC classification.
LENS (Lens, 2024).

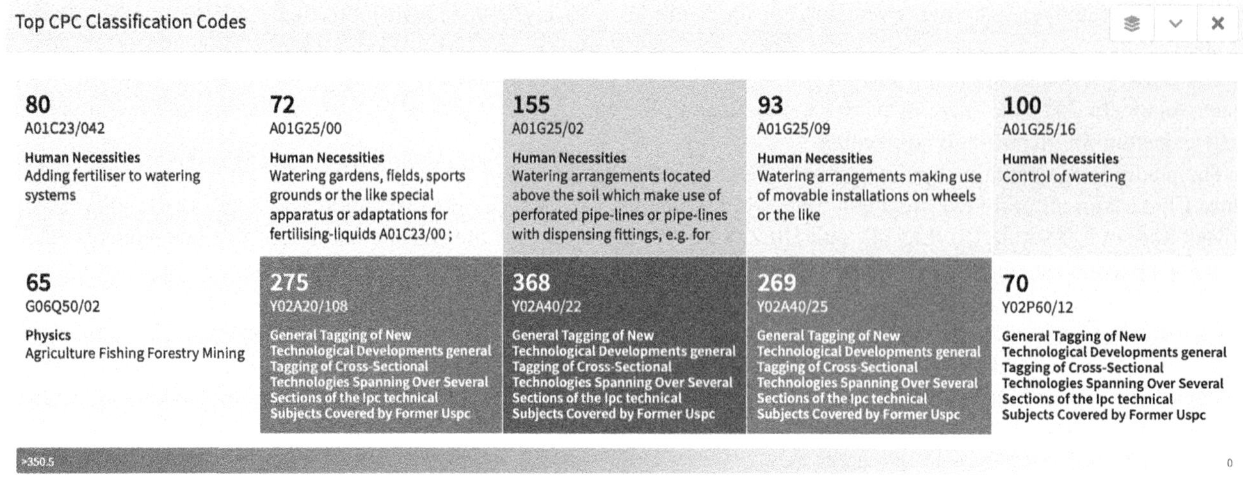

FIGURE 12.12 Grouping of agricultural water patents by CRC classification.
LENS (Lens, 2024).

The largest number of patents in the water efficiency sector (Y02 class) concern the following areas:

1. "Improving land use, improving water use or availability, controlling erosion" (368 patents).
2. "Water conservation; efficient water supply; efficient water use; rainwater harvesting" (275 patents).
3. "Adaptation technologies in agriculture, forestry, livestock, or agro-alimentary; greenhouse technology, e.g., cooling systems" (269 patents).

Class A01 is represented by the following subclasses: "Watering gardens, fields, sports grounds, or the like, special apparatus or adaptations for fertilizing liquids; nozzles or outlets, spraying apparatus; watering arrangements located above the soil which make use of perforated pipe-lines or pipe-lines with dispensing fittings, e.g., for drip irrigation" (155 patents); "Control of watering" (100 patents).

Table 12.5 lists the top 5 cited patents related to energy and water use in agriculture.

As indicated by the patent titles, the majority of patents in the field of energy usage pertain to various technologies that harness solar energy for agricultural production processes. Concerning water use, the focus is primarily on controlling water within irrigation systems and enhancing decision-making processes related to irrigation needs.

12.9 Policy perspective for enhancing water–energy use efficiency toward carbon footprint management

Several macroeconomic factors influence energy efficiency in agriculture. According to the study by Liu et al. (2021), the most significant macro factor is agriculture's share in GDP. An increase in this share leads to a decline in energy efficiency within the sector. Additionally, the level of urbanization and the growth of GDP per capita negatively affect agricultural energy efficiency. Conversely, a higher share of oil in the total energy consumed in agriculture has a positive impact on energy efficiency, as diesel and gasoline vehicles are generally more fuel-efficient and productive (Liu et al., 2021). Therefore to increase the efficiency of energy and water use in agriculture and reduce the footprint, it is also important to transform the relevant policies.

Grasping the connections between water, energy, and C is crucial for effective resource use and policy formulation (Meng et al., 2019). Policies should aim to minimize resource use, shift toward clean energy solutions, and control emissions to address these interconnected issues (Liu et al., 2023). Table 12.6 outlines the primary approaches for evaluating the interdependencies among water, energy, and C.

Water and energy are essential and interconnected resources, and their generation and distribution within the electricity sector lead to C emissions and impact climate change.

TABLE 12.5 Top-5 most cited patents by patent count (both energy and water).

Patent title	Description (type; family; jurisdiction; legal status; publishing date)
Energy and Agriculture	
"Aerial Solar Agricultural Irrigation, Energy Generation, Hydro Conservation with Beneficiary Sharing for Relieving Poverty, Protecting Animals, Wildlife and the Environment Autonomous Apparatus" (US 2021/0294287 A1)	Patent Application; 2s/4ex; US; active; 2021
"A Multifunctional Automatic Drying Room for Agricultural Products with Coupling Heat Supply of Solar Energy and Heat Pump" (AU 2020/101226 A4)	Limited Patent; 1s/1ex; AU; inactive; 2020
"Novel agricultural greenhouse system and solar energy storage and supply system" (CN 110268882A)	Patent Application; 1s/1ex; CN; discontinued; 2019
"Integrated flexible self-charging power source for agricultural environment energy collection and preparation method" (CN 110739879A)	Patent Application; 5s/5ex; CN, WO, US; active; 2020
"Novel agricultural energy-saving fertilizer distributor" (CN 211792885 U)	Limited Patent; 1s/1ex; CN; inactive; 2020
Water and Agriculture	
"Irrigation district agricultural irrigation water demand decision-making method and system" (CN 110689173A)	Patent Application; 1s/1ex; CN; pending; 2020
"Facility agriculture intelligent water and fertilizer integrated control system and control method" (CN 110187688A)	Patent Application; 1s/1ex; CN; discontinued; 2019
"Agricultural irrigation water demand prediction method" (CN 110580657A)	Patent Application; 7s/7ex; JP, US, CN, WO; active; 2019
"Agricultural irrigation water conservancy monitoring system" (CN 109863870A)	Patent Application; 2s/2ex; CN; active; 2019
"Water Saving Apparatus for Use in Agricultural Apparatus, and Method for Water Saving Apparatus" (WO 2022/094954 A1)	Patent Application; 1s/1ex; WO; pending; 2020

Source: LENS (Lens, 2024).

12.9.1 Water policy to increase water use efficiency: the case of China

Policy measures are essential for creating an enabling environment for improving WUE (Azad and Ancev, 2014; Shah et al., 2023). Governments and policymakers can promote efficient water use through incentives, regulations, and capacity-building initiatives. For instance, applying water-pricing mechanisms that mirror the actual cost of water can incentivize farmers to use water more judiciously. Another policy measure is the water rights trading policy.

Water rights trading policy is an innovative approach to managing water resources, especially in areas experiencing water shortages (D'Odorico et al., 2020; Fang and Zhang, 2020). This policy permits the buying and selling of water rights, facilitating a more efficient allocation of water resources among users. By creating a market for water rights, water trading encourages users to optimize their water use, thereby improving overall WUE. This subsection explores the concept of water trading policy and how it affects WUE in agriculture, with a specific focus on China.

Water trading policy operates on the principle that water rights can be transferred from one user to another. This market-based approach incentivizes water conservation by allowing users who save water to sell their surplus to others who need it. In agriculture, where water is a critical input, this policy can lead to significant improvements in WUE. Farmers who adopt water-saving technologies and practices can benefit financially by selling their excess water rights, while those who need more water can purchase it, ensuring that water is allocated to its most productive uses.

China provides a compelling case study for the implementation of water trading policy (Pan et al., 2023; Zhang et al., 2021). The country faces severe water scarcity, with water availability per capita being considerably lower than the global average. To address this challenge, the Chinese government has implemented pilot projects for trading water rights in various provinces, including Henan, Ningxia, Jiangxi, Hubei, Inner Mongolia, Gansu, and Guangdong. These pilot projects aim to optimize water resource management through market mechanisms, thereby enhancing WUE in agriculture.

TABLE 12.6 Methods for assessing water–energy–carbon interrelationship.

Method	Characteristic
Materials flow analysis (MFA)	It is a systematic method used to track and quantify the flow of materials and energy within a system, often focusing on the life cycle of resources from extraction to disposal. It aims to account for the amount of material and energy entering, circulating within, and leaving a system, typically at a regional, national, or global scale. MFA can be used to study a wide range of systems, including industrial processes, agricultural systems, or environmental processes, such as water, energy, and C flows. In the context of water–energy–C interrelationships, MFA helps to identify the flow of water, energy, and C through various stages of a system, revealing the connections between these resources. By understanding how water and energy are used and how they interact with C emissions throughout different processes, MFA provides valuable insights into the sustainability of systems and the impact of different sectors on the environment.
Life cycle assessment (LCA)	It is a method used to evaluate the environmental impacts of a product, process, or service from its inception to its disposal, often referred to as a product's "cradle-to-grave" analysis. LCA involves the comprehensive evaluation of all stages in a product's life cycle, including raw material extraction, manufacturing, distribution, use, and disposal, with a focus on the consumption of resources and the emissions generated. When applied to the water–energy–C interrelationship, LCA helps to assess the total environmental impacts of processes or systems, including water use, energy consumption, and C emissions. It quantifies and assesses the environmental effects associated with resource flows throughout a product or service's life cycle, making it valuable for understanding the environmental trade-offs between water, energy, and C usage.
Regional Energy & Water Supply Scenarios model (REWSS)	It is a tool used to assess the environmental impacts of energy and water resources within a specific geographic region. This model helps evaluate the energy and water use patterns across different sectors, considering both direct and indirect effects, and can predict the consequences of various future scenarios. The REWSS model is typically used to simulate the long-term effects of different strategies, such as changes in energy generation methods or water usage policies, on the regional environment. By modeling scenarios that address water and energy supply in conjunction, REWSS focuses on providing a comprehensive view of how these two resources are interlinked. It accounts for the pressures of water use in energy production, the impact of energy production on water availability (e.g., water withdrawals for cooling in power plants), and the environmental impacts related to GHG emissions, water consumption, and water stress. Additionally, it uses methods such as Monte Carlo simulations to incorporate uncertainties and variability into its predictions.
Hydrological rainfall-runoff-routing model (LISFLOOD) and power system model (Dispa-SET)	The LISFLOOD is a modeling tool that simulates the hydrological processes of rainfall, runoff, and river flow routing. It is designed to predict the flow of water across a given watershed or river basin by considering factors such as precipitation, soil moisture, evaporation, and topography. The model is often used for flood forecasting, water resource management, and land-use planning. It simulates the movement of water from precipitation through the landscape and into water bodies such as rivers, lakes, or reservoirs. The Dispa-SET model is a simulation tool used to optimize the operation of power systems, with a focus on balancing energy production and consumption while minimizing costs and emissions. Dispa-SET incorporates the dynamics of energy production, grid distribution, and storage systems to simulate how different energy resources (e.g., renewable, nuclear, fossil fuels) can be dispatched to meet electricity demand. The model also integrates various environmental factors, including the effects of C emissions, renewable energy potential, and the integration of energy storage. When combined, LISFLOOD and Dispa-SET can be used to analyze the interrelationship between water, energy, and C in a region or power system. The two models work together by simulating water availability (via LISFLOOD) and electricity generation and consumption (via Dispa-SET), which helps assess the

(Continued)

TABLE 12.6 (Continued)

Method	Characteristic
	impacts of water availability on energy generation, especially in regions where hydropower is a significant energy source. By integrating these two models, the combined approach provides a comprehensive analysis of the water–energy–C nexus, evaluating the effects of water resources on energy systems and the environmental impacts of energy production.
Competitive Markov Decision Process model (CMDP)	The Competitive Markov Decision Process (CMDP) model is a sophisticated tool used for decision-making in competitive environments. It is based on the principles of Markov Decision Processes (MDP), which are mathematical frameworks used for modeling decision-making where outcomes are partly random and partly under the control of a decision-maker. In CMDP, multiple decision-makers (agents) compete or cooperate, and each agent aims to maximize its own utility or minimize its costs while interacting with other agents who are also optimizing their own objectives. In the context of water–energy–C interrelationships, CMDP is applied to simulate and optimize the interactions between different stakeholders (such as utilities, governments, and private entities) within water and energy systems. It can be used to model how water resources are allocated for energy production (especially hydropower), how energy consumption affects water use (for example, in cooling processes or irrigation), and how C emissions impact both systems. CMDP helps in identifying strategies that balance the competing needs of water, energy, and C mitigation across various sectors and decision-makers.
Meta-system architecture model (MA)	It is an integrated modeling approach that allows the analysis of material and energy flows within interconnected systems at a large scale. It specifically focuses on the evaluation of key systems such as national electricity grids, water supply systems, and sewage networks. The MA model is used to assess how these complex systems interact with one another, considering parameters such as water, energy, environment, wastewater, and material flows. The model is designed to capture the interdependencies between various infrastructure components and offers insights into how changes in one sector (e.g., energy generation) can affect other sectors (e.g., water management and wastewater treatment). In the context of water–energy–C interrelationships, the MA model provides a holistic framework for understanding how water resources are utilized in energy production, how energy demand impacts water availability, and how both sectors contribute to C emissions. By modeling the entire metasystem, MA helps identify opportunities for optimizing resource use, reducing emissions, and improving sustainability across different sectors.
Multiregional Input-Output Analysis (MRIO)	It takes into account both direct and indirect consumption. This allows it to analyze the virtual water embedded in energy exchanges and track the hidden water flows between sectors. However, managing and obtaining data on the water–energy–C footprint within a specific network presents significant challenges.
Ecological Network Analysis (ENA)	It is a systems-based approach used to study and assess the flow of materials, energy, and information within ecological systems. Originally developed in ecological sciences, ENA has been adapted to study human-engineered systems, including the interconnections between environmental resources such as water, energy, and C. ENA is particularly valuable in assessing how different components of a system (e.g., energy production, water usage, C emissions) interact with one another and affect the overall functioning and sustainability of the system. ENA uses network-based models to analyze material and energy flows, which helps identify critical pathways, bottlenecks, and resource dependencies. The key principle behind ENA is that any system, whether ecological or socio-economic, can be analyzed as a network of interconnected components. In the case of water–energy–C interrelationships, ENA is used to understand the flows of water in energy production, the energy required for water treatment and distribution, and the associated C emissions throughout the processes. In this method, nodes in

(Continued)

TABLE 12.6 (Continued)

Method	Characteristic
	the network represent different processes or sectors, while links show how materials and energy flow between them. By analyzing these flows, ENA can help quantify the efficiency and sustainability of systems, highlighting areas where improvements in resource use or reductions in C emissions can be made.
Social Network Analysis (SNA)	It is a method used to study the relationships and interactions between entities, referred to as "actors" in the network. Traditionally, SNA has been applied to human social structures, examining how individuals or organizations are connected and how information, resources, or influence flows through these networks. However, SNA can also be adapted to study material and energy flows within environmental systems, such as the interactions between water, energy, and C in different sectors or regions. In the context of water–energy–C interrelationships, SNA analyzes how different systems, such as water supply, energy generation, and C emissions, are linked and influence each other. It examines the flow of resources (e.g., water used for energy generation), the interdependencies between sectors (e.g., water needed for hydropower plants), and the feedback loops that might affect C emissions or energy consumption. By representing these systems as networks, SNA helps uncover the underlying structures that drive the use and movement of resources, providing insights into how changes in one part of the system might affect others.
Mixed-unit Multi-Regional Input – Output Analysis (Mixed-unit MRIO)	It is an advanced analytical method used to assess the flow of materials, energy, and environmental impacts across different regions, industries, and sectors. This method integrates data from multiple regions and accounts for various units of measurement, thus providing a more nuanced and detailed analysis of interregional economic relationships and environmental impacts, including water, energy, and C emissions. MRIO typically combines input-output (IO) tables from different countries or regions to trace the flow of goods and services and their environmental implications. The "mixed-unit" aspect refers to the ability of the MRIO model to handle multiple units of measurement, such as tons, kilowatt-hours, or cubic meters, making it adaptable for cross-sector and cross-regional analysis. In the context of water–energy–C interrelationships, MRIO models can help assess the interconnectedness of water, energy, and C emissions by accounting for both direct and indirect flows. For example, MRIO can track the embodied energy or water used in the production of goods and services across different sectors, and it can evaluate the C emissions associated with the use of energy and water resources in each sector. This comprehensive approach allows for a better understanding of how regional and global supply chains contribute to the water–energy–C nexus.
Urban Water to Air Model (UWtoA)	It is a modeling approach designed to simulate the interactions and flow of energy, water, and air pollutants within urban water systems. This model specifically focuses on the relationship between energy consumption, water management, and the emission of air pollutants within urban settings, which often face complex challenges related to water supply, wastewater treatment, energy usage, and air quality. The UWtoA model helps assess how urban water systems (e.g., water treatment, distribution, wastewater treatment) contribute to the overall environmental footprint of cities, considering not only water consumption but also energy use and emissions of pollutants. By focusing on the energy required to treat and distribute water, as well as the associated air quality impacts from energy consumption (such as CO_2 emissions or other air pollutants), the model offers a holistic view of the water–energy–C nexus in urban areas.
Other correlation analysis methods	Such as gray correlation analysis, structural path analysis, and data envelopment analysis, among others, examine the interrelationships of regional water, energy, and C. These methods analyze the coupling of these elements by simultaneously minimizing their consumption across these sectors.

Source: Liu et al. (2023) and Ghodrati et al. (2023).

One of the key advantages of China's water trading policy is its ability to support the adoption of water-saving irrigation technologies (Zhang et al., 2021). Conventional irrigation techniques, such as flooding irrigation, are highly inefficient, leading to substantial water losses through evaporation and runoff. By contrast, modern irrigation technologies like drip and sprinkler systems provide water straight to the plant roots, reducing losses and ensuring that water is used more efficiently. The financial incentives provided by water trading encourage farmers to invest in these technologies, as they can sell any water they save through improved irrigation practices.

The effect of water rights trading policy on the efficiency of water use in China has been significant (Pan et al., 2023; Yang et al., 2024). It incentivizes farmers to invest in efficient irrigation technologies. Studies have shown that the implementation of water-saving irrigation methods has increased in regions where the policy is implemented. For instance, in regions like Ningxia and Henan, the introduction of water trading has spurred the adoption of modern irrigation technologies. Farmers in these areas have increasingly turned to drip irrigation systems. This shift not only enhances WUE but also boosts crop yields and agricultural productivity.

Moreover, water trading policy has also contributed to improved food security in China (Zhang et al., 2021). By promoting efficient water use, the policy helps ensure that sufficient water is available for agricultural production, even in times of scarcity. This is particularly important in a country where agriculture accounts for a substantial amount of water consumption. By optimizing water allocation, water trading helps maintain stable food production levels, thereby supporting food security.

12.10 Future of water–energy use efficiency for carbon footprint management and overall sustainability

To evaluate the future potential effects of current trends in water and energy usage on human well-being, we examine the link between economic development and emission levels (Prasad et al., 2021). This analysis is based on a model derived from the Environmental Kuznets Curve (EKC). The EKC is a theory that explains the connection between economic development and environmental harm. It proposes that in the early stages of economic growth, environmental degradation typically rises, but once a certain income per capita threshold is reached, the trend reverses, leading to improvements in environmental quality. When plotted, this creates an inverted U-shaped curve that represents environmental degradation against per capita income (Fig. 12.13) and is divided into three distinct phases.

1. Early stage of development: Preindustrial economies $\Longrightarrow$ Rising environmental degradation. In the initial phases of economic growth, industries expand, and countries prioritize economic output, often at the expense of environmental protection. During this stage, pollution, deforestation, and resource depletion tend to increase as countries focus on industrialization, infrastructure development, and urbanization.
2. Turning point: peak of environmental degradation. As countries grow wealthier and reach higher levels of income per capita, they tend to become more aware of environmental issues. The turning point is when economic development reaches a threshold and societies begin to demand better environmental standards. Governments begin to implement policies that prioritize sustainability, cleaner technologies, and pollution control.
3. Later stage of development: Postindustrial economies (service economy) $\Longrightarrow$ Decreasing environmental degradation. In the advanced stages of development, nations tend to allocate resources toward cleaner technologies, implement stricter environmental laws, and transition to more sustainable methods. Consequently, environmental quality improves, leading to a decrease in pollution, better resource efficiency, and greater investment in renewable energy and eco-friendly technologies.

So, the EKC drivers are:

1. Technological advancements—as countries develop, they have more access to cleaner and more efficient technologies that reduce the environmental impact of production.
2. Environmental awareness—wealthier populations tend to have greater awareness and political support for environmental protection, leading to stronger regulations.
3. Structural economic changes as economies mature, they often shift from heavily polluting industries (e.g., manufacturing) to less resource-intensive sectors (e.g., services and technology).
4. Institutional strengthening (as countries develop, institutions that enforce environmental regulations, such as environmental agencies, tend to become more effective and better resourced).

In our opinion, the EKC also has a number of shortcomings and unaccounted points. The EKC does not apply uniformly to all environmental problems. While it may describe trends for certain pollutants (CO_2, etc.), it may not hold true for

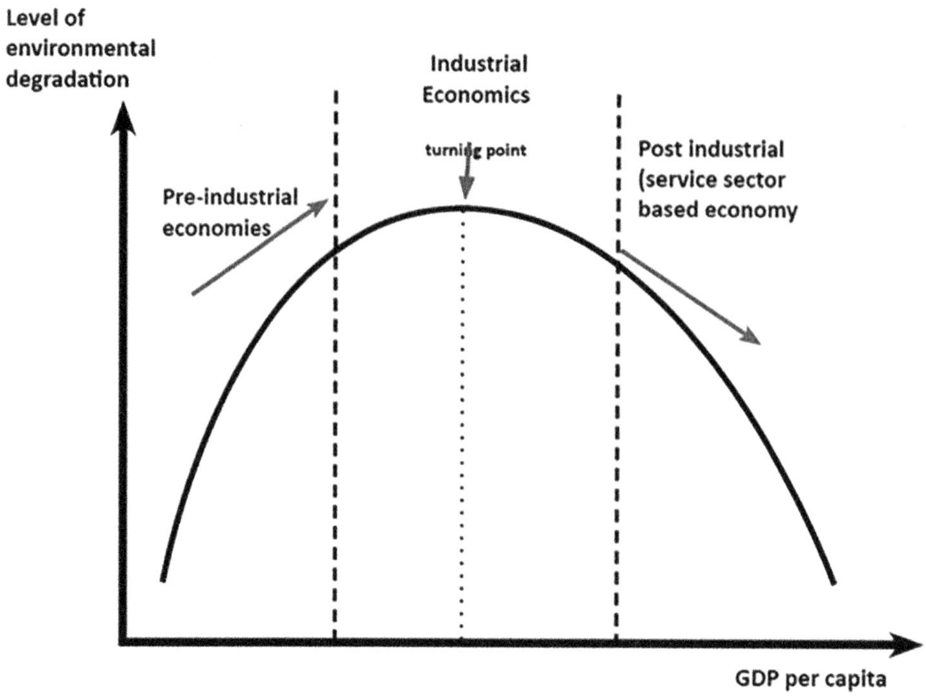

FIGURE 12.13 Environmental Kuznets Curve.
Based on Leal and Marques (2022), Sarkodie and Strezov (2018), and Stern (2017).

issues such as biodiversity loss, climate change, or resource depletion. Some types of environmental degradation, such as biodiversity loss or deforestation, may be irreversible even if income increases.

The EKC may give a misleading impression that economic growth alone will lead to environmental improvement. Wealthier countries often outsource polluting industries to poorer nations, effectively reducing their domestic pollution but contributing to global environmental degradation. Moreover, the EKC does not take into account the time lags. The curve assumes that countries can "grow out" of pollution, but the timeline for this transition can be long, and the environmental damage done during the early stages of development may be significant and hard to reverse.

For assessing the connection between GDP per capita and the volume of pollutant emissions, we estimate the following model (Eq. 12.21):

$$\ln CO2_{it} = \beta_0 + \beta_1 \times \ln GDP_{it} + \beta_2 \times \ln GDP_{it}^2 + \beta_3 \times \ln ENER_{it} + \beta_4 \times \ln ISL_{it} + \varepsilon_{it}, \tag{12.21}$$

where i: denotes the countries considered ($i = 1,\ldots,38$); t are the years ($t = 2010,\ldots,2020$); β: model parameters that measure the effects of a change in independent variable in the period t for the i-th country.

The research implements an analysis of 38 European countries for the period 2010–2020 using such indicators as CO_2 emissions (CO_2), GDP per capita, current US$ (GDP), primary energy consumption, TWh (ENER), and inequality indicator–income share held by the lowest 10% (ISL). Since the relationship between economic development (GDP) and emissions is nonlinear, we also took the square of GDP.

Descriptive characteristics of the indicators are presented in Table 12.7.

The modeling results are presented in Table 12.8.

As we can see, an increase in the level of GDP in developing countries leads to a stronger increase in emissions, while for highly developed countries that are willing to invest in reducing emissions, such a decrease will be much smaller.

CO_2 emissions increase as energy consumption rises primarily because a significant portion of the world's energy is generated from fossil fuels such as coal, oil, and natural gas. These fuels contain C, and when they are burned to produce energy, they release CO_2 as a byproduct. Although renewable energy sources like solar, wind, and hydropower are growing, fossil fuels still account for a large share of global energy production. This means that as energy consumption

TABLE 12.7 Descriptive characteristics of the indicators.

Characteristic	CO_2	GDP	ENER	ISL
Min.	2.511	2125	15,864.63	0.800
1st Qu.	4.454	13,062	16,771.26	2.400
Median	5.938	23,424	17,225.64	3.000
Mean	6.518	32,159	17,142.48	2.972
3rd Qu.	7.734	47,284	17,523.85	3.500
Max.	21.756	123,679	18,282.51	4.500

Source: Own processing.

TABLE 12.8 The modeling results of the relationship between economic development and CO_2 emissions.

Balanced panel: n = 38, T = 11, N = 418					
Residuals:					
Min.	1st Qu.	Median	3rd Qu.	Max.	
−0.73828	−0.26092	−0.00351	0.23192	0.92774	
Coefficients:					
	Estimate	Std. Error	t-value	Pr(>ltl)	
(Intercept)	−96.00419	16.28134	−5.897	7.72e−09	***
log(GDP)	29.85496	5.02166	5.945	5.88e−09	***
I(log(GDP)^2)	−3.04276	0.51393	-5.921	6.76e−09	***
log(ENER)	0.10322	0.01743	5.922	6.70e−09	***
log(ISL)	0.27576	0.06523	4.227	2.91e−05	***

Signif. codes: 0 '***' 0.001 '**' 0.01 '*' 0.05 '.' 0.1 ' ' 1.
Residual standard error: 0.3353 on 413 degrees of freedom.
Multiple R-squared: 0.249, Adjusted R-squared: 0.2418.
F-statistic: 34.24 on 4 and 413 DF, p value: < 2.2e−16.
Source: Own processing.

increases, more fossil fuels are burned, resulting in higher CO_2 emissions. In summary, increased energy consumption leads to increased CO_2 emissions, but switching to clean and renewable energy sources can help reduce this dependence.

The results obtained show that as the income of the poorest 10% of the population increases, the level of CO_2 emissions will also increase. Increasing income allows the poorest segments of the population to increase their consumption of goods and services, which is associated with increased energy use. Households may start using electrical appliances, cars, and air conditioning, which all increase energy consumption and, by extension, emissions. With more income, poor households often move from energy sources like biomass (wood or charcoal) to electricity or fossil fuels (gasoline, natural gas), which can be more CO_2-intensive, especially in regions where electricity generation relies heavily on coal or natural gas. As incomes rise, people tend to adopt consumption patterns more similar to those of wealthier populations. This may involve more energy-intensive activities such as buying consumer electronics, traveling more frequently (especially by air or car), or moving into bigger homes that require more energy for heating, cooling, and lighting. Higher income levels also tend to lead to changes in diet, such as increased meat consumption, which has a higher C footprint due to the energy-intensive nature of livestock farming compared to plant-based diets.

Moreover, despite the growth of CO_2 emissions, increasing the incomes of the poorest 10% of the population is important in terms of improving the quality of life and social justice. While the poorest 10% gaining more income does lead to an increase in emissions, their overall contribution is still relatively small compared to wealthier segments of the population, whose consumption is far more energy- and C-intensive. The rise in emissions could be mitigated if the increase in income is accompanied by access to cleaner energy sources, such as solar or wind power, or by policies promoting energy efficiency.

12.11 Conclusion

Summarizing the results of the conducted research on the efficiency of the use of energy and water in agriculture and measures to reduce the associated ecological footprint, several key conclusions emerge. The energy efficiency concept, referring to the measurement of agricultural output per unit of energy consumed, should become the central principle in decision-making regarding the use of resources, technologies, and practices. Existing criteria, such as the energy ratio, productivity, and specific energy, can serve as reliable tools to support decision-making and implement effective practices. This approach should also be applied to water consumption, with a focus on the concurrent use of indicators of water efficiency and water productivity, aiming to produce more protein with less water.

The analysis revealed that agriculture has a significant and increasing C footprint, particularly at the farm level and within the pre and postproduction phases between 2000 and 2021. Simultaneously, the global ecological footprint of all agricultural activities at the farm level is growing. The water footprint of agriculture comprises both green and blue components, and this footprint is substantial, as the industry consumes approximately 70% of the world's freshwater and is responsible for 78% of ocean and freshwater eutrophication.

In agriculture, the use of water and energy is closely interrelated, jointly contributing to the industry's significant environmental impact. Water performance in irrigation systems is often linked to the intensive use of energy resources, which exacerbates climate change. Consequently, it is crucial to implement measures that minimize excessive energy consumption. Such measures include modifications in soil management, tillage, weed control, seed selection, AF, and general farm management. Additionally, modern digital technologies can significantly transform agrotechnological practices, promoting higher energy and water efficiency. These technologies include systems for monitoring and controlling environmental parameters, precision agriculture, and the regulation of fertilizer and seed application. Ultimately, agricultural practices must find a middle ground between the needs of plants, overall economic goals, and available resources, including water and energy. Furthermore, the implementation of these measures must be systemic, addressing all stages of production and areas of the value chain, and should be preceded by proper training for farmers.

Equally important are measures related to the rational spatial planning of production. This can be achieved by optimizing field geometry, planting arrangements, using shading devices, and carefully planning irrigation technologies. Institutional factors, including access to knowledge, investment, and political regulation of resource prices and usage, are also critical to improving the efficiency of energy and water use in agriculture.

Contemporary research predominantly focuses on water use. The number of patents in this area is more than four times higher than the number of patents related to energy use. Nonetheless, the number of new patents has generally decreased in recent years, following a peak in 2021. Notably, research in energy use tends to exhibit a higher level of implementation. China leads the field, accounting for 91%–92% of global share of patent applications. Most patents, in both energy and water use, focus on CCM and adaptation. In the energy sector, these innovations primarily involve the use of renewable (solar) energy. In contrast, in the water sector, they address technologies for determining irrigation needs and controlling irrigation processes, which are also more energy-efficient. This aligns with current trends in scientific research. Ultimately, policymakers can utilize the EKC framework to align economic growth with environmental safeguards by encouraging cleaner technologies from the outset of development, implementing proactive ecological policies to minimize damage before reaching the turning point on the EKC, and promoting SD practices that simultaneously support economic growth and ecological preservation.

References

Alharbi, S., Felemban, A., Abdelrahim, A., Al-Dakhil, M., 2024. Agricultural and technology-based strategies to improve water-use efficiency in arid and semiarid areas. Water 16 (13), 1842. Available at https://doi.org/10.3390/w16131842.

An, Y., Li, Q., Zhang, L., 2022. Managing agricultural water use in a changing climate in China. Sustainable Production and Consumption 33, 978–990. Available at https://doi.org/10.1016/j.spc.2022.08.023.

Azad, M.A.S., Ancev, T., 2014. Measuring environmental efficiency of agricultural water use: a Luenberger environmental indicator. Journal of Environmental Management 145, 314–320. Available at https://doi.org/10.1016/j.jenvman.2014.05.037.

Bajan, B., Mrówczyńska-Kamińska, A., Poczta, W., 2020. Economic energy efficiency of food production systems. Energies 13 (21), 5826. Available at https://doi.org/10.3390/en13215826.

Banerjee, A., Meena, R.S., Jhariya, M.K., Yadav, D.K., 2020. Agroecological Footprints Management for Sustainable Food System. Springer, Singapore, pp. 1–514. Available at https://doi.org/10.1007/978-981-15-9496-0.

Banerjee, A., Jhariya, M.K., Bargali, S.S., Palit, D., 2023. Ecorestoration for Sustainability. Wiley-Scrivener, USA, pp. 1–441.

Banerjee, A., Jhariya, M.K., Raj, A., Mechergui, T., 2024. Ecosystem management: climate change and global sustainability—an introduction. Ecosystem Management: Climate Change and Sustainability. wiley, India, pp. 1–21. Available at https://doi.org/10.1002/9781394231249.ch1.

Bar-On, Y.M., Phillips, R., Milo, R., 2018. The biomass distribution on Earth. Proceedings of the National Academy of Sciences 115 (25), 6506–6511. Available at https://doi.org/10.1073/pnas.1711842115.

Benedek, A., Rokicki, T., Szeberényi, A., 2023. Bibliometric evaluation of energy efficiency in agriculture. Energies 16 (16), 5942. Available at https://doi.org/10.3390/en16165942.

Brown, G.O., 2002. The history of the Darcy–Weisbach equation for pipe flow resistance. Proceedings of the Environmental and Water Resources History. American Society of Civil Engineers (ASCE), pp. 34–43. Available at https://doi.org/10.1061/40650(2003)4.

Bwambale, E., Abagale, F.K., Anornu, G.K., 2022. Smart irrigation monitoring and control strategies for improving water use efficiency in precision agriculture: a review. Agricultural Water Management 260, 107324. Available at https://doi.org/10.1016/j.agwat.2021.107324.

Cai, W., Ullah, S., Yan, L., Lin, Y., 2021. Remote sensing of ecosystem water use efficiency: a review of direct and indirect estimation methods. Remote Sensing 13 (12), 2393. Available at https://doi.org/10.3390/rs13122393.

Chandra, M.S., Naresh, R.K., Dhaliwal, S.S., Rajput, P., Harish, J., Yadav, S., Kumar, R., 2020. Re-designing energy-efficient and intensively managed cereal based cropping systems for reducing the environmental footprints of production in north-west india: a review. International Journal of Environment and Climate Change 342–361. Available at https://doi.org/10.9734/ijecc/2020/v10i1230310.

Chen, G., Kong, X., Gan, Y., Zhang, R., Feng, F., Yu, A., Zhao, C., Wan, S., Chai, Q., 2018. Enhancing the systems productivity and water use efficiency through coordinated soil water sharing and compensation in strip-intercropping. Scientific Reports 8 (1). Available at https://doi.org/10.1038/s41598-018-28612-6.

Du, T., Kang, S., Sun, J., Zhang, X., Zhang, J., 2010. An improved water use efficiency of cereals under temporal and spatial deficit irrigation in north China. Agricultural Water Management 97 (1), 66–74. Available at https://doi.org/10.1016/j.agwat.2009.08.011.

Dycian, Y., 2022. The Carbon Footprint of Water. Water Intelligence. Available at. https://wint.ai/blog/the-carbon-footprint-of-water/.

D'Odorico, P., Chiarelli, D.D., Rosa, L., Bini, A., Zilberman, D., Rulli, M.C., 2020. The global value of water in agriculture. Proceedings of the National Academy of Sciences of the United States of America. 117 (36), 21985–21993. Available at https://doi.org/10.1073/pnas.2005835117.

EEA, 2024. Energy efficiency. European Environment Agency. Available at https://www.eea.europa.eu/en/topics/in-depth/energy-efficiency.

EESI, 2024. Energy efficiency. Environmental and Energy Study Institute. Available at https://www.eesi.org/topics/energy-efficiency/description.

Enerdata, 2023. Global Energy Transition Statistics. Available at https://yearbook.enerdata.net/.

Eskandari, H., 2023. Improving energy efficiency in agronomical systems. Iranian Journal of Energy and Environment 14 (3), 221–227. Available at https://doi.org/10.5829/ijee.2023.14.03.04.

Fang, L., Zhang, L., 2020. Does the trading of water rights encourage technology improvement and agricultural water conservation? Agricultural Water Management 233, 106097. Available at https://doi.org/10.1016/j.agwat.2020.106097.

Fan, J., Lu, X., Gu, S., Guo, X., 2020a. Improving nutrient and water use efficiencies using water-drip irrigation and fertilization technology in Northeast China. Agricultural Water Manage. 241, 106352.

Fan, X., Zhang, W., Chen, W., Chen, B., 2020b. Land–water–energy nexus in agricultural management for greenhouse gas mitigation. Applied Energy 265, 114796. Available at https://doi.org/10.1016/j.apenergy.2020.114796.

FAO, 2011. Energy-smart food for people and climate. Issue Paper (p. 67). Available at https://openknowledge.fao.org/server/api/core/bitstreams/b9f2c018-1a88-429e-a1fe-d59d9c938b59/content.

Food and Agriculture Organization of the United Nations, 2011. The State of the World's Land and Water Resources for Food and Agriculture: Managing Systems at Risk, first ed. Routledge, London. Available at https://doi.org/10.4324/9780203142837.

FAO UN, 2024. Emissions total. FAOSTAT. Available at https://www.fao.org/faostat/en/#data/GT.

Farooq, M., Hussain, M., Ul-Allah, S., Siddique, K.H.M., 2019. Physiological and agronomic approaches for improving water-use efficiency in crop plants. Agricultural Water Management 219, 95–108. Available at https://doi.org/10.1016/j.agwat.2019.04.010.

Ghodrati, S., Kargari, N., Farsad, F., Javid, A.H., Kani, A.H., 2023. Assessing the water–energy–carbon nexus (WECN) in combined cycle power plants in Iran. International Journal of Environmental Science and Technology 20 (3), 2649–2672. Available at https://doi.org/10.1007/s13762-022-04749-w.

IEA, 2024. Efficiency policy momentum builds, but global energy intensity progress slows. International Energy Agency. Available at https://www.iea.org/energy-system/energy-efficiency-and-demand/energy-efficiency.

Jhariya, M.K., Meena, R.S., Banerjee, A., Kumar, S., Raj, A., 2023. Agroforestry for Carbon and Ecosystem Management. Elsevier, India, pp. 1–434. Available at https://doi.org/10.1016/C2021-0-02916-3.

Jhariya, M.K., Meena, R.S., Banerjee, A., Meena, S.N., 2022. Natural Resources Conservation and Advances for Sustainability. Elsevier. Available at https://doi.org/10.1016/C2019-0-03763-6.

Kilemo, D.B., 2022. The review of water use efficiency and water productivity metrics and their role in sustainable water resources management. OALib 09 (01), 1–21. Available at https://doi.org/10.4236/oalib.1107075.

Konapala, G., Mishra, A.K., Wada, Y., Mann, M.E., 2020. Climate change will affect global water availability through compounding changes in seasonal precipitation and evaporation. Nature Communications 11 (1). Available at https://doi.org/10.1038/s41467-020-16757-w.

Kourtis, M.A., Batistatos, M., Xylouris, G., Oikonomakis, A., Santorinaios, D., Zarakovitis, C., Chochliouros, I., 2023. Energy efficiency in agriculture through tokenization of 5G and edge applications. Energies 16 (13). Available at https://doi.org/10.3390/en16135182.

Kumar, S., Meena, R.S., Jhariya, M.K., 2020. Resources Use Efficiency in Agriculture. Springer Nature, Singapore, pp. 760. Available at https://doi.org/10.1007/978-981-15-6953-1.

Kumar, S., Meena, R.S., Sheoran, P., Jhariya, M.K., 2024. Regenerative Agriculture for Sustainable Food Systems. Springer Science+Business Media, India, pp. 1–493. Available at https://doi.org/10.1007/978-981-97-6691-8.

Leal, P.H., Marques, A.C., 2022. The evolution of the environmental Kuznets curve hypothesis assessment: a literature review under a critical analysis perspective. Heliyon 8 (11). Available at https://doi.org/10.1016/j.heliyon.2022.e11521.

Lens. (2024). Lens patent database. https://www.lens.org/.

Liu, J., Wang, H., Rahman, S., Sriboonchitta, S., 2021. Energy efficiency, energy conservation and determinants in the agricultural sector in emerging economies. Agriculture 11 (8), 773. Available at https://doi.org/10.3390/agriculture11080773.

Liu, X., Vu, D., Perera, S.C., Wang, G., Xiong, R., 2023. Nexus between water-energy-carbon footprint network: multiregional input-output and coupling coordination degree analysis. Journal of Cleaner Production 430, 139639. Available at https://doi.org/10.1016/j.jclepro.2023.139639.

Meena, R.S., Pradhan, G., 2023. Industrial garbage-derived biocompost enhances soil organic carbon fractions, CO_2 biosequestration, potential carbon credits and sustainability index in a rice-wheat ecosystem. Environmental Research 235, 116525. Available at https://doi.org/10.1016/j.envres.2023.116525.

Meena, R.S., Yadav, A., Kumar, S., Jhariya, M.K., Jatav, S.S., 2022. Agriculture ecosystem models for CO_2 sequestration, improving soil physicochemical properties, and restoring degraded land. Ecological Engineering 176. Available at https://doi.org/10.1016/j.ecoleng.2022.106546.

Meena, R.S., Pradhan, G., Kumar, S., Lal, R., 2023. Using industrial wastes for rice-wheat cropping and food-energy-carbon-water-economic nexus to the sustainable food system. Renewable and Sustainable Energy Reviews 187, 113756. Available at https://doi.org/10.1016/j.rser.2023.113756.

Meena, R.S., Pradhan, G., Nalani, Kumar, S., Singh, A.K., Verma, S.K., Mina, K.K., Kumar, S., Chaturvedi, R.K., 2024a. Diversified cropping modules designed for soil restoration, CO_2 sequestration, and generating carbon credits. Land Degradation and Development 35 (14), 4314–4329. Available at https://doi.org/10.1002/ldr.5224.

Meena, R.S., Pradhan, G., Singh, K., Kumar, S., Singh, A.K., Shashidhar, K.S., Mina, K.K., Rao, C.S., 2024b. Agriculture models for restoring degraded land to enhance CO_2 biosequestration and carbon credits in the Vindhyan region of India. Science of the Total Environment 929. Available at https://doi.org/10.1016/j.scitotenv.2024.172661.

Meena, R.S., Lal, R., Kumar, S., Pradhan, G., Srinivasarao, C., Singh, A.K., Pathak, H., Abhilash, P.C., Kumar, A., Sharma, S.K., Jat, M.L., Singh, S., 2024c. Potential of Indian agriculture for capturing atmospheric CO_2 and monetizing carbon credits to the farmers: an overview and policy framework. Advances in Agronomy 188, 101–206. Available at https://doi.org/10.1016/bs.agron.2024.06.001.

Meena, R.S., Pradhan, G., Kumar, S., 2025. Energy flow, eco-efficiency, and economic circulation with recycled industrial waste compost application in wheat and subsequent rice farming. Science of the Total Environment 967. Available at https://doi.org/10.1016/j.scitotenv.2025.178779.

Meena, R.S., Kumar, S., Pradhan, G., Kumar Jhariya, M., Srinivasarao, C., Meena, R.S., Pradhan, G.S., Kumar, S., Jhariya, M.K., Rao, C.S., 2026. Carbon-Negative Agriculture. Springer Nature, Singapore, pp. 1–500. Available at https://doi.org/10.1007/978-981-95-1483-0.

Meng, F., Liu, G., Liang, S., Su, M., Yang, Z., 2019. Critical review of the energy-water-carbon nexus in cities. Energy 171, 1017–1032. Available at https://doi.org/10.1016/j.energy.2019.01.048.

Mushtaq, S., Maraseni, T.N., Maroulis, J., Hafeez, M., 2009. Energy and water tradeoffs in enhancing food security: a selective international assessment. Energy Policy 37 (9), 3635–3644. Available at https://doi.org/10.1016/j.enpol.2009.04.030.

Pan, J., Peng, J., Yang, X., Xuan, S., 2023. Does water rights trading promote resources utilisation efficiency and green growth? Evidence from China's resources trading policy. Resources Policy 86, 104235. Available at https://doi.org/10.1016/j.resourpol.2023.104235.

Pimentel, C., 2022. Agronomic practices to improve water use efficiency. Environmental Sciences and Ecology: Current Research (ESECR 3 (8), 1–4. Available at https://doi.org/10.54026/esecr/1081.

Platis, D.P., Anagnostopoulos, C.D., Tsaboula, A.D., Menexes, G.C., Kalburtji, K.L., Mamolos, A.P., 2019. Energy analysis, and carbon and water footprint for environmentally friendly farming practices in agroecosystems and agroforestry. Sustainability (Switzerland) 11 (6), 1664. Available at https://doi.org/10.3390/su11061664.

Poore, J., Nemecek, T., 2018. Reducing food's environmental impacts through producers and consumers. Science 360 (6392), 987–992. Available at https://doi.org/10.1126/science.aaq0216.

Pradhan, G., Meena, R.S., 2023a. Interaction impact of biocompost on nutrient dynamics and relations with soil biota, carbon fractions index, societal value of CO_2 equivalent and ecosystem services in the wheat-rice farming. Chemosphere 339. Available at https://doi.org/10.1016/j.chemosphere.2023.139695.

Pradhan, G., Meena, R.S., 2023b. Utilizing waste compost to improve the atmospheric CO_2 capturing in the rice-wheat cropping system and energy-cum-carbon credit auditing for a circular economy. Science of the Total Environment 892. Available at https://doi.org/10.1016/j.scitotenv.2023.164572.

Pradhan, G., Meena, R.S., Kumar, S., Lal, R., 2023. Utilizing industrial wastes as compost in wheat-rice production to improve the above and below-ground ecosystem services. Agriculture, Ecosystems & Environment 358, 108704. Available at https://doi.org/10.1016/j.agee.2023.108704.

Prasad, R., Jhariya, M.K., Banerje, A., 2021. Advances in Sustainable Development and Management of Environmental and Natural Resources: Economic Outlook and Opinions, 2-volume set. Apple Academic Press, New York, pp. 1–428. Available at https://doi.org/10.1201/9781003187455.

Qin, J., Duan, W., Zou, S., Chen, Y., Huang, W., Rosa, L., 2024. Global energy use and carbon emissions from irrigated agriculture. Nature Communications 15, 3084. Available at https://doi.org/10.1038/s41467-024-47383-5.

Ritchie, H., Rosado, P., Roser, M., 2024. Energy Mix. Our World in Data. Available at https://ourworldindata.org/energy-mix.

Roy, O., Meena, R.S., Kumar, S., Jhariya, M.K., Pradhan, G., 2022. Assessment of land use systems for CO_2 sequestration, carbon credit potential, and income security in Vindhyan region, India. Land Degradation and Development 33 (4), 670–682. Available at https://doi.org/10.1002/ldr.4181.

Roy, D., Ghosh, T., Roy, A., Gogoi, M., Alam, K., Mitra, R., 2023. Enhancing agricultural water productivity of crops at field scale. In: Ghosh, A., Mahajan, A., Langangmeilu, G., Ghosh, A., Soujanya, K. (Eds.), Futuristic Trends in Agriculture Engineering & Food Sciences Volume 2 Book 9. Iterative International Publishers, pp. 376–390. Available at https://doi.org/10.58532/V2BS9CH29.

Sarkodie, S.A., Strezov, V., 2018. Empirical study of the environmental kuznets curve and environmental sustainability curve hypothesis for Australia, China, Ghana and USA. Journal of Cleaner Production 201, 98–110. Available at https://doi.org/10.1016/j.jclepro.2018.08.039.

Shah, H., Hao, G., Yasmeen, R., Yan, H., Shen, J., Lu, Y., 2023. Role of China's agricultural water policy reforms and production technology heterogeneity on agriculture water usage efficiency and total factor productivity change. Agriculture Water Management 287, 108429.

Stern, D.I., 2017. The environmental Kuznets curve after 25 years. Journal of Bioeconomics 19 (1), 7–28. Available at https://doi.org/10.1007/s10818-017-9243-1.

UNEP, 2024. As shortages mount, countries hunt for novel sources of water. Available at https://www.unep.org/news-and-stories/story/shortages-mount-countries-hunt-novel-sources-water.

Wani, P.P., Mate, P., Chande, K., Humane, A.N., Bele, K., Ramteke, P.R., Bajad, H., 2020. Management interventions for enhancing agricultural production by improving water use efficiency. International Journal of Chemical Studies 8 (6), 1478–1481. Available at https://doi.org/10.22271/chemi.2020.v8.i6u.10971.

Wysokiński, M., Domagała, J., Gromada, A., Golonko, M., Trębska, P., 2020. Economic and energy efficiency of agriculture. Agricultural Economics (Zemědělská ekonomika) 66 (8), 355–364. Available at https://doi.org/10.17221/170/2020-agricecon.

Yang, Y., Huang, Z., He, W., Li, J., Yuan, L., 2024. Can water rights trading improve water resources use efficiency: empirical evidence from China's pilot program. Environment, Development and Sustainability. Available at https://doi.org/10.1007/s10668-024-04569-7.

Yu, L., Zhao, X., Gao, X., Siddique, K.H.M., 2020. Improving/maintaining water-use efficiency and yield of wheat by deficit irrigation: a global meta-analysis. Agricultural Water Management 228. Available at https://doi.org/10.1016/j.agwat.2019.105906.

Zhang, H., Zhou, Q., Zhang, C., 2021. Evaluation of agricultural water-saving effects in the context of water rights trading: an empirical study from China's water rights pilots. Journal of Cleaner Production 313, 127725. Available at https://doi.org/10.1016/j.jclepro.2021.127725.

Chapter 13

Application of artificial intelligence to evaluate carbon credits in agriculture

Suchismita Satapathy[1], Krishna Kant Mina[2], Priyanka Saha[3], Shashi Shekhar[4], Manoj Kumar Jhariya[5], Arnab Banerjee[6], Gourisankar Pradhan[1], Sanjeev Kumar[1], and Ram Swaroop Meena[1]

[1]*Department of Agronomy, Institute of Agricultural Sciences, Banaras Hindu University, Varanasi, Uttar Pradesh, India,* [2]*Department of Soil Science and Agricultural Chemistry, Institute of Agricultural Sciences, Banaras Hindu University, Varanasi, Uttar Pradesh, India,* [3]*ICAR Krishi Vigyan Kendra, West Garo Hills, ICAR RC NEH Region, Umiam, Meghalaya, India,* [4]*Department of Agricultural Engineering, Institute of Agricultural Sciences, Banaras Hindu University, Varanasi, Uttar Pradesh, India,* [5]*Department of Farm Forestry, Sant Gahira Guru Vishwavidyalaya, Ambikapur, Chhattisgarh, India,* [6]*School of Studies in Environmental Science, Pt. Ravishankar Shukla University, Raipur, Chhattisgarh, India*

Highlights

- Addressing climate change requires smart agricultural transformation.
- Economic gain needs to be incorporated in the form of carbon credits.
- Technological intervention in terms of machine learning has huge potential.
- Application of AI would create a new era of agriculture.
- Machine learning can be suitably used for assessing carbon credits.

13.1 Introduction

Climate change is a crucial issue and must be addressed promptly. According to McPherson and Kendall (2014), cutting down trees for farming, infrastructure, or other purposes releases carbon (C) stored in the trees, leading to emissions. Ultimately, the planet's surface is heated by potent greenhouse gases (GHGs) that trap sunlight in the atmosphere (El Zein et al., 2015). This brings in the concept of C credit farming, which has become a key climate change mitigation strategy (Raj et al., 2025). In this approach, carbon dioxide (CO_2) is sequestered from the atmosphere through agricultural or land management practices that enhance C capture and storage (CCS) in soils, plants, or trees (van der Gaast et al., 2018). This method involves obtaining C credits, marketable licenses, or certificates that permit the emission of a specific amount of CO_2 or other GHGs. The demand for C credits generates a market for innovative technologies and practices that enable carbon sequestration (C_{seq}) and emissions reduction (Meena et al., 2024). This promotes the development and implementation of green technologies, including CCS, renewable energy, and low-C transportation, thereby advancing progress toward a more sustainable and climate-resilient future (Bui et al., 2018). C credit farming offers financial incentives for landowners to adopt environmentally sustainable practices by participating in various C credit programs that help to mitigate climate change by reducing overall GHG emissions (Raina et al., 2024). Furthermore, it can support sustainable land management practices and enhance soil health and biodiversity. C credits generated from agricultural practices can be used by corporations and industries to offset their C emissions, enabling them to meet regulatory requirements, achieve sustainability goals, and improve their environmental profiles by investing in initiatives that remove or reduce CO_2 from the atmosphere (Gupta and Rakshit, 2023). Individuals or organizations can earn C credits by removing or preventing C emissions through actions such as tree planting, sustainable agriculture, or land restoration. Farmers are being educated on the benefits of C-farming to ensure equitable advantages for all participants. C credit prices per ton in 2023 ranged from US $40 to 80, with demand expected to increase 15-fold by 2030. As a result, the market value of C credits may rise significantly, making them increasingly viable to sell. Indian state and federal governments

Carbon Incentives, Plans, and Policies in Agriculture. DOI: https://doi.org/10.1016/B978-0-443-33557-0.00013-6

have aligned natural, regenerative, agroforestry, and organic farming schemes to encourage farmers' participation in C credit programs (Jhariya et al., 2022, 2023).

Considering the growing importance of C credits in agriculture, various artificial intelligence (AI) tools have emerged. AI has become a widely discussed topic, with ongoing discourse about its potential, challenges, and risks across different sectors. Although interest in AI has surged in recent years, the field has been recognized for over 50 years, and it has undergone numerous developmental shifts. Recently, advanced technologies such as machine learning (ML), AI, and deep learning have been increasingly adopted to extract and utilize information related to C credits, benefiting farmers.

13.2 Relationship between climate change and carbon credits

Climate change is considered one of the foremost challenges confronting mankind in the 21st century. It arises from the greenhouse effect, which sustains Earth's average temperature of 15°C, facilitating the existence of life. The American Meteorological Society defines climate change as a systematic alteration in the long-term statistics of climate factors (including temperature, precipitation, pressure, or winds) maintained over several decades or more (National Research Council, Division on Earth, Life Studies, Board on Atmospheric Sciences, Committee on Understanding, Monitoring Abrupt Climate Change, & Its Impacts, 2013).

GHGs include CO_2, nitrous oxide (N_2O), sulfur hexafluoride, hydrofluorocarbons, and perfluorocarbons. Emissions are believed to have peaked in the last three decades (1983–2012), leading to progressively higher surface temperatures on Earth in each of the preceding decades since 1850. The excessive concentration of these gases in the atmosphere results in climate change. The influence of human activity on the climate system is apparent. The open combustion of biomass accounts for approximately 37% of global black C emissions, resulting in annual emissions of 2760 Gg (Bond et al., 2013).

According to the Indian Ministry of New and Renewable Energy, India generates around 500 million tons of agricultural residue each year, of which an average of 92 million tons are burned annually. Although the quantity incinerated constitutes about 18.4% of the total agricultural residue generated, it remains a significant source of concern regarding GHG emissions and climate change.

Additionally, according to the Report on Crop Residue Burning (2021), approximately 66% of all GHGs produced from the combustion of crop residues are CO_2. The Kyoto Protocol, approved in 1997, aimed to minimize global GHG emissions and address climate change by introducing the concepts of C credits and a C market. Furthermore, a C credit is a repayment mechanism used by governments or entities that emit excessive GHGs (Lokuge and Anders, 2022). These credits can be bought or sold in international markets at specific prices, depending on various conditions. This prompted the establishment of the international C market, facilitating C trading among parties to promote environmentally favorable and sustainable policies regarding global C emissions (Zhou and Li, 2019). The incineration of crop residues in the Indian agricultural sector significantly contributes to CO_2 emissions and results in substantial losses of CO_2 credits in the global C market. In summary, India may generate significant foreign exchange through C-trading if stubble burning is eliminated.

13.3 Origins of the carbon market

C markets originated from the need to address climate change and reduce GHG emissions, with the concept gaining traction through international agreements and the establishment of trading mechanisms. The Kyoto Protocol in 1997 was a landmark moment, introducing emissions trading and other flexible mechanisms such as the Clean Development Mechanism (CDM) and Joint Implementation. These mechanisms allowed countries and companies to trade C credits, essentially giving them the right to emit a certain amount of GHGs (Paul and Nair, 2024).

C markets can be broadly categorized into compliance (or regulated) markets and voluntary markets. Compliance markets are established by governments or multigovernment bodies to regulate C emissions, while voluntary markets allow companies and individuals to offset their emissions by purchasing C credits.

13.3.1 Voluntary carbon markets

Voluntary carbon markets (VCMs) differ from compliance markets, as they involve private firms that voluntarily choose to either offset their C emissions or promote sustainability goals in the absence of governmental mandates (Battocletti et al., 2024). Voluntarily traded C credits predominantly finance Reduced Emissions from Deforestation and Forest Degradation initiatives in developing nations, as well as C_{seq} and removal projects related to agriculture, renewable energy, waste management, and other sectors (Ebeling and Yasué, 2008).

The creation of a legal framework for VCMs is essential to prevent greenwashing, double-counting, or C credit leakage. Such a framework would ensure the uniform quality of C credits, uphold market integrity, and provide assurance to purchasers (Chawla et al., 2023).

13.3.2 Carbon trading

C trading is a regulatory mechanism used to manage CO_2 emissions through economic incentives, wherein a central authority establishes a limit on the permissible amount of CO_2 emissions. Various entities—such as companies, individuals, or other organizations—are allocated emission licenses. They must hold a corresponding number of allowances (or credits) that authorize them to emit a designated quantity. The total number of allowances and credits must not exceed the cap, thereby constraining overall emissions to that threshold (Murray et al., 2009).

Companies that require more emission allowances must acquire credits from entities that emit less or possess surplus C credits. The buyer incurs a cost for excess emissions, while the seller is compensated for reducing emissions. In recent years, this exchange has led to the development of C markets (Tian et al., 2016).

13.3.3 Scenarios of carbon credit value in different Asian countries

The evolution of the global C market has garnered significant interest. The C Forward Asia conference, held on March 7 and 8 in Singapore, convened industry experts and scholars to examine current market conditions and future trends. Delegates from Singapore, Malaysia, China, South Korea, and the European Union (EU) focused on two primary topics: the advancement of C markets in the Asia-Pacific region and the prospects of the global voluntary C market.

In contrast to the EU Emissions Trading System, Southeast Asia has established its regulations for C pricing. An examination of Southeast Asia's C pricing methods indicates that each nation employs distinct strategies (Ewing, 2022). Among all the South Asian countries, Indonesia—a leading global greenhouse gas emitter—is the first nation in the region to implement an Emissions Trading System.

Many Southeast Asian countries prefer voluntary C markets, which primarily emphasize C offsetting, over government-led, mandated C markets. For instance, Malaysia's Bursa Carbon Exchange and Thailand's FTIX Carbon Trading Platform commenced operations last year. Singapore's C tax framework and C credit trading exhibit greater maturity, characterized by comprehensive regulations and strong market liquidity (Li et al., 2023).

In 2021 the estimated value of the voluntary C market surpassed USD 2 billion. Despite a 50% decline in C credit trading volume in the subsequent year, the overall transaction value remained comparable, signifying an increase in the price per C unit (Fig. 13.1). This tendency is chiefly attributable to the growing focus on premium C credits.

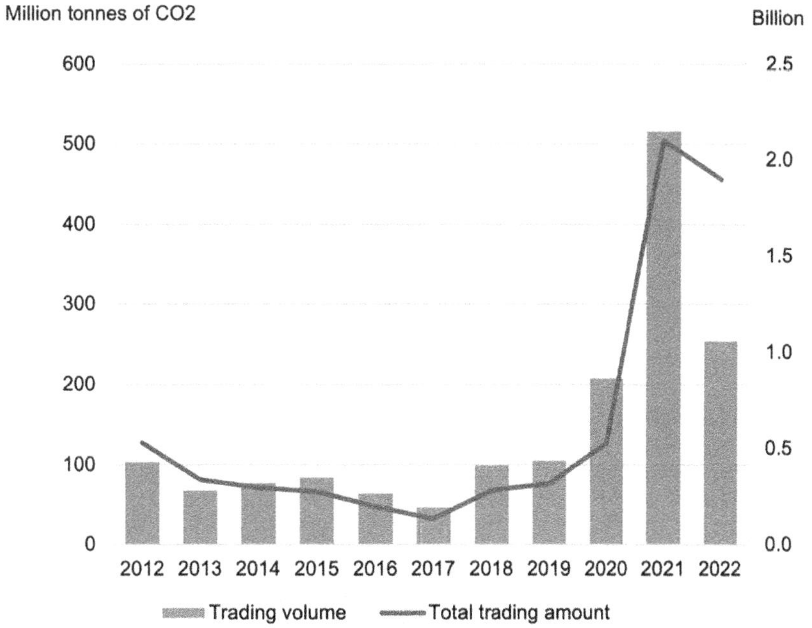

FIGURE 13.1 Global voluntary carbon market trading volume and transaction value (2024 Carbon Forward Asia: In-depth analysis).

13.4 Current challenges in carbon credit assessment

13.4.1 Source-specific challenges

Despite the potential of C credit assessment, several challenges hinder the quality of C credits. Understanding the type of production source is a crucial step that can assist organizations in making informed decisions when selecting C credits to mitigate their C emissions (Bayon et al., 2012).

Nature-based C credits originate from afforestation and forest management. C credits derived from afforestation help mitigate climate change, conserve natural resources, and provide protection against droughts, heat waves, and floods (Moomaw et al., 2019). These nature-based C credits are often considered more valuable than renewable energy credits, as they focus on removing C from the atmosphere rather than merely reducing emissions (Johnstone, 2023). Similar to nature-based C credits, negative emissions technologies aim to eliminate C (Meena et al., 2026). These include direct air capture, soil C capture, and bioenergy CCS, serving as alternatives to afforestation (Fawzy et al., 2020).

13.4.2 Location-specific challenges

The location of projects designated for generating C credits is another crucial aspect, encompassing two key factors: international and local (Khan, 2024). Currently, C projects are predominantly located in underdeveloped nations, while multinational companies and affluent countries acquire these C credits at lower prices. This approach is theoretically beneficial for both the demand and host countries (Blum, 2020).

A further concern relates to regulation and quality control, given the absence of a uniform global framework and consistent local government monitoring of these projects. Verifying the legitimacy of C offsetting initiatives becomes complex and challenging due to issues that may undermine the value of C credits. These include double-counting, project mismanagement or failure, creative accounting, and other such practices (Kreibich and Hermwille, 2021). Therefore, emphasis and priority should be given to locally generated C credits to support the achievement of C neutrality.

13.4.3 Timeframe affecting carbon credit quality

C credits are inherently time-sensitive, and a major challenge lies in ensuring their availability and long-term quality. Identifying these characteristics is essential to understanding the actual impact of the issued credits (Maddali et al., 2015).

For instance, some organizations promote their products as C-neutral. However, the issuance of C credits used for offsetting may not occur for 2 years or more (Northern, 2011), indicating that these credits are not currently available for immediate emission offsetting, thereby rendering them invalid or of low quality.

13.4.4 Fidelity affecting carbon credit

The fidelity dimension evaluates elements such as the trustworthiness of a C credit and its actual performance (Khuzairi et al., 2024). The dependability of the entire process is characterized by the reliability of C credits. However, the majority of issued credits are susceptible to the risks of double issuance and double counting, which undermine their credibility. These risks are not new (Kreibich and Obergassel, 2016). Double issuance refers to the allocation of a single C credit to multiple entities, even though the emissions reduction or removal is counted only once, resulting in double counting (Juvonen, 2021).

13.5 Role of artificial intelligence with carbon emissions

AI in agriculture can reduce C emissions from farming practices and rejuvenate the overall sector (Mor et al., 2021). AI-driven C footprint prediction systems evaluate data from various sources—such as meteorological forecasts, soil conditions, and agricultural outputs—using ML algorithms. This technology can accurately calculate the C footprints of farms by analyzing the collected data (Frackiewicz, 2023).

A bibliometric analysis of research conducted in this field from 2000 to 2021 highlights the role of AI in sustainable agriculture. The implementation of AI in agriculture can lower C footprints by providing farmers with accurate assessments of their emissions and enhancing the efficiency of farming practices.

13.6 Artificial intelligence application for assessment of carbon credit value

Climate change is an urgent challenge that necessitates new strategies to mitigate C emissions. To address this, AI and blockchain technologies have the potential to transform C credit trading (Baklaga, 2024). AI, or machine intelligence, is a powerful technological force represented through ML algorithms, predictive analytics, and decision-making systems. Typically, the development of AI models follows a centralized path, encompassing data collection, training, and deployment (Shandilya et al., 2024).

AI-driven predictive models and ML algorithms can interact with blockchain systems to enhance resource allocation, stimulate economic growth, and reduce environmental damage (Chauhan and Sahoo, 2024). AI plays a crucial role in decentralized C markets by improving emission forecasting, credit pricing, and trading strategies. ML algorithms enable the analysis of large datasets, the identification of patterns, and adaptability to market dynamics (Baklaga, 2024).

According to Paramesha et al. (2024), the integration of AI enhances agility and accuracy, enabling stakeholders to make informed decisions based on real-time data. Meanwhile, blockchain technology supports AI by collecting and securing data from various sources, such as clients, servers, and networks.

The convergence of blockchain and AI aims to resolve challenges such as inefficient processes, lack of transparency, and the risk of double-counting, ultimately creating a more efficient, transparent, and adaptable ecosystem for C trading. There are various AI methods and tools available for assessing the value of C credits.

13.6.1 Remote sensing and satellite imagery

Remote sensing technologies, especially satellite-based approaches, are essential for estimating C stocks by delivering precise, regular, and accurate data (Issa et al., 2020). These methodologies help to evaluate agricultural ecosystem health, land-use changes, and the potential for C_{seq} (Ponce-Hernandez et al., 2004). Remote sensing data is primarily used to monitor land-use and land-cover changes to estimate C stocks (Goetz and Dubayah, 2011), which is important for rural and urban planning, natural resource management, and environmental stewardship (Kumar et al., 2021).

Conservation studies reveal land-use and land-cover (LULC) changes, and LULC maps enable quick and accurate assessments of green vegetation and soil C content (Khachoo et al., 2024). This information allows farmers and governments to allocate land efficiently, rotate crops, and manage land sustainably (Nehzak et al., 2022). Satellite data is a valuable resource for LULC classification (Steinhausen et al., 2018), and Earth Observation datasets are often used to collect LULC and land-use/land-cover change (LULCC) data.

Remote sensing technology can also estimate tree heights, diameters, and volumes—key factors in assessing C stocks (White et al., 2016)—which are crucial for climate change research and C_{seq} programs. Puliti et al. found that multispectral satellite data improved biomass assessment. Additionally, by combining Landsat-derived forest loss data with extensive aerial laser scanning data, Breidenbach et al. (2021) enhanced the accuracy of C stock estimates.

13.6.2 Predictive analysis

Predictive analytics is widely acknowledged as a crucial instrument in C_{seq}, offering valuable insights for sustainable investment (Hamdan et al., 2024). By utilizing advanced data analytics techniques and various AI algorithms, predictive analytics enables the forecasting of C credit values in agriculture. It helps to mitigate climate-related risks, including physical risks such as extreme weather events and transition risks such as rising CO_2 concentrations in the atmosphere. It incorporates diverse data sources, including climate models, satellite imagery, socioeconomic indicators, and financial data (Choi and Kim, 2023). Furthermore, predictive analytics provides actionable insights by analyzing historical trends and future projections to optimize the value of C credits. Various ML models are used to estimate C prices and trading volumes in C markets.

13.7 Potential artificial intelligence tools and techniques for carbon credit assessment in agriculture

13.7.1 Machine learning

ML is a field of artificial intelligence that uses and/or creates algorithms and statistical models to learn from data and improve task performance without explicit programming (Helm et al., 2020). Based on their learning methods, ML algorithms can be categorized as supervised, unsupervised, or reinforcement learning (Sarker, 2021). Furthermore, within these broad categories, ML can be further classified according to the transformation method employed—into linear and

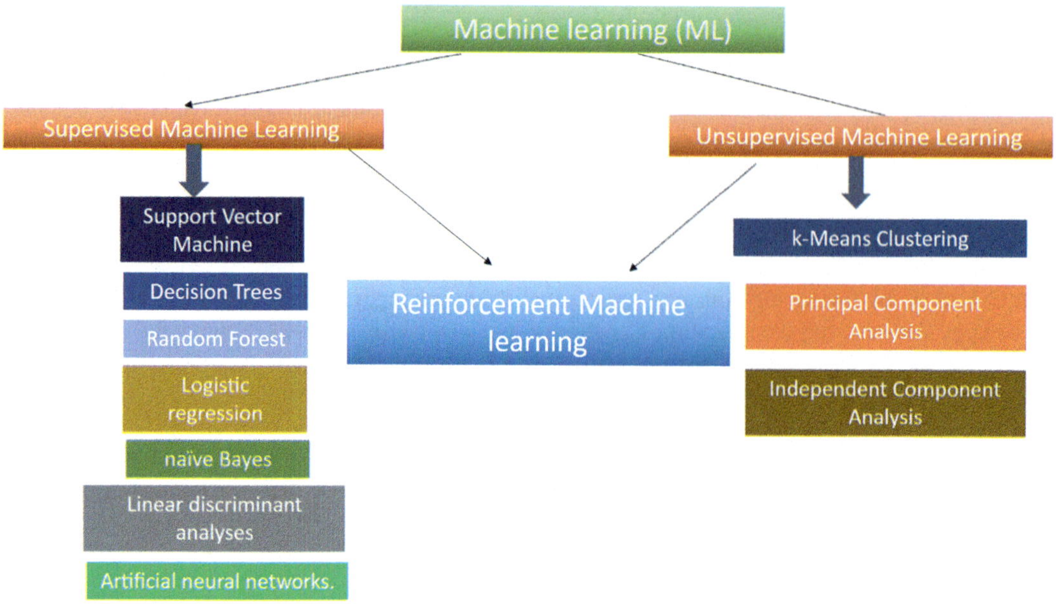

FIGURE 13.2 Diagrammatic classification of various algorithms used in ML (Mishra and Mishra, 2023) (*ML*, machine learning).

nonlinear regression techniques (Maulud and Abdulazeez, 2020). This technique is widely applicable in the agricultural field to assess various parameters.

Agricultural soils can function as both sources and sinks of GHGs, particularly CO_2. The adoption of climate-smart farming practices, along with variations in soil properties and environmental factors, contributes to the reduction of CO_2 emissions (McNunn et al., 2020). However, challenges such as food security, population growth, and climate change require more innovative agricultural approaches to mitigate CO_2 emissions.

These models, however, come with limitations, including the need for skilled users with agro-environmental knowledge, requirements for calibration and validation, and variations in input parameters based on site-specific conditions. ML algorithms (Fig. 13.2) serve as an alternative to biophysical models for simulating and predicting CO_2 emissions and environmental events. ML models have been effectively applied in agriculture for image recognition and classification, land cover and crop type categorization, and forecasting future trends (Hamrani et al., 2020).

ML is also extensively used in crop monitoring. Algorithms can analyze remote sensing data—including satellite imagery, spectral analysis, and other imaging techniques—to predict and monitor yield, quality, CO_2 emissions, and C_{seq} (Omia et al., 2023).

13.7.2 Natural language processing

European agriculture faces global challenges arising from human activity, climate change, and the ongoing conflict between Russia and Ukraine. Global temperatures are projected to rise by at least 1.5°C, resulting in severe weather events and climatic phenomena across Southern Europe (IPCC Intergovernmental Panel on Climate Change, 2022). This will lead to droughts and heatwaves, threatening biodiversity, human health, and agricultural systems. Additionally, the conflict exacerbates the energy crisis, raising prices and jeopardizing food security and farmers' livelihoods (Kiryluk-Dryjska et al., 2020). In this context, European authorities must adapt to these emerging realities.

Natural language processing (NLP), a data analytics tool widely used across sectors such as tourism, banking, and healthcare (Bermeo-Almeida et al., 2019), holds potential for this adaptation. However, its application in agri-food research remains limited. NLP can extract insights from user-generated content on platforms such as Twitter/X, capturing consumer perceptions about products and their sourcing and manufacturing processes (Ramos-Sandoval and Beltran, 2021). This capability can support the development of innovative agri-food policies aligned with societal demands and values. Consequently, NLP can aid policymaking, forecasting, trend analysis, and collaborative policy development—including the formulation of various C credit generation policies in agriculture (Patel and Patel, 2020).

NLP is a domain within computer science, AI, and linguistics that focuses on enabling computers to understand human language (Srivastava et al., 2022). One specific tool in this domain is the Statistical Morphological Analyzer, which uses

ML to predict the morphological structure of words, including features such as lemma, gender, number, and case. Techniques such as sentiment analysis and opinion mining have gained considerable attention in academic research (Stieglitz et al., 2018).

Named Entity Recognition (NER) is an automated technique for analyzing C credit documentation. It efficiently detects and classifies key terms like "C_{seq}," "soil organic carbon (SOC)," and "greenhouse gas (GHG) reduction" in the context of climate change and environmental issues (Toprak and Turan, 2023). NER is part of the broader task of information extraction, identifying and categorizing named entities such as people, organizations, locations, and dates from unstructured text (Nasar et al., 2022). It aids in extracting and organizing information on climate change, environmental policies, and scientific research by identifying key phrases and topics (Weston et al., 2019). In the context of C_{seq}, NER helps recognize and classify references to processes such as atmospheric CO_2 capture and storage. It can also identify mentions of SOC, which is vital for soil health and C_{seq} (Nayak et al., 2019). Furthermore, NER plays a significant role in identifying and improving the classification of references to GHG mitigation, a key goal in climate action.

NLP offers several valuable features for agricultural applications, including sentiment analysis of farmer feedback, machine translation for cross-language communication, and text summarization to extract key insights from large datasets. Additionally, NLP can be used for crop identification, feature extraction from agricultural data, NER of crops and diseases, and data processing to extract meaningful information (see Fig. 13.3).

13.7.3 Internet of Things

Agriculture 4.0 represents the most recent stage in agricultural development, utilizing technologies such as the Internet of Agricultural Things (IoAT), AI, big data analytics, robotics, and precision agriculture (Singh et al., 2021). Technological integration is expected to play a crucial role in addressing global food production challenges, maintaining resource sustainability, and adapting to the demands of climate change (Kakamoukas et al., 2021). Agriculture 4.0, through the use of the Internet of Things (IoT), promotes sustainable farming and C neutrality, enhances environmental governance, and helps mitigate climate change (Dayıoğlu and Turker, 2021). The integration of IoT, big data, and ML supports the development of comprehensive system frameworks for monitoring agricultural C footprints (Singh et al., 2021). This includes the collection, analysis, and translation of data into actionable policy recommendations.

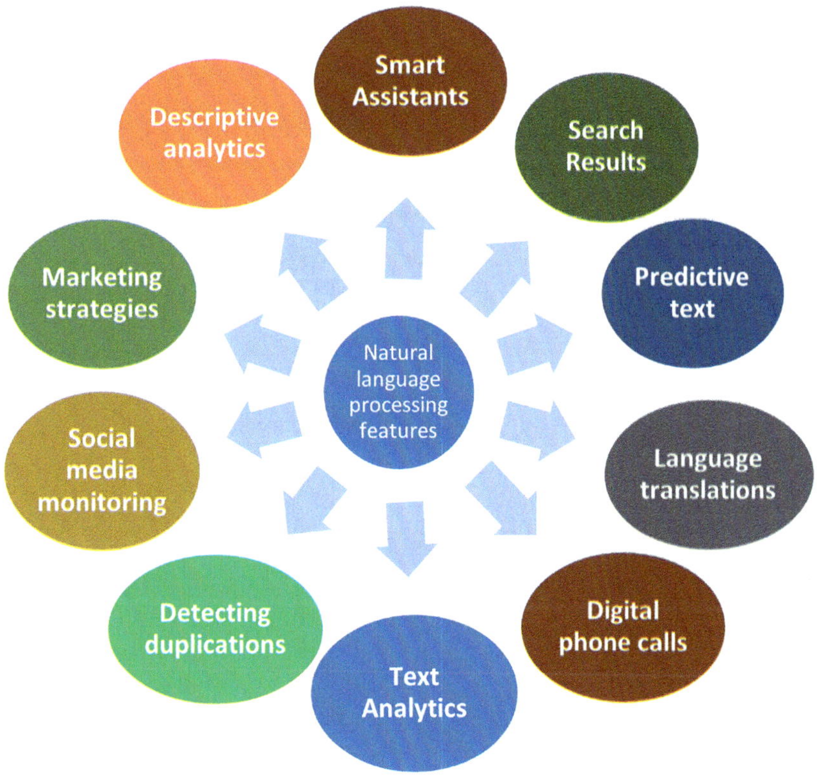

FIGURE 13.3 Various features of natural language processing.

13.7.3.1 *Frameworks of the Internet of Things in agriculture*

IoAT utilizes modern agricultural methods and enhances the IoT framework for data-driven insights and agricultural automation (Ojha et al., 2021). The sensory layer (perception layer) requires sensors to measure soil moisture, temperature, humidity, and meteorological factors (Shaikh et al., 2022). Real-time data from soil moisture sensors helps to optimize irrigation schedules (Adak et al., 2023). This layer includes infrared gas analyzers for CO_2 and methane detection (Ma et al., 2024), normalized difference vegetation index sensors for crop health assessment, and weather stations for microclimatic data collection.

Agricultural equipment, including tractors, is increasingly equipped with sensors and actuators to facilitate automation (Fig. 13.4). GPS-guided tractors can now perform tasks such as planting and fertilizing with exceptional precision, using GPS sensors and actuators (Vellingiri et al., 2024). Unmanned aerial vehicles and drones enhance these capabilities by capturing airborne imagery and data across large areas, enabling the assessment of crop health (Barbedo, 2019; Olson and Anderson, 2021).

Communication among these components at the networking layer is crucial. However, describing this communication as "seamless" may be inaccurate. Data from these sensors is transmitted to the cloud for further analysis at the application layer. This process requires reliable infrastructure, which may not always be available in all agricultural settings. Ultimately, at the application layer, IoAT supports solutions such as precision agriculture, predictive maintenance, crop monitoring, and supply chain optimization.

IoT and AI have enhanced sustainable agriculture by optimizing resource utilization and improving environmental outcomes. (Prabhat and Singh, 2021) found that IoT-enabled precision irrigation systems, using real-time soil moisture and meteorological data, reduced water consumption by 30% and increased agricultural yields. AI-driven digital soil-mapping algorithms, such as Perennial's, have reduced the need for physical soil samples by 40%, improving C_{seq} estimates and enabling targeted interventions for regenerative farming practices (Mehta, 2024; Kumar et al., 2024). Moreover, Big Data analytics in agriculture has enhanced crop production forecasts by up to 25% and optimized nitrogen fertilizer use, resulting in cost reductions of approximately 15% (Weraikat et al., 2024) (Table 13.1).

13.7.4 Geographic information systems

SOC is an essential factor in assessing soil quality, particularly in agricultural regions (Tajik et al., 2020), and it affects soil's physical, chemical, and biological properties (Milne et al., 2015; Zeraatpisheh et al., 2021, 2022). The water-retention capacity and precipitation infiltration influenced by SOC enhance moisture availability in the landscape

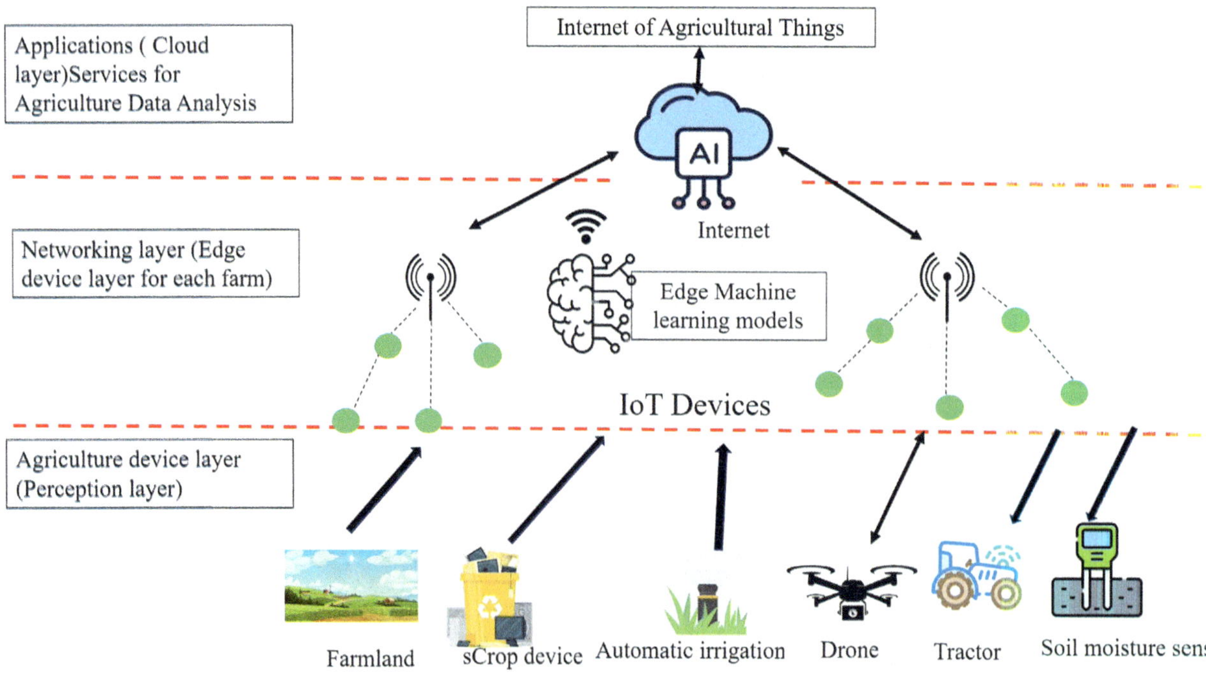

FIGURE 13.4 Architecture of an Internet of Agricultural Things (IoAT) (Ahmed and Shakoor, 2025; Udutalapally et al., 2021).

TABLE 13.1 Examples of some potential AI tools widely used for the generation of C credits.

Serial no.	AI tools	Utilization	Reference
1	Pachama	Use a combination of satellite data, ML, and dynamic baseline methodologies, ensuring high-quality credits are issued.	https://pachama.com/
2	Indigo Ag	It employs a hybrid approach, combining soil sampling and modeling, to help farmers generate agricultural C credits.	https://www.indigoag.com/carbon-credits
3	Planet Labs	This employs satellite images and GIS tools to facilitate the creation and validation of C credits, specifically concerning forest C monitoring and C offset initiatives.	https://www.planet.com/products/forest-carbon/
4	Silvia Terra, also now known as Natural Capital Exchange (NCX)	This is a climate tech company that facilitates the creation and trading of C credits from forest and agricultural lands.	https://ncx.com/

AI, artificial intelligence; *GIS*, geographic information systems.

FIGURE 13.5 Top five countries by C emissions in billion tons.

(Hartemink and McSweeney, 2014). Overall, SOC is a crucial factor in sustainable agriculture (Johnston et al., 2009). Understanding the key elements and roles of SOC, along with its precise measurement, is vital in addressing soil degradation (Ramifehiarivo et al., 2017). This knowledge also contributes to improved agricultural productivity (Jat et al., 2019). The accurate and comprehensive geographical distribution of SOC is essential for various land-use processes.

Geographic information systems (GIS) provide the spatial analysis needed to understand the geographical distribution of C emissions, C sinks, and C offset initiatives. By integrating location data with environmental parameters, GIS offers insights into the variability of C emissions and reductions across different geographies (Mansourihanis et al., 2025). There are various ways in which GIS supports the development of a sustainable C credit policy in agriculture.

13.7.4.1 Spatial analysis of carbon emissions

GIS facilitates the visualization of C emissions at global, national, or local levels, pinpointing emission hotspots that require focused action (Fig. 13.5). This geographical analysis can integrate data from diverse sources, including satellite imagery, industrial monitoring systems, and transportation networks, to provide comprehensive emission maps (Nagavi et al., 2024). For example, GIS can delineate areas with high deforestation rates or urban locations with significant emissions from transportation and industrial activities. Policymakers can use this data to implement more localized C reduction initiatives (Dalvi et al., 2006).

13.7.4.2 Locating carbon offset projects

GIS is essential for determining appropriate sites for C offset initiatives, including afforestation, wind energy facilities, and solar power installations. In reforestation initiatives, GIS can assist in pinpointing deforested regions with significant potential for C_{seq} by evaluating climate data, soil quality, and land-use trends to identify the most suitable locations for tree-planting programs (Bilal et al., 2024). Furthermore, GIS can delineate existing C offset initiatives, offering information on their geographical distribution and proximity to emission sources. This spatial intelligence is essential for ensuring the effectiveness of C offset schemes and their alignment with regional requirements (Gupta, 2024).

13.7.4.3 Tracking carbon credit ownership

GIS can facilitate the monitoring of C credit ownership and transactions by integrating blockchain technology, enabling the creation of a transparent and immutable record of C credit exchanges (Khanna and Maheshwari, 2023). This ensures that credits are not redundantly counted and that the complete lifecycle of a C credit—from inception to retirement—is thoroughly documented.

The integration of AI and GIS offers a powerful resource for managing C-credits. While AI excels at analyzing large datasets, identifying trends, and generating forecasts, GIS provides the spatial framework that makes these insights actionable (Safavi, 2024). Together, these technologies optimize every phase of the C credit process, including early emission monitoring, project implementation, and credit trading (Avwioroko, 2024). By integrating AI algorithms with GIS maps, organizations can monitor emissions levels and C-reduction initiatives in real time. AI can analyze satellite data to detect deforestation or emissions from industrial facilities, while GIS can visualize these occurrences, offering a clear depiction of areas requiring intervention (Garg, 2024).

This integration enables precise, real-time reporting—crucial for validating C credits. Various AI tools, including GIS, can collaborate to improve the design and execution of C offset initiatives by evaluating historical data and future trends to recommend the most suitable project types (e.g., wind, solar, or reforestation) (Kosgei, 2009).

The combination of AI, GIS, and blockchain technologies can bring unmatched transparency to C markets. GIS offers a spatial view of where C credits are generated and traded, AI ensures accurate tracking and verification; and blockchain securely records these transactions—mitigating fraud and ensuring full credit accountability (Baklaga, 2024).

13.8 Future of the carbon market

The global C credit market is valued at USD 531.2 billion in 2024 and is projected to reach USD 7126.6 billion by 2034, exhibiting a compound annual growth rate (CAGR) of 29.8% during the forecast period from 2025 to 2034. A C credit is a transferable permit or certificate that authorizes its holder to emit 1 ton of CO_2 or an equivalent quantity of another GHG.

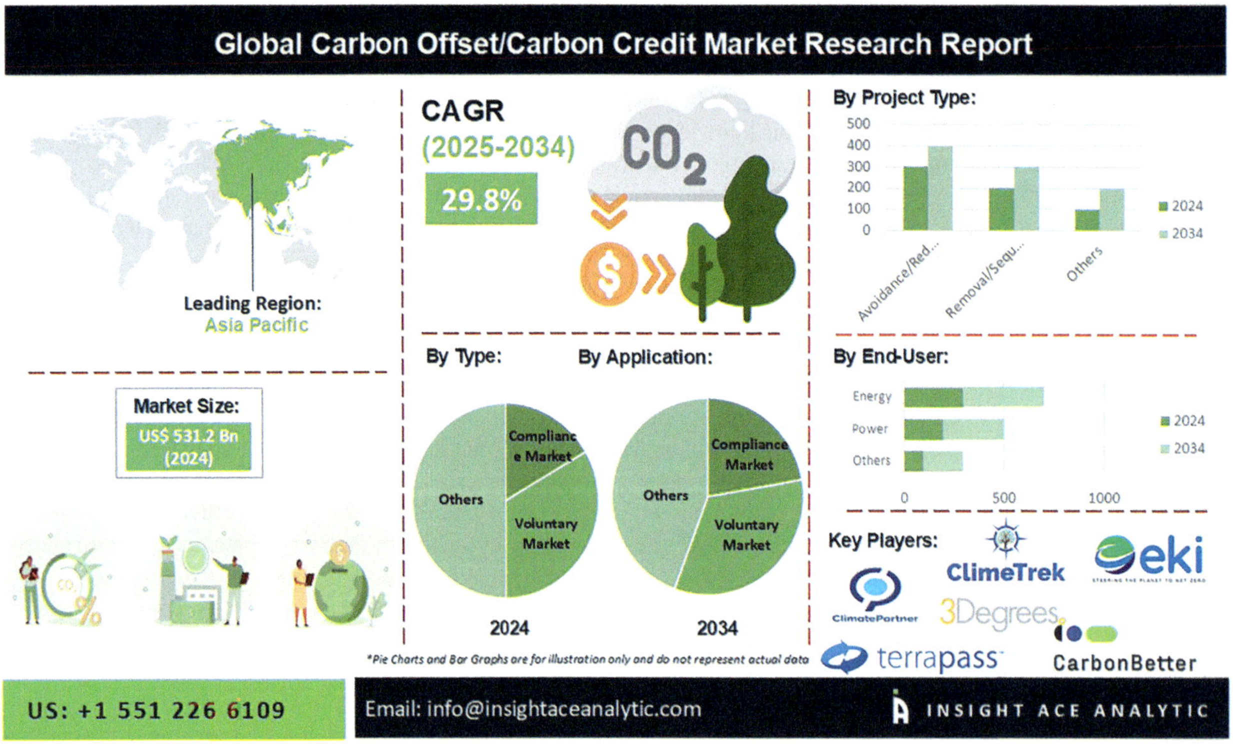

FIGURE 13.6 Ref: Global carbon credit market report, 2025. Map lines delineate study areas and do not necessarily depict accepted national boundaries.

It serves as compensation for individuals or entities that emit harmful gases. As global climate initiatives progress, C markets are becoming increasingly crucial in achieving net-zero GHG emissions (Fig. 13.6).

C capture solutions are at the forefront of combating global warming through the development and implementation of novel, scalable technologies that both prevent CO_2 emissions and enable the extraction of previously emitted CO_2. Numerous international, national, and subnational market systems are being established worldwide. The growing diversity of C markets necessitates a strong regulatory framework to prevent issues such as the double-counting of activities.

Innovative funding structures—such as public–private partnerships and microfinance programs that provide affordable access to long-term AI solutions—can help to address this challenge. A three-pronged strategy is essential:

1. The first prong focuses on delivering necessary support for inputs required for AI deployment in agriculture;
2. The second emphasizes the rapid and practical implementation of the latest AI solutions, including various techniques at the farm level, along with adequate training;
3. The third aims to provide AI tools to small farms through collective, cooperative, or public–private partnership models, similar to crop insurance in India, to achieve truly sustainable agriculture with minimal C footprints.

13.9 Conclusion

Climate change has rendered conventional agricultural knowledge obsolete, costly, and labor-intensive, accompanied by rudimentary decision-making methodologies. Historically, agricultural decisions were made by farmers and their advisors based on experience or memory, often influenced by intuition, which led to inaccuracies, particularly in determining the correct price for produce and in addressing discrepancies between supply and demand. Consequently, the ambition to provide "green" quality food for the growing population may not be fulfilled. The return on investment in Indian agriculture is expected to increase due to enhanced cost efficiency, reduced inputs, and higher outputs per unit of land. Advanced AI in precision agriculture can significantly decrease C emissions while improving crop production, resulting in more resilient and stable agricultural practices in India. To ensure the successful implementation of sustainable AI solutions in Indian agriculture, certain challenges must be addressed. A major issue is the lack of information and technical proficiency among farmers and agricultural professionals. Educating farmers on the benefits and applications of sustainable AI technologies—through targeted training initiatives and awareness campaigns—can help to overcome this barrier.

References

Adak, S., Mandal, N., Maity, P.P., Mukhopadhyay, A., Gunturi, A., 2023. Digital soil health: unveiling the potential of soil health monitoring. Food Science Reports 4 (9), 96–103.

Ahmed, N., Shakoor, N., 2025. Advancing agriculture through IoT, Big Data, and AI: a review of smart technologies enabling sustainability. Smart Agricultural Technology 10, 100848. Available at https://doi.org/10.1016/j.atech.2025.100848.

Avwioroko, A., 2024. The integration of smart grid technology with carbon credit trading systems: benefits, challenges, and future directions. Engineering Science & Technology Journal 4 (2), 33–45. Available at https://doi.org/10.51594/estj.v4i2.1287.

Baklaga, L., 2024. Synergizing AI and Blockchain: innovations in decentralized carbon markets for emission reduction through intelligent carbon credit trading. Journal of Computer Science and Technology Studies 6 (2), 111–120. Available at https://doi.org/10.32996/jcsts.2024.6.2.13.

Barbedo, J.G.A., 2019. A review on the use of unmanned aerial vehicles and imaging sensors for monitoring and assessing plant stresses. Drones 3 (2), 1–27. Available at https://doi.org/10.3390/drones3020040.

Battocletti, V., Enriques, L., Romano, A., 2024. The voluntary carbon market: market failures and policy implications. University of Colorado – Law Review 95, 519. Available at https://doi.org/10.4337/9781035350636.00013. Sustainability, Due Diligence and Value Chain Governance. Edward Elgar Publishing Ltd., Netherlands Edward Elgar Publishing Ltd., Netherlands (2025), 86–104.

Bayon, R., Hawn, A., Hamilton, K., 2012. Voluntary Carbon Markets: An International Business Guide to What They are and How They Work. Routledge.

Bermeo-Almeida, O., del Cioppo-Morstadt, J., Cardenas-Rodriguez, M., Cabezas-Cabezas, R., Bazán-Vera, W., 2019. Sentiment analysis in social networks for agricultural pests. In: ICT for agriculture and environment: Second International Conference, CITAMA 2019, Guayaquil, Ecuador, January 22-25, 2019, Proceedings 2 (pp. 122-129). Springer International Publishing.

Bilal, H., Okonkwo, E.C., Al-Ansari, T., 2024. A pathway to 25% reduction in greenhouse gas emission by 2030 using GIS-based multi-criteria decision making for renewables, date palm (phoenix dactylifera), and carbon credits: a case for Qatar. Frontiers in Energy Research 12. Available at https://doi.org/10.3389/fenrg.2024.1272993.

Blum, M., 2020. The legitimation of contested carbon markets after Paris–empirical insights from market stakeholders. Journal of Environmental Policy and Planning 22 (2), 226–238. Available at https://doi.org/10.1080/1523908X.2019.1697658.

Bond, T.C., Doherty, S.J., Fahey, D.W., Forster, P.M., Berntsen, T., DeAngelo, B.J., et al., 2013. Bounding the role of black carbon in the climate system: a scientific assessment. Journal of Geophysical Research: Atmospheres 118 (11), 5380–5552. Available at https://doi.org/10.1002/jgrd.50171.

Breidenbach, J., Ivanovs, J., Kangas, A., Nord-Larsen, T., Nilsson, M., Astrup, R., 2021. Improving living biomass C-stock loss estimates by combining optical satellite, airborne laser scanning, and NFI data. Canadian Journal of Forest Research 51 (10), 1472–1485. Available at https://doi.org/10.1139/cjfr-2020-0518.

Bui, M., Adjiman, C.S., Bardow, A., Anthony, E.J., Boston, A., Brown, S., et al., 2018. Carbon capture and storage (CCS): the way forward. Energy and Environmental Science 11 (5), 1062–1176. Available at https://doi.org/10.1039/c7ee02342a.

Chauhan, M., Sahoo, D.R., 2024. Towards a greener tomorrow: exploring the potential of AI, Blockchain, and IoT in sustainable development. Nature Environment and Pollution Technology 23 (2), 1105–1113. Available at https://doi.org/10.46488/NEPT.2024.v23i02.044.

Chawla, V., Sethuraman, D., Ashok, T., Kitts, J., 2023. The role of the voluntary carbon market in scaling impact globally - an analysis of recent trends. National Law School Business Law Review 9 (2), 109–135. Available at https://doi.org/10.55496/toar9864.

Choi, I., Kim, W.C., 2023. Enhancing financial literacy in South Korea: integrating AI and data visualization to understand financial instruments' interdependencies. Societal Impacts 1 (1-2), 100024. Available at https://doi.org/10.1016/j.socimp.2023.100024.

Dalvi, M., Beig, G., Patil, U., Kaginalkar, A., Sharma, C., Mitra, A.P., 2006. A GIS based methodology for gridding of large-scale emission inventories: application to carbon-monoxide emissions over Indian region. Atmospheric Environment 40 (16), 2995–3007. Available at https://doi.org/10.1016/j.atmosenv.2006.01.013.

Dayıoğlu, M.A., Turker, U., 2021. Digital transformation for sustainable future-agriculture 4.0: a review. Journal of Agricultural Science 27 (4), 373–399.

Ebeling, J., Yasué, M., 2008. Generating carbon finance through avoided deforestation and its potential to create climatic, conservation and human development benefits. Philosophical Transactions of the Royal Society B: Biological Sciences 363 (1498), 1917–1924. Available at https://doi.org/10.1098/rstb.2007.0029.

El Zein, A.L., Chehayeb, N.A., El Zein, A.L., Chehayeb, N.A., 2015. The effect of greenhouse gases on earth's temperature. International Journal of Environmental Monitoring and Analysis 3 (2), 74.

Ewing, J., 2022. Roadmap to a Northeast Asian Carbon Market. Asia Society.

Fawzy, S., Osman, A.I., Doran, J., Rooney, D.W., 2020. Strategies for mitigation of climate change: a review. Environmental Chemistry Letters 18 (6), 2069–2094. Available at https://doi.org/10.1007/s10311-020-01059-w.

Frackiewicz, M., 2023. The use of artificial intelligence in predicting and managing carbon footprint in agriculture [Online]. Available from: https://ts2.space/en/the-use-of-artificial-intelligence-in-predicting-and-managing-carbon-footprint-in-agriculture/.

Garg, P.K., 2024. Geospatial approaches for environmental pollution mapping, analysis and mitigation. Spatial Intelligence for a Greener Planet: AI Applications in Environmental Pollution Mapping, Analysis and Mitigation. CRC Press, India, pp. 1–19.

Goetz, S., Dubayah, R., 2011. Advances in remote sensing technology and implications for measuring and monitoring forest carbon stocks and change. Carbon Management 2 (3), 231–244. Available at https://doi.org/10.4155/cmt.11.18.

Gupta, K., 2024. Carbon credits and offsetting: navigating legal frameworks, innovative solutions, and controversies. International Journal of Multidisciplinary Research 6 (2), 1–12.

Gupta, H., Rakshit, D., 2023. Assessing the demand for carbon credits from the most polluting and hard-to-abate sectors in India. Journal of Cleaner Production 425, 138825. Available at https://doi.org/10.1016/j.jclepro.2023.138825.

Hamdan, A., Ibekwe, K.I., Etukudoh, E.A., Umoh, A.A., Ilojianya, V.I., 2024. AI and machine learning in climate change research: a review of predictive models and environmental impact. World Journal of Advanced Research and Reviews 21 (1), 1999–2008. Available at https://doi.org/10.30574/wjarr.2024.21.1.0257.

Hamrani, A., Akbarzadeh, A., Madramootoo, C.A., 2020. Machine learning for predicting greenhouse gas emissions from agricultural soils. Science of the Total Environment 741. Available at https://doi.org/10.1016/j.scitotenv.2020.140338.

Hartemink, A.E., McSweeney, K., 2014. Soil carbon. In: Hartemink, A.E., McSweeney, K. (Eds.), Soil carbon. Springer International Publishing, pp. 7–16.

Helm, J.M., Swiergosz, A.M., Haeberle, H.S., Karnuta, J.M., Schaffer, J.L., Krebs, V.E., et al., 2020. Machine learning and artificial intelligence: definitions, applications, and future directions. Current Reviews in Musculoskeletal Medicine 13 (1), 69–76. Available at https://doi.org/10.1007/s12178-020-09600-8.

IPCC Intergovernmental Panel on Climate Change, 2022. Climate change 2022: impacts, adaptation, and vulnerability. In: Pörtner, H.O., Roberts, D.C., Tignor, M., Poloczanska, E.S., Mintenbeck, K., Alegría, A. (Eds.), Contribution of Working Group II to the Sixth Assessment Report of the Intergovernmental Panel on Climate Change. Cambridge University Press.

Issa, S., Dahy, B., Ksiksi, T., Saleous, N., 2020. A review of terrestrial carbon assessment methods using geo-spatial technologies with emphasis on arid lands. Remote Sensing 12 (12), 2008. Available at https://doi.org/10.3390/rs12122008.

Jat, H.S., Datta, A., Choudhary, M., Sharma, P.C., Yadav, A.K., Choudhary, V., et al., 2019. Climate smart agriculture practices improve soil organic carbon pools, biological properties and crop productivity in cereal-based systems of North-West India. CATENA 181, 104059. Available at https://doi.org/10.1016/j.catena.2019.05.005.

Jhariya, M.K., Meena, R.S., Banerjee, A., Meena, S.N., 2022. Natural Resources Conservation and Advances for Sustainability. Elsevier. Available at https://doi.org/10.1016/C2019-0-03763-6.

Jhariya, M.K., Meena, R.S., Banerjee, A., Kumar, S., Raj, A., 2023. Agroforestry for Carbon and Ecosystem Management Agroforestry for Carbon and Ecosystem Management. Elsevier, India, pp. 1–434.

Johnston, A.E., Poulton, P.R., Coleman, K., 2009. Chapter 1 soil organic matter. Its importance in sustainable agriculture and carbon dioxide fluxes. Advances in Agronomy 101, 1–57. Available at https://doi.org/10.1016/S0065-2113(08)00801-8.

Johnstone, I., 2023. Nature-based solutions and the voluntary carbon market: opportunities and limits. Handbook of Nature-Based Solutions to Mitigation and Adaptation to Climate Change. Springer International Publishing, Cham, pp. 1–23.

Juvonen, J., 2021. Corporate carbon neutrality claims and the legal effect of double counting. Institute for International Finance 26.

Kakamoukas, G., Sarigiannidis, P., Maropoulos, A., Lagkas, T., Zaralis, K., Karaiskou, C., 2021. Towards climate smart farming—a reference architecture for integrated farming systems. Telecom 2 (1), 52–74. Available at https://doi.org/10.3390/telecom2010005.

Khachoo, Y.H., Cutugno, M., Robustelli, U., Pugliano, G., 2024. Impact of land use and land cover (LULC) changes on carbon stocks and economic implications in calabria using Google Earth Engine (GEE). Sensors 24 (17). Available at https://doi.org/10.3390/s24175836.

Khan, A.A., 2024. Empowering cities a framework for utilizing carbon credits in local area planning (Doctoral dissertation, SPA Bhopal).

Khanna, A., Maheshwari, P., 2023. Blockchain-powered NFTs: A paradigm shift in carbon credit transactions for traceability, transparency, and accountability. European, Mediterranean, and Middle Eastern Conference on Information Systems. Springer Nature Switzerland, Cham, pp. 75–87.

Khuzairi, N.M.S., Cob, Z.C., Dass, A.S., 2024. A carbon credit quality assessment and verification model towards carbon neutrality. Innovation and Technological Advances for Sustainability. CRC Press, pp. 532–540.

Kiryluk-Dryjska, E., Beba, P., Poczta, W., 2020. Local determinants of the Common Agricultural Policy rural development funds' distribution in Poland and their spatial implications. Journal of Rural Studies 74, 201–209. Available at https://doi.org/10.1016/j.jrurstud.2020.01.018.

Kosgei, J., 2009. Implementing Web GIS for Monitoring Carbon Sequestration in Sustainable Agroforestry Projects. Armacost Library, University of Redlands.. Available at https://doi.org/10.26716/redlands/master/2009.8.

Kreibich, N., Obergassel, W., 2016. Carbon Markets After Paris- How to Account for the Transfer of Mitigation Results? Wuppertal Institute for Climate, Environment and Energy, Wuppertal, Germany.

Kreibich, N., Hermwille, L., 2021. Caught in between: credibility and feasibility of the voluntary carbon market post-2020. Climate Policy 21 (7), 939–957. Available at https://doi.org/10.1080/14693062.2021.1948384.

Kumar, S., Meena, R.S., Sheoran, S., Jangir, C.K., Jhariya, M.K., Banerjee, A., et al., 2022. Remote sensing applications for natural resources and agriculture management. Natural Resources Conservation and Advances for Sustainability. Elsevier, India, pp. 91–135. Available at https://doi.org/10.1016/B978-0-12-822976-7.00012-0.

Kumar, S., Meena, R.S., Sheoran, P., Jhariya, M.K., 2024. Regenerative Agriculture for Sustainable Food Systems. Springer Science+Business Media, India, pp. 1–493.

Li, Y., Wang, Y., Chong, D., Xu, Z., Li, L., Hu, Y., 2023. Carbon taxation in Singapore's semiconductor sector: a mini-review on GHG emission metrics and reporting. Carbon Research 2 (1). Available at https://doi.org/10.1007/s44246-023-00082-0.

Lokuge, N., Anders, S., 2022. Carbon-credit systems in agriculture: a review of literature. The School of Public Policy Publications 15.

Ma, W., Ji, X., Ding, L., Yang, S.X., Guo, K., Li, Q., 2024. Automatic monitoring methods for greenhouse and hazardous gases emitted from ruminant production systems: a review. Sensors 24 (13). Available at https://doi.org/10.3390/s24134423.

Maddali, V., Tularam, G.A., Glynn, P., 2015. Economic and time-sensitive issues surrounding CCS: a policy analysis. Environmental Science and Technology 49 (15), 8959–8968. Available at https://doi.org/10.1021/acs.est.5b00839.

Mansourihanis, O., Maghsoodi Tilaki, M.J., Kookhaei, T., Zaroujtaghi, A., Sheikhfarshi, S., Abdoli, N., 2025. Integrating geospatial intelligence and spatio-temporal modeling for monitoring tourism-related carbon emissions in the United States. Management of Environmental Quality 36 (1), 107–129. Available at https://doi.org/10.1108/MEQ-04-2024-0156.

Maulud, D., Abdulazeez, A.M., 2020. A review on linear regression comprehensive in machine learning. Journal of Applied Science and Technology Trends 1 (2), 140–147. Available at https://doi.org/10.38094/jastt1457.

McNunn, G., Karlen, D.L., Salas, W., Rice, C.W., Mueller, S., Muth, D., et al., 2020. Climate smart agriculture opportunities for mitigating soil greenhouse gas emissions across the U.S. Corn-Belt. Journal of Cleaner Production 268. Available at https://doi.org/10.1016/j.jclepro.2020.122240.

McPherson, E.G., Kendall, A., 2014. A life cycle carbon dioxide inventory of the Million Trees Los Angeles program. International Journal of Life Cycle Assessment 19 (9), 1653–1665. Available at https://doi.org/10.1007/s11367-014-0772-8.

Meena, R.S., Kumar, S., Pradhan, G., Kumar Jhariya, M., Srinivasarao, C., Meena, R.S., et al., 2025. Carbon-Negative Agriculture. Springer Nature, Singapore, pp. 1–500. Available at https://doi.org/10.1007/978-981-95-1483-0.

Mehta, A., 2024. From field to the cloud: how AI is helping regenerative agriculture to grow. Ethical Corporation Magazine. Thomson Reuters.

Milne, E., Banwart, S.A., Noellemeyer, E., Abson, D.J., Ballabio, C., Bampa, F., et al., 2015. Soil carbon, multiple benefits. Environmental Development 13, 33–38. Available at https://doi.org/10.1016/j.envdev.2014.11.005.

Mishra, H., Mishra, D., 2023. Artificial intelligence and machine learning in agriculture: transforming farming systems. Research Trends in Agricultural Sciences 1, 1–16.

Moomaw, W.R., Masino, S.A., Faison, E.K., 2019. Intact forests in the United States: proforestation mitigates climate change and serves the greatest good. Frontiers in Forests and Global Change 2. Available at https://doi.org/10.3389/ffgc.2019.00027.

Mor, S., Madan, S., Prasad, K.D., 2021. Artificial intelligence and carbon footprints: roadmap for Indian agriculture. Strategic Change 30 (3), 269–280. Available at https://doi.org/10.1002/jsc.2409.

Murray, B.C., Newell, R.G., Pizer, W.A., 2009. Balancing cost and emissions certainty: an allowance reserve for cap-and-trade. Review of Environmental Economics and Policy 3 (1), 84–103. Available at https://doi.org/10.1093/reep/ren016.

Nagavi, J.C., Shukla, B.K., Bhati, A., Rai, A., Verma, S., 2024. Harnessing geospatial technology for sustainable development: a multifaceted analysis of current practices and future prospects. Sustainable Development and Geospatial Technology: Volume 1: Foundations and Innovations. Springer Nature Switzerland, Cham, pp. 147–170.

Nasar, Z., Jaffry, S.W., Malik, M.K., 2022. Named entity recognition and relation extraction: state-of-the-art. ACM Computing Surveys 54 (1). Available at https://doi.org/10.1145/3445965.

National Research Council, Division on Earth, Life Studies, Board on Atmospheric Sciences, Committee on Understanding, Monitoring Abrupt Climate Change, & Its Impacts, 2013. Abrupt Impacts of Climate Change: Anticipating Surprises. National Academies Press (2013).

Nayak, A.K., Rahman, M.M., Naidu, R., Dhal, B., Swain, C.K., Nayak, A.D., et al., 2019. Current and emerging methodologies for estimating carbon sequestration in agricultural soils: a review. Science of the Total Environment 665, 890–912. Available at https://doi.org/10.1016/j.scitotenv.2019.02.125.

Nehzak, H.K., Aghaei, M., Mostafazadeh, R., Rabiei-Dastjerdi, H., 2022. Assessment of machine learning algorithms in land use classification. Computers in Earth and Environmental Sciences. Elsevier, pp. 97–104.

Northern, G., 2011. Greenwashing the organic label: abusive green marketing in an increasingly eco friendly marketplace. Journal of Food Law Policy 7, 101–134.

Ojha, T., Misra, S., Raghuwanshi, N.S., 2021. Internet of things for agricultural applications: the state of the art. IEEE Internet of Things Journal 8 (14), 10973–10997. Available at https://doi.org/10.1109/JIOT.2021.3051418.

Olson, D., Anderson, J., 2021. Review on unmanned aerial vehicles, remote sensors, imagery processing, and their applications in agriculture. Agronomy Journal 113 (2), 971–992. Available at https://doi.org/10.1002/agj2.20595.

Omia, E., Bae, H., Park, E., Kim, M.S., Baek, I., Kabenge, I., et al., 2023. Remote sensing in field crop monitoring: a comprehensive review of sensor systems, data analyses and recent advances. Remote Sensing 15 (2). Available at https://doi.org/10.3390/rs15020354.

Paramesha, M., Rane, N., Rane, J., 2024. Big data analytics, artificial intelligence, machine learning, internet of things, and blockchain for enhanced business intelligence. SSRN Electronic Journal. Available at https://doi.org/10.2139/ssrn.4855856.

Patel, M.A.K., Patel, B.R., 2020. Sentiment analysis for social media: an overview and open challenges. International Journal of Computer Applications 1 (1), 3.

Paul, A., Nair, R., 2024. Carbon emission trading–a critical analysis.

Ponce-Hernandez, R., Koohafkan, P., Antoine, J., 2004. Assessing carbon stocks and modelling win-win scenarios of carbon sequestration through land-use changes (Vol. 1). Food & Agriculture Org.

Prabhat, P.K., Singh, K.K., 2021. Advancing smart agriculture: a study on iot-enabled preci sion irrigation systems for sustainable water management. International Journal of Scientific Research in Science and Technology. Available at https://doi.org/10.32628/IJSRST.

Raina, N., Zavalloni, M., Viaggi, D., 2024. Incentive mechanisms of carbon farming contracts: a systematic mapping study. Journal of Environmental Management 352, 120126. Available at https://doi.org/10.1016/j.jenvman.2024.120126.

Raj, A., Jhariya, M.K., Meena, R.S., 2025. Agroforestry for Monetising Carbon Credits. Springer Nature, Switzerland.

Ramifehiarivo, N., Brossard, M., Grinand, C., Andriamananjara, A., Razafimbelo, T., Rasolohery, A., et al., 2017. Mapping soil organic carbon on a national scale: towards an improved and updated map of Madagascar. Geoderma Regional 9, 29–38. Available at https://doi.org/10.1016/j.geodrs.2016.12.002.

Ramos-Sandoval, R., Beltran, C.S., 2021. Using Twitter data for consumer research: the Peruvian's consumer perspective on short food supply chains. In 2021 IEEE Sciences and Humanities International Research Conference (SHIRCON) (pp. 1-4). IEEE.

Report on Crop Residue Burning: Challenges and Solutions [Internet]. Indian Council of Food and Agriculture (ICFA). [cited 2021 Feb 16]. Available from: http://icfa.org.in/assets/doc/report s/Crop_Residue_Burning.pdf.

Safavi, S., 2024. Leveraging GIS for environmental planning. Global Challenges in Environmental and Climate Change 22, 308.

Sarker, I.H., 2021. Machine learning: algorithms, real-world applications and research directions. SN Computer Science 2 (3). Available at https://doi.org/10.1007/s42979-021-00592-x.

Shaikh, F.K., Karim, S., Zeadally, S., Nebhen, J., 2022. Recent trends in internet-of-things-enabled sensor technologies for smart agriculture. IEEE Internet of Things Journal 9 (23), 23583–23598. Available at https://doi.org/10.1109/JIOT.2022.3210154.

Shandilya, S.K., Datta, A., Kartik, Y., Nagar, A., 2024. Role of artificial intelligence and machine learning. Digital Resilience: Navigating Disruption and Safeguarding Data Privacy. Springer Nature Switzerland, Cham, pp. 313–399.

Singh, R.K., Berkvens, R., Weyn, M., 2021. AgriFusion: an architecture for IoT and emerging technologies based on a precision agriculture survey. IEEE Access 9, 136253–136283. Available at https://doi.org/10.1109/ACCESS.2021.3116814.

Srivastava, R., Bharti, P.K., Verma, P., 2022. Comparative analysis of lexicon and machine learning approach for sentiment analysis. International Journal of Advanced Computer Science and Applications 13 (3), 71–77. Available at https://doi.org/10.14569/IJACSA.2022.0130312.

Steinhausen, M.J., Wagner, P.D., Narasimhan, B., Waske, B., 2018. Combining Sentinel-1 and Sentinel-2 data for improved land use and land cover mapping of monsoon regions. International Journal of Applied Earth Observation and Geoinformation 73, 595–604. Available at https://doi.org/10.1016/j.jag.2018.08.011.

Stieglitz, S., Mirbabaie, M., Ross, B., Neuberger, C., 2018. Social media analytics – challenges in topic discovery, data collection, and data preparation. International Journal of Information Management 39, 156–168. Available at https://doi.org/10.1016/j.ijinfomgt.2017.12.002.

Tajik, S., Ayoubi, S., Zeraatpisheh, M., 2020. Digital mapping of soil organic carbon using ensemble learning model in Mollisols of Hyrcanian forests, northern Iran. Geoderma Regional 20, e00256. Available at https://doi.org/10.1016/j.geodrs.2020.e00256.

Tian, Y., Akimov, A., Roca, E., Wong, V., 2016. Does the carbon market help or hurt the stock price of electricity companies? Further evidence from the European context. Journal of Cleaner Production 112, 1619–1626. Available at https://doi.org/10.1016/j.jclepro.2015.07.028.

Toprak, A., Turan, M., 2023. Enhanced named entity recognition algorithm for financial document verification. Journal of Supercomputing 79 (17), 19431–19451. Available at https://doi.org/10.1007/s11227-023-05371-4.

Udutalapally, V., Mohanty, S.P., Pallagani, V., Khandelwal, V., 2021. SCrop: a novel device for sustainable automatic disease prediction, crop selection, and irrigation in internet-of-agro-things for smart agriculture. IEEE Sensors Journal 21 (16), 17525–17538. Available at https://doi.org/10.1109/JSEN.2020.3032438.

van der Gaast, W., Sikkema, R., Vohrer, M., 2018. The contribution of forest carbon credit projects to addressing the climate change challenge. Climate Policy 18 (1), 42–48. Available at https://doi.org/10.1080/14693062.2016.1242056.

Vellingiri, A., Kokila, R., Nisha, P., Kumar, M., Chinnusamy, S., Boopathi, S., 2024. Harnessing GPS, sensors, and drones to minimize environmental impact: precision agriculture. Designing Sustainable Internet of Things Solutions for Smart Industries. IGI Global, India, pp. 77–108.

Weraikat, D., Šorič, K., Žagar, M., Sokač, M., 2024. Data analytics in agriculture: enhancing decision-making for crop yield optimization and sustainable practices. Sustainability 16 (17), 7331. Available at https://doi.org/10.3390/su16177331.

Weston, L., Tshitoyan, V., Dagdelen, J., Kononova, O., Trewartha, A., Persson, K.A., et al., 2019. Named entity recognition and normalization applied to large-scale information extraction from the materials science literature. Journal of Chemical Information and Modeling 59 (9), 3692–3702. Available at https://doi.org/10.1021/acs.jcim.9b00470.

White, J.C., Coops, N.C., Wulder, M.A., Vastaranta, M., Hilker, T., Tompalski, P., 2016. Remote sensing technologies for enhancing forest inventories: a review. Canadian Journal of Remote Sensing 42 (5), 619–641. Available at https://doi.org/10.1080/07038992.2016.1207484.

Zeraatpisheh, M., Ayoubi, S., Mirbagheri, Z., Mosaddeghi, M.R., Xu, M., 2021. Spatial prediction of soil aggregate stability and soil organic carbon in aggregate fractions using machine learning algorithms and environmental variables. Geoderma Regional 27, e00440. Available at https://doi.org/10.1016/j.geodrs.2021.e00440.

Zeraatpisheh, M., Garosi, Y., Reza Owliaie, H., Ayoubi, S., Taghizadeh-Mehrjardi, R., Scholten, T., et al., 2022. Improving the spatial prediction of soil organic carbon using environmental covariates selection: a comparison of a group of environmental covariates. CATENA 208, 105723. Available at https://doi.org/10.1016/j.catena.2021.105723.

Zhou, K., Li, Y., 2019. Carbon finance and carbon market in China: progress and challenges. Journal of Cleaner Production 214, 536–549. Available at https://doi.org/10.1016/j.jclepro.2018.12.298.

Chapter 14

Carbon sequestration potential of fruit trees

Niharika[1], Suddhasuchi Das[2], Chanderkant[1], and Ram Swaroop Meena[3]

[1]*Dr Yashwant Singh Parmar University of Horticulture and Forestry, Nauni, Solan, Himachal Pradesh, India,* [2]*Malda Krishi Vigyan Kendra, Uttar Banga Krishi Vishwavidyalaya, Ratua, Malda, West Bengal, India,* [3]*Department of Agronomy, Institute of Agricultural Sciences, Banaras Hindu University, Varanasi, Uttar Pradesh, India*

Highlights

- Incorporation of higher plants in agroecosystems enhances the profits.
- Fruit tree species can serve a dual purpose of being productive and gaining carbon credits.
- Screening of suitable fruit tree species with compatibility with the crops is a prime concern.
- Fruit trees help toward low-energy input, higher carbon capture, and resilience practices.

14.1 Introduction

Agricultural landscapes, particularly those involving perennials such as fruit trees, offer scalable opportunities for carbon sequestration (C_{seq}) (Peng et al., 2020). As global climate targets, such as those outlined in the Paris Agreement, call for net-zero emissions by mid-century, C_{seq} approaches must go beyond afforestation and industrial C capture. Simultaneously, reducing reliance on external inputs such as synthetic fertilizers and irrigation through energy-efficient practices becomes critical for achieving sustainable production (Meena et al., 2025). Fruit trees, through their shading, deep root systems, and microclimatic buffering, reduce evapotranspiration losses and can significantly lower on-farm energy consumption. Beyond C_{seq}, fruit trees also contribute to energy efficiency. Their shade reduces soil temperature and water evaporation, conserving irrigation water and lowering the energy required for pumping. Studies on integrated orchards in Argentina found that energy inputs for irrigation, fertilization, and pesticide application were 15%–30% lower in mixed-species orchards compared to monocultures. In tropical regions, shading by tree canopies also moderates heat stress on crops and livestock, enhancing overall farm resilience. Silvopastoral systems, where fruit trees are grown alongside fodder grasses or livestock, demonstrate improved energy balance and increased biodiversity, further contributing to the system's sustainability (Ioannidou et al., 2025).

Human-induced climate change remains one of the most pressing environmental challenges of the 21st century. Atmospheric carbon dioxide (CO_2) levels have increased from preindustrial concentrations of approximately 280 ppm to over 420 ppm in 2023, primarily due to widespread deforestation, fossil fuel combustion, and unsustainable agricultural practices (Sharma et al., 2021a; Meena et al., 2024b). Agriculture accounts for nearly 23% of global greenhouse gas (GHG) emissions, transforming land-use systems crucial in climate change mitigation (CCM) strategies. Among several nature-based solutions (NBS), perennial fruit tree systems have emerged as highly promising for their dual roles in C_{seq} and sustainable food production (Gelaye and Mesfin, 2024). Fruit tree-based systems, especially when integrated into agroforestry (AF) and regenerative agriculture, not only absorb atmospheric CO_2 but also store it in both biomass and soils over the long-term (Kumar et al., 2024; Ioannidou et al., 2025; Meena et al., 2024a). In contrast to annual crops, which are short-lived and leave land bare between planting cycles, fruit trees maintain continuous canopy cover and root activity throughout the year, enhancing their potential for long-term C capture. Furthermore, these systems contribute to improved energy efficiency by modifying microclimates, reducing irrigation needs, and stabilizing ecosystem services (Fig. 14.1).

Carbon Incentives, Plans, and Policies in Agriculture. DOI: https://doi.org/10.1016/B978-0-443-33557-0.00009-4

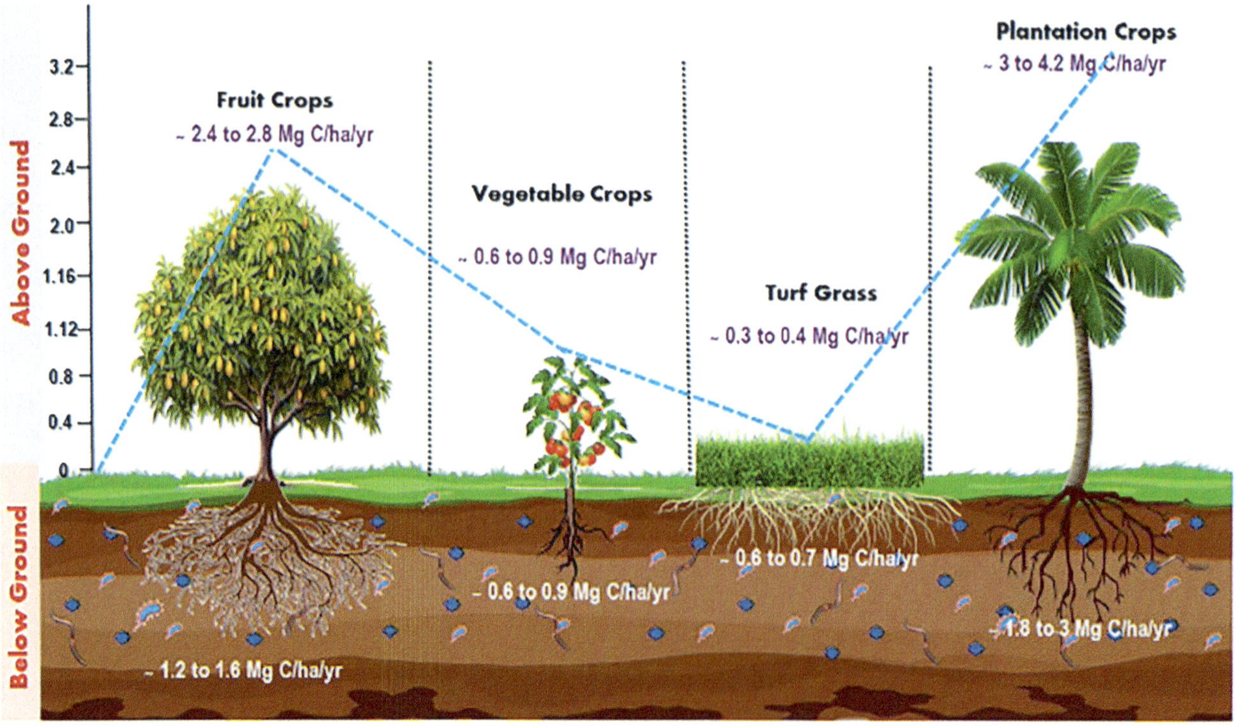

FIGURE 14.1 Trends of C_{seq} in fruit crops (Ilakiya et al., 2024). C_{seq}, carbon sequestration.
Source: Ilakiya, T., Parameswari, E., Swarnapriya, R., Yazhini, G., Kalaiselvi, P., Davamani, V. et al., 2024. Unlocking the carbon sequestration potential of horticultural crops. C. 10 (3), 65. https://doi.org/10.3390/c10030065.

Given the urgent need for scalable, sustainable C sinks and the cobenefits offered by perennial systems, fruit trees are a critical component of climate-resilient agriculture. Their ability to combine food production with long-term C storage and improved energy efficiency makes them uniquely valuable in both mitigation and adaptation strategies (Jhariya et al., 2023, 2022; Meena et al., 2024c). This chapter aims to provide an in-depth analysis of the C_{seq} potential and energy efficiency of fruit tree systems across climatic zones. It reviews the latest scientific evidence, identifies key influencing factors, and suggests best practices for maximizing climate benefits while maintaining productivity. By bridging ecological theory and practical application, the chapter serves as a reference for researchers, policymakers, and practitioners in sustainable horticulture and CCM.

14.2 Role of fruit trees in different climatic zones

Fruit trees are cultivated across a broad range of climatic zones, including temperate, subtemperate, tropical, and subtropical regions. Each of these zones presents distinct dynamics in C accumulation and energy efficiency (Roussos, 2024). In temperate regions, deciduous fruit trees like apple, pear, cherry, and peach store substantial biomass in trunks and branches that remain intact for decades. Cooler climates promote slower decomposition of litterfall, increasing soil organic carbon (SOC) accumulation. Subtemperate systems, which lie between temperate and tropical belts, often support stone fruits and nut trees with similar sequestration dynamics, but with faster C cycling due to warmer temperatures and increased microbial activity (Ioannidou et al., 2025). Tropical and subtropical regions host species like mango, guava, citrus, and banana. These trees grow rapidly, produce dense canopies, and exhibit high root turnover, resulting in significant below-ground C input (Meena and Pradhan, 2023; Kumar and Sinha, 2022). Tropical climates also support year-round photosynthesis, enhancing net primary productivity and C uptake. In Ethiopia, fruit-based agroforestry systems (AFs) were reported to store more than 40% of their total C in above-ground biomass alone (Gelaye and Mesfin, 2024).

14.3 Role of fruit trees for a sustainable environment

Fruit tree systems contribute substantially to environmental sustainability by delivering ecosystem services, enhancing biodiversity, stabilizing soils, improving water regulation and nutrient cycling, and supporting climate-smart livelihoods. These multifunctional landscapes align with the three pillars of sustainability—environmental, social, and economic—while addressing pressing global challenges such as biodiversity loss, soil degradation, and climate change (Freitas and Silva, 2022; Meena et al., 2023). Fruit orchards integrated into AF or silvopastoral systems provide a permanent canopy cover, enriching soil ecosystems with organic matter (OM) and promoting healthy microbial communities, which in turn sustain water retention, nutrient turnover, and C_{seq} (Freitas and Silva, 2022; Peng et al., 2020).

Traditional and restored orchards also host rich biodiversity: a study in urban Prague found that restored fruit orchards supported diverse taxa—including lichens, butterflies, orthopterans, and birds—with greater species richness than abandoned sites, demonstrating how management improves ecological quality even in fragmented landscapes (Rada et al., 2022). This biodiversity in orchards supports pollination, pest control, and resilience to ecological disturbance (Rada et al., 2022). On soil and ecosystem function, fruit trees foster soil aggregation through root networks and leaf litter input, reduce erosion, and enhance soil structure. Chestnut orchards, for instance, demonstrate deep rooting and high soil water retention, reduced leaching and GHG emissions, and increased abundance of soil invertebrates—showcasing how perennial fruit systems improve soil health while sequestering C. Plant Growth-Promoting Bacteria applied in fruit orchards enhance nutrient cycling, microbial diversity, root growth, and soil biodiversity, all of which are central to environmental sustainability (Freitas and Silva, 2022). Fruit trees further enable water regulation and microclimate moderation. In AFs across semiarid to humid climates, canopy shade reduces soil evaporation and moderates temperature extremes. Surveys across Iran indicate that alley cropping in fruit orchards has improved water-use efficiency, stabilized soil moisture, and enhanced livelihoods in climate-vulnerable zones—underscoring fruit tree systems as a valuable climate adaptation asset (Reyes et al., 2023). Moreover, fruit-tree systems integrate effectively into agroecological landscapes. A global review confirms that AF—where fruit trees coexist with crops or livestock—enhances ecosystem services, boosts C stocks, supports biodiversity, and promotes food security (Azra et al., 2022; Montanaro et al., 2021). AF adoption increased resilience to floods and drought while improving income and C retention (Aggarwal et al., 2024). The synthesis of sustainable practices, biodiversity, and C storage in these systems underscores their unique environmental value (Banerjee et al., 2024). Fruit orchards also support urban sustainability. Urban fruit orchards—managed community orchards on vacant public land—enhance food access, improve green space, support pollinator habitat, and reduce urban heat islands (Rada et al., 2022). These green urban orchards contribute to soil regeneration and create shared social spaces for education, community cohesion, and environmental stewardship.

Ultimately, fruit systems promote sustainable livelihoods by reducing reliance on synthetic agrochemicals and regenerative practices. Fruit production offers economic incentives for smallholder farmers to maintain canopy cover and soil health while providing nutritious food (Freitas and Silva, 2022). AF fruit systems reduce pressure on natural forests by supplying locally produced tree-derived products, thereby aiding forest conservation (Azra et al., 2022). When deployed at scale, these multifunctional systems generate socio-ecological feedback loops: as biodiversity and soil health improve, farmers' incentive to conserve trees grows—forming a self-reinforcing pathway toward sustainability.

14.4 Understanding carbon sequestration in fruit tree systems

C_{seq} refers to the biological, chemical, or physical processes through which CO_2 is removed from the atmosphere and stored in a stable form. In fruit tree systems, this includes C stored in the trunk, branches, leaves, roots, litterfall, and soil organic matter (SOM) (Sharma et al., 2021a). Compared to annual cropping systems, perennial fruit crops often have higher C densities due to their larger biomass and extended lifespans (Gelaye and Mesfin, 2024). Moreover, fruit trees support microbial communities and soil aggregates that promote the stabilization of SOC (Peng et al., 2020). A recent study in China on apple orchards reported sequestration rates of up to 14 t C ha^{-1} yr^{-1} in mature orchards, with most C stored in above-ground biomass and root systems. Similarly, guava orchards in India have shown total C stocks ranging from 106 to 114 t C ha^{-1}, underscoring the relevance of tropical fruit systems in C budgeting (Kumar and Sinha, 2022).

14.5 Biomass allocation and carbon stability

C_{seq} efficiency depends not just on how much C is captured, but also on how long it remains stored. Woody tissues, such as trunks and branches, act as long-term C sinks, while leaves and fruits represent short-term C pools. The stability of SOC is further influenced by interactions between root systems, microbial biomass, and soil mineral fractions (Peng et al.,

2020). Moreover, species-specific traits, such as wood density, root architecture, and growth form, determine overall C accumulation and turnover rates. Interestingly, a new mechanism of inorganic C_{seq} has been discovered in some tropical fruit trees, such as figs (*Ficus* spp.), where calcium oxalate crystals formed in plant tissues are eventually converted into calcium carbonate, resulting in stable mineral C in woody tissues. This oxalate–carbonate pathway offers an additional form of long-term sequestration unique to certain fruit trees.

14.6 Relationship between carbon sequestration and fruit trees

Fruit tree systems, perennial, woody agricultural environments, play a critical role in capturing atmospheric CO_2 and storing it in stable biomass and soil compartments, clearly outperforming conventional annual cropping in long-term C_{seq} (Fig. 14.2).

14.6.1 Carbon capture pathways and biomass allocation

Fruit trees assimilate CO_2 through photosynthesis, distributing the resulting C into structural biomass: trunks, branches, roots, leaves, and fruit residues (Sharma et al., 2021b; Pradhan and Meena, 2023a). While leaves and fruit biomass are replenished annually, woody tissues serve as long-term C reservoirs, gradually accumulating over decades in mature orchards. Fallen leaves, root exudates, and decaying fine roots enrich SOC pools—enabling up to moderate stabilization in the topsoil under perennial cover, unlike annual croplands, which often disrupt SOC via tillage (Sharma et al., 2021b) (Figs. 14.3 and 14.4).

14.6.2 Quantitative sequestration estimates

Recent empirical studies and reviews demonstrate that fruit-tree-based systems can sequester approximately 2–3 t C ha^{-1} yr^{-1}, considerably higher than the ~0.3–0.5 t C ha^{-1} yr^{-1} typically observed in annual cropping systems. For example, Sharma et al. (2021a) report that orchards contribute significantly to C pools globally, especially when integrated into AF practices (Montanaro et al., 2021; Sharma et al., 2021a) (Table 14.1).

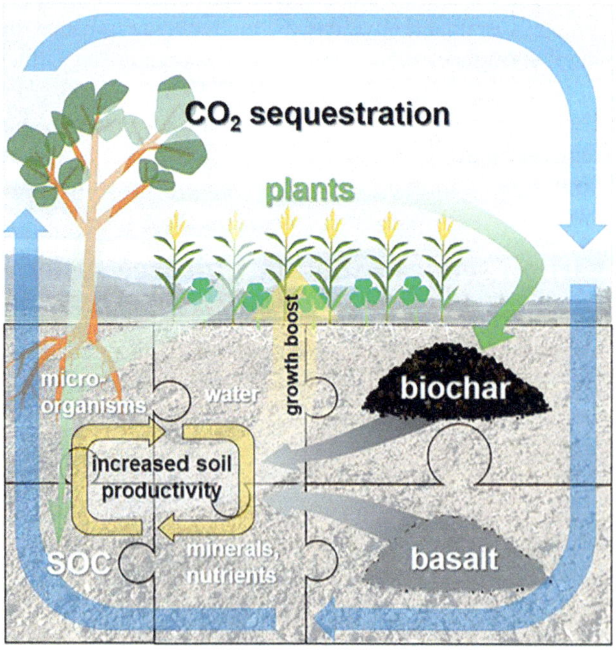

FIGURE 14.2 Process of C_{seq} in plants. C_{seq}, carbon sequestration.
Source: Ilakiya, T., Parameswari, E., Swarnapriya, R., Yazhini, G., Kalaiselvi, P., Davamani, V. et al., 2024. Unlocking the carbon sequestration potential of horticultural crops. C. 10 (3), 65. https://doi.org/10.3390/c10030065.

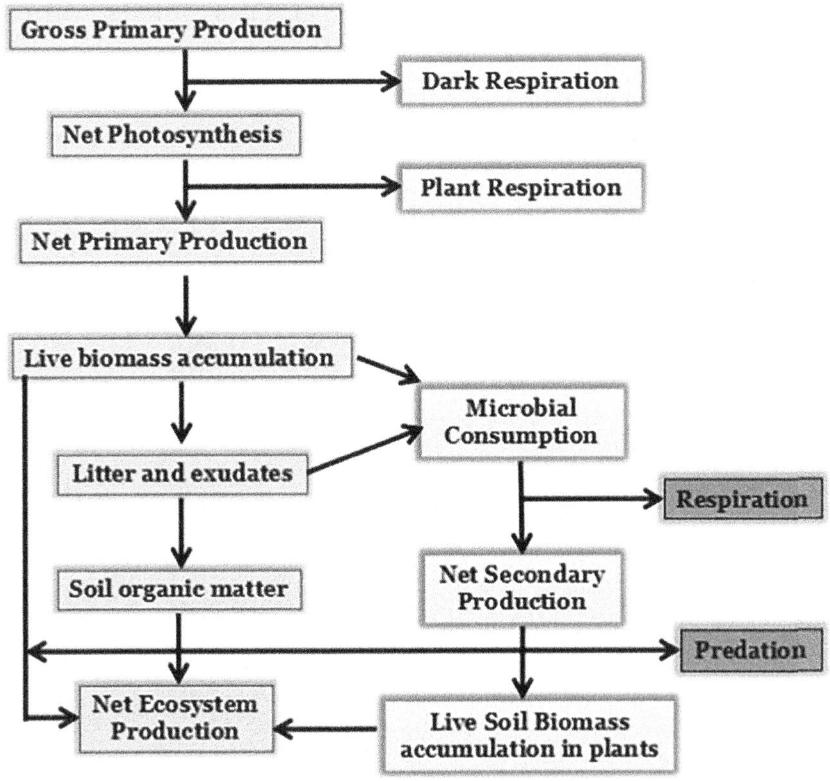

FIGURE 14.3 Mechanism of C_{seq}. C_{seq}, carbon sequestration.
Source: Ilakiya, T., Parameswari, E., Swarnapriya, R., Yazhini, G., Kalaiselvi, P., Davamani, V. et al., 2024. Unlocking the carbon sequestration potential of horticultural crops. C. 10 (3), 65. https://doi.org/10.3390/c10030065.

14.7 Factors influencing sequestration efficacy

14.7.1 Several factors determine carbon storage outcomes

1. **Species traits**: Trees with high wood density, extensive root systems, or nitrogen-fixing/arbuscular mycorrhizae (AM) mycorrhizal associations enhance C stability and SOC accumulation, with AM-associated species producing up to 10%–22% higher mineral soil C stocks (Peng et al., 2020) (Fig. 14.5).
2. **Orchard design**: Mixed-species orchards support complementary growth patterns, greater litter diversity, and more stable C dynamics than monocultures (Ioannidou et al., 2025; Montanaro et al., 2021). Stand age and growth phase: Most C accumulation occurs during midlife; younger orchards sequester progressively more as biomass develops.
3. **Climatic zone comparisons:** Fruit tree C dynamics differ across climatic zones:
4. **Temperate regions** (e.g., apple, pear, cherry): Exhibit substantial woody biomass storage, with slower litter decomposition contributing to stable SOC buildup (Sharma et al., 2021b).
5. **Sub-temperate zones**: Support fast-growing species with moderate C turnover between above- and below-ground pools.
6. **Tropical and subtropical regions** (e.g., mango, avocado): Demonstrate rapid growth, dense canopy, and high root turnover rates. For instance, smallholder tropical systems in Kenya report mango and avocado above-ground C stocks averaging ~10 Mg C ha^{-1}, underscoring the C_{seq} potential of home gardens and orchards. While rapid turnover in warmer soils may reduce SOC persistence, above-ground biomass accumulation remains substantial in tropical systems.
7. **Soil carbon distribution and stability:** Richer SOC under fruit orchards often extends deeper than under annual crops, as perennial roots penetrate beyond plow layers. Species with nitrogen-fixing capacity or AM associations commonly show 6%–10% higher SOC stocks in mineral soil layers compared to nonassociated trees (Peng et al., 2020). SOC is further stabilized by aggregate formation and microbial by-products such as glomalin proteins.

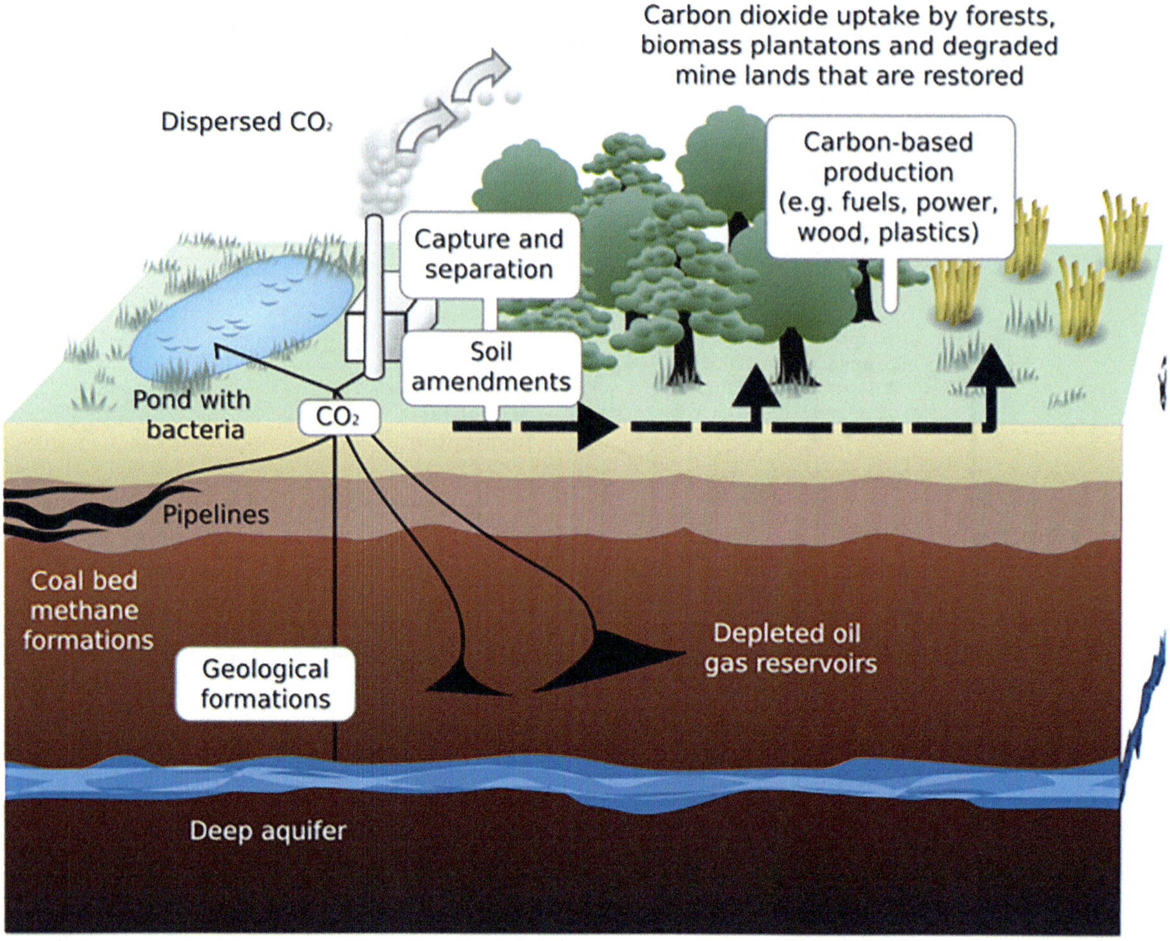

FIGURE 14.4 Geologic and biologic C_{seq} (CCS Explained, 2020). C_{seq}, carbon sequestration.
Source: CCS Explained, 2020. UKUKCCSRC. Archived from the original on 28 June 2020. Retrieved 27 June 2020.

TABLE 14.1 System type annual sequestration (t C ha^{-1} yr^{-1}).

Mixed-species fruit orchard	**2.0–3.0**
Monoculture orchard	1.0–1.5 (age/species dependent)
Annual cropping system	0.3–0.5

Source: Sharma et al. (2021b) and Montanaro et al. (2021).

8. **Lifecycle carbon balance:** Initial GHG emissions during orchard establishment—stemming from nursery operations, tillage, fertilization, and planting—may account for up to 30%–40% of total lifecycle emissions. However, mature orchards often become net C sinks as biomass accumulation surpasses upfront emissions. For example, life cycle analyses in Mediterranean mandarin orchards indicate that above-ground biomass storage offsets approximately 25%–30% of initial C costs (Ioannidou et al., 2025).
9. **Integration with agroforestry and multilateral systems:** Fruit orchards integrated into AF or silvo-pasture systems enhance C_{seq} by combining trees with crops or pasture. Such systems store more C than monoculture cropping, reduce

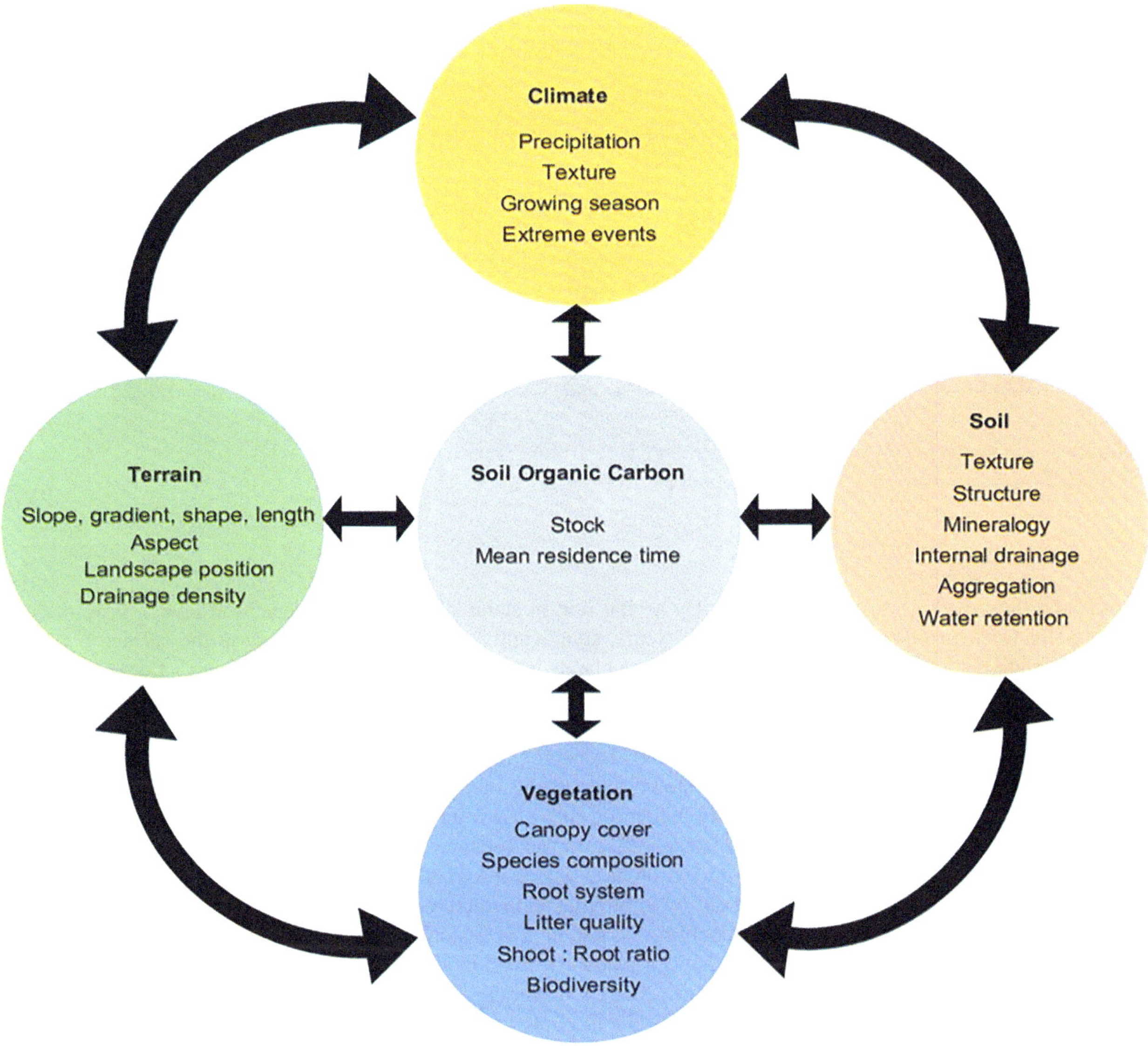

FIGURE 14.5 Factors determining C storage outcomes.
Illustrated by the author.

GHG emissions from machinery, and support biodiversity and ecosystem resilience (Montanaro et al., 2021; Ioannidou et al., 2025).

10. **Emerging findings: mineral carbon pathways:** Recent research suggests fruit trees, particularly fig species, may deposit calcium oxalate within wood tissues that mineralize into stable calcium carbonate—effectively locking C into inorganic forms that persist beyond tree lifespans. While still under investigation, this oxalate–carbonate pathway presents an additional mechanism for C permanence in fruit tree biomass.

14.8 Carbon sequestration in soil and trees

C_{seq} in fruit tree systems occurs through two major pools: above-ground tree biomass and SOC. As perennial woody crops, fruit trees continuously capture atmospheric CO_2 through photosynthesis and store it in their trunks, branches, leaves, and roots over decades, making them superior C sinks compared to annual cropping systems (Sharma et al., 2021a). Above-ground biomass acts as a long-term C reservoir, especially as trees mature and accumulate dense wood; meanwhile, the root system contributes directly to SOC by exuding organic compounds, producing root litter, and fostering microbial interactions (Peng et al., 2020; Pradhan and Meena, 2023b). The microbial decomposition of root

TABLE 14.2 Carbon sequestration pathways in fruit tree systems.

Component	Mechanism of C_{seq}	References
Above-ground biomass	CO_2 is fixed in wood, leaves, branches, and fruits. Long-term C storage in stable woody tissue	Sharma et al. (2021a)
Root biomass and exudates	C is transferred to soil via roots, and microbial activity enhances soil OM and deeper C layers	Peng et al. (2020)
Soil organic carbon (SOC)	Litterfall, microbial decomposition, and root turnover. Stable soil C pools in both surface and deep layers	
Biochar application	Charcoal addition from orchard waste increases SOC, improves structure, and reduces GHG emissions.	
Agroforestry/Silvopasture	Integrated systems with fruit trees and livestock/crops, higher sequestration due to system diversity and synergy	Montanaro et al. (2021)
Inorganic C (e.g., fig spp.)	Oxalate–carbonate conversion in tree tissues. Permanent mineral C_{seq}	

C_{seq}, carbon sequestration; *OM*, organic matter; *GHG*, greenhouse gases.

biomass and the action of mycorrhizal fungi play a vital role in stabilizing this C in deeper soil layers. Fruit orchards, exceptionally when maintained for more than 20 years, significantly enhance SOC not only at the surface but also in deeper horizons (30–90 cm), outperforming adjacent croplands in multiple regions, including China's Loess Plateau. For example, Wang et al. reported up to 190% higher SOC in 15–20-year-old apple orchards than in neighboring farmland. Additionally, the vertical movement of C is more effective in perennial systems due to the deeper root networks, which increase the stability and longevity of C storage, thereby reducing the risk of rapid mineralization. Climate zones influence sequestration rates; temperate orchards tend to slowly but steadily build woody biomass and SOC due to slower litter decomposition, whereas tropical fruit trees like mango or guava grow and accumulate biomass rapidly. However, high microbial activity in these zones can lead to faster SOC turnover (Gelaye and Mesfin, 2024). To further stabilize SOC, amendments such as biochar derived from pruned orchard wood or agricultural residues are proving effective. Studies have shown that biochar not only enhances SOC by improving soil structure and microbial activity but also reduces GHG emissions from soil. Lifecycle assessments confirm that although emissions are generated during orchard establishment (nursery activities, tillage, and fertilizer use), they are gradually offset by the C_{seq} gains in both biomass and soil over the productive lifespan of the orchard (Ioannidou et al., 2025). On average, fruit orchards can sequester between 2 and 3 tons of C per hectare annually when both tree biomass and soil pools are considered, with higher values reported in mixed-species or AF-based systems (Montanaro et al., 2021). Integrating fruit trees into silvo-pastoral or AF models can further enhance C_{seq} by improving biodiversity, reducing erosion, and lowering synthetic input use. However, caution is needed: not all SOC increases imply effective mitigation if turnover is high or if management leads to emissions elsewhere. The most stable systems are those that combine deep-rooted species, permanent soil cover, organic amendments, and species with beneficial microbial associations. One emerging area of interest is the oxalate–carbonate pathway found in fig species, where calcium carbonate formation within tree tissues could represent an additional, inorganic, and stable form of C storage. This process, though still under research, indicates that some fruit trees may store C not only in organic forms but also in minerals, offering an extra layer of permanence. Ultimately, the synergy between C stored in biomass and the long-term stabilization of soil C makes fruit trees a powerful tool in CCM, especially when supported with suitable management strategies, species selection, and landscape integration (Table 14.2).

14.9 Factors affecting carbon sequestration in fruit crops

C_{seq} in fruit crops is not uniform. Still, it is influenced by a suite of interrelated factors: tree species and physiological traits, orchard design and diversity, management practices, soil and climate conditions, and agro-ecological integration.

Tree species and traits are fundamental. Species with dense wood, deep root systems, and slower turnover of leaves and branches store C more stably in both biomass and soil (Peng et al., 2020). Mycorrhizal associations—especially AM—and nitrogen-fixing species significantly enhance soil aggregation and SOC; a global meta-analysis shows that AM-associated and N-fixing trees can have 6%–22% higher mineral soil C stocks than nonassociated species (Peng et al., 2020).

TABLE 14.3 Key factors influencing C_{seq} in fruit crops.

Factor	Impact mechanism	Sequestration effect
Species traits (wood density, rooting depth, AM/N-fixing)	Stabilizes biomass and enhances SOC via root–microbe interactions, 6%–22% increase in SOC	Greater biomass retention
Orchard design and diversity	Mixed-species cover improves litter input and rooting profiles	Higher sequestration than monocultures
Management (organic, no-till, and cover cropping)	Reduced soil disturbance; enhanced soil fertility and structure	Improved SOC longevity and lower emissions
Tree age and growth phase	Biomass accumulation increases with age	Cumulative sequestration in mature orchards
Soil and climate conditions	Affects decomposition, moisture, and root penetration	Temperate: slow decay; Tropical: high turnover but rapid growth
Agroforestry integration	Synergies with crops/livestock improve ecosystem resilience	Greater C storage and cobenefits
Novel pathways (e.g., oxalate–carbonate)	Inorganic C locked in woody tissues	Added permanence beyond organic biomass

C_{seq}, carbon sequestration; *SOC*, soil organic carbon; *AM*, arbuscular mycorrhizae.

Allometric equations developed for fruit tree species—such as mango and avocado—enable accurate above-ground biomass estimates to inform C stock assessments.

Orchard design and species diversity matter: mixed-species orchards, including combinations of pome, stone, nut, and tropical fruit trees, outperform monocultures for C storage and ecosystem stability (Ioannidou et al., 2025; Montanaro et al., 2021). These designs diversify canopy structures, litterfall types, root depths, and microbial interactions, enhancing both biomass accumulation and soil C retention. Management practices, such as fertilization, tillage, irrigation, pruning, and crop cover, significantly affect C_{seq}. Integrated and organic systems—using compost, minimal tillage, cover crops, and optimized nutrient inputs—can boost SOC and reduce practitioner emissions. Conservation tillage, mulching, and no-till reduce soil disruption and preserve aggregates and microbial biomass, all contributing to improved SOC levels. Conversely, intensive inorganic fertilization without adequate organic inputs can reduce C retention (Pradhan et al., 2023).

Tree age and biomass accumulation stage influence sequestration trajectories. Younger fruit trees sequester less C initially; rates accelerate as trees reach mid-growth and generally plateau in maturity. Long-lived fruit orchards may continue to accumulate biomass—and thus C—for decades, with cumulative benefits outweighing the initial establishment costs (Sharma et al., 2021a). Soil type and climate are crucial contextual factors. Soil texture and moisture influence root penetration and microbial activity; sandy soils may support deeper C deposition but risk faster SOC turnover, while clay soils foster aggregation and C protection (Peng et al., 2020). Climate zones affect both biomass production and decomposition rates: temperate regions slow litter decay, enhancing SOC accumulation, whereas tropics support rapid biomass growth but also faster microbial decomposition (Gelaye and Mesfin, 2024). Integration into AF or silvopastoral systems amplifies C benefits. Fruit trees intercropped with fodder grasses or annual crops stabilize soils, increase biodiversity, reduce synthetic inputs, and sequester more C than isolated orchards (Montanaro et al., 2021; Ioannidou et al., 2025). Such systems also moderate microclimate extremes, enhancing both productivity and resilience.

Lastly, innovative mechanisms, such as the oxalate–carbonate pathway in particular fig species, may contribute inorganic, long-term C_{seq} beyond biomass and SOC. While still emerging, this pathway illustrates how species traits can unlock new C storage avenues (Table 14.3).

14.10 Carbon capturing and long-residence in woody, leaf, fruit tissues

C_{seq} in fruit tree systems depends not only on how much CO_2 is fixed, but also on where and how long C remains stored. Different tree organs—woody tissues, leaves, fruits, and roots—have distinct C residence times and ecological functions that influence overall sequestration potential. This section explores these pathways, highlighting examples from key species such as apples, mangoes, and avocados. Fruit trees assimilate CO_2 through photosynthesis, channeling C into different organs. Woody biomass—including trunks and branches—serves as the longest-term C reservoir, with stored C lifespan often spanning several decades (Sharma et al., 2021a). Allometric studies in apple trees indicate that C accumulation during orchard maturity (approx. 15–20 years) peaks, storing over 200 t C ha^{-1} in mature orchards. While the initial years (under ~8 years) may generate more emissions than sink capacity, orchards become net C sinks as trunks and branches accumulate C more rapidly than short-lived tissues turn over. Also, net ecosystem C balance in agricultural systems maximizes or minimizes the C level by cultivation over a passage of time (Antar et al., 2021) as presented in Fig. 14.5.

Leaves and fruit, in contrast, represent short-term C pools. Leaves turn over seasonally, returning most of their C to soil via litterfall within one year. Fruits, harvested or abscissed, remove C from trees, though if residues and pruning are returned to soil, some of that C may be retained as SOC. Apple orchard studies have estimated that leaf litter decomposition contributes significantly to SOC dynamics (Sharma et al., 2021a), but represents only a small portion (<30%) of annual net C accumulation.

Roots and fine roots sequester C both directly via biomass accumulation and indirectly through exudation and turnover. While fine root biomass represents a minor fraction of total tree biomass (<1%), turnover rates may account for 30%–50% of annual net primary productivity, significantly contributing to SOC over time (Sharma et al., 2021a). The exudates and decomposition of fine roots feed soil microbial pools and support stable C aggregation (Peng et al., 2020). Quantitative species-specific data illustrate these processes clearly. In Kenyan smallholder systems, mango (*Mangifera indica*) trees were found to store on average 10.5 ± 2.9 Mg C ha^{-1} in above-ground biomass, and avocado (*Persea americana*) trees 9.7 ± 2.5 Mg C ha^{-1}, indicating substantial woody C even in relatively young plantings. Longan (*Dimocarpus longan*) orchards in Thailand showed average above-ground C stocks of 1.0 Mg C ha^{-1} in Edo species and 0.86 Mg C ha^{-1} in Paungtong species under specific density and age conditions, illustrating the influence of tree size and age on woody C storage. Temporal dynamics are critical: younger trees (under ~8 years) have nursery and establishment emissions. By middle age (10–20 years), above-ground net primary productivity peaks, woody C accumulates fastest, and orchards transition into net C sinks. As orchards age beyond ~18 years, net sequestration capability stabilizes or declines due to reduced growth rates, but long-lived woody C remains sequestered.

Inorganic C pathways are emerging as additional mechanisms of long-term storage. Particular fig species (*Ficus spp.*) have been shown in recent studies to form calcium oxalate crystals that potentially mineralize into calcium carbonate inside woody tissues—offering mineral C permanence beyond organic storage. Although the research is preliminary, this mechanism may significantly augment C residence time in woody biomass (Fig. 14.6 and Table 14.4).

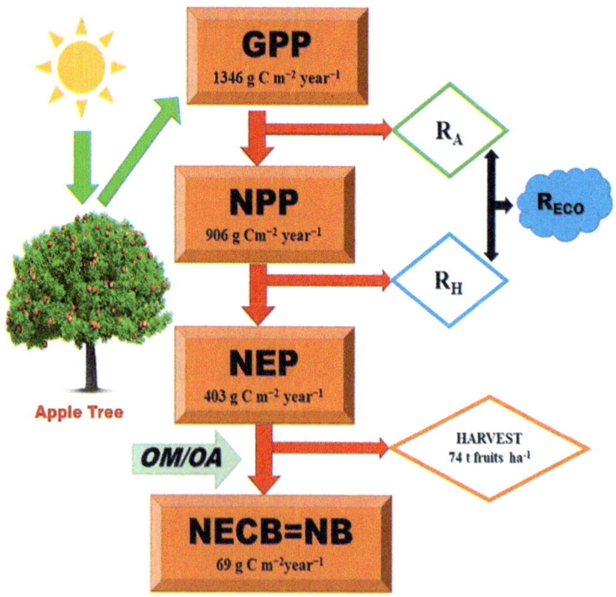

FIGURE 14.6 Relationship between fruit trees and C_{seq}. C_{seq}, carbon sequestration.
Compiled from Sharma, R., Pradhan, L., Kumari, M., Bhattacharya, P. (2021b). Assessment of Carbon Sequestration Potential of Tree Species in Amity University Campus Noida. Environmental Sciences Proceedings, 3:52.

TABLE 14.4 Carbon pools and residence in fruit trees.

Tissue type	Residence time	Function and C flow	Example
Woody biomass	Decades to centuries	Trunks, branches store long-term structural C	Apple: >200 t C ha^{-1} in mature orchards
Leaves	1 year	Photosynthate is fixed and reintroduced to the soil via litterfall	Apple leaf turnover <30% of annual NPP (Sharma et al., 2021a)
Fruit	Harvested annually	C removed; residues may re-enter SOC if returned	Mango/Avocado: harvest impacts but biomass is heavy
Fine roots and exudates	Seasonal turnover but recurring	Inject C into the soil directly and support microbial SOC build-up	Apple: fine root turnover up to 50% NPP (Sharma et al., 2021a)
Inorganic (oxalate–carbonate)	Potentially centuries or more	Mineral C storage inside Ficus woody tissue	*Ficus species*: oxalate–carbonate mechanism

NPP, Net primary productivity; *SOC*, soil organic carbon.

C captured in fruit tree systems is partitioned across multiple pools—with woody biomass providing long-term permanence, leaves and fruit functioning as short-cycle reservoirs, and roots contributing to persistent soil C. Species such as apple, mango, avocado, and longan illustrate how tree traits (e.g., wood density, size, and age) and management influence the scale and retention of C. Emerging discoveries of inorganic C pathways in some species may further enhance sequestration potential. Understanding these dynamics enables strategic species selection and management practices aimed at maximizing long-term C storage in fruit tree systems.

14.11 Environmental elements affecting fruit carbon stocks

Fruit orchards' C storage capacity is strongly modulated by environmental conditions—including microclimate, soil moisture and temperature, microbial activity, and orchard floor management—all of which influence both biomass accumulation and SOC stability. Canopy shading from tree foliage reduces soil surface temperature and limits evaporation, thereby conserving water and moderating microbial respiration rates (Sharma et al., 2021a). This creates favorable moisture retention under trees, especially in dry zones. Moreover, soil moisture and temperature regimes play a central role: in apple orchards on China's Loess Plateau, studies using isotope-based water balance found that tree age correlated negatively with soil evaporation—older trees with higher leaf area index reduced soil evaporation by shading, while cooler soil slowed OM breakdown (Huang et al., 2025). Similarly, in subtropical fruit systems like grapefruit, mango, and litchi, evapotranspiration partitioning shows that soil+understory evaporation can constitute 30%–35% of total evapotranspiration, meaning that orchard floor management—such as mulching or cover cropping—can substantially diminish nonbeneficial water losses and enhance SOC (Li et al., 2024). Soil microbial communities are essential in converting root exudates and litter into stable C; thus, soils with rich microbial biomass, especially in systems with organic amendments or reduced tillage, retain more SOC (Sharma et al., 2021b). In mango–citrus orchards in Oman, application of compost and ground cover significantly improved microbial diversity, nutrient cycling, SOM, and fruit yields (Janke et al., 2024). Soil texture and chemistry are also influential: clay-rich soils promote aggregate formation that protects OM, whereas sandy or acidic soils may limit root development, microbial diversity, and C stabilization. Micronutrient status (e.g., magnesium and calcium) regulates microbial function and enzymatic activity, affecting SOC formation (Peng et al., 2020). Abiotic stressors, such as prolonged drought or extreme heat, suppress photosynthesis and root exudation, reducing C inputs to soils; without canopy cover or water-conserving practices, drought also accelerates SOC loss (Huang et al., 2025). Finally, agroecological integration—for instance, AF or silvopasture—enhances environmental buffering and improves C stocks through diverse canopy structure, reduced soil disturbance, and enhanced soil health (Montanaro et al., 2021). In well-managed orchards, these factors combine to elevate SOC retention, maximize biomass growth, and ensure long-term C residence across tissues and soil. Understanding and optimizing these environmental drivers through strategic orchard design, mulch or cover strategies, and water management is essential to unlocking the full sequestration potential of fruit tree systems in different climatic contexts (Table 14.5).

TABLE 14.5 Environmental factors influencing C stocks in fruit orchards.

Environmental factor	Mechanism	Effect on C stocks	Source
Canopy shading/Microclimate	Reduces soil temperature and evaporation	Conserves moisture and slows SOC decomposition	Sharma et al. (2021b)
Soil moisture and temperature	Influences microbial activity and OM turnover	Cooler, moist soils retain SOC longer	Huang et al. (2025)
Orchard floor and evapotranspiration	Soil + understory evaporation contributes 30%–35% of ET	Mulching/cover cropping reduces nonbeneficial losses	Li et al. (2024)
Soil microbial community	Converts plant inputs into stable SOC	Vital for SOC stabilization	Janke et al. (2024)
Soil texture and nutrients	Affects aggregation, root growth, and microbial function	Clay, balanced pH, and micronutrients support SOC retention	Peng et al. (2020)
Abiotic stress (heat, drought)	Limits photosynthesis and C input	Reduces both biomass and SOC, increases decomposition risk	Huang et al. (2025)
Agroforestry/Silvipasture	Introduces species diversity and understorey vegetation	Enhances C stocks, biodiversity, and system resilience	Montanaro et al. (2021)

SOC, soil organic carbon; *OM*, organic matter.

14.12 Carbon stock potential of fruit trees

Fruit trees contribute significantly to climate mitigation by combining long-term C_{seq} in biomass and soil with ecosystem resilience, food security, and socio-economic benefits. Through photosynthesis, fruit trees lock C into woody tissues and root systems and enhance SOC via litterfall and root exudates; when integrated within AFs, these contributions can reach 2–3 t C ha^{-1} yr^{-1}, markedly exceeding the rates of annual cropping systems (Sharma et al., 2021a; Gelaye and Mesfin, 2024). In Ethiopia, multiple fruit tree species—in mixed-species home gardens—accounted for more than 40% of above- and below-ground C storage, underscoring the importance of diversity (Gelaye and Mesfin, 2024). Beyond C capture, fruit-tree AFs deliver climate adaptation services by regulating microclimates, reducing soil erosion, and buffering temperature extremes. For example, silvopasture models and mixed orchards improve moisture retention and reduce evaporation, enhancing resilience under drought-prone conditions (Montanaro et al., 2021; CIFOR-ICRAF, 2023). Policies in Kenya and Rwanda that support the household-level planting of fruit trees have simultaneously boosted nutrition security, livelihoods, and climate resilience (Muthuri et al., 2023). From a climate finance perspective, these community-based efforts lay the groundwork for C credit access, though standardizing monitoring, reporting, and verification remains a challenge (Muthuri et al., 2023; Raj et al., 2025). On the scientific frontier, recent research has discovered that fig species such as *Ficus wakefieldii*, *F. natalensis*, and *F. glumosa*—native to Kenya—sequester inorganic C by forming calcium carbonate in woody tissues via the oxalate–carbonate pathway, creating a mineral C sink that persists even after tree death (Egan, 2025; ScienceDaily, 2025). While further validation is needed for global AF species, the discovery points to the potential selection of fruit tree species that offer dual-mode C capture—both organic and inorganic. Lifecycle assessments indicate that once establishment emissions (e.g., nursery operations, planting, and fertilization) are factored in, well-managed fruit orchards ultimately become net C sinks over time, especially when incorporating mixed-species design, reduced tillage, organic amendments, and renewable irrigation systems (Sharma et al., 2021a; Montanaro et al., 2021). Fruit trees also contribute to CCM indirectly by reducing pressure on natural forests—through fruit production, they reduce reliance on clearing woodlands for fuel or timber (Gelaye and Mesfin, 2024). Simultaneously, they foster biodiversity and ecosystem services, serving as habitat for pollinators and beneficial fauna, enriching soil fertility, and enhancing water cycling (Montanaro et al., 2021; CIFOR-ICRAF, 2023). In urban settings, community orchards and tree gardens serve as urban cooling infrastructure, sequester C, and promote social cohesion and access to nutritious food (CIFOR-ICRAF, 2023). These multifaceted roles position fruit tree systems as robust nature-based climate strategies—bridging mitigation, adaptation, ecosystem health, and rural development. To fully

TABLE 14.6 C stock potential of fruit trees in different C pools.

C pools in the fruit tree ecosystems	Level of C stock
Store C	Photosynthesis converts CO_2 into glucose and oxygen, while trees store C in biomass and SOM. Fruit plants sequester CO_2differently with varying C stocks, such as mango and jackfruit.
Above and below-ground biomass	An Indonesian study finds that fruit tree biomass has a high C_{seq} capacity, with an average sequestration of 951.1356 tons per hectare and 191.5545 tons per hectare. Coffee plants contribute 4.4 Gg.
Litter biomass	Litter has a low C storage capacity and low C_{seq} potential; however, above-ground litter input increases SOC pools and microbial biomass, thereby benefiting sequestration despite decomposition.
Soil C	Soil is the largest terrestrial C sink, storing OM. Reducing loss involves minimal disturbance, clay particle management, and OC addition.

Source: Gelaye, T., Mesfin, D., 2024. Soil organic carbon dynamics in Ethiopian fruit-based agroforestry systems. Agroecology and Sustainable Food Systems. 48 (2), 123–142. https://doi.org/10.1080/21683565.2023.2294544.

TABLE 14.7 C_{seq} potential of various fruit crops.

Crop	Crop details	Above-ground	Below-ground C	Total C_{seq}	References
Mango	15 years old	47.61 t/ha	7.17 t/ha	54.78 t/ha	Janiola and Marin (2016)
Citrus	17 years old	149.53 kg/tree	36.72 kg/tree	186.25 kg/tree	Kimball et al. (2007)
Guava	15 years old	43.11 kg/tree	11.21 kg/tree	54.31 kg/tree	Shinde et al. (2020)
Jamun	15 years old	62.54 kg/tree	16.26 kg/tree	78.79 kg/tree	Shinde et al. (2020)
Aonla	14 years old	8.98 t/ha	1.89 t/ha	10.88 t/ha	Wankhede et al. (2018)
Wood Apple	10 years old	0.31 t/tree	0.08 t/tree	0.39 t/tree	Shrestha and Malla (2016)
Rambutan	12 years old	13.4 t/ha	2.01 t/ha	15.41 t/ha	Janiola and Marin (2016)
Litchi	14 years old	0.41 t/tree	0.09 t/tree	0.50 t/tree	Shrestha and Malla (2016)
Banana	25 trees	15.04 kg/tree	2.26 kg/tree	17.30 kg/tree	Sharma et al. (2021b)
Grape	1631 vines	8.2 Mg/ha	4.1 Mg/ha	12.3 Mg/ha	Morandé et al. (2017)
Ber	37 trees	0.36 t/ha	0.07 t/ha	0.43 t/ha	Maiti (2015)
Date Palm	77 trees	14.33 Kg/tree	2.15 kg/tree	16.48 kg/tree	Ganeshamurthy et al. (2019)
Custard Apple	1462 trees	35.95 t/ha	9.35 t/ha	45.30 t/ha	Vishnu and Patil (2016)
Apple	2632 trees	7.65 t/ha	4.73 t/ha	12.91 t/ha	Scandellari et al. (2016)

harness their potential, it is essential to mainstream fruit tree systems within climate-smart agriculture frameworks, promote species suitable for both OC and inorganic C retention, and integrate them into national C accounting and incentive programs (Table 14.6).

14.13 Carbon sequestration potential of different fruit crops

Regardless of soil moisture levels, fruit trees are likely to exhibit higher rates of photosynthetic activity and biomass production. The increased amounts of C are therefore most likely stored in fruit tree branches and woody trunks. According to Patil and Kumar (2017), fruit-based systems are considered self-sustaining because they can harness solar energy at different heights, enabling the efficient use of soil resources with higher cropping intensities. The main crop, intercrop, and filler crop are the main components of this system. They are spread out across the production system area at three different levels. The primary crop's fruit species have a long juvenile and productive phase and a large canopy size. Filler crops are characterized by fruit species with short stature, precocity, and prolific bearing. They comprise the middle layer of the multitier system, where economic productivity is achieved. The remaining vacant spaces are used to grow the lowest tiers of the multitier system. The trees' capacity for abundant flowering and fruiting enhances the removal of CO_2 from the atmosphere, and the CO_2 is stored in significant amounts as cellulose. Santol (32 years old), mango (15 years old), and rambutan (12 years old) have C storage potentials of roughly 203.62, 122.34, and 112.17 t/ha, respectively (Janiola and Marin, 2016). China's nectarine orchards store between 13 and 15 tons of C per hectare (Wang et al., 2015). About 258.005 Mt of C were sequestered from the entire Indian mango plantation (Ganeshamurthy et al., 2019) (Tables 14.7 and 14.8).

TABLE 14.8 Average C footprint values of fruit crops cultivated worldwide.

Type of fruit	Mean value of C_{seq}	References
Papaya	0.3 kg (0.67 lb) of CO_2e/kg	Choinova (2023)
Jackfruit	0.9 kg CO_2e/kg	
Guava	0.15 kg CO_2e/kg	
Passionfruit	0.77 kg CO_2e/kg	
Dragonfruit	0.9 kg CO_2e/kg	
Banana	0.21 kg (0.48 lb)/kg	
Mango	0.21 kg (0.46 lbs) CO_2e/kg	Choinova (2023)
Avocado	0.19 kg CO_2e/kg	
Pineapple	0.09 kg (0.20 lb) of CO_2e/kg	
Orange, Lemons, and Limes	0.3 kg (0.66 lbs) CO_2e/kg	
Apples	0.24 kg (0.53 lbs) CO_2e/kg	
Plum	0.4 kg (0.88 lbs) CO_2e/kg	
Peach	0.17 kg (0.38 lbs) CO_2e/kg	
Pomegranate	0.39 kg (0.87 lb) CO_2e/kg	Choinova (2023)
Cherry	0.584 kg CO_2e/kg	
Apricot	0.16 kg (0.36 lb) CO_2e/kg	
Kiwifruits	0.9 kg CO_2e/kg	
Grape	0.64 kg (1.42 lbs) CO_2e/kg	
Pear	0.34 kg CO_2e/kg	
Strawberry	0.39 kg (0.88 lb) CO_2e/kg	

14.14 Conclusion

Fruit trees sit at the intersection of ecological restoration, CCM, and sustainable agriculture. As this chapter has shown, their role goes well beyond food production. Through their ability of C_{seq}, improve soil health, conserve water, support biodiversity, and stabilize ecosystems, fruit trees become vital allies in the global effort for environmental sustainability. In both rural and urban areas, fruit-based agro-ecosystems provide multiple benefits that address urgent environmental challenges. Their perennial nature enables long-term C storage in biomass and soil, while their integration with crops or livestock supports diverse and resilient farming systems. In arid and semiarid regions, fruit trees help buffer microclimates and improve soil moisture retention, thereby mitigating the effects of climate extremes. In urban settings, they contribute to green infrastructure, cool cityscapes, and foster social cohesion through community-based orchards. Importantly, fruit trees exemplify a regenerative approach to agriculture—one that balances productivity with environmental care. Including them in agroecological and AFs not only conserves natural resources but also enhances the livelihoods of farmers and rural communities. By promoting circular nutrient flows, reducing dependence on synthetic inputs, and preserving ecosystem services, fruit tree systems closely align with global goals for climate resilience, biodiversity protection, and sustainable development. The evidence presented throughout this chapter underscores the importance of integrating fruit trees into policy and practice. From smallholder farms to urban green zones and institutional landscapes, there is significant potential to expand fruit tree-based systems as NBS. Achieving this requires strategic investments in research, capacity development, and inclusive land-use planning that emphasizes multifunctionality and long-term sustainability. In summary, fruit trees are more than just components of orchard economies—they are living infrastructure that supports ecological stability, enriches agro-ecosystems, and guides us toward a more sustainable, climate-resilient future. Recognizing their multifunctional roles and using them thoughtfully across diverse agro-ecological contexts is essential for building sustainable landscapes that nourish both people and the planet.

References

Aggarwal, R., Sharma, R., Singh, A., 2024. Integrating fruit trees for resilient agroforestry systems in Bangladesh. Agroforestry Systems 98 (3), 245–258. Available at https://doi.org/10.1007/s10457-023-01025-y.

Antar, M., Lyu, D., Nazari, M., Shah, A., Zhou, X., Smith, D.L., 2021. Biomass for a sustainable bioeconomy: an overview of world biomass production and utilization. Renewable and Sustainable Energy Reviews 139. Available at https://doi.org/10.1016/j.rser.2020.110691.

Azra, M.N., Noor, M.I.M., Eales, J., Sung, Y.Y., Ghaffar, M.A., 2022. What evidence exists for the impact of climate change on the physiology and behaviour of important aquaculture marine crustacean species in Asia? A systematic map protocol. Environmental Evidence 11, 9. Available at https://doi.org/10.1186/s13750-022-00263-1.

Banerjee, A., Jhariya, M.K., Raj, A., Mechergui, T., 2024. Ecosystem Management: Climate Change and Sustainability. Wiley, India, pp. 1–669.

CCS Explained, 2020. UKUKCCSRC. Archived from the original on 28 June 2020. Retrieved 27 June 2020.

Choinova, V., 2023. Identifying Data for Horticulture Categories to Propose a Standard Carbon Footprint Model (Thesis). South Eastern Finland, University of Applied Sciences. Available at https://www.theseus.fi/bitstream/handle/10024/802190/Viliuia_Choinova.pdf?sequence=2.

CIFOR-ICRAF, 2023. Fruit trees for climate resilience and nutrition: technical brief. Nairobi, Kenya. Available from: https://www.worldagroforestry.org/publication.

Egan, A., 2025. Ficus species mineralize carbon in woody tissues via the oxalate-carbonate pathway. Nature Ecology & Evolution 9 (2), 114–121. Available at https://doi.org/10.1038/s41559-025-02491-z.

Freitas, M.D., Silva, A.C., 2022. Plant growth-promoting bacteria in fruit orchards: soil health and nutrient dynamics. Soil Biology & Biochemistry 158, 108303. Available at https://doi.org/10.1016/j.soilbio.2021.108303.

Ganeshamurthy, A.N., Ravindra, V., Rupa, T.R., 2019. Carbon sequestration potential of mango orchards in India. Current Science 117 (12), 2006. Available at 10.18520/cs/v117/i12/2006-2013.

Ganeshamurthy, A.N., Ravindra, V., Rupa, T.R., Bhatt, R.M., 2019. Carbon sequestration potential of mango orchards in the tropical hot and humid climate of Konkan region, India. Current Science 116 (8), 1417–1423. Available at https://doi.org/10.18520/cs/v116/i8/1417-1423.

Gelaye, T., Mesfin, D., 2024. Soil organic carbon dynamics in Ethiopian fruit-based agroforestry systems. Agroecology and Sustainable Food Systems. 48 (2), 123–142. Available at https://doi.org/10.1080/21683565.2023.2294544.

Huang, J., Li, X., Chen, Y., Huang, J., Li, X., Chen, Y., 2025. Soil carbon responses to drought and warming in perennial horticultural systems. Global Change Biology 31 (5), 3201–3215. Available at https://doi.org/10.1111/gcb.16230.

Ilakiya, T., Parameswari, E., Swarnapriya, R., Yazhini, G., Kalaiselvi, P., Davamani, V., et al., 2024. Unlocking the carbon sequestration potential of horticultural crops. C 10 (3), 65. Available at https://doi.org/10.3390/c10030065.

Ioannidou, S., Litskas, V., Vogiatzakis, I., Ioannidou, S., Litskas, V., Vogiatzakis, I., 2025. CO_2 sequestration in mixed Mediterranean fruit orchards: allometric approaches. Agronomy 15 (3), 423. Available at https://doi.org/10.3390/agronomy15030423.

Janiola, M.D.C., Marin, A., 2016. Carbon sequestration potential of fruit tree plantations in Southern Philippines. Journal of Biodiversity and Environmental Sciences 8, 164–174.

Janke, J.R., Brown, M.E., Smith, T., Janke, J.R., Brown, M.E., Smith, T., 2024. Soil microbial pathways in fruit orchards under organic management. Applied Soil Ecology 190, 104150. Available at https://doi.org/10.1016/j.apsoil.2023.104150.

Jhariya, M.K., Meena, R.S., Banerjee, A., Meena, S.N., 2022. Natural Resources Conservation and Advances for Sustainability. Elsevier. Available at https://doi.org/10.1016/C2019-0-03763-6.

Jhariya, M.K., Meena, R.S., Banerjee, A., Kumar, S., Raj, A., 2023. Agroforestry for Carbon and Ecosystem Management. Academic Press, pp. 1–434. Available at 10.1016/C2021-0-02916-3.

Kimball, B.A., Idso, S.B., Johnson, S., Rillig, M.C., 2007. Seventeen years of carbon dioxide enrichment of sour orange trees: final results. Global Change Biology 13 (10), 2171–2183. Available at https://doi.org/10.1111/j.1365-2486.2007.01430.x.

Kumar, N., Sinha, V., 2022. Role of fruit-based agroforestry in rural sustainability and food security. Sustainability 14 (10), 5895. Available at https://doi.org/10.3390/su14105895.

Kumar, S., Meena, R.S., Sheoran, P., Jhariya, M.K., 2024. Regenerative Agriculture for Sustainable Food Systems. Springer, Singapore, pp. 1–493. Available at https://doi.org/10.1007/978-981-97-6691-8.

Li, H., Wang, Y., Zhao, Q., 2024. Evapotranspiration dynamics under orchard groundcovers. Agriculture Water Manage 278, 108012. Available at https://doi.org/10.1016/j.agwat.2022.108012.

Maiti, S.K., 2015. Carbon Sequestration Study at Eco-restoration Sites of BCCL. Centre of Minning Environment, Dhanbad, Jharkhand, pp. 1–136 (2015), 1–136.

Meena, R.S., Pradhan, G., 2023. Industrial garbage-derived biocompost enhances soil organic carbon fractions, CO_2 biosequestration, potential carbon credits and sustainability index in a rice-wheat ecosystem. Environmental Research 235, 116525. Available at https://doi.org/10.1016/j.envres.2023.116525.

Meena, R.S., Pradhan, G., Kumar, S., Lal, R., 2023. Using industrial wastes for rice-wheat cropping and food-energy-carbon-water-economic nexus to the sustainable food system. Renewable and Sustainable Energy Reviews 187, 113756. Available at https://doi.org/10.1016/j.rser.2023.113756.

Meena, R.S., Pradhan, G., Singh, K., Kumar, S., Singh, A.K., Shashidhar, K.S., et al., 2024a. Agriculture models for restoring degraded land to enhance CO_2 biosequestration and carbon credits in the Vindhyan region of India. Science of the Total Environment 929. Available at https://doi.org/10.1016/j.scitotenv.2024.172661.

Meena, R.S., Pradhan, G., Nalani, Kumar, S., Singh, A.K., Verma, S.K., et al., 2024b. Diversified cropping modules designed for soil restoration, CO_2 sequestration, and generating carbon credits. Land Degradation and Development 35 (14), 4314–4329. Available at https://doi.org/10.1002/ldr.5224.

Meena, R.S., Lal, R., Kumar, S., Pradhan, G., Srinivasarao, C., Singh, A.K., et al., 2024c. Potential of Indian agriculture for capturing atmospheric CO_2 and monetizing carbon credits to the farmers: an overview and policy framework. Advances in Agronomy 188, 101–206. Available at https://doi.org/10.1016/bs.agron.2024.06.001.

Meena, R.S., Pradhan, G., Kumar, S., 2025. Energy flow, eco-efficiency, and economic circulation with recycled industrial waste compost application in wheat and subsequent rice farming. Science of the Total Environment 967. Available at https://doi.org/10.1016/j.scitotenv.2025.178779.

Montanaro, G., Amato, D., Briglia, N., Russo, C., Nuzzo, V., 2021. Carbon fluxes in sustainable tree crops: field, ecosystem and global dimension. Sustainability 13 (16), 8750. Available at https://doi.org/10.3390/su13168750.

Morandé, J.A., Stockert, C.M., Liles, G.C., Williams, J.N., Smart, D.R., Viers, J.H., 2017. From berries to blocks: carbon stock quantification of a California vineyard. Carbon Balance and Management 12 (1). Available at https://doi.org/10.1186/s13021-017-0071-3.

Muthuri, C.W., Kimaro, A., Sileshi, G., 2023. Enhancing carbon sequestration through fruit tree planting initiatives in East Africa. ICRAF Working Paper. Nairobi, Kenya.

Patil, P., Kumar, A.K., 2017. Biological carbon sequestration through fruit crops (perennial crops-natural "sponges" for absorbing carbon dioxide from atmosphere). Plant Archives 17 (2), 1041–1046.

Peng, Y., Liu, B., Zhang, D., 2020. Soil nutrient cycling and microbial function in subtropical orchards. Soil Tillage Research 204, 104729. Available at https://doi.org/10.1016/j.still.2020.104729.

Pradhan, G., Meena, R.S., 2023a. Interaction impact of biocompost on nutrient dynamics and relations with soil biota, carbon fractions index, societal value of CO_2 equivalent and ecosystem services in the wheat-rice farming. Chemosphere 339. Available at https://doi.org/10.1016/j.chemosphere.2023.139695.

Pradhan, G., Meena, R.S., 2023b. Utilizing waste compost to improve the atmospheric CO_2 capturing in the rice-wheat cropping system and energy-cum-carbon credit auditing for a circular economy. Science of the Total Environment 892. Available at https://doi.org/10.1016/j.scitotenv.2023.164572.

Pradhan, G., Meena, R.S., Kumar, S., Lal, R., 2023. Utilizing industrial wastes as compost in wheat-rice production to improve the above and below-ground ecosystem services. Agriculture, Ecosystems & Environment 358, 108704. Available at https://doi.org/10.1016/j.agee.2023.108704.

Rada, P., Kalusová, V., Tropek, R., 2022. Biodiversity in restored and abandoned fruit orchards in urban landscapes. Urban Forestry Urban Greening 68, 127481. Available at https://doi.org/10.1016/j.ufug.2021.127481.

Raj, A., Jhariya, M.K., Meena, R.S., 2025. Agroforestry for Monetising Carbon Credits. Springer Nature, Switzerland.

Reyes, J., Karimi, M., Noor, A., 2023. Orchard-based climate mitigation strategies in semi-arid Iran. Journal of Agroecology 13 (1), 88–101. Available at https://doi.org/10.1177/2345678923120012.

Roussos, P.A., 2024. Climate change challenges in temperate and sub-tropical fruit tree cultivation. Encyclopedia 4 (1), 558–582. Available at https://doi.org/10.3390/encyclopedia4010036.

Scandellari, F., Caruso, G., Liguori, G., Meggio, F., Palese Assunta, M., Zanotelli, D., et al., 2016. A survey of carbon sequestration potential of orchards and vineyards in Italy. European Journal of Horticultural Science 81 (2), 106–114. Available at https://doi.org/10.17660/eJHS.2016/81.2.4.

ScienceDaily, 2025. Fruit trees discovered to trap carbon in mineral form. Available from: https://www.sciencedaily.com/releases/2025/03/250304092319.htm.

Sharma, R., Pradhan, L., Kumari, M., Bhattacharya, P., 2021. Assessment of carbon sequestration potential of tree species in Amity University Campus Noida. Environmental Sciences Proceedings 3 (1), 52.https://doi.org/10.3390/IECF2020-08075.

Sharma, S., Rana, V.S., Prasad, H., Lakra, J., Sharma, U., Sharma, R., et al., 2021b. Sequestration potential of tropical fruit trees under different agro-climatic zones. Frontiers Envionmental Sciences 9, 700768. Available at https://doi.org/10.3389/fenvs.2021.700768.

Shinde, V.V., Maheswarappa, H.P., Ghavale, S.L., Sumitha, S., Wankhede, S.M., Haldankar, P.M., 2020. Productivity and carbon sequestration potential of coconut-based cropping system as influenced by integrated nutrient management practices. Journal of Plantation Crops 48 (2), 103–110. Available at https://doi.org/10.25081/jpc.2020.v48.i2.6368.

Shrestha, G., Malla, G., 2016. Estimation of atmospheric carbon sequestration by fruit plants in mid-western Terai region Nepal. Nepalese Journal of Agricultural Science 14, 211–215.

Vishnu, P.R., Patil, S.S., 2016. Carbon storage and sequestration by trees in and around University Campus of Aurangabad City, Maharashtra. International Journal of Innovative Research in Science, Engineering and Technology 5, 5459–5468. Available at 10.15680/IJIRSET.2016.0504179.

Wang, Y.X., Weng, B.Q., Ye, J., Zhong, Z.M., Huang, Y.B., 2015. Carbon sequestration in a nectarine orchard as affected by green manure in China. European Journal of Horticultural Science 80 (5), 208–215. Available at https://doi.org/10.17660/ejhs.2015/80.5.2.

Wankhede, S., Saini, M.K., Kothari, S.L., Bala, N., Singh, G., Gour, V.S., 2018. Evaluation of carbon sequestration potential in amla (Emblica officinalis Gaertn.) orchards in semi-arid region of India. Proceedings of the National Academy of Sciences India Section B - Biological Sciences 88 (4), 1655–1660. Available at https://doi.org/10.1007/s40011-017-0917-1.

Section IV

Methodological approaches for carbon incentives

Chapter 15

Methodological approaches for calculating carbon credits from soils

Riziki Mwadalu[1], Mary Gathara[1], Mercy Rutto[1], Abey M. Abdi[1], Luciana R. Chappa[2], Emmanuely Z. Nungula[3], and Harun I. Gitari[4]

[1]*Central Highlands Eco-Region Research Programme, Kenya Forestry Research Institute, Nairobi, Kenya,* [2]*Department of Agricultural Sciences, College of Agriculture, Mwalimu Nyerere University of Agriculture and Technology, Butiama, Tanzania,* [3]*Department of Environment and Sustainable Development, Mzumbe University, Morogoro, Tanzania,* [4]*Department of Agricultural Science and Technology, School of Agriculture and Environmental Sciences, Kenyatta University, Nairobi, Kenya*

Highlights

- Agricultural lands cover approximately 35% of Earth's land area.
- Carbon neutrality in agriculture can be attained through increasing soil organic carbon stocks.
- Carbon sequestration potential of agricultural soils is approximately 1.2 billion tons of carbon annually.
- The soil carbon pool is about 4.5 times that of biotic pools (560 Pg) and 3.3 times the atmospheric pool (760 Pg).

15.1 Introduction

Globally, agricultural activities have significantly contributed to greenhouse gas (GHG) emissions, thus contributing to climate change. The farm sector also bears the brunt of climate change through erratic weather patterns, which have reduced productivity (Raj et al., 2024; Gitari et al., 2024; Meena et al., 2026; Nungula et al., 2024). Agriculture-based emissions account for 11% of GHGs, largely from nitrous oxide and methane from fertilizers and livestock, respectively. Biological processes are the leading causes of agricultural emissions (Paul et al., 2023). The main carbon (C) sink in agricultural systems is soil, which constitutes the main C pool, storing three times the aboveground biomass and twice the atmosphere. Studies have shown that most agricultural soils have lower C stocks compared to soils under other land uses, such as forests. Loss of SOC (soil organic carbon) in these ecosystems is attributed to intensive tillage that emits carbon dioxide (CO_2) into the atmosphere (Meena et al., 2025). As such, C farming has the potential to reduce losses. In 2021, agricultural land was estimated at about 4.8 billion hectares, accounting for over 30% of the terrestrial landmass (Kizeková et al., 2024). The agricultural lands by 2021 comprised 1.6 billion hectares of croplands and 3.2 billion hectares of permanent grasslands. High carbon dioxide removal in agricultural soils can be achieved through enhancing SOC. Agricultural land accounts for 31% of habitable landmass globally. Slight increases in SOC in agricultural can significantly mitigate climate change. C neutrality in agriculture can be attained by increasing SOC stocks, which involves the removal of CO_2 from the atmosphere. Increasing SOC in agricultural landscapes requires changes in management practices (Paul et al., 2023; Meena et al., 2024b).

The contribution of soil to various ecosystems has gained traction due to its carbon sequestration (C_{seq}) and storage potential (Abdullahi et al., 2018; Banerjee et al., 2024; Meena et al., 2024a). Studies have reported that mineralization of approximately 10% of the global SOC pool is equivalent to about three decades of emissions from anthropogenic activities. Global trends show that the largest C pool of nearly 1500 gigatons (Gt) is found in the first 1 m of the soil profile. Soil degradation has been massive and accounts for 24% of global soil and nearly half of the agricultural soils. The C_{seq} potential of agricultural soils is approximately 1.2 billion tons of C annually. Soil further provides approximately 75% of terrestrial C; therefore, projects aimed at increasing soil C storage, such as afforestation and sustainable land management, are crucial (Zeraatpishe and Khormali, 2012). Sequestration of SOC has majorly contributed to climate

Carbon Incentives, Plans, and Policies in Agriculture. DOI: https://doi.org/10.1016/B978-0-443-33557-0.00004-5

change mitigation (CCM) (Powlson and Galdos, 2023; Meena et al., 2024c). However, C dynamics in soils are influenced by numerous land-use and land management practices such as tillage and intensive agriculture (Joensuu et al., 2021). Accurate quantification of SOC is crucial for C accounting, which could earn farmers C credits.

15.1.1 Soil carbon pools in agricultural production

The soil C pool is crucial in the global C cycle. However, soils across the world have lost about 55–90 Pg (Pg = petagram = 1 billion tons =1* 10^{15}) of C, mainly due to land-use changes from natural ecosystems to agricultural ecosystems (Lal, 2004, 2006). The soil C pool comprises two major components: soil inorganic carbon (SIC) and SOC. The SOC pools comprise active humus and inert C in charcoal, while the SIC pools include carbonate minerals and essential C (Lal, 2007; Lal et al., 2019). Globally, the soil C pool for 1 m depth is approximated at 1550 Pg of SOC and 950 Pg of SIC (Batjes, 1996). Soil C consists of 4.5 times the biotic pools (560 Pg) and 3.3 times the atmospheric pool (760 Pg). Apart from the degradation of soil and water qualities, large quantities of soil C pools lost from degraded ecosystems and agricultural soils are released as CO or methane (Lal, 2006) back to the atmosphere.

The ability of the soil C pool to act as a source or a sink of the atmospheric pool is highly dependent on the land use and management (Lal, 2007). The SOC pools consist of two major components, namely, the inert (recalcitrant) component, which is mainly not involved in the soil mineralization process and is dependent on climate, soil type, landscape position, and history of land use; and (2) the active (labile) component, which is mainly dependent on the management practices. The SOC changes resulting from shifts in land use and management are mainly driven by dynamics in the active fraction. Links exist between the concentration of the active fraction of SOC and soil quality, particularly in degraded soils of the tropics and subtropics. This has led to a 60%–80% loss of their SOC pools due to soil degradation by subsistence farming (Lal, 2006; Lal et al., 2019).

Agriculture is among the main terrestrial biomes and covers >35% of the earth's total land mass and occupies about 1500 million hectares of the total land area (Asner et al., 2004; Lal et al., 2015). Conversion of natural environments to agricultural lands depletes C pools due to lower return of biomass and losses of C pools through erosion, leaching, and mineralization, thereafter leading to variations in moisture regimes and temperatures (Meena and Pradhan, 2023; Lal et al., 2015). Most soils have C pools that are way below their potential levels, mainly due to excessive C depletion by land use systems that are based on destructive C soil fertility management practices (Lal, 2007; Meena et al., 2023).

The main factors leading to the depletion of SOC pools and emission of CO_2 are soil erosion and plowing. Therefore practices such as the use of crop residues in the rotation cycle and the utilization of integrated pest management can help to restore SOC pools in farmland ecosystems (Kumar et al., 2020). In addition, restoring eroded soils and controlling erosion can be critical strategies for improving and maintaining SOC pools in agricultural ecosystems (Lal, 2007; Lal et al., 2019).

Undisturbed soils have high contents of OC pools, and the properties and magnitude of these SOC pools are usually dependent on soil properties (soil water retention, soil type, clay content, and nutrient reserves), soil profiles, landscape orientation, terrain features, rainfall (more in wet regions), and temperature (more in cold climates) (Lal, 2007). In natural vegetation, the SOC pool ranges between 40 and 400 Mg C/hectares (Post and Kwon, 2000). The conversion of undisturbed natural ecosystems to agricultural farmland rapidly depletes existing OC pools. The degree of depletion may account for 50%–75% over 5–20 years after clearing vegetation in the soils of tropical regions, in comparison to 25%–50% over 20–50 years in temperate climate soils (Lal, 2004; Lal et al., 2019).

Croplands contain about 25%–75% less C pools than other undisturbed ecosystems. If managed well, agricultural soils have massive potential in offsetting emissions and C_{seq}; therefore, soil re-carbonization is a critical strategy toward CCM and adaptation (Lal et al., 2015). C pool depletion in agricultural ecosystems is usually higher for those practices that do not reuse agricultural wastes (straw and crop residues) and other biosystems that usually utilize the use of organic amendments and mulch farming. In addition, no-tillage agricultural practices in seedbed preparation generally result in lower SOC depletion when compared to plow-based agricultural ecosystems (Lal, 2007). Agricultural ecosystems, where all C inputs are balanced through the application of recommended inputs, usually have lower SOC depletion rates when compared to subsistence farming, which has fertility mining and other C-extracting characteristics in which the C inputs are less than the outputs (Lal, 2007), as shown in Table 15.1.

15.2 Importance of the soil carbon pool in agroecosystems

Agriculturalists in the tropics are not mainly disturbed by the depletion of soil C pools but mainly by decreased yields due to degradation of OC pools. SOC is not a plant nutrient and is not critical to plant growth per se, and its benefits can

TABLE 15.1 Potential impacts of reduced soil organic carbon in agricultural production.

Potential impacts of reduced soil organic carbon in agricultural production	References
Reduced SOC pools in agricultural farmlands have been found mostly in unfertilized soils in comparison to soils that are fertilized, and there is a buildup in the SOC pool in soils that were applied with recommended rates of fertilizers compared with manure applications.	Ghosh et al. (2003), Lal et al. (2019)
Legume and millet rotation with 30 kg N/ ha in Niger, sustained crop yields and maintained high levels in the SOC pool.	Lal (2006)
In Kenya, cultivated soils have lost their SOC pool by 50–75 Mg C ha^{-1}	Briggs (1995)
In Senegal, SOC levels in the top 0–10 cm layer decreased from 28 g/kg to 10 g/kg after more than 850 years of subsistence farming, which had negative impacts on soil quality and the potential of soils to resist droughts.	Emerson (1995), Siband (1974)
In Kenya, a study conducted on red clay soils proved that yields of beans and maize ranged from 1.4 Mg/ha/y when crop residues were removed and without fertilizer inputs to up to 6.0 Mg/ha/y when crop residues were retained and fertilizers and manure applied.	Lal (2006)
The C depletion rates will usually be high in soils when C inputs into agricultural ecosystems (such as the return of biomass and crop wastes and the use of farm inputs) are less than C outputs (mainly leaching, mineralization, or erosion).	Lal (2007), Lal et al. (2019)
Excessive depletion of OC pools can result in poor soil health, nutrient and elemental imbalance, poor water balance due to high runoff and evaporation, and reduced soil biodiversity, mainly impacting soil microfauna such as earthworms.	Lal (2004, 2006)
Reduced soil quality will decrease net agricultural productivity, which will, in turn, cause reduced availability and usefulness of the crop residues returned to the soils, further worsening the depletion of C and SOC pools.	Lal (2006, 2007)

be mainly substituted by the application of external inputs such as fertilizer, though at certain levels (Lal, 2006; Pradhan and Meena, 2023a). However, there is a direct connection between improved crop outputs and enhanced OC in soil, mainly through increased residues returned to soil and net biomass (Reilly and Fuglie, 1998). It has been publicized that C pools in the tropical soils can be improved through the combination of manuring, use of agroforestry systems (AFs), nontillage farming, N fertilization, residue retention, and incorporation of legumes and grass in the rotation cycles (Pal, 2003; Devi et al., 2024; Jhariya et al., 2023). The importance of soil C pools is shown in Table 15.2.

15.3 Assessment of carbon credits in the soil ecosystem

The assessment of C credits in soil ecosystems plays a significant role in global strategies for implementing CCM. Soil is a significant C sink, storing atmospheric CO_2 as SOC, which can be quantified and traded as C credits (Ellis and Paustian, 2024; Meena et al., 2023; Pradhan and Meena, 2023b). This potential has gained increased attention recently due to C_{seq} being recognized in international climate frameworks, particularly under the Paris Agreement (IPCC, 2019). New research from 2022 to 2023 implies the increasing importance of intricate soil C projects in addressing both climate change and food security challenges (Friedlingstein et al., 2022).

Soils have the main biogeochemical active C pool on Earth. Estimates of SOC range from 500 to 3000 Gt of C, as demonstrated by Paul et al. (2023), a larger amount than that stored in the atmosphere and terrestrial biomass combined. Studies have shown that most emissions occur in topsoil layers due to losses from decomposition. As such, emphasis should be placed on replenishing C directly back to the soils using suitable farming practices to enhance the circular economy. Sustainable agricultural practices and systems, such as agroforestry (AF), no-tillage farming, and organic soil amendments, can significantly enhance soil C storage (Jhariya et al., 2022). Farmers find that higher SOC amounts have plenty of benefits. One major benefit of SOC is improved soil fertility and increased resilience against climate-related losses in yield and yield components. Good agricultural management changes, although with increased associated costs, increase SOC levels (Paul et al., 2023; Pradhan et al., 2023).

TABLE 15.2 Importance of soil carbon pools in agroecosystems.

Importance of soil carbon pools in agroecosystems	Magnitude of the importance to agricultural ecosystems	References
Increases crop yields	• Enhancing the OC pools of eroded soils would improve crop yields by increasing available water, enhancing nutrient supply, and improving soil physical properties such as structure. • A rise in SOC pool concentration can cause equal amounts of crop yields with decreased levels of fertilizer, mainly due to improved fertilizer-use efficiency and reduced leaching of nitrates. • Experiments in Canada have shown that a reduction in OC pools in soils by 1 Mg/ha in the top 0–8 cm soil layer reduced yields of wheat between 40 and 20 kg/ha in different locations. • Wheat yield reduced by 26.5 kg/ha for every 1 Mg/ha reduction in OC pools at the top 0–50 cm of the soil layer in North Dakota, USA. • In Russia, improvement of OC contents by 1% led to enhanced yields of cereals by 1 Mg/ha, equal to 42 kg/ha/y of soil humus for soil bulk density of 1.2 Mg m^{-3} at a depth of 20 cm. • Yields of wheat and other agricultural crops in the UK improved with an enhancement in OC contents from 0.7% to 0.9%. • In Pampas, Argentina, the yields of wheat were related to SOC concentration less than 17.5 g/kg and the reduction of soil organic matter (SOM) by 1 Mg decreased wheat yields by up to 40 kg/ha. The authors recommended practices such as nontillage soil management and the combination of crop residues in the rotation cycles. • In India, the reuse of crop wastes and manure improved moisture contents, increased SOC levels, and enhanced yields of pearl millet by 0.1–0.2 Mg/ha. • 25–40 million Mg/y can enhance food production in the growing economies due to increased soil qualities by reducing degradation processes and enhancing the SOC pool. • Retention of straws, mulching, biosolids, and the utilization of manure are critical in improving and maintaining the OC pools in the tropical soils.	Lal (2004, 2006, 2007), Lal et al. (2015, 2019), Aggarwal et al. (1997), Lal (2006), Larney et al. (2000), Lal (2006), Ganzhara (1998), Díaz-Zorita and Grosso (2000), Aggarwal et al. (1997), Aggarwal et al. (1997)
Enhances plant available water	• A positive relationship exists between plant-available moisture capacity, OC pools, and the potential of plants to withstand harsh conditions such as drought. • The available soil moisture content improves by 1–10 g for each 1 g enhancement in organic matter (OM) content in the soil. The improvement seems small, but it can help sustain plant growth for 5–10 days without rainfall. • Some studies have pointed to a slight or no impact of OC pools in the soils on accessible plant water capacity.	Emerson (1995), Lal et al. (2019), Lal (2004, 2007)
Improves nutrient supplies and soil biota	• The preservation of SOC pools would enhance the biological activities of soil microorganisms and increase nutrient supplies. • Soil biota are critical to reversing soil degradation, and their roles in ecosystem restoration cannot be overlooked. • The SOC pool drives biological activities and is the main foundation of nutrients and energy for living organisms in the soil. • Soils with depleted SOC pools usually have a slow response to fertilizers and other inputs as compared to soils with enhanced qualities.	Lavelle et al. (1997), Lal (2006, 2007), Powlson et al. (2001)
Increases cation exchange capacities	• Increasing OC pools in the soil enhances cation exchange capacity. • Reduction in 1 g/kg of OC pool decreased effective cation exchange capacity by about 4.3 mmol/kg, mainly in soils with low activity clays (like in the Sahel region), decreasing the potential of such soils to retain nutrients. • A significant positive relation exists between the SOC and exchangeable bases and removable potassium in the ferrosols of Tasmania, Australia.	Lavelle et al. (1997), Bationo and Mokwunye (1991), Cotching et al. (2002)
Improves soil physical properties	• Enhanced OC pools in the soil improve soil aggregation, texture, and structure, causing soils to be less susceptible to compaction, crusting, and soil erosion.	Díaz-Zorita and Grosso (2000), Lal et al. (2019)

(Continued)

TABLE 15.2 (Continued)

Importance of soil carbon pools in agroecosystems	Magnitude of the importance to agricultural ecosystems	References
Improves the C_{seq} potential of soils	• Improved SOC pools at 1 Mg C/ha/y over 500 million ha can cause sequestration potential of 0.5 Pg C/y, which is essential in offsetting the emissions from fossil fuels. • C seizure by the C pools in agricultural soils is critical in the stabilization of GHGs in the atmosphere. • C_{seq} potential of agricultural soils is between 0.4 and 1.2 Pg C. This can be enhanced through proper identification and utilization of soil and crop-specific management practices.	Lal (2004, 2007), Lal et al. (2015, 2019)

The SOC stock often accumulates over varying time periods (Blakemore, 2024) and is categorized as follows:

- Mineral soils consisting of <17% SOC, which stores humic C due to N deficiency.
- Biomass, which is mostly recycled through decomposition processes within short time periods.
- Fossil fuels formed from geological anoxic accumulation of SOM.
- Seasonally frozen soil, which is mainly referred to as permafrost (contains permafrost peatlands).
- Oxygen-deficient nonpermafrost peat originating from waterlogging.
- Dissolved organic carbon or sediments washed from the watersheds.

SOC forms a major component of ecosystems. It is a dependable indicator of soil quality, which affects a wide range of soil physico-chemical properties (Ellis and Paustian, 2024). The SOC in the soil originates from different sources, such as biomass inputs that contain OM from plant roots, residues, and microbial biomass, and are major contributors to soil C pools. Soil aggregation also plays a key role in C stabilization through preventing OM from microbial decomposition (Friedlingstein et al., 2022; Six et al., 1999). Further, microbial activities drive decomposition of OM and influence SOC stabilization (Blakemore, 2024; Ellis and Paustian, 2024).

C credits represent the sequestration of one metric ton of CO_2 equivalent (CO_2e) and are tradeable in compliance and voluntary C markets (Lorenz et al., 2019). Soil C credits are particularly important in the agriculture, forestry, and land-use (AFOLU) sector, which is gaining momentum as a key area for CCM (Badgery et al., 2021). Recent updates in C market standards are encouraging the expansion of soil C projects, especially in developing regions. C offset certifiers are systems that report and track C credit data for projects, including credits produced, nature of ownership, sales, sales generated, and retirement (Liu, 2022). The plan or "protocol" reviewed often by the certifiers ensures that project developers are executing against key criteria of being real, completion, consistency, transparency, accuracy in the conservation, leakage, additionality, permanence, and ownership. Despite this, country parties to the Paris Agreement are yet to finalize rules for market-based cooperation under Article 6.

15.3.1 Methodologies for assessing soil carbon credits

15.3.1.1 Direct measurement

Direct soil sampling and laboratory analysis provide the most accurate method for quantification of soil C stocks. This method involves analyzing SOC using techniques like loss on ignition using dry combustion (Nelson and Sommers, 2018). While costly, direct measurement ensures precise quantification of changes in C stocks.

15.3.1.2 Remote sensing and modeling

Recent advancements in remote sensing, coupled with machine learning and artificial intelligence, have revolutionized the assessment of soil C (Dai et al., 2025). Complimentary cost-effective soil spectroscopy holds substantial potential for tackling this challenge (Friedlingstein et al., 2022). Whereas previous research has focused on predicting SOC using mid-infrared spectroscopy technology, other technologies, such as visible and near-infrared spectroscopy, remain unexplored, especially for tropical soils. The use of AI in C modeling has shown promise in reducing the cost and complexity of monitoring projects (Friedlingstein et al., 2022).

15.3.1.3 Carbon accounting standards

C accounting standards, such as the Gold Standard and the Verified Carbon Standard (VCS), offer frameworks for measuring and verifying soil C projects (Plan Vivo Foundation, 2023). Recent updates have introduced new protocols for smallholder farmers, which are designed to ensure fair access to C markets and equitable benefit sharing (Liu, 2022). C accounting standards provide frameworks for measurement, reporting, and verifying (MRV) soil C_{seq} projects. Key standards include:

- VCS: A leading voluntary standard for agricultural and land-use C projects.
- The Gold Standard: Known for its strong environmental and social safeguards, the gold standard supports soil C projects with a focus on sustainable development outcomes.
- Climate Action Reserve (CAR): Provides methodologies for North American agricultural and forestry projects, including soil C_{seq} protocols (CAR, 2020). Each standard includes methodologies for establishing baselines, assessing additionality, and ensuring permanence and leakage avoidance to guarantee the integrity of C credits.

15.3.2 Challenges in soil carbon credit assessment

15.3.2.1 Permanence and leakage

Ensuring the permanence of sequestered C is a key challenge for soil C projects. C stored in soils can be rapidly lost due to changes in land management or natural events (Kumar et al., 2024). Leakage, where C_{seq} in one area leads to emissions elsewhere, also poses significant challenges. Several nature-based mitigation pathways rely on changing the status of ecological and economic systems. The impact of these interventions has the potential of being traded as C credits in the existing C accounting framework, which become offsets when used to substitute for other mitigation actions (Liu, 2022). Leakage occurs when the intended effects of an intervention fall outside the accounting boundary used to track mitigation. Inaccuracies in C accounting make these beyond-boundary effects difficult to quantify (Dai et al., 2025; Filewod and McCarney, 2023).

15.3.2.2 Baseline establishment and additionality

Accurately establishing baselines and proving additionality are essential to the credibility of C credits. New frameworks are addressing these challenges by using advanced remote sensing and machine learning to create more dynamic and reliable baselines. This approach still falls short in accounting for factors such as surface soil moisture, roughness, and vegetation cover, which can result in less reliable estimations. However, ongoing advancements in the systems aim to address these limitations (Misch and Wingender, 2024). Ensuring additionality means demonstrating that the C benefits of a project would not have occurred without its intervention (Filewod and McCarney, 2023; Misch and Wingender, 2024).

15.3.2.3 Monitoring, verification, and reporting

Accurately measuring and verifying C credits from forestry and agricultural activities is often challenging and expensive. While analyzing soil samples to measure SOC remains the most precise method for estimating soil C storage, it is too costly and time-intensive for widespread application (Wongpiyabovorn et al., 2023). Technological advancements in remote sensing and AI are improving the accuracy and affordability of monitoring systems for soil C_{seq}. Recent studies emphasize the importance of continuous monitoring to ensure project integrity (Batjes et al., 2024; Misch and Wingender, 2024).

15.3.3 Opportunities and future directions

Advances in AI, machine learning, and big data are expanding opportunities for scaling up soil C assessments (Batjes et al., 2024). Schemes for payment for ecosystem services (PES) are helping to incentivize sustainable land management practices that enhance C_{seq}, especially in regions with high potential for C storage. Emerging research highlights the cobenefits of soil C_{seq} projects, such as improved soil health, biodiversity conservation, and enhanced food security. There has been growing interest in integrating soil C_{seq} into nationally determined contributions (NDCs) under the Paris Agreement (Wiese et al., 2021).

15.4 Assessment of carbon credits in agroecosystems

Anthropogenic activities worldwide have contributed to GHG emissions, making the earth warmer by 1.1°C since the preindustrial period (Cariappa and Krishna, 2025). In total, 76% of GHG emissions are a result of electricity, heat, transport, and industrial activities, while 24% comes from agriculture (Cariappa and Krishna, 2025; Sapkota et al., 2021). Climate change impacts on agroecosystems are a complex challenge; thus, there is a need to curb GHG emissions through conservation agriculture (CA) technologies. Estimation of CA technology adopters and nonadopters fails to account for statistics of the level of adoption, thus leading to bias (Cariappa et al., 2024). Increasing crop and animal production through agricultural interventions at the farm production stage can reduce GHG emissions (Mukherji et al., 2023). This could lead to win–win actions across the agroecosystem when considering C-credits trading.

C-credit is a solution to CCM through incentives for reducing GHG in the atmosphere (Kumar, 2023). C credits in agroecosystems are earned by implementing sustainable practices that sequester C. For example, in India, when a farmer achieves a reduction of 1.5 tons of CO_2 per hectare, it is equivalent to 1 ton of C credit worth US $10 at market value (Kumar, 2023). The quality of C credits is a critical factor that can impact their value. These are additional benefits beyond C_{seq}, such as sustainable agriculture (Mukherji et al., 2023). In Italy, agroecosystem projects convert conventional farmlands to agroecological practices. These practices reduce GHG emissions and also promote soil health and biodiversity and provide a new form of income to small holders. Such agroecology-based projects invest in C credits by supporting local communities to preserve important cultural landscapes and promoting sustainable farming practices, thus reducing GHG emissions (Rudd et al., 2022; Crippa et al., 2021). The projects ensure that the C credits issued accurately represent the CO_2 sequestered through the use of remote sensing and soil sampling techniques. Remote sensing techniques help in detecting land use (LU) changes and vegetation cover, estimating CO_2 sequestered with a high level of accuracy (Rudd et al., 2022).

Several factors are considered in certification standards of C credits, such as the location and the type of project (CMW, 2020). With climate change concerns and GHG emission reductions, shelterbelts made of planted trees and shrub species in agricultural landscapes are included in the assessment of agroecosystem C credit. These are rows established within crop fields, surrounding farmyards, infrastructures for livestock, and riparian buffer for protection against wind and soil erosion (Rudd et al., 2022). A study in Saskatchewan (SK), Canada, provided details on a 1 km-long farm shelterbelt, where production and transportation stages for 1000 shrub shelter belt seedlings were estimated to generate 2200 kg of CO_2 emissions regardless of species. All phases of the shelterbelt life cycle were assessed; these included primary sources of GHG emission, such as use for heating and lighting during seedling growth, and from transportation of the seedlings (Rudd et al., 2022). This research used two tools for inventory analysis and decision-making, where assessment was based on total emissions by the various inputs and uses (Rudd et al., 2022).

Other programs assessing C credits within an agroecosystem and paying farmers to capture CO_2 through cover crops include Bayer, Land O'Lakes, and Cargill in the climate agenda for America. Cover crops are increasing in reputation for climate change C credit accounting (Cariappa et al., 2024). In 2021 an American program named Land O'Lakes paid $4 million to farmers to grow cover crops, capturing 200,000 metric tons of C from the atmosphere (Crippa et al., 2021). In India, other practices recommended for reducing GHG emissions and C_{seq} in soil include zero tillage and continuous practice of CA. In India, CA implementation for 20 years in a research and development project reduced GHG gases by >50%, amounting to 85.5 million tons of CO_2 emissions yearly (Cariappa et al., 2024).

Agriculture C credits in Sub-Saharan Africa offer an innovative approach to overcoming persistent financing difficulties. By transforming environmental benefits into tradable commodities, this has contributed to low GHG emissions while improving biodiversity conservation (Mukherji et al., 2023). In Sub-Saharan Africa, one C credit is equivalent to one metric ton of CO_2 equivalent (1 t CO_2 e). Thus C-credit projects in this region support ways of reducing GHG emissions while allowing other entities to trade their rights to emit CO_2. In consultation with farmers and relevant stakeholders to understand needs, priorities, and challenges of a community, project developers collect baseline data on current land use and environmental conditions (Crippa et al., 2021). Potential environmental impacts on biodiversity, soil health, and water quality are assessed; a cost-benefit analysis of the project is conducted; and financial feasibility is evaluated. For long-term sustainability, the projects cooperate with adaptive management strategies and ensure proposed C_{seq} practices are integrated with existing agricultural activities. In Kenya, the top five prioritized technologies in agriculture for GHG emission reduction are biogas production, solar home systems, solar dryers, afforestation and reforestation, and the use of wind energy (Mukherji et al., 2023; Filewod and McCarney, 2023).

15.5 Potential benefits, utility, and application of the carbon credit schemes in agriculture

In agriculture, several types of C market projects offer farmers offsets on reduced GHG emissions through C credits. These projects promote low-emission agricultural practices such as CA as an innovative approach to reduce GHG emissions in tradable commodities. The GHGs causing global warming and climate change by trapping heat in the atmosphere are CO_2, methane (CH_4), and nitrous oxide (N_2O). Environmental benefits through GHG emissions reduction include improved biodiversity conservation, enhanced soil health, and enhanced nutrient cycling (Miharza et al., 2023).

Improved livelihood is one of the benefits of C credit facilitation from certified projects that reduce GHG emissions. There are opportunities for farmers to earn C credit income through C removal activities. Therefore with C credit support, farmers can improve their livelihoods by accessing resources such as irrigation and shade trees to protect their farms from the effects of climate change (Miharza et al., 2023). AF is a common practice in agriculture as a C removal activity. Advantages of AF practice include income generation, improved main crop yields, protection of from risks pests and diseases, extreme weather, and soil fertility, and finally prevention of biodiversity loss (Oviantari et al., 2021). In addition, there is potential to increase C stock in the field and reduce GHG emissions.

Integrating trees and shrubs with crops strategically in AF practice has multiple benefits, such as enhanced soil fertility, improved soil moisture, and increased biodiversity (Schilling et al., 2023). In the forest and agricultural sectors, C credit projects must develop a written memorandum of agreement with farmers to hold C stocks for at least 40 years. Benefits attained from the C credit project can be in kind or in cash. During Free Prior Informed Consent, payment under the C credit mechanism is usually agreed upon between the project parties and communities. Cash benefits include household payments, community funds, budget transfers, seed capital for business groups, de-risking instruments for micro-credit, loans, and savings to reduce loan costs (Schilling et al., 2023). In-kind benefits include employment, training, equipment, and community infrastructure development project implementation with beneficial income (Commercial Agriculture for Smallholders and Agribusiness, 2022).

In Africa, C farming projects contribute to CCM through strengthening food security, reducing poverty, and the vulnerability of smallholder farmers (Schilling et al., 2023). In Africa, Verra and Gold Standard C projects invest with smallholder farmers in mitigating emissions while delivering adaptation benefits through carefully designed nature-based solutions, which account for 94% of farms globally and 18% of the world's agricultural land (Commercial Agriculture for Smallholders and Agribusiness, 2022). These nature-based solutions include AF and improved land, production, and waste management practices that have the potential to deliver adaptation goals. In Ghana, C_{seq} can potentially benefit by planting 1 hectare of cocoa with appropriate and suitable shade tree species and cover crops with the potential to enhance soil fertility; this could remove an average of 2 t CO_2e/ha/annum during the first 20 years of tree growth (Gunamantha et al., 2021). Therefore if this were scaled up to a larger area, the overall annual C_{seq} potential would be high for each of the major nature-based conservation technologies. Key benefits to the farmer would include improved climate resilience and soil health and increased fodder and fuel availability (Commercial Agriculture for Smallholders and Agribusiness, 2023).

In Kenya, the BioCarbon Fund of the World Bank did a prefeasibility study on the Kenya smallholder coffee C credit project to identify cropping systems with high economic potential for sustainable agricultural land management. This C project at Komothai, Kiambu County aims to reduce climate change vulnerability through restoring sustainable coffee farming practices. The C project has a cooperative of 9000 smallholder farmers, of whom 50% are coffee farmers and 50% are subsistence agriculture farmers, aiming to restore 7200 hectares in the first phase and 10,000 hectares in the second phase. During the first phase of the project, approximately 3.5 t CO_2/ha/Yr will be mitigated across the total project area through the adoption of Sustainable Agriculture Land Management (SALM) (Seeberg-Elverfeldt, 2010). This C credit project at Komothai will lead to increased potential income, coffee yields, and resilience to climate change in agricultural production systems.

15.6 Methodological advances in carbon credit assessment of soils

Soil C finance schemes aim to provide incentives for farmers and landowners to transition from business-as-usual to alternative management strategies. Reductions or removal of soil-based CO_2e units are purchased or financed to offset C emissions (British Society of Soil Science, 2023). Schemes often combine measurements of SOC stocks with the prediction of derived GHG emission reductions (FAO, 2019). Soil C assessments can be conducted through modeling, especially when estimation or extrapolation of C stock changes is the main goal. The complexity of the models used depends on the context of the desired results, factoring in soil type, climate, and management. Models used for C estimation are categorized into the following:

- Level 1 models: These models are used to predict changes in C stock. It is often used when access to data-based factors is sufficient. However, this model has limited accuracy resulting from its simplicity (British Society of Soil Science, 2023).

TABLE 15.3 Procedures for trading soil carbon credits.

Procedure	Description	References
Monitoring	Required sampling and measurement of soil C stocks every 3–5 years to establish the changes from the baseline due to management in the soil C project.	Jat et al. (2022), FAO (2019)
Reporting	Production of a monitoring report with the necessary information about the soil C project over a specified monitoring period. The report includes C changes and is submitted to the organization issuing C credits.	
Verification	This is a process whereby skilled and qualified personnel examine the monitoring report. Verification can also be done independently by the C scheme.	

- Level 2 models: These models are mainly utilized in instances where regional data on SOC emission factors are unavailable. These factors include humification coefficients. In this model, data on plant biomass inputs and environmental factors influencing C losses are required (British Society of Soil Science, 2023).
- Level 3 models: These models are used when the main aim is to combine feedback from numerous processes on soil–plant–atmospheric relationships and dynamics of SOC. These models are used to assess various effects on agricultural management, crops, and soils. The models also estimate the impacts of climate change feedbacks on crop productivity and SOC dynamics (British Society of Soil Science, 2023). They are also used to estimate the trade-offs between environmental indicators and SOC change.

The main procedures for trading soil C credits are as outlined in Table 15.3.

15.7 Soil carbon management for enhancing soil carbon credit in agriculture

Soils play a pivotal role in CCM by regulating atmospheric GHGs. Global soils are dominant nature-based C sinks holding more C than other terrestrial habitats. Changing management practices in agricultural systems has the potential to increase SOC stocks and reduce GHG emissions (British Society of Soil Science, 2023). Soil C_{seq} is estimated at 0.8–1.5 Gt C/year from the atmosphere. Management practices aimed at reducing CO_2 in the atmosphere by increasing SOC storage require verification, including periodic soil sampling to estimate changes in SOC stocks over a specified period (FAO, 2019). The SOC pool in the first 1 m depth contains approximately twice the atmospheric C. Therefore the quantity of C should reflect a balance between inputs and release associated with microbial activities (Harbo et al., 2022). Table 15.4 presents the SOC management practices adopted by most soil C projects globally.

15.7.1 Calculating soil carbon change

Soil C stock changes are calculated by estimating changes in soil C levels achieved after project registration compared with baseline C stocks, as shown in Fig. 15.1 (Australian Government, 2021). C credits are obtained only for the C_{seq} as a result of the C project.

15.8 R&D toward soil carbon credit assessment

There are limited approaches for calculating C storage and its CO_2 sequestration potential (Oldfield et al., 2022). A lack of data on the spatial and temporal patterns of SOC under different management is also a major impediment to C credit assessments. Soils in managed ecosystems have lower SOC than in natural environments. Factors affecting C_{seq} in soils that should be factored when assessing soil C include: abiotic factors such as clay content, structural stability, soil moisture, landscape position, mineralogy, and temperature regimes, and biotic factors such as climate and soil microbial activities. Research geared toward soil C assessment is still limited, and data required for location-specific C credit quantification are not consistent. Methodologies for C credit assessments ought to be standardized to ensure accuracy across different regions and circumstances. Fig. 15.2 shows techniques required for monitoring, reporting, and verification of C stocks in soil for C credit assessment.

TABLE 15.4 SOC management practices for carbon enhancement.

SOC management practices	Adoption constraints	Trade-offs	References
Sustainable grazing	Inadequate pastureland for grazing, poverty, and scarcity of fodder	Disproportionate grazing, inadequate regulations, and community perceptions	Jat et al. (2022)
Data-driven land use plans coupled with sustainable soil management	Lack of implementation of relevant policies	Competing land uses with urbanization offering a viable alternative	Jat et al. (2022), FAO (2019)
Minimum tillage	Inadequate knowledge and skills; lack of farming equipment, community perceptions, misconceptions, and inadequate locally adapted packages	Community perceptions on conventional tillage and inadequate incentives for eco-friendly services	Joensuu et al. (2021)
Recycling and retention of residues on farms	Other economic uses agro-based wastes; inadequate knowledge and capacity	Residue removed or burned	Joensuu et al. (2021)
Biochar utilization	Availability of feedstock; cost of biochar production; efficiency of biochar for crop production	Inadequate knowledge and capacity for biochar adoption	Jat et al. (2022)
Balanced use of nutrients	Inadequate capacity, inaccessibility of soil amendments, and high cost of inputs	Inefficient use of nutrients	Barbato and Strong (2023)
Nitrogen application based on crop needs	Inadequate capacity, easily accessible fertilizer from subsidies by governments	Poor nutrient use efficiency	Barbato and Strong (2023)
Controlled irrigation	Inadequate irrigation infrastructure	Ineffective water management; increased GHG emissions from waterlogged soils	Jat et al. (2022)
Improved fallows	Inadequate incentives for fallow management; small landholding by farmers	Bare fallow	Barbato and Strong (2023)
Deep-rooted crop varieties	Inadequate research and breeding for suitable crops	Use of shallow-rooted crops available	Jat et al. (2022)
Crop rotation	Inadequate capacity for sustainable crop rotation, lack of incentives	Suboptimal crop rotation	Jat et al. (2022)

SOC, soil organic carbon.

The number of samples and sampling strategy will be dependent on:

- Variability in the area
- Level of carbon stocks to be detected
- Current land use

IPCC guidelines should be followed with soil sampling done to depths of 0–30 cm, 30–60 cm, and beyond. Analytical methods for quantification of SOC (%) and bulk density (g cm^{-3}) should be accurate

15.8.1 Quantification of carbon credits in soil carbon projects

Soil C projects aim to store additional C in agricultural soils, thereby reducing CO_2 levels in the atmosphere (Harbo et al., 2022). Approaches of measuring soil C for C projects include:

- Measurement-only approach, where the SOC of the sampled area is averaged (Australian Government, 2021)
- A hybrid approach that estimates results using a combination of soil C model estimates and soil core measurements (Australian Government, 2021).

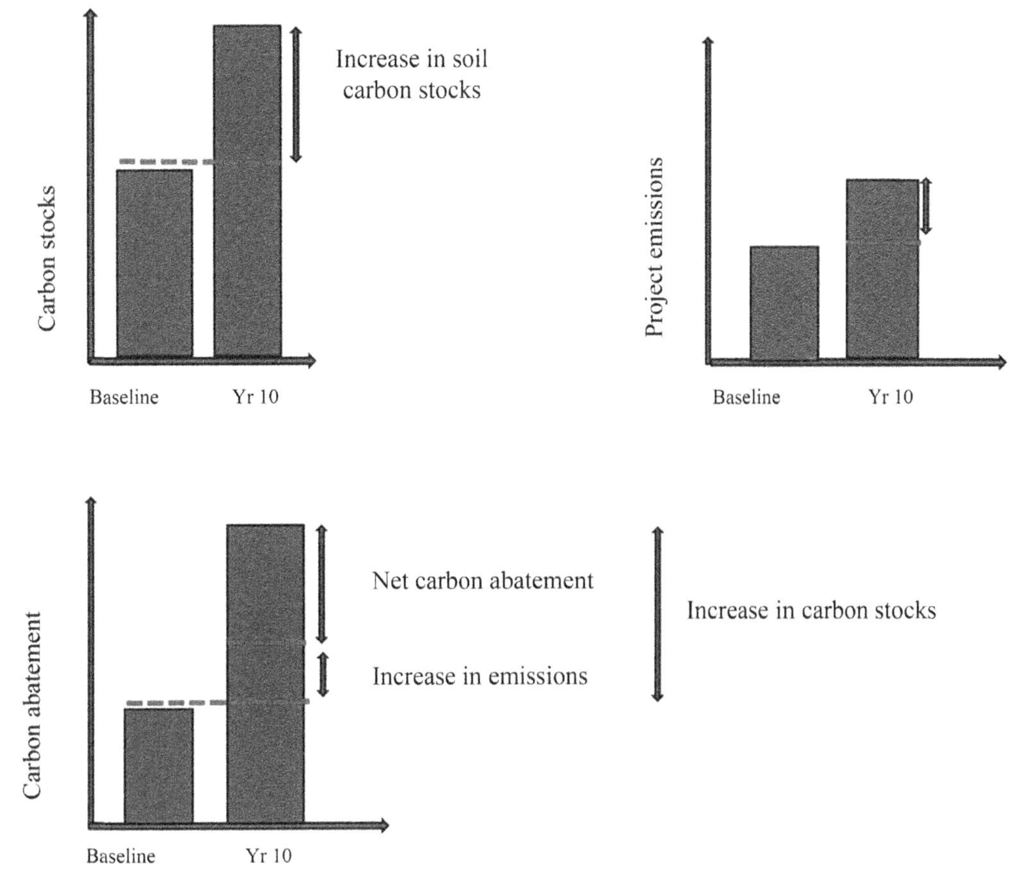

FIGURE 15.1 Process of estimating soil carbon change (Australian Government, 2021).

To quantify C credits resulting from C projects, there is a need to have a baseline, which will be used to evaluate changes resulting from the management activities of the soil C projects (Oldfield et al., 2022). Parameters to consider when quantifying C credits for C markets are highlighted in Fig. 15.3.

15.8.2 Process of participating in carbon projects: the case of Australia

Numerous approaches exist for registering C credit projects globally. These approaches are region- and country-specific. However, all soil C projects aim to sequester CO_2 for CCM. The Australian government, for example, has developed comprehensive guidelines for soil C projects. The 2021 guidelines are an improvement of the 2018 guidelines, which outlined procedures for registering C projects for C credits (Australian Government, 2021). For efficient and sustainable implementation of soil C projects, the steps highlighted in Fig. 15.4 should be followed.

15.8.3 Project lifecycle for soil carbon credits

Soil C projects undergo numerous steps during their life cycle. Table 15.5 highlights the main steps from planning to reporting and receiving C credits of the specified project. The whole process of coming up with a C project is data-driven and requires an accurate estimation of the C_{seq} potential of the proposed project (FAO, 2019). The minimum requirements for SOC measurement are outlined below:

- SOC should be done within a specific soil depth
- SOC of fine earth (2 mm size)
- Coarse mineral fraction content (>2 mm)
- Soil bulk density

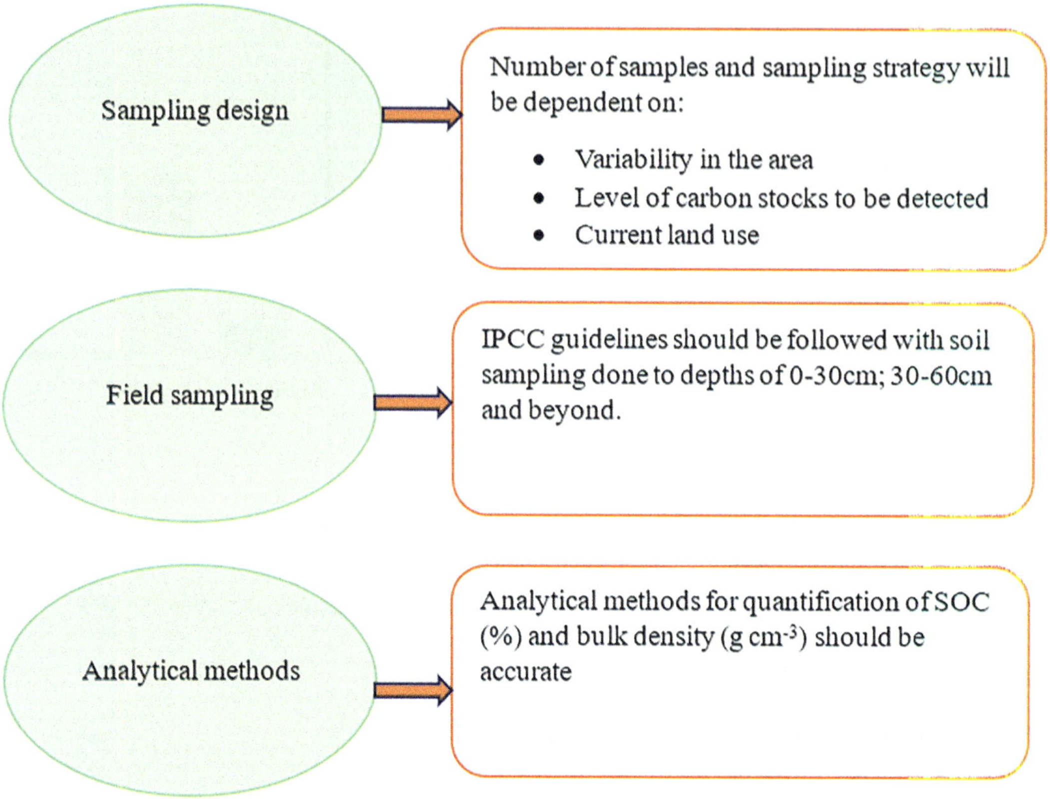

FIGURE 15.2 Techniques required for monitoring, reporting, and verification of carbon stocks in soil.
Source: Oldfield, E.E., Eagle, A.J., Rubin, R.L., Rudek, J., Sanderman, J., Gordon, D.R., 2022. Crediting agricultural soil carbon sequestration. Science. 375 (6586), 1222–1225. https://doi.org/10.1126/science.abl7991.

- Soil sampling depth of at least 30 cm
- All samples should be georeferenced

15.9 Policy perspectives on soil carbon credit systems in agroecosystems

Soil C credit systems have gained momentum as part of broader efforts to CCM through SLM. Agroecosystems, with their vast potential for C_{seq}, offer significant opportunities for integrating soil C credits into global C markets. This chapter explores the policy landscape shaping soil C credit systems in agroecosystems, focusing on recent developments in international frameworks, national policies, and market-based mechanisms.

With the growing urgency to achieve the targets of the Paris Agreement, governments, international organizations, and the private sector are increasingly aligning their strategies to leverage soil C_{seq}. The emphasis is on enhancing soil health, improving agricultural productivity, and addressing food security while reducing GHG emissions (IPCC, 2023).

15.9.1 International climate frameworks and soil carbon credits

Global climate frameworks, particularly the Paris Agreement, highlight the importance of C_{seq} in soils as a critical component of mitigation efforts. Countries are encouraged to include soil C_{seq} in their NDCs, promoting sustainable agricultural practices such as AF, cover cropping, and conservation tillage (IPCC, 2019).

In 2021 the COP26 Glasgow Pact emphasized the need for land-based solutions, including soil C projects, to accelerate climate action. To add to this effort, the United Nations Framework Convention on Climate Change has been developing guidelines for C accounting in the AFOLU sector, including methodologies for quantifying and verifying soil C_{seq}. Soil C credits are increasingly recognized as a mechanism that can support national climate goals while providing additional environmental and social cobenefits.

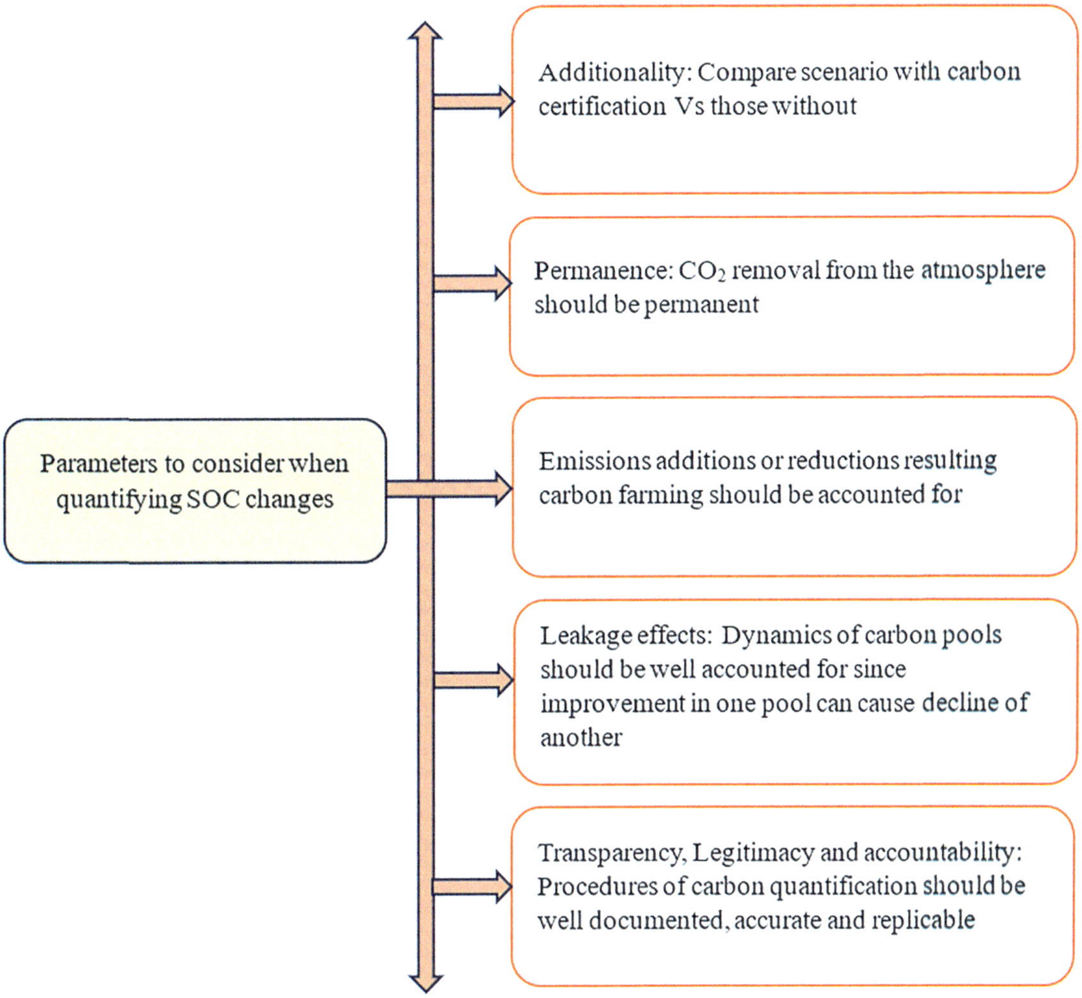

FIGURE 15.3 Parameters to consider when quantifying carbon credits for carbon markets (Oldfield et al., 2022).

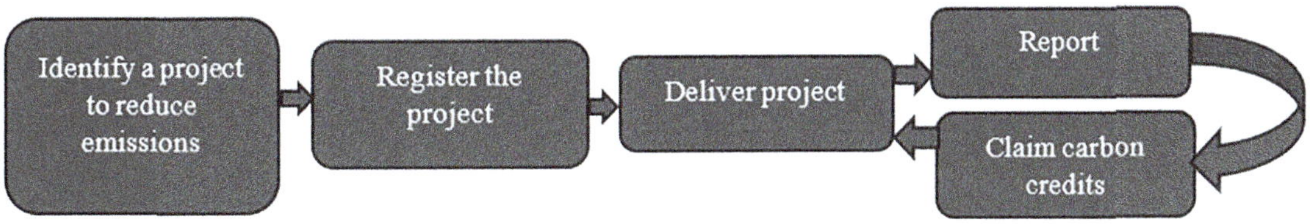

FIGURE 15.4 Process of participating in carbon projects (Australian Government, 2021).

15.9.2 National policies supporting soil carbon credit systems

Several countries have integrated soil C_{seq} into their national climate policies. In the United States, the Growing Climate Solutions Act of 2021 developed a framework to enable farmers and landowners to participate in C credit markets by promoting climate-smart agriculture practices that support soil C_{seq}. The act facilitates the certification of C credits and aims to provide clear pathways for agricultural stakeholders to access C markets.

Similarly, Australia has been a pioneer in soil C credit systems through its carbon farming initiative (CFI). The initiative provides financial incentives for farmers to adopt on-farm activities that sequester C in soils, such as rotational grazing and organic soil amendments. Australia's CFI is part of the broader Emissions Reduction Fund, which allows landholders to generate C credits that can be sold in domestic and international markets. In the European Union (EU), the common

TABLE 15.5 Project lifecycle for soil carbon credits (Australian Government, 2021).

Stage	Activity
Planning	• Check general eligibility • Management activities • Key concepts • Decide on project proponents • Map project area • Prepare land management strategy • Estimate project returns
Registration	• Prepare project registration requirements • Calculate abatement estimates • Apply for registration
Deliver project	• Conduct baseline sampling • Determine baseline project emissions • Carry out eligible land management activities • Comply with prohibited activities
Report and credit	• Calculate C abatement • Report on your project and earn credit • Conduct project audit • Continue with management activities

agricultural policy has integrated soil C_{seq} into its climate action strategies. The EU's Farm to Fork Strategy, launched in 2020, promotes sustainable food systems, with an emphasis on increasing soil C storage through agroecological practices (European Commission, 2021). Additionally, the European Green Deal seeks to advance C farming initiatives across member states, providing incentives for farmers to adopt practices that enhance soil health and mitigate GHG emissions (Wiese et al., 2021).

15.9.3 Market-based mechanisms and carbon trading

The rise of voluntary C markets has opened new avenues for soil C projects. In 2022 global C markets saw a surge in demand for high-quality C credits, including those generated from soil C_{seq} in agroecosystems (Badgery et al., 2021). Market platforms such as the VCS and the Gold Standard have developed protocols for soil C credits, offering pathways for landholders to monetize their C_{seq} efforts.

However, several challenges remain in scaling up soil C credits in voluntary markets. One key issue is ensuring the additionality of projects—demonstrating that the C_{seq} can only occur when C credits provide the incentives. Additionally, maintaining the permanence of sequestered C and preventing leakage are critical concerns that policymakers and standard setters are working to address. Emerging markets such as Plan Vivo and the CAR are focusing on smallholder farmers in developing countries, ensuring that C finance reaches the communities most affected by climate change. These systems aim to provide transparent, accessible, and equitable platforms for generating and trading C credits (Plan Vivo Foundation, 2023).

15.9.4 Incentives for farmers and landholders

Policymakers are increasingly recognizing the role of financial incentives in encouraging farmers to participate in C credit systems. PES schemes, which reward farmers for adopting practices that increase soil C_{seq}, are being implemented across various regions. For example, in Kenya, the Vi Agroforestry Project has successfully implemented soil C_{seq} practices in AFs, allowing smallholder farmers to generate C credits and access global markets (Nyong and Martin, 2019). In the United States, C credit buyers such as Indigo Agriculture and Nori are collaborating with farmers to promote regenerative agriculture practices that increase soil C_{seq}. These market-based approaches provide farmers with new revenue streams while encouraging them to adopt climate-friendly practices.

15.9.5 Challenges and barriers to scaling soil carbon credits

15.9.5.1 Measurement, reporting, and verification

A major challenge in scaling soil C credit systems is the complexity of accurately MRV C_{seq} in soils. Advances in remote sensing, AI, and machine learning have improved the accuracy and scalability of MRV systems, but significant data gaps and uncertainties remain. Policymakers are working to standardize MRV protocols, ensuring that soil C credits are credible and comparable across different markets and regions (Batjes et al., 2024).

15.9.5.2 Permanence and additionality

Enhancing the permanence of sequestered C and additionality in soil C projects remains a critical concern. C stored in soils can be re-released due to land-use changes or climate-related events, such as droughts or floods (Liu, 2022; Wiese et al., 2021). Policymakers are addressing these risks by introducing buffer pools and long-term monitoring requirements for soil C projects.

15.9.5.3 Equity and access for smallholder farmers

While the potential of soil C credits is significant, ensuring equitable access for smallholder farmers remains a key challenge. Many smallholders lack the resources or technical expertise to engage with C markets, raising concerns about the inclusivity of these systems. Policy interventions are needed to provide support in terms of capacity building, financing, and infrastructure development (Kiptot and Franzel, 2012).

15.9.6 Future directions for policy and practice

Policymakers must continue to refine soil C credit systems to ensure that they are transparent, inclusive, and effective in CCM. Scaling up soil C_{seq} in agroecosystems requires a coordinated effort that includes improving MRV systems, addressing equity concerns, and integrating soil C credits into broader agricultural and climate policies (Friedlingstein et al., 2022).

In 2024 there is growing momentum to incorporae soil C_{seq} into national climate action plans, with countries such as Brazil, India, and Indonesia leading efforts to integrate soil C into their NDCs (Wiese et al., 2021). The potential for C credits to drive both environmental and socio-economic benefits makes this an area of significant policy interest for moving forward.

15.10 Future thrust

There is growing momentum for incorporating soil C_{seq} into national climate action plans. Therefore methodological advancement is required to ensure SOC stocks are accurately estimated to ensure the validity of soil C projects. There is a need for standardization of soil C estimation methods globally; thus, more research should be undertaken to ensure this is achieved.

15.11 Conclusion

Soils are a major C sink globally. The C_{seq} potential of agricultural soils is approximately 1.2 billion tons of C annually. Based on the foregoing, countries and individuals have begun tapping into C credits from soil C projects. These projects have mainly focused on SLUM practices that enhance C storage. However, a major impediment lies in the availability of reliable and replicable methodologies for the estimation of changes in SOC stocks resulting from the proposed soil C projects. There is, therefore, a need for research geared toward the standardization of these methodologies while factoring in country-specific circumstances.

References

Abdullahi, A.C., Siwar, C., Shaharudin, M.I., Anizan, I., 2018. Carbon sequestration in soils: the opportunities and challenges. In: Agarwal, R.K. (Ed.), Carbon Capture, Utilization and Sequestration. InTech. Available at https://doi.org/10.5772/intechopen.79347.

Aggarwal, R.K., Praveen-Kumar, Power, J.F., 1997. Use of crop residue and manure to conserve water and enhance nutrient availability and pearl millet yields in an arid tropical region. Soil and Tillage Research 41 (1-2), 43–51. Available at https://doi.org/10.1016/s0167-1987(96)01082-3.

Asner, G.P., Elmore, A.J., Olander, L.P., Martin, R.E., Harris, T., 2004. Grazing systems, ecosystem responses, and global change. Annual Review of Environment and Resources 29, 261–299. Available at https://doi.org/10.1146/annurev.energy.29.062403.102142.

Australian Government, 2021. Understanding Your Soil Carbon Project - Simple method guide. Australian Government. Available at https://cer.gov.au/node/3906.

Badgery, W., Murphy, B., Cowie, A., Orgill, S., Rawson, A., Simmons, A., et al., 2021. Soil carbon market-based instrument pilot-the sequestration of soil organic carbon for the purpose of obtaining carbon credits. Soil Research 59 (1), 12–23. Available at https://doi.org/10.1071/SR19331.

Banerjee, A., Jhariya, M.K., Raj, A., Mechergui, T., 2024. Ecosystem Management: Climate Change and Sustainability. wiley, India, pp. 1–669.

Barbato, C.T., Strong, A.L., 2023. Farmer perspectives on carbon markets incentivizing agricultural soil carbon sequestration. npj Climate Action 2 (1). Available at https://doi.org/10.1038/s44168-023-00055-4.

Bationo, A., Mokwunye, A.U., 1991. Role of manures and crop residue in alleviating soil fertility constraints to crop production: with special reference to the Sahelian and Sudanian zones of West Africa. Fertilizer Research 29 (1), 117–125. Available at https://doi.org/10.1007/bf01048993.

BATJES, N.H., 1996. Total carbon and nitrogen in the soils of the world. European Journal of Soil Science 47 (2), 151–163. Available at https://doi.org/10.1111/j.1365-2389.1996.tb01386.x.

Batjes, N.H., Ceschia, E., Heuvelink, G.B.M., Demenois, J., le Maire, G., Cardinael, R., et al., 2024. Towards a modular, multi-ecosystem monitoring, reporting and verification (MRV) framework for soil organic carbon stock change assessment. Carbon Management 15 (1). Available at https://doi.org/10.1080/17583004.2024.2410812.

Blakemore, R.J., 2024. Biomass refined: 99% of organic carbon in soils. Available from: https://www.preprints.org/manuscript/202409.2383.

Briggs, J., 1995. More People, Less Erosion: Environmental Recovery in Kenya, by M. Tiffen, M. Mortimore, and F. Gichuki. Transactions of the Institute of British Geographers 20 (4), 520. Available at https://doi.org/10.2307/622984.

British Society of Soil Science, 2023. Guidance Note for Land Managers Soil Carbon: What are carbon stocks and how can they be measured? (land use for net zero, pp. 1–10). British Society of Soil Science. Available from: http://www.soils.org.uk/.

Cariappa, A.A.G., Krishna, V.V., 2025. Carbon farming in India: are the existing projects inclusive, additional, and permanent? Climate Policy 25 (5), 756–771. Available at https://doi.org/10.1080/14693062.2024.2416497.

CAR, 2020. Climate Action Reserve 2020 Annual Report. Los Angeles, California. http://www.climateactionreserve.org/.

Cariappa, A.A.G., Konath, N.C., Sapkota, T.B., Krishna, V.V., 2024. Evaluating the potential and eligibility of conservation agriculture practices for carbon credits. Scientific Reports 14 (1). Available at https://doi.org/10.1038/s41598-024-59262-6.

CMW, 2020. CMW-English-Carbon-Markets-101-The-Ultimate-Guide-To-Market-Based-Climate-Mechanisms-Final-2020-Web.Pdf—Google Search. Available from: https://www.google.com/search?q=CMW-English-Carbon-Markets-101-The-Ultimate-Guide-To-Market-Based-Climate-Mechanisms-Final-2020-Web.Pdf&rlz=1C1GCEA_enKE1132KE1132&oq=CMW-English-Carbon-Markets-101-The-Ultimate-Guide-To-Market-Based-Climate-Mechanisms-Final-2020-Web.Pdf&gs_lcrp=EgZjaHJvbWUyBggAEEUYOdIBCDM1OTBqMGo3qAIIsAIB&sourceid=chrome&ie=UTF-8.

Commercial Agriculture for Smallholders and Agribusiness. (2023). CASA research recommendations digest for donors. https://casaprogramme.com/wp-content/uploads/2023/07/CASA-research-recommendations-digest-for-donors.pdf.

Commercial Agriculture for Smallholders and Agribusiness. (2022). Carbon finance for smallholder farmers and agribusinesses-Analytical briefing on agroforestry solutions. https://casaprogramme.com/wp-content/uploads/2024/06/TechnoServe-December-2022-Carbon-Finance-for-Smallholder-Farmers-and-Agribusinesses-Analytical-briefing-on-agroforestry-solutions-CASA.pdf.

Cotching, W.E., Hawkins, K., Sparrow, L.A., McCorkell, B.E., Rowley, W., 2002. Crop yields and soil properties on eroded slopes of red ferrosols in north-west Tasmania. Australian Journal of Soil Research 40 (4), 625–642. Available at https://doi.org/10.1071/sr01062.

Crippa, M., Solazzo, E., Guizzardi, D., Monforti-Ferrario, F., Tubiello, F.N., Leip, A., 2021. Food systems are responsible for a third of global anthropogenic GHG emissions. Nature Food 2 (3), 198–209. Available at https://doi.org/10.1038/s43016-021-00225-9.

Dai, L., Wang, Z., Zhuo, Z., Ma, Y., Shi, Z., Chen, S., 2025. Prediction of soil organic carbon fractions in tropical cropland using a regional visible and near-infrared spectral library and machine learning. Soil and Tillage Research 245, 106297. Available at https://doi.org/10.1016/j.still.2024.106297.

Devi, A., Jhariya, M.K., Raj, A., Banerjee, A., Singh, K.P., Singh, B., 2024. Acacia nilotica: a promising species for soil sustainability. Land and Environmental Management through Forestry. Wiley, India, pp. 339–353.

Díaz-Zorita, M., Grosso, G.A., 2000. Effect of soil texture, organic carbon and water retention on the compactability of soils from the Argentinean pampas. Soil and Tillage Research 54 (1-2), 121–126. Available at https://doi.org/10.1016/s0167-1987(00)00089-1.

Ellis, E., Paustian, K., 2024. Importance of on-farm research for validating process-based models of climate-smart agriculture. Carbon Balance and Management 19 (1). Available at https://doi.org/10.1186/s13021-024-00260-6.

Emerson, W.W., 1995. Water-retention, organic-C and soil texture. Australian Journal of Soil Research 33 (2), 241–251. Available at https://doi.org/10.1071/sr9950241.

European Commission, 2021. Commission staff working document; sustainable carbon cycles—carbon farming. European Commission. Available from: https://ec.europa.eu/eip/agriculture/en/eip-agri-projects/projects/operational-groups.

FAO, 2019. Measuring and Modelling Soil Carbon Stocks and Stock Changes in Livestock Production Systems: Guidelines for Assessment (Version 1), 1st ed. FAO.

Filewod, B., McCarney, G., 2023. Avoiding carbon leakage from nature-based offsets by design. One Earth 6 (7), 790–802. Available at https://doi.org/10.1016/j.oneear.2023.05.024.

Friedlingstein, P., O'sullivan, M., Jones, M.W., Andrew, R.M., Gregor, L., Hauck, J., et al., 2022. Global carbon budget 2022. Earth System Science Data 14 (11), 4811–4900. Available at https://doi.org/10.5194/essd-14-4811-2022.

Ganzhara, N.F., 1998. Humus, soil properties, and yield. https://www.cabidigitallibrary.org/doi/full/10.5555/19991906052.

Ghosh, P.K., Dayal, D., Mandal, K.G., Wanjari, R.H., Hati, K.M., 2003. Optimization of fertilizer schedules in fallow and groundnut-based cropping systems and an assessment of system sustainability. Field Crops Research 80 (2), 83–98. Available at https://doi.org/10.1016/s0378-4290(02)00170-3.

Gitari, H.H., Nungula, E.Z., Chappa, L.R., Raza, M.A., Ranjan, S., Sow, S., et al., 2024. Agroforestry for Climate Security. In: Raj, A., Jhariya, M.K., Banerjee, A., Jha, R.K., Singh, K.P. (Eds.), Agroforestry, 1st ed. Wiley, pp. 319–344. Available at https://doi.org/10.1002/9781394231164.ch11.

Gunamantha, I.M., Sudiana, I.K., Sastrawidana, D.K., Suryaputra, I.N.G.A., Oviantari, M.V., 2021. The evaluation of soil fertility status of open space in campus area and their suitability for tropical fruits production. Journal of Soil Science and Environmental Management 12, 78–85. Available at https://doi.org/10.5897/jssem2021.0872.

Harbo, L.S., Olesen, J.E., Liang, Z., Christensen, B.T., Elsgaard, L., 2022. Estimating organic carbon stocks of mineral soils in Denmark: impact of bulk density and content of rock fragments. Geoderma Regional 30. Available at https://doi.org/10.1016/j.geodrs.2022.e00560.

Intergovernmental Panel on Climate Change (IPCC), 2023. Climate Change 2022 – Impacts, Adaptation and Vulnerability: Working Group II Contribution to the Sixth Assessment Report of the Intergovernmental Panel on Climate Change. Cambridge University Press. Available at http://doi.org/10.1017/9781009325844.

IPCC, 2019. Climate change and land: an IPCC special report on climate change, desertification, land degradation, sustainable land management, food security, and greenhouse gas fluxes in terrestrial ecosystems. Available from: www.ipcc.ch.

Jat, M.L., Chakraborty, D., Ladha, J.K., Parihar, C.M., Datta, A., Mandal, B., et al., 2022. Carbon sequestration potential, challenges, and strategies towards climate action in smallholder agricultural systems of South Asia. Crop and Environment 1 (1), 86–101. Available at https://doi.org/10.1016/j.crope.2022.03.005.

Jhariya, M.K., Meena, R.S., Banerjee, A., Meena, S.N., 2022. Natural Resources Conservation and Advances for Sustainability. Elsevier. Available at https://doi.org/10.1016/C2019-0-03763-6.

Jhariya, M.K., Meena, R.S., Banerjee, A., Kumar, S., Raj, A., 2023. Agroforestry for Carbon and Ecosystem Management. Elsevier, India, pp. 1–434.

Joensuu, K., Rimhanen, K., Heusala, H., Saarinen, M., Usva, K., Leinonen, I., et al., 2021. Challenges in using soil carbon modelling in LCA of agricultural products—the devil is in the detail. The International Journal of Life Cycle Assessment 26 (9), 1764–1778. Available at https://doi.org/10.1007/s11367-021-01967-1.

Kiptot, E., Franzel, S., 2012. Gender and agroforestry in Africa: a review of women's participation. Agroforestry Systems 84 (1), 35–58. Available at https://doi.org/10.1007/s10457-011-9419-y.

Kizeková, M., Kanianska, R., Jančová, Ľ., Čunderlík, J., Dugátová, Z., 2024. Carbon and nitrogen stocks in agricultural soils under different natural conditions and management in slovakia. Land 13 (2), 179. Available at https://doi.org/10.3390/land13020179.

Kumar, R., 2023. Impact of climate change and biological factors on agriculture in western himalaya- people's perception. Indian Journal of Ecology. Available at https://doi.org/10.55362/IJE/2023/4128.

Kumar, S., Meena, R.S., Jhariya, M.K., 2020. Resources Use Efficiency in Agriculture. Springer. Available at https://doi.org/10.1007/978-981-15-6953-1.

Kumar, S., Meena, R.S., Sheoran, P., Jhariya, M.K., 2024. Regenerative Agriculture for Sustainable Food Systems. Springer Science+Business Media, India, pp. 1–493.

Lal, R., 2004. Soil carbon sequestration impacts on global climate change and food security. Science 304 (5677), 1623–1627. Available at https://doi.org/10.1126/science.1097396.

Lal, R., 2006. Enhancing crop yields in the developing countries through restoration of the soil organic carbon pool in agricultural lands. Land Degradation & Development 17 (2), 197–209. Available at https://doi.org/10.1002/ldr.696.

Lal, R., 2007. Carbon management in agricultural soils. Mitigation and Adaptation Strategies for Global Change 12 (2), 303–322. Available at https://doi.org/10.1007/s11027-006-9036-7.

Lal, R., Negassa, W., Lorenz, K., 2015. Carbon sequestration in soil. Current Opinion in Environmental Sustainability 15, 79–86. Available at https://doi.org/10.1016/j.cosust.2015.09.002.

Lal, R., Kimble, J., Follett, R., 2019. Land use and soil C pools in terrestrial ecosystems. Management of Carbon Sequestration in Soil. CRC Press, pp. 1–10.

Larney, F.J., Janzen, H.H., Olson, B.M., Lindwall, C.W., 2000. Soil quality and productivity responses to simulated erosion and restorative amendments. Canadian Journal of Soil Science 80 (3), 515–522. Available at https://doi.org/10.4141/S99-100.

Lavelle, P., Bignell, D., Lepage, M., Wolters, V., Roger, P., Ineson, P., et al., 1997. Soil function in a changing world: the role of invertebrate ecosystem engineers. European Journal of Soil Biology 33 (4), 159–193.

Liu, S., 2022. Restoring Carbon Sinks in Agriculture through Carbon Credits. Sustainable Finance Initiative: Stanford, CA, USA. https://sfi.stanford.edu/sites/sfi/files/media/file/restoring_carbon_sinks_in_agriculture_through_carbon_credits_whitepaper_0_0_0.pdf.

Lorenz, K., Lal, R., Ehlers, K., 2019. Soil organic carbon stock as an indicator for monitoring land and soil degradation in relation to United Nations' Sustainable Development Goals. Land Degradation and Development 30 (7), 824–838. Available at https://doi.org/10.1002/ldr.3270.

Meena, R.S., Pradhan, G., 2023. Industrial garbage-derived biocompost enhances soil organic carbon fractions, CO_2 biosequestration, potential carbon credits and sustainability index in a rice-wheat ecosystem. Environmental Research 235, 116525. Available at https://doi.org/10.1016/j.envres.2023.116525.

Meena, R.S., Pradhan, G., Kumar, S., Lal, R., 2023. Using industrial wastes for rice-wheat cropping and food-energy-carbon-water-economic nexus to the sustainable food system. Renewable and Sustainable Energy Reviews 187, 113756. Available at https://doi.org/10.1016/j.rser.2023.113756.

Meena, R.S., Pradhan, G., Singh, K., Kumar, S., Singh, A.K., Shashidhar, K.S., et al., 2024a. Agriculture models for restoring degraded land to enhance CO_2 biosequestration and carbon credits in the Vindhyan region of India. Science of the Total Environment 929. Available at https://doi.org/10.1016/j.scitotenv.2024.172661.

Meena, R.S., Pradhan, G., Nalani, Kumar, S., Singh, A.K., Verma, S.K., et al., 2024b. Diversified cropping modules designed for soil restoration, CO_2 sequestration, and generating carbon credits. Land Degradation and Development 35 (14), 4314–4329. Available at https://doi.org/10.1002/ldr.5224.

Meena, R.S., Lal, R., Kumar, S., Pradhan, G., Srinivasarao, C., Singh, A.K., et al., 2024c. Potential of Indian agriculture for capturing atmospheric CO_2 and monetizing carbon credits to the farmers: an overview and policy framework. Advances in Agronomy 188, 101–206. Available at https://doi.org/10.1016/bs.agron.2024.06.001.

Meena, R.S., Pradhan, G., Kumar, S., 2025. Energy flow, eco-efficiency, and economic circulation with recycled industrial waste compost application in wheat and subsequent rice farming. Science of the Total Environment 967. Available at https://doi.org/10.1016/j.scitotenv.2025.178779.

Meena, R.S., Kumar, S., Pradhan, G., Kumar Jhariya, M., Srinivasarao, C., Meena, R.S., et al., 2026. Carbon-negative Agriculture. Springer Nature, Singapore, pp. 1–500. Available at http://doi.org/10.1007/978-981-95-1483-0.

Misch, F., Wingender, P., 2024. Revisiting carbon leakage. Energy Economics 140, 107786. Available at https://doi.org/10.1016/j.eneco.2024.107786.

Miharza, T., Wijayanto, N., Roshetko, J.M., Siregar, I.Z., 2023. Carbon stocks and footprints of smallholder cacao systems in Polewali Mandar, West Sulawesi. Frontiers in Environmental Science 11. Available at https://doi.org/10.3389/fenvs.2023.680984.

Mukherji, A., Arndt, C., Arango, J., Flintan, F.E., Derera, J., Francesconi, W., et al., 2023. Achieving agricultural breakthrough: a deep dive into seven technological areas. Available from: https://hdl.handle.net/10568/131852.

Nelson, D.W., Sommers, L.E., 2018. Total Carbon, Organic Carbon, and Organic Matter. Wiley, pp. 961–1010. Available at http://doi.org/10.2136/sssabookser5.3.c34.

Nungula, E.Z., Chappa, L.R., Ranjan, S., Sow, S., Alnemari, A.M., Seleiman, M.F., et al., 2024. Ecosystem services through agroforestry systems and its sustainability. Agroforestry. Wiley, pp. 223–254. Available at http://doi.org/10.1002/9781394231164.ch8.

Nyong, P.A., Martin, N.T., 2019. Enhancing agricultural sustainability and productivity under changing climate conditions through improved agroforestry practices in smallholder farming systems in Sub-Saharan Africa. African Journal of Agricultural Research 14 (7), 379–388. Available at https://doi.org/10.5897/ajar2018.12972.

Oldfield, E.E., Eagle, A.J., Rubin, R.L., Rudek, J., Sanderman, J., Gordon, D.R., 2022. Crediting agricultural soil carbon sequestration. Science 375 (6586), 1222–1225. Available at https://doi.org/10.1126/science.abl7991.

Oviantari, M. V., Sudiana, I. K., Sastrawidana, I. D. W. K., Suryaputra, I. A. N. G., & Gunamantha, I. M. (2021). Contribution of Tropical Fruit Plants and Soil Properties to the Potential of Carbon Sequestration in Open Land Utilization for Mixed Plantations: 4th International Conference on Innovative Research Across Disciplines (ICIRAD 2021), Singaraja, Indonesia. Available at https://doi.org/10.2991/assehr.k.211222.007.

Pal, M.S., 2003. Alternative crop production strategies for rice-wheat cropping systems in the Indo-Gangetic plains of India. Australian Journal of Experimental Agriculture 43 (6), 605–615. Available at https://doi.org/10.1071/ea01197.

Paul, C., Bartkowski, B., Dönmez, C., Don, A., Mayer, S., Steffens, M., et al., 2023. Carbon farming: are soil carbon certificates a suitable tool for climate change mitigation? Journal of Environmental Management 330, 117142. Available at https://doi.org/10.1016/j.jenvman.2022.117142.

Plan Vivo Foundation, 2023. Annual Report 2022-2023. Plan Vivo Foundation. Available from: www.planvivo.org.

Post, W.M., Kwon, K.C., 2000. Soil carbon sequestration and land-use change: processes and potential. Global Change Biology 6 (3), 317–327. Available at https://doi.org/10.1046/j.1365-2486.2000.00308.x.

Powlson, D.S., Galdos, M.V., 2023. Challenging claimed benefits of soil carbon sequestration for mitigating climate change and increasing crop yields: Heresy or sober realism? Global Change Biology 29 (9), 2381–2383. Available at https://doi.org/10.1111/gcb.16640.

Powlson, D.S., Hirsch, P.R., Brookes, P.C., 2001. The role of soil microorganisms in soil organic matter conservation in the tropics. Nutrient Cycling in Agroecosystems 61 (1-2), 41–51. Available at https://doi.org/10.1023/A:1013338028454.

Pradhan, G., Meena, R.S., 2023a. Interaction impact of biocompost on nutrient dynamics and relations with soil biota, carbon fractions index, societal value of CO_2 equivalent and ecosystem services in the wheat-rice farming. Chemosphere 339. Available at https://doi.org/10.1016/j.chemosphere.2023.139695.

Pradhan, G., Meena, R.S., 2023b. Utilizing waste compost to improve the atmospheric CO_2 capturing in the rice-wheat cropping system and energy-cum-carbon credit auditing for a circular economy. Science of the Total Environment 892. Available at https://doi.org/10.1016/j.scitotenv.2023.164572.

Pradhan, G., Meena, R.S., Kumar, S., Lal, R., 2023. Utilizing industrial wastes as compost in wheat-rice production to improve the above and below-ground ecosystem services. Agriculture, Ecosystems & Environment 358, 108704. Available at https://doi.org/10.1016/j.agee.2023.108704.

Raj, A., Jhariya, M.K., Banerjee, A., Jha, R.K., Singh, K.P., Sow, S., et al., 2024. Agroforestry for climate security. Agroforestry. Wiley, pp. 319–344. Available at http://doi.org/10.1002/9781394231164.ch11.

Reilly, J.M., Fuglie, K.O., 1998. Future yield growth in field crops: what evidence exists? Soil and Tillage Research 47 (3-4), 275–290. Available at https://doi.org/10.1016/S0167-1987(98)00116-0.

Rudd, L., Belcher, K., Kulshreshtha, S., Laroque, C., Bentham, M., 2022. Life cycle assessment of carbon dioxide emissions from shelterbelt seedling production and transportation in Saskatchewan, Canada. American Journal of Agriculture and Forestry 10 (3), 85. Available at https://doi.org/10.11648/j.ajaf.20221003.11.

Sapkota, T.B., Jat, M.L., Rana, D.S., Khatri-Chhetri, A., Jat, H.S., Bijarniya, D., et al., 2021. Crop nutrient management using Nutrient Expert improves yield, increases farmers' income and reduces greenhouse gas emissions. Scientific Reports 11 (1). Available at https://doi.org/10.1038/s41598-020-79883-x.

Schilling, F., Baumüller, H., Ecuru, J., & von Braun, J. (2023). Carbon farming in Africa: Opportunities and challenges for engaging smallholder farmers [Report]. Center for Development Research (ZEF). Available at https://doi.org/10.48565/bonndoc-122.

Seeberg-Elverfeldt, C. (2010). Carbon finance possibilities for agriculture, forestry and other land use projects in a smallholder context (pp. 23). Rome: FAO.

Siband, P.-L., 1974. Evolution des caractères et de la fertilité d'un sol rouge de Casamance. https://www.sidalc.net/search/Record/dig-cirad-fr-421134/Description.

Six, J., Elliott, E.T., Paustian, K., 1999. Aggregate and soil organic matter dynamics under conventional and no-tillage systems. Soil Science Society of America Journal 63 (5), 1350–1358. Available at https://doi.org/10.2136/sssaj1999.6351350x.

Wiese, L., Wollenberg, E., Alcántara-Shivapatham, V., Richards, M., Shelton, S., Hönle, S.E., et al., 2021. Countries' commitments to soil organic carbon in nationally determined contributions. Climate Policy 21 (8), 1005–1019. Available at https://doi.org/10.1080/14693062.2021.1969883.

Wongpiyabovorn, O., Plastina, A., Crespi, J.M., 2023. Challenges to voluntary Ag carbon markets. Applied Economic Perspectives and Policy 45 (2), 1154–1167. Available at https://doi.org/10.1002/aepp.13254.

Zeraatpishe, M., Khormali, F., 2012. Carbon stock and mineral factors controlling soil organic carbon in a climatic gradient, Golestan province. Journal of Soil Science and Plant Nutrition 12 (ahead). Available at https://doi.org/10.4067/s0718-95162012005000022 0–0.

Chapter 16

Carbon credit estimation in croplands

Victor V. Ngaiza[1], Emmanuely Z. Nungula[2], Norbert J. Ngowi[2], Luciana R. Chappa[3], Rogerio P. Soratto[4], and Harun I. Gitari[5]

[1]*School of Life Sciences and Bio-engineering, The Nelson Mandela African Institution of Sciences and Technology, Arusha, Tanzania,* [2]*Department of Environment and Sustainable Development, Mzumbe University, Morogoro, Tanzania,* [3]*Department of Agricultural Sciences, College of Agriculture, Mwalimu Nyerere University of Agriculture and Technology, Butiama, Tanzania,* [4]*Department of Crop Science, College of Agricultural Sciences, São Paulo State University (UNESP), Botucatu, Brazil,* [5]*Department of Agricultural Science and Technology, School of Agriculture and Environmental Sciences, Kenyatta University, Nairobi, Kenya*

Highlights

- Agroecosystem appears to be a potential carbon sink.
- Agroecosystem performs a dual function of carbon sequestration and climate change mitigation.
- A comprehensive methodology for carbon credit assessment in an agroecosystem is required.
- Suitable cultivation practices provide an enhancement of carbon credits from the agroecosystem.

16.1 Introduction

Carbon (C) credit estimation involves quantifying the carbon dioxide (CO_2) emissions that can be reduced through specific agricultural practices or land management strategies (Meena et al., 2025b). The financial gain for sustainable practices can be generated in the C markets by selling the determined quantities of CO_2 (Salma et al., 2024; Velvizhi et al., 2023). Enejison et al. (2022) demonstrate that C credits have become an indispensable instrument for promoting sustainable farming practices, particularly when strategizing for climate change mitigation (CCM). C credits, then, are a market-based mechanism for encouraging reductions in greenhouse gas (GHG) emissions by enabling enterprises to purchase and sell credits obtained from confirmed emissions reductions (Hannay, 2023; Raj et al., 2025; Meena et al., 2024c).

According to recent reports, the agricultural sector accounts for 10%–12% of GHG emissions, with croplands accounting for a major share of this total. According to the Intergovernmental Panel on Climate Change (IPCC), sustainable agricultural practices can capture up to 1.5 billion tons of carbon dioxide equivalent (CO_2e) annually, making them a key priority for climate action. According to research by Smith et al. (2015), applying conservation agricultural technologies might sequester an additional 0.4–1.2 gigatons of CO_2e per year by 2030. Similarly, the 4-per-mille initiative of the French Minister of Agriculture at the UN Climate Change Conference (COP21) in Paris (2015) indicates that, through sustainable soil and land management practices, the soil organic carbon (SOC) stocks are expected to increase by 0.4% per year (Tiefenbacher et al., 2021).

Assessment of C credits in croplands offers an opportunity for agricultural transformation and facilitates understanding and improvement of the carbon sequestration (C_{seq}) capacity of agricultural methods that contribute to CCM (Tiefenbacher et al., 2021; Meena et al., 2025b; Kiran et al., 2023; Meena et al., 2024a). Additionally, Xie et al. (2024) indicated that several techniques are used to estimate C credits in croplands, such as measuring SOC stocks, assessing land-use changes, and using models to forecast how various agricultural practices affect C_{seq} (Kumar et al., 2024).

Several soil management techniques, including reduced tillage and cover cropping, can raise SOC by 0.2%–1.0% annually, increasing croplands' potential for C credits (Oscar Kisaka et al., 2023). However, since C financing firms offer incentives to farmers and encourage the shift toward sustainable agriculture, they can be economically advantageous for

Carbon Incentives, Plans, and Policies in Agriculture. DOI: https://doi.org/10.1016/B978-0-443-33557-0.00015-X

farmers. For instance, according to the World Bank, additional C credits worth \$10–\$50 per ton of CO_2e might be obtained through the C market, contingent on legal frameworks and market conditions (Thapa, 2019).

16.2 Carbon and cropland ecosystems

Croplands cover about 12% of the total Earth's crust, and this part serves as both a source of GHG emissions and a C sink. Cover crops, reduced tillage, and agroforestry (AF) can all contribute to storing considerable amounts of CO_2 in cropland soil and plant biomass (Jhariya et al., 2022, 2023; Meena et al., 2024b). Studies indicate that sustainable agricultural methods may increase SOC storage, storing up to 1.5 billion tons of CO_2e globally per annum. Such capacity to store C is crucial for CCM while maintaining agricultural productivity (Sah et al., 2024; Meena and Pradhan, 2023; Dahal and Bajracharya, 2010). Fig. 16.1 illustrates the variables that determine SOC storage in temperate regions.

C dynamics in cropland ecosystems are explained by processes such as sequestration, storage, and release within agricultural ecosystems (Lorenz and Lal, 2012). The key process for sequestration is photosynthesis, a process whereby atmospheric CO_2 is trapped and converted into organic matter (OM) by plants (Nasar et al., 2021). The accumulated OM is converted into SOC after the plant dies and gets decomposed (Kumar and Das, 2014; Nanda et al., 2016; Nyawade et al., 2019; Girijaveni et al., 2023; Ngaiza et al., 2025). Several farming practices, such as traditional tillage, can alter soil structure and cause stored C to be released into the atmosphere (Nazir et al., 2024; Poeplau and Don, 2015). However, several strategies aid in retaining C in the soil. Some of these techniques include cover crop planting and reduced tillage. Furthermore, environmental factors, including meteorological conditions and soil type, influence dynamics by influencing plant development and microbial activity (Hussain et al., 2021). Table 16.1 displays the processes that affect the C dynamics in soil.

Their biological and physiological traits govern the photosynthetic ability of plants, and as a result, different crops were found to have varying capacities in contributing to biomass and storage of C in the soil (Dignac et al., 2017; Mathew et al., 2017) (Table 16.2). It was revealed that perennial crops are more capable of absorbing and storing C than annuals. They are distinguished by their longer growing seasons and deeper roots. Legumes, which fix nitrogen, have also been shown to contribute more C storage (Toensmeier, 2016; Vico and Brunsell, 2018; Devi et al., 2024). The anticipated capacity for C_{seq} for various crop varieties is shown in Table 16.2.

Croplands can, however contribute to GHG emissions, especially when conventional farming methods are used. For example, the use of excessive fertilizer, conventional tillage, and land can release methane (CH_4), nitrous oxide (N_2O), and CO_2 into the atmosphere (Kraus et al., 2016; Naz et al., 2021; Sahu and Arya, 2024; Nungula et al., 2025c). The IPCC identifies agriculture as a potential contributor to GHG emissions (10%–12%). This proportion emphasizes the importance of implementing sustainable measures to offset such effects (Tongwane and Moeletsi, 2018; Nsabiyeze et al., 2024; Kamyab et al., 2024). Farmers can cut emissions while enhancing land resilience and fertility by adopting measures that increase soil health and C_{seq}.

Improving C_{seq} in croplands has significant positive effects on the economy and ecology (Prasad et al., 2021a,b). By selling C credits produced through sustainable practices, farmers can not only increase their revenue but also contribute to achieving broader climate goals (Tao et al., 2019; Morgan et al., 2010). Moreover, Smith et al. (2015) and Garbach et al. (2014) highlight that robust agricultural ecosystems support biodiversity and offer vital functions, including water

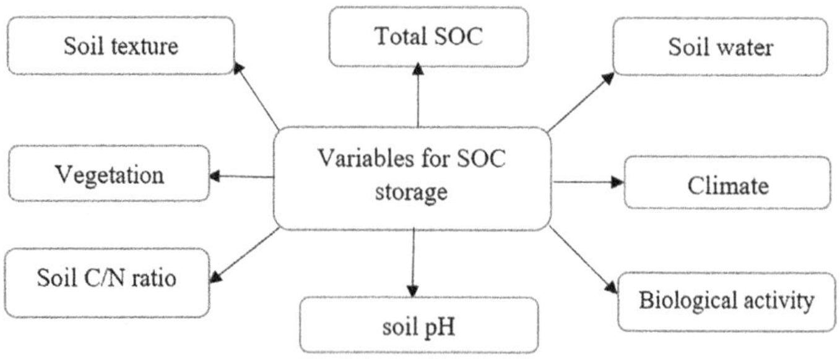

FIGURE 16.1 Key predictor variables for soil organic carbon storage in temperate regions.
Modified from Tiefenbacher, A., Sandén, T., Haslmayr, H.P., Miloczki, J., Wenzel, W., Spiegel, H., 2021. Optimizing carbon sequestration in croplands: a synthesis. Agronomy. 11 (5), 882. https://doi.org/10.3390/agronomy11050882

TABLE 16.1 Processes that determine the carbon dynamics in the cropland ecosystem.

Process	Explanation	Effect on carbon	References
Photosynthesis	Conversion of CO_2 into plant biomass	C sink	Reddy et al. (2010), Paul and Foyer (2001), Wu and Wu (2022)
Crop management	Practices like reduced tillage enhance SOM	Enhanced C sink	Jarecki and Lal (2003), Hussain et al. (2021)
Soil respiration	Release of CO_2 from soil and plant roots	C source	Shi et al. (2020), Watts et al. (2021), Criscuoli et al. (2024)
Change in land use	Conversion of forests to croplands	Increased CO_2	Koga et al. (2020), Azadi et al. (2021)
Decomposition	Breakdown of organic matter	C source	Prescott and Vesterdal (2021), Bhattacharyya et al. (2022)

filtration and nutrient cycling. Understanding the importance of croplands in the C cycle is becoming increasingly crucial as the world's food consumption rises. Croplands benefit the environment, and agricultural productivity can be guaranteed through implementing sustainable management practices (Beyer et al., 2022; Bithas and Latinopoulos, 2021).

16.3 Importance of carbon financing for cropland ecosystems

It is increasingly acknowledged that cropland ecosystems can benefit from C financing as a means to combat climate change and promote sustainable farming methods (Smith et al., 2020). C financing encourages landowners and farmers to implement techniques that improve soil and vegetation retention of C (Raina et al., 2024; Marks, 2019; Lokuge and Anders, 2022). According to Marks (2019) and Xu et al. (2024), if C is sold at $160/ton of CO_2e, C_{seq} on agricultural land may generate an extra $375 billion in revenue by 2050. The money raised is used to reduce GHG emissions and encourage the adoption of techniques that enhance soil health and agricultural productivity. C financing is recognized to significantly assist poor-resource farmers, particularly in developing countries with limited access to capital (Gujba et al., 2012; Odhong' et al., 2019).

By using sustainable methods, farmers can monetize the C they sequester through C financing mechanisms, such as C credits and payments for ecosystem services (Maitra et al., 2023). SOC levels can be considerably raised by methods such as AF, reduced tillage, and cover cropping (Raina et al., 2024; Henderson et al., 2022; Nungula et al., 2025b; Meena and Pradhan, 2023). Globally, these activities have been shown to sequester between 0.7 and 2.5 gigatons of CO_2e annually. Farmers who participate in C markets can receive compensation for their C_{seq} efforts, which can help offset the costs of implementing these techniques and create a new source of income (Raina et al., 2024; Henderson et al., 2022; Von Unger and Emmer, 2018).

By providing financial resources and technical support, C financing programs can enable these farmers to implement sustainable practices that improve productivity and climate change resilience (Chirambo, 2019; Osumba et al., 2021). By effectively restoring millions of hectares of degraded land, for instance, the Farmer Managed Natural Regeneration program in Africa has improved agricultural output and C stocks while also benefiting local communities economically (Weston et al., 2015).

16.4 Methodological assessment of carbon credits in croplands

A crucial field of study that focuses on how agricultural practices might be measured, validated, and sold in C markets is the methodological evaluation of C credits in croplands (Dupla et al., 2024). To ensure that C credits are reliable and effective in minimizing GHG emissions, they must be generated from cropland techniques that employ sound methodologies (Everhart, 2023; Lokuge and Anders, 2022; Pradhan and Meena, 2023a). Finding suitable practices, measuring C_{seq}, confirming outcomes, and registering credits for trading are usually some of the crucial elements in the process. To ensure dependability and transparency in the C market, each of these procedures must adhere to specific guidelines (Everhart, 2023; Lessmann et al., 2022).

TABLE 16.2 Different types of crops and their carbon sequestration potential.

Crop type	Examples	Carbon sequestration potential	Root depth (m)	Description	References
Tree crops	Almond, chestnut	Very high	1–3	Significantly above and below-ground storage	Siarudin et al. (2021), Yadav et al. (2022)
Perennial crops	Switchgrass, Bermuda grass	High	1–3	Deep roots enhance SOM	Sainju et al. (2017), Mapfumo et al. (2023)
Legumes	Clover and Alfalfa	Moderate to high	0.5–1.5	Improves soil health	Boniface (2023)
Oilseeds	Canola, sunflower	Moderate	0.5–1	promotes soil structure and fertility	Kumar and Kunhamu (2022)
Cover crop	Wheat, rye, hairy vetch	Moderate	0.5–1	protects the soil and adds biomass	Moore (2022), Mirsky et al. (2013)
Cereal crops	Wheat, barley	Moderate	0.5–1	Benefit from sustainable practices	Bhardwaj et al. (2019), Alam et al. (2019), Meena et al. (2020)

Verra's VM0042 is a well-known approach for evaluating C credits in croplands that emphasizes better agricultural land management. The additional criteria outlined in this methodology ensure that the C reductions achieved exceed what would occur in a typical business scenario (Nayak et al., 2019; Croft et al., 2012; Brummitt et al., 2024). In line with this, Cariappa et al. (2024) indicated that practices such as no-till farming, cover cropping, and rotation of crops can qualify for C credits if they meet the additionality condition, meaning farmers must adopt these practices specifically to generate C credits rather than as part of their standard operations.

The 50th model, with 83 Pg C and a median of 59.1 Mg C/ha, was used in an experiment to assess SOC content in topsoil (0–30 cm) in 1410 hectares of croplands worldwide. The results revealed significant concentrations of additional SOC storage in ecoregions with high precultivation densities, such as flooded grasslands, savannas, and mixed woods. According to the results, the possible additional SOC storage is 29 Pg C (the 75th and 50th percentile differences), with shrublands, mixed forests, temperate zones, and broadleaf and mixed forests accounting for 59% of the total potential additional topsoil SOC storage. Table 16.3 displays all topsoil (0–30 cm) SOC stocks for croplands worldwide, categorized by ecoregion.

16.4.1 Baseline assessment

Baseline assessment is a crucial step in quantifying C credits. It aims at establishing knowledge of the existing C and GHG emissions after adopting C_{seq} practices. The established value is used as a standard for computing the net C benefits from the implemented practices. Fig. 16.2 illustrates the objectives for conducting the baseline assessment (Henry et al., 2023; Roberts and Thibaudeau, 2024; Paustian et al., 2019).

TABLE 16.3 The soil organic carbon (0-30 cm depth) in Pg1 C for global croplands by ecoregions.

Global cropland	Cropland area (km^2)	Current (q50)	Intermediate (q75)	Technical maximum (q50)
Tropical and subtropical moist broadleaf forests	1,771,950	9.87	13.37	17.48
Temperate grasslands, savannas, and shrublands	2,777,674	20.33	26.58	33.91
Temperate broadleaf and mixed forests	4,178,195	29.47	40.38	54.03
Mediterranean forests, woodlands, and scrub	859,799	4.38	5.84	7.58
Deserts and Xeric shrublands	1,052,258	3.23	4.45	5.88
Tropical and subtropical dry broadleaf forests	1,314,051	4.89	6.71	9.25
Tropical and subtropical grasslands, savannas, and shrublands	1,262,423	4.92	6.91	9.53
Boreal forests/Taiga	174,224	1.64	2.33	3.06
Temperate conifer forests	173,098	1.15	1.54	2.02
Flooded grasslands and Savannas	145,205	0.97	1.41	1.94
Montane grasslands and shrublands	214,602	1.29	1.73	2.18
Tropical and subtropical coniferous forests	43,780	0.27	0.37	0.48
Mangroves	31,880	0.20	0.27	0.35
Tundra	0.05	0.07	0.09	10,712
Global total	**14,009,852**	**82.68**	**111.97**	**147.75**

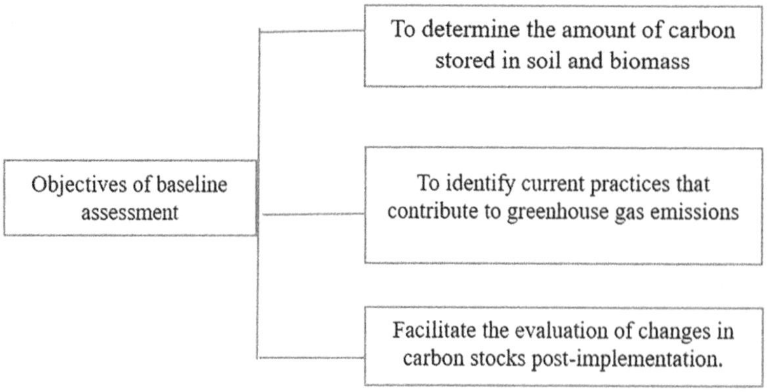

FIGURE 16.2 Objectives of baseline assessment.

Various methodologies are employed to perform a baseline assessment. These methodologies help to collect data on soil C content, biomass, and historical land use, each offering unique advantages and insights (Ciais et al., 2014). Similarly, Phelan et al. (2024) and Kamyab et al. (2024) indicated that several methodologies can also be combined to guarantee stakeholders a comprehensive understanding of C dynamics in croplands, thereby facilitating informed decision-making and effective participation in C markets. Table 16.4 presents various methodologies employed in baseline assessments, along with their associated advantages and disadvantages.

16.4.2 Verification of results

The assessment results obtained are verified to ensure their accuracy and reliability for the claimed C reductions (Chen and Lin, 2021; Gupta et al., 2012). Third-party verifiers play a significant role in this stage, conducting audits and assessments to confirm that the C credits generated meet the established standards. For instance, organizations such as the Gold Standard and Verra provide frameworks for verification that include rigorous protocols for data collection and analysis. This independent verification process is essential for maintaining the integrity of C markets and building trust among stakeholders, including farmers, investors, and regulatory bodies (Nardelli, 2012; Pradhan and Meena, 2023b).

16.4.3 Registration of credits for trading

C registries are used to record C credits, which track their issuance and trading. These registries ensure that each credit is unique and that double-counting does not occur. For example, the Agoro Carbon Alliance and Indigo are among the project developers that work with farmers to generate and register C credits, providing a streamlined process for farmers to participate in C markets (Santier, 2023; Michaelowa et al., 2019; Mwadalu et al., 2025a). As C markets continue to evolve, the methodologies for assessing C credits in croplands will also need to adapt, incorporating advancements in technology and data analytics to enhance accuracy and efficiency (Kyriakarakos et al., 2024).

16.5 Screening of suitable cultivation practices for enhanced carbon credits

Screening of suitable cultivation practices for enhanced C credits is important not only for promoting sustainable agriculture but also for addressing climate change (Piñeiro et al., 2020). The agricultural practices that farmers can adopt to enhance soil C_{seq}, thereby generating C credits that can be traded in C markets, are discussed in detail below:

16.5.1 No-tilling

No-till farming reduces soil disturbance, thereby preserving soil structure and enhancing C storage (Jayaraman et al., 2024; Oscar Kisaka et al., 2023). According to research, using no-till methods, one can raise SOC by 0.1–0.3 metric tons per hectare per year, depending on the area and soil type (Li et al., 2020). Topsoil (0–10 cm) with no or minimal tillage has a higher SOC content than conventional tillage. Tillage procedures have little effect on SOC when soil depth increases (>10 cm) (Beniston et al., 2015; Mazzoncini et al., 2016; Francaviglia et al., 2019). Minimal soil disturbance increases aggregate stability and thus reduces erosion. Mazzoncini et al. (2016) identified higher amounts of topsoil SOC in

TABLE 16.4 **Techniques used in the measurement of soil carbon.**

Technique	Description	Advantage	Disadvantage	References
Soil sampling	Physical collection of soil samples	Direct measurement	Labor-intensive, spatially limited	Stolbovoy et al. (2005), Schumacher (2002), Wotherspoon et al. (2015)
Remote sensing	Satellite imagery analysis	Broad coverage	Limited soil depth	Kuhnert et al. (2022), Zhao (2024), Odebiri et al. (2021)
Modeling data	Simulation of C processes	Predictive power	Requires calibration	Zhang et al. (2024), Dotto et al. (2018), Stevens et al. (2013)

nontilled land than under conventional plowing of 30/35 cm. Despite the opportunities presented for no-till practices in C credit generation, there are several challenges, as shown in Table 16.5

16.5.2 Cover cropping

This method is promising in improving C credits. Farmers can increase C_{seq}, prevent erosion, and improve soil health by cover cropping during the off-season. An experiment performed in the Midwest United States revealed that cover crops could sequester an additional 0.5–1.0 metric tons of CO_2 equivalent ha^{-1} $year^{-1}$ (Clark et al., 2019). This approach not only adds to C credits but also increases biodiversity and improves nitrogen cycling in agricultural systems. Furthermore, including legumes as cover crops may reduce the need for synthetic fertilizers, resulting in lower emissions of GHGs associated with fertilizer manufacturing and application (Moukanni et al., 2022; Gitari et al., 2019). Fig. 16.3 displays the effects of growing cover crops on improving soil health and increasing SOC.

According to Acharya et al. (2024), soil physiochemical properties like soil texture, porosity, soil pH, EC, total CEC, total N, and available phosphorus tend to influence the particulate organic C, while the mineral-associated organic C is influenced by the microbial community structure and size (Fig. 16.4).

TABLE 16.5 Challenges of no-till practices for carbon credit generation.

Challenge	References
High initial cost	Somasundaram et al. (2020), Lokuge and Anders (2022)
Lack of standard methods for measurement and verification	Dupla et al. (2024), Brummitt et al. (2024)
Market volatility: Fluctuation of the C credit price	Joseph (2002), Raina et al. (2024)
Long-term commitment and uncertainty	Pudasaini et al. (2025), Thompson et al. (2022)

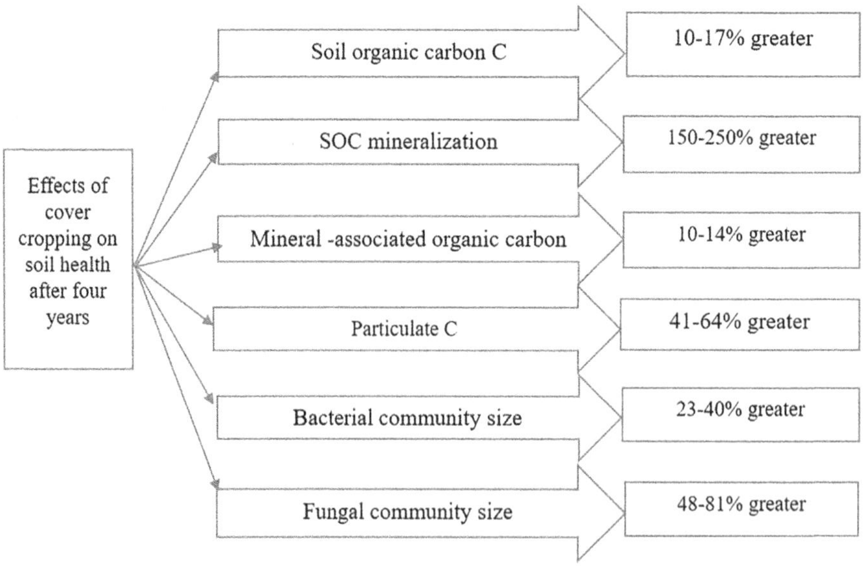

FIGURE 16.3 Effects of growing cover crops on soil health after four years.
Modified from Acharya, P., Ghimire, R., Acosta-Martínez, V., 2024. Cover crop-mediated soil carbon storage and soil health in semi-arid irrigated cropping systems. Agriculture, Ecosystems & Environment. 361, 108813. https://doi.org/10.1016/j.agee.2023.108813.

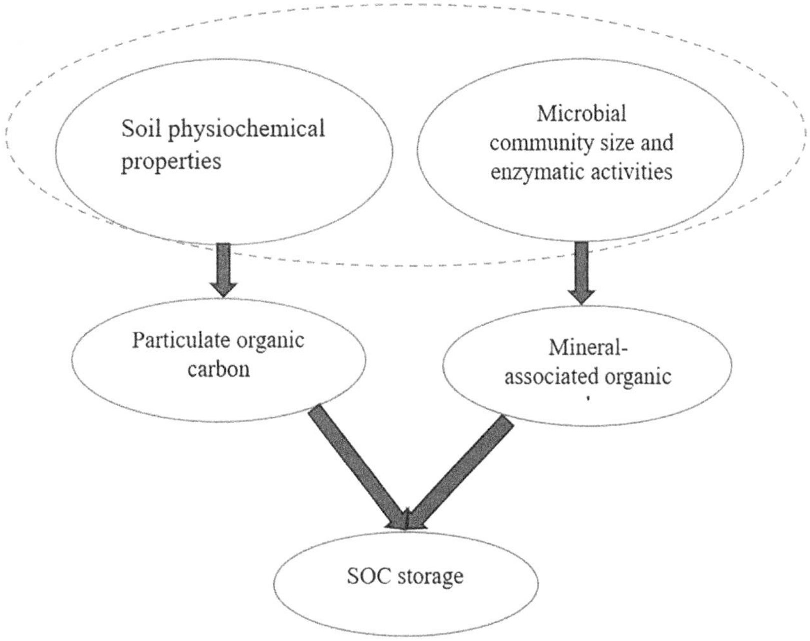

FIGURE 16.4 Influence of physiochemical properties and microbial community on SOC storage (*SOC*, soil organic carbon).
Modified from Acharya, P., Ghimire, R., Acosta-Martínez, V., 2024. Cover crop-mediated soil carbon storage and soil health in semi-arid irrigated cropping systems. Agriculture, Ecosystems & Environment. 361, 108813. https://doi.org/10.1016/j.agee.2023.108813.

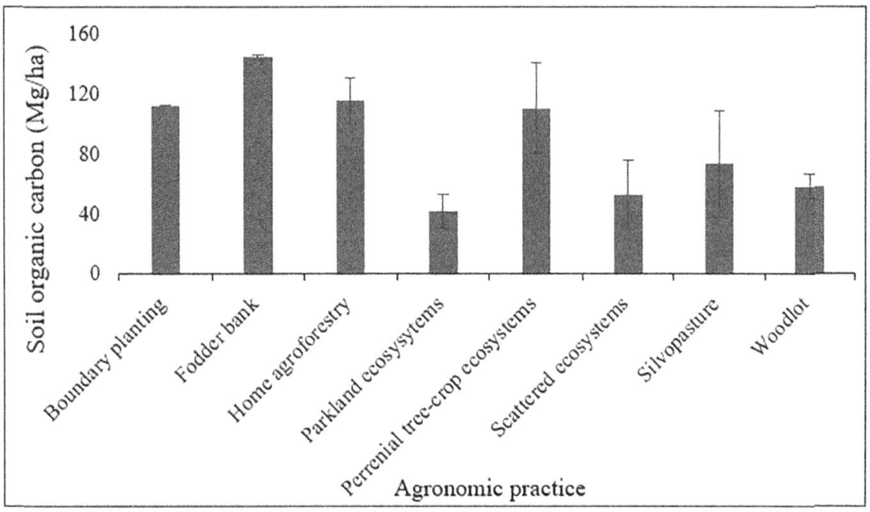

FIGURE 16.5 Estimates of SOC (0–60 cm depth, Mg C ha^{-1}) in key agroforestry in East Africa (*SOC*, soil organic carbon).

16.5.3 Agroforestry

AF combines crop farming and tree planting, another successful strategy for increasing C credits (Raj et al., 2024). This approach helps sequester C in trees and soil, thereby enhancing climatic resilience and increasing biodiversity and soil fertility (Cheruto et al., 2025). In one instance, research in Brazil found that agroforestry systems (AFs) may sequester up to 10 pounds of CO_2-equivalent per hectare per year, thereby significantly contributing to C credit generation (Nair et al., 2021). Tree incorporation into agricultural landscapes can also provide shade and alleviate water stress for crops, hence increasing productivity. Fig. 16.5 indicates the quantity of SOC accumulated under different agronomic practices.

16.5.4 Adoption of improved nutrient management strategies

This involves precision agricultural tactics that maximize fertilizer use based on soil health and crop requirements and can help to reduce nitrous oxide emissions, a strong GHG. Research has shown that implementing precision nutrient management can reduce nitrous oxide emissions by up to 30% while maintaining crop yields (Robertson and Grace, 2019). Furthermore, Lokuge and Anders (2022) highlighted the dual benefits of reducing emissions while enhancing soil C storage, making improved nutrient management a valuable practice for farmers seeking to participate in C credit markets.

16.5.5 Crop rotation

Compared to the monoculture system, crop rotation has a greater impact on C storage and increases biomass accumulation in the topsoil. For example, Poeplau and Don (2015) reported large SOC stocks in the field diversity of crop cycles, including cover crops, perennials, and forages. A diverse crop rotation, in addition to increasing SOC stocks, may improve soil aggregate stability, enhance microbial diversity, and even promote organic C in the subsoil via deep-rooting. Deep-rooted crops are essential in the storage of SOC because their roots can store up to 2.3 times more C than the biomass above-ground (Tiefenbacher et al., 2021; Morari et al., 2019). Tiefenbacher et al. (2021) noted that the SOC increase is most pronounced in the topsoil and diminishes with increasing soil depth. Agricultural topsoil in rotational crop systems has higher C_{seq} potential than in monocropping systems (Loges et al., 2018; Pradhan et al., 2023). Poeplau and Don (2015) reported an increment in C_{seq} for (0–20/30 cm) by 216 ± 117 kg C ha^{-1} y^{-1}. Similarly, Minasny et al. (2017) reported an increment of 360 kg C ha^{-1} y^{-1} in clay and 590 kg C ha^{-1} y^{-1} in the loamy texture of topsoil (0–20 cm) after 35 years of crop rotation.

16.5.6 Biochar application

Biochar is a highly recalcitrant material, so it offers a long-lasting sink for C when applied to the soil. Various studies, such as Liu et al. (2016) and Maestrini et al. (2015), revealed a significant increase in the amount of C in agricultural fields, in both topsoil and subsoil, following the incorporation of biochar. For evaluating the C_{seq}, the biochar's recalcitrance is commonly used as a deciding factor. For instance, various estimates, indicate that biochar incorporation has a C_{seq} of 6 ± 5.14 Mg C ha^{-1} y^{-1} (Paustian et al., 2016; Majumder et al., 2019; Woolf et al., 2010). Fig. 16.6 displays the biochar C credit market system.

16.5.7 Organic farming

Organic farming systems offer a C_{seq} potential of 287 ± 102 kg C ha^{-1} y^{-1} as opposed to conventional farming (Autret et al., 2019; Gattinger et al., 2012; Müller et al., 2016; Freibauer et al., 2004). This C_{seq} in organic farming is linked to

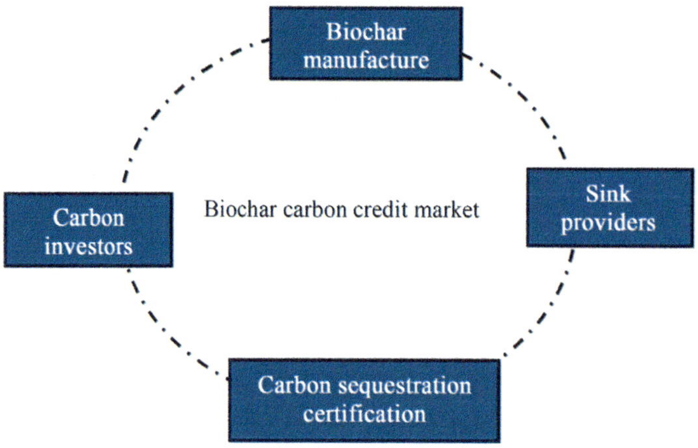

FIGURE 16.6 Biochar carbon credit market system.
Modified from Salma, A., Fryda, L., Djelal, H., 2024. Biochar: a key player in carbon credits and climate mitigation. Resources. 13 (2), 31. https://doi.org/10.3390/resources13020031.

measures such as organic fertilizer application, cover cropping, and the rotation of a range of crops with legumes (Blanco-Canqui et al., 2017; Müller et al., 2016). Cattle manure has a reduced C_{seq} potential (160 kg C ha^{-1} y^{-1}) compared to alfalfa green manure (180 kg C ha^{-1} y^{-1}) (Blanco-Canqui et al., 2017). Although principles of organic farming (diversified crop rotation, organic fertilizer application, and the use of cover crops) can also be used in conventional farming, organic farming is more widespread. Additionally, increasing C_{seq} potential, organic farming enhances biodiversity and reduces water contamination with nutrients (eutrophication) (Müller et al., 2016).

Additionally, there is less consumption of primary energy than in conventional farming. This is because organic farming is characterized by minimal or no application of artificial inputs such as industrial chemical pesticides and easily soluble inorganic fertilizers. Moreover, the sowing and management of crops under organic farming are associated with fewer emissions of CO_2 than in conventional farming (Bellarby et al., 2008).

16.5.8 Water table management

Artificially applying water to plants increases water availability, enhancing primary output, particularly during dry conditions. Increased production correspondingly increases the organic SOC stored in the soils, but it also accelerates C and N mineralization, potentially reducing the importance of irrigation for SOC (Trost et al., 2013; Zhou et al., 2016). Irrigation practice has a C_{seq} ranging from 50 to 100 kg C ha^{-1} y^{-1}. However, in 25–30 m soil depth, SOC decreased by −144 kg C ha^{-1} y^{-1} (Meena et al., 2020; Ngaiza et al., 2025; Baumgarten et al., 2014; Nungula et al., 2025c; Chappa et al., 2025b; Mwadalu et al., 2025a).

16.5.9 Planting of catch crops

Catch crops, commonly referred to as green manure crops, are grown after harvesting the main crop (mostly the legume crop) or sometimes sown under the main crops. Yin et al. (2023), Adetunji et al. (2020), Koehler-Cole and Elmore (2020), and Cottney et al. (2022) provided examples of catchy crops, which include cruciferous, rye, legume, and grass varieties. The purpose of planting catchy crops is to ensure arable land is covered, which also increases the time for C assimilation (Chahal et al., 2020).

Additional gains of planting catchy crops include reduced soil erosion, preventing weed pressure, and minimization of nitrate and other nutrient loss (Crézé and Horwath, 2021; Pasket, 2023). Planting catchy crops in agricultural topsoil (0–25/30 cm) has been shown to increase SOC stocks, with an average annual C_{seq} potential of 403 ± 142 kg C ha^{-1} y^{-1} (De Notaris et al., 2021; Petito et al., 2024; Jian et al., 2020). The research conducted by Tiefenbacher et al. (2021) in Denmark found that rotating catchy crops raised SOC stock in topsoil (0–25 cm) by 210 kg C ha^{-1} y^{-1}. A systematic review of 131 global research studies revealed that catchy crops have an average C_{seq} rate of 560 kg C ha^{-1} y^{-1} (Jian et al., 2020).

16.6 Potential economic gains through carbon credits in agroecosystems

The elimination of CO_2 from the air through agricultural methods creates and defines C credits, allowing farmers to sell these credits in C markets. This not only generates extra income for farmers but also encourages the adoption of sustainable techniques that improve soil health and biodiversity (Jhariya et al., 2021a,b). For example, AF, the use of cover plants, and no-till farming can considerably enhance SOC levels, leading to the production of C credits that can be traded for profit (Li et al., 2020).

A notable example of economic gain through C credits is the implementation of cover cropping in the Midwestern United States. Farmers adopting cover crops can sequester between 0.5 and 1.0 metric tons of CO_2 equivalent ha^{-1} $year^{-1}$, translating to potential earnings of approximately \$10–\$25 per ton of C credit sold in optional markets (Clark et al., 2019). The practice enhances soil fertility and reduces soil erosion, also enhancing agroecosystem resilience. As farmers increasingly recognize the financial benefits of C credits, the adoption of cover cropping is expected to rise, further amplifying its positive effect on both farm income and the environment.

AFs present significant opportunities for generating C credit. By integrating trees into agricultural landscapes, farmers can enhance C_{seq} while simultaneously enhancing the biodiversity and soil health. For example, a study in Brazil found that AFs could sequester up to 10 metric tons of CO_2-equivalent/ha/year, providing substantial economic returns through C credits (Nair et al., 2021). Additionally, AF can offer farmers diversified income sources from timber, fruits, and nuts, thereby increasing their economic resilience against market fluctuations in traditional crops.

Sustainable agriculture methods in India are being investigated as a potential source of C credits. Adopting conservation agriculture practices can dramatically increase farmers' C credit profits. For instance, farmers in Punjab can earn approximately $30 per hectare through the implementation of zero tillage and reduced fertilizer use, which meets the criteria for C credit generation under Verra's methodologies (Singh et al., 2020). Table 16.6 indicates the potential earnings from C credits across different regions.

C credits encourage farms to embrace sustainable practices while also helping to minimize GHG emissions and battle changes in climate (Nsabiyeze et al., 2024; Marks, 2019; Nungula et al., 2025a).

The economic potential of C credits in agroecosystems is further enhanced by the growing demand for sustainable products in global markets. As consumers increasingly seek environmentally friendly options, farmers who engage in C farming can market their products as sustainably produced, potentially commanding higher prices (Khurana et al., 2024; Jha et al., 2011; Chappa et al., 2025b). This trend is supported by the establishment of various C credit programs and registries that facilitate the trading of C credits, making it easier for farmers to participate in these markets and realize the economic benefits of their sustainable practices (Robertson and Grace, 2019).

Furthermore, the integration of technology in monitoring and reporting can significantly improve the efficiency and accuracy of C-credit generation in agroecosystems. Innovations such as remote sensing, machine learning, and blockchain technology enable precise tracking of C_{seq} and verification processes. For instance, a recent study highlighted how satellite imagery could be used to monitor changes in land use and C stocks, providing an efficient way to examine the effect of sustainable practices (Oldfield et al., 2022). Leveraging these technologies, farmers can enhance their credibility in C markets, attract more investment, and ultimately increase their economic gains through C credits (Marks, 2019; Mwadalu et al., 2025a).

16.7 Research and development toward cropland carbon credit assessment

16.7.1 Development of methodologies for various agricultural practices

For example, researchers have revealed that CA practices such as cover cropping, crop rotation, and no-till farming can significantly increase SOC levels (Alhammad et al., 2023). Research conducted in India indicates that farmers adopting CA practices can sequester between 1.23 and 1.97 metric tons of CO_2 equivalent per hectare annually, translating to potential earnings of $16–$30 per hectare from C credits, depending on the region (Singh et al., 2020). This highlights the economic viability of sustainable practices and the importance of developing accurate assessment methods to facilitate farmers' participation in C markets.

TABLE 16.6 Potential earnings from carbon credits from different regions.

Region	C credit earned (USD ha^{-1})	Additional benefits	References
California, USA	40	Enhanced water conservation and biodiversity	Zhang (2023)
Australia	35	Reduced erosion and soil fertility enhancement	White (2022), Bradshaw et al. (2013)
Punjab, India	30	Improved crop yield and marketability	Khurana et al. (2024), Cariappa et al. (2024)
Indonesia	28	Enhanced biodiversity conservation	Nawari et al. (2021), Pusparini et al. (2023)
Brazil	25	Minimal deforestation	Moutinho et al. (2009), Alves and Diniz (2022), Barros et al. (2023)
Eastern Europe	22	Improved soil structure and crop yields	Springer (2003)
Bihar, India	18	Reduced the cost of farm inputs and promoted the fertility of the soil	Khurana et al. (2024), Meena et al. (2021)

16.7.2 Integration of technology in carbon credit assessment

Innovations such as remote sensing (RS), machine learning, and blockchain technology are being explored to ensure C monitoring, and they have proved to be more efficient and accurate. For instance, RS can provide instantaneous data on land-use changes and C stock variations, enabling more precise calculations of C credits generated from specific agricultural practices. A recent study demonstrated how satellite imagery could be utilized to monitor changes in soil C levels, thereby enhancing the credibility of C credit claims (Oldfield et al., 2022).

16.7.3 Socio-economic impacts of carbon credit systems on farming communities

Understanding how C markets can benefit smallholder farmers is fundamental to promoting the widespread shift to sustainable practices. C credit earnings can significantly contribute to farmers' incomes, particularly in regions where traditional agricultural practices are economically challenging. For example, farmers in Punjab, India, can earn substantial income from C credits by implementing practices that reduce nitrogen fertilizer use and promote zero tillage, which also aligns with national climate goals (Sow et al., 2024).

Lastly, the establishment of robust regulatory frameworks and standards is vital for the successful implementation of C credit programs. Research and development are important in informing policy decisions and creating guidelines that guarantee the credibility and effectiveness of C markets. For instance, the development of standards by organizations like Verra and the Gold Standard helps to establish clear criteria for C credit generation, ensuring that projects are not only environmentally beneficial but also socially equitable (Enejison et al., 2022; Espenan, 2023).

16.8 Policy perspectives cropland carbon credit systems

The policy perspective on cropland C credit systems is increasingly important as governments and organizations seek to address climate change while promoting sustainable agricultural practices. Credit systems enable farmers to generate income from C_{seq} in their soils through sustainable farming practices. These credits can then be sold in C markets, providing an additional revenue stream for farmers while incentivizing environmentally friendly practices. A good example is the European Union's Emissions Trading System, which has begun to explore the inclusion of agricultural emissions and C_{seq} in its framework, which could significantly enhance the economic viability of sustainable farming practices across Europe (European Commission, 2021).

Internationally, the Kyoto Protocol and the Paris Agreement have laid the groundwork for C credit systems by establishing frameworks for emissions reductions and C trading. Under these agreements, countries can implement national policies that promote C_{seq} in agriculture. For example, New Zealand has developed a unique system that allows farmers to earn C credits through afforestation and reforestation projects, which are then sold in both domestic and international markets (NZMPI–New Zealand Ministry for Primary Industries, 2021). This approach incentivizes sustainable land use and helps to meet national GHG emissions minimization goals.

However, the effectiveness of cropland C credit systems is contingent upon robust monitoring, reporting, and verification frameworks. Policymakers need to ensure that C credit is economically sound and effective. Cutting-edge technologies such as RS and soil C modeling are being integrated into these systems to enhance accuracy and transparency. For instance, the use of satellite imagery to monitor land-use changes and soil health can provide reliable data for assessing C credits, thereby increasing trust among C market stakeholders (Oldfield et al., 2022; Nungula et al., 2024).

For a successful C credit system, it is essential to ensure that local communities and other stakeholders are effectively engaged in the design and implementation. Policymakers need to consider the social dimensions of C credit programs to ensure that they do not disproportionately benefit large agribusinesses at the expense of smallholder farmers. Inclusive policies that provide training and resources to smaller farms can enhance participation and equity within C markets (Moges et al., 2020; Mwadalu et al., 2025a).

Moreover, educating farmers about the benefits of C credit systems is essential for fostering the widespread adoption of sustainable practices (Kandel et al., 2022). Extension services and outreach programs can play a critical role in informing farmers about the potential financial and environmental benefits of participating in C markets. For instance, pilot programs in regions such as the Midwest United States have shown that when farmers are provided with clear information and support on C credit generation, they are more likely to adopt practices that enhance C_{seq} (Fargione et al., 2018).

16.9 Conclusion

C credit estimation in farmland is a viable business approach that not only helps farmers financially but also contributes significantly to the advancement of sustainable agricultural practices and the fight against climate change. Accurate assessment methodologies should be used to encourage farmers to embrace C credit-generating practices that improve soil health while simultaneously enhancing biodiversity. However, clear rules and supportive regulations are required that encourage farmers to adopt practices that maximize C storage. There is an urgent need to educate farmers about the C credit industry and the importance of adopting sustainable agricultural methods. A collaborative strategy involving agricultural stakeholders, scientists, policymakers, and local people will be critical for harnessing the full potential of C credits. This involvement will accelerate collaborative efforts to have a significant impact on C_{seq} by establishing a mutually beneficial condition. Mitigating climate change while assuring sustainable food production and safeguarding our natural resources for future generations will therefore be achieved.

References

Acharya, P., Ghimire, R., Acosta-Martínez, V., 2024. Cover crop-mediated soil carbon storage and soil health in semi-arid irrigated cropping systems. Agriculture, Ecosystems & Environment 361, 108813. Available at https://doi.org/10.1016/j.agee.2023.108813.

Adetunji, A.T., Ncube, B., Mulidzi, R., Lewu, F.B., 2020. Management impact and benefit of cover crops on soil quality: a review. Soil and Tillage Research 204. Available at https://doi.org/10.1016/j.still.2020.104717.

Alam, M.K., Bell, R.W., Biswas, W.K., 2019. Increases in soil sequestered carbon under conservation agriculture cropping decrease the estimated greenhouse gas emissions of wetland rice using life cycle assessment. Journal of Cleaner Production 224, 72–87. Available at https://doi.org/10.1016/j.jclepro.2019.03.215.

Alhammad, B.A., Roy, D.K., Ranjan, S., Padhan, S.R., Sow, S., Nath, D., et al., 2023. Conservation tillage and weed management influencing weed dynamics, crop performance, soil properties, and profitability in a rice–wheat–greengram system in the eastern indo-gangetic plain. Agronomy 13 (7). Available at https://doi.org/10.3390/agronomy13071953.

Alves, V.D.P., Diniz, M.B., 2022. Reducing carbon emissions from avoided deforestation in the Brazilian Amazon: an approach based on the Business-as-Usual (BAU) scenario. Revista de Gestão Ambiental e Sustentabilidade (GeAS) 11 (1).

Autret, B., Beaudoin, N., Rakotovololona, L., Bertrand, M., Grandeau, G., Gréhan, E., et al., 2019. Can alternative cropping systems mitigate nitrogen losses and improve GHG balance? Results from a 19-yr experiment in Northern France. Geoderma 342, 20–33. Available at https://doi.org/10.1016/j.geoderma.2019.01.039.

Azadi, H., Taheri, F., Burkart, S., Mahmoudi, H., De Maeyer, P., Witlox, F., 2021. Impact of agricultural land conversion on climate change. Environment, Development and Sustainability 23 (3), 3187–3198. Available at https://doi.org/10.1007/s10668-020-00712-2.

Barros, FdeV., Lewis, K., Robertson, A.D., Pennington, R.T., Hill, T.C., Matthews, C., et al., 2023. Cost-effective restoration for carbon sequestration across Brazil's biomes. Science of the Total Environment 876, 162600. Available at https://doi.org/10.1016/j.scitotenv.2023.162600.

Baumgarten, A., Geitner, C., Haslmayr, H.P., Zechmeister-Boltenstern, S., Bohner, A., Blum, W.E. et al., 2014. Kapitel 5: Der Einfluss des Klimawandels auf die Pedosphäre.

Bellarby, J., Foereid, B., Hastings, A.F.S.J., Smith, P., 2008. Cool farming: climate impacts of agriculture and mitigation potential. https://www.researchgate.net/publication/237224382_Cool_Farming_Climate_Impacts_of_Agriculture_and_Mitigation_Potential

Beniston, J.W., Shipitalo, M.J., Lal, R., Dayton, E.A., Hopkins, D.W., Jones, F., et al., 2015. Carbon and macronutrient losses during accelerated erosion under different tillage and residue management. European Journal of Soil Science 66 (1), 218–225. Available at https://doi.org/10.1111/ejss.12205.

Beyer, R.M., Hua, F., Martin, P.A., Manica, A., Rademacher, T., 2022. Relocating croplands could drastically reduce the environmental impacts of global food production. Communications Earth and Environment 3 (1). Available at https://doi.org/10.1038/s43247-022-00360-6.

Bhardwaj, A.K., Rajwar, D., Mandal, U.K., Ahamad, S., Kaphaliya, B., Minhas, P.S., et al., 2019. Impact of carbon inputs on soil carbon fractionation, sequestration and biological responses under major nutrient management practices for rice-wheat cropping systems. Scientific Reports 9 (1). Available at https://doi.org/10.1038/s41598-019-45534-z.

Bhattacharyya, S.S., Ros, G.H., Furtak, K., Iqbal, H.M., Parra-Saldívar, R., 2022. Soil carbon sequestration–an interplay between soil microbial community and soil organic matter dynamics. Science of the Total Environment 815, 152928.

Bithas, K., Latinopoulos, D., 2021. Managing tree-crops for climate mitigation. An economic evaluation trading-off carbon sequestration with market goods. Sustainable Production and Consumption 27, 667–678. Available at https://doi.org/10.1016/j.spc.2021.01.033.

Blanco-Canqui, H., Francis, C.A., Galusha, T.D., 2017. Does organic farming accumulate carbon in deeper soil profiles in the long term? Geoderma 288, 213–221. Available at https://doi.org/10.1016/j.geoderma.2016.10.031.

Boniface, H.S., 2023. Quantifying the impacts of climate-smart farming practices for improved management and long-term carbon storage (Master's thesis, University of Maryland, College Park).

Bradshaw, C.J.A., Bowman, D.M.J.S., Bond, N.R., Murphy, B.P., Moore, A.D., Fordham, D.A., et al., 2013. Brave new green world - consequences of a carbon economy for the conservation of Australian biodiversity. Biological Conservation 161, 71–90. Available at https://doi.org/10.1016/j.biocon.2013.02.012.

Brummitt, C.D., Mathers, C.A., Keating, R.A., O'Leary, K., Easter, M., Friedl, M.A., et al., 2024. Solutions and insights for agricultural monitoring, reporting, and verification (MRV) from three consecutive issuances of soil carbon credits. Journal of Environmental Management 369. Available at https://doi.org/10.1016/j.jenvman.2024.122284.

Cariappa, A.A.G., Konath, N.C., Sapkota, T.B., Krishna, V.V., 2024. Evaluating the potential and eligibility of conservation agriculture practices for carbon credits. Scientific Reports 14 (1). Available at https://doi.org/10.1038/s41598-024-59262-6.

Chahal, I., Vyn, R.J., Mayers, D., Van Eerd, L.L., 2020. Cumulative impact of cover crops on soil carbon sequestration and profitability in a temperate humid climate. Scientific Reports 10 (1). Available at https://doi.org/10.1038/s41598-020-70224-6.

Chappa, L.R., Nungula, E.Z., Ngaiza, V.V., Mwadalu, R., Makwinja, Y.H., Dlamini, J.C., et al., 2025. The challenges and solutions for carbon credits. In: Raj, A., Jhariya, M.K., Meena, R.S. (Eds.), Carbon Trading in Agroforestry Under Changing Climate. Springer Nature, Cham, pp. 525–549. Available at https://doi.org/10.1007/978-3-031-97151-8_20.

Chen, X., Lin, B., 2021. Towards carbon neutrality by implementing carbon emissions trading scheme: policy evaluation in China. Energy Policy 157, 112510. Available at https://doi.org/10.1016/j.enpol.2021.112510.

Cheruto, G., Nungula, E.Z., Nyawira, L., Chappa, L.R., Kahuthia-Gathu, R., Mwadalu, R., et al., 2025. Agroforestry tree species: Acacia tortilis (Vachellia tortilis), biology, importance, agroforestry production, and biotechnology application. In: Kalia, R.K., Pathak, R. (Eds.), Tree Biology and Biotechnology. Springer Nature, Singapore, pp. 145–161. Available at https://doi.org/10.1007/978-981-96-0002-1_10.

Chirambo, D., 2019. Climate finance as a catalyst for initiating an african green revolution to enhance drought resilience: policy prospects and challenges. In: Leal Filho, W. (Ed.), Handbook of Climate Change Resilience. Springer, Cham, pp. 1843–1857. Available at https://doi.org/10.1007/978-3-319-93336-8_105.

Ciais, P., Dolman, A.J., Bombelli, A., Duren, R., Peregon, A., Rayner, P.J., et al., 2014. Current systematic carbon-cycle observations and the need for implementing a policy-relevant carbon observing system. Biogeosciences 11 (13), 3547–3602. Available at https://doi.org/10.5194/bg-11-3547-2014.

Clark, A.J., et al., 2019. Cover crops increase soil carbon and nitrogen. Agronomy J 111 (5), 2185–2196. Available at https://doi.org/10.2134/agronj2019.01.0030.

Cottney, P., Black, L., Williams, P., White, E., 2022. How cover crop sowing date impacts upon their growth, nutrient assimilation and the yield of the subsequent commercial crop. Agronomy 12 (2), 369. Available at https://doi.org/10.3390/agronomy12020369.

Criscuoli, I., Martelli, A., Falconi, I., Galioto, F., Lasorella, M.V., Maurino, S., et al., 2024. Lessons learned from existing carbon removal methodologies for agricultural soils to drive European Union policies. European Journal of Soil Science 75 (5). Available at https://doi.org/10.1111/ejss.13577.

Croft, H., Kuhn, N.J., Anderson, K., 2012. On the use of remote sensing techniques for monitoring spatio-temporal soil organic carbon dynamics in agricultural systems. CATENA 94, 64–74. Available at https://doi.org/10.1016/j.catena.2012.01.001.

Crézé, C.M., Horwath, W.R., 2021. Cover cropping: a malleable solution for sustainable agriculture? Meta-analysis of ecosystem service frameworks in perennial systems. Agronomy 11 (5). Available at https://doi.org/10.3390/agronomy11050862.

Dahal, N., Bajracharya, R.M., 2010. Prospects of soil organic carbon sequestration: implications for Nepal's mountain agriculture. Journal of Forest Livelihood 9 (1), 45–56.

Devi, A., Jhariya, M.K., Raj, A., Banerjee, A., Singh, K.P., Singh, B., 2024. Acacia nilotica: a promising species for soil sustainability. Land and Environmental Management through Forestry. wiley, India, pp. 339–353. Available at https://doi.org/10.1002/9781119910527.ch14.

Dignac, M.F., Derrien, D., Barré, P., Barot, S., Cécillon, L., Chenu, C., et al., 2017. Increasing soil carbon storage: mechanisms, effects of agricultural practices and proxies. A review. Agronomy for Sustainable Development 37 (2). Available at https://doi.org/10.1007/s13593-017-0421-2.

Dotto, A.C., Dalmolin, R.S.D., ten Caten, A., Grunwald, S., 2018. A systematic study on the application of scatter-corrective and spectral-derivative preprocessing for multivariate prediction of soil organic carbon by Vis-NIR spectra. Geoderma 314, 262–274. Available at https://doi.org/10.1016/j.geoderma.2017.11.006.

Dupla, X., Bonvin, E., Deluz, C., Lugassy, L., Verrecchia, E., Baveye, P.C., et al., 2024. Are soil carbon credits empty promises? Shortcomings of current soil carbon quantification methodologies and improvement avenues. Soil Use and Management 40 (3). Available at https://doi.org/10.1111/sum.13092.

Enejison, M., Ejide, O., Nemanic, C., 2022. Blocks and credits: a sustainability lens on blockchain technology in voluntary carbon markets.

Espenan, N.P., 2023. Improving voluntary carbon markets through standardization and blockchain technology. Wyoming Law Review 23 (1), 141–177. Available at https://doi.org/10.59643/1942-9916.1473.

European Commission, 2021. Fit for 55: Delivering the EU's 2030 Climate Target on the Way to Climate Neutrality. Available from: https://ec.europa.eu/clima/policies/eu-climate-action/fit-for-55_en.

Everhart, S., 2023. Growing carbon credits: strengthening the agricultural sector's participation in voluntary carbon markets through law and policy. NYU Env't LJ 31, 65.

Fargione, J.E., Bassett, S., Boucher, T., Bridgham, S.D., Conant, R.T., Cook-Patton, S.C., et al., 2018. Natural climate solutions for the United States. Science Advances 4 (11). Available at https://doi.org/10.1126/sciadv.aat1869.

Francaviglia, R., Di Bene, C., Farina, R., Salvati, L., Vicente-Vicente, J.L., 2019. Assessing "4 per 1000" soil organic carbon storage rates under Mediterranean climate: a comprehensive data analysis. Mitigation and Adaptation Strategies for Global Change 24 (5), 795–818. Available at https://doi.org/10.1007/s11027-018-9832-x.

Freibauer, A., Rounsevell, M.D.A., Smith, P., Verhagen, J., 2004. Carbon sequestration in the agricultural soils of Europe. Geoderma 122 (1), 1–23. Available at https://doi.org/10.1016/j.geoderma.2004.01.021.

Garbach, K., Milder, J.C., Montenegro, M., Karp, D.S., DeClerck, F.A.J., 2014. Biodiversity and ecosystem services in agroecosystems. In: Van Alfen, N.K. (Ed.), Encyclopedia of Agriculture and Food Systems. Elsevier BV, pp. 21–40. Available at https://doi.org/10.1016/b978-0-444-52512-3.00013-9.

Gattinger, A., Muller, A., Haeni, M., Skinner, C., Fliessbach, A., Buchmann, N., et al., 2012. Enhanced top soil carbon stocks under organic farming. Proceedings of the National Academy of Sciences of the United States of America 109 (44), 18226–18231. Available at https://doi.org/10.1073/pnas.1209429109.

Girijaveni, V., Reddy, K.S., Prasad, J.V.N.S., Singh, V.K., 2023. Carbon footprint and sustainability of different agricultural production systems in climate change scenario. Handbook of Energy Management in Agriculture. Springer Nature, India, pp. 77–100. Available at https://doi.org/10.1007/978-981-19-7736-7_6-1.

Gitari, H.I., Gachene, C.K.K., Karanja, N.N., Kamau, S., Nyawade, S., Schulte-Geldermann, E., 2019. Potato-legume intercropping on a sloping terrain and its effects on soil physico-chemical properties. Plant and Soil 438, 447–460. Available at https://doi.org/10.1007/s11104-019-04036-7.

Gujba, H., Thorne, S., Mulugetta, Y., Rai, K., Sokona, Y., 2012. Financing low carbon energy access in Africa. Energy Policy 47 (1), 71–78. Available at https://doi.org/10.1016/j.enpol.2012.03.071.

Gupta, A., Lövbrand, E., Turnhout, E., Vijge, M.J., 2012. In pursuit of carbon accountability: The politics of REDD+ measuring, reporting and verification systems. Current Opinion in Environmental Sustainability 4 (6), 726–731. Available at https://doi.org/10.1016/j.cosust.2012.10.004.

Hannay, J.A., 2023. Private carbon credit initiatives in the agricultural sector: investigating motivations and understanding their effects (Master's thesis, University of Waterloo).

Henderson, B., Lankoski, J., Flynn, E., Sykes, A., Payen, F.T., MacLeod, M., 2022. Soil carbon sequestration by agriculture: policy options.

Henry, B., Dalal, R., Harrison, M.T., Keating, B., 2023. Creating frameworks to foster soil carbon sequestration.

Hussain, S., Hussain, S., Guo, R., Sarwar, M., Ren, X., Krstic, D., et al., 2021. Carbon sequestration to avoid soil degradation: a review on the role of conservation tillage. Plants 10 (10), 2001. Available at https://doi.org/10.3390/plants10102001.

Jarecki, M.K., Lal, R., 2003. Crop management for soil carbon sequestration. Critical Reviews in Plant Sciences 22 (6), 471–502. Available at https://doi.org/10.1080/713608318.

Jayaraman, S., Naorem, A., Dalal, R.C., Sinha, N.K., Srinivasa Rao, C., Lal, R., et al., 2024. No-till farming and climate change mitigation: lessons learnt from long-term no-till experiments and future perspectives. Advances in Agronomy 187, 21–107. Available at https://doi.org/10.1016/bs.agron.2024.05.005.

Jhariya, M.K., Banerjee, A., Swaroop Meena, R., Kumar, S., Raj, A., 2021. Sustainable Intensification for Agroecosystem Services and Management. Springer, Singapore, pp. 1–870. Available at https://doi.org/10.1007/978-981-16-3207-5.

Jhariya, M.K., Meena, R.S., Banerjee, A., 2021b. Ecological Intensification of Natural Resources for Sustainable Agriculture. Springer Singapore, India, pp. 1–655. Available at https://doi.org/10.1007/978-981-33-4203-3.

Jhariya, M.K., Meena, R.S., Banerjee, A., Meena, S.N., 2022. Natural Resources Conservation and Advances for Sustainability. Elsevier. Available at https://doi.org/10.1016/C2019-0-03763-6.

Jhariya, M.K., Meena, R.S., Banerjee, A., Kumar, S., Raj, A., 2023. Agroforestry for Carbon and Ecosystem Management. Elsevier, India, pp. 1–434. Available at https://doi.org/10.1016/C2021-0-02916-3.

Jha, S., Bacon, C.M., Philpott, S.M., Rice, R.A., Méndez, V.E., Läderach, P., 2011. A review of ecosystem services, farmer livelihoods, and value chains in shade coffee agroecosystems. In: Campbell, W., Lopez Ortiz, S. (Eds.), Integrating Agriculture, Conservation and Ecotourism: Examples from the Field. Issues in Agroecology – Present Status and Future Prospectus, vol 1. Springer, Dordrecht, pp. 141–208. Available at https://doi.org/10.1007/978-94-007-1309-3_4.

Jian, J., Du, X., Reiter, M.S., Stewart, R.D., 2020. A meta-analysis of global cropland soil carbon changes due to cover cropping. Soil Biology and Biochemistry 143. Available at https://doi.org/10.1016/j.soilbio.2020.107735.

Joseph, S.R., 2002. An economic analysis of the influence of carbon credits on farm management practices (Doctoral dissertation, University of Guelph).

Kamyab, H., SaberiKamarposhti, M., Hashim, H., Yusuf, M., 2024. Carbon dynamics in agricultural greenhouse gas emissions and removals: a comprehensive review. Carbon Letters 34 (1), 265–289. Available at https://doi.org/10.1007/s42823-023-00647-4.

Kandel, M., Anghileri, D., Alare, R.S., Lovett, P.N., Agaba, G., Addoah, T., et al., 2022. Farmers' perspectives and context are key for the success and sustainability of farmer-managed natural regeneration (FMNR) in northeastern Ghana. World Development 158. Available at https://doi.org/10.1016/j.worlddev.2022.106014.

Khurana, A., Kajale, D., Cariappa, A.A.G., Krishna, V.V., 2024. Shaping India's climate future: a perspective on harnessing carbon credits from agriculture. Outlook on Agriculture 53 (2), 113–130. Available at https://doi.org/10.1177/00307270241240778.

Kiran, K.K., Pal, S., Chand, P., Kandpal, A., 2023. Carbon sequestration potential of sustainable agricultural practices to mitigate climate change in Indian agriculture: a meta-analysis. Sustainable Production and Consumption 35, 697–708. Available at https://doi.org/10.1016/j.spc.2022.12.015.

Koehler-Cole, K., Elmore, R.W., 2020. Seeding rates and productivity of broadcast interseeded cover crops. Agronomy 10 (11). Available at https://doi.org/10.3390/agronomy10111723.

Koga, N., Shimoda, S., Shirato, Y., Kusaba, T., Shima, T., Niimi, H. et al., 2020. Assessing changes in soil carbon stocks after land use conversion from forest land.

Kraus, D., Weller, S., Klatt, S., Santabárbara, I., Haas, E., Wassmann, R., et al., 2016. How well can we assess impacts of agricultural land management changes on the total greenhouse gas balance (CO_2, CH_4 and N_2O) of tropical rice-cropping systems with a biogeochemical model? Agriculture, Ecosystems and Environment 224, 104–115. Available at https://doi.org/10.1016/j.agee.2016.03.037.

Kuhnert, M., Vetter, S., Smith, P., 2022. Measuring and monitoring soil carbon sequestration. In: Rumpel, C. (Ed.), Understanding and Fostering Soil Carbon Sequestration. Burleigh Dodds Science Publishing. Available at https://shop.bdspublishing.com/store/bds/detail/product/3-190-9781801467063.

Kumar, K., Das, D., 2014. Carbon dioxide sequestration by biological processes. Transformation and Utilization of Carbon Dioxide. Springer, Berlin, Heidelberg, pp. 303–334.

Kumar, B.M., Kunhamu, T.K., 2022. Nature-based solutions in agriculture: A review of the coconut (*Cocos nucifera* L.)-based farming systems in Kerala, "the Land of Coconut Trees. Nature-Based Solutions 2, 100012. Available at https://doi.org/10.1016/j.nbsj.2022.100012.

Kumar, S., Meena, R.S., Sheoran, P., Jhariya, M.K., 2024. Regenerative Agriculture for Sustainable Food Systems. Springer, Singapore, pp. 1–493. Available at https://doi.org/10.1007/978-981-97-6691-8.

Kyriakarakos, G., Petropoulos, T., Marinoudi, V., Berruto, R., Bochtis, D., 2024. Carbon farming: bridging technology development with policy goals. Sustainability 16 (5), 1903. Available at https://doi.org/10.3390/su16051903.

Lessmann, M., Ros, G.H., Young, M.D., de Vries, W., 2022. Global variation in soil carbon sequestration potential through improved cropland management. Global Change Biology 28 (3), 1162–1177. Available at https://doi.org/10.1111/gcb.15954.

Li, H., et al., 2020. Calibration of DNDC model for nitrate leaching from an intensively cultivated region of Northern China. Geoderma 223-225, 108–118. Available at .

Liu, Z., Dugan, B., Masiello, C.A., Barnes, R.T., Gallagher, M.E., Gonnermann, H., 2016. Impacts of biochar concentration and particle size on hydraulic conductivity and DOC leaching of biochar-sand mixtures. Journal of Hydrology 533, 461–472. Available at https://doi.org/10.1016/j.jhydrol.2015.12.007.

Loges, R., Bunne, I., Reinsch, T., Malisch, C., Kluß, C., Herrmann, A., et al., 2018. Forage production in rotational systems generates similar yields compared to maize monocultures but improves soil carbon stocks. European Journal of Agronomy 97, 11–19. Available at https://doi.org/10.1016/j.eja.2018.04.010.

Lokuge, N., Anders, S., 2022. Carbon-credit systems in agriculture: a review of literature. The School Public Policy Publications 15.

Lorenz, K., Lal, R., 2012. Cropland soil carbon dynamics. In: Lal, R., Lorenz, K., Hüttl, R., Schneider, B., von Braun, J. (Eds.), Recarbonization of the Biosphere: Ecosystems and the Global Carbon Cycle. Springer, Dordrecht, pp. 303–346. Available at https://doi.org/10.1007/978-94-007-4159-1_14.

Maestrini, B., Nannipieri, P., Abiven, S., 2015. A meta-analysis on pyrogenic organic matter induced priming effect. GCB Bioenergy 7 (4), 577–590. Available at https://doi.org/10.1111/gcbb.12194.

Maitra, S., Sahoo, U., Sairam, M., Gitari, H.I., Rezaei-Chiyaneh, E., Battaglia, M.L., et al., 2023. Cultivating sustainability: a comprehensive review on intercropping in a changing climate. Research on Crops 24 (4), 702–715. Available at https://doi.org/10.31830/2348-7542.2023.ROC-1020.

Majumder, S., Neogi, S., Dutta, T., Powel, M.A., Banik, P., 2019. The impact of biochar on soil carbon sequestration: meta-analytical approach to evaluating environmental and economic advantages. Journal of Environmental Management 250, 109466. Available at https://doi.org/10.1016/j.jenvman.2019.109466.

Mapfumo, E., Baron, V.S., Lemke, R., Naeth, M.A., Chanasyk, D.S., Dick, C., 2023. Management impacts on organic carbon under continuous perennial grass, perennial grass-legume mixture, and annual cereals on a thick Black Chernozemic soil. Canadian Journal of Soil Science 104 (1), 108–123. Available at https://doi.org/10.1139/cjss-2023-0041.

Marks, A.B., 2019. Carbon farming our way out of climate change. Denver Law Review 97, 497.

Mathew, I., Shimelis, H., Mutema, M., Chaplot, V., 2017. What crop type for atmospheric carbon sequestration: results from a global data analysis. Agriculture, Ecosystems and Environment 243, 34–46. Available at https://doi.org/10.1016/j.agee.2017.04.008.

Mazzoncini, M., Antichi, D., Di Bene, C., Risaliti, R., Petri, M., Bonari, E., 2016. Soil carbon and nitrogen changes after 28 years of no-tillage management under Mediterranean conditions. European Journal of Agronomy 77, 156–165. Available at https://doi.org/10.1016/j.eja.2016.02.011.

Meena, R.S., Pradhan, G., 2023. Industrial garbage-derived biocompost enhances soil organic carbon fractions, CO_2 biosequestration, potential carbon credits and sustainability index in a rice-wheat ecosystem. Environmental Research 235, 116525. Available at https://doi.org/10.1016/j.envres.2023.116525.

Meena, R.S., Kumar, S., Yadav, G.S., 2020. Soil carbon sequestration in crop production. In: Meena, R. (Ed.), Nutrient Dynamics for Sustainable Crop Production. Springer, Singapore, pp. 1–39. Available at https://doi.org/10.1007/978-981-13-8660-2_1.

Meena, R.S., Kumar, S., Sheoran, S., Jhariya, M.K., Bhatt, R., Yadav, G.S., et al., 2021. Soil organic carbon restoration in India: Programs, policies, and thrust areas. Soil Organic Matter and Feeding the Future: Environmental and Agronomic Impacts. CRC Press, India, pp. 305–338. Available at https://doi.org/10.1201/9781003102762-13.

Meena, R.S., Pradhan, G., Singh, K., Kumar, S., Singh, A.K., Shashidhar, K.S., et al., 2024a. Agriculture models for restoring degraded land to enhance CO_2 biosequestration and carbon credits in the Vindhyan region of India. Science of the Total Environment 929. Available at https://doi.org/10.1016/j.scitotenv.2024.172661.

Meena, R.S., Lal, R., Kumar, S., Pradhan, G., Srinivasarao, C., Singh, A.K., et al., 2024b. Potential of Indian agriculture for capturing atmospheric CO_2 and monetizing carbon credits to the farmers: an overview and policy framework. Advances in Agronomy 188, 101–206. Available at https://doi.org/10.1016/bs.agron.2024.06.001.

Meena, R.S., Pradhan, G., Nalani, Kumar, S., Singh, A.K., Verma, S.K., et al., 2024c. Diversified cropping modules designed for soil restoration, CO_2 sequestration, and generating carbon credits. Land Degradation and Development 35 (14), 4314–4329. Available at https://doi.org/10.1002/ldr.5224.

Meena, R.S., Pradhan, G., Kumar, S., 2025b. Energy flow, eco-efficiency, and economic circulation with recycled industrial waste compost application in wheat and subsequent rice farming. Science of the Total Environment 967. Available at https://doi.org/10.1016/j.scitotenv.2025.178779.

Michaelowa, A., Shishlov, I., Hoch, S., Bofill, P., Espelage, A.,2019. Overview and comparison of existing carbon crediting schemes. https://www.researchgate.net/publication/336599849_Overview_and_comparison_of_existing_carbon_crediting_schemes

Minasny, B., Malone, B.P., McBratney, A.B., Angers, D.A., Arrouays, D., Chambers, A., et al., 2017. Soil carbon 4 per mille. Geoderma 292, 59–86. Available at https://doi.org/10.1016/j.geoderma.2017.01.002.

Mirsky, S.B., Ryan, M.R., Teasdale, J.R., Curran, W.S., Reberg-Horton, C.S., Spargo, J.T., et al., 2013. Overcoming weed management challenges in cover crop-based organic rotational no-till soybean production in the Eastern United States. Weed Technology 27, 193–203. Available at https://doi.org/10.1614/WT-D-12-00078.1.

Moges, A., et al., 2020. Community-based carbon credit initiatives in East Africa: challenges and opportunities. Environmental Science & Policy 113, 186–197. Available at https://doi.org/10.1016/j.envsci.2020.07.015.

Moore, E.A., 2022. Cover crops for semi-arid crop production (Doctoral dissertation, University of Wyoming).

Morari, F., Berti, A., Dal Ferro, N., Piccoli, I., 2019. Deep carbon sequestration in cropping systems. In: Lal, R., Francaviglia, R. (Eds.), Sustainable Agriculture Reviews 29. Sustainable Agriculture Reviews, vol 29. Springer, Cham, pp. 33–65. Available at https://doi.org/10.1007/978-3-030-26265-5_2.

Morgan, J.A., Follett, R.F., Allen, L.H., Del Grosso, S., Derner, J.D., Dijkstra, F., et al., 2010. Carbon sequestration in agricultural lands of the United States. Journal of Soil and Water Conservation 65 (1). Available at https://doi.org/10.2489/jswc.65.1.6A 6A.

Moukanni, N., Brewer, K.M., Gaudin, A.C.M., O'Geen, A.T., 2022. Optimizing carbon sequestration through cover cropping in mediterranean agroecosystems: synthesis of mechanisms and implications for management. Frontiers in Agronomy 4. Available at https://doi.org/10.3389/fagro.2022.844166.

Moutinho, P., Cenamo, M.C., Moreira, P.F., 2009. Reducing Carbon Emissions By Slowing Deforestation: REDD Initiatives in Brazil Avoided Deforestation: Prospects for Mitigating Climate Change. Brazil Routledge Taylor & Francis Group, Brazil, pp. 90–109. Available at https://doi.org/10.4324/9780203880999.

Mwadalu, R., Abdi, A.M., Rutto, M., Chappa, L.R., Nungula, E.Z., Rezaei-Chiyaneh, E., et al., 2025a. Potential of agroforestry for enhancing Africa's access to carbon markets: a review. In: Raj, A., Jhariya, M.K., Meena, R.S. (Eds.), Carbon Trading in Agroforestry Under Changing Climate. Springer Nature, Switzerland.

Müller, A., Bautze, L., Meier, M., Gattinger, A., Gall, E., Chatzinikolaou, E., et al., 2016. Organic farming, climate change mitigation and beyond. Reducing the Environmental Impacts of EU Agriculture.

Nair, P.K.R., et al., 2021. Agroforestry systems in a changing climate: challenges and opportunities. Climate Change 167 (3), 49. Available at https://doi.org/10.1007/s10584-021-03186-5.

Nanda, S., Reddy, S.N., Mitra, S.K., Kozinski, J.A., 2016. The progressive routes for carbon capture and sequestration. Energy Science and Engineering 4 (2), 99–122. Available at https://doi.org/10.1002/ese3.117.

Nardelli, C., 2012. Trust, but verify: ensuring the accuracy of carbon credits registered under the clean-development mechanism. George Washington Journal of Energy and Environmental Law 3, 204.

Nasar, J., Khan, W., Khan, M.Z., Gitari, H.I., Gbolayori, J.F., Moussa, A.A., et al., 2021. Photosynthetic activities and photosynthetic nitrogen use efficiency of maize crop under different planting patterns and nitrogen fertilization. Journal of Soil Science and Plant Nutrition 21 (3), 2274–2284. Available at https://doi.org/10.1007/s42729-021-00520-1.

Nawari, Syahza, A., Siregar, Y.I., 2021. Community-based mangrove forest management as ecosystem services provider for reducing CO_2 emissions with carbon credit system in Bengkalis District, Riau, Indonesia. Journal of Physics: Conference Series 2049 (1). Available at https://doi.org/10.1088/1742-6596/2049/1/012074.

Nayak, A.K., Rahman, M.M., Naidu, R., Dhal, B., Swain, C.K., Nayak, A.D., et al., 2019. Current and emerging methodologies for estimating carbon sequestration in agricultural soils: a review. Science of the Total Environment 665, 890–912. Available at https://doi.org/10.1016/j.scitotenv.2019.02.125.

Nazir, M.J., Li, G., Nazir, M.M., Zulfiqar, F., Siddique, K.H.M., Iqbal, B., et al., 2024. Harnessing soil carbon sequestration to address climate change challenges in agriculture. Soil and Tillage Research 237. Available at https://doi.org/10.1016/j.still.2023.105959.

Naz, M., Akhtar, K., Khan, A., Nizamani, G.S., 2021. Agricultural practices can reduce soil greenhouse gas emissions: challenges and future perspectives. Engineering Tolerance in Crop Plants Against Abiotic Stress. CRC Press, China, pp. 263–274. Available at https://doi.org/10.1201/9781003160717-13.

Ngaiza, V.V., Nungula, E.Z., Chappa, L.R., Mwadalu, R., Nyambele, K.A., Shankar, T., et al., 2025. Biotechnological approaches to fluoride remediation. In: Sharma, K. (Ed.), Fluorides in Drinking Water. Environmental Science and Engineering. Springer, Cham, pp. 163–188. Available at https://doi.org/10.1007/978-3-031-77247-4_7.

De Notaris, C., Jensen, J.L., Olesen, J.E., Stumpf da Silva, T., Rasmussen, J., Panagea, I., et al., 2021. Long-term soil quality effects of soil and crop management in organic and conventional arable cropping systems. Geoderma 403. Available at https://doi.org/10.1016/j.geoderma.2021.115383.

Nsabiyeze, A., Ma, R., Li, J., Luo, H., Zhao, Q., Tomka, J., et al., 2024. Tackling climate change in agriculture: a global evaluation of the effectiveness of carbon emission reduction policies. Journal of Cleaner Production 468, 142973. Available at https://doi.org/10.1016/j.jclepro.2024.142973.

Nungula, E.Z., Mugwe, J., Massawe, B.H.J., Seleiman, M.F., Ali, N., Gitari, H.I., 2024. GIS-AHP based approach in land evaluation and suitability assessment for sunflower (*Helianthus annus*) production. Cogent Food and Agriculture 10 (1). Available at https://doi.org/10.1080/23311932.2024.2309831.

Nungula, E.Z., Chappa, L.R., Ngaiza, V.V., Msangi, G.S., Dlamini, J.C., Saeed, A., et al., 2025a. Climate change and carbon emission trading. In: Raj, A., Jhariya, M.K., Meena, R.S. (Eds.), Carbon Trading in Agroforestry Under Changing Climate. Springer Nature, Switzerland.

Nungula, E.Z., Chappa, L.R., Ngaiza, V.V., Saeed, A., Ranjan, S., Sow, S., et al., 2025b. Agroforestry for monetising carbon credits. In: Raj, A., Jhariya, M.K., Meena, R.S. (Eds.), Carbon Trading in Agroforestry Under Changing Climate. Springer Nature, Switzerland.

Nungula, E.Z., Ngaiza, V.V., Chappa, L.R., Mwadalu, R., Nyambele, K.A., Uddin, S., et al., 2025. Fluoride in groundwater: causes, implications and mitigation measures. In: Sharma, K. (Ed.), Fluorides in Drinking Water. Environmental Science and Engineering. Springer, Cham, pp. 283–303. Available at https://doi.org/10.1007/978-3-031-77247-4_11.

Nyawade, S.O., Karanja, N.N., Gachene, C.K.K., Gitari, H.I., Schulte-Geldermann, E., Parker, M.L., 2019. Short-term dynamics of soil organic matter fractions and microbial activity in smallholder potato-legume intercropping systems. Applied Soil Ecology 142, 123–135. Available at https://doi.org/10.1016/j.apsoil.2019.04.015.

NZMPI–New Zealand Ministry for Primary Industries, 2021. Emissions trading scheme. Available from: https://www.mpi.govt.nz/our-work/climate-change/emissions-trading-scheme/.

Odebiri, O., Odindi, J., Mutanga, O., 2021. Basic and deep learning models in remote sensing of soil organic carbon estimation: a brief review. International Journal of Applied Earth Observation and Geoinformation 102, 102389. Available at https://doi.org/10.1016/j.jag.2021.102389.

Odhong', C., Wilkes, A., van Dijk, S., Vorlaufer, M., Ndonga, S., Sing'ora, B., et al., 2019. Financing large-scale mitigation by smallholder farmers: what roles for public climate finance? Frontiers in Sustainable Food Systems 3. Available at https://doi.org/10.3389/fsufs.2019.00003.

Oldfield, E.E., Eagle, A.J., Rubin, R.L., Rudek, J., Sanderman, J., Gordon, D.R., 2022. Crediting agricultural soil carbon sequestration. Science (New York, N.Y.) 375 (6586), 1222–1225. Available at https://doi.org/10.1126/science.abl7991.

Oscar Kisaka, M., Shisanya, C., Cournac, L., Raphael Manlay, J., Gitari, H., Muriuki, J., 2023. Integrating no-tillage with agroforestry augments soil quality indicators in Kenya's dry-land agroecosystems. Soil and Tillage Research 227. Available at https://doi.org/10.1016/j.still.2022.105586.

Osumba, J.J.L., Recha, J.W., Oroma, G.W., 2021. Transforming agricultural extension service delivery through innovative bottom–up climate-resilient agribusiness farmer field schools. Sustainability 13 (7), 3938. Available at https://doi.org/10.3390/su13073938.

Pasket, A.M., 2023. Seasonal dynamics and impact of P fertilizer treatments and cover crops on soil health indicators in a no-till corn-soybean system (Doctoral dissertation, Kansas State University).

Paul, M.J., Foyer, C.H., 2001. Sink regulation of photosynthesis. Journal of Experimental Botany 52 (360), 1383–1400. Available at https://doi.org/10.1093/jexbot/52.360.1383.

Paustian, K., Lehmann, J., Ogle, S., Reay, D., Robertson, G.P., Smith, P., 2016. Climate-smart soils. Nature 532 (7597), 49–57. Available at https://doi.org/10.1038/nature17174.

Paustian, K., Collier, S., Baldock, J., Burgess, R., Creque, J., DeLonge, M., et al., 2019. Quantifying carbon for agricultural soil management: from the current status toward a global soil information system. Carbon Management 10 (6), 567–587. Available at https://doi.org/10.1080/17583004.2019.1633231.

Petito, M., Cantalamessa, S., Pagnani, G., Pisante, M., 2024. Modelling and mapping soil organic carbon in annual cropland under different farm management systems in the Apulia region of Southern Italy. Soil and Tillage Research 235, 105916. Available at https://doi.org/10.1016/j.still.2023.105916.

Phelan, L., Chapman, P.J., Ziv, G., 2024. The emerging global agricultural soil carbon market: the case for reconciling farmers' expectations with the demands of the market. Environmental Development 49, 100941. Available at https://doi.org/10.1016/j.envdev.2023.100941.

Piñeiro, V., Arias, J., Dürr, J., Elverdin, P., Ibáñez, A.M., Kinengyere, A., et al., 2020. A scoping review on incentives for adoption of sustainable agricultural practices and their outcomes. Nature Sustainability 3 (10), 809–820. Available at https://doi.org/10.1038/s41893-020-00617-y.

Poeplau, C., Don, A., 2015. Carbon sequestration in agricultural soils via cultivation of cover crops - a meta-analysis. Agriculture, Ecosystems and Environment 200, 33–41. Available at https://doi.org/10.1016/j.agee.2014.10.024.

Pradhan, G., Meena, R.S., 2023a. Utilizing waste compost to improve the atmospheric CO_2 capturing in the rice-wheat cropping system and energy-cum-carbon credit auditing for a circular economy. Science of the Total Environment 892. Available at https://doi.org/10.1016/j.scitotenv.2023.164572.

Pradhan, G., Meena, R.S., 2023b. Interaction impact of biocompost on nutrient dynamics and relations with soil biota, carbon fractions index, societal value of CO_2 equivalent and ecosystem services in the wheat-rice farming. Chemosphere 339. Available at https://doi.org/10.1016/j.chemosphere.2023.139695.

Pradhan, G., Meena, R.S., Kumar, S., Lal, R., 2023. Utilizing industrial wastes as compost in wheat-rice production to improve the above and below-ground ecosystem services. Agriculture, Ecosystems & Environment 358, 108704. Available at https://doi.org/10.1016/j.agee.2023.108704.

Prasad, R., Jhariya, M.K., Banerje, A., 2021. Advances in Sustainable Development and Management of Environmental and Natural Resources: Economic Outlook and Opinions: Volume 11. Apple Academic Press, India, pp. 1–428. Available at https://doi.org/10.1201/9781003187455.

Prasad, R., Jhariya, M.K., Banerjee, A., 2021b. Advances in sustainable development and management of environmental and natural resources: economic outlook and opinions Volume II, pp. 1–428. https://doi.org/10.1201/9781003187455.

Prescott, C.E., Vesterdal, L., 2021. Decomposition and transformations along the continuum from litter to soil organic matter in forest soils. Forest Ecology and Management 498, 119522. Available at https://doi.org/10.1016/j.foreco.2021.119522.

Pudasaini, K., Bhattarai, T., Rolfe, J., 2025. Exploring the barriers to farmer participation in soil carbon projects under the Australian Carbon Credit Unit Scheme. Australasian Journal of Environmental Management 32 (2), 97–115. Available at https://doi.org/10.1080/14486563.2024.2393246.

Pusparini, W., Cahyana, A., Grantham, H.S., Maxwell, S., Soto-Navarro, C., Macdonald, D.W., 2023. A bolder conservation future for Indonesia by prioritising biodiversity, carbon and unique ecosystems in Sulawesi. Scientific Reports 13 (1). Available at https://doi.org/10.1038/s41598-022-21536-2.

Raina, N., Zavalloni, M., Viaggi, D., 2024. Incentive mechanisms of carbon farming contracts: a systematic mapping study. Journal of Environmental Management 352, 120126. Available at https://doi.org/10.1016/j.jenvman.2024.120126.

Raj, A., Jhariya, M.K., Meena, R.S., 2025. Agroforestry for Monetising Carbon Credits. Springer Nature, Switzerland.

Raj, A., Jhariya, M.K., Banerjee, A., Jha, R.K., Singh, K.P., Alnemari, A.M., et al., 2024. Outlooks on major agroforestry systems. Agroforestry. Wiley, pp. 21–48. Available at https://doi.org/10.1002/9781394231164.ch2.

Reddy, A.R., Rasineni, G.K., Raghavendra, A.S., 2010. The impact of global elevated CO_2 concentration on photosynthesis and plant productivity. Current Science 99 (1), 46–57.

Robertson, G.P., Grace, P.R., 2019. Nitrous oxide emissions from soils. Nature Sustainability 2 (9), 929–940. Available at https://doi.org/10.1038/s41893-019-0286-2.

Roberts, M., Thibaudeau, P., 2024. Soil carbon sequestration in building life cycle assessment: offsetting measure or site impact. IOP Conference Series: Earth and Environmental Science 1363 (1), 012061. Available at https://doi.org/10.1088/1755-1315/1363/1/012061.

Sahu, S., Arya, S.K., 2024. Mitigating greenhouse gas emissions in agriculture and food processing through sustainable management practices for climate change. Advances and Technology Development in Greenhouse Gases: Emission, Capture and Conversion. Elsevier, pp. 71–96.

Sah, P.K., Paudel, S., Oli, B., Shrestha, A., 2024. Greenhouse gases (GHG) emissions from agricultural soil: a review. SSRN, Nepal SSRN. Available from: https://www.ssrn.com/index.cfm/en/. doi: 10.2139/ssrn.4826174.

Sainju, U.M., Allen, B.L., Lenssen, A.W., Mikha, M., 2017. Root and soil total carbon and nitrogen under bioenergy perennial grasses with various nitrogen rates. Biomass and Bioenergy 107, 326–334. Available at https://doi.org/10.1016/j.biombioe.2017.10.021.

Salma, A., Fryda, L., Djelal, H., 2024. Biochar: a key player in carbon credits and climate mitigation. Resources 13 (2), 31. Available at https://doi.org/10.3390/resources13020031.

Santier, T., 2023. The next priorities of the voluntary carbon market for mass adoption: the need for new technologies, carbon policy frameworks, and a meta registry (Doctoral dissertation, Université Paris-Saclay).

Schumacher, B.A., 2002. Methods for Determining Total Organic Carbon (TOC) in Soils and Sediments. US Environmental Protection Agency, Office of Research and Development, Ecological Risk Assessment Support Center, Washington, DC, pp. 1–23.

Shi, P., Qin, Y., Liu, Q., Zhu, T., Li, Z., Li, P., et al., 2020. Soil respiration and response of carbon source changes to vegetation restoration in the Loess Plateau, China. Science of the Total Environment 707, 135507. Available at https://doi.org/10.1016/j.scitotenv.2019.135507.

Siarudin, M., Rahman, S.A., Artati, Y., Indrajaya, Y., Narulita, S., Ardha, M.J., et al., 2021. Carbon sequestration potential of agroforestry systems in degraded landscapes in west Java, Indonesia. Forests 12 (6). Available at https://doi.org/10.3390/f12060714.

Singh, S., et al., 2020. Economic viability of conservation agriculture in Punjab, India: a carbon credit perspective. Journal of Cleaner Production 244, 118594. Available at https://doi.org/10.1016/j.jclepro.2019.118594.

Smith, P., Cotrufo, M.F., Rumpel, C., Paustian, K., Kuikman, P.J., Elliott, J.A., et al., 2015. Biogeochemical cycles and biodiversity as key drivers of ecosystem services provided by soils. SOIL 1 (2), 665–685. Available at https://doi.org/10.5194/soil-1-665-2015.

Smith, P., Calvin, K., Nkem, J., Campbell, D., Cherubini, F., Grassi, G., et al., 2020. Which practices co-deliver food security, climate change mitigation and adaptation, and combat land degradation and desertification? Global Change Biology 26 (3), 1532–1575. Available at https://doi.org/10.1111/gcb.14878.

Somasundaram, J., Sinha, N.K., Dalal, R.C., Lal, R., Mohanty, M., Naorem, A.K., et al., 2020. No-till farming and conservation agriculture in South Asia – issues, challenges, prospects and benefits. Critical Reviews in Plant Sciences 39 (3), 236–279. Available at https://doi.org/10.1080/07352689.2020.1782069.

Sow, S., Ranjan, S., Kumar, N., Nilanjaya, Gitari, H., Dayal, P., et al., 2024. Sustainable fodder production in South Asia through silvopastoral systems. Current Science 126 (10), 1217–1224. Available at https://doi.org/10.18520/cs/v126/i10/1217-1224.

Springer, U., 2003. The market for tradable GHG permits under the Kyoto Protocol: a survey of model studies. Energy Economics 25 (5), 527–551. Available at https://doi.org/10.1016/s0140-9883(02)00103-2.

Stevens, Nocita, Tóth, Montanarella, van Wesemael, 2013. Prediction of soil organic carbon at the European scale by visible and near-infrared reflectance spectroscopy. PLoS One 8 (6).

Stolbovoy, V., Montanarella, L., Filippi, N., Jones, A., Gallego, J., Grassi, G., 2005. Soil sampling protocol to certify the changes of organic carbon stock in mineral soils of European Union. Office for Official Publications of the European Communities. Available at https://esdac.jrc.ec.europa.eu/ESDB_Archive/eusoils_docs/other/EUR21576_2.pdf.

Tao, F., Palosuo, T., Valkama, E., Mäkipää, R., 2019. Cropland soils in China have a large potential for carbon sequestration based on literature survey. Soil and Tillage Research 186, 70–78. Available at https://doi.org/10.1016/j.still.2018.10.009.

Thapa, S., 2019. Sharing the carbon revenues: mainstreaming carbon and credit markets for renewable energy technologies in Nepal.

Thompson, N.M., Hughes, M.N., Nuworsu, E.K., Reeling, C.J., Armstrong, S.D., Mintert, J.R., et al., 2022. Opportunities and challenges associated with "carbon farming" for US row-crop producers. Choices (New York, N.Y.) 37 (3), 1–10.

Tiefenbacher, A., Sandén, T., Haslmayr, H.P., Miloczki, J., Wenzel, W., Spiegel, H., 2021. Optimizing carbon sequestration in croplands: a synthesis. Agronomy 11 (5). Available at https://doi.org/10.3390/agronomy11050882.

Toensmeier, E., 2016. The Carbon Farming Solution: A Global Toolkit of Perennial Crops and Regenerative Agriculture Practices for Climate Change Mitigation and Food Security. Chelsea Green Publishing.

Tongwane, M.I., Moeletsi, M.E., 2018. A review of greenhouse gas emissions from the agriculture sector in Africa. Agricultural Systems 166, 124–134. Available at https://doi.org/10.1016/j.agsy.2018.08.011.

Trost, B., Prochnow, A., Drastig, K., Meyer-Aurich, A., Ellmer, F., Baumecker, M., 2013. Irrigation, soil organic carbon and N_2O emissions. A review. Agronomy for Sustainable Development 33 (4), 733–749. Available at https://doi.org/10.1007/s13593-013-0134-0.

Von Unger, M., Emmer, I., 2018. Carbon Market Incentives to Conserve, Restore and Enhance Soil Carbon. Silvestrum and The Nature Conservancy, Arlington.

Velvizhi, G., Nair, R., Goswami, C., Arumugam, S.K., Shetti, N.P., Aminabhavi, T.M., 2023. Carbon credit reduction: a techno-economic analysis of "drop-in" fuel production. Environmental Pollution 316. Available at https://doi.org/10.1016/j.envpol.2022.120507.

Vico, G., Brunsell, N.A., 2018. Tradeoffs between water requirements and yield stability in annual vs. perennial crops. Advances in Water Resources 112, 189–202. Available at https://doi.org/10.1016/j.advwatres.2017.12.014.

Watts, J.D., Natali, S.M., Minions, C., Risk, D., Arndt, K., Zona, D., et al., 2021. Soil respiration strongly offsets carbon uptake in Alaska and Northwest Canada. Environmental Research Letters 16 (8). Available at https://doi.org/10.1088/1748-9326/ac1222.

Weston, P., Hong, R., Kaboré, C., Kull, C.A., 2015. Farmer-managed natural regeneration enhances rural livelihoods in dryland West Africa. Islands Environmental Management 55 (6), 1402–1417. Available at https://doi.org/10.1007/s00267-015-0469-1.

White, R.E., 2022. The role of soil carbon sequestration as a climate change mitigation strategy: an Australian case study. Soil Systems 6 (2). Available at https://doi.org/10.3390/soilsystems6020046.

Woolf, D., Amonette, J.E., Street-Perrott, F.A., Lehmann, J., Joseph, S., 2010. Sustainable biochar to mitigate global climate change. Nature Communications 1 (5). Available at https://doi.org/10.1038/ncomms1053.

Wotherspoon, A., Voroney, R.P., Thevathasan, N.V., Gordon, A.M., 2015. Comparison of three methods for measurement of soil organic carbon. Communications in Soil Science and Plant Analysis 46 (sup1), 362–374. Available at https://doi.org/10.1080/00103624.2014.989111.

Wu, Y., Wu, Y., 2022. The increase in the karstification–photosynthesis coupled carbon sink and its implication for carbon neutrality. Agronomy 12 (9), 2147. Available at https://doi.org/10.3390/agronomy12092147.

Xie, E., Chen, J., Peng, Y., Yan, G., Zhao, Y., 2024. Historical and future dynamics of cropland soil organic carbon stocks in an intensive human-impacted area of southeastern China. Agriculture, Ecosystems & Environment 372, 109098. Available at https://doi.org/10.1016/j.agee.2024.109098.

Xu, Y., Smith, P., Qin, Z., 2024. Sustainable bioenergy contributes to cost-effective climate change mitigation in China. iScience 27 (7). Available at https://doi.org/10.1016/j.isci.2024.110232.

Yadav, V.S., Yadav, S.S., Gupta, S.R., Meena, R.S., Lal, R., Sheoran, N.S., et al., 2022. Carbon sequestration potential and CO_2 fluxes in a tropical forest ecosystem. Ecological Engineering 176. Available at https://doi.org/10.1016/j.ecoleng.2022.106541.

Yin, L., Tao, F., Chen, Y., Wang, Y., Ciais, P., Smith, P., 2023. Novel cropping-system strategies in China can increase plant protein with higher economic value but lower greenhouse gas emissions and water use. One Earth 6 (5), 560–572. Available at https://doi.org/10.1016/j.oneear.2023.04.010.

Zhang, Z., 2023. Mitigating investment risks in nature-based solutions: a strategic approach towards sustainable project implementation (Doctoral dissertation, Massachusetts Institute of Technology).

Zhang, L., Heuvelink, G.B.M., Mulder, V.L., Chen, S., Deng, X., Yang, L., 2024. Using process-oriented model output to enhance machine learning-based soil organic carbon prediction in space and time. Science of the Total Environment 922. Available at https://doi.org/10.1016/j.scitotenv.2024.170778.

Zhao, T., 2024. Digital soil mapping of soil chemical and biological properties in Australian sugarcane growing districts (Doctoral dissertation, UNSW Sydney).

Zhou, X., Zhou, L., Nie, Y., Fu, Y., Du, Z., Shao, J., et al., 2016. Similar responses of soil carbon storage to drought and irrigation in terrestrial ecosystems but with contrasting mechanisms: a meta-analysis. Agriculture, Ecosystems and Environment 228, 70–81. Available at https://doi.org/10.1016/j.agee.2016.04.030.

Chapter 17

Carbon credit estimation in agroforestry systems

Anjana J. Atapattu, Tharindu D. Nuwarapaksha, D.M.N.S. Dissanayaka, and Shashi S. Udumann
Agronomy Division, Coconut Research Institute, Lunuwila, Sri Lanka

Highlights

- Agroforestry offers significant potential for carbon sequestration and carbon credit generation.
- Accurate assessment methodologies for carbon stock estimation in agroforestry systems are crucial.
- Scaling up agroforestry practices can contribute to carbon-neutrality and negative-emissions goals.

17.1 Introduction

17.1.1 Definition and components of agroforestry systems

Agroforestry (AF) is defined as a form of land use that involves the practice of growing trees and shrubs besides cultivable land for farming, hence integrating agricultural and forestry practices for better land utilization and productivity (Nuwarapaksha et al., 2024b; Raj et al., 2019). The very essence of AF is the intentional integration of woody perennials, cited as trees and shrubs, together with crops and livestock within the same parcel of land (Fig. 17.1). This integration is intended to increase biological diversification efforts, promote soil quality, and enhance abiotic stress resistance in farming systems (Nuwarapaksha et al., 2024b; Rolo et al., 2023; Meena et al., 2025b). According to the structure and functions, there is a possibility of dividing agroforestry systems (AFs) into some components. Silvopasture is the practice of growing trees in combination with pastures with stock-grazing, the trees act as a provider of shade and shelter to the animals, while grazing also improves nutrient cycling (Nuwarapaksha et al., 2023b; "Gene" Garrett et al., 2015). Alley cropping involves growing crops in rows in between rows of trees or shrubs to reduce wind damage as well as provide protection to the crops (Wolz and DeLucia, 2018). Shelter belts are planted rows of trees or shrubs used to guard crops, soil, and livestock from the effects of excessively strong winds to moderate the climate. Forest farming is the practice of growing desirable crops, such as herbs and fungi, in forests while exploiting the favorable climatic conditions (Karjalainen et al., 2010). Riparian zones are narrow zones of vegetation established along water bodies that slow flow, minimize erosion, allow water purification, and support wildlife (Broadmeadow and Nisbet, 2004; Kujur et al., 2021, 2022). By integrating all these varied components, AFs produce complex landscapes that are economical, ecological, and sociable, thereby enhancing the sustainability and robustness of agricultural practices (Jhariya et al., 2022, 2023; Meena et al., 2024c).

17.1.2 Agroforestry's role in sustainability

As a sustainable agriculture practice, AF can make a very positive contribution to more than one sustainable development goal at the same time (Atapattu et al., 2024). AF helps to increase food security since there are different ways and more than one type of product or service that can be obtained from the garden, as well as food sources (Nuwarapaksha et al., 2023c; Taye Gifawesen et al., 2020). A commercial tree in AFs can fix nitrogen (N), add organic matter (OM), and enhance nutrient cycling in the soils (Dissanayaka et al., 2023b; Nuwarapaksha et al., 2023a; Meena et al., 2024a). The use of leguminous cover crops in an AFs can improve the soil nutrient status through the N-fixation properties of legumes, while at the same time offering many stand-alone functions of erosion control and species diversification

Carbon Incentives, Plans, and Policies in Agriculture. DOI: https://doi.org/10.1016/B978-0-443-33557-0.00001-X

FIGURE 17.1 Components of the agroforestry system.
Figure by authors.

TABLE 17.1 Principles of social sustainability in sustainable agriculture perspectives.

Dimension	Description
Quality of life	Promotes the overall welfare of individuals and communities by meeting basic needs and addressing agricultural challenges, such as enhancing food security and safety, with a commitment to ongoing improvement.
Equity	Addresses the equitable promotion of a sustainable global environment, particularly in the context of alleviating poverty through the adoption of self-sustaining agricultural methods.
Diversification	Promotes the expansion of agricultural product variety and encourages active community participation with both public and private stakeholders in shaping and implementing sustainable agricultural systems.

Source: Authors' compilation based on data from Salleh and Harun (2013).

(Dissanayaka et al., 2022, 2024; Udumann et al., 2023). They also have strong root systems that contribute to water conservation and reduce soil erosion. The land becomes more resistant to drought and other forms of weather (Dissanayaka et al., 2025). However, AF also enhances abiotic factors such as creating ecological niches for several organisms, biological control of pests, and the use of minimal inorganic inputs such as chemical fertilizers. This comes in smart in times of climate change, whereby trees and shrubs can store some of the C, preventing climate change and ensuring long-term sustainability (Bogale and Bekele, 2023; Meena et al., 2024b). In the context of food production, AF addresses the issue of land productivity as it enables several crops and livestock to be produced at once, and if one type fails in the market, others may be able to succeed (Mbow et al., 2014;Udumann et al., 2025). It ensured that farmers access other sources of income through the sale of timber, fruits, nut-producing trees, and other tree products, thus breaking the cycle of dependence on a single source of income. In the social sphere, it contributes to the food security of local communities, offers employment among growers, and creates awareness of sustainable practices. Table 17.1 outlines the principles of social sustainability in relation to sustainable agriculture in AFs (Salleh and Harun, 2013). In other words, AF is a system that is friendly with ecological concepts and an effective way of addressing hurdles related to farming, being environmentally sustainable while benefiting the farmer (Tsonkova et al., 2018; Kumar et al., 2024).

17.1.3 The potential of agroforestry in climate change mitigation

AF remains one of the main approaches to combating climate change because it helps to implement the process of C_{seq} to improve the stability of ecosystems and reduce the emission of GHG (Abbas et al., 2017; Banerjee et al., 2024). Including trees and shrubs in farming systems may go a long way in removing carbon dioxide (CO_2) from the atmosphere through what is referred to as C storage. Forest plantations capture CO_2 from the air during photosynthesis, integrate it into the trunks, branches, and foliage of trees within AFs, and deposit it through the root systems and the deposition of OM into the soil (Patel et al., 2023). This plays an important role in achieving an overall decrease in the share of GHGs in the atmosphere. AFs also influence the physical characteristics of the soil, and thus the capacity of the soil to C_{seq} is also affected (Lorenz and Lal, 2014). Alley cropping, silvopasture, and forest farming improve soil organic matter (SOM), enhance soil micro-organisms, and therefore improve C_{seq} (Shrestha et al., 2018). At the same time, these systems also minimize the levels of soil erosion and degradation, retaining the soils' capability to C_{seq} in their structure (Atapattu et al., 2017; Nuwarapaksha et al., 2022; Meena and Pradhan, 2023). Since AF entails diversification of ecosystems, it enhances their ability to withstand the effects of climate change. The integration of a diverse plant population forms part of the AFs and support system, making ecosystems more suitable and less vulnerable to violent, destructive forces such as floods and droughts (Wilson and Lovell, 2016). Besides, it improves the C store and offers shelter to a variety of animals, thus benefiting the state of the environment.

The AF can also lead to the mitigation of the release of GHG emissions from the process of agriculture. Trees and shrubs used as windbreaks help in curbing evapotranspiration, by which energy-using irrigation systems are not ought to be used (Thevs et al., 2019). Under economic incentives, there is prospective income from the timber, fruits, nuts, and other products from trees, hence encouraging farmers to embrace sustainable farming (Prasad et al., 2021; Atapattu et al., 2025a). This adoption results in other changes of practices on land and the use of agricultural inputs that can also influence climate change over the long term (Gebrehiwot and Van Der Veen, 2013; Meena and Pradhan, 2023). Finally, the prospects of AF as one of the viable solutions to climate change are to serve as C sinks, improve the quality of soil, increase the level of ecological stability, and decrease the emission of GHGs while maintaining food production and preserving the living standards of rural communities (Dissanayaka et al., 2023c; Meena et al., 2023). This chapter aims to identify the methodologies, challenges, and opportunities for the assessment of C credits in AF and the systems that are used as climate change mitigation (CCM) practices, while highlighting the need for efficient methods of C credit assessment, as well as discussing the prospects for future research and policy impacts on the development of AF.

17.2 Carbon sequestration potential of agroforestry

17.2.1 Understanding carbon sequestration in agroforestry

Integrating the trees and shrubs into the agriculture production system, known as AF, provides a viable option for massive C_{seq}, an important procedure in combating climatic variations (Ghale et al., 2022; Pradhan and Meena, 2023a). C_{seq} occurs through the C cycle, and Fig. 17.2 illustrates the movement of CO_2 within different natural and anthropogenic systems, including photosynthesis, respiration, industrial emissions, and fossil fuel usage. C_{seq} means the process of collecting CO_2 from the atmosphere and stocks in biomass (plants and trees) and soils (Lal, 2005). AFs promote this process by taking advantage of trees' natural ability to C_{seq} and their relationship with crops and livestock. AF C_{seq} mainly through the perennial trees and shrubs that are required for the livelihood of people in various parts of the globe (Sahoo and Wani, 2019). These plants take in CO_2 during the process of photosynthesis, and it is then fixed as organic carbon (OC) in the biomass trunks, branches, leaves, and roots of the plants. The trees meant for AF are long-lived structures in comparison with annuals and, thus, are able to C_{seq} more than once. The complex system of roots the trees work by depositing OM to the soil, hence enhancing the soil C_{seq} (Dissanayaka et al., 2025; Hoosbeek et al., 2018). Alley cropping, silvopasture, and forest farming are practices in which plantations are composed of complex layers to increase productivity and stability of the ecosystem. Such systems are favorable for the health of soils because they enhance the content of OM, diminish soil erosion, and intensify nutrient cycling. Better health of the soil implies that a given soil can sequester more C in both organic and inorganic matrices (Rahman et al., 2019; Meena et al., 2023a). Biochar is one of the stable forms of OC that can be applied to the soil. When applied to soil, they are observed to increase the water retention capacity of the soil, which in turn lowers water use demand and drought magnitude in crop production systems (Dissanayaka et al., 2023a,b,c,d; Nuwarapaksha et al., 2024a). Furthermore, trees decrease the temperature of the soil and reduce rates of evaporation, therefore, increasing water content in the soil, which in turn improves plant growth and accounts for increased C storage. Another positive effect of applying AF within the agricultural land is the reduction of deforestation and land degradation (Gupta et al., 2020). Since the management of AF provides alternative sources of income and

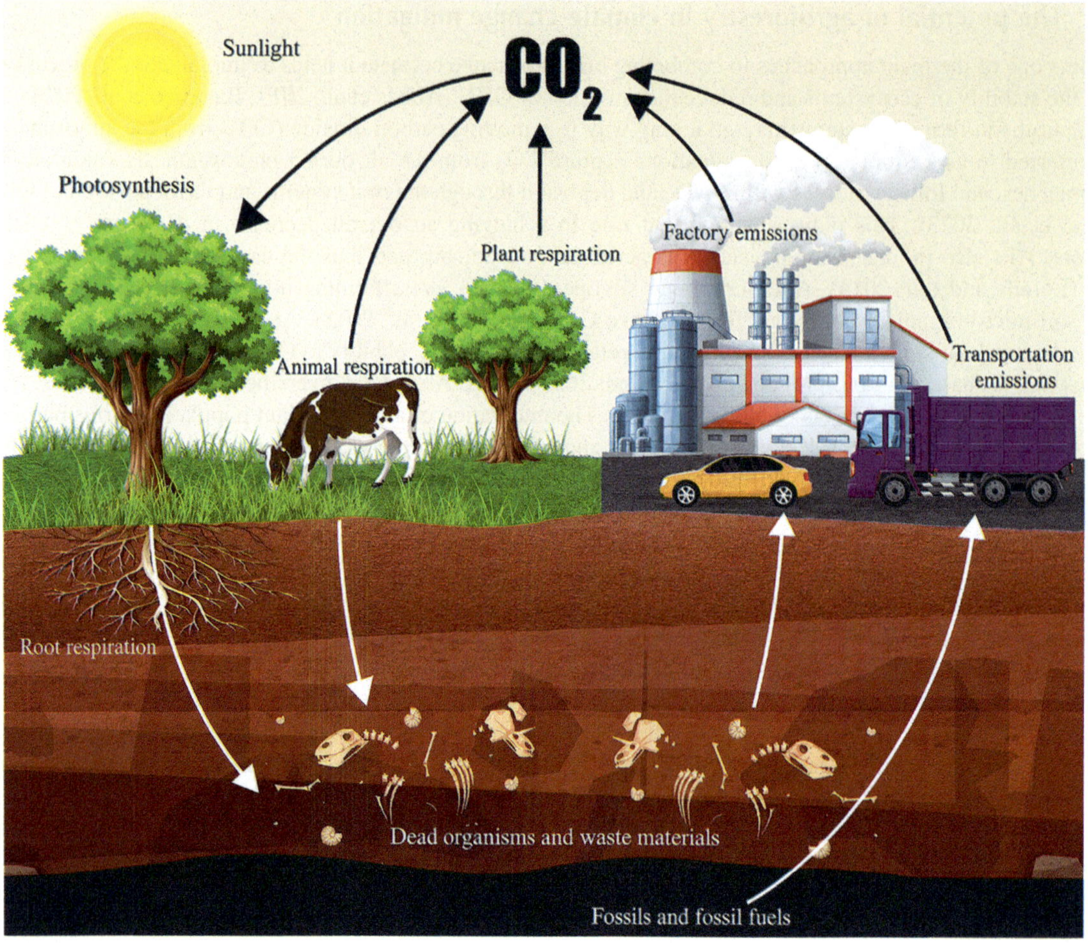

FIGURE 17.2 Carbon cycle.
Figure by authors.

resources, such as timber, fruits, and fodder, to the people, it helps reduce the exploitation of natural forests and regulates the utilization of land (Raj et al., 2021). This, in turn, protects genetic variation and preserves the ecosystem, which helps build claims to profitability and CCM.

17.2.2 Factors influencing carbon sequestration rates

The C_{seq} rates depend on a lot of factors that depend on biology, environment, and management practices within the AFs (Ramachandran Nair et al., 2009). Species have to be selected well, with fast-growing and long-lived species storing more C because of their greater growth rates. Tree density and arrangement significantly impact C_{seq}, and dense plantations generally result in higher C stocks (Biber et al., 2020). The type of soil, its depth, and initial C level decide how much C can be sequestered; the soils that are highly structured, containing clay, also tend to store more C. Climate influences include temperature and rainfall, affecting the rate of plant growth and decomposition of SOM, respectively (Sierra et al., 2015). Working practices such as pruning, thinning, and harvesting rotations are said to directly impact C_{seq}, while operations that reduce soil turnover are said to be the key to building SOC. Age of the AFs is also a consideration since the rates of sequestration are usually higher in the early years of the AFs and progressively reduce with time. Cropping or raising animals alters the productivity of the entire system as well as C flow. Nutrient application and water availability can improve plant growth rate, biomass, and C storage, but with high levels of nutrient and water inputs, they may lead to emissions (Li et al., 2009). Baseline soil C stock and the scope for further C storage are defined by the land-use history of the site. The physical environment, in terms of area ecology and land usage, may alter local climate and system productivity, as well as the rate of C storage in AFs (Williams et al., 2018).

TABLE 17.2 Soil carbon stocks in major agroforestry systems under an ecological-based approach.

Agroforestry system	Soil carbon stock ($Mg/C/ha^{-1}$)
Humid lowlands	
Alley cropping systems	10–25
Woodlots	61–75
Improved fallow lands	123–149
Home gardens	108–119
Tree intercropping	27–78
Silvopasture	96–173
Shaded perennial cropping systems	21–235
Tropical highlands	
Silvopastoral systems	132–173
Shaded perennial systems	21–97
Arid and semiarid lowlands	
Live fences	24
Grazing lands	27–33
Fodder – tree-dominated	25–80
Crop dominated	20–70
Forage banks	33
Fuelwood dominated	30–90

Source: Authors' compilation based on data from ?Nair et al. (2009).

17.2.3 Agroforestry compared to other agricultural systems

Analyzing the adequate sequestration capacity of various land-use systems, it was found that AFs are likely to sequester a much higher amount of C due to their unique principle of incorporating trees with crops or livestock (Pandey, 2002; Yang et al., 2025). These systems generally store higher C stock compared to conventional agricultural systems due to the presence of higher biomass in both shoot and root parts. This C stock is two to sixteen times higher than that stored by monoculture cropping systems in temperate lands and up to eight times higher in the tropics (Nadège et al., 2019). Trees planted may contribute to the SOC pool through inputs such as litter, root decomposition, and better soil aggregation. Relative to intensive farming models, AFs are characterized by low rates of soil disturbance, thus reducing CO_2 emissions (Dissanayaka et al., 2025; Jhariya et al., 2021a,b). According to Table 17.2, C stocks are comparatively higher in silvopastoral systems than in some pastoral systems, especially in the tropics (Nair et al., 2009; Varsha et al., 2019). Therefore, natural forests are capable of storing more C than both plantations and AFs. They provide a basis for balancing C stocks and agricultural productivity, useful in regions where the restoration of natural forests is not possible (Banerjee et al., 2023). An example is urban forestry systems, which, while effective in city environments, do not have as high sequestration rates of elements compared to rural AFs because of physical restraints and urban stresses (Taylor and Lovell, 2021). AF presents enhanced long-term C_{seq} over bioenergy crop systems due to its constant standing biomass.

17.3 Methodologies for carbon stock assessment in agroforestry

17.3.1 Field measurement techniques

Evaluating C stocks in AFs and related investigations requires field measurements. These methods include direct quantification of C stocks in the form of above- and below-ground biomass (live and dead) and SOC or litter accumulation

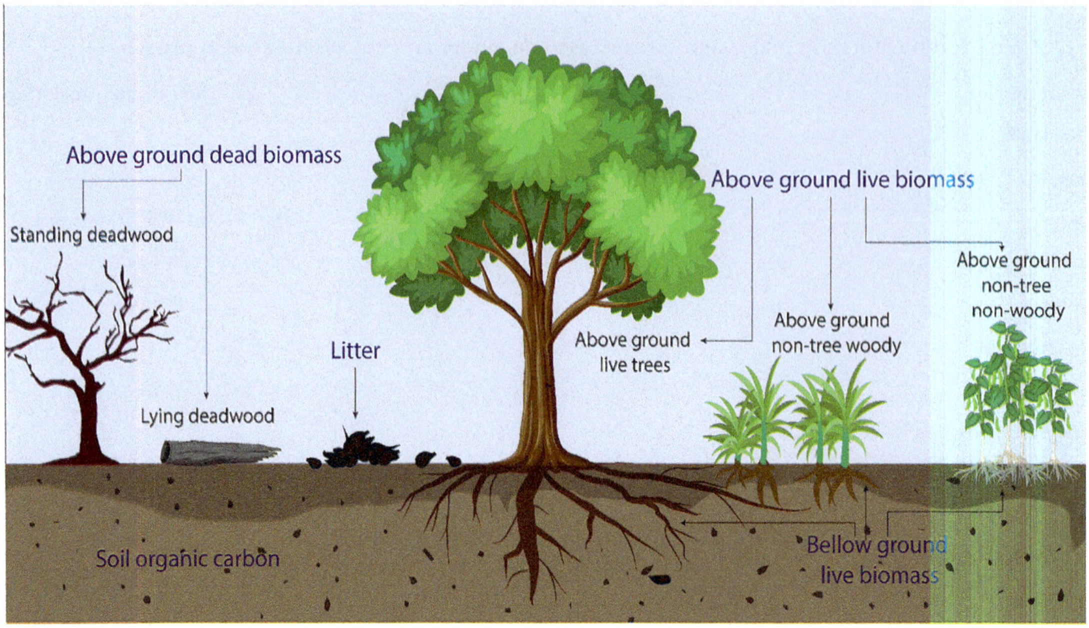

FIGURE 17.3 Quantification of components of biomass and soil carbon stock.

(Fig. 17.3) (Sohrabi et al., 2016). Above-ground biomass is the largest and often easily measured C pool, as it includes all the living plant parts above the ground (Ma et al., 2024). This consists of trees, shrubs, and those herbaceous plants, too. C content is often assessed based on tree size data and using the allometric equations or biomass conversion factors (Nuwarapaksha et al., 2025a). Another essential but less recognized C storage is the below-ground biomass, referring mainly to root systems (Forfora et al., 2024). It is often determined with the help of root-to-shoot ratios or through direct measurement methods, which include soil coring or pit digging. SOC is the major C pool in AF and many times has a higher percentage than the trees' C stock (Raju Singh et al., 2020). It is quantified in several degrees up to 1 m and involves the use of soil sampling and subjecting it to analysis, such as dry combustion or wet oxidation. Another C pool includes litter and deadwood deposits. Litter can be described as dead plant parts, such as fallen leaves, twigs, and other OM found on the ground of a forest (Krishna and Mohan, 2017). Deadwood includes both living and recently dead standing trees as well as snags and down woody debris. These are generally assumed by the use of plot-based sampling techniques. Processed wood obtained from the AF also serves as the C pool in the long run, as the wood is harvested for various uses (Besar et al., 2020). This consists of timber, fuelwood, and other wood-based products that have been produced from the system but are still holding C. Foliage, including crops or pastures under silvopastoral systems, is a component of C stocks, which is often estimated by biomass surveys and estimation (Feliciano et al., 2018; Pradhan and Meena, 2023b). In multispecies AF, the C content of shrubs and herbs in the understory layer is usually estimated individually. Microbial biomass C is less in quantity but has significantly more turnover and is occasionally incorporated into stock figures (Atapattu et al., 2025b; Don et al., 2013). These different C pools help give a complete picture of the C stock in the AFs. It enables further distinction between various management practices and contributes to the creation of approaches to enhance C stock.

17.3.2 Remote sensing applications

Remote sensing methods have thus gained prominence in assessing C stock for AFs, as they provide accurate inventory methods in addition to field surveys (Nuwarapaksha et al., 2025b; Sharma et al., 2023). These techniques utilize satellite, airborne, and unmanned aerial system data to predict various aspects related to C stocks across extensive regions. Medium or coarse-scale data, such as satellite data from Landsat or Sentinel-2, is used to determine vegetation indices, for example, the normalized difference vegetation index (NDVI) (Nuwarapaksha et al., 2026; Xia and Zhang, 2022). These indices are related to biomass, and when combined with field data, one can estimate the above-ground C stocks. Multitemporal data, for instance, facilitates monitoring of vegetation cover and the stock of C over a set period. AF C stock with the help of Light Detection and Ranging (LiDAR) has done a paradigm shift (Chowdhary et al., 2019). They provide accurate three-dimensional descriptive characteristics of the vegetation, such as height, canopy density, and crown size of trees. These estimations can be done with relatively high accuracy and literally add to the field measurements and allometric equations. Synthetic aperture radar (SAR) is more effective in the tropical part of the world, where cloud formation hinders the use of optical satellites. SAR can operate in the cloud and can derive structural and biomass information of vegetation in very dense vegetation cover. Aerial remote sensing using unmanned aerial vehicles is gaining increasing attention due to the technology's ability to provide high-resolution images and 3D data at a lower cost (Zhang and Zhu, 2023). The song is most valuable when highly accurate assessments of small areas of AF are required. Advanced machine learning techniques and artificial intelligence are also being integrated into the analysis of remotely sensed information to increase the efficiency of C stock assessment and apply approaches for automated monitoring of large land areas.

17.3.3 Modeling approaches for carbon estimation

C estimation modeling strategies for AFs include C amounts and exchange projections that utilize mathematical and statistical modeling (Marçal et al., 2021). These models combine multiple forms of data and information about the environment to estimate probable levels of C uptake and stock at micro, meso, and macro levels and across various time horizons. Process-based models simulate the physiological processes of C uptake, allocation, and release in AFs. Those models, including CO_2FIX and FORECAST, take photosynthesis, respiration, and decomposition to elaborate on C storage (Abbas et al., 2017; Jose et al., 2023). Sometimes, they involve climate data, soil requirements, and management practices for the successful functioning of the model. Empirical models, on the other hand, utilize statistical correlations between easily quantifiable variables (such as diameter and height) and C pools (Coomes et al., 2012). These include the allometric equations and biomass expansion factors, which are the most commonly used predictors for estimating above-ground biomass and C stock from field measurements. Spatial models at a broader scale include InVEST and CENTURY, which incorporate data to estimate the C content as well as C_{seq} capacity at the landscape level (Lahiji et al., 2020; Pennock and Frick, 2001). The most suitable application of these models is to assess the optimal land-use type for C stocks in AFs. In C estimation, random forests and neural networks are the most used machine learning techniques at present. These techniques are useful since flexibility in trend estimation is allowed, and model prediction accuracy may be enhanced, especially for nonlinear associations between environmental factors and C stocks. Integrated models utilize features of both process-based approaches and empirical models, thereby benefiting from the strengths of each (Makela et al., 2000). The growth equations in the 3-PG model combine physiological-based functions to express growth; parameters are then calibrated based on empirical data. It is worth mentioning that Bayesian models are becoming increasingly popular due to their capacity to deal with C uncertainty and incorporate knowledge (Almeida et al., 2010). These models remain useful for long-term monitoring because they can be updated as soon as new data is received.

17.4 Carbon credit mechanisms and frameworks

17.4.1 Overview of existing carbon credit systems

Carbon credit systems have been defined as being market-based for efficient lowering of GHG emissions, whereby entities can buy, sell, and trade credits that represent the ability to emit CO_2 or an equivalent quantity of any GHGs (Ulucak et al., 2019; Raj et al., 2025; Pradhan et al., 2023). The notable international systems are the Clean Development Mechanism (CDM) by the United Nations (UN) and the European Union Emissions Trading System and the voluntary C markets that target anyone attempting to balance emissions (Verde and Borghesi, 2022). Other significant systems are REDD+, Gold Standard, Verified Carbon Standard (VCS), American Carbon Registry, and Climate Action Reserve (Pan et al., 2022). These frameworks differ in the methods and the kind of projects, but all of them are focused on generating the mechanism of economic promotion of emission reductions. Some of the hurdles include how to achieve additionality, avoid double-counting, and protect the environment. Nevertheless, C credit procedures remain relevant and are quite useful in the present global climate change initiatives, providing certain options for companies to achieve emission requirements and fund various emission reduction schemes globally (Perdan and Azapagic, 2011).

17.4.2 Agroforestry in voluntary and compliance carbon markets

Agroforestry, which is a combination of trees and crops or livestock production practices, has come out as a major player in both voluntary and compliance C credit mechanisms and frameworks (Porro et al., 2012). This approach creates a win–win chance to store C while giving numerous ecological and socio-economic advantages and makes it more appealing for C credits. In voluntary C markets, there has been an increase in interest in AF projects (Lee et al., 2016). These markets, resulting from corporate and individual purchases of emission reductions, demand AF because of benefits other than C stock. Some standards, such as the VCS and gold standard, have developed certain methodologies that can be used on AF projects, which facilitate measurement and certification of C credits (Lee et al., 2018). They are commonly implemented with smallholder farmers in least developed countries, using sustainable land management and tree planting to enhance C stocks and C_{seq} while at the same time supporting local livelihoods. There are various compliance markets, such as CDM or national emissions trading schemes, in which AF has been less integrated (Bryan et al., 2010). However, they are gradually starting to realize its benefits. A few countries have now incorporated AF in their nationally determined contributions (NDCs) under the Paris Agreement, which could serve to enhance its addition to compliance markets (Gupta and Dube, 2018). The adoption of AF in these markets may enhance the global extent of C storage and foster sustainable rural growth.

17.4.3 Incentives for adopting agroforestry practices

Stipends that apply to AFs under C credit mechanisms and frameworks aim to persuade landowners and farmers to incorporate trees in their fields, thereby raising C stock and improving other environmental benefits (Montagnini and Nair, 2004). These incentives are intended to address the first kind of barriers, which include the expensive start-up and the long-term nature of tree investments. Economic motivations comprise the main motives where C credits are a definite new source of revenue generation (Wunder, 2007). Because trees and soils fix C, the farmers are enslaved to own the credits and sell them in voluntary or compliance markets. Certain programs provide for the direct provision of payments or grants to support the development of AFs, thereby alleviating the cost of implementation. Tax incentives are another tool, with some jurisdictions offering tax breaks or deductions for AF investments (Ramos-Gerena, 2024). Some of these may include the possibility of cutting property taxes or income tax for selling C credits. Technical support and capacity development are two nonsubsidizing tools. Most of the programs offer free or low-cost technical assistance and education to farmers for the effective adoption and establishment of AFs. Cobenefit recognition is gradually popularized, and some frameworks allow charging higher prices for the C credits from the projects that provide the extra nonclimate benefits, such as biodiversity protection or water-use efficiency enhancement (Onaindia et al., 2013). Other policies, such as good land-use policies and the integration of AF in the national climate change plans can foster the increased practice of AF. Finally, performance-based incentives include what has to be done for market access incentives, wherein endorsement programs for sustaining AF goods and services could bolster economic returns through potentially accessing relatively better market prices to farmers who implement these practices (Liu and Chuang, 2023).

17.5 Challenges and concerns in carbon credit assessment in agroforestry

17.5.1 Methodological challenges in carbon estimation

There are still some major difficulties in C estimation in AFs that pose methodological challenges for C credits' evaluation. These systems are complex because they involve tree cropping and sometimes animals, and getting the C flows into a format that can be accurately measured and modeled is difficult (Nair, 2011). When it comes to the C estimation, the conventional techniques established for monoculture forestry or agricultural lands do not work well for AFs. The first major issue involves sampling all C pools, including above- and below-ground plant components, soil C, and litter (Petrokofsky et al., 2012). All these components operate differently, in AF enterprises as opposed to simple agriculture or silviculture. The areal distribution of AF plots is irregular, so agreement between sample plot data and data obtained for larger areas is not always accurate (Gonçalves et al., 2019). There is also the issue of temporal heterogeneity, because the rates of C_{seq} vary with stand age, crop cycle, and management practices. Therefore, creating simple measurement techniques that consider this complexity and can be used across such purposes and regions is rather a challenge. Moreover, using remote sensing data incorporated with ground observations, which has potential, is still far from an accurate measurement at large scales (Glenn et al., 2007). These methodological challenges also influence the methods of C credits' enumeration and the credibility of AF-based C offset projects in the market. Addressing these challenges requires cross-sectoral synthesis, cutting-edge approaches, and methods that can either characterize the spatial patterns of various facets of AFs or reach the rigor that the C market requires.

17.5.2 Variability and uncertainty in agroforestry systems

The inherent variability of the AFs adds a very unpredictable characteristic to the C credit calculations. This variation can be due to climatic differences, the type of soil, the kind of vegetation, and the ways in which resources must be utilized (Wienhold and Goulao, 2023). In contrast to the monoculture systems, AF incorporates several types of species that grow and sequester C and have different life spans. Thus while promoting the overall protection of ecosystems, it poses challenges in calculating C stock. Concerning analogous areas, disparities in C_{seq} result from differences in tree stocking, crops, and the choice of management practices undertaken by farmers (Plaza-Bonilla et al., 2015). Another issue is temporal variability, which means that C stocks tend to vary with the changes in season, crops, and trees' growth cycles. The capacity to sequester C in the long run is rather questionable since the effects of climate change on productivity and stability of ecosystems are still unpredictable (Plaza-Bonilla et al., 2015). Future change in C_{seq} in AFs is conditional because of extreme social events, changes in temperature and precipitation regimes, and pests and diseases. Furthermore, the distribution of C in above- and below-ground biomass also differs based on species selection and site conditions, and therefore, the creation of fixed C credit rules may not be an easy task (Henry et al., 2009). These intrinsic variations and risks not only increase the imprecision of C credit estimation but also question the permanence and enhancement of C stocks in AF projects. These issues require robust evaluation procedures that can effectively handle the complexity of risk assessment for AFs, as well as management techniques that incorporate adjustable C credits.

17.5.3 Monitoring, reporting, and verification issues

The application of C credit programs in AFs depends more on monitoring, reporting, and verification mechanisms. However, such processes are accompanied by several issues that can weaken the trustworthiness and effectiveness of C offset projects (van Kooten and de Vries, 2013). There is one major concern, namely, optimizing low-cost monitoring procedures suitable for small-scale AF initiatives. It is common to find that many AF interventions target growers in the developing world, who may not have easy access to state-of-the-art technology or significant experience. It is vital to develop monitoring procedures that are both precise and applicable to these settings. Remote sensing technologies are potentially useful, but sometimes, it is necessary to go on the ground to sample, and they do not adequately capture the variability of AFs (Nair, 2011). Another area of difficulty is reporting practices, which include challenges in achieving comparability and consistency of eports across different projects and countries. Such a combination often experiences challenges due to variation in reporting formats and measures taken by various organizations. Additionally, many smallholders may possess limited abilities to keep records or engage in complex record-keeping metrics, which may lead to their exclusion or lack of access to C markets. Verification also tends to present some challenges, such as the requirement of a third party to verify C credits to increase their credibility (Su et al., 2024). Nonetheless, the costs of promoting verification can be very expensive, depending on the scale of the projects, which hinders potential participants from participating. Establishing the additionality and permanence of C stocks in the constantly changing dynamic systems of AF is a tedious process that might be costly. Another key problem of the monitoring, reporting, and verification processes is data quality, consistency, and transparency (Tang et al., 2018). This chapter identifies accuracy, currency, and accessibility of data as key aspects significant for trust in the C credit programs. Solving these monitoring, reporting, and verification issues entails the use of strategies that are both robust and realistic; possibilities could include involving communities in monitoring, applying relatively simple techniques where smallholders are involved, and using new technologies that improve efficiency without compromising accuracy (Perosa et al., 2023).

17.6 Research and development towards carbon credits and agroforestry

17.6.1 Improving estimation methodologies

The process of quantitative estimation and valuation of C stocks in terms of C credits within AFs has a significant level of inherent assumptions in current methods that need to be addressed. Current techniques are generally applied arbitrarily and enforced in stereotypical approaches, which may not reflect specific practices of AF. Subsequent studies need to concentrate on the enhancement of more sophisticated and protocol-specific prediction models (Jin et al., 2024). This could include utilizing LiDAR and hyperspectral imaging in addition to other key technologies as a machine learning technique to enhance the accuracy of above-ground biomass measurements. Unconventional technologies such as soil sequestration cannot be efficiently characterized using existing techniques and, thus, should employ new methods such as portable soil spectroscopy and isotopic analysis of roots and SOC (Nayak et al., 2019). More long-term research should be devoted to attempting to understand how C evolves in various AFs, not only based on climatic differences or types of soil and vegetation but also by changing land use practices and techniques (Verchot et al., 2007). More research should accompany solutions to the problem of quantifying avoided emissions and leakage in AF projects. This involves developing sound reference states and refining models that reveal the impacts of intervention on C cycles in AF land and its vicinity. Lastly, the hope is to generate a better approach for an affordable and more sound mechanism to estimate the potential of C credits from AFs, helping to increase the quality of these credits and to expand their use on global C markets (Nambiar, 2021).

17.6.2 Developing decision support tools for agroforestry planning

Strategic decision support tools are the most important to enhance and improve the planning of the AF action and C_{seq} (Abbas et al., 2017). These tools should take into consideration multiple values, such as ecological, economic, and social aspects, to enable farmers, policymakers, and project developers to make the right decisions. Thus future research should aim to develop easily navigated software tools based on Geographic Information Systems mapping, climate simulation, and economic modeling. Such tools could enable the generation of possible AF situations and the assessment of Cseq capacities, as well as the overall profitable yields in succession (Achmad et al., 2022). The integration of machine learning algorithms could enable these tools to learn and refine their recommendations based on actual field data from implemented AF projects. This will enable one to get even more precise recommendations for each region. Further study is also required to design modules for other valuable services that can be procured from AF besides the C stock, which includes biological diversity, soil quality, and water control (Gitz et al., 2022). Another potential area is the inclusion of risk analysis tools that enable users to prevent issues like the spread of pests, changes in markets, or adverse effects of climate change, among others (Szyniszewska et al., 2024). Such tools should also be compatible with the purpose to be easily applied at different tiers, ranging from farm-level planning to AF at the landscape level. It would enhance both smallholder agriculture and the national and global climate change policy, thus connecting the grassroots initiatives and international climate change policy.

17.6.3 Breeding high carbon-sequestering tree and crop species

This approach holds a promising avenue for extending the potential of AFs in contributing to improved C_{seq} that comes with breeding high C_{seq} tree and crop species (Puri and Nair, 2004). This chapter's direction is research that seeks to come up with plant varieties that will reap both the farmer and the soil and biomass C stocks. The trees should be the subject of study, and this should include the ability to detect and improve the characteristics that help trees to grow faster and achieve higher wood density, as well as possess wide root systems (Hodge et al., 2009). This could include seeking to identify the genes that underpin these traits and using techniques, including marker-assisted selection, to speed up the process of bringing forward improved types. Likewise, for crop breeding, intervention should focus on those characteristics that can increase the biomass production and enhance the allocation of C below-ground (Comas et al., 2013). This could involve the generation of varieties with more root–shoot ratio characteristics or with characteristics that produce more of the reserve OM, which takes longer to decompose and thus can provide more long-term soil C storage. This will be a crucial aspect of this research, and it will seek to employ varieties that sequester high levels of C to also have climate resilience and pest and disease tolerance, as well as be well-suited for the different agro-ecological regions (Ghosh et al., 2023). This will require a multidisciplinary effort involving plant breeders and geneticists, agronomists and soil scientists, and specialists in AF. Furthermore, breeding strategies that involve farmer participation may help develop varieties that cater to both the needs of C_{seq} and farmers (Himanen et al., 2016). Capacity should also be designed with an additional focus on identifying the trade-offs or synergies between C and other ecosystem services or agricultural yields in AFs for the development of multiple-benefit designs.

17.7 Policy perspectives and scaling up agroforestry

17.7.1 Integration with climate-smart agriculture policies

The recommendation for the enhancement of climate-smart agriculture (CSA) policies is that AF should be included since it provides solutions to climate change, food insecurity, and sustainable agriculture (Sileshi et al., 2023). This integration needs a long-term approach to ensure that AFs are in line with the overall CSA development. Specifically, policymakers should understand the role of AF as one of the components of the CSA approach, focusing on the opportunities related to C stocks, climate change adaptation, and CCM (Zerssa et al., 2021). This means establishing a formulation of policies that promote the practice of AF by offering bonuses for service production that include payments for the ecosystem services or C credits. Further studies, surveys, and investigations are needed to determine the extent of the interaction between AF and other CSA techniques in order to provide policy recommendations (Ntawuruhunga et al., 2023). This may involve analyzing how AF supports conservation agriculture, integrated soil fertility management, and water-smart practices and how they can be integrated profitably and sustainably. They should also incorporate measures to facilitate AF development by clearing hurdles such as rights of use to land, availability of quality planting materials, and market for AF products. Having extensible support from institutional services and the right extension agency that is suited for providing extension services to AFs will also be important (Berthet and Hickey, 2018). Moreover, the inclusion of AF at even broader levels of climate policy can occur through the inclusion of AFs into national mitigation and adaptation plans, also known as NDCs under the Paris Agreement (Dhyani et al., 2021). This has the implication of creating solid monitoring, reporting, and verification systems for AF and climate mitigation contributions. Effective integration will require synchronization of policies of the agricultural, forestry, and environment ministries and agencies while enabling full climate-smart opportunities in AFs.

17.7.2 Strategies for scaling up agroforestry practices

A way forward is to promote commercial implementation of the practices to enhance the extent and impacts of their C-storing potential as well as sustainable use of the land (Dignac et al., 2017). This implies addressing strategic and technical inputs, as well as socio-economic and policy constraints, to facilitate societal change, supporting wider uptake. The first is the promotion of context-adapted AF models that are ecologically suitable, farmer-friendly, and market-required (Wienhold and Goulao, 2023). This entails participatory research as well as demonstration initiatives aimed at illustrating the practical applicability of, as well as the added values arising from, integrated AFs. Another consideration is that tree improvement through the use of quality germplasm and planting materials should be enhanced. Promoting decentralized operations that involve the stewardship of nurseries and seed banks, as well as blind knowledge delivery programs in tree production and maintenance, would help farmers practice AF (Groninger, 2012). That is why such financial forms as grants, tax incentives, subsidies, and other sources of financing contribute significantly to scaling up. This encompasses innovative options to support AF activities, such as blended finance, impact investing, and C finance schemes, which can provide both initial capital and long-term revenues for AF projects. Developing and diversifying markets for its products is important, and sustaining value chains for AF products is crucial to the economy (Leakey et al., 2005). This includes working on postharvest handling, particularly in processing, enhancing marketing channels, and supporting accreditation programs for AF products to secure better prices. Intersecting engagement at different levels of systems strengthening is also imperative. Education and awareness raising involve providing farmers, extension agents, local institutions, senior policymakers, financial institutions, and other stakeholders with knowledge on the ability of AFs and the capability of training them on the various practices to be adopted when undertaking AFs (Rois-Díaz et al., 2018).

17.7.3 Contribution to carbon neutrality and negative emissions goals

Current studies show that AF has a heightened potential of contributing to the goals of C neutrality and negative emissions through the effective implementation of a C neutrality and negative emissions goals strategy on large areas of agricultural land (Chen et al., 2022; Meena et al., 2025a). Again, concerning the capture of CO_2 from the atmosphere, amalgamation of AF has the capacity to capture a large volume of CO_2, potentially in the range of billions of tons per year. This may be particularly helpful in addressing the climate situation, as it enables negative CO_2 emissions, that is, forcibly removing CO_2 from the air with substantial by-products that benefit nutrition and rural income (Praveen and Sharma, 2019). To harness this potential, the actual and potential rates of C_{seq} in different AFs have to be scientifically determined along the ecological gradient (Marsden et al., 2020). This also entails enhancing better approaches to quantifying both the above-ground and below-ground C stocks in diverse AF environments. It is necessary to include AF in national and international C markets and standards. This entails setting up guide rails for the quantitative assessment, documentation, and certification of C stocks in AFs for purposes of marketing them under C assets and accessing climate funds (Jacobi et al., 2017). The policy frameworks should therefore provide for the multiple values of AF besides the mitigation of C, such as the conservation of the biological diversity, soil health, and the enhancement of the climate (Meena et al., 2022; Roy et al., 2022). It will, in this regard, contribute to the realization of the overall sustainable development objectives of AF. Longitudinal assessment and quantitative modeling of the AF sector are crucial to extrapolating the level of future contribution toward C neutrality objectives (Dupraz et al., 2019). This also encompasses aspects such as the premature effects of some risks, like climate change, on the growth of the trees and C_{seq}.

17.8 Conclusion

This chapter aims to present the interactions between AFs and C credit estimation for C offsets, including their potential and associated issues. Trees, where they are intercropped with crops and pastures in AF, increase biomass and soil C stocks, and this is often more than the levels attained in pure agricultural systems. But the estimates of the C stock in these AFs are still tentative, as the quantification of this C_{seq} more often than not is hampered by the variability and complexity of the AF geography. Developments in the assessment of field measurements, remote sensing technologies, and modeling procedures have enhanced the evaluation of C own in AFs. However, methodological limitations still occur, especially regarding the identification of all C pools and the spatial and temporal differentiation. Therefore, the integration of AF with the C credit systems and actions in both the voluntary and compliance markets should hold ample opportunities for possible pathways to promote and expand the practices. But challenges such as monitoring, reporting, and verification have to be solved prior to the credibility and efficiency of projects based on AF to provide C offsets. More research should be pursued to refine the estimation procedures, build decision-making tools, and develop elite genotypes to produce more C_{seq}. Furthermore, the synergy of AF with CSA policies and the need to come up with a roadmap on how to enhance the practices in achieving the desired C-neutral and negative emissions are critical. As the world faces the problem of climate change in the future, AF appears to be a solution complex enough to deal with both CCM and adaptation, as well as the development agenda. The potential of Cseq in the AFs and analyzing the problems in the estimation of C credits open up scope for further exploration of sustainable management of land resources and climate change.

References

Abbas, F., Hammad, H.M., Fahad, S., Cerdà, A., Rizwan, M., Farhad, W., et al., 2017. Agroforestry: a sustainable environmental practice for carbon sequestration under the climate change scenarios—a review. Environmental Science and Pollution Research 24 (12), 11177–11191. Available at https://doi.org/10.1007/s11356-017-8687-0.

Achmad, B., Sanudin, Siarudin, M., Widiyanto, A., Diniyati, D., Sudomo, A., et al., 2022. Traditional subsistence farming of smallholder agroforestry systems in Indonesia: a review. Sustainability 14 (14), 8631. Available at https://doi.org/10.3390/su14148631.

Almeida, A.C., Siggins, A., Batista, T.R., Beadle, C., Fonseca, S., Loos, R., 2010. Mapping the effect of spatial and temporal variation in climate and soils on *Eucalyptus* plantation production with 3-PG, a process-based growth model. Forest Ecology and Management 259 (9), 1730–1740. Available at https://doi.org/10.1016/j.foreco.2009.10.008.

Atapattu, A.A.A.J., Senarathne, S.H.S., Raveendra, S.A.S.T., Egodawatte, W.C.P., Mensah, S., 2017. Effect of short term agroforestry systems on soil quality in marginal coconut lands in Sri Lanka. Agricultural Research Journal 54 (3), 324. Available at https://doi.org/10.5958/2395-146X.2017.00060.6.

Atapattu, A.J., Ranasinghe, C.S., Nuwarapaksha, T.D., Udumann, S.S., Dissanayaka, N.S., 2024. Sustainable agriculture and sustainable development goals (SDGs). Emerging Technologies and Marketing Strategies for Sustainable Agriculture. IGI Global, Sri Lanka, pp. 1–27. Available at https://doi.org/10.4018/979-8-3693-4864-2.ch001.

Atapattu, Anjana J., Nuwarapaksha, T.D., Dissanayaka, D.M.N.S., Udumann, S.S., 2025a. Ecofriendly Frontiers, (Eds.) O. O. Babalola et al. Springer Nature, Switzerland, pp. 3–38. Available at 10.1007/978-3-031-98700-7_1.

Atapattu, A.J., Nuwarapaksha, T.D., Udumann, S.S., Dissanayaka, N.S., 2025b. Integrated Farming Systems: A Holistic Approach to Sustainable Agriculture. In: Babu, S., Singh, R., Rathore, S.S., Das, A., Singh, V.K. (Eds.), Agricultural Diversification for Sustainable Food Production. Sustainability Sciences in Asia and Africa. Springer, Singapore, pp. 89–127. Available at https://doi.org/10.1007/978-981-97-7517-0_4.

Banerjee, A., Jhariya, M.K., Bargali, S.S., Palit, D., 2023. Ecorestoration for Sustainability. Wiley-Scrivener, USA, pp. 1–441.

Banerjee, A., Jhariya, M.K., Raj, A., Mechergui, T., 2024. Ecosystem management: climate change and global sustainability—an introduction. Ecosystem Management: Climate Change and Sustainability. wiley, India, pp. 1–21. Available at https://doi.org/10.1002/9781394231249.ch1.

Berthet, E.T., Hickey, G.M., 2018. Organizing collective innovation in support of sustainable agro-ecosystems: the role of network management. Agricultural Systems 165, 44–54. Available at https://doi.org/10.1016/j.agsy.2018.05.016.

Besar, N.A., Suardi, H., Phua, M.H., James, D., Mokhtar, M.B., Ahmed, M.F., 2020. Carbon stock and sequestration potential of an agroforestry system in Sabah, Malaysia. Forests 11 (2). Available at https://doi.org/10.3390/f11020210.

Biber, P., Felton, A., Nieuwenhuis, M., Lindbladh, M., Black, K., Bahýl', J., et al., 2020. Forest biodiversity, carbon sequestration, and wood production: modeling synergies and trade-offs for ten forest landscapes across Europe. Frontiers in Ecology and Evolution 8. Available at https://doi.org/10.3389/fevo.2020.547696.

Bogale, G.A., Bekele, S.E., 2023. Sustainability of agroforestry practices and their resilience to climate change adaptation and mitigation in sub-Saharan Africa: a review. Ekológia (Bratislava) 42 (2), 179–192. Available at https://doi.org/10.2478/eko-2023-0021.

Broadmeadow, S., Nisbet, T.R., 2004. The effects of riparian forest management on the freshwater environment: a literature review of best management practice. Hydrology and Earth System Sciences 8 (3), 286–305. Available at https://doi.org/10.5194/hess-8-286-2004.

Bryan, E., Akpalu, W., Yesuf, M., Ringler, C., 2010. Global carbon markets: opportunities for sub-Saharan Africa in agriculture and forestry. Climate and Development 2 (4), 309–331. Available at https://doi.org/10.3763/cdev.2010.0057.

Chen, L., Msigwa, G., Yang, M., Osman, A.I., Fawzy, S., Rooney, D.W., et al., 2022. Strategies to achieve a carbon neutral society: a review. Environmental Chemistry Letters 20 (4), 2277–2310. Available at https://doi.org/10.1007/s10311-022-01435-8.

Chowdhary, G., Gazzola, M., Krishnan, G., Soman, C., Lovell, S., 2019. Soft robotics as an enabling technology for agroforestry practice and research. Sustainability 11 (23), 6751. Available at https://doi.org/10.3390/su11236751.

Comas, L.H., Becker, S.R., Cruz, V.M.V., Byrne, P.F., Dierig, D.A., 2013. Root traits contributing to plant productivity under drought. Frontiers in Plant Science 4. Available at https://doi.org/10.3389/fpls.2013.00442.

Coomes, D.A., Holdaway, R.J., Kobe, R.K., Lines, E.R., Allen, R.B., 2012. A general integrative framework for modelling woody biomass production and carbon sequestration rates in forests. Journal of Ecology 100 (1), 42–64. Available at https://doi.org/10.1111/j.1365-2745.2011.01920.x.

Dhyani, S., Murthy, I.K., Kadaverugu, R., Dasgupta, R., Kumar, M., Gadpayle, K.A., 2021. Agroforestry to achieve global climate adaptation and mitigation targets: are South Asian countries sufficiently prepared? Forests 12 (3), 303. Available at https://doi.org/10.3390/f12030303.

Dignac, M.F., Derrien, D., Barré, P., Barot, S., Cécillon, L., Chenu, C., et al., 2017. Increasing soil carbon storage: mechanisms, effects of agricultural practices and proxies. A review. Agronomy for Sustainable Development 37 (2). Available at https://doi.org/10.1007/s13593-017-0421-2.

Dissanayaka, D.M.N.S., Nuwarapaksha, T.D., Udumann, S.S., Dissanayake, D.K.R.P.L., Atapattu, A.J., 2022. A sustainable way of increasing productivity of coconut cultivation using cover crops: a review. Circular Agricultural Systems. 2. Available at https://doi.org/10.48130/CAS-2022-0007.

Dissanayaka, D.M.N.S., Dissanayake, L., Dassanayake, S.D., Udumann, S.S., Keerthisinghe, J.P., Jayalath, N., et al., 2023a. Enhancing sustainable agriculture through king coconut husk ash: investigating optimal processing parameters for high potassium content and efficient waste management. IECAG 2023. MDPI, Basel Switzerland, pp. 17. Available at https://doi.org/10.3390/IECAG2023-15802.

Dissanayaka, D.M.N.S., Dissanayake, D.K.R.P.L., Udumann, S.S., Nuwarapaksha, T.D., Atapattu, A.J., 2023b. Agroforestry—a key tool in the climate-smart agriculture context: a review on coconut cultivation in Sri Lanka. Frontiers in Agronomy 5. Available at https://doi.org/10.3389/fagro.2023.1162750.

Dissanayaka, D.M.N.S., Udumann, S.S., Nuwarapaksha, T.D., Atapattu, A.J., 2023c. Agroforestry: an avenue for resilient and productive farming through integrated crops and livestock production. In: Kiba, D.I. (Ed.), Transitioning to Zero Hunger. MDPI Books, Basel, Switzerland, pp. 115–136. Available at https://doi.org/10.3390/books978-3-03897-863-3-5.

Dissanayaka, D.M.N.S., Udumann, S.S., Nuwarapaksha, T.D., Atapattu, A.J., 2023d. Effects of pyrolysis temperature on chemical composition of coconut-husk biochar for agricultural applications: a characterization study. Technology in Agronomy 3. Available at https://doi.org/10.48130/TIA-2023-0013.

Dissanayaka, D.M.N.S., Udumann, S.S., Nuwarapaksha, T.D., Atapattu, A.J., 2024. Harnessing the potential of Mucuna cover cropping: a comprehensive review of its agronomic and environmental benefits. Circular Agricultural Systems. 4. Available at https://doi.org/10.48130/cas-0024-0001.

Dissanayaka, N.S., Nimanthika, N., Nuwarapaksha, T.D., Udumann, S.S., Atapattu, A.J., 2025. Agroforestry and climate-smart agriculture. Agroforestry for a Climate-Smart Future. IGI Global, Sri Lanka, pp. 173–199. Available at https://doi.org/10.4018/979-8-3693-8282-0.ch006.

Dissanayaka, Nuwandhya S., Udumann, Shashi S., Nuwarapaksha, Tharindu D., Atapattu, Anjana J., 2025. Microbial partnerships in agriculture: boosting crop health and productivity. Circular Agricultural Systems 5. Available at 10.48130/cas-0025-0011.

Dissanayake, D.K.R.P.L., Udumann, S.S., Dissanayaka, D.M.N.S., Nuwarapaksha, T.D., Atapattu, A.J., 2023a. Effect of biochar application rate on macronutrient retention and leaching in two coconut growing soils. Technology in Agronomy 3. Available at https://doi.org/10.48130/TIA-2023-0005.

Dissanayake, D.K.R.P.L., Dissanayaka, D.M.N.S., Udumann, S.S., Nuwarapaksha, T.D., Atapattu, A.J., 2023b. Is biochar a promising soil amendment to enhance perennial crop yield and soil quality in the tropics. Technology in Agronomy 3. Available at https://doi.org/10.48130/TIA-2023-0004.

Don, A., Rödenbeck, C., Gleixner, G., 2013. Unexpected control of soil carbon turnover by soil carbon concentration. Environmental Chemistry Letters 11 (4), 407–413. Available at https://doi.org/10.1007/s10311-013-0433-3.

Dupraz, C., Wolz, K., Lecomte, I., Talbot, G., Vincent, G., Mulia, R., et al., 2019. Hi-sAFe: a 3D agroforestry model for integrating dynamic tree–crop interactions. Sustainability 11 (8), 2293. Available at https://doi.org/10.3390/su11082293.

Feliciano, D., Ledo, A., Hillier, J., Nayak, D.R., 2018. Which agroforestry options give the greatest soil and above ground carbon benefits in different world regions? Agriculture, Ecosystems and Environment 254, 117–129. Available at https://doi.org/10.1016/j.agee.2017.11.032.

Forfora, N., Azuaje, I., Vivas, K.A., Vera, R.E., Brito, A., Venditti, R., et al., 2024. Evaluating biomass sustainability: why below-ground carbon sequestration matters. Journal of Cleaner Production 439. Available at https://doi.org/10.1016/j.jclepro.2024.140677.

Gebrehiwot, T., Van Der Veen, A., 2013. Farm level adaptation to climate change: the case of farmer's in the ethiopian highlands. Environmental Management 52 (1), 29–44. Available at https://doi.org/10.1007/s00267-013-0039-3.

"Gene" Garrett, H.E., Brauer, D., Clason, T.R., 2015. Silvopastoral practices. North American Agroforestry: An Integrated Science and Practice. Wiley, pp. 105–131. Available at https://doi.org/10.2134/2009.northamericanagroforestry.2ed.c6.

Ghale, B., Mitra, E., Sodhi, H.S., Verma, A.K., Kumar, S., 2022. Carbon sequestration potential of agroforestry systems and its potential in climate change mitigation. Water, Air, and Soil Pollution 233 (7). Available at https://doi.org/10.1007/s11270-022-05689-4.

Ghosh, S., Sarkar, S., Garai, S., Roy, A., Saha, S., Dey, S., et al., 2023. Climate change and global crop production: an inclusive insight. Climate-Resilient Agriculture. 2, 1–34. Available at https://doi.org/10.1007/978-3-031-37428-9_1.

Gitz, V., Xu, J., Lu, Y., Springgay, E., Animon, I., Mubarak, R.K.A., et al., 2022. Trees as hotspots: using forests, trees, and agroforestry to foster diverse sustainable landscapes. Circular Agricultural Systems 2. Available at https://doi.org/10.48130/CAS-2022-0004.

Glenn, E.P., Huete, A.R., Nagler, P.L., Hirschboeck, K.K., Brown, P., 2007. Integrating remote sensing and ground methods to estimate evapotranspiration. Critical Reviews in Plant Sciences 26 (3), 139–168. Available at https://doi.org/10.1080/07352680701402503.

Gonçalves, A.C., Sousa, A.M.O., Mesquita, P., 2019. Functions for aboveground biomass estimation derived from satellite images data in Mediterranean agroforestry systems. Agroforestry Systems 93 (4), 1485–1500. Available at https://doi.org/10.1007/s10457-018-0252-4.

Groninger, J.W., 2012. Reforestation strategies amid social instability: lessons from Afghanistan. Environmental Management 49 (4), 833–845. Available at https://doi.org/10.1007/s00267-012-9817-6.

Gupta, H., Dube, L.C., 2018. Addressing biodiversity in climate change discourse: paris mechanisms hold more promise. International Forestry Review 20 (1), 104–114. Available at https://doi.org/10.1505/146554818822824282.

Gupta, S.R., Dagar, J.C., Teketay, D., 2020. Agroforestry for rehabilitation of degraded landscapes: achieving livelihood and environmental security. In: Dagar, J.C., Gupta, S.R., Teketay, D. (Eds.), Agroforestry for Degraded Landscapes: Recent Advances and Emerging Challenges-Vol.1. Springer, Singapore, pp. 23–68. Available at https://doi.org/10.1007/978-981-15-4136-0_2.

Henry, M., Tittonell, P., Manlay, R.J., Bernoux, M., Albrecht, A., Vanlauwe, B., 2009. Biodiversity, carbon stocks and sequestration potential in aboveground biomass in smallholder farming systems of western Kenya. Agriculture, Ecosystems & Environment 129 (1-3), 238–252. Available at https://doi.org/10.1016/j.agee.2008.09.006.

Himanen, S., Mäkinen, H., Rimhanen, K., Savikko, R., 2016. Engaging farmers in climate change adaptation planning: assessing intercropping as a means to support farm adaptive capacity. Agriculture 6 (3), 34. Available at https://doi.org/10.3390/agriculture6030034.

Hodge, A., Berta, G., Doussan, C., Merchan, F., Crespi, M., 2009. Plant root growth, architecture and function. Plant and Soil 321 (1-2), 153–187. Available at https://doi.org/10.1007/s11104-009-9929-9.

Hoosbeek, M.R., Remme, R.P., Rusch, G.M., 2018. Trees enhance soil carbon sequestration and nutrient cycling in a silvopastoral system in south-western Nicaragua. Agroforestry Systems 92 (2), 263–273. Available at https://doi.org/10.1007/s10457-016-0049-2.

Jacobi, J., Rist, S., Altieri, M.A., 2017. Incentives and disincentives for diversified agroforestry systems from different actors' perspectives in Bolivia. International Journal of Agricultural Sustainability 15 (4), 365–379. Available at https://doi.org/10.1080/14735903.2017.1332140.

Jhariya, M.K., Meena, R.S., Banerjee, A., 2021. Ecological Intensification of Natural Resources for Sustainable Agriculture. Springer, Singapore, pp. 1–655. Available at https://doi.org/10.1007/978-981-33-4203-3.

Jhariya, M.K., Banerjee, A., Swaroop Meena, R., Kumar, S., Raj, A., 2021. Sustainable Intensification for Agroecosystem Services and Management. Springer, Singapore, pp. 1–870. Available at https://doi.org/10.1007/978-981-16-3207-5.

Jhariya, M.K., Meena, R.S., Banerjee, A., Meena, S.N., 2022. Natural Resources Conservation and Advances for Sustainability. Elsevier. Available at https://doi.org/10.1016/C2019-0-03763-6.

Jhariya, M.K., Meena, R.S., Banerjee, A., Kumar, S., Raj, A., 2023. Agroforestry for Carbon and Ecosystem Management Agroforestry for Carbon and Ecosystem Management. Elsevier, India, pp. 1–434. Available at https://doi.org/10.1016/C2021-0-02916-3.

Jin, Y., Sharifi, A., Li, Z., Chen, S., Zeng, S., Zhao, S., 2024. Carbon emission prediction models: a review. Science of the Total Environment 927, 172319. Available at https://doi.org/10.1016/j.scitotenv.2024.172319.

Jose, K., Bandopadhyay, A., Arya, A., Chaturvedi, R.K., 2023. Forest ecosystem modeling for policy planning: a review. Ecosystem and Species Habitat Modeling for Conservation and Restoration. Springer Nature, India, pp. 439–457. Available at https://doi.org/10.1007/978-981-99-0131-9_24.

Karjalainen, E., Sarjala, T., Raitio, H., 2010. Promoting human health through forests: overview and major challenges. Environmental Health and Preventive Medicine 15 (1), 1–8. Available at https://doi.org/10.1007/s12199-008-0069-2.

Krishna, M.P., Mohan, M., 2017. Litter decomposition in forest ecosystems: a review. Energy, Ecology and Environment 2 (4), 236–249. Available at https://doi.org/10.1007/s40974-017-0064-9.

Kujur, E., Jhariya, M.K., Yadav, D.K., Banerjee, A., 2021. Biomass, carbon storage and CO_2 mitigation potential of two riparian sites of northern Chhattisgarh, India. Environmental Earth Sciences 80 (21). Available at https://doi.org/10.1007/s12665-021-10019-8.

Kujur, E., Jhariya, M.K., Yadav, D.K., Banerjee, A., 2022. Phytosociological attributes and regeneration potential of riparian vegetation in Northern Chhattisgarh, India. Environment, Development and Sustainability 24 (2), 2861–2886. Available at https://doi.org/10.1007/s10668-021-01557-z.

Kumar, S., Meena, R.S., Sheoran, P., Jhariya, M.K., 2024. Regenerative Agriculture for Sustainable Food Systems. Springer Science+Business Media, India, pp. 1–493. Available at https://doi.org/10.1007/978-981-97-6691-8.

Lahiji, R.N., Dinan, N.M., Liaghati, H., Ghaffarzadeh, H., Vafaeinejad, A., 2020. Scenario-based estimation of catchment carbon storage: linking multi-objective land allocation with InVEST model in a mixed agriculture-forest landscape. Frontiers of Earth Science 14 (3), 637–646. Available at https://doi.org/10.1007/s11707-020-0825-1.

Lal, R., 2005. Forest soils and carbon sequestration. Forest Ecology and Management 220 (1-3), 242–258. Available at https://doi.org/10.1016/j.foreco.2005.08.015.

Leakey, R.R.B., Tchoundjeu, Z., Schreckenberg, K., Shackleton, S.E., Shackleton, C.M., 2005. Agroforestry tree products (AFTPs): targeting poverty reduction and enhanced livelihoods. International Journal of Agricultural Sustainability 3 (1), 1–23. Available at https://doi.org/10.1080/14735903.2005.9684741.

Lee, J., Ingalls, M., Erickson, J.D., Wollenberg, E., 2016. Bridging organizations in agricultural carbon markets and poverty alleviation: an analysis of pro-Poor carbon market projects in East Africa. Global Environmental Change 39, 98–107. Available at https://doi.org/10.1016/j.gloenvcha.2016.04.015.

Lee, D.H., Kim, Dh, Kim, Si, 2018. Characteristics of forest carbon credit transactions in the voluntary carbon market. Climate Policy 18 (2), 235–245. Available at https://doi.org/10.1080/14693062.2016.1277682.

Li, S.X., Wang, Z.H., Malhi, S.S., Li, S.Q., Gao, Y.J., Tian, X.H., 2009. Chapter 7 nutrient and water management effects on crop production, and nutrient and water use efficiency in dryland areas of China. Advances in Agronomy 102, 223–265. Available at https://doi.org/10.1016/S0065-2113(09)01007-4.

Liu, W.Y., Chuang, Y.L., 2023. Assessing the incentives and financial compensation of agroforestry considering the uncertainty of price and yield. Ecological Indicators 146. Available at https://doi.org/10.1016/j.ecolind.2022.109753.

Lorenz, K., Lal, R., 2014. Soil organic carbon sequestration in agroforestry systems. A review. Agronomy for Sustainable Development 34 (2), 443–454. Available at https://doi.org/10.1007/s13593-014-0212-y.

Ma, T., Zhang, C., Ji, L., Zuo, Z., Beckline, M., Hu, Y., et al., 2024. Development of forest aboveground biomass estimation, its problems and future solutions: a review. Ecological Indicators 159, 111653. Available at https://doi.org/10.1016/j.ecolind.2024.111653.

Makela, A., Landsberg, J., Ek, A.R., Burk, T.E., Ter-Mikaelian, M., Agren, G.I., et al., 2000. Process-based models for forest ecosystem management: current state of the art and challenges for practical implementation. Tree Physiology 20 (5-6), 289–298. Available at https://doi.org/10.1093/treephys/20.5-6.289.

Marçal, M.F.M., de Souza, Z.M., Tavares, R.L.M., Farhate, C.V.V., Oliveira, S.R.M., Galindo, F.S., 2021. Predictive models to estimate carbon stocks in agroforestry systems. Forests 12 (9). Available at https://doi.org/10.3390/f12091240.

Marsden, C., Martin-Chave, A., Cortet, J., Hedde, M., Capowiez, Y., 2020. How agroforestry systems influence soil fauna and their functions - a review. Plant and Soil 453 (1-2), 29–44. Available at https://doi.org/10.1007/s11104-019-04322-4.

Mbow, C., Van Noordwijk, M., Luedeling, E., Neufeldt, H., Minang, P.A., Kowero, G., 2014. Agroforestry solutions to address food security and climate change challenges in Africa. Current Opinion in Environmental Sustainability 6 (1), 61–67. Available at https://doi.org/10.1016/j.cosust.2013.10.014.

Meena, R.S., Pradhan, G., 2023. Industrial garbage-derived biocompost enhances soil organic carbon fractions, CO_2 biosequestration, potential carbon credits and sustainability index in a rice-wheat ecosystem. Environmental Research 235, 116525. Available at https://doi.org/10.1016/j.envres.2023.116525.

Meena, R.S., Pradhan, G., Kumar, S., Lal, R., 2023a. Using industrial wastes for rice-wheat cropping and food-energy-carbon-water-economic nexus to the sustainable food system. Renewable and Sustainable Energy Reviews 187, 113756. Available at https://doi.org/10.1016/j.rser.2023.113756.

Meena, R.S., Yadav, A., Kumar, S., Jhariya, M.K., Jatav, S.S., 2022. Agriculture ecosystem models for CO_2 sequestration, improving soil physicochemical properties, and restoring degraded land. Ecological Engineering 176. Available at https://doi.org/10.1016/j.ecoleng.2022.106546.

Meena, R.S., Pradhan, G., Singh, K., Kumar, S., Singh, A.K., Shashidhar, K.S., et al., 2024a. Agriculture models for restoring degraded land to enhance CO_2 biosequestration and carbon credits in the Vindhyan region of India. Science of the Total Environment 929. Available at https://doi.org/10.1016/j.scitotenv.2024.172661.

Meena, R.S., Lal, R., Kumar, S., Pradhan, G., Srinivasarao, C., Singh, A.K., et al., 2024b. Potential of Indian agriculture for capturing atmospheric CO_2 and monetizing carbon credits to the farmers: an overview and policy framework. Advances in Agronomy 188, 101–206. Available at https://doi.org/10.1016/bs.agron.2024.06.001.

Meena, R.S., Pradhan, G., Nalani, Kumar, S., Singh, A.K., Verma, S.K., et al., 2024c. Diversified cropping modules designed for soil restoration, CO_2 sequestration, and generating carbon credits. Land Degradation and Development 35 (14), 4314–4329. Available at https://doi.org/10.1002/ldr.5224.

Meena, R.S., Pradhan, G.S., Kumar, S., Jhariya, M.K., Rao, C.S., 2025a. Carbon-Negative Agriculture. Springer, Singapore, pp. 1–500. Available at https://doi.org/10.1007/978-981-95-1483-0.

Meena, R.S., Pradhan, G., Kumar, S., 2025b. Energy flow, eco-efficiency, and economic circulation with recycled industrial waste compost application in wheat and subsequent rice farming. Science of the Total Environment 967. Available at https://doi.org/10.1016/j.scitotenv.2025.178779.

Montagnini, F., Nair, P.K.R., 2004. Carbon sequestration: an underexploited environmental benefit of agroforestry systems. In: Nair, P.K.R., Rao, M.R., Buck, L.E. (Eds.), ew Vistas in Agroforestry. Advances in Agroforestry, vol 1. Springer, Dordrecht, pp. 281–295. Available at https://doi.org/10.1007/978-94-017-2424-1_20.

Nadège, M.T., Louis, Z., Cédric, C.D., Louis-Paul, K.B., Funwi, F.P., Ingrid, T.T., et al., 2019. Carbon storage potential of cacao agroforestry systems of different age and management intensity. Climate and Development 11 (7), 543–554. Available at https://doi.org/10.1080/17565529.2018.1456895.

Nair, P.K.R., 2011. Methodological Challenges in Estimating Carbon Sequestration Potential of Agroforestry Systems. In: Kumar, B., Nair, P, (Eds.), Carbon Sequestration Potential of Agroforestry Systems. Advances in Agroforestry, vol 8. Springer, Dordrecht, pp. 3–16. Available at https://doi.org/10.1007/978-94-007-1630-8_1.

Nair, P.K.R., Nair, V.D., Kumar, B.M., Haile, S.G., 2009. Soil carbon sequestration in tropical agroforestry systems: a feasibility appraisal. Environmental Science and Policy 12 (8), 1099–1111. Available at https://doi.org/10.1016/j.envsci.2009.01.010.

Nambiar, E.K.S., 2021. Small forest growers in tropical landscapes should be embraced as partners for Green-growth: increase wood supply, restore land, reduce poverty, and mitigate climate change. Trees, Forests and People 6, 100154. Available at https://doi.org/10.1016/j.tfp.2021.100154.

Nayak, A.K., Rahman, M.M., Naidu, R., Dhal, B., Swain, C.K., Nayak, A.D., et al., 2019. Current and emerging methodologies for estimating carbon sequestration in agricultural soils: a review. Science of the Total Environment 665, 890–912. Available at https://doi.org/10.1016/j.scitotenv.2019.02.125.

Ntawuruhunga, D., Ngowi, E.E., Mangi, H.O., Salanga, R.J., Shikuku, K.M., 2023. Climate-smart agroforestry systems and practices: a systematic review of what works, what doesn't work, and why. Forest Policy and Economics 150. Available at https://doi.org/10.1016/j.forpol.2023.102937.

Nuwarapaksha, T.D., Udumann, S.S., Dissanayaka, D.M.N.S., Dissanayake, D.K.R.P.L., Atapattu, A.J., 2022. Coconut based multiple cropping systems: an analytical review in Sri Lankan coconut cultivations. Circular Agricultural Systems 2. Available at https://doi.org/10.48130/CAS-2022-0008.

Nuwarapaksha, T.D., Dissanayaka, N.S., Udumann, S.S., Atapattu, A.J., 2023a. Gliricidia as a beneficial crop in resource-limiting agroforestry systems in Sri Lanka. Indian Journal of Agroforestry 25 (1), 12–18.

Nuwarapaksha, T.D., Rajapaksha, U.N., Ekanayake, J., Weerasooriya, S.A., Atapattu, A.J., 2023b. Exploring the economic viability of integrating Jamnapari goat into underutilized pastures under coconut cultivations in coconut research institute, Sri Lanka. IECAG 2023. MDPI, Basel Switzerland, pp. 27. Available at https://doi.org/10.3390/IECAG2023-14990.

Nuwarapaksha, T.D., Udumann, S.S., Dissanayaka, N.S., Atapattu, A.J., 2023c. Coconut-based livestock farming: a sustainable approach to enhancing food security in Sri Lanka. In: Kiba, D.I. (Ed.), Transitioning to zero hunger. MDPI Books, Basel, Switzerland, pp. 197–213. Available at https://doi.org/10.3390/books978-3-03897-863-3-8.

Nuwarapaksha, T.D., Dissanayaka, D.M.N.S., Udumann, S.S., Vinujan, S., Atapattu, A.J., 2024a. Exploring the impact of pyrolysis temperature on nutrient composition of *Gliricidia sepium* biochar: a comprehensive study. Technology in Agronomy 4 (1). Available at https://doi.org/10.48130/tia-0024-0014 0–0.

Nuwarapaksha, T.D., Udumann, S.S., Atapattu, A.J., 2024b. Fostering food and nutritional security through agroforestry practices. Agroforestry. wiley, Sri Lanka, pp. 285–318. Available at https://doi.org/10.1002/9781394231164.ch10.

Nuwarapaksha, T.D., Dilhan, R.M.C.P., Udumann, S.S., Ranasinghe, C.S., Egodawatta, W.C.P., Dissanayaka, N.S., et al., 2025a. Estimation of the total carbon sequestration potential of coconut-gliricidia mixed cropping systems in Sri Lanka. Carbon Footprints 4 (1). Available at https://doi.org/10.20517/cf.2024.31.

Nuwarapaksha, T.D., Udumann, S.S., Dissanayaka, N.S., Atapattu, A.J., 2025b. Future prospects and emerging trends in agroforestry research. Agroforestry for a Climate-Smart Future. IGI Global, Sri Lanka, pp. 497–518. Available at https://doi.org/10.4018/979-8-3693-8282-0.ch017.

Nuwarapaksha, Tharindu D., Udumann, Shashi S., Dissanayaka, Nuwandhya S., Atapattu, Anjana J., 2026. Climate change impacts and adaptation strategies in coconut plantations: integrating remote sensing and real-time monitoring. Frontiers in Climate 8, 1762364.. Available at 10.3389/fclim.2026.1762364.

Onaindia, M., Fernández de Manuel, B., Madariaga, I., Rodríguez-Loinaz, G., 2013. Co-benefits and trade-offs between biodiversity, carbon storage and water flow regulation. Forest Ecology and Management 289, 1–9. Available at https://doi.org/10.1016/j.foreco.2012.10.010.

Pan, C., Shrestha, A., Innes, J.L., Zhou, G., Li, N., Li, J., et al., 2022. Key challenges and approaches to addressing barriers in forest carbon offset projects. Journal of Forestry Research 33 (4), 1109–1122. Available at https://doi.org/10.1007/s11676-022-01488-z.

Pandey, D., 2002. Carbon sequestration in agroforestry systems. Climate Policy 2 (4), 367–377. Available at https://doi.org/10.1016/s1469-3062(02)00025-6.

Patel, R., Patel, B., Jaison, J.M., Dash, B., Mukherjee, S., 2023. Impact of agroforestry ecosystem on carbon sequestration potential and climate change. Agricultural Soil Sustainability and Carbon Management. Elsevier, India, pp. 269–297. Available at https://doi.org/10.1016/B978-0-323-95911-7.00002-5.

Pennock, D.J., Frick, A.H., 2001. The role of field studies in landscape-scale applications of process models: an example of soil redistribution and soil organic carbon modeling using CENTURY. Soil and Tillage Research 58 (3-4), 183–191. Available at https://doi.org/10.1016/s0167-1987(00)00167-7.

Perdan, S., Azapagic, A., 2011. Carbon trading: current schemes and future developments. Energy Policy 39 (10), 6040–6054. Available at https://doi.org/10.1016/j.enpol.2011.07.003.

Perosa, B., Newton, P., da Silva, R.F.B., 2023. A monitoring, reporting and verification system for low carbon agriculture: a case study from Brazil. Environmental Science and Policy 140, 286–296. Available at https://doi.org/10.1016/j.envsci.2022.12.006.

Petrokofsky, G., Kanamaru, H., Achard, F., Goetz, S.J., Joosten, H., Holmgren, P., et al., 2012. Comparison of methods for measuring and assessing carbon stocks and carbon stock changes in terrestrial carbon pools. How do the accuracy and precision of current methods compare? A systematic review protocol. Environmental Evidence 1 (1). Available at https://doi.org/10.1186/2047-2382-1-6.

Plaza-Bonilla, D., Arrúe, J.L., Cantero-Martínez, C., Fanlo, R., Iglesias, A., Álvaro-Fuentes, J., 2015. Carbon management in dryland agricultural systems. A review. Agronomy for Sustainable Development 35 (4), 1319–1334. Available at https://doi.org/10.1007/s13593-015-0326-x.

Porro, R., Miller, R.P., Tito, M.R., Donovan, J.A., Vivan, J.L., Trancoso, R., et al., 2012. Agroforestry in the Amazon region: a pathway for balancing conservation and development. In: Nair, P., Garrity, D. (Eds.), Agroforestry - The Future of Global Land Use. Advances in Agroforestry, vol 9. Springer, Dordrecht, pp. 391–428. Available at https://doi.org/10.1007/978-94-007-4676-3_20.

Pradhan, G., Meena, R.S., 2023a. Interaction impact of biocompost on nutrient dynamics and relations with soil biota, carbon fractions index, societal value of CO_2 equivalent and ecosystem services in the wheat-rice farming. Chemosphere 339, 139695. Available at https://doi.org/10.1016/j.chemosphere.2023.139695.

Pradhan, G., Meena, R.S., 2023b. Utilizing waste compost to improve the atmospheric CO_2 capturing in the rice-wheat cropping system and energy-cum-carbon credit auditing for a circular economy. Science of the Total Environment 892. Available at https://doi.org/10.1016/j.scitotenv.2023.164572.

Pradhan, G., Meena, R.S., Kumar, S., Lal, R., 2023. Utilizing industrial wastes as compost in wheat-rice production to improve the above and below-ground ecosystem services. Agriculture, Ecosystems & Environment 358, 108704. Available at https://doi.org/10.1016/j.agee.2023.108704.

Prasad, R., Jhariya, M.K., Banerje, A., 2021. Advances in Sustainable Development and Management of Environmental and Natural Resources: Economic Outlook and Opinions, 2-volume set. Apple Academic Press, New York, pp. 1–428. Available at https://doi.org/10.1201/9781003187455.

Praveen, B., Sharma, P., 2019. A review of literature on climate change and its impacts on agriculture productivity. Journal of Public Affairs 19 (4). Available at https://doi.org/10.1002/pa.1960.

Puri, S., Nair, P.K.R., 2004. Agroforestry research for development in India: 25 Years of experiences of a national program. Agroforestry Systems 61-62 (1-3), 437–452. Available at https://doi.org/10.1023/B:AGFO.0000029014.66729.e0.

Rahman, G.K.M.M., Rahman, M.M., Alam, M.S., Kamal, M.Z., Mashuk, H.A., Datta, R., et al., 2019. Biochar and organic amendments for sustainable soil carbon and soil health. In: Datta, R., Meena, R., Pathan, S., Ceccherini, M. (Eds.), Carbon and Nitrogen Cycling in Soil. Springer, Singapore, pp. 45–85. Available at https://doi.org/10.1007/978-981-13-7264-3_3.

Raj, A., Jhariya, M.K., Yadav, D.K., Banerjee, A., Meena, R.S., 2019. Agroforestry: a holistic approach for agricultural sustainability. Sustainable agriculture, forest and environmental management. Springer, pp. 101–131. Available at https://doi.org/10.1007/978-981-13-6830-1_4.

Raj, A., Jhariya, M.K., Banerjee, A., Meena, R.S., Nema, S., Khan, N., et al., 2021. Agroforestry a model for ecological sustainability. Natural Resources Conservation and Advances for Sustainability. Elsevier, India, pp. 289–307. Available at https://doi.org/10.1016/B978-0-12-822976-7.00002-8.

Raj, A., Jhariya, M.K., Meena, R.S., 2025. Agroforestry for Monetising Carbon Credits. Springer Nature, Cham. Available at https://doi.org/10.1007/978-3-031-97151-8.

Raju Singh, N., Kumar, D., Rao, K.K., Bhatt, B.P., 2020. Agroforestry: soil organic carbon and its carbon sequestration potential. Climate change and agroforestry systems. Apple Academic Press, pp. 119–142. Available at https://doi.org/10.1201/9780429286759-5.

Ramachandran Nair, P.K., Mohan Kumar, B., Nair, V.D., 2009. Agroforestry as a strategy for carbon sequestration. Journal of Plant Nutrition and Soil Science 172 (1), 10–23. Available at https://doi.org/10.1002/jpln.200800030.

Ramos-Gerena, C.E., 2024. Regulating belonging: contradictions in Puerto Rico's agricultural land-use policies. Land use Policy 145, 107279. Available at https://doi.org/10.1016/j.landusepol.2024.107279.

Rois-Díaz, M., Lovric, N., Lovric, M., Ferreiro-Domínguez, N., Mosquera-Losada, M.R., den Herder, M., et al., 2018. Farmers' reasoning behind the uptake of agroforestry practices: evidence from multiple case-studies across Europe. Agroforestry Systems 92 (4), 811–828. Available at https://doi.org/10.1007/s10457-017-0139-9.

Rolo, V., Rivest, D., Maillard, É., Moreno, G., 2023. Agroforestry potential for adaptation to climate change: a soil-based perspective. Soil Use and Management 39 (3), 1006–1032. Available at https://doi.org/10.1111/sum.12932.

Roy, O., Meena, R.S., Kumar, S., Jhariya, M.K., Pradhan, G., 2022. Assessment of land use systems for CO_2 sequestration, carbon credit potential, and income security in Vindhyan region, India. Land Degradation and Development 33 (4), 670–682. Available at https://doi.org/10.1002/ldr.4181.

Sahoo, G., Wani, A.M., 2019. Multifunctional agroforestry systems in India for livelihoods. Annals of Horticulture 12 (2), 139. Available at https://doi.org/10.5958/0976-4623.2019.00022.7.

Salleh, A.M., Harun, N.Z., 2013. The environmental benefits of agroforestry systems in relation to social sustainability. 301–317. Available at https://www.researchgate.net/publication/304259524_The_Environmental_Benefits_of_Agroforestry_Systems_in_relation_to_Social_Sustainability.

Sharma, P., Bhardwaj, D.R., Singh, M.K., Nigam, R., Pala, N.A., Kumar, A., et al., 2023. Geospatial technology in agroforestry: status, prospects, and constraints. Environmental Science and Pollution Research 30 (55), 116459–116487. Available at https://doi.org/10.1007/s11356-022-20305-y.

Shrestha, B.M., Chang, S.X., Bork, E.W., Carlyle, C.N., 2018. Enrichment planting and soil amendments enhance carbon sequestration and reduce greenhouse gas emissions in agroforestry systems: a review. Forests 9 (6), 369. Available at https://doi.org/10.3390/f9060369.

Sierra, C.A., Trumbore, S.E., Davidson, E.A., Vicca, S., Janssens, I., 2015. Sensitivity of decomposition rates of soil organic matter with respect to simultaneous changes in temperature and moisture. Journal of Advances in Modeling Earth Systems 7 (1), 335–356. Available at https://doi.org/10.1002/2014MS000358.

Sileshi, G.W., Dagar, J.C., Nath, A.J., Kuntashula, E., 2023. Agroforestry as a climate-smart agriculture: strategic interventions, current practices and policies. In: Dagar, J.C., Gupta, S.R., Sileshi, G.W. (Eds.), Agroforestry for Sustainable Intensification of Agriculture in Asia and Africa. Sustainability Sciences in Asia and Africa. Springer, Singapore, pp. 589–640. Available at https://doi.org/10.1007/978-981-19-4602-8_18.

Sohrabi, H., Bakhtiarvand-Bakhtiari, S., Ahmadi, K., 2016. Above- and below-ground biomass and carbon stocks of different tree plantations in central Iran. Journal of Arid Land 8 (1), 138–145. Available at https://doi.org/10.1007/s40333-015-0087-z.

Su, M., Zhao, R., Jiang, J., Zhao, J., Wang, M., Zha, D., et al., 2024. A blockchain system supporting cross-border data protection and consistency verification in unified global carbon emissions trading framework. Journal of Cleaner Production 448, 141693. Available at https://doi.org/10.1016/j.jclepro.2024.141693.

Szyniszewska, A.M., Akrivou, A., Björklund, N., Boberg, J., Bradshaw, C., Damus, M., et al., 2024. Beyond the present: how climate change is relevant to pest risk analysis. EPPO Bulletin 54 (1), 20–37. Available at https://doi.org/10.1111/epp.12986.

Tang, R., Guo, W., Oudenes, M., Li, P., Wang, J., Tang, J., et al., 2018. Key challenges for the establishment of the monitoring, reporting and verification (MRV) system in China's national carbon emissions trading market. Climate Policy 18 (1), 106–121. Available at https://doi.org/10.1080/14693062.2018.1454882.

Taye Gifawesen, S., Tola, F.K., Duguma, M.S., 2020. Review on role of home garden agroforestry practices to improve livelihood of small scale farmers and climate change adaptation and mitigation. Journal of Plant Sciences 8 (5), 134. Available at https://doi.org/10.11648/j.jps.20200805.15.

Taylor, J.R., Lovell, S.T., 2021. Designing multifunctional urban agroforestry with people in mind. Urban Agriculture and Regional Food Systems 6 (1). Available at https://doi.org/10.1002/uar2.20016.

Thevs, N., Gombert, A., Strenge, E., Lleshi, R., Aliev, K., Emileva, B., 2019. Tree wind breaks in central Asia and their effects on agricultural water consumption. Land 8 (11), 167. Available at https://doi.org/10.3390/land8110167.

Tsonkova, P., Mirck, J., Böhm, C., Fütz, B., 2018. Addressing farmer-perceptions and legal constraints to promote agroforestry in Germany. Agroforestry Systems 92 (4), 1091–1103. Available at https://doi.org/10.1007/s10457-018-0228-4.

Udumann, S.S., Dissanayaka, N.S., Nuwarapaksha, T.D., Thelwadana, E.P., Atapattu, A.J., 2023. Assessing the growth potential of Sunn hemp (*Crotalaria juncea* L.) as a cover crop for major coconut-growing soils. Trends in Horticulture. 6 (2), 3579. Available at https://doi.org/10.24294/th.v6i2.3579.

Udumann, Shashi S., Dissanayaka, Nuwandhya S., Nuwarapaksha, Tharindu D., Atapattu, Anjana J., 2025. Ecological Solutions to Agricultural Land Degradation, Eds. Babu, A. Das, S. S. Rathore, R. Springer, Singapore, pp. 117–143. Available at 10.1007/978-981-96-3392-0_5.

Ulucak, R., Yücel, A.G., Koçak, E., 2019. The process of sustainability. Environmental Kuznets Curve (EKC). Elsevier, pp. 37–53. Available at https://doi.org/10.1016/B978-0-12-816797-7.00005-9.

van Kooten, G.C., de Vries, F.P., 2013. Carbon Offsets. In: Shogren, J.F. (Ed.), Encyclopedia of Energy, Natural Resource, and Environmental Economics, volume 1. Elsevier Science, pp. 6–8. Available at https://doi.org/10.1016/B978-0-12-375067-9.00165-0.

Varsha, K.M., Raj, A.K., Kurien, E.K., Bastin, B., Kunhamu, T.K., Pradeep, K.P., 2019. High density silvopasture systems for quality forage production and carbon sequestration in humid tropics of Southern India. Agroforestry Systems 93 (1), 185–198. Available at https://doi.org/10.1007/s10457-016-0059-0.

Verchot, L.V., Van Noordwijk, M., Kandji, S., Tomich, T., Ong, C., Albrecht, A., et al., 2007. Climate change: linking adaptation and mitigation through agroforestry. Mitigation and Adaptation Strategies for Global Change 12 (5), 901–918. Available at https://doi.org/10.1007/s11027-007-9105-6.

Verde, S.F., Borghesi, S., 2022. The international dimension of the EU emissions trading system: bringing the pieces together. Environmental and Resource Economics 83 (1), 23–46. Available at https://doi.org/10.1007/s10640-022-00705-x.

Wienhold, K., Goulao, L.F., 2023. The embedded agroecology of coffee agroforestry: a contextualized review of smallholder farmers' adoption and resistance. Sustainability 15 (8), 6827. Available at https://doi.org/10.3390/su15086827.

Williams, D.R., Phalan, B., Feniuk, C., Green, R.E., Balmford, A., 2018. Carbon storage and land-use strategies in agricultural landscapes across three continents. Current Biology 28 (15), 2500. Available at https://doi.org/10.1016/j.cub.2018.05.087.

Wilson, M., Lovell, S., 2016. Agroforestry—The next step in sustainable and resilient agriculture. Sustainability 8 (6), 574. Available at https://doi.org/10.3390/su8060574.

Wolz, K.J., DeLucia, E.H., 2018. Alley cropping: global patterns of species composition and function. Agriculture, Ecosystems and Environment 252, 61–68. Available at https://doi.org/10.1016/j.agee.2017.10.005.

Wunder, S., 2007. The efficiency of payments for environmental services in tropical conservation. Conservation Biology 21 (1), 48–58. Available at https://doi.org/10.1111/j.1523-1739.2006.00559.x.

Xia, C., Zhang, Y., 2022. Comparison of the use of Landsat 8, Sentinel-2, and Gaofen-2 images for mapping soil pH in Dehui, northeastern China. Ecological Informatics 70, 101705. Available at https://doi.org/10.1016/j.ecoinf.2022.101705.

Yang, C., Dossa, G.G.O., Atapattu, A.J., Xia, S., Yang, X., 2025. Soil fauna explain the plant apparency effect on litter decomposition in tropical rainforests. CATENA 257, 109157. Available at 10.1016/j.catena.2025.109157.

Zerssa, G., Feyssa, D., Kim, D.-G., Eichler-Löbermann, B., 2021. Challenges of smallholder farming in ethiopia and opportunities by adopting climate-smart agriculture. Agriculture 11 (3), 192. Available at https://doi.org/10.3390/agriculture11030192.

Zhang, Z., Zhu, L., 2023. A review on unmanned aerial vehicle remote sensing: platforms, sensors, data processing methods, and applications. Drones 7 (6), 398. Available at https://doi.org/10.3390/drones7060398.

Chapter 18

Carbon credit estimation from grassland ecosystems

Emmanuely Z. Nungula[1], Norbert J. Ngowi[1], Luciana R. Chappa[2], Victor V. Ngaiza[3], Stephen Mwakabonga[1], Paisteras H. Ilomo[4], Kesia A. Nyambele[5], Riziki Mwadalu[6], and Harun I. Gitari[7]

[1]*Department of Environment and Sustainable Development, Mzumbe University, Morogoro, Tanzania,* [2]*Department of Agricultural Sciences, College of Agriculture, Mwalimu Nyerere University of Agriculture and Technology, Butiama, Tanzania,* [3]*School of Life Sciences and Bio-engineering, The Nelson Mandela African Institution of Sciences and Technology, Arusha, Tanzania,* [4]*Department Engineering Management Studies, Mzumbe University, Morogoro, Tanzania,* [5]*Faculty of Agribiology, Food and Natural Resources, Czech University of Life Sciences, Prague, Czech Republic,* [6]*Central Highlands Eco-Region Research Programme, Kenya Forestry Research Institute, Nairobi, Kenya,* [7]*Department of Agricultural Science and Technology, School of Agriculture and Environmental Sciences, Kenyatta University, Nairobi, Kenya*

Highlights

- Grassland provides substantial ecological services for the well-being of humankind.
- Carbon credits can be an additional benefit generated from the grasslands.
- The grassland ecosystem makes a significant contribution to the carbon sink and the soil carbon pool.

18.1 Introduction

Grasslands comprise around 3.5 billion hectares, equal to 26 of the global land (Zhou et al., 2018). They also hold 20% of the world's carbon (C) soil stores. These grassland resources are essential to human food and fodder production. Although developed crops have replaced about 25% of the world's natural grasslands, meadows still provide a sizable amount of the world's milk (27%) and meat (23%). The livestock sector relies primarily on these grasslands for its operations, supporting one-third of the world's livestock population and providing employment for nearly 1 billion people (Yang et al., 2019a). According to Piao et al. (2022a), the livestock industry in certain developing nations even contributes between 50% and 80% of GDP. However, these grasslands are currently under pressure from intensive grazing, particularly in developing countries, which are susceptible to climate change but are still anticipated to provide a significant amount of meat and milk in the years to come. Overgrazing has already caused 7.5% of the grasslands globally to deteriorate as a result of previous practices (Chappa et al., 2024).

Grasslands are particularly important to people (Lange et al., 2019; Huang et al., 2021). Due to their extensive distribution in both African savannas and Eurasian steppes, they account for approximately 40% of the global land coverage (Bahram et al., 2018; Bai and Cotrufo, 2022). C cycling between plants, microorganisms, soils, and the atmosphere underpins many of the diverse ecological functions that grasslands provide, including sustaining high biodiversity, controlling climate, and biogeochemical cycling (Meena et al., 2025b; Cavender-Bares et al., 2022). Globally, grasslands contribute up to 34% of the C stock, making them significant even though their C pool is lower than that of forests (Deng et al., 2021; Gitari et al., 2024). Like forests, the variation in C uptake from photosynthetic processes and losses from respiration determines the C balance of grasslands.

However, grasslands may be subject to more intense solar radiation and larger extreme weather events due to their comparatively lower stature and sparser vegetation, compared to forests (Nungula et al., 2024). Therefore abiotic processes are more important in controlling the grassland C balance (Cheruto et al., 2025; Goldstein et al., 2020). Global warming will alter many of the abiotic controls of the grassland C balance, and land-use conversion will cause more anthropogenic disturbance in grasslands, which may lead to the decline of soil organic carbon (SOC), which cannot be

Carbon Incentives, Plans, and Policies in Agriculture. DOI: https://doi.org/10.1016/B978-0-443-33557-0.00010-0

recovered on timescales (Fuhlendorf et al., 2018; Guo et al., 2021; Meena et al., 2024b). Therefore it is crucial to understand the driving forces behind the grassland carbon cycle and estimate the amount of C that grasslands sequester (Huang et al., 2018). Hence, the major objective of this chapter is to evaluate the potential of the grassland ecosystem for the generation of C credit that may provide additional benefits and ecological services for the community stakeholders.

18.2 Grassland ecosystems and the soil carbon pool

Grasslands offer numerous cultural benefits, support food production, and serve as habitats for biodiversity (Li et al., 2020). But also offer a C storage pool of more than 90% belowground and about 34% of the terrestrial C stock that is essential for soil carbon sequestration (C_{seq}) (Li et al., 2018; Jia et al., 2021; Meena et al., 2024c). Due to anthropogenic disturbances, such as overgrazing and land-use change, grasslands are particularly vulnerable to degradation (Iqbal, 2021). Grasslands around the world have seen notable reductions in biodiversity and ecological functioning, which has led to less SOC storage (Dixon et al., 2014; Bright et al., 2015; Meena et al., 2024a).

Grasslands store between 525 and 634 Pg C worldwide. Although over 80% of the grassland C pool is found in the soil as plant C (roots, other belowground biomass) and SOC, grassland biomass C makes up a minor portion of the system's overall C storage, in contrast to forests (Huang et al., 2021). Crucially, soil C can endure for years or perhaps millennia in contrast to the transient plant biomass C in grasslands (Lyu et al., 2023). Thus the most important factor influencing grassland's ability to trap C and slow down climate change is its soil characteristics and processes (Fig. 18.1) (Buisson et al., 2022). One of the key elements affecting grassland productivity and, consequently, soil C stores is plant diversity (Angst et al., 2021; IPCC, 2021; Day et al., 2019). Diversity has a greater impact on SOC accumulation over time, similar to biomass productivity (Lyu et al., 2023; Meena and Pradhan, 2023; Charles, 2018; Jhariya et al., 2022, 2023).

SOC accumulation in species-rich plant communities may be due to enhanced soil microbial activity and abundance in response to a wider range of plant inputs (Xiong et al., 2020; Nachshon, 2018). The soil's capacity to store C is greatly impacted by the ability of soil microorganisms to mineralize plant C inputs or decompose soil organic matter (SOM) (Meena et al., 2022; Conant, 2010). The interaction between plants and soil microbes accelerates decomposition activities that enable soil C storage. Because grasslands generate high-quality litter with a lower concentration of lignin and a lower C to nitrogen ratio, bacteria are drawn to these areas and take control of the decomposition processes (Li et al., 2018; Cui et al., 2024; Meena and Pradhan, 2023). Grasslands enable the fastest decomposition because they have the lowest fungal-to-bacterial ratio (Li et al., 2018; Duffy et al., 2021). Additionally, the superior quality of grassland plant litter encourages

FIGURE 18.1 Grassland carbon emissions and the sinking cycle.

the efficiency of microbial C consumption. In grasslands, microbial breakdown accounts for three times as much SOC (Angst et al., 2021; Wu et al., 2021).

Soil C_{seq} is dependent on the soil's ability to hold and stabilize organic carbon (OC) as well as the amount of plant C (Tang et al., 2018). Microbial products make up a major portion of SOC in the development and preservation of SOM (Fang et al., 2018; Piao et al., 2022b). Finer-textured soils can sequester more OC due to their increased surface area, which also protects C from further microbial degradation and stabilization with soil minerals (Buisson et al., 2022). Grasslands frequently exhibit proportionately lower levels of partially digested OM due to the interaction between soil texture and the quick microbial turnover of plant litter (Yang et al., 2022). However, grasslands in arid areas have young soils with a coarse texture in addition to low plant production and microbial activity (Deng et al., 2018). Therefore compared to grasslands in moister climes, arid grasslands often have reduced SOC storage (An et al., 2019).

Therefore interactions between climate, plant, soil microorganisms, and soil characteristics all have a significant impact on SOC storage. SOC is typically considered to be the majority of the terrestrial C pool, while soil inorganic carbon (SIC) is often overlooked (Gutknecht et al., 2012; Pradhan and Meena, 2023a). The majority of grasslands are found in drylands, where over 90% of soil C stocks may be inorganic (Hopkinson et al., 2020). The distribution pattern of SIC concentrations differs from that of SOC, and they tend to increase with depth (Evaristo and McDonnell, 2019). The breakdown and leaching of soil carbonate are restricted in regions with significant potential evapotranspiration and inadequate precipitation, which are home to major SIC stocks (Zhou et al., 2021).

Environmental conditions and soil properties play a key role in regulating carbonate precipitation. In arid regions, as soils dry and water is flushed through, more atmospheric CO_2 can dissolve and later precipitate as carbonate (Dixon et al., 2014). Soil pH is another crucial factor controlling carbonate formation, as it determines the balance of dissolved inorganic C species (Bright et al., 2015; Lian et al., 2021). However, when soil pH drops, carbonate dissolves, and essential cations are leached, leading to irreversible SIC loss. Grassland topsoils worldwide have experienced SIC depletion due to increased soil acidity resulting from the use of synthetic fertilizers(Liu et al., 2018). Current estimates suggest that SIC loss from fertilizer and soil acidification results in C emissions of 0.273 petagrams (Pg) of C annually, roughly 30% of the 0.9 Pg C released each year from land-use changes (Berdugo et al., 2020; Gitari et al., 2024).

18.3 Importance of carbon credit assessment of grassland ecosystems

18.3.1 Increasing the ability to conserve natural grasslands

The majority of natural grasslands that have not been disturbed are C-neutral, but changes in climate cause grasslands to become C sources. For instance, alpine grasslands and tundra, particularly those in permafrost regions, may experience significant C loss as a result of climate change (Dass et al., 2018; Lyu et al., 2023). In addition to causing global grassland degradation, overgrazing and agricultural practices in wild grasslands significantly lower SOC storage (Deng et al., 2018; Yang et al., 2019b). Thus maintaining global grassland C neutrality requires protecting the presently intact natural grasslands. Grasslands have been improved by the application of techniques, including limiting the entry of cattle into environmentally sensitive regions and using controlled fire to lessen woody encroachment (Raza et al., 2020). Additionally, the integration of grasslands with herbivores improves ecosystem C persistence by encouraging herbaceous vegetation with low albedo and high root biomass (Huang et al., 2021). Natural grassland conservation must strike a balance between ecological, economic, and policy goals because human societies rely heavily on the services provided by grassland ecosystems (Day et al., 2019).

18.3.2 Increasing our knowledge of the grassland carbon cycle processes

Climate variability significantly impacts abiotic processes that define the grassland C cycle, leading to the loss of both SOC and SIC (Dixon et al., 2014; Raza et al., 2020). According to recent research, abiotic processes and how they interact with microbial and plant activity have the potential to significantly alter the C and nutrition cycles (Evaristo and McDonnell, 2019; Pradhan and Meena, 2023b). More significantly, it is extremely unknown how abiotic processes will react to changes in the world (Tang et al., 2018; Nachshon, 2018). Climate change, for instance, may influence the frequency of strong winds and precipitation regimes, which accelerate erosion (Rocci et al., 2021; Bright et al., 2015). However, the deposition of soil particles may enhance C storage in deposition locations, even as erosion may cause a rapid loss of C from the damaged sites (Nair et al., 2009). Additionally, soil deposition contributes more N and other mineral elements (Lian et al., 2021; Berdugo et al., 2022).

18.3.3 Creating long-term plans for restoring degraded grasslands

There is compelling evidence that managing degraded grasslands sustainably could improve soil C storage quickly and aid in reducing the effects of climate change (Chen et al., 2018; Tang et al., 2018). Tree planting in damaged grassland should never be mistaken for grassland restoration (Deng et al., 2018). Furthermore, the mix of plant communities may influence ecosystem resilience to climate change (Acharya et al., 2012; Kumar et al., 2024; Banerjee et al., 2023, 2024). Since various grassland species respond differently to changes in precipitation and have varying degrees of drought conditions (Levy et al., 2007). Since species vary in the amount of mycorrhizal C inputs, which determine SOC storage, the diversity of species significantly contributes to soil C storage (Ghosh and Mahanta, 2014; Pradhan et al., 2023c).

Therefore screening for species that have complementary roles within the plant community, robust tolerance to climatic variations, and appropriate leaf and root features could speed up C_{seq} during grassland restoration (Rai et al., 2013; Lange et al., 2019). Furthermore, a complex canopy structure might increase the system's resistance to wind erosion and the vegetation's capacity to absorb dust deposition (Xiong et al., 2020; Dixon et al., 2014). It's crucial to evaluate which management techniques may impact the grassland cycle and energy balance, which usually respond to the local climate (Deng et al., 2018). Thus the integration of both biological and abiotic processes involved in grassland C dynamics should be considered when creating sustainable grassland restoration projects (Dass et al., 2018).

18.3.4 Increasing the precision of estimates of grassland carbon sinks

Grassland ecosystems' role in the global C sink is still up for debate (Conant, 2010). The significant regional variation in grassland C stocks and the disparity in research methodologies used for grassland C sink estimation are the primary causes of the uncertainty surrounding these estimations (Zhou et al., 2021). Combining flux monitoring, ground inventory, and model simulation could accomplish this. Moreover, contemporary remote sensing methods like laser scanning, digital photogrammetry, and hyperspectral imaging can accurately measure the distribution of soil C pools and can track grassland dynamics on a broad regional scale (Li et al., 2022; Chapin et al., 2011).

18.4 Methodology for estimation of carbon credits in grassland ecosystems

Since they store 20%–30% of the world's SOC, grasslands, which make up around 40% of the planet's land area, are essential C sinks. Grasslands store C mostly in their soil and root systems, which makes them resistant to disturbances like fire, in contrast to forests, which store C above ground (Chen et al., 2018). Some of the largest glasslands in the world include Africa's savannas, the American Great Plains, and Mongolian and Chinese lands in Asia and South America. The study by Liu et al. (2023) in Inner Mongolia, China, analyzed the grassland C sink capacity from the year 2000 to 19. The study employed Net Ecosystem Productivity, Sen +MK (Mann-Kendall), trend test method, and soil respiratory model to estimate the C sources/sinking capacity of the grassland (Chapin et al., 2011). Estimation of C credit in a grassland ecosystem involves the collection and analysis of geospatial data. Where data from the specific geographical area are monitored and collected over a certain period and then analyzed to provide an estimated amount of C_{seq} in that area (Lyu et al., 2023).

18.4.1 Defining the study area

This is the first step in the process where the targeted grassland ecosystem to be studied is identified according to various factors, the key factors are the grassland type, the temperature of the area, and other weather data of the location, the land usage history, and the land management practices in that area (Lyu et al., 2023).

18.4.2 Collection of geospatial data

Geospatial data are types of data that are collected in a specific geographical location at a specific period; these data include remote sensing data, meteorological data, and field investigation data (Lyu et al., 2023). The data include SOC, above-ground, and below-ground biomass (Zhou et al., 2021).

18.4.3 Calculating net ecosystem productivity

Net ecosystem productivity (is employed to measure the size of ecosystem C at the regional scale (Chang et al., 2021). The discrepancy between soil microorganisms' respiratory consumption and net primary productivity is known as net ecosystem productivity (NEP). To have more accurate data, the NEP trend of a certain time is evaluated with the help of various methods, such as the Sen+ Mann-Kendall trend test method. From that trend, the estimated capacity of a given grassland ecosystem to capture C is obtained, which is then traded as the C credit (Zhou et al., 2018).

18.5 Grassland management practices toward enhanced carbon credits

Various studies highlighted several management actions for enhancing C credit (Raj et al., 2025; Meena et al., 2025a). Dass et al. (2018) proposed four practices for enhancing grassland C_{seq}; the first is having regional priority areas, which are areas that will be managed strategically for C_{seq} purposes. Second is dynamic monitoring and management; due to frequent changes in regional and global standards and human activity, there is a need to frequently update the C trading protocol to cope with the changes (Lavallee et al., 2020). Also, control of grazing intensity, extreme grazing leads to land degradation, which affects the land's capacity to sink C (Zhou et al., 2021), and the fourth is the implementation of ecological compensation. Another important practice is to mitigate grassland degradation is essential for C credit enhancement. This is because grassland degradation reduces the soil's C_{seq} capability (Cui et al., 2024).

18.5.1 Grasslands as carbon sinks

It has been estimated that by 2030, grassland soils might trap between 0.2 and 0.8 Gt CO_2 annually on a worldwide scale (An et al., 2019). AF practices, such as silvopastoral systems, extend the growing season, create more diverse niches for water and nutrient uptake, and, when N-fixing species are included, enhance soil fertility, all contributing to increased C absorption (Table 18.1). C is stored both in tree biomass and, often, in the soil when silvopastoral systems are implemented in suitable environments (Singh and Gill, 2014). Converting 630 million hectares of underperforming or degraded land to silvopastoral systems could sequester up to 0.59 teragrams (Tag) of C annually by 2040, while better management of existing agroforestry systems (AFS) could capture an additional 0.012 Tag of C per year. C_{seq} capacity within AFS has been found to range between 12 and 228 megagrams per hectare (Mg/ha) (Ghosh and Mahanta, 2014). According to Kirby and Potvin (2007), AFS is thought to have a greater capacity to trap C than field crops or pastures. This theory is predicated on the idea that planting trees in grassland and croplands will increase both the net above ground and below ground C_{seq} (Haile et al., 2008).

18.5.2 Restoring damaged areas and planting grasses on fertile land

Restoring degraded lands increases C inputs and C_{seq} while improving production in regions with low herbage productivity. It is now widely acknowledged that improved grazing management practices, such as silvopasture, can enhance sequestration in nondegraded grasslands, increase feed production, and more effectively utilize land resources (Yang et al., 2022). Grasslands have the same potential for sequestration as croplands, as they comprise a significant portion of the Earth's surface (Levy et al., 2007). In addition to increasing animal feed production, management techniques may also improve soil C stocks, which would help soils store more C from the atmosphere. Improved management techniques include irrigation, fertilization, rigorous grazing management, and planting beneficial legumes and forage grasses (Rai et al., 2013).

TABLE 18.1 Silvopastoral systems' impact on soil carbon.

AFS	Years	SOC (Mg ha^{-1})	Soil depth (cm)
Silvopastoral system	19	0–100	173
Fodder bank	8–40	0–100	33.4
Agrisilviculture + field crops	5–1016	0–60	27.4
Gliricidia sepium + maize (*Zea mays*)		0–200	123
Alley cropping system + maize and bean	6-9	0–40	1.62
Agroforestry	11	0–45	95.89
Silvopastoral system	10–16	0–100	132
Tree-based pastures + bahiagrass	10	0–125	6.9–24.2

AFS, agroforestry systems.

18.5.3 Grazing management

It is well established that properly managed grazing enhances N cycling in grassland ecosystems and promotes the establishment of herbaceous plants. Compared to ungrazed prairies/exclosures, grazed mixed-grass areas revealed higher photosynthesis. The grazing process has a significant impact on the rate of turnover or disintegration of the plant community's above-ground component. Shoot turnover was found to be 39% and 36% under light and high grazing, respectively, while it was 28% in ungrazed exclosures (Derner and Schuman, 2007). It was determined that animal traffic may accelerate the pace of decomposition, soil integration, and physical breakdown of standing dead plant matter and litter (Tang et al., 2018). If grazing intensities are reduced or if grazing is postponed during the period when forage species are developing at their fastest rate, grazing management may result in less C removal (Mengistu and Mekuriaw, 2014). Therefore C inputs and C stocks can be raised by sustainable grazing management without necessarily lowering fodder output. Restoring productive forage species through grazing management can also increase C inputs.

18.5.4 Fertilizer management

Applying fertilizer increased the amount of litter produced, which improved the soil's capacity to store C. In pasture grassland, the total organic carbon (TOC) of the soil significantly increased with the application of 40 kg N/ha (Lian et al., 2021). Compared to the natural grassland, the rate of TOC formation was 1.5 times higher. OC storage was also improved by applying additional nutrients where they were lacking (Yang et al., 2019a; Raza et al., 2024). However, the advantages of enhanced soil C_{seq} must be contrasted with the costs of fertilizer application (Grünzweig et al., 2022). In a similar vein, applying water or irrigation can improve nitrogen and water balances, increasing carbon inputs and plant output.

18.5.5 Fire management

By altering the density or invasion of woody species, fire management in some grasslands also affects the concentration of C stored in biomass. Charcoal, a type of C that is extremely resistant to decomposition, is produced when biomass is burned (Lange et al., 2019). This can account for a sizable portion of the C retained in some grassland soils. After 10 years, it was discovered that grazing and annual burning on the tall grass prairie increased soil C storage by 2.2 Mg/ha (Zhou et al., 2021).

18.5.6 Planting of improved forage species

Increased biomass production can result from the introduction or promotion of improved forage species that are more suited to the change in climate, with drought-resistant, grazing-resilient characteristics, and able to improve soil fertility. Ultimately, increasing productivity results in higher C inputs and C_{seq}. *Macroptilium lathyroides* was found to raise TOC in natural grassland by 1.29 times when compared to the original grassland. Increased soil TOC and mixed biomass production were also observed in legume-incorporated grassland (Ghosh and Mahanta, 2014) (Table 18.2).

The restoration of the degraded land was greatly aided by the silvipasture method. The relative effectiveness of eight distinct tree species as a component of a silvopasture based on *Cenchrus ciliaris* on herbage production was evaluated in comparison to *Cenchrus ciliaris* as a pasture (Table 18.3). The performance of the understory plants under various tree types was impacted by photosynthetically active radiation that reached the ground. The percentage of relative radiation infiltration through tree species ranged from 58% to 78%, with *Eucalyptus* sp. having the highest percentage and *Acacia nilotica* having the lowest (Nair et al., 2009). The TOC concentration of the surface soil was influenced by variations in biomass production and litterfall under different plant species. *Leucaena leucocephala* increased the TOC by 3.4 times above the original levels within the same period (Murphy-Bokern, 2009).

It has been noted that SOM degrades extremely quickly under various arable land-use schemes. Reduced inputs of OM, greater crop residue decomposability, and tillage effects that diminish the amount of physical protection against decomposition are all responsible for a large portion of this loss in SOC (Acharya et al., 2012). On agricultural soils, incorporating grass into the rotation cycle can decrease soil disturbance from tillage and boost production return OM, which contributes to C_{seq} (Yang et al., 2019b; Tang et al., 2018).

According to research conducted in France, C_{seq} averaged 44 g C msquare^{-1} year^{-1} over 20 years at soil depths ranging from 0 to 30 cm. In the northeastern United States, temperate pastures are quite productive and may serve as important C sinks (Stevens et al., 2022). However, a comparatively high amount of biomass is removed from these pastures either as hay or by grazing animals consuming it. It has been shown that the overuse of biomass generated in grasslands prevents them from becoming C sinks. The C_{seq} capacity of managed grasslands in the United States was shown to vary from 10 to 90 g C msquare^{-1} year^{-1}, depending on the extent of modification. According to the Kyoto Protocol on Climate Change

TABLE 18.2 Impact of range legumes of natural grassland on soil organic carbon.

Treatments	FDM (Mg ha^{-1})	TOC (g kg^{-1})	SOC (g kg^{-1}yr^{-1})
Clitoria ternatea	4.4	7.47	0.64
Alysicarapus rugosus	4.2	7.55	0.67
Natural grassland	3.3	7.78	0.74
Macroptilium atropurpureum	4.1	7.99	0.81
Stizolobium deeringianum	4	8.1	0.85
Stylosanthes humilis	4	8.22	0.89
Glycine javanica	3.8	8.58	1.01
Mimosa invisa	3.7	8.91	1.12
Vigna luteola	4.2	9.15	1.2
Atylosia scarabaeoides	4.1	9.22	1.22
Desmodium tortusum	4.2	9.72	1.39
Dolichos lablab	4.7	10.07	1.51
Stylosanthes guianensis	4.2	10.53	1.66
Macroptilium thyroids	4.9	11.05	1.83

DFM, dry forage matter; *TOC*, total organic carbon; *SOC*, soil organic carbon.

TABLE 18.3 Influence of tree species on TOC content of the soil in a grassland field.

Tree	DFY (Mg/ha)	Leaf litters (Mg/ha)	TOC (g/kgsoil)	RI
Initial	2	3.87	6.91	2.26
	-	5.95	6.91	2.26
Open (without tree)	1.5	5.34	7.18	2.35
Acacia tortilis	0.7	4.36	7.44	2.43
Albizzia procera	2.3	4.21	8.37	2.73
Dichrostachys nutan Hardwikia binata Eucalypytus sp.	2	5.46	9.04	2.96
Albizzia lebbeck	3.4	5.63	9.31	3.04
Leucaena leucocephala	3.2	5.28	10.37	3.39

DFY, dry forage yield; *TOC*, total organic carbon; *RI*, relative increase.

for Europe, the C_{seq} for arable land conversion to grasslands was 144 g C/m^2/year, whereas the C_{seq} for existing grasslands was 52 g C msquare^{-1} year^{-1} (IPCC, 2021). The projected sequestration in Switzerland was 40.1 g C m^{-2} year^{-1}, whereas the source in Portugal was 4.5 g C m^{-2} year^{-1} (Wu et al., 2021).

18.6 R&D toward carbon credits and grassland ecosystems

More research is needed to understand how future shifts in these abiotic processes and their interactions with biotic processes and one another may affect grassland carbon dynamics, as well as the scales at which these processes operate (Wu et al., 2021; Raza et al., 2020). A clearer understanding of the role abiotic processes play in grassland C cycling

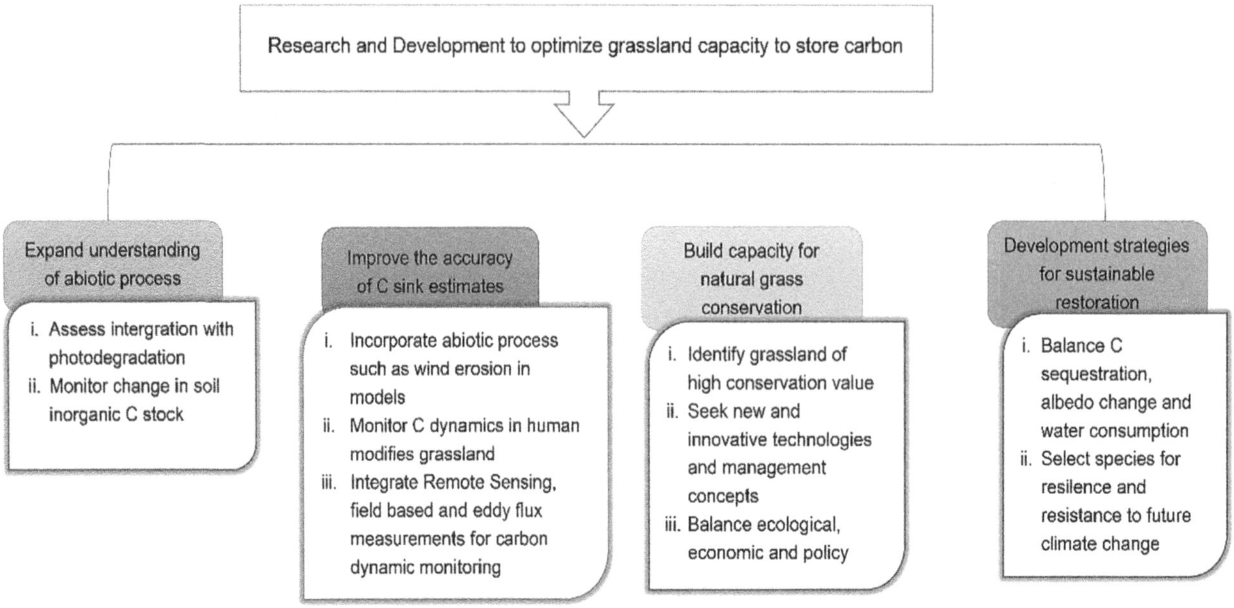

FIGURE 18.2 Essentials to consider to optimize grasslands' capacity to store carbon.

would provide a stronger scientific foundation for land management policies aimed at maintaining or enhancing C stocks (Li et al., 2018). Significant uncertainty remains about the extent to which grassland ecosystems contribute to the global C sink (Zhou et al., 2024). This uncertainty largely stems from the substantial regional variability in grassland C stocks and the differences in research techniques used to estimate C storage (Li et al., 2022). As a result, determining the C sink capacity of grasslands across different regions remains a major challenge, despite numerous studies measuring C stored in soil and vegetation at the site level (Li et al., 2022; Xia and Wan, 2008).

There is a lot of ambiguity when the results are extrapolated to the regional level (100). Because inventory studies or eddy covariance measurements rarely measure it, the lateral movement of C by wind cannot be reliably calculated in Earth System Models (98, 100). Additionally, the majority of C flow monitoring data originates from areas with low anthropogenic disturbance (Piao et al., 2022a). There is a need to integrate several methods to enhance the measurements of lateral C transfer and SOC and SIC dynamics, particularly in regions where human activity has significantly disrupted the environment. Moreover, according to Li et al. (2022), to optimize grasslands' capacity to store C, it is essential to increase knowledge of abiotic processes, enhance the precision of C sink estimations, increase the ability to conserve natural grasslands, and create sustainable restoration plans Fig. 18.2.

18.7 Policy perspectives for enhancing carbon credit in grassland ecosystems

To address climate mitigation, environmental conservation, food insecurity, and poverty alleviation, national and international policy must acknowledge the significance of grasslands in C_{seq} (Zamanian et al., 2024). The absence of grassland representation at the national and international policy levels is a significant element in the degradation of grasslands. Since ecosystems such as freshwater, forests, croplands, and oceans are the scientific focus of global sustainability strategies, grasslands are frequently undervalued (Wilcox et al., 2017). Although the exact causes of this disregard for grasslands in sustainability strategy are unknown, historical circumstances and false beliefs about them are to blame (Xiong et al., 2020).

Furthermore, natural grasslands are generally located in areas where soil and climate conditions only offer few economic benefits to land users, although their economic relevance for the provision of ecosystem services has been fully acknowledged (Xia and Wan, 2008; Zamanian et al., 2021). It's also possible that this prejudice stems from the fact that forests were the first places in the world to be converted to farmland, which helped people better grasp how deforestation affects biodiversity and society (Li et al., 2022). Grasslands are often overlooked, despite being recognized in some national and international environmental laws. However, international conversions, such as global frameworks like the sustainable development goals (SDGs), aim to prevent and reverse ecological degradation; yet, they do not specifically focus on grasslands (Cotrufo et al., 2019). Unlike forests, oceans, and coastal habitats, which are frequently included in targets and indicators, grasslands are absent from the SDGs (Zhou et al., 2021).

Grasslands receive far less attention than forests, which play a key role in the European Biodiversity Strategy for 2020. Although the EU's Common Agricultural Policy includes measures to protect grasslands, their effectiveness has been widely criticized (Yang et al., 2019a). Additionally, unintended consequences of national and international environmental and socioeconomic policies have intensified pressures on grasslands. C_{seq} efforts that prioritize tree planting, such as reducing emissions from deforestation and forest degradation (REDD+), increasingly target natural grasslands, often leading to their degradation. Over the past 25 years, China has heavily invested in afforestation, adding trees to previously treeless areas (Bardgett et al., 2021).

However, there are growing concerns that planting water-intensive trees in dry and semiarid grasslands is accelerating grassland degradation and reducing plant diversity. Afforestation, particularly when trees are planted on natural grasslands, can lead to nutrient depletion, soil acidification, and soil C loss, making these areas more susceptible to C loss from fires (Compagnoni et al., 2021). Additionally, while healthy, restored grasslands can store comparable amounts of OC to forests, most of this C is stored underground. In contrast, initiatives like REDD+ primarily focus on C captured in above-ground tree biomass, overlooking grasslands' significant below-ground C storage potential (Yang et al., 2019b).

Also, evidence indicates that grasslands may act as more durable C sinks than forests. When trees are planted on natural grasslands, afforestation can cause nutrient depletion, soil acidification, and C loss from the soil, increasing vulnerability to fire-related C emissions (Bengtsson et al., 2019; Lyu et al., 2023). Moreover, while healthy, restored grasslands can store similar amounts of SOC as forests, most of this C is held underground. If grasslands are to be valued and successfully maintained, global and regional policies must change to recognize their significance as hotspots for biodiversity and providers of several ecosystem services, as well as to establish objectives for their preservation and sustainable management (Berdugo et al., 2022).

18.8 Conclusion

Conclusively, grasslands' contribution to C neutrality hinges on our ability to preserve natural grasslands and rehabilitate degraded ones. Future climate change may make grassland vegetation less resilient to natural disasters. Strategies for grassland management, restoration, and conservation must take into account how to support the functioning of the grassland ecosystem today and its capacity to adapt to future climate change. To guarantee the grassland is robust and resistant to future climatic changes given that grasslands around the world serve as nature-based solutions to attaining C neutrality, cooperation between scientists, landowners, and policymakers is therefore necessary.

References

Acharya, B.S., Rasmussen, J., Eriksen, J., 2012. Grassland carbon sequestration and emissions following cultivation in a mixed crop rotation. Agriculture, Ecosystems and Environment 153, 33–39. Available at https://doi.org/10.1016/j.agee.2012.03.001.

An, H., Wu, X., Zhang, Y., Tang, Z., 2019. Effects of land-use change on soil inorganic carbon: a meta-analysis. Geoderma 353, 273–282. Available at https://doi.org/10.1016/j.geoderma.2019.07.008.

Angst, G., Mueller, K.E., Nierop, K.G., Simpson, M.J., 2021. Plant-or microbial-derived? A review on the molecular composition of stabilized soil organic matter. Soil Biology and Biochemistry 156, 108189.

Bahram, M., Hildebrand, F., Forslund, S.K., Anderson, J.L., Soudzilovskaia, N.A., Bodegom, P.M., et al., 2018. Structure and function of the global topsoil microbiome. Nature 560 (7717), 233–237. Available at https://doi.org/10.1038/s41586-018-0386-6.

Bai, Y., Cotrufo, M.F., 2022. Grassland soil carbon sequestration: current understanding, challenges, and solutions. Science (New York, N.Y.) 377 (6606), 603–608. Available at https://doi.org/10.1126/science.abo2380.

Banerjee, A., Jhariya, M.K., Bargali, S.S., Palit, D., 2023. Ecorestoration for Sustainability. Wiley-Scrivener, USA, pp. 1–441.

Banerjee, A., Jhariya, M.K., Raj, A., Mechergui, T., 2024. Ecosystem management: Climate change and global sustainability—an introduction. Ecosystem Management: Climate Change and Sustainability. Wiley, India, pp. 1–21. Available at http://doi.org/10.1002/9781394231249.ch1.

Bardgett, R.D., Bullock, J.M., Lavorel, S., Manning, P., Schaffner, U., Ostle, N., et al., 2021. Combatting global grassland degradation. Nature Reviews Earth and Environment 2 (10), 720–735. Available at https://doi.org/10.1038/s43017-021-00207-2.

Bengtsson, J., Bullock, J.M., Egoh, B., Everson, C., Everson, T., O'Connor, T., et al., 2019. Grasslands—more important for ecosystem services than you might think. Ecosphere 10 (2). Available at https://doi.org/10.1002/ecs2.2582.

Berdugo, M., Delgado-Baquerizo, M., Soliveres, S., Hernández-Clemente, R., Zhao, Y., Gaitán, J.J., et al., 2020. Global ecosystem thresholds driven by aridity. Science (New York, N.Y.) 367 (6479), 787–790. Available at https://doi.org/10.1126/science.aay5958.

Berdugo, M., Vidiella, B., Solé, R.V., Maestre, F.T., 2022. Ecological mechanisms underlying aridity thresholds in global drylands. Functional Ecology 36 (1), 4–23. Available at https://doi.org/10.1111/1365-2435.13962.

Bright, R.M., Zhao, K., Jackson, R.B., Cherubini, F., 2015. Quantifying surface albedo and other direct biogeophysical climate forcings of forestry activities. Global Change Biology 21 (9), 3246–3266. Available at https://doi.org/10.1111/gcb.12951.

Buisson, E., Archibald, S., Fidelis, A., Suding, K.N., 2022. Ancient grasslands guide ambitious goals in grassland restoration. Science (New York, N.Y.) 377 (6606), 594–598. Available at https://doi.org/10.1126/science.abo4605.

Cavender-Bares, J., Schweiger, A.K., Gamon, J.A., Gholizadeh, H., Helzer, K., Lapadat, C., et al., 2022. Remotely detected aboveground plant function predicts belowground processes in two prairie diversity experiments. Ecological Monographs 92 (1). Available at https://doi.org/10.1002/ecm.1488.

Chang, J., Ciais, P., Gasser, T., Smith, P., Herrero, M., Havlík, P., et al., 2021. Climate warming from managed grasslands cancels the cooling effect of carbon sinks in sparsely grazed and natural grasslands. Nature Communications 12 (1). Available at https://doi.org/10.1038/s41467-020-20406-7.

Chapin III, F.S., Matson, P.A., Vitousek, P., 2011. Principles of Terrestrial Ecosystem Ecology. Springer Science & Business Media.

Chappa, L.R., Nungula, E.Z., Makwinja, Y.H., Ranjan, S., Sow, S., Alnemari, A.M., et al., 2024. Outlooks on major agroforestry systems. Agroforestry. Wiley, Sweden, pp. 21–48. Available at http://doi.org/10.1002/9781394231164.ch2.

Charles, J. (2018). Bay Nature Magazine: How Dirt and Ranchers Could Save the World. Available from: https://baynature.org/article/the-grass-really-is-greener/.

Chen, S., Wang, W., Xu, W., Wang, Y., Wan, H., Chen, D., et al., 2018. Plant diversity enhances productivity and soil carbon storage. Proceedings of the National Academy of Sciences 115 (16), 4027–4032. Available at https://doi.org/10.1073/pnas.1700298114.

Cheruto, G., Nungula, E.Z., Nyawira, L., Chappa, L.R., Kahuthia-Gathu, R., Mwadalu, R., et al., 2025. Agroforestry tree species: Acacia tortilis, biology, importance, agroforestry production, and biotechnology application. Tree Biology and Biotechnology. Springer Nature, Kenya, pp. 145–161. Available at http://doi.org/10.1007/978-981-96-0002-1_10.

Compagnoni, A., Levin, S., Childs, D.Z., Harpole, S., Paniw, M., Römer, G., et al., 2021. Herbaceous perennial plants with short generation time have stronger responses to climate anomalies than those with longer generation time. Nature Communications 12 (1). Available at https://doi.org/10.1038/s41467-021-21977-9.

Conant, R.T., 2010. Challenges and Opportunities for Carbon Sequestration in Grassland Systems, Vol. 9. FAO, Rome, Italy.

Cotrufo, M.F., Ranalli, M.G., Haddix, M.L., Six, J., Lugato, E., 2019. Soil carbon storage informed by particulate and mineral-associated organic matter. Nature Geoscience 12 (12), 989–994. Available at https://doi.org/10.1038/s41561-019-0484-6.

Cui, Z., Xin, J., Yang, X., Dang, Y., Lin, C., Ma, Z., et al., 2024. Contribution of glomalin-related soil protein to soil organic carbon following grassland degradation and restoration: a case from alpine meadow of qinghai–tibet plateau. Land 13 (12), 2223. Available at https://doi.org/10.3390/land13122223.

Dass, P., Houlton, B.Z., Wang, Y., Warlind, D., 2018. Grasslands may be more reliable carbon sinks than forests in California. Environmental Research Letters 13 (7), 074027. Available at https://doi.org/10.1088/1748-9326/aacb39.

Day, T.A., Bliss, M.S., Placek, S.K., Tomes, A.R., Guénon, R., 2019. Thermal abiotic emission of CO_2 and CH4 from leaf litter and its significance in a photodegradation assessment. Ecosphere 10 (5). Available at https://doi.org/10.1002/ecs2.2745.

Deng, M., Liu, L., Jiang, L., Liu, W., Wang, X., Li, S., et al., 2018. Ecosystem scale trade-off in nitrogen acquisition pathways. Nature Ecology & Evolution 2 (11), 1724–1734. Available at https://doi.org/10.1038/s41559-018-0677-1.

Deng, M., Li, P., Wang, Z., Guo, L., Wu, Y., Huang, J., et al., 2021. Drought and salinization stress induced by stand development alters mineral element cycling in a larch plantation. Journal of Geophysical Research: Biogeosciences 126 (3). Available at https://doi.org/10.1029/2020JG005906.

Derner, J.D., Schuman, G.E., 2007. Carbon sequestration and rangelands: a synthesis of land management and precipitation effects. Journal of Soil and Water Conservation 62 (2), 77–85. Available at https://doi.org/10.1080/00224561.2007.12435927.

Dixon, A.P., Faber-Langendoen, D., Josse, C., Morrison, J., Loucks, C.J., Ebach, M., 2014. Distribution mapping of world grassland types. Journal of Biogeography 41 (11), 2003–2019. Available at https://doi.org/10.1111/jbi.12381.

Duffy, K.A., Schwalm, C.R., Arcus, V.L., Koch, G.W., Liang, L.L., Schipper, L.A., 2021. How close are we to the temperature tipping point of the terrestrial biosphere? Science Advances 7 (3). Available at https://doi.org/10.1126/sciadv.aay1052.

Evaristo, J., McDonnell, J.J., 2019. Global analysis of streamflow response to forest management. Nature 570 (7762), 455–461. Available at https://doi.org/10.1038/s41586-019-1306-0.

Fang, J., Yu, G., Liu, L., Hu, S., Chapin, F.S., 2018. Climate change, human impacts, and carbon sequestration in China. Proceedings of the National Academy of Sciences 115 (16), 4015–4020. Available at https://doi.org/10.1073/pnas.1700304115.

Fuhlendorf, S.D., Davis, C.A., Elmore, R.D., Goodman, L.E., Hamilton, R.G., 2018. Perspectives on grassland conservation efforts: should we rewild to the past or conserve for the future? Philosophical Transactions of the Royal Society B: Biological Sciences 373 (1761). Available at https://doi.org/10.1098/rstb.2017.0438.

Ghosh, P.K., Mahanta, S.K., 2014. Carbon sequestration in grassland systems. Range Management and Agroforestry 35 (2), 173–181.

Gitari, H.H., Nungula, E.Z., Chappa, L.R., Raza, M.A., Ranjan, S., Sow, S., et al., 2024. Agroforestry for climate security. Agroforestry. Wiley, Kenya, pp. 319–344. Available at http://doi.org/10.1002/9781394231164.ch11.

Goldstein, A., Turner, W.R., Spawn, S.A., Anderson-Teixeira, K.J., Cook-Patton, S., Fargione, J., et al., 2020. Protecting irrecoverable carbon in earth's ecosystems. Nature Climate Change 10 (4), 287–295. Available at https://doi.org/10.1038/s41558-020-0738-8.

Grünzweig, J.M., De Boeck, H.J., Rey, A., Santos, M.J., Adam, O., Bahn, M., et al., 2022. Dryland mechanisms could widely control ecosystem functioning in a drier and warmer world. Nature Ecology and Evolution 6 (8), 1064–1076. Available at https://doi.org/10.1038/s41559-022-01779-y.

Guo, L., Deng, M., Yang, S., Liu, W., Wang, X., Wang, J., et al., 2021. The coordination between leaf and fine root litter decomposition and the difference in their controlling factors. Global Ecology and Biogeography 30 (11), 2286–2296. Available at https://doi.org/10.1111/geb.13384.

Gutknecht, J.L.M., Field, C.B., Balser, T.C., 2012. Microbial communities and their responses to simulated global change fluctuate greatly over multiple years. Global Change Biology 18 (7), 2256–2269. Available at https://doi.org/10.1111/j.1365-2486.2012.02686.x.

Haile, S.G., Nair, P.K.R., Nair, V.D., 2008. Carbon storage of different soil-size fractions in Florida silvopastoral systems. Journal of Environmental Quality 37 (5), 1789–1797. Available at https://doi.org/10.2134/jeq2007.0509.

Hopkinson, P., Hammond, M., Bartolome, J.W., Macaulay, L., 2020. Using consecutive prescribed fires to reduce shrub encroachment in grassland by increasing shrub mortality. Restoration Ecology 28 (4), 850–858. Available at https://doi.org/10.1111/rec.13138.

Huang, Y., Chen, Y., Castro-Izaguirre, N., Baruffol, M., Brezzi, M., Lang, A., et al., 2018. Impacts of species richness on productivity in a large-scale subtropical forest experiment. Science (New York, N.Y.) 362 (6410), 80–83. Available at https://doi.org/10.1126/science.aat6405.

Huang, J., Liu, W., Pan, S., Wang, Z., Yang, S., Jia, Z., et al., 2021. Divergent contributions of living roots to turnover of different soil organic carbon pools and their links to plant traits. Functional Ecology 35 (12), 2821–2830. Available at https://doi.org/10.1111/1365-2435.13934.

IPCC, 2021. Climate change: The physical science basis. Contribution of Working Group I to the Sixth Assessment Report of the Intergovernmental Panel On Climate Change. Cambridge University Press, Cambridge.

Iqbal, M.A., 2021. Grasses and grassland-new perspectives. IntechOpen eBooks. Available at https://doi.org/10.5772/intechopen.95204.

Jhariya, M.K., Meena, R.S., Banerjee, A., Meena, S.N., 2022. Natural Resources Conservation and Advances for Sustainability. Elsevier. Available at https://doi.org/10.1016/C2019-0-03763-6 ISBN:9780128229767.

Jhariya, M.K., Meena, R.S., Banerjee, A., Kumar, S., Raj, A., 2023. Agroforestry for Carbon and Ecosystem Management Agroforestry for Carbon and Ecosystem Management. Elsevier, India, pp. 1–434. Available at http://doi.org/10.1016/C2021-0-02916-3.

Jia, Z., Li, P., Wu, Y., Yang, S., Wang, C., Wang, B., et al., 2021. Deepened snow cover alters biotic and abiotic controls on nitrogen loss during non-growing season in temperate grasslands. Biology and Fertility of Soils 57 (2), 165–177. Available at https://doi.org/10.1007/s00374-020-01514-4.

Kirby, K.R., Potvin, C., 2007. Variation in carbon storage among tree species: implications for the management of a small-scale carbon sink project. Forest Ecology and Management 246 (2-3), 208–221. Available at https://doi.org/10.1016/j.foreco.2007.03.072.

Kumar, S., Meena, R.S., Sheoran, P., Jhariya, M.K., 2024. Regenerative Agriculture for Sustainable Food Systems. Springer Science+Business Media, India, pp. 1–493. Available at http://doi.org/10.1007/978-981-97-6691-8.

Lange, M., Koller-France, E., Hildebrandt, A., Oelmann, Y., Wilcke, W., Gleixner, G., 2019. How plant diversity impacts the coupled water, nutrient and carbon cycles. Advances in Ecological Research 61, 185–219. Available at https://doi.org/10.1016/bs.aecr.2019.06.005.

Lavallee, JM, Soong, JL, Cotrufo, MF, 2020. Conceptualizing soil organic matter into particulate and mineral-associated forms to address global change in the 21st century. Global Change Biology 26 (1), 261–273. Available at https://doi.org/10.1111/gcb.14859.

Levy, P.E., Mobbs, D.C., Jones, S.K., Milne, R., Campbell, C., Sutton, M.A., 2007. Simulation of fluxes of greenhouse gases from European grasslands using the DNDC model. Agriculture, Ecosystems & Environment 121 (1-2), 186–192. Available at https://doi.org/10.1016/j.agee.2006.12.019.

Li, P., Sayer, E.J., Jia, Z., Liu, W., Wu, Y., Yang, S., et al., 2020. Deepened winter snow cover enhances net ecosystem exchange and stabilizes plant community composition and productivity in a temperate grassland. Global Change Biology 26 (5), 3015–3027. Available at https://doi.org/10.1111/gcb.15051.

Li, P., Liu, L., Wang, J., Wang, Z., Wang, X., Bai, Y., et al., 2018. Wind erosion enhanced by land use changes significantly reduces ecosystem carbon storage and carbon sequestration potentials in semiarid grasslands. Land Degradation and Development 29 (10), 3469–3478. Available at https://doi.org/10.1002/ldr.3118.

Li, P., Sayer, E.J., Jia, Z., Wu, Y., Deng, M., Wang, X., et al., 2022. Deepened snow cover mitigates soil carbon loss from intensive land-use in a semi-arid temperate grassland. Functional Ecology 36 (3), 635–645. Available at https://doi.org/10.1111/1365-2435.13958.

Lian, X., Piao, S., Chen, A., Huntingford, C., Fu, B., Li, L.Z.X., et al., 2021. Multifaceted characteristics of dryland aridity changes in a warming world. Nature Reviews Earth and Environment 2 (4), 232–250. Available at https://doi.org/10.1038/s43017-021-00144-0.

Liu, Weixing, Qiao, Chunlian, Yang, Sen, Bai, Wenming, Liu, Lingli, 2018. Microbial carbon use efficiency and priming effect regulate soil carbon storage under nitrogen deposition by slowing soil organic matter decomposition. Geoderma 332, 44. Available at https://doi.org/10.1016/j.geoderma.2018.07.008.

Liu, L., Sayer, E.J., Deng, M., Li, P., Liu, W., Wang, X., et al., 2023. The grassland carbon cycle: mechanisms, responses to global changes, and potential contribution to carbon neutrality. Fundamental Research 3 (2), 209–218. Available at https://doi.org/10.1016/j.fmre.2022.09.028.

Lyu, X., Li, X., Wang, K., Zhang, C., Dang, D., Dou, H., et al., 2023. Strengthening grassland carbon source and sink management to enhance its contribution to regional carbon neutrality. Ecological Indicators 152, 110341. Available at https://doi.org/10.1016/j.ecolind.2023.110341.

Meena, R.S., Yadav, A., Kumar, S., Jhariya, M.K., Jatav, S.S., 2022. Agriculture ecosystem models for CO_2 sequestration, improving soil physicochemical properties, and restoring degraded land. Ecological Engineering 176. Available at https://doi.org/10.1016/j.ecoleng.2022.106546.

Meena, R.S., Pradhan, G., 2023. Industrial garbage-derived biocompost enhances soil organic carbon fractions, CO_2 biosequestration, potential carbon credits and sustainability index in a rice-wheat ecosystem. Environmental Research 235, 116525. Available at https://doi.org/10.1016/j.envres.2023.116525.

Meena, R.S., Lal, R., Kumar, S., Pradhan, G., Srinivasarao, C., Singh, A.K., et al., 2024a. Potential of Indian agriculture for capturing atmospheric CO_2 and monetizing carbon credits to the farmers: An overview and policy framework. Advances in Agronomy 188, 101–206. Available at https://doi.org/10.1016/bs.agron.2024.06.001.

Meena, R.S., Pradhan, G., Nalani, Kumar, S., Singh, A.K., Verma, S.K., et al., 2024b. Diversified cropping modules designed for soil restoration, CO_2 sequestration, and generating carbon credits. Land Degradation and Development 35 (14), 4314–4329. Available at https://doi.org/10.1002/ldr.5224.

Meena, R.S., Pradhan, G., Singh, K., Kumar, S., Singh, A.K., Shashidhar, K.S., et al., 2024c. Agriculture models for restoring degraded land to enhance CO_2 biosequestration and carbon credits in the Vindhyan region of India. Science of the Total Environment 929. Available at https://doi.org/10.1016/j.scitotenv.2024.172661.

Meena, R.S., Pradhan, G.S., Kumar, S., Jhariya, M.K., Rao, C.S., 2025a. Carbon-Negative Agriculture. Springer, pp. 1–500. Available at https://link.springer.com/book/9789819514823.

Meena, R.S., Pradhan, G., Kumar, S., 2025b. Energy flow, eco-efficiency, and economic circulation with recycled industrial waste compost application in wheat and subsequent rice farming. Science of the Total Environment 967. Available at https://doi.org/10.1016/j.scitotenv.2025.178779.

Mengistu, A., Mekuriaw, S., 2014. Challenges and opportunities for carbon sequestration in grassland system: a review. International Journal of Environmental Engineering and Natural Resources 1, 1–12.

Murphy-Bokern, D., 2009. The environmental performance of native grassland based livestock production systems. Murphy-Bokern Konzepte, Lindinweg, Lohne, Germany.

Nachshon, U., 2018. Cropland Soil Salinization and associated hydrology: trends, processes and examples. Water 10 (8), 1030. Available at https://doi.org/10.3390/w10081030.

Nair, P.K.R., Kumar, B.M., Nair, V.D., 2009. Agroforestry as a strategy for carbon sequestration. Journal of Plant Nutrition and Soil Science 172 (1), 10–23. Available at https://doi.org/10.1002/jpln.200800030.

Nungula, E.Z., Chappa, L.R., Ranjan, S., Sow, S., Alnemari, A.M., Seleiman, M.F., et al., 2024. Ecosystem services through agroforestry systems and its sustainability. Agroforestry. Wiley, Tanzania, pp. 223–254. Available at http://doi.org/10.1002/9781394231164.ch8.

Piao, S., He, Y., Wang, X., Chen, F., 2022a. Estimation of China's terrestrial ecosystem carbon sink: methods, progress and prospects. Science China Earth Sciences 65 (4), 641–651. Available at https://doi.org/10.1007/s11430-021-9892-6.

Piao, S., Yue, C., Ding, J., Guo, Z., 2022b. Perspectives on the role of terrestrial ecosystems in the 'carbon neutrality' strategy. Science China Earth Sciences 65 (6), 1178–1186. Available at https://doi.org/10.1007/s11430-022-9926-6.

Pradhan, G., Meena, R.S., 2023a. Interaction impact of biocompost on nutrient dynamics and relations with soil biota, carbon fractions index, societal value of CO_2 equivalent and ecosystem services in the wheat-rice farming. Chemosphere 339, 139695. Available at https://doi.org/10.1016/j.chemosphere.2023.139695.

Pradhan, G., Meena, R.S., 2023b. Utilizing waste compost to improve the atmospheric CO_2 capturing in the rice-wheat cropping system and energy-cum-carbon credit auditing for a circular economy. Science of the Total Environment 892. Available at https://doi.org/10.1016/j.scitotenv.2023.164572.

Pradhan, G., Meena, R.S., Kumar, S., Lal, R., 2023c. Utilizing industrial wastes as compost in wheat-rice production to improve the above and below-ground ecosystem services. Agriculture, Ecosystems & Environment 358, 108704. Available at https://doi.org/10.1016/j.agee.2023.108704.

Rai, A.K., Ghosh, P.K., Ram, S.N., Singh, S., Kumar, S., Mahanta, S.K., et al., 2013. Carbon sequestration in forage based land use systems. Indian Grassland and Fodder Research Institute, Jhansi. pp. 1-52.

Raj, A., Jhariya, M.K., Meena, R.S., 2025. Agroforestry for Monetising Carbon Credits. Springer.

Raza, S., Miao, N., Wang, P., Ju, X., Chen, Z., Zhou, J., et al., 2020. Dramatic loss of inorganic carbon by nitrogen-induced soil acidification in Chinese croplands. Global Change Biology 26 (6), 3738–3751. Available at https://doi.org/10.1111/gcb.15101.

Raza, S., Irshad, A., Margenot, A., Zamanian, K., Li, N., Ullah, S., et al., 2024. Inorganic carbon is overlooked in global soil carbon research: a bibliometric analysis. Geoderma 443. Available at https://doi.org/10.1016/j.geoderma.2024.116831.

Rocci, K.S., Lavallee, J.M., Stewart, C.E., Cotrufo, M.F., 2021. Soil organic carbon response to global environmental change depends on its distribution between mineral-associated and particulate organic matter: a meta-analysis. Science of the Total Environment 793. Available at https://doi.org/10.1016/j.scitotenv.2021.148569.

Singh, B., Gill, R.I.S., 2014. Carbon sequestration and nutrient removal by some tree species in an agrisilviculture system in Punjab, India. Range Management and Agroforestry 35 (1), 107–114.

Stevens, N., Bond, W., Feurdean, A., Lehmann, C.E.R., 2022. Grassy ecosystems in the anthropocene. Annual Review of Environment and Resources 47, 261–289. Available at https://doi.org/10.1146/annurev-environ-112420-015211.

Tang, X., Zhao, X., Bai, Y., Tang, Z., Wang, W., Zhao, Y., et al., 2018. Carbon pools in China's terrestrial ecosystems: new estimates based on an intensive field survey. Proceedings of the National Academy of Sciences of the United States of America 115 (16), 4021–4026. Available at https://doi.org/10.1073/pnas.1700291115.

Wilcox, KR, Shi, Z, Gherardi, LA, Lemoine, NP, Koerner, SE, Hoover, DL, Bork, E, Byrne, KM, Cahill J, Jr, Collins, SL, Evans, S, Gilgen, AK, Holub, P, Jiang, L, Knapp, AK, LeCain, D, Liang, J, Garcia-alacios P, P, Peñuelas, J, Pockman, WT, Smith, MD, un S, S, White, SR, Yahdjian, L, Zhu, K, Luo, Y, et al., 2017. Asymmetric responses of primary productivity to precipitation extremes: A synthesis of grassland precipitation manipulation experiments. Global Change Biology 23 (10), 4376–4385. Available at https://doi.org/10.1111/gcb.13706.

Wu, Z., Bi, J., Gao, Y., 2021. Drivers and environmental impacts of vegetation greening in a semi-arid region of northwest china since 2000. Remote Sensing 13 (21), 4246. Available at https://doi.org/10.3390/rs13214246.

Xia, J., Wan, S., 2008. Global response patterns of terrestrial plant species to nitrogen addition. New Phytologist 179 (2), 428–439. Available at https://doi.org/10.1111/j.1469-8137.2008.02488.x.

Xiong, D., Wei, C.Z., Wubs, E.R.J., Veen, G.F., Liang, W., Wang, X., et al., 2020. Nonlinear responses of soil nematode community composition to increasing aridity. Global Ecology and Biogeography 29 (1), 117–126. Available at https://doi.org/10.1111/geb.13013.

Yang, S., Liu, W., Qiao, C., Wang, J., Deng, M., Zhang, B., et al., 2019a. The decline in plant biodiversity slows down soil carbon turnover under increasing nitrogen deposition in a temperate steppe. Functional Ecology 33 (7), 1362–1372. Available at https://doi.org/10.1111/1365-2435.13338.

Yang, Y., Tilman, D., Furey, G., Lehman, C., 2019b. Soil carbon sequestration accelerated by restoration of grassland biodiversity. Nature Communications 10 (1). Available at https://doi.org/10.1038/s41467-019-08636-w.

Yang, Y., Shi, Y., Sun, W., Chang, J., Zhu, J., Chen, L., et al., 2022. Terrestrial carbon sinks in China and around the world and their contribution to carbon neutrality. Science China Life Sciences 65 (5), 861–895. Available at https://doi.org/10.1007/s11427-021-2045-5.

Zamanian, K., Zhou, J., Kuzyakov, Y., 2021. Soil carbonates: the unaccounted, irrecoverable carbon source. Geoderma 384, 114817. Available at https://doi.org/10.1016/j.geoderma.2020.114817.

Zamanian, M., Golzardi, F., Gitari, H., Nungula, E., Salehi, F., Heydarzadeh, S., 2024. Enhancing forage nutritional value in Persian clover (*Trifolium resupinatum*) and crimson clover (*Trifolium incarnatum*) through intercropping and optimized seeding rate. Cogent Food & Agriculture 10 (1), 2410459. Available at https://doi.org/10.1080/23311932.2024.2410459.

Zhou, C., Xia, H., Yang, T., Zhang, Z., Zheng, G., 2024. Grassland degradation affected vegetation carbon density but not soil carbon density. BMC Plant Biology 24 (1). Available at https://doi.org/10.1186/s12870-024-05409-6.

Zhou, W., Yang, H., Zhou, L., Chen, Y., Huang, L., Ju, W., 2018. Dynamics of grassland carbon sequestration and its coupling relation with hydrothermal factor of Inner Mongolia. Ecological Indicators 95, 1–11. Available at https://doi.org/10.1016/j.ecolind.2018.07.008.

Zhou, W., Li, H., Xie, L., Nie, X., Wang, Z., Du, Z., et al., 2021. Remote sensing inversion of grassland aboveground biomass based on high accuracy surface modeling. Ecological Indicators 121, 107215. Available at https://doi.org/10.1016/j.ecolind.2020.107215.

Chapter 19

Energy conservation approaches to generate carbon credits in the agroecosystems

Hari Harjanto Setiawan and Rizal Akbar Aldyan
National Research and Innovation Agency (BRIN), Research Center for Social Welfare, Villages and Connectivity, Jakarta, Indonesia

Highlights

- Mina Padi technique integrates rice cultivation and fish farming, which enhances biodiversity and overall sustainability.
- Carbon credits enable farmer participation in carbon markets for additional income.
- Policy and collaborative approaches are required for effective adoption of carbon credits.
- Technological intervention optimizes resource use and resilience.
- Carbon credit mechanism diversifies income, strengthens local economies, and enhances food security in rural communities.

19.1 Introduction

Current development is a transformation process that optimizes the utilization of natural resource potential. The emergence of this sustainable dimension is motivated by the decline of human wisdom and saving humanity for the future due to the ineffectiveness of human resources in managing natural resources as well as possible and the use of technology for various development efforts (Jhariya et al., 2022, 2023). Polluting technology, excessive consumption, inharmonious development policies, rapid population growth, and poverty issues contribute to a decline in environmental quality. Problems in agricultural and environmental development are critical land issues that significantly impact agricultural productivity and effective land management (Meena et al., 2025a). This study examines the energy conservation approach to generate carbon (C) credits in agroecosystems.

An agroecosystem is a system that is applied to advance world agriculture. A system that involves natural and social factors and creates harmony in life (Raj et al., 2021). An agroecosystem is an environmental system that has been modified and managed by humans for specific purposes. An agroecosystem can be interpreted as a plant and animal community interacting with its environment. Humans have modified it to produce food, fiber, fuel, and other products for human consumption and processing. An agroecosystem is an environmental system that was modified and managed by humans for the benefit of their lives. Unlike natural ecosystems, agroecosystems utilize energy sources derived from organic materials and incorporate other sources that humans, such as fertilizers, pesticides, and various technologies, have developed. The results of these modifications then cause visible differences in the ecosystem. Agroecosystems tend to be dominated by a single variety, while natural ecosystems that are still pure certainly exhibit greater diversity.

The application of agroecosystems can be useful for protecting, restoring, and improving agricultural and food systems in the face of various climate shocks and pressures. Moreover, the application of agroecosystems also plays a role in combating C problems in peatlands of oil palm agroccosystcms, which are crucial in maintaining global climate balance. C reserves above ground level come from vegetation formed in terrestrial ecosystems. We can find out based on the biomass value of the vegetation. Agroecosystems can answer global issues related to C (Banerjee et al., 2021; Meena et al., 2025b). Developing sustainable and environmentally friendly agroecosystems holds great potential but also presents

Carbon Incentives, Plans, and Policies in Agriculture. DOI: https://doi.org/10.1016/B978-0-443-33557-0.00005-7

a significant challenge. Agroecosystems modify ecosystems to meet the needs of life. Various changes in the natural system will certainly affect and impact the social system. The agricultural industry continues to adapt to future challenges, including climate change. One innovative solution that has emerged is a C credit program for rice cultivation. This program enables rice farmers to generate additional income by adopting more sustainable practices, such as drip irrigation. Rice mina can significantly reduce methane emissions, a greenhouse gas (GHG) contributing to global warming (Shine et al., 2005; Meena et al., 2024b). This C credit program benefits not only the environment but also the farmers. Farmers can increase yields and reduce production costs by switching to more sustainable practices (Raj et al., 2025). The program is still in its early stages, but it has enormous potential. If many rice farmers participate, this program can significantly contribute to combating climate change and improving farmers' welfare.

Amidst the challenging business world and the need to hunt for innovative solutions for rice cultivation. C credits, essentially permits that represent reduced GHG emissions, are generated when farmers switch to more sustainable practices (Meena et al., 2024c). We must recognize that to meet global food needs sustainably, innovative approaches are needed. This technology aims to realize more sustainable, efficient, and resilient agriculture in the face of climate change (Kumar et al., 2024). As the agricultural sector faces serious challenges from climate change and increasing global food demand, Rice-fish farming reflects a growing trend in the industry. This effort shows the potential for innovative solutions to address food security and environmental issues (Mushtaq et al., 2025; Meena et al., 2024a). Some argue that such programs allow pollution in one place and offset it in another. Others would extend such programs to real global emissions reductions. The agricultural sector's ability to balance profitability with sound environmental management will be crucial in addressing this pressing issue. Joint efforts with other industry players can potentially shape a sustainable ecosystem. However, ongoing monitoring and refinement of these programs will be essential to ensure continued effectiveness in combating climate change.

Energy is a fundamental component in agroecosystems, determining agricultural productivity and environmental sustainability. Agroecosystems require various energy inputs, including solar energy, human labor, machinery, fertilizers, and irrigation systems. The energy utilization efficiency in these systems directly influences crop yields, resource conservation, and overall sustainability (Iqbal et al., 2018). Efficient energy utilization in agroecosystems involves optimizing input–output ratios to maximize production while minimizing waste (Meena and Pradhan, 2023). Sustainable farming techniques such as precision agriculture, organic farming, and integrated pest management contribute to improved energy efficiency. Additionally, reducing dependency on fossil fuels by incorporating renewable energy sources, such as solar and biogas, significantly lowers the energy footprint of agricultural systems. The footprint of agroecosystems is a critical measure of environmental impact. Agricultural activities, particularly conventional farming, contribute to GHG emissions through excessive chemical fertilizers, intensive mechanization, and water-intensive irrigation (Meena and Pradhan, 2023). Implementing conservation agriculture, agroforestry (AF), and circular farming models can help to mitigate energy footprints and enhance environmental sustainability (Rayner and Morgan, 2018).

Energy conservation in agroecosystems focuses on reducing energy consumption while maintaining or improving productivity. Several approaches were adopted to achieve this goal, emphasizing sustainable practices and innovative technologies (Mishra et al., 2014). One effective approach is conservation tillage, which minimizes soil disturbance, reduces fuel use, and preserves soil organic matter (SOM). Crop rotation and intercropping enhance soil fertility and nutrient cycling, reducing the need for synthetic fertilizers and associated energy expenditures (Amin et al., 2021). Precision farming techniques, such as global positioning system-guided machinery and automated irrigation systems, optimize resource use, further contributing to energy conservation (Mishra et al., 2014). Agroecological practices, including organic farming and permaculture, integrate natural processes to sustain agricultural productivity with minimal external energy inputs (Prasad and Mangipudi, 2022; Pradhan and Meena, 2023a). Utilizing biological pest control instead of chemical pesticides and adopting composting and biofertilizers significantly reduces energy-intensive inputs (Iqbal et al., 2018). Moreover, efficient water management systems, such as drip irrigation and rainwater harvesting, contribute to energy conservation by reducing the reliance on electrically powered irrigation pumps.

Technological advancements have introduced various energy-saving solutions in agroecosystems, improving efficiency and reducing carbon footprints. Innovative farming technologies, including the Internet of Things, AI-based monitoring, and automated machinery, enhance resource utilization, reducing energy waste and production costs (Iqbal et al., 2018). Renewable energy solutions, such as solar-powered irrigation, wind energy for grain drying, and bioenergy production from agricultural waste, significantly lower fossil fuel dependency (Rayner and Morgan, 2018). Biogas digesters, which convert organic farm waste into usable energy, provide a sustainable alternative for on-farm energy needs while contributing to waste reduction and nutrient recycling (Ng and Chan, 2016).

19.2 The rationale of the study

19.2.1 Energy conservation approaches

Conservation is the regulation of humans' use of the biosphere so that sustainable results are obtained for the present generation by maintaining the potential for the needs of future generations (Ariani et al., 2021). The definition of conservation is an effort or action to maintain the existence of something continuously, both in terms of quantity and in quality. The purpose of conservation is to strive to realize the sustainability of natural resources and the balance of their ecosystems so that they can better support efforts to improve community welfare and the quality of human life. Conservation of natural resources and their ecosystems is the responsibility and activity of the government and society. The usefulness of natural resource conservation is realized by maintaining the condition of nature and its environment, which involves conserving the conservation area to prevent damage (Dong et al., 2024).

Avoiding disasters caused by changes in nature, which means disturbances experienced by flora/fauna and their ecosystems in particular, and natural resources in general, that cause changes in the form of damage or a decrease in the number and quality of these natural resources. Avoiding rare and nonrare living things from extinction means that disturbances that cause a decrease in the number and quality of living things, if left unchecked without any control efforts, will result in these living things becoming extinct or even extinct altogether. Thus conservation efforts are efforts to maintain and preserve flora and fauna (Al-Qadami et al., 2024). Able to realize environmental balance, both micro and macro, meaning that in the ecosystem, there is a close relationship between living things and between living things and their environment (Table 19.1).

19.2.2 Agroecosystems

An agroecosystem is a community of plants and animals that are related to their environment (both physical and chemical) that humans have modified to produce food, feed, fiber, firewood, and other products (Gomes de Sá et al., 2024). Another definition of an agroecosystem is that it is a form of ecosystem built by humans with the aim of producing agricultural production to meet human needs. The concept of an agroecosystem is an ecological system found in a farming environment, which is usually a natural system that occurs after being formed by humans. In other words, an agroecosystem is a system where certain living things are cultivated, including all the organisms that live in it and the materials that interact with them. To make it easier to understand, agricultural land can be interpreted broadly, including production forests with industrial crop commodities (Hutan Tanaman Industri [HTI]), livestock areas with grazing land, and fish ponds. An agroecosystem is a system applied to advance world agriculture. A system that involves natural and social factors and creates harmony in life (Ajah et al., 2014).

Two systems are the main components in agroecosystems: (1) Social systems are social components such as demographics, social organizations, economics, political institutions, and belief systems. These components will certainly affect the character, durability, stability, and level of progress of the ecosystem that is formed. (2) Natural ecosystems, including biophysical factors such as soil, water, climate, light, plants, animals, and so on. Natural factors in this case also

TABLE 19.1 Energy inputs and outputs in agroecosystems.

Type of energy input	Source	Efficiency rating	Impact on C footprint	Sources
Solar energy	Solar panels, photosynthesis	High	Positive	Iqbal et al. (2018)
Human labor	Manual farming activities	Medium	Neutral	Mishra et al. (2014)
Machinery	Tractors, harvesters	Low	Negative	Iqbal et al. (2018)
Chemical fertilizers	Synthetic nitrogen fertilizers	Low	Negative	Prasad and Mangipudi (2022)
Organic fertilizers	Compost, biofertilizers	High	Positive	Ng and Chan (2016)
Irrigation systems	Drip irrigation, groundwater pumps	Medium	Neutral	Mishra et al. (2014)
Renewable energy	Biogas, wind energy	High	Positive	Rayner and Morgan (2018)

include the presence of pests, parasites, first-level consumers (herbivores), and various interactions, such as competition and symbiosis between components (Nunes et al., 2024; Pradhan and Meena, 2023b).

Traditional agroecosystem practices in Indonesia are also relatively diverse. Examples include shifting cultivation systems, traditional rainfed rice fields, home gardens, and traditional shaded coffee and cocoa systems. The following are examples in Indonesia: (1) Intensive agroecosystems, agroecosystems dominated by single open plants, highly dependent on minerals carried by water as food (therefore at a certain stage also require water structure intervention), have a relatively stable balance, and tend to overcome population pressures in a centralized manner. Examples of agroecosystem models like this are found in Java. (2) Extensive agroecosystems have a fairly high level of plant diversity, are closed, the circulation of nutrients that support the system occurs through life mechanisms (biotic), have a fragile level of balance, and tend to overcome population pressures by spreading. Examples of agroecosystem models like this are slash-and-burn fields outside Java (Suhardono et al., 2025).

19.3 Materials and methods

The method of writing this book chapter is based on research data and a literature review described using qualitative descriptive analysis. Discussion materials are obtained from various reputable publication sources and research activities related to several programs in this article. Substantive discussion focuses on the relationship between several sources on agroecosystems implemented through various development programs in Indonesia. Analysis to support the agroecosystem development process concerns the production of agricultural and fishery products through rice-fish farming. This chapter describes the agroecosystem scheme through rice-fish farming. The community is experiencing social and economic changes in the agricultural sector. This further illustrates the adjustment of society in facing the agricultural industry 5.0. Data collection was carried out through documentation studies, observations, and interviews.

Primary data were obtained through observation and interviews. Twenty-five informants were selected purposively. The purpose of the interviews with officials at the Ministry of Agriculture was to obtain information on how the agricultural sector developed the rice-fish farming system scheme. In July 2023, the observation and in-depth interview process began. In-depth interviews lasted approximately sixty to ninety minutes. Farmers have shared their experiences with the scheme, and agricultural extension workers have provided information on policies and programs designed to protect farmers from the threats associated with climate change induced by industrial transformation. Thematic analysis was employed to identify patterns within the collected data. The data collection process can be seen in Fig. 19.1.

19.4 Results and discussion

19.4.1 The rice–fish farming system and climate change

The expanse of rice plants in the rice fields is planted with rice, tilapia, and catfish in the exact location as the rice. The rice-fish farming model is farming while raising fish on the same land. Some of his colleagues were initially unsure that rice-fish farming would be effective and profitable for farmers. After careful consideration, they finally decided to implement it. Unlike conventional rice planting, the rice-fish farming process begins by plowing the rice fields and making a 4x1-meter pond with a fish path in the middle. Then, compost and rice are planted with rice. After one or two weeks of planting rice, the rice fields are irrigated, and fish seeds are put in. "Per 1000 square meters, 200 kilograms of compost and 8 kilograms of urea fertilizer are required."

It turns out that farmers benefit from the rice-fish farming system, specifically that the rice becomes more fertile, the use of chemical fertilizers is reduced, and farmers' income increases from fish farming. In the conventional way, 10–15 rice shoots planted produce 160–170 grains of rice.

Meanwhile, 40 rice shoots produce 215 rice grains with the rice-fish system. Rice plants are also not attacked by rats and brown plant hoppers (insects) because the plants are submerged in water. However, bird attacks are still a threat to farmers. While fish do not require much feed, they need to be supported by clean irrigation. Every 1000 meters of rice and fish can produce up to 10 quintals or more. Meanwhile, from fish, approximately 9000 fish can be obtained from 4000 square meters. A farmer can harvest fish measuring 4–7 fish/kilogram every 2.5 months, with a selling price of Rp. 18,000.

The price of tilapia is also stable and tends to increase, unlike catfish and gourami. At the end of 2014, he harvested 8000 fish from 9000 spread fish seeds. The advantages of rice-fish farming are minimal labor, abundant agricultural products, and fish harvests as added value. Rice-fish farming increases rice and fish self-sufficiency and will help Indonesia realize food sovereignty. Tilapia is chosen in the rice-fish farming system because it is not easily attacked by disease, the price continues to rise, and demand is increasing. According to him, 1500 seeds can produce three quintals

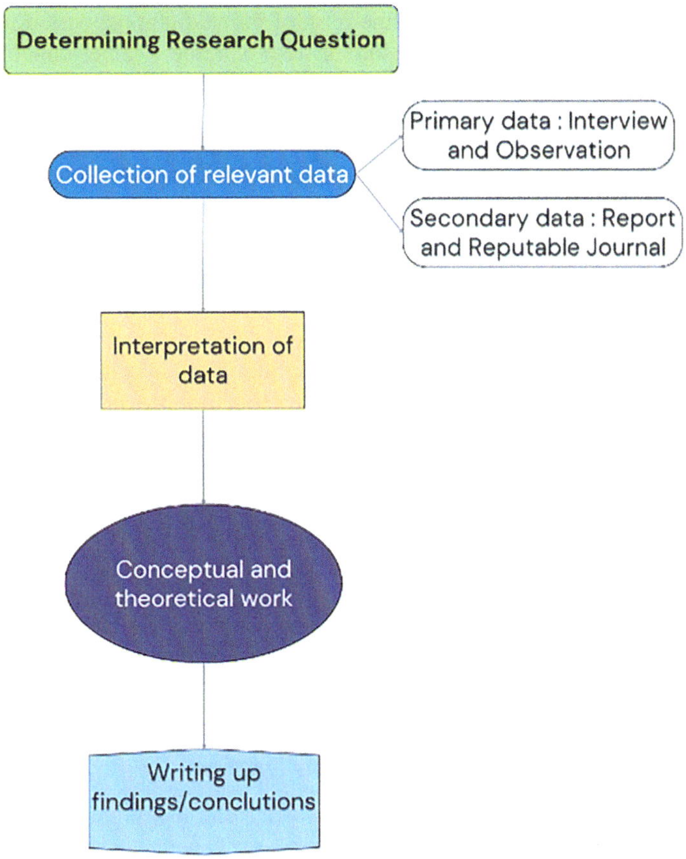

FIGURE 19.1 Research data collection process.

FIGURE 19.2 An agroecosystem with the rice-fish farming system.

of fish. However, a rice system that is profitable for farmers requires agricultural land with abundant water, as shown in Fig. 19.2.

Among the many technologies tested on agricultural land to improve environmental quality in anticipation of climate anomalies, one promising technology is rice-fish farming. Rice-fish farming has been developed in Indonesia for a century. According to the research of the Director General of Aquaculture, Ministry of Marine Affairs and Fisheries in 2011, rice-fish farming is an integrated cultivation that can increase farmers' income in the form of increased production of up to 10%, increase the diversity of agricultural products in the form of fish, increase soil and water fertility with reduced fertilizer use by 30% and reduce pests and diseases in the form of brown plant hoppers. Rice-fish farming is considered one of the solutions for addressing low productivity resulting from extreme weather conditions.

From the publication of the Director General of Aquaculture entitled "The Role of rice fish farming: Reducing Methane Gas Emissions (CH_4) in the Air as an Anticipation of Climate Anomalies," it is stated that rice-fish farming can fertilize land from fish waste, which helps to accelerate environmental improvement. The rice-fish farming pattern can also reduce methane gas released from fertilizer residues. The report states that climate change, as a global phenomenon, is triggered by human activities, especially those related to fossil fuels, natural processes, and land conversion activities, including agricultural and livestock activities. Agriculture and livestock contribute 8.05% of GHG emissions emitted into the atmosphere. The impact of global warming on the agricultural and farming sectors includes the shifting of seasons and changes in rainfall patterns, which lead to delays in planting or harvesting seasons and failures due to flooding, landslides, and drought. Rice cultivation produces the most methane gas, namely 2.57 tons/year. Geographically, 21.2% of the methane gas is contributed by rice cultivation land from West Java, 20.9% from East Java, and 15.9% from Central Java.

Through rice-fish farming, soil fertility in rice fields can increase because fish and food waste function as fertilizer. Fish waste contains various nutrients, thereby reducing the need for inorganic fertilizers by 30%. Fish also reduce the growth of other plants that are competitors with rice in terms of using nutrients, thereby reducing the cost of weeding out wild plants. Therefore rice-fish farming must be supported by the selection of rice varieties. Using superior and adaptive varieties to integrate farming practices will reduce chemical fertilizer inputs. The report also stated that the rice-fish farming system increases oxygen in the water, thus having a positive impact on fish growth. The development of the rice-fish farming system can bring several benefits, namely saving the environment from GHG emissions, fulfilling the need for environmentally friendly organic fertilizers, and supporting the achievement of fisheries production targets of up to 35.5%. Therefore the rice-fish farming system requires support from government programs to conserve the environment, particularly in addressing global warming.

19.4.2 The social and economic impacts of rice–fish farming

Rice-fish farming is an integrated farming system that combines rice and fish production and is increasingly popular in various regions of Indonesia. This innovative approach offers economic benefits and enhances food security, as well as the livelihoods of local communities. Rice-fish farming offers a sustainable solution to address the social and economic challenges farmers face by utilizing the rice field ecosystem as a habitat for fish. One of the main advantages of implementing the rice-fish farming system is increased agricultural productivity. Research data shows that this method can produce a rice harvest of 6–7.5 tons per hectare, an increase of around 10%–20% compared to traditional farming techniques (Dian Prama Dewi et al., 2023). In addition to rice, farmers can also harvest up to 1–2 tons of fish per hectare, providing a dual source of income (Dian Prama Dewi et al., 2023; Kriska et al., 2022).

In addition to generating economic benefits, integrated fish farming helps to reduce dependence on pesticides and chemical fertilizers through natural control of pests and weeds (Merliana et al., 2021). This approach not only supports environmental sustainability but also reduces production costs. Diversification of crops in the rice-fish farming system helps to stabilize farmers' incomes and reduces economic risks due to fluctuations in the price of a single commodity (Intan Christian et al., 2022).

The rice-fish farming system has been demonstrated to enhance the availability of both rice and fish, thereby strengthening local food distribution. The combination of rice as a source of carbohydrates and fish as a source of protein provides significant nutritional benefits, encouraging a more balanced diet for the community (Sumiarsih et al., 2019). The harvest surplus allows farmers to sell or trade their products, positively impacting the local economy (Junaidi et al., 2023). In addition, this system contributes to developing local markets for rice and fish commodities, creating new business opportunities and supporting economic growth in local communities (Intan Christian et al., 2022). Thus rice-fish farming meets local food needs and increases farmers' competitiveness in the market.

Socially, the implementation of rice-fish farming encourages cooperation and solidarity among farmers. The formation of farmer groups plays a vital role in sharing information and disseminating best practices related to this system (Kriska et al., 2022; Intan Christian et al., 2022). Interactions within these groups strengthen social networks, improve farmers' technical skills, and open access to broader resources (Shalihah et al., 2022). Furthermore, the rice-fish farming system encourages the active participation of farmers in the decision-making process, creating a greater sense of ownership over their farming practices (Intan Christian et al., 2022). This approach empowers individual farmers and strengthens social cohesion within communities. The rice-fish farming approach has significant economic and social impacts on local communities in Indonesia. By increasing agricultural productivity, strengthening food security, and encouraging community collaboration, this system serves as a sustainable model for improving the welfare of farmers and their

TABLE 19.2 Social and economic benefits of rice-fish farming.

Benefit type	Description	Impact level	Sources
Economic benefits	Increased farmer income, dual crop production	High	Kriska et al. (2022)
Social benefits	Community cooperation. Knowledge sharing, job creation	Medium	Junaidi et al. (2023)
Environmental benefits	Reduced methane emissions, improved biodiversity	High	Merliana et al. (2021)
Food security	Increased rice and fish availability	High	Sumiarsih et al. (2019), Intan Christian et al. (2022)

families. In the future, wider adoption of the rice-fish farming system has the potential to support national food security and community- based economic growth significantly (Table 19.2).

19.4.3 Comparison of rice–fish farming agroecosystems in Indonesia and other countries

Agroecosystems that integrate rice and fish, particularly in Indonesia and Southeast Asia, are crucial to achieving sustainable agriculture. This integrated system, known as rice-fish farming, not only increases food security but also supports environmental conservation. With strong historical roots, this practice continues to evolve, adapting to various ecological and socio-economic conditions in the regions where it is applied. Rice-fish farming is recognized as one of Indonesia's most important agricultural heritage systems due to its extraordinary ecological and economic benefits. This system not only improves the quality of rice and fish harvests but also strengthens biodiversity and supports sustainable aquaculture development (Wang et al., 2022). The synergy between rice and fish allows for more efficient resource utilization. Fish help to control pests and recycle soil nutrients, reducing the need for chemical fertilizers and pesticides (Xie et al., 2011). This symbiotic relationship increases the yield and production quality, providing significant economic benefits to farmers (Yang et al., 2024).

Other Southeast Asian countries, such as Thailand and Vietnam, also practice rice-fish farming with various local adaptations. In Thailand, government policies significantly influence rice production, indirectly influencing the adoption of this integrated farming system (Laiprakobsup, 2019). In Vietnam, the integration of rice-fish farming reflects an effort to balance traditional and modern farming approaches with an increasing focus on sustainability and ecosystem preservation (Pradadimara, 2015). However, amid its benefits, the expansion of aquaculture in Southeast Asia has created challenges. Competition for land between fish farms and traditional rice fields has raised concerns about food security and environmental impacts. Aquaculture farms often replace high-value agricultural land, posing serious challenges to the food system. However, innovations such as planting tall- stemmed rice in aquaculture farms offer promising solutions, reducing GHG emissions and improving land use efficiency (Li et al., 2023a).

In addition to economic benefits, the rice-fish system provides significant ecological contributions. Research shows that this system can increase soil fertility and improve water quality, strengthening the overall health of the agroecosystem (Li et al., 2018). Integrating aquaculture into rice farming also helps increase resilience to climate change, as more diverse systems tend to be more resilient to environmental changes (Yang et al., 2024). The rice-fish farming agroecosystem is one example of a relevant and effective sustainable agricultural practice, both in Indonesia and in the Southeast Asian region. By optimizing the synergy between rice and fish, this system addresses the challenge of food security and supports ecological balance and economic stability. Rice-fish farming is a crucial foundation for policy development and further research, accelerating the transformation of agriculture toward sustainability in this region (Table 19.3).

19.4.4 New technologies in the rice–fish farming agroecosystem

The rice-fish farming agroecosystem, with its unique characteristics in agricultural practices and environmental conditions, is currently undergoing a transformation in the application of new technologies and digitalization. These changes are increasing productivity and resilience to the challenges of climate change and natural resource degradation. One of the critical innovations in managing the rice-fish farming agroecosystem is the application of climate-smart agriculture (CSA). CSA technology, including superior crop varieties, composting, and AF, has been shown to increase food security, adapt agricultural systems to climate change, and reduce GHG emissions. For example, crop rotation and

TABLE 19.3 Comparison of rice-fish farming models in different countries.

Country	Farming Model	Environmental Impact	Economic Benefit	Sources
Indonesia	Traditional rice-fish farming	Medium	High	Xie et al. (2011)
Thailand	Government-supported rice-fish integration	Low	Medium	Laiprakobsup (2019)
Vietnam	Commercial rice-aquaculture	High	High	Wang et al. (2022)
China	Intensive rice-fish farming	High	High	Yang et al. (2024)
India	Small-scale organic rice-fish farming	Low	Medium	Li et al. (2023a)

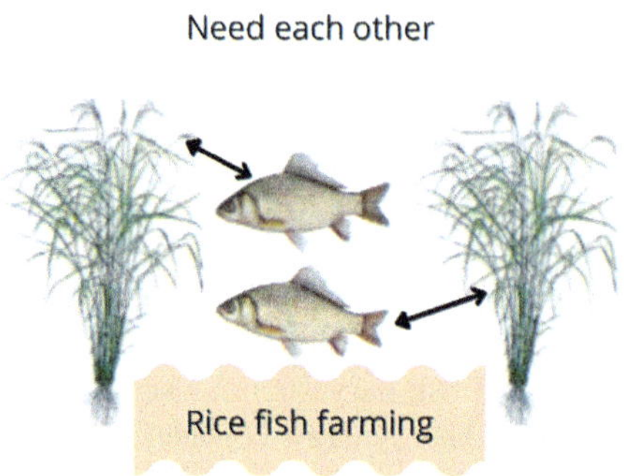

FIGURE 19.3 Mutual need between fish and rice plants.

soil and water conservation practices not only strengthen food security but also reduce the negative impacts of climate change (Teklu et al., 2022). These approaches are particularly relevant in regions facing significant challenges due to resource degradation, such as in Ethiopia (Submitter et al., 2021).

Furthermore, integrating precision agricultural technologies is key to optimizing resource use and crop management. Precision agriculture utilizes advanced technologies, including global navigation satellite systems, remote sensing, and data analytics, to enhance decision-making processes in agriculture (Rayhan Shaheb et al., 2022). On the other hand, the application of machine learning through wireless sensor networks enables the collection and analysis of real-time data. This is essential for maintaining crop health and soil conditions, which ultimately supports increased agroecosystem productivity and sustainability (Mekonnen et al., 2019; Murugamani et al., 2022).

Other innovative practices, such as conservation tillage and the use of grimaces, have proven to be effective in improving soil quality and enhancing food security (Mgolozeli et al., 2020; Sadiq et al., 2024). These techniques not only maintain soil moisture and SOM but also contribute to reducing GHG emissions from agricultural activities (Sadiq et al., 2024). Additionally, AF practices have sought to enhance biodiversity and ecosystem function, thereby increasing the resilience of agroecosystems to climate change (Visscher et al., 2024). The mutual need between fish and rice plants can be seen in Fig. 19.3.

Digitization has also become a significant transformational element in the rice-fish farming agroecosystem. Applying blockchain technology in smart farming, for example, can increase transparency and efficiency in the supply chain, thereby strengthening the overall destruction of agricultural practices. Furthermore, using automation and robotics in agricultural operations can reduce labor costs while increasing operational efficiency. This is a very relevant solution considering the challenges of labor shortages due to the impacts of climate change (Kumareswaran and Jayasinghe, 2022). Thus the rice-fish farming agroecosystem is now at the intersection of traditional practices and modern technological innovations. The integration of CSA, precision technology, and digitalization not only addresses the challenges of climate change but also opens up opportunities for a more sustainable and resilient agricultural future. Farmers can increase productivity, ensure food security, and support the restoration of essential ecosystem functions through these innovations.

19.4.5 Carbon market mechanisms

The rice-fish farming system, which combines traditional rice cultivation with aquaculture, offers a unique opportunity for Indonesian farmers to generate C credits by implementing sustainable agricultural practices. As a relevant approach to climate change mitigation (CCM), this mechanism helps farmers monetize their efforts to reduce GHG emissions and significantly increase agricultural productivity. Rice-fish farming integrates rice cultivation with fisheries, supporting biodiversity and improving soil quality. By adopting practices such as minimal tillage, organic fertilizers, and integrated pest management, farmers can increase the land's capacity to absorb C. This approach not only reduces C emissions but also increases the farming system's resilience to climate change's impacts. C credits generated from these practices can be traded on the C market, creating an additional source of income for farmers (Benjamin, 2013).

To access the C market, farmers must implement specific practices that comply with C credit certification standards. This process involves documenting the farming methods used, measuring the level of carbon sequestration (C_{seq}), and adhering to monitoring protocols established by the certification body. The involvement of local cooperatives or agricultural organizations is important in facilitating farmers' access to these markets while also providing the necessary training and technical support (Ajah et al., 2014). For example, cooperatives can assist farmers in providing funding, training, and technical support to help them adopt sustainable practices that qualify for C credits (Sarah et al., 2024).

On the other hand, access to agricultural credit is also a key component in the success of these initiatives. The credit provides opportunities for farmers to invest in sustainable technologies and practices that increase productivity and the land's capacity to absorb C (Ammani, 2012). Research indicates that access to formal credit has a positive impact on agricultural productivity, while encouraging the adoption of innovations that generate more C credits (Nordjo and Adjasi, 2019). However, the disparity in credit access between small and large holders remains a challenge. Farmers who are financially better off are often better equipped to invest in needed technologies, while smallholder farmers risk being left behind in these C market mechanisms (Karlan et al., 2014).

The C market mechanism based on the rice-fish farming system opens up strategic opportunities for farmers to contribute to CCM through sustainable agricultural practices. With access to agricultural credit supported and cooperatives strengthened, farmers can increase productivity while gaining economic benefits from the C market. Integrating financial support with the implementation of sustainable agriculture not only strengthens farmers' economic resilience but also optimizes their contribution to reducing GHG emissions.

19.4.6 The role of government policies in supporting energy conservation and the rice–fish farming system

Indonesian government policies play a central role in supporting energy conservation and the development of the rice-fish farming system as an integrated approach to rice farming that focuses on sustainability. The Indonesian government has implemented various strategic policies to enhance energy conservation, which is crucial for economic growth and environmental sustainability. For example, the National Energy Policy was designed to diversify the energy mix by encouraging renewable energy sources and reducing dependence on fossil fuels. This step aims to support long- term economic growth while protecting the environment (Fitriana et al., 2024; Yudha and Tjahjono, 2019).

Energy conservation in Indonesia is not just about reducing energy consumption but also about encouraging the creation of a culture of sustainability within the community and the industrial sector. The government recognizes the importance of public education and awareness raising in achieving this goal. Initiatives such as energy-saving campaigns have been implemented in various sectors, including manufacturing and agriculture, pillars of the national economy (Jaelani et al., 2020). In addition, integrating Islamic economic principles into energy management strategies provides a moral basis for saving energy to realize sustainable development (Jaelani et al., 2020).

The rice-fish farming system exemplifies the synergy between energy conservation and agricultural productivity. This integrated farming approach enables farmers to grow rice and fish simultaneously, thereby optimizing land use and enhancing overall productivity (Merliana et al., 2021). The dual output of this system not only increases farmers' income but also contributes to resource efficiency, which aligns with the government's goal for sustainable agricultural practices (Kriska et al., 2022). The successful implementation of the rice-fish farming system is highly dependent on the support of local farmer groups and agricultural extension officers. These stakeholders play a crucial role in enabling farmers to adopt best practices and maximize the benefits of rice and fish production (Kriska et al., 2022; Tarigan et al., 2023). This approach highlights the potential of integrated farming systems to advance sustainable agriculture and improve rural livelihoods.

In the rice-fish farming system context, government policies play a significant role in supporting the implementation of environmentally friendly agricultural techniques. This system integrates aquaculture with rice farming, increasing productivity and reducing the use of water and chemical inputs. This policy aligns with sustainability goals, as the rice-fish farming system supports increased biodiversity and resource efficiency (Basuki et al., 2022). The successful implementation of this system depends on a collaborative governance model involving various stakeholders, including local communities (Kusumawardhani et al., 2022). Indonesia's commitment to international agreements such as the Paris Agreement further strengthens the government's efforts to reduce GHG emissions and promote sustainable practices in various sectors, including energy and agriculture (Ariani et al., 2021). This is reflected in policies focusing on increasing energy efficiency and using renewable energy sources, which are key elements in CCM (Wahyudi, 2024). The government is also critical in facilitating stakeholder collaboration and ensuring policies are designed, communicated, and implemented effectively to achieve optimal outcomes (Berdej et al., 2016). Overall, Indonesian government policies play a crucial role in promoting energy conservation and the development of the rice-fish farming system. The government is laying the foundation for a sustainable future by strengthening the regulatory framework that supports sustainable practices, raising public awareness, and fostering stakeholder collaboration. These steps aim to balance economic growth with responsible environmental management.

19.4.7 The role of collaboration between stakeholders

Implementing rice-fish farming practices, namely the integration of rice and fish farming, has excellent potential to increase productivity and sustainability in the agricultural sector. However, this practice's successful implementation and expansion require close synergy between various stakeholders, such as farmers, government, nongovernmental organizations (NGOs), and the private sector. Each party has a strategic role in overcoming various challenges and creating an ecosystem that supports innovation and sustainability in agriculture. First, collaboration between farmers and local governments is a crucial foundation for building a support system that encourages the adoption of rice-fish farming practices. Local governments can provide various facilities, such as training, infrastructure, and financial assistance, which are vital for small farmers who often face limited resources (Wu and Li, 2024). For example, the establishment of agricultural business units under the auspices of the village government has proven effective in providing business capital and marketing support, thereby improving the standard of living of farmers (Habib and Kiswanto, 2024). This collaborative approach ensures that farmers' needs are adequately met, promoting the successful implementation of sustainable agriculture.

NGOs are also important as a liaison between farmers and other stakeholders, such as the government and the private sector. As "intermediary organizations," NGOs often facilitate cross- sector communication and collaboration (Wu and Zhang, 2013). Additionally, through targeted training and resource provision, NGOs assist farmers in enhancing their skills, particularly in fish farming practices that necessitate specialized technical knowledge (Fontana and Pisalyaput, 2023). Experiences from various sectors demonstrate that collaborations between farmers and NGOs can yield significant environmental and economic benefits, underscoring the importance of these partnerships in promoting agricultural innovation (Fontana and Pisalyaput, 2023).

The private sector also makes a significant contribution to this collaborative framework. Companies can provide broader market access and innovative technologies to improve efficiency and productivity. For example, collaboration between farmers and private companies can build supply chains responsive to market dynamics, thereby increasing farmers' incomes and the sustainability of their businesses (Li et al., 2023b). Additionally, the private sector can support research and development that focuses on optimizing farming practices tailored to local characteristics (Ren and Cui, 2024).

Integrating these various forms of collaboration can create a holistic approach to agricultural development. Multistakeholder engagement can improve the effectiveness of farm policies and programs, ensuring their relevance and suitability to the real needs and challenges farmers face. This approach is particularly relevant in areas where agricultural practices remain fragmented and coordinated collaboration is necessary to support productivity and long-term sustainability. The synergy between farmers, government, NGOs, and the private sector is key to the successful implementation of rice-fish farming. Effective collaboration not only provides the necessary support and resources but also creates a conducive environment for innovation and sustainability. By leveraging each party's unique strengths, the challenges faced by farmers can be addressed, and sustainable agricultural practices can be more widely implemented to improve community welfare.

19.4.8 Multidisciplinary approach in the rice–fish farming agroecosystem

A multidisciplinary approach that holistically integrates ecology, economics, and sociology can significantly strengthen energy conservation efforts in rice-fish agroecosystems. This combination enables a deeper understanding of the complex dynamics that occur in agroecosystems, thereby fostering the development of sustainable practices that are ecologically harmonious, economically competitive, and socially inclusive. From an ecological perspective, agroecological principles highlight the importance of biodiversity and environmental interactions in supporting agroecosystem sustainability. Crop diversification, for example, creates complex ecological networks, increases energy flows, and strengthens nutrient cycles in agricultural systems (Nicholls and Altieri, 2016). This diversity also facilitates symbiotic relationships between plants, insects, and microorganisms, which in turn increases the resilience of agroecosystems against pest and disease threats (Ratnadass et al., 2012). In addition, managing ecological processes such as intercropping to optimize nutrient absorption and natural control of pest populations can reduce dependence on external inputs, including chemical fertilizers and pesticides, thereby significantly saving energy (Tixier et al., 2013). By utilizing this ecological approach, agroecosystems become more productive and sustainable, preserving natural resources for future generations (Nicholls et al., 2017; Pradhan et al., 2023c).

From an economic perspective, transitioning to sustainable agroecosystems requires transforming traditional agricultural practices by adopting innovative economic models. Eco-economic clubs can support the simultaneous production of agricultural products and ecological services, encouraging collaboration between farmers and other stakeholders (Omer, 2023). This collaborative framework allows for the sharing of resources and knowledge, ultimately increasing the efficiency of energy conservation practices and economic competitiveness. In addition, economic analysis in agroecological planning can identify cost-effective strategies that still accommodate farmers' financial needs (Velten et al., 2021). Thus integrating these economic aspects ensures that energy conservation solutions are environmentally friendly and provide sustainable financial returns. From a sociological perspective, the success of energy conservation in agroecosystems is highly dependent on the underlying social dynamics.

Trust and collaboration between farmers, government agencies, and local communities are key foundations for implementing sustainable practices (Norton et al., 2020). Involving local communities in planning and decision-making processes ensures that energy conservation strategies align with local cultural values and are socially just, increasing their acceptability and effectiveness. Furthermore, educational programs that raise awareness of the importance of energy-efficient practices can empower communities to adopt sustainable approaches in their agricultural activities (Armendáriz et al., 2024). By considering this social dimension, a multidisciplinary approach can create inclusive strategies supporting energy conservation while strengthening stakeholder relationships. Integrating ecological, economic, and sociological perspectives in one interdisciplinary approach can provide more comprehensive and effective energy conservation solutions for rice-fish farming agroecosystems. By strengthening biodiversity, creating innovative economic collaborations, and empowering local communities, this approach can revolutionize agricultural practices to be more energy-efficient, sustainable, and resilient in the future.

19.4.9 Research limitations

Because each country's characteristics and policies differ, this study cannot be generalized to all countries. Due to the varying and unstable socio-economic conditions of people around the world, this research cannot be used as a benchmark for measurement. However, the results of this research can be used to increase the resilience of agricultural communities to Industry 5.0 financing. Therefore local communities from various countries with various characteristics must assess the condition of their communities. This program can be implemented only if the conditions in Indonesia are the same. Due to the short observation time in this study, the results cannot be applied to certain situations.

19.4.10 Recommendations

Enhancing energy conservation in agroecosystems requires integrating ecological, economic, and social perspectives. Leveraging precision agriculture and blockchain technologies can optimize resources and facilitate C credit access. Policies should incentivize sustainable practices and foster multistakeholder collaboration, addressing barriers for small farmers by simplifying certification and providing financial support. Expanding traditional systems like rice-fish farming by integrating local wisdom with modern methods and supporting R&D for regional adaptation is essential. Robust monitoring systems and education programs are needed to build capacity, while international collaboration can drive funding and knowledge exchange, promoting global sustainability in agroecosystems.

19.5 Conclusion

This study examines the integration of energy conservation approaches in agroecosystems, with a focus on the innovative "rice-fish farming" system and its ability to address environmental, economic, and social challenges in agriculture.

By combining traditional practices with advanced technologies, the rice-fish farming system exemplifies a sustainable model that optimizes resource utilization, reduces GHG emissions, and increases agricultural productivity. The study emphasizes the ecological benefits of integrating rice and fish farming. This approach promotes biodiversity, reduces the need for chemical inputs, and improves soil quality. Economically, the dual production system increases farmers' income and opens pathways for participation in C credit markets. Socially, the system fosters stakeholder collaboration, empowers local communities, and promotes equitable distribution of resources.

Key strategies highlighted include precision agriculture, CSA technologies, and blockchain-based systems that simplify operations and increase transparency. Government policies that emphasize sustainability and collaboration are crucial in supporting these initiatives. However, challenges such as gaps in access to finance and market integration for smallholder farmers require targeted interventions. The study advocates a multidisciplinary approach that combines ecological, economic, and sociological perspectives. This integration ensures energy-efficient, resilient, and inclusive agroecosystem practices aligned with global sustainability goals. The findings underscore the transformative potential of aligning indigenous knowledge with modern innovations to create balanced synergies between environmental health, economic viability, and community wellbeing. This study lays the groundwork for further research and policy development to refine agroecosystem practices and expand their application across ecological and socio-economic contexts. The recommendations offer practical pathways to scale up sustainable agricultural innovation, ultimately contributing to the broader goals of food security, climate resilience, and sustainable development.

Acknowledgemnts

Both authors made significant contributions to the completion of this chapter. We thank the Village Research Center for Social Welfare and Connectivity, as well as the National Research and Innovation Agency (BRIN), for their support of this research.

References

Ajah, E.A., Itam, K.O., Asuquo, A.I., 2014. Analysis of cooperative societies effectiveness in credit delivery to agricultural enterprises in Calabar Municipality of Cross River State, Nigeria. Greener Journal of Agricultural Sciences 4 (8), 354–361. Available at https://doi.org/10.15580/GJAS.2014.8.080914323.

Al-Qadami, E., Mohammad Razi, M.A., Ideris, M.F.M., Mahamud, M., Shah, S.M.H., Pu, J.H., et al., 2024. Comprehensive biodiversity assessment of flora and fauna along the Kedah coastline: implications for conservation and sustainable management. Results in Engineering 24, 103120. Available at https://doi.org/10.1016/j.rineng.2024.103120.

Amin, M.N., Asaduzzaman, M., Kabir, A., Snigdha, S.S., Hossain, M.S., 2021. Lessons from local indigenous climate adaptation practices: perceptions and evidence from coastal Bangladesh. Routledge, Australia Local Environment 26 (8), 967–984. Available at https://doi.org/10.1080/13549839.2021.1937970.

Ammani, A.A., 2012. An investigation into the relationship between agricultural production and formal credit supply in Nigeria. International Journal of Agriculture and Forestry 2 (1), 46–52. Available at https://doi.org/10.5923/j.ijaf.20120201.08.

Ariani, R., Hamzani, A.I., Rahayu, K., 2021. Acceleration of clean energy use based on the 2015 Paris Agreement. Journal of Energy Engineering and Thermodynamics (21), 1–10. Available at https://doi.org/10.55529/jeet.21.1.10.

Armendáriz, M.B.-H.C., Rodríguez, F.D.J.G., Aguirre, G.M.Q., Moreno, M.M., Cabrera, Á.F.C., Yah, M.Y.U., 2024. Entomofauna associated with italian pumpkin cultivation "cucurbita pepo/gray zucchini" under agroecological system in campeche, Mexico. Journal of Agricultural Sciences Research 4 (2), 2–8. Available at https://doi.org/10.22533/at.ed.973422408022.

Banerjee, A., Jhariya, M.K., Yadav, S.K., Khan, N., Raj, A., Meena, R.S., et al., 2021. Climate change vulnerability and agroecosystem services. Sustainable Intensification for Agroecosystem Services and Management. Springer, Singapore, pp. 163–195. Available at https://doi.org/10.1007/978-981-16-3207-5_6.

Basuki, T.M., Nugroho, H.Y.S.H., Indrajaya, Y., Pramono, I.B., Nugroho, N.P., Supangat, A.B., et al., 2022. Improvement of integrated watershed management in Indonesia for mitigation and adaptation to climate change: a review. Sustainability 14 (16), 9997. Available at https://doi.org/10.3390/su14169997.

Benjamin, E.O., 2013. Credit risk modelling and sustainable agriculture: asset evaluation and rural carbon revenue. Journal of Sustainable Finance & Investment 3 (1), 57–69. Available at https://doi.org/10.1080/20430795.2013.765382.

Berdej, S.M., Armitage, D.R., Gao, Z.-K., 2016. Bridging organizations drive effective governance outcomes for conservation of Indonesia's marine systems. PLoS One 11 (1), e0147142. Available at https://doi.org/10.1371/journal.pone.0147142.

Dian Prama Dewi, G., Mursitama, T.N., Noerlina, D.N., Utama, S.A., 2023. Abrori, Investigating the impact of Mina Padi cultivation technology on rice productivity: attaining the Sustainable Development Goals 2.3. E3S Web of Conferences 426, 01113. Available at https://doi.org/10.1051/e3sconf/202342601113.

Dong, Z., Zhou, Z., Ananzeh, M., Hoang, K.N., Shamansurova, Z., Luong, T.A., 2024. Exploring the asymmetric association between fintech, clean energy, climate policy, natural resource conservations and environmental quality. A post-COVID perspective from Asian countries. Resources Policy 88, 104489. Available at https://doi.org/10.1016/j.resourpol.2023.104489.

Fitriana, I., Hadiyanto, Warsito, B., Himawan, E., Santosa, J., 2024. The optimization of power generation mix to achieve net zero emission pathway in indonesia without specific time target. International Journal of Sustainable Energy Planning and Management 41, 5–19. Available at https://doi.org/10.54337/ijsepm.8263.

Fontana, E., Pisalyaput, N., 2023. Understanding the importance of farmer–NGO collaboration for sustainability and business strategy: evidence from the coffee supply chain. Business Strategy and the Environment 32 (6), 2715–2735. Available at https://doi.org/10.1002/bse.3266.

Gomes de Sá, É.F.G., Silva, T.A.A., Cordeiro-Estrela, P., 2024. The curious case of small mammal community in a rice-Pantanal agroecosystem of Brazil: a tale of multiple diversity metrics. Ecological Indicators 163, 112028. Available at https://doi.org/10.1016/j.ecolind.2024.112028.

Habib, F., Kiswanto, E., 2024. Village-owned enterprise strategy in improving the economic welfare of sustainable farmers. Populasi 32 (1), 61. Available at https://doi.org/10.22146/jp.97556.

Intan Christian, A., Partini, P., Andarwati, S., 2022. Peran Aktor dalam Perubahan Sosial Ekonomi Petani Pasca Penerapan Mina Padi. Prospect: Jurnal Pemberdayaan Masyarakat 1 (1), 29–39. Available at https://doi.org/10.55381/jpm.v1i1.1.

Iqbal, Q., Hassan, S.H., Akhtar, S., Khan, S., 2018. Employee's green behavior for environmental sustainability: a case of banking sector in Pakistan. World Journal of Science, Technology and Sustainable Development 15 (2), 118. Available at https://doi.org/10.1108/wjstsd-08-2017-0025.

Jaelani, A., Layaman, L., Fatmasari, D., Salikin, A.D., Dz, A.S., 2020. Energy conservation and energy management for industry in indonesia in islamic economic perspective. International Journal of Energy Economics and Policy 10 (3), 239–249. Available at https://doi.org/10.32479/ijeep.9061.

Jhariya, M.K., Meena, R.S., Banerjee, A., Meena, S.N., 2022. Natural Resources Conservation and Advances for Sustainability. Elsevier. Available at https://doi.org/10.1016/C2019-0-03763-6.

Jhariya, M.K., Meena, R.S., Banerjee, A., Kumar, S., Raj, A., 2023. Agroforestry for Carbon and Ecosystem Management. Elsevier, India, pp. 1–434. Available at http://doi.org/10.1016/C2021-0-02916-3.

Junaidi, Y., Rachmi Indahsari, S., Ahmad Adi, S., 2023. Aspek Keberlanjutan Program Mina Padi Di Desa Sungai Rebo: Pelajaran Dari CSR Pt Kilang Pertamina Internasional RU III Plaju. Journal of Comprehensive Science (JCS) 2 (8), 1431–1442. Available at https://doi.org/10.59188/jcs.v2i8.497.

Karlan, D., Osei, R., Osei-Akoto, I., Udry, C., 2014. Agricultural decisions after relaxing credit and risk constraints* The Quarterly Journal of Economics 129 (2), 597–652. Available at https://doi.org/10.1093/qje/qju002.

Kriska, M., Harsoyo, H., Sulami, R.P., Putra, N.T.W., Kusuma, Y., 2022. Keberlanjutan usaha tani mina padi di kalurahan sumberagung kapanewon moyudan daerah istimewa yogyakarta. SEPA: Jurnal Sosial Ekonomi Pertanian dan Agribisnis 18 (2), 225. Available at https://doi.org/10.20961/sepa.v18i2.51481.

Kumar, S., Meena, R.S., Sheoran, P., Jhariya, M.K., 2024. Regenerative Agriculture for Sustainable Food Systems. Springer Science+Business Media, India, pp. 1–493. Available at http://doi.org/10.1007/978-981-97-6691-8.

Kumareswaran, K., Jayasinghe, G.Y., 2022. Systematic review on ensuring the global food security and covid-19 pandemic resilient food systems: towards accomplishing sustainable development goals targets. Discover Sustainability 3 (1). Available at https://doi.org/10.1007/s43621-022-00096-5.

Kusumawardhani, D., Asmara, A.Y., Purwadi, Pradana, A.W., Indriasari, D.T., 2022. A collaborative governance model for bioethanol research and innovation ecosystem. IOP Conference Series: Earth and Environmental Science 1105 (1), 012042. Available at https://doi.org/10.1088/1755-1315/1105/1/012042.

Laiprakobsup, T., 2019. The policy effect of government assistance on the rice production in Southeast Asia: comparative case studies of Thailand, Vietnam, and the Philippines. Development Studies Research 6 (1), 1–12. Available at https://doi.org/10.1080/21665095.2019.1568900.

Li, Q., Xu, L., Xu, L., Qian, Y., Jiao, Y., Bi, Y., et al., 2018. Influence of consecutive integrated rice–crayfish culture on phosphorus fertility of paddy soils. Land Degradation & Development 29 (10), 3413–3422. Available at https://doi.org/10.1002/ldr.3107.

Li, F., Qian, H., Yang, T., Wang, M., Fang, F., Jiang, Y., et al., 2023a. Higher food yields and lower greenhouse gas emissions from aquaculture ponds with high-stalk rice planted. Environmental Science & Technology 57 (33), 12270–12279. Available at https://doi.org/10.1021/acs.est.3c02667.

Li, W., Liu, C., Yang, Q., You, Y., Zhuo, Z., Zuo, X., 2023b. Factors influencing farmers' vertical collaboration in the agri-chain guided by leading enterprises: a study of the table grape industry in China. Agriculture 13 (10), 1915. Available at https://doi.org/10.3390/agriculture13101915.

Meena, R.S., Pradhan, G., 2023. Industrial garbage-derived biocompost enhances soil organic carbon fractions, CO2 biosequestration, potential carbon credits and sustainability index in a rice-wheat ecosystem. Environmental Research 235, 116525. Available at https://doi.org/10.1016/j.envres.2023.116525.

Meena, R.S., Lal, R., Kumar, S., Pradhan, G., Srinivasarao, C., Singh, A.K., et al., 2024a. Potential of Indian agriculture for capturing atmospheric CO2 and monetizing carbon credits to the farmers: an overview and policy framework. Advances in Agronomy 188, 101–206. Available at https://doi.org/10.1016/bs.agron.2024.06.001.

Meena, R.S., Pradhan, G., Nalani, Kumar, S., Singh, A.K., Verma, S.K., et al., 2024b. Diversified cropping modules designed for soil restoration, CO2 sequestration, and generating carbon credits. Land Degradation and Development 35 (14), 4314–4329. Available at https://doi.org/10.1002/ldr.5224.

Meena, R.S., Pradhan, G., Singh, K., Kumar, S., Singh, A.K., Shashidhar, K.S., et al., 2024c. Agriculture models for restoring degraded land to enhance CO2 biosequestration and carbon credits in the Vindhyan region of India. Science of the Total Environment 929. Available at https://doi.org/10.1016/j.scitotenv.2024.172661.

Meena, R.S., Pradhan, G.S., Kumar, S., Jhariya, M.K., Rao, C.S., 2025a. Carbon-Negative Agriculture. Springer, pp. 1–500. Available at https://link.springer.com/book/9789819514823.

Meena, R.S., Pradhan, G., Kumar, S., 2025b. Energy flow, eco-efficiency, and economic circulation with recycled industrial waste compost application in wheat and subsequent rice farming. Science of the Total Environment 967. Available at https://doi.org/10.1016/j.scitotenv.2025.178779.

Mekonnen, Y., Namuduri, S., Burton, L., Sarwat, A., Bhansali, S., 2019. Review—machine learning techniques in wireless sensor network based precision agriculture. Journal of the Electrochemical Society 167 (3), 37522. Available at https://doi.org/10.1149/2.0222003jes.

Merliana, A., Mulyatno Setiawan, B., Budiraharjo, K., 2021. Economic efficiency of regular rice farming and mina padi production factors. SOCA: Jurnal Sosial, Ekonomi Pertanian 15 (3), 531. Available at https://doi.org/10.24843/soca.2021.v15.i03.p10.

Mgolozeli, S., Nciizah, A.D., Wakindiki, I.I.C., Mudau, F.N., 2020. Innovative pro-smallholder farmers' permanent mulch for better soil quality and food security under conservation agriculture. Agronomy 10 (4), 605. Available at https://doi.org/10.3390/agronomy10040605.

Mishra, D., Akman, I., Mishra, A., 2014. Theory of reasoned action application for green information technology acceptance. Computers in Human Behavior 36, 29–40. Available at https://doi.org/10.1016/j.chb.2014.03.030.

Murugamani, C., Shitharth, S., Hemalatha, S., Kshirsagar, P.R., Riyazuddin, K., Naveed, Q.N., et al., 2022. Machine learning technique for precision agriculture applications in 5G-based Internet of Things. Wireless Communications and Mobile Computing 2022 (1). Available at https://doi.org/10.1155/2022/6534238.

Mushtaq, M.W., Qadiri, S.S.N., Khan, I.A., Mattoo, A., Shakeel, U., Shabir, A., 2025. Breeding technology for sustainable aquaculture and food security: current advancements and future prospects. In: Sundaray, J.K., Rather, M.A., Ahmad, I., Amin, A. (Eds.), Food Security, Nutrition and Sustainability Through Aquaculture Technologies. Springer, Cham, pp. 95–105. Available at https://doi.org/10.1007/978-3-031-75830-0_6.

Ng, Y.S., Chan, D.J.C., 2016. Wastewater phytoremediation by *Salvinia molesta*. Journal of Water Process Engineering 15, 107. Available at https://doi.org/10.1016/j.jwpe.2016.08.006.

Nicholls, C.I., Altieri, M.A., Vazquez, L., 2017. Agroecological principles for the conversion of farming systems. Agroecological Practices For Sustainable Agriculture: Principles, Applications, And Making The Transition. World Scientific Publishing Co, United States, pp. 1–18. Available at http://doi.org/10.1142/9781786343062_0001.

Nicholls, C.I., Altieri, M.A., 2016. Agroecology: principles for the conversion and redesign of farming systems. J. Ecosystem Ecography 01 (s5). Available at https://doi.org/10.4172/2157-7625.S5-010.

Nordjo, R.E., Adjasi, C.K.D., 2019. The impact of credit on productivity of smallholder farmers in Ghana. Agricultural Finance Review 80 (1), 91–109. Available at https://doi.org/10.1108/afr-10-2018-0096.

Norton, D.A., Suryaningrum, F., Buckley, H.L., Case, B.S., Cochrane, C.H., Forbes, A.S., et al., 2020. Achieving win-win outcomes for pastoral farming and biodiversity conservation in New Zealand. New Zealand Journal of Ecology 44 (2). Available at https://doi.org/10.20417/nzjecol.44.15.

Nunes, A.L.P., et al., 2024. Chapter 12 - Role of *Pseudomonas fluorescens* secondary metabolites in agroecosystem applications. In: Abd-Elsalam, K.A., Mohamed, H.I. (Eds.), Bacterial Secondary Metabolites: Synthesis and Applications in Agroecosystem. Elsevier, pp. 211–220. Available at https://doi.org/10.1016/B978-0-323-95251-4.00008-9.

Omer, A., 2023. A club model of nature-smart agriculture for biodiversity, climate, and productivity enhancements. Integrated Environmental Assessment and Management 19 (2), 412–421. Available at https://doi.org/10.1002/ieam.4641.

Pradadimara, D., 2015. Rice in colonial and post colonial southeast asia: a food regime analysis. Paramita: Historical Studies Journal 25 (1). Available at https://doi.org/10.15294/paramita.v25i1.3418.

Pradhan, G., Meena, R.S., 2023a. Interaction impact of biocompost on nutrient dynamics and relations with soil biota, carbon fractions index, societal value of CO2 equivalent and ecosystem services in the wheat-rice farming. Chemosphere 339. Available at https://doi.org/10.1016/j.chemosphere.2023.139695.

Pradhan, G., Meena, R.S., 2023b. Utilizing waste compost to improve the atmospheric CO2 capturing in the rice-wheat cropping system and energy-cum-carbon credit auditing for a circular economy. Science of the Total Environment 892. Available at https://doi.org/10.1016/j.scitotenv.2023.164572.

Pradhan, G., Meena, R.S., Kumar, S., Lal, R., 2023c. Utilizing industrial wastes as compost in wheat-rice production to improve the above and below-ground ecosystem services. Agriculture, Ecosystems & Environment 358, 108704. Available at https://doi.org/10.1016/j.agee.2023.108704.

Prasad, K.D.V., Mangipudi, M.R., 2022. The effect of green human resource management on employee eco-friendly behaviour, employee environment-friendly response and organizational sustainability: an empirical study concerning IT industry in Hyderabad Metro. Global Journal of Management and Business Research 22 (4), 31–42. Available at https://doi.org/10.34257/gjmbravol22is4pg31.

Raj, A., Jhariya, M.K., Banerjee, A., Khan, N., Meena, R.S., Oraon, P.R., et al., 2021. Agroecosystem service management and environmental sustainability. Sustainable Intensification for Agroecosystem Services and Management. Springer, Singapore, pp. 379–402. Available at http://doi.org/10.1007/978-981-16-3207-5_12.

Raj, A., Jhariya, M.K., Meena, R.S., 2025. Agroforestry for Monetising Carbon Credits. Springer Nature, Switzerland.

Ratnadass, A., Fernandes, P., Avelino, J., Habib, R., 2012. Plant species diversity for sustainable management of crop pests and diseases in agroecosystems: a review. Agronomy for Sustainable Development 32 (1), 273–303. Available at https://doi.org/10.1007/s13593-011-0022-4.

Rayhan Shaheb, M., Sarker, A., A. Shearer, S., 2022. Precision Agriculture for Sustainable Soil and Crop Management. In: Aide, M., Braden, I. (Eds.), Soil Science - Emerging Technologies, Global Perspectives and Applications. IntechOpen. Available at https://doi.org/10.5772/intechopen.101759.

Rayner, J., Morgan, D., 2018. An empirical study of 'green' workplace behaviours: ability, motivation and opportunity. Asia Pacific Journal of Human Resources 56 (1), 56–78. Available at https://doi.org/10.1111/1744-7941.12151.

Ren, G., Cui, X., 2024. The government–farmer cooperation mechanism and its implementation path to realize the goals of optimizing grain planting structure. Land 13 (3), 358. Available at https://doi.org/10.3390/land13030358.

Sadiq, M., Rahim, N., Tahir, M.M., Alasmari, A., Alqahtani, M.M., Albogami, A., et al., 2024. Conservation tillage: a way to improve yield and soil properties and decrease global warming potential in spring wheat agroecosystems. Frontiers in Microbiology 15. Available at https://doi.org/10.3389/fmicb.2024.1356426.

Sarah, E.N.W.A., Theophilus Miebi, G.B.I.G.B.I., Governor Ekene, O.Y.I.T.A., Veronica U, I.K.E.N.G.A., 2024. Assessing the efficacy of cooperative societies in agricultural credit delivery: a comprehensive review in Nigeria. Magna Scientia Advanced Research and Reviews 10 (1), 132–142. Available at https://doi.org/10.30574/msarr.2024.10.1.0001.

Shalihah, M., Sugihardjo, Rusdiyana, E., 2022. Mina Padi farmers' adapting strategy in dealing with the climate change. IOP Conference Series: Earth and Environmental Science 1105 (1), 012047. Available at https://doi.org/10.1088/1755-1315/1105/1/012047.

Shine, K.P., Fuglestvedt, J.S., Hailemariam, K., Stuber, N., 2005. Alternatives to the Global Warming Potential for comparing climate impacts of emissions of greenhouse gases. Climatic Change 68 (3), 281–302. Available at https://doi.org/10.1007/s10584-005-1146-9.

Submitter, F.E.E.M.R.P.S., Bazzana, D., Foltz, J., Zhang, Y., 2021. Impact of climate smart agriculture on food security: an agent-based analysis. SSRN Electronic Journal. Available at https://doi.org/10.2139/ssrn.3857718.

Suhardono, S., Lee, C.-H., Suryawan, I.W.K., 2025. Valuation of marine integrated disaster management amidst global warming in Southern Coast of Java, Indonesia. Marine Pollution Bulletin 211, 117446. Available at https://doi.org/10.1016/j.marpolbul.2024.117446.

Sumiarsih, E., Kausar, K., Adriman, A., Eddiwan, E., Hasibuan, I.F., 2019. Penerapan minapadi untuk mendukung ketahanan pangan dalam meningkatkan kesejahteraan masyarakat Desa Koto Benai Kabupaten Kuantan Singingi. Unri Conference Series: Community Engagement 1, 304–309. Available at https://doi.org/10.31258/unricsce.1.304-309.

Tarigan, W.P.L., Suyanto, S., Paidi, P., Wilujeng, I., Tarigan, C.U., 2023. Analysis of Mina Padi Innovation Village based on local wisdom as a learning resource assisted by biological simulation applications. Agricultural and Environmental Education 2 (1), em003. Available at https://doi.org/10.29333/agrenvedu/13088.

Teklu, A., Simane, B., Bezabih, M., 2022. Effectiveness of climate-smart agriculture innovations in smallholder agriculture system in Ethiopia. Sustainability 14 (23), 16143. Available at https://doi.org/10.3390/su142316143.

Tixier, P., Duyck, P.-F., Côte, F.-X., Caron-Lormier, G., Malézieux, E., 2013. Food web-based simulation for agroecology. Agronomy for Sustainable Development 33 (4), 663–670. Available at https://doi.org/10.1007/s13593-013-0139-8.

Velten, S., Jager, N.W., Newig, J., 2021. Success of collaboration for sustainable agriculture: a case study meta-analysis. Environment, Development and Sustainability 23 (10), 14619–14641. Available at https://doi.org/10.1007/s10668-021-01261-y.

Visscher, A.M., Meli, P., Fonte, S.J., Bonari, G., Zerbe, S., Wellstein, C., 2024. Agroforestry enhances biological activity, diversity and soil-based ecosystem functions in mountain agroecosystems of Latin America: a meta-analysis. Global Change Biology 30 (1). Available at https://doi.org/10.1111/gcb.17036.

Wahyudi, H., 2024. The relationship between the renewable energy and CO2 emissions to the indonesian economy. International Journal of Energy Economics and Policy 14 (3), 349–357. Available at https://doi.org/10.32479/ijeep.15903.

Wang, E., Zhou, Y., Liang, Y., Ling, F., Xue, X., He, X., et al., 2022. Rice flowering improves the muscle nutrient, intestinal microbiota diversity, and liver metabolism profiles of tilapia (*Oreochromis niloticus*) in rice-fish symbiosis. Microbiome 10 (1). Available at https://doi.org/10.1186/s40168-022-01433-6.

Wu, B., Zhang, L., 2013. Farmer innovation diffusion via network building: a case of winter greenhouse diffusion in China. Agriculture and Human Values 30 (4), 641–651. Available at https://doi.org/10.1007/s10460-013-9438-6.

Wu, S., Li, S., 2024. Collaboration to address the challenges faced by smallholders in practicing organic agriculture: a case study of the organic sorghum industry in Zunyi City, China. Agriculture 14 (5), 726. Available at https://doi.org/10.3390/agriculture14050726.

Xie, J., Hu, L., Tang, J., Wu, X., Li, N., Yuan, Y., et al., 2011. Ecological mechanisms underlying the sustainability of the agricultural heritage rice–fish coculture system. Proceedings of the National Academy of Sciences 108 (50). Available at https://doi.org/10.1073/pnas.1111043108.

Yang, M., Li, Z., Shao, L., Zhao, X., Chu, J., Yu, F., 2024. Retraction Notice: Impact of different modes of comprehensive rice field planting and aquaculture systems in paddy fields on rice yield, quality, and economic benefits. Polish Journal of Environmental Studies 34 (2), 1415–1423. Available at https://doi.org/10.15244/pjoes/187139.

Yudha, S.W., Tjahjono, B., 2019. Stakeholder mapping and analysis of the renewable energy industry in Indonesia. Energies 12 (4). Available at https://doi.org/10.3390/en12040602.

Chapter 20

Greenhouse gas and carbon equivalent emissions: calculation and quantification

Keivan Arastou and Shervin Jamshidi
Department of Civil Engineering, University of Isfahan, Isfahan, Iran

Highlights

- Agricultural practices and changing LULC contribute to C emissions.
- SWAT-C can integrate the C cycle with large-scale water basin modeling.
- SWAT-C enables policymakers to account for Cseq by changing LULC.
- LCIA converts the SWAT-C outcomes into quantified environmental impacts globally.
- Integrated SWAT-C, LCIA, and C footprint facilitates inland C calculations.

20.1 Introduction

20.1.1 Background and definition

Greenhouse gases (GHGs), such as water vapor, carbon dioxide (CO_2), methane (CH_4), nitrous oxide (N_2O), and ozone (O_3), can partially absorb infrared radiation. Consequently, GHGs can keep the heat and energy level required for all beings constant (Buder, 2020; Easterbrook, 2016). Among the above GHGs, CO_2 is known as a pivotal concern. Despite its small atmospheric concentration, CO_2 is often linked to the phenomenon of "global warming." This correlation is mostly because the main sources of increasing CO_2 are anthropogenic and extensive globally. Since global warming establishes the base of the "climate change" phenomenon, predicting the variations of CO_2 concentration, along with other GHGs, such as N_2O and CH_4, is currently an issue of primary concern (Intergovernmental Panel on Climate Change (IPCC), 2014; Meena et al., 2025b).

Increasing GHGs originate from anthropogenic sources, such as industrial activities, transportation, energy production, and agriculture (Meena et al., 2024a). Accurately calculating and quantifying these emissions is challenging, but it is essential for developing effective mitigation strategies. Carbon footprint (CF) can measure the GHG emissions (as tonnes of CO_2 equivalents: CO_2 eq.) emitted from a specific activity (e.g., agriculture). Similar to other footprint indices, CF can encompass both direct emissions from controlled and monitored activities and indirect emissions from the use of energy from various sources, such as electricity or fossil fuels. This chapter highlights the potential of CF as the primary GHG emissions indicator for use in integrated calculation methodologies that measure carbon (C) and GHG emissions. This indicator can provide a unified framework and overview, with a focus on CO_2, and explain the significance of C equivalent emissions. Here, CF quantifies GHG emissions to aid in emission management and mitigation. It involves GHG accounting at each life cycle stage of a product or activity (Pandey and Agrawal, 2014). Key resources include: (1) GHG protocol (WRI/WBCSD): Basis for most guidelines, including ISO 14064 and PAS 2050, (2) ISO standards: ISO 14064 for organizational and project emissions, ISO 14040 and 14044 for life cycle assessment (LCA), (3) PAS 2050 (BSI): Assesses life cycle GHG emissions of goods and services, (4) IPCC guidelines: Categorizes GHG sources into energy, industrial processes, agriculture, forestry, and waste (ISO, 2006a; ISO, 2006b; ISO, 2006c; Specification, 2008). The framework for accounting CF includes selecting GHG, setting scopes and boundaries (spatial and temporal), collecting emission data, and calculating the footprint as the final step.

Agriculture significantly contributes to GHG emissions through various activities, such as using organic or chemical fertilizers, deforestation, soil erosion, or product transportation. The latter happens when the cultivated food or plantation

Carbon Incentives, Plans, and Policies in Agriculture. DOI: https://doi.org/10.1016/B978-0-443-33557-0.00024-0

seeds are often exported to/imported from long distances (Wakeland et al., 2012). Yet, it should be noted that agriculture can play a dual role in GHG emissions: First, it acts as a C sink where the plantations adsorb and settle C into the ground. On the contrary, using fertilizers and energy, deforestation changing LULC, and rotting the remnants of harvested plants can gradually emit significant amounts of N_2O, CH_4, and CO_2 (Kim, 2015; Meena et al., 2024b). The agricultural sector is one major contributor to GHG emissions, accounting for 52% of global anthropogenic CH_4 and 84% of N_2O emissions. Direct CO_2 emissions from agriculture are relatively small, primarily resulting from on-farm energy use and the production and transportation of agricultural products (Painkra et al., 2025). Efforts to mitigate agricultural GHG emissions are crucial for climate change policies. The net CO_2 emissions from agriculture are spatially variable, influenced by factors such as soil characteristics and farming practices (Roy et al., 2022). Understanding these local variations is essential for accurately estimating GHG mitigation potential during cultivation. Food contributes 10%–30% of the CF, with a larger share in lower-income households (Owusu et al., 2023). Production accounts for 68% of food-related emissions, while transportation makes up 5%. These emissions primarily consist of CO_2, N_2O, and CH_4, resulting from agricultural activities.

Livestock production, rice cultivation, and the use of chemical fertilizers are key sources, releasing substantial amounts of all potent GHGs (CO_2, CH_4, and N_2O). In general, agriculture is responsible for 10%–12% of the total anthropogenic emissions of GHGs in the world, including CO_2, CH_4, and N_2O (Benbi, 2013). Here, emissions from the agricultural sector grow at an average annual rate of 2.2% (Ramachandra and Mahapatra, 2015). It should be noted that meat products have relatively higher CFs per calorie compared to grains or vegetables due to the inefficient conversion of plant energy to animal energy and CH_4 emissions from manure management and enteric fermentation in ruminants (Cedrez et al, 2023; Meena et al., 2024c).

A study on measuring the direct GHG emissions from cultivating rice, wheat, and maize found that rice has relatively the highest CF, about 2.1 kg CO_2 per kg rice produced, followed by wheat with an average rate of 1.26 kg CO_2 emission per kg wheat production yield. Maize also has 0.95 kg CO_2 emission per its kg weight (Cheng et al., 2015a; Zhang et al., 2017). The total CF includes emissions from farming, manufacturing, processing, and transportation of inputs such as fertilizer, electricity, chemicals, and machinery. Notably, wheat's emissions are predominantly from the mentioned activities (86%), followed by maize (63%) and rice (50%) (Jaiswal and Agrawal, 2020). Differences in CF across regions are attributed to variations in nutrient requirements, irrigation practices, and farming methods. For instance, in paddy fields during cultivation, the average daily discharged nutrient loads (pollution export coefficients) range between 45.3 (phosphorus-based) and 56.2 gr/day (nitrogen-based) as a matter of fertilizer application, leaching, and wash-out rates from the farms, irrigation and precipitation rates, and farming practices like tillage (Jamshidi et al., 2024a). However, these studies revealed that whenever direct water consumption in agriculture or its pollution discharge is considered, water and nutrient cycles are typically emphasized, and the soil C cycles in agriculture are usually ignored or less frequently addressed. Since agricultural practices are regionally variable, as they follow local traditions and are reliant on economic, climate, and crop conditions, the GHG emission models considering both nutrient and C cycles must be validated and refined for the specific areas studied (Roudbari et al., 2023; Meena and Pradhan, 2023). This kind of calibration requires field-based data and experiments in advance.

Direct measurement is the most reliable method for evaluating soil-borne GHG emissions and cycling due to their sensitivity to environmental conditions (Pandey and Agrawal, 2014; Meena et al., 2023). Therefore implementing advanced agricultural methods can help to reduce CO_2 emissions and restore soil organic carbon (SOC) (Nazir et al., 2024). However, implementing all of the required practices in the study is time-consuming, costly, and case-specific. These practices are also very variable in basics. For example, some recent key strategies introduced rehabilitating riparian buffer zones and using filter strips in the area between irrigated farmlands and water bodies (Jamshidi and Naderi, 2023b; Pongratz et al., 2021). In addition, practicing reduced tillage with cover crops and using best management practices (BMPs) in farms and LULCs were also recommended (Jamshidi et al., 2020). BMPs aim to reduce environmental impact through various methods. Vegetation filter strips, constructed wetlands, and hydroponic vegetation can effectively trap sediments and reduce nutrients, mainly total nitrogen (TN) and phosphorus (TP) (Jamshidi and Naderi, 2023a). Ponds, pits, and fencing stabilize sediments and control stormwater runoff. Terracing and tillage modification reduce erosion, while artificial aeration and pressurized drip irrigation enhance dissolved oxygen and manage nutrient levels. Fertilizer reduction and water recycling further improve efficiency, though chemical additives are not recommended due to potential hazardous effects. Developing local markets for trading both water (fresh or reclaimed water) and pollutants, for example, nutrients and C, simultaneously is a new trend that emphasizes using economic incentives for motivating farmers toward a more sustainable, self-organized, and cooperative option in farming (Imani et al., 2017; Jamshidi et al., 2014). This approach includes trading between wastewater treatment plants and farmers, yielding significant environmental and economic benefits without penalties or the need for changing LULC (Souri et al., 2025). Improving nutrient cycling and reducing chemical fertilizer application through composting and manure application is another implementation approach. However, assessing these practices with their local-global impacts requires a comprehensive modeling framework (Kumar et al., 2024; Meena et al., 2025a).

This chapter deals with the details of calculation methods, including the use of emission factors, and advanced simulation tools and models, such as the Soil and Water Assessment Tool for carbon (SWAT-C). SWAT-C integrates C cycling processes with soil erosion models to estimate C losses and transformations in basins (Zhang, 2018). Therefore SWAT-C is a new tool that can estimate total C emissions from farmlands on a basin scale. The SWAT-C model is used to simulate CO_2 and N_2O fluxes in diverse upland ecosystems. The model's C module has been tested at 16 eddy covariance flux towers (Yang and Zhang, 2016). The chapter aims to introduce SWAT-C for simulating long-term sediment and eroded C. This research is expected to enhance understanding of the lateral movement of C from cropland to water bodies, supporting future development and applications of the model to better understand C cycling across terrestrial and aquatic ecosystems.

Additionally, this chapter covers the life cycle impact assessment (LCIA) method, which categorizes environmental impacts, such as climate change and global warming, to provide a holistic view of an agricultural production's environmental footprint (Acero et al., 2016; Pradhan and Meena, 2023a). It addresses the significant contributions of agriculture to GHG emissions. LCA is used to account for GHG emissions and environmental impacts in agriculture, but there is no consensus on including soil C changes (Brandão et al., 2013). CF, a single-impact LCA, considers global warming potential (GWP) as the only impact category. The GWP measures the impact of gases over a 100-year period. CO_2 has a GWP of 1.0 (base/reference), CH_4 is 21 (up to 25), and N_2O is 298, indicating CH_4 and N_2O are much more potent than CO_2 (Ramachandra and Mahapatra, 2015).

Furthermore, this chapter introduced tools for developing reliable GHG inventories and covers international standards and protocols for GHG reporting, such as the GHG Protocol and ISO 14064, ensuring transparency and accuracy (Weng and Boehmer, 2006). Through real-world case studies, this chapter illustrates best practices and emerging trends in C emission measurement and reporting, making it an invaluable resource for environmental professionals, policymakers, and researchers dedicated to addressing these challenges.

Understanding the C cycle requires knowledge of GHGs and their categories. These gases are categorized based on their sources and impacts on the atmosphere, such as natural processes, industrial activities, and agricultural practices. Additionally, these parameters are considered in LCIA by methods such as the ReCiPe, which evaluates environmental impacts at both midpoint and endpoint levels, focusing on areas such as human health, ecosystem quality, and resource scarcity (Huijbregts et al., 2016). In the following sections, a detailed explanation of C cycles, with an introduction and classification of each pollutant and its impacts, is presented.

20.1.2 Carbon cycle and sequestration

The C cycle can be observed in several areas beyond terrestrial and aquatic ecosystems (Bockheim, 2010), as illustrated in (Fig. 20.1). The atmosphere plays a crucial role by storing and exchanging CO_2 with other reservoirs (Bachu, 2008). CO_2 is released into the atmosphere through processes such as respiration, the combustion of fossil fuels, and volcanic eruptions, and is absorbed by plants during photosynthesis and by the oceans through dissolution (Trabalka, 1985). The geosphere, which includes the Earth's crust and mantle, stores C in rocks and sediments, cycling it through processes such as weathering, volcanic activity, and the formation of fossil fuels (Ranaweera et al., 2023). The pedosphere, or soil layer, stores organic carbon (OC) from decomposed plant and animal matter. Microbial activity breaks down this organic matter (OM) and releases CO_2 back into the atmosphere. Soil management practices can significantly impact the amount of C stored in soils (Swift, 2001). The cryosphere, which encompasses ice caps, glaciers, and permafrost, also plays a crucial role in the C cycle. OC trapped in permafrost can be released as CO_2 and CH_4 when the permafrost thaws due to rising temperature, contributing to atmospheric GHG concentrations (Pradhan and Meena, 2023b; Ping et al., 2015). Understanding the C cycle in these various areas is essential for developing comprehensive strategies for climate change mitigation (CCM) and managing C emissions effectively.

Human activities have significantly altered the global C cycle over decades, leading to increased atmospheric CO_2 (1.46 W/m^2 since 1850) and temperature (0.15°C/decade since 1976) (Lal, 2004a). Natural processes alone cannot absorb all the excess CO_2 (Goeppert et al., 2012). Oceans and terrestrial ecosystems act as C sinks, but their capacity is limited and influenced by various factors (Falkowski et al., 2000). Secondary mechanisms and phenomena, such as global warming, can reduce the efficiency of these sinks (Luo, 2007). Understanding these changes is crucial for developing policies to mitigate future impacts. Key points of the C cycle include:

1. *Human impact:* Human activities, such as burning fossil fuels and deforestation, have significantly increased atmospheric CO_2 levels, disrupting the natural C cycle (Ayres et al., 1994).

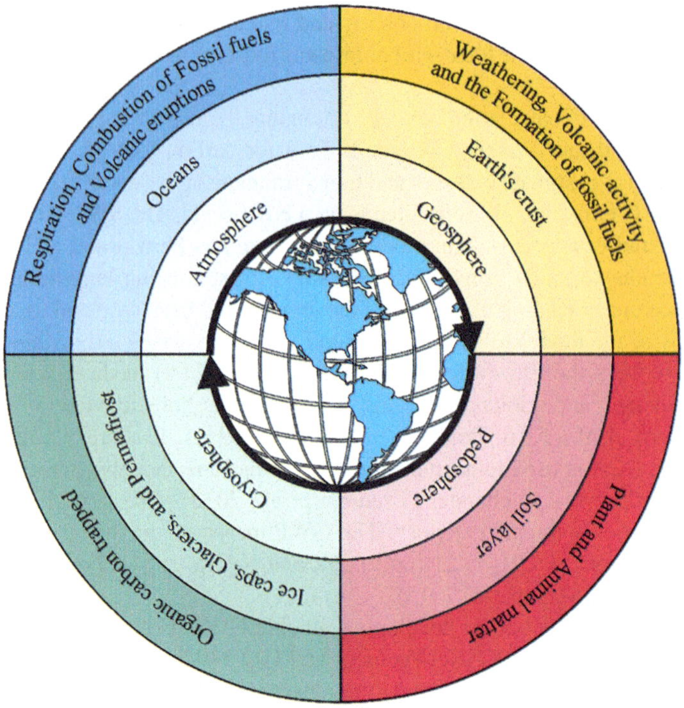

FIGURE 20.1 Carbon cycle areas by diffusion and absorption environment (by authors).

2. *Natural processes:* While natural processes can slow the increase of atmospheric CO_2, they cannot fully absorb the excess CO_2 produced by human activities (Fred Singer, 2008).
   - *Carbon reservoirs:* The oceans and terrestrial ecosystems act as C sinks, absorbing CO_2 from the atmosphere. However, their capacity to absorb CO_2 is limited and influenced by various factors, including temperature and nutrient availability (Keenan and Williams, 2018).
   - *Feedback mechanisms:* The efficiency of C sinks is affected by secondary mechanisms. For example, global warming can reduce the ocean's ability to absorb CO_2 by increasing water stratification and weakening algal activities (Heinze et al., 2015).
   - *Biogeochemical cycles:* The C cycle interacts with other biogeochemical cycles, such as the nitrogen and phosphorus cycles. Human activities have also altered these cycles, further complicating the C cycle dynamics (Ciais et al., 2014).
   - *Future projections:* Given current trends in energy use and population growth, atmospheric CO_2 levels are expected to continue rising. This will have significant implications for climate and ecosystem stability (Levy et al., 2004).

The Earth's layers significantly contribute to carbon sequestration (C_{seq}) through soil and geological formations (Lal, 2008). The detailed potential for soil organic C_{seq} in agricultural and forestry ecosystems within temperate cool climates is presented in (Table 20.1).

Soil absorbs and stores C via plant photosynthesis and decomposition, while sedimentary rocks store C through chemical weathering (Horwath, 2007). These processes reduce atmospheric CO_2 and aid in CCM. Soil C_{seq} offers a way to address climate change by revitalizing soils, boosting biomass production, purifying water, and compensating for fossil fuel emissions, as shown in (Fig. 20.2). Embracing these techniques is essential for tackling environmental challenges (Nazir et al., 2024). Soil C_{seq}, involving a global soil C pool of 2500 petagrams (Pg), could offset 0.4–1.2 Pg of C per year from fossil fuels, equivalent to 5%–15% of global emissions (Lal, 2004b; Petersen et al., 2013). C_{seq} is variable across different countries depending on climatic conditions, soil type, and other factors. For example, in the top 20 cm of soil depth, the annual soil C_{seq} rates are approximately 0.62, 0.23, 0.45, and 0.78 tons/ha in China, India, the USA, and Australia, respectively (Minasny et al., 2017). However, these rates can vary in different regions of each country.

Accurate quantification of C cycling is also essential for developing effective management strategies to reduce C emissions. Current estimations indicate a missing terrestrial C sink of 1.7–2.6 Pg of C per year for the 1980s and 1990s, respectively (IPCC, 2013). This missing sink is comparable to other major components of the global C budget, such as the net ocean uptake of approximately 2.3 Pg C per year and C emissions of about 1.0 Pg C per year from changing LULC

TABLE 20.1 Soil organic carbon sequestration in agricultural and forestry ecosystems (Lal, 2004a).

Land-use land-cover	Carbon sequestration rate (kg C/ha/year)
1. Agriculture and forestry	
a. Cropland	200–600
b. Pasture and grassland	200–400
c. Forest	100–200
2. Land use conversion	
a. Restoring severely degraded soils	200–800
b. Mineral soil reclamation	200–500
c. Wetland restoration	500-1000
3. Urban area	
a. Urban forestry	200–500
b. Recreational area	400–600

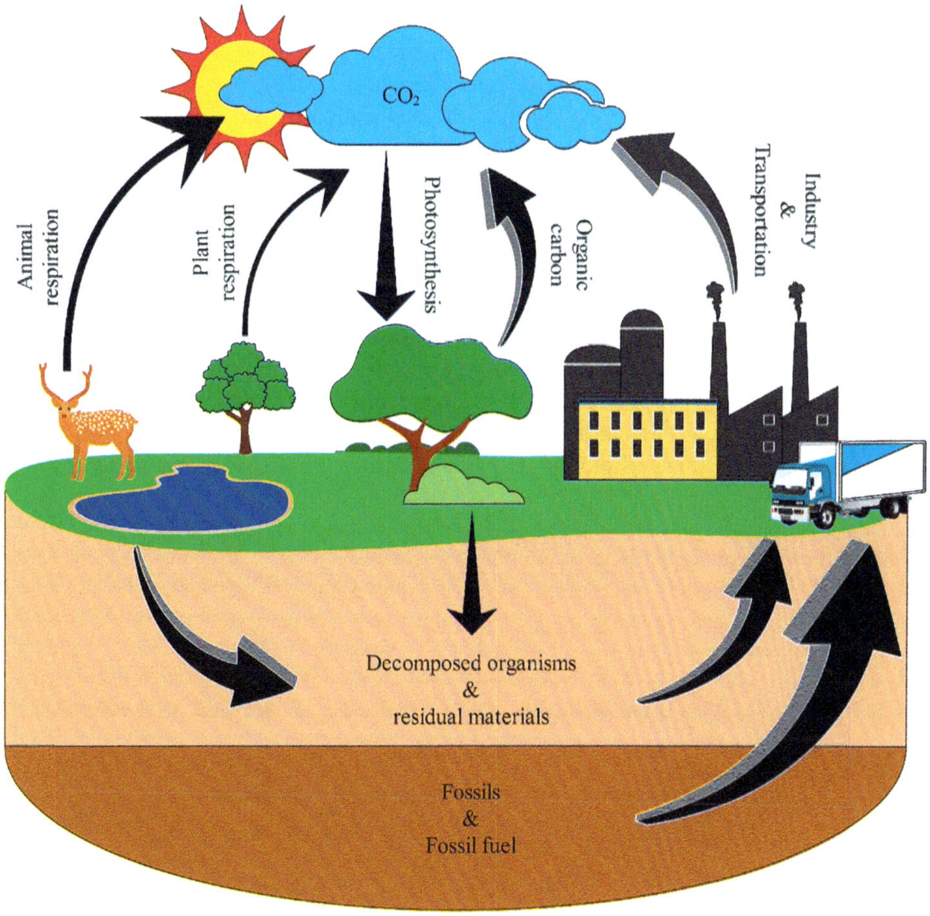

FIGURE 20.2 Carbon cycle with arrows indicating the movement of carbon between reservoirs (by authors).

(Ciais et al., 2014). Soil erosion processes may contribute to this missing sink, potentially inducing a global C sink of 0.12–1.5 Pg C per year (Quinton et al., 2010; Smith et al., 2001). Moreover, soil erosion redistributes SOC and other nutrients, with global estimates of SOC erosion fluxes ranging from 0.5 to 6.0 Pg C per year (Müller-Nedebock and Chaplot, 2015). Modern agricultural practices can restore SOC and slow CO_2 emissions. Soil C_{seq} offers a holistic approach to combat climate change by rejuvenating soils, enhancing biomass, purifying water bodies, and offsetting fossil fuel emissions.

20.1.3 Greenhouse gas classification and impacts

GHG plays a crucial role in the Earth's climate system by trapping heat in the atmosphere (Kweku et al., 2018). Understanding these gases and their sources is essential for addressing climate challenges, particularly in the context of agriculture, as given in Table 20.2, which is both a significant source and a potential storage of GHGs (Paustian et al., 2001).

The primary GHGs include CO_2, CH_4, N_2O, and fluorinated gases, along with other notable contributors such as ozone (O_3) and water vapor (H_2O), as introduced in the following:

1. *Carbon dioxide:* In agriculture, CO_2 is primarily released through soil cultivation, burning of crop residues, and the use of fossil fuels in farm machinery (West and Marland, 2002). It is a major contributor to the greenhouse effect, leading to global warming and climate change, which in turn affect crop yields and alter growing seasons. Additionally, deforestation for agricultural expansion also releases significant amounts of CO_2 (Raihan et al., 2022). Practices such as no-till farming and reforestation can help to mitigate CO_2 emissions by enhancing C_{seq} in soils and vegetation (Kane, 2015).
2. *Methane:* CH_4 is emitted from enteric fermentation in ruminant animals (like cows and sheep), rice paddies, and the decomposition of OM in anaerobic conditions (Broucek, 2014). It is a potent GHG with a global warming potential much higher than CO_2, significantly contributing to the greenhouse effect and climate change (Jain et al., 2015; Pradhan et al., 2023). CH_4 emissions are also associated with manure management practices and the use of organic fertilizers (Chadwick et al., 2011). Implementing better manure management techniques, such as anaerobic digestion, and optimizing livestock diets can reduce CH_4 emissions (Yao et al., 2020).
3. *Nitrous oxide:* N_2O emissions in agriculture come from the application of chemical fertilizers, manure management, and the cultivation of nitrogen-fixing crops (Snyder et al., 2009). It has a high global warming potential and contributes to the depletion of the ozone layer, affecting soil health and water quality (Smith, 1997). The overuse of nitrogen-based fertilizers can lead to increased N_2O emissions, which also contribute to eutrophication in water bodies (Abrol et al., 2007). Precision farming techniques, such as controlled-release fertilizers and nitrification inhibitors, can help to reduce N_2O emissions (Shoji et al., 2001).
4. *Fluorinated gases:* This category includes gases such as chlorofluorocarbons, hydrochlorofluorocarbons, hydrofluorocarbons, sulfur hexafluoride, and nitrogen trifluoride (Kamyab et al., 2024; Lynch et al., 2021; Singh et al., 2023). Though not commonly linked to agriculture, these gases can be emitted during certain agricultural activities, particularly those involving freezing processes and pesticide application (Weinstein and Davison, 2004). They have a very high global warming potential and can remain in the atmosphere for a long time, significantly contributing to climate change (Sovacool et al., 2021). Fluorinated gases are often used in cooling systems and as propellants in

TABLE 20.2 Key greenhouse gases and their major agricultural sources (Paustian et al., 2001).

GHG	Major agricultural sources
Carbon dioxide	Soil cultivation, crop residue burning, fossil fuel use, deforestation
Methane	Enteric fermentation, rice paddy fields, organic matter decomposition, manure management
Nitrous oxide	Chemical fertilizers, manure, nitrogen-fixing crops
Fluorinated gases	Refrigerants, cooling systems, pesticide applications
Ozone	Pesticides, fertilizers, VOCs, and NO_x reactions in sunlight
Water vapor	Irrigation, deforestation (indirect influences)

agricultural sprays (Sicard and Baker, 2020). Transitioning to alternative refrigerants with lower global warming potential and improving the efficiency of cooling systems can help to mitigate emissions of fluorinated gases.

5. *Ozone:* While O_3 in the stratosphere protects life on Earth by blocking harmful ultraviolet radiation, ground-level ozone is a GHG that can be harmful to crops and human health (Pleijel, 2000). It is formed by chemical reactions between volatile organic compounds and nitrogen oxides in the presence of sunlight (Swamy et al., 2012). Agricultural activities, such as the use of pesticides and fertilizers, can contribute to the formation of ground-level ozone.
6. *Water vapor:* H_2O is the most abundant GHG and plays a crucial role in the natural greenhouse effect (Dalal and Allen, 2008). However, its concentration in the atmosphere is primarily controlled by natural processes, such as evaporation and condensation. Human activities, such as irrigation and deforestation, can indirectly influence water vapor levels by altering local climates and hydrological cycles.

Understanding the various GHGs and their impacts is essential for developing effective strategies to mitigate climate change. By addressing the sources and effects of these gases, particularly in agriculture, as shown in (Fig. 20.3) (Ourworldindata, 2024a), it is possible to reduce their emissions and enhance the sustainability of agricultural practices.

Human land use covers over three-quarters of the available land surface (Luyssaert et al., 2014). Changes in land uses and their management, known as LULC changes, can significantly affect the local to global climate. Understanding these effects is crucial for identifying the drivers of climate change and evaluating the impact of humans. LULC changes toward a sustainable vegetated area with greener applications can help to mitigate global warming and adapt local climates to reduce climate change impacts (Pongratz et al., 2021). It is especially important in socioeconomic scenarios aligned with the Paris Agreement, which require large-scale CO_2 removal through methods such as afforestation, biomass plantations with C capture and storage, farmers' cooperation, and social education and awareness about the cause-and-effect relationship between sustainable agriculture and climate change adaptation (Babaeian et al., 2023). Additionally, non-GHG effects of LULC changes, such as energy and water exchanges between land and atmosphere, are increasingly recognized as significant.

Land uses and changes also play a critical role in the C cycle by influencing C storage and release in terrestrial ecosystems, as shown in (Fig. 20.4). Managed forests (red part in Fig. 20.3) and managed wetlands (purple part in Fig. 20.4) are connected to croplands (blue part in Fig. 20.4) in two ways: by agricultural re-growth and conversion, respectively. This land use provides an impact on the C cycle by either sequestering C in vegetation and soil or releasing it through agricultural activities. These processes play a crucial role in regulating atmospheric C levels and CCM. LULC changes also significantly affect SOC dynamics and net GHG emissions, such as annually sequestering approximately 100 kg of C per hectare when converting from conventional to organic soybean production (Petersen et al., 2013). Significant SOC losses are reported for LULC changes from forest and grassland to cultivated land, while the reverse usually increases SOC (Shrestha et al., 2014). Changing soil C varies interannually due to land use and climate interactions (Zhang et al., 2010). Many LCA studies ignore soil C dynamics, highlighting the need for standardized methods to account for soil C in agricultural LCAs (Arzoumanidis et al., 2014; Petersen et al., 2013).

Data on global GHG emissions from agriculture and land use change are presented in (Fig. 20.5), focusing on three dominant key gases: N_2O, CH_4, and CO_2 (Ourworldindata, 2024b). Over the past two decades (from 2000 to 2022), the

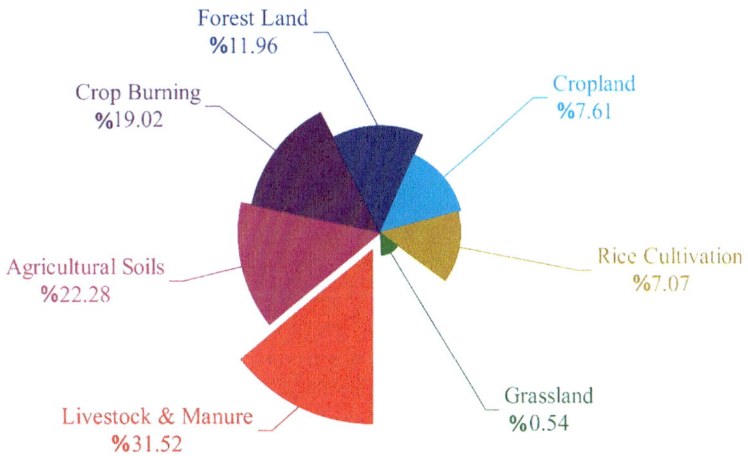

FIGURE 20.3 Global GHG emissions by agriculture, forestry, and land use (Ourworldindata, 2024a). *GHG*, greenhouse gases.

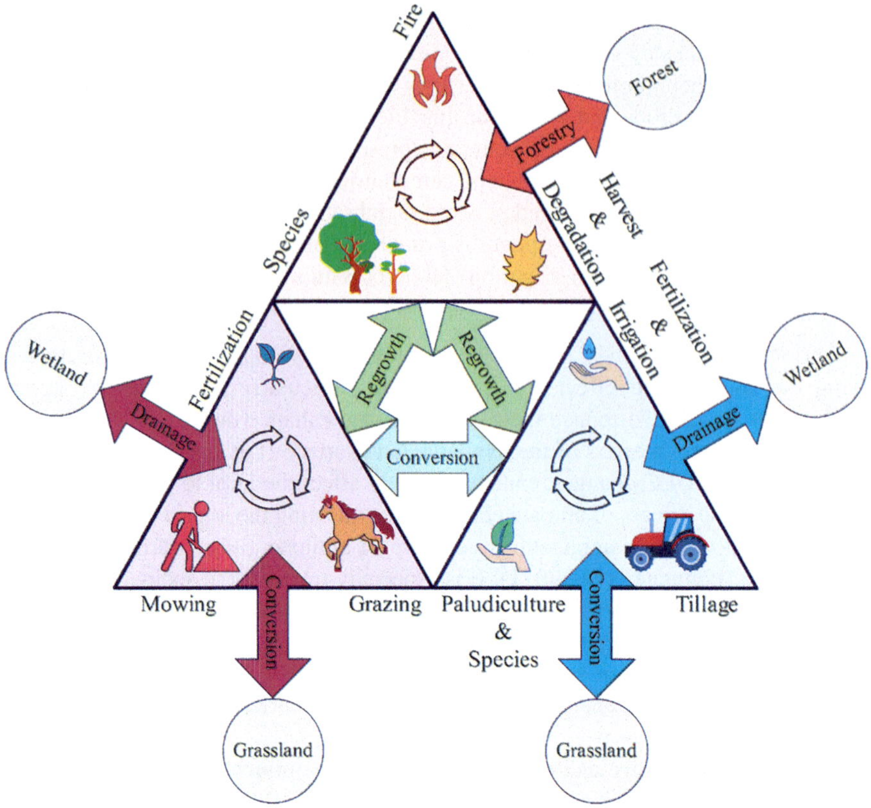

FIGURE 20.4 Land-use practices, including related land-cover changes and land management (by authors).

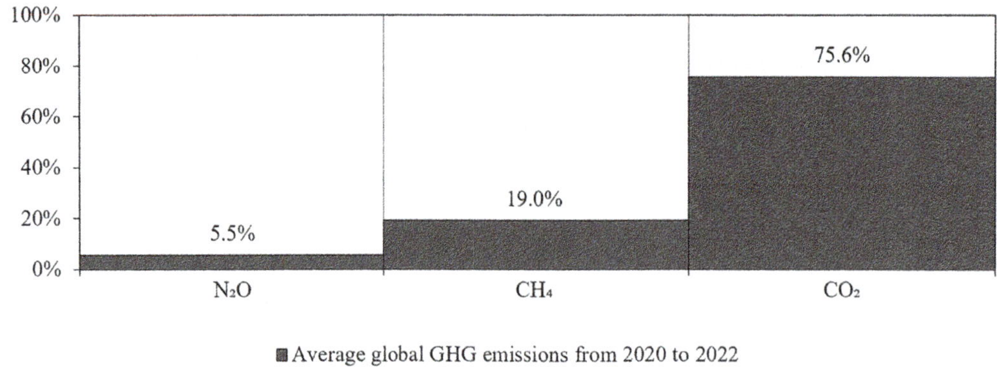

FIGURE 20.5 Global GHG emissions from all sources, including agriculture and land-use change, 2020–2022 (Ourworldindata, 2024b). *GHG*, greenhouse gases.

percentages of these gases have fluctuated, reflecting changes in agricultural practices, land use, and environmental policies. N_2O emissions, primarily from synthetic fertilizers and manure management, have shown slight variations, generally averaging around 5.5%, indicating the ongoing use of nitrogen-based fertilizers and the importance of managing agricultural soils to reduce emissions. CH_4 emissions, mainly from enteric fermentation in livestock and paddies, have remained relatively stable, averaging around 19.0%, suggesting consistent livestock populations and rice cultivation practices, and highlighting the need for improved CH_4 management strategies in agriculture. CO_2 emissions, resulting from soil cultivation, deforestation, and the use of fossil fuels in farm machinery, have also shown minor fluctuations, averaging around 75%, underscoring the significant impact of agricultural activities on CO_2 emissions and the importance of sustainable land management practices to mitigate these emissions. Overall, (Fig. 20.4) illustrates the critical role of agriculture and land use in global GHG emissions and the need for continued efforts to adopt sustainable practices to reduce these emissions.

According to the results, CO_2 constitutes a significant part of GHG emissions. Therefore, calculating C equivalent emissions using methods such as SWAT-C or LCIA is essential to assess and manage the environmental impact of agricultural practices accurately.

20.1.4 International rules and standards

According to the reports from the Conference of the Parties on its 23rd session held in Bonn (6–18 November 2017) and the 27th session held in Sharm el-Sheikh (6–20 November 2022), comprehensive visions have been published on C control in agriculture, which can be considered by applications such as SWAT-C. These reports propose various strategies for reducing C emissions in the agricultural sector (UNFCCC Secretariat, 2018; Sherif, 2023):

1. Enhancing soil C, health, and fertility in various systems, along with effective water management, is crucial.
2. Optimizing nutrient use and manure management is essential for sustainable agriculture.
3. Climate change's socioeconomic impacts and food security challenges in agriculture are critical.
4. Existing tools for adaptation assessment need refinement. Developing country-specific tools and sharing best practices is vital. Science, technology, and capacity-building are key to effective data collection and assessment.
5. Addressing soil and water management issues contextually can boost productivity, food security, and C_{seq}.
6. Effective soil and nutrient management, including organic fertilizers and improved manure management, is fundamental for climate-resilient, sustainable food production and global food security.

The above mainstream policies and regulations imply that reducing GHGs from agriculture and climate change adaptation should not be an obstacle to food production and jeopardize food security. However, farmers are not allowed to use chemical fertilizers or emit GHGs from farmlands freely, and this puts their lives at risk. Therefore this kind of environment-food nexus thinking emphasizes the development, assessment, and implementation of engineered solutions to achieve greener and more sustainable agricultural production.

We know that global emissions pathways to limit warming to 1.5°C or well below 2°C, as per the Paris Agreement, require significant reductions in agricultural GHG (CH_4 and N_2O) and achieving net zero CO_2 (Leahy et al., 2020). Without reducing agricultural emissions, more rapid cuts in CO_2 emissions would be necessary, jeopardizing the 1.5°C target. Current models suggest pricing agricultural emissions as a solution. Nevertheless, countries are more or less reluctant to implement mandatory C pricing for agriculture, and consequently, they postpone applying stringent climate policies to this sector (Leahy et al., 2020). Environmental and climate policies can incentivize the production of nutritious foods, but trade-offs may undermine their affordability (FAO; IFAD; UNICEF; WFP; WHO, 2022). Addressing environmental externalities employing pricing mechanisms and economic levers, such as C taxes and credits, could alter food prices, particularly when policy-makers encounter several uncertainties (technical, economical, and social) and the risk of decision-making is high. Under these conditions, practical implementation is challenging and requires both global agreements and socio-hydrological modeling approaches. It is obvious that effective mitigation of agricultural GHG requires realistic policy pathways and comprehensive efforts beyond government price policies, addressing scalability and efficacy challenges.

One of the policies that can support these efforts is the adoption of international standards for GHG quantification. ISO 14064 is a global standard developed by the International Organization for Standardization (ISO) for the quantification, reporting, and verification of GHG emissions. Completed in March 2006, it consists of three parts (Wintergreen and Delaney, 2007):

1. Part one focuses on the quantification and reporting of GHG emissions and removals at the organizational level, providing guidelines for creating GHG inventories, including setting boundaries, quantifying emissions, and reporting.
2. Part two addresses the quantification and reporting of GHG emission reductions from specific project activities.
3. Part three establishes a process for the validation and verification of GHG assertions, including organizational inventories, outlining principles and procedures for independent verification.

Key aspects include principles regarding the relevance, completeness, consistency, accuracy, and transparency of measured or estimated data, defining organizational and operational boundaries, a detailed quantification process, and comprehensive reporting requirements. ISO 14064 is compatible with the GHG Protocol, providing a complementary framework for managing GHG emissions. Both private and public sectors use it to develop credible and consistent GHG inventories, serving as a foundation for policymakers to build GHG mitigation programs and helping organizations improve the consistency and credibility of their GHG reporting. Additionally, ISO 14064 supports the understanding and management of the C cycle, which is crucial for addressing climate change and ensuring sustainable environmental practices.

20.1.5 Related environmental indicators

20.1.5.1 Environmental performance index

In evaluating the impact of CO_2 emissions, it is essential to consider multiple criteria to gain a comprehensive understanding of environmental performance. The environmental performance index (EPI) provides a framework for this evaluation, focusing on three key criteria: climate change, environmental health, and ecosystem vitality, introduced as follows:

1. *Climate change:* This criterion focuses on the efforts and effectiveness of land in CCM. It includes indicators such as GHG emissions, renewable energy adoption, and policies aimed at reducing CF. The goal is to assess how well a country is managing its contributions to global climate change and its progress towards international climate targets.
2. *Environmental health:* This criterion measures the impact of environmental factors on human health. This includes air quality, access to clean water and sanitation, and exposure to harmful pollutants. The criterion assesses how effectively a land protects its population from environmental risks that can lead to health issues, thereby ensuring a healthier living environment.
3. *Ecosystem vitality:* Ecosystem vitality assesses the health and sustainability of natural ecosystems. It includes indicators related to biodiversity, habitat protection, sustainable agriculture, and forest management. This criterion assesses a country's ability to conserve its natural resources and biodiversity, thereby ensuring the long-term sustainability of its ecosystems.

These criteria reflect a comprehensive approach for evaluating environmental performance, highlighting the importance of addressing climate change, protecting human health, and maintaining ecosystem health.

For assessing the impacts of CO_2 emissions, three criteria can be weighted based on the EPI. According to (Fig. 20.6), these impact criteria have evolved over different periods in response to global conditions. For instance, climate change and its associated effects, such as emissions of CO_2, N_2O, and CH_4, were previously categorized under ecosystem vitality (before 2021). However, the EPI 2022 introduced climate change as a separate category with a weighting of 38% among the other two main classes. In 2024, the weightings for the three criteria: climate change, environmental health, and ecosystem vitality are 30%, 25%, and 45%, respectively, in which the weight of climate impact is alleviated, and changes reflect a shift in the management priorities for these issues.

The framework for climate change organizes various components and their respective contributions to overall GHG emissions, shown in (Fig. 20.7) (EPI, 2024). CO_2 is the largest contributor, accounting for 27% of emissions. CH_4 follows with 10%, while N_2O contributes 3%. Fluorinated gases (F) add another 7%, and black C accounts for 5%. Land use, LULC, and forestry contribute 3% to the emissions. Additionally, GHG emission intensity and GHG emission per capita both stand at 20% individually. Projections for GHG emissions in 2050 and budget use are estimated at 5%. This structured approach helps in understanding the relative impact of each subparameter based on climate change, facilitating targeted strategies for mitigation and adaptation.

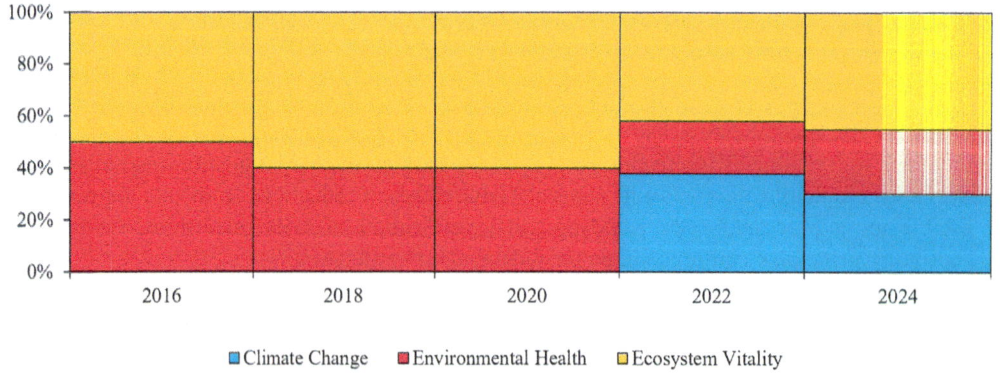

FIGURE 20.6 Three main policy objectives and their weights in the environmental performance index (EPI) during the time (Chandrasekharan et al., 2020; EPI, 2024; Hsu et al., 2016).

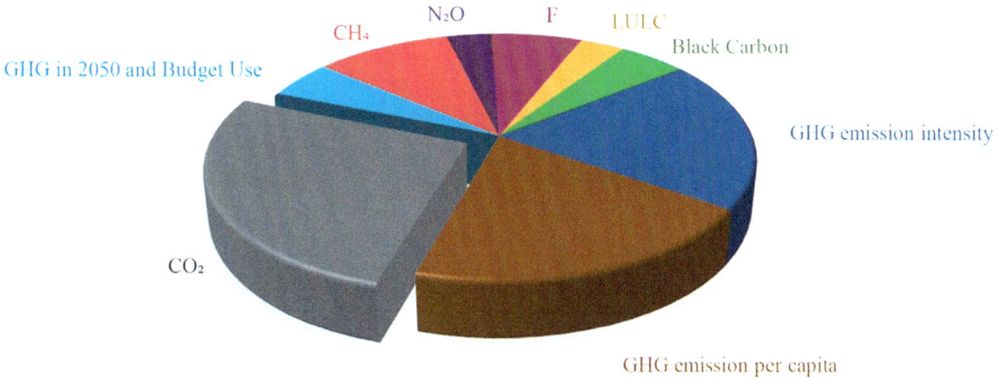

FIGURE 20.7 Climate change category of the environmental performance index (EPI) framework 2024 (EPI, 2024).

EPI subparameters are not only important for annual national reporting but also recent literature has referred to their proportional weights for similarly weighting the main LCIA categories (Jamshidi and Naderi, 2023a; Jamshidi et al., 2024a). Here, the LCA endpoint coefficients driven by ReCiPe used weightings based on EPI to accumulate the quantified equivalent damages into a unified single value of the impaired environment.

20.1.5.2 Carbon footprint

Calculating CF involves estimating the total amount of GHG, primarily CO_2, that is produced through activities (Pandey et al., 2011). Soil plays a crucial role in C_{seq}, which is the process of capturing and storing atmospheric CO_2 (Lal, 2008). Over 100 years, CH_4 has a GWP of about 28–36 times that of CO_2 (Zhao et al., 2020). This means that 1 kilogram of CH_4 is equivalent to 28–36 kilograms of CO_2. N_2O has a GWP of approximately 298 times that of CO_2 over the same period (Gao et al., 2014a). Therefore 1 kilogram of N_2O is equivalent to 298 kilograms of CO_2. By definition, the GWP of CO_2 is 1, so 1 kilogram of CO_2 is equivalent to 1 kilogram of CO_2. For example, if a farm emits 2 kilograms of CH_4 and 1 kilogram of N_2O, the total CO_2 equivalent emissions would be calculated as follows:

$$CH_4\text{: } 2\text{ kg} \times 28\text{–}36 = 56\text{–}72\text{ kg } CO_2 \text{ equivalent}$$

$$N_2O\text{: } 1\text{ kg} \times 298 = 298\text{ kg } CO_2 \text{ equivalent}$$

Thus the total emissions would be between 354 and 370 kilograms of CO_2 equivalent. CF is determined by summing the GWP of each product (i) (Pandey and Agrawal, 2014). Therefore, CF based on the CO_2 equivalent is considered as given in Eq. (20.1) (Cheng et al., 2015b):

$$CF = \sum_{i=1}^{n} [(CH_4 \times 25) + (N_2O \times 298) + (CO_2 \times 1)]_i \quad (20.1)$$

The final representation of the CF of product for agricultural systems can be expressed as either spatial (area) or yield-scaled CF, using the following Eq. (20.2) (Cheng et al., 2015a):

$$CF_P = \frac{CF}{Yield} \quad (20.2)$$

Accurate CF calculation requires comprehensive data on material quantities, activities, and emission factors throughout a product's life cycle (Gao et al., 2014b). This ensures all inputs, outputs, and waste are accounted for without omissions.

20.2 Greenhouse gas equivalent assessment methodologies

20.2.1 Soil and water assessment tool-carbon

The SWAT-C is an improved version of the SWAT model, specifically designed to simulate C dynamics within watersheds (Fig. 20.8) (Zhang et al., 2013). It focuses on soil organic matter (SOM) and C cycling, making it valuable for assessing the impacts of land management practices on C_{seq} and CO_2 emissions. SWAT-C incorporates terrestrial C algorithms from models like CENTURY (Zhang et al., 2013).

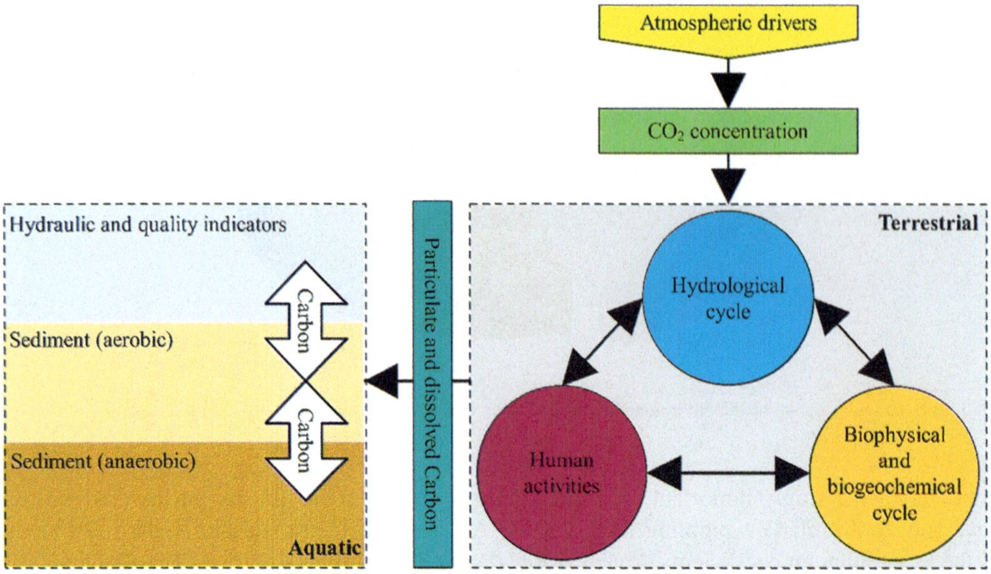

FIGURE 20.8 Integrating terrestrial and aquatic sciences in the SWAT-C model (by authors). *SWAT-C*, soil and water assessment tool for carbon.

SWAT-C represents SOM with five pools: metabolic litter (easily decomposable), structural litter (recalcitrant), active microbial biomass (short turnover), slow humus (medium turnover), and passive humus (long turnover) (Zhang et al., 2013). The decomposition and transformation of these SOM pools are influenced by factors such as soil temperature, moisture, tillage, oxygen availability, and soil texture. This model also simulates plant growth processes using the principles of radiation-use efficiency, where daily gains in plant biomass are affected by environmental factors such as vapor pressure deficits, atmospheric CO_2 concentrations, water, and nutrient availability.

Additionally, SWAT-C calculates eroded SOC by multiplying the amount of eroded soil by the concentration of SOC and its enrichment ratio. Smaller particles (e.g., clay) are more easily transported to streams, resulting in a sediment load with a higher proportion of clay-sized particles. Since SOM attaches to clay particles, the sediment load has a higher concentration of OC than the soil surface layer. In the SWAT-C model, the enrichment ratio is estimated using the sediment concentration in surface runoff (Zhang, 2018).

This helps in understanding the transportation and loss of C within the watershed during different farm practices or variations in basin LULC. In general, SWAT-C can provide a comprehensive framework, besides water and nutrient simulation at the basin scale by the SWAT model, for modeling C dynamics, helping to develop strategies for enhancing C_{seq} and reducing CO_2 emissions.

The C cycle in the soil is influenced by three key layers: (1) the surface layer (organic layer), (2) the topsoil, and (3) the subsoil layers as shown in (Fig. 20.9) Each layer plays a distinct role as detailed in (Table 20.3) in the production and emission of GHG, influenced by various factors such as SOM composition, temperature, and water dynamics (Ontl and Schulte, 2012).

In the organic layer of the soil, only active SOM, such as metabolic litter and microbial biomass in lignin or nonlignin structures, is present, as shown in (Fig. 20.10) (Zhang et al., 2013). The SWAT-C model includes a residue pool consisting of lignin, nonlignin, and metabolic residue, each with its own decomposition rate. Approximately 10 millimeters (mm) below the surface, in the organic layer, there are both active SOM and passive structures such as slow humus and passive humus. Residue dynamics occur at the soil surface and in the top 10 mm layer. SOM is simulated as microbial, slow, and passive pools, each with distinct turnover rates. The microbial pool is present in all soil layers, while the slow and passive pools are absent in the top 10 mm. The decomposition of residue and mineralization of SOM in SWAT-C depend on factors such as lignin content, soil temperature, texture, moisture, tillage effects, and oxygen content. These components, along with temperature and water, play a significant role in the production and emission of CO_2. The primary factor driving the increased transfer of this gas in the organic layer is surface runoff. Interestingly, the type of soil affects the movement of CO_2. An increase in clay soil facilitates the transfer of CO_2 to other layers, while granular soils, such as sand and gravel, promote the transfer of CO_2 within the layer. In the topsoil layer, which lies below the organic layer, both active and passive SOM are present. Lateral flows are the main factor in the increased transfer of CO_2 from this layer to other layers. This transfer continues in the subsoil layer until the underlying water flow carries away the CO_2.

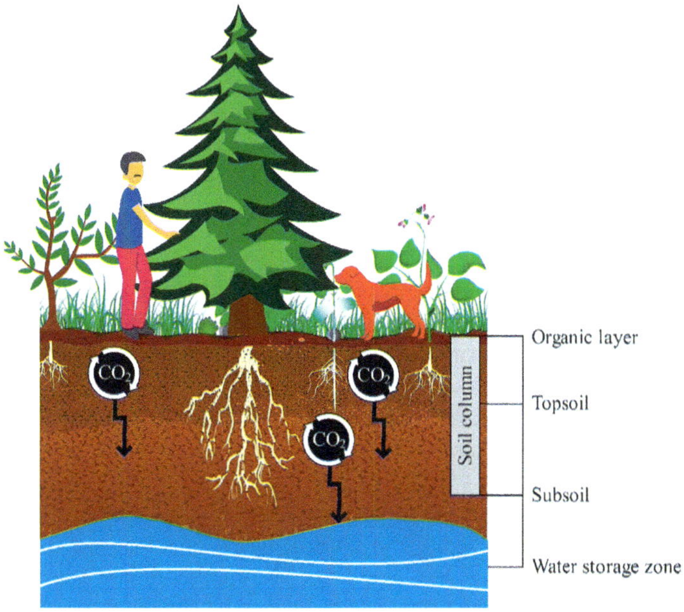

FIGURE 20.9 The role of soil layers in the carbon cycle and CO_2 emission dynamics (by authors).

TABLE 20.3 Soil layers and their role in carbon dynamics (Ontl and Schulte, 2012).

Soil layer	SOM composition	Key factors
Organic layer	Active SOM (metabolic, microbial)	Surface runoff drives CO_2 transfer
Topsoil	Active and passive SOM	Lateral flows move CO_2 between layers
Subsoil	Passive SOM	CO_2 is transported by water flow

SOM, soil organic matter.

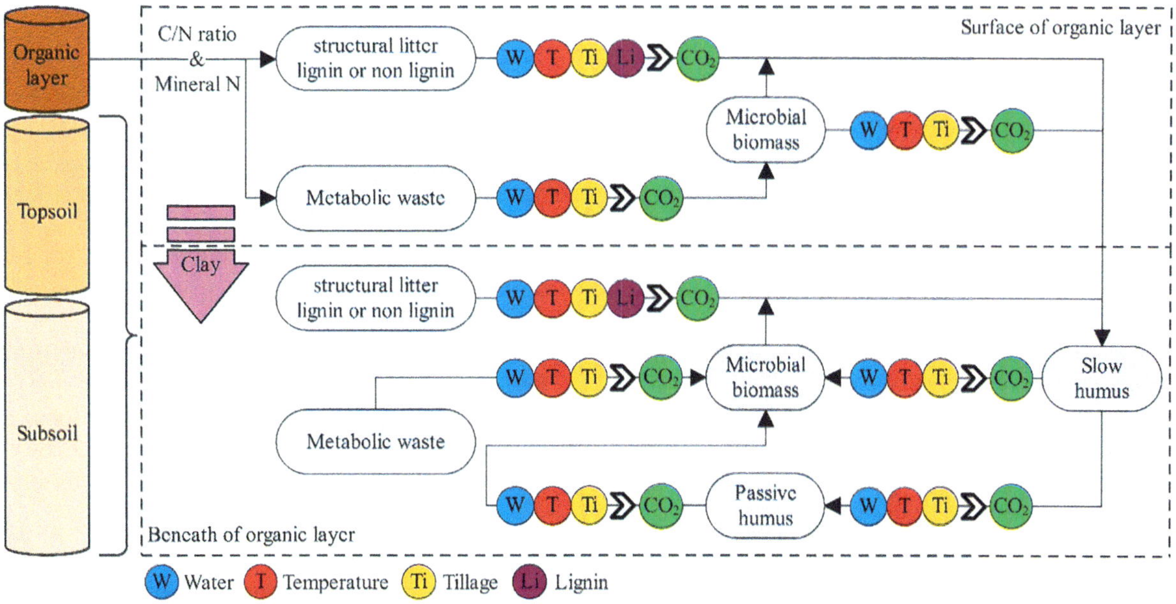

FIGURE 20.10 Schematic of SOM dynamics simulated in the SWAT-C model (by authors). *SOM*, soil organic matter; *SWAT-C*, soil and water assessment tool for carbon.

The model calculates residue composition and lignin content based on plant age, and all decomposition processes result in the transformation of C and nitrogen, ultimately releasing CO_2 (Haney et al., 2018). The SWAT-C model successfully simulates daily CO_2 fluxes except during rewetting periods. SWAT-C was tested by comparing simulated results to corn and soybean crop yields across the U.S. Midwest, showing good performance in predicting long-term average crop production where detailed management data were available (Zhang et al., 2013). SWAT-C effectively models C dynamics through detailed simulation of residue decomposition and SOM mineralization, influenced by various environmental and management factors.

The SWAT-C model simulates how C moves and changes within a basin. Here is a simple overview of how it works (Zhang et al., 2013):

1. *Residue pools:* SWAT-C includes different plant residues, such as lignin, nonlignin, and metabolic residue, each breaking down at different rates. These residues are found at the soil surface and in the top 10 mm layer (organic layer).
2. *SOM pools:* SOM is divided into three types: microbial, slow, and passive pools, each with its own turnover rate. The microbial pool is present in all soil layers, while the slow and passive pools are not found in the top 10 mm.
3. *Decomposition and mineralization:* The breakdown of residues and SOM depends on factors like lignin content, soil temperature, texture, moisture, tillage effects, and oxygen levels. These processes transform carbo and nitrogen, ultimately releasing CO_2.
4. *Carbon fluxes:* SWAT-C simulates daily CO_2 emissions, considering various environmental and management factors. The model calculates residue composition and lignin content based on plant age, and all decomposition processes contribute to C dynamics.
5. *Simulation and validation:* SWAT-C has been tested by comparing its simulated results to crop yields across different regions, such as corn and soybeans. This helps ensure the model's accuracy in predicting long-term average crop production and C dynamics.

By integrating these components, the SWAT-C model effectively models the C cycle within a basin, providing insights into how different factors influence C storage and emissions. This model is adept at simulating the dynamics of water, sediment, and agricultural chemicals within extensive and intricate watershed systems (Zhang, 2018; Zhang et al., 2013). One of its primary functions is to estimate both eroded C yield and sediment yield. Eroded C yield pertains to the C lost from soil due to erosion. When soil particles are dislodged and carried by water or wind, they transport OC along with them. SWAT-C can model this process and provide estimates of C loss in kilograms per hectare (kg/ha), which is essential for understanding the effects of soil erosion on C cycling and storage within ecosystems. Sediment yield quantifies the total sediment transported from land surfaces to water bodies, including soil particles and associated nutrients and OM. SWAT-C calculates sediment yield in kg/ha, assisting in the evaluation of soil erosion's impact on water quality and land productivity. By integrating C cycling with soil erosion processes, SWAT-C offers significant insights into how various land management practices and environmental conditions influence C and sediment losses. This is crucial for developing strategies to reduce soil erosion and enhance C_{seq} in soils. Another output can be changed to SOC, which indicates the level of C fixation in soil layers. This means we have C transferred through sediment and absorbed into the soil. This process can be related to GHG dynamics, as increased soil C_{seq} can help to mitigate GHG emissions.

20.2.2 Life cycle impact assessment

LCIA is indeed the third phase of the LCA methodology (Hauschild, 2019). This phase involves evaluating the environmental impacts of a product or service throughout its entire life cycle, such as climate change and toxicity. The ISO standard for LCA is ISO 14040:2006 (Arvanitoyannis, 2008; ISO, 2006a). This standard outlines the principles and framework for conducting LCAs. The four main stages of an LCA are Goal and Scope Definition, which establishes the purpose and boundaries of the assessment; life cycle inventory (LCI), which compiles an inventory of relevant inputs and outputs; LCIA, which evaluates the potential environmental impacts associated with the inputs and outputs identified in the LCI; and Interpretation, which analyzes the results to make informed decisions and recommendations (Finnveden et al., 2009; ISO, 2006a).

This text compares four widely used LCIA methods: Centrum voor Milieukunde Leiden (CML), ReCiPe 2016, Environmental Footprint (EF), and TRACI (Tool for the Reduction and Assessment of Chemicals and Other Environmental Impacts) (Golsteijn and Vieira, 2020; Frontera et al., 2020), highlighting their approaches, focus areas, and applications for GHG impact assessment as detailed in (Table 20.4) (Agency, 2024; Bare et al., 2012; European Commission, 2024; Guinée, 2002; Huijbregts et al., 2016). Each method offers unique insights and is suited for different types of environmental assessments as follows:

1. *CML:* This method is a midpoint-oriented LCIA method developed by Leiden University (Guinée, 2002). It focuses on specific environmental impacts such as global warming, acidification, and eutrophication. By using characterization

factors, CML translates emissions into potential environmental impacts. This method is commonly used in Europe and is particularly suitable for studies that require detailed midpoint impact categories.

2. *ReCiPe:* This method is a comprehensive LCIA method that provides both midpoint and endpoint assessments (Huijbregts et al., 2016). Midpoint categories are similar to those in CML, focusing on specific environmental impacts. Endpoint categories, however, translate these impacts into broader areas of protection, such as human health, ecosystem quality, and resource scarcity. ReCiPe links midpoint impacts to endpoint damages, offering a holistic assessment. It is used globally and is suitable for studies needing detailed and comprehensive impact assessments.
3. *EF:* This method is part of the European Commission's initiative to measure and communicate the environmental performance of products and organizations (European Commission, 2024; Golsteijn and Vieira, 2020). It includes both product EF and organization EF methods. EF aims to provide a standardized approach for assessing environmental impacts across various categories, such as climate change, resource use, and ecosystem quality. EF is a midpoint method and is suitable for comprehensive environmental performance assessments.
4. *TRACI:* This method is developed by the EPA TRACI is used for assessing the environmental impacts of chemicals and other factors (Agency, 2024; Bare et al., 2012). It provides characterization factors for various impact categories, including ozone depletion, climate change, acidification, eutrophication, smog formation, human health impacts, and ecotoxicity. TRACI is widely used in the United States for LCIA, industrial ecology, and sustainability metrics.

The LCIA method is recommended for its ability to use various footprints, such as CF, biodiversity, and EF, to account for cumulative environmental impacts (Jamshidi and Naderi, 2023a). Development of social LCIA indices, integrating social factors such as farmers' living conditions, employment, social fairness, and public commitment to sustainability into environmental assessments. Mentioned applications, such as ReCiPe, in environmental assessments emphasize the importance of using a variety of midpoint indices for a comprehensive evaluation of environmental damage. These factors introduced in (Table 20.4), which are used in the calculation of LCIA, are explained below:

1. *The midpoint coefficient (M_C):* M_C for climate change was commonly represented by the GWP. GWP measured in ReCiPe the additional radiative forcing over time (20, 100, or 1000-year time horizon) caused by emitting 1 kg of a GHG compared to the same effect from 1 kg of CO_2 (Huijbregts et al., 2016). This resulted in a GWP value expressed in kg CO_2 equivalents per kg of GHG. The latest Intergovernmental Panel on Climate Change (IPCC) report in 2013 provided GWP values for 20 years (individualist), 100 years (Hierarchist), and 1000 years (egalitarian) directly.

TABLE 20.4 Comparison of LCIA Methods CML, ReCiPe, EF, and TRACI factors for GHG (Agency, 2024; Bare et al., 2012; European Commission, 2024; Guinée, 2002; Huijbregts et al., 2016).

Substance	Method	Midpoint coefficient (M_C)	Midpoint to endpoint (E_C)	Midpoint (M_P)	Endpoint (E_P)
		kg CO_2 eq./kg GHG	Species.year/kg CO_2 eq.	kg CO_2 eq./person	Species.year/person
CO_2	CML	1.0	-	8.46E+03	-
CH_4		21			
N_2O		310			
CO_2	ReCiPe	1.0	2.80E-09	7.99E+03	2.24E-05
CH_4		34			
N_2O		298			
CO_2	EF	1.0			
CH_4		29.8	-	7.55E+03	-
N_2O		273			
CO_2	TRACI	1.0	-	-	-
CH_4		25			
N_2O		298			

CML, Centrum voor Milieukunde Leiden; *EF*, Environmental Footprint; *GHG*, Greenhouse gases; *LCIA*, life cycle impact assessment; *TRACI*, Tool for the Reduction and Assessment of Chemicals and Other Environmental Impacts.

Under the individualist perspective, there is complete societal adjustment, resulting in minimal effects on human health and low biodiversity decline. In contrast, the hierarchical viewpoint involves moderate societal adjustment, leading to average effects on human health and moderate biodiversity decline. Meanwhile, the egalitarian approach assumes no societal adjustment, which results in significant effects on human health and high biodiversity decline (Goldstein et al., 2016). The GWP for the egalitarian time horizon was calculated differently. Table 20.4 presented M_C values derived from the hierarchist perspective for CO_2, NH_4, and N_2O.

2. *The midpoint to endpoint (E_C):* E_C factors for human health damage, terrestrial ecosystem damage, and freshwater ecosystem damage, along with the three cultural perspectives, were analyzed (Huijbregts et al., 2016). The initial step in the E_C model involved converting time-integrated radiative forcing into time-integrated temperature increase. For human health damage due to climate change, the increased risks of diseases such as malnutrition, malaria, and diarrhea, as well as higher flood risks, resulted in additional health damage measured in disability-adjusted life years (DALY). These effects did not impact all regions equally. Table 20.4 highlighted E_C values derived from terrestrial ecosystem damage caused by CO_2, NH_4, and N_2O emissions, primarily from agricultural activities.
3. *The midpoint characterization factor (M_P):* M_P is a measure used in LCIA to evaluate the potential environmental impacts of a product or process at an intermediate stage in the cause-effect chain per person (Huijbregts et al., 2016). These individualistic factors focus on specific environmental mechanisms, such as GWP, acidification potential, or eutrophication potential. Quantifying these impacts at a midpoint allows for a more detailed understanding of how different emissions and resource uses contribute to broader environmental issues. For instance, the GWP measures the impact of GHSs on climate change by comparing their radiative forcing relative to CO_2. M_P factors are essential for identifying and mitigating specific environmental impacts before they translate into broader endpoint damages.
4. *The endpoint characterization factor (E_P):* E_P is a measure used to assess the ultimate impacts of environmental changes on areas of protection, such as human health, ecosystem quality, and resource availability (Huijbregts et al., 2016). Unlike midpoint factors, which focus on specific environmental mechanisms, endpoint factors provide a broader perspective by linking these mechanisms to tangible outcomes. For example, in the context of climate change, endpoint characterization factors might quantify the impact on human health in terms of DALY, the effect on ecosystem quality in terms of species loss, or the depletion of resources in terms of future availability. These factors are crucial for comprehensive environmental impact assessments, as they help policymakers and stakeholders understand the full scope of potential damage and make informed decisions to mitigate adverse effects.

Subsequently, to assess the equivalent amount of CO_2 emitted from the quantity (mass) of substance (i) GHG emitted, this equivalent amount is determined using the CML, ReCiPe, EF, and TRACI methods as outlined in Eq. (20.3).

$$CO_2 \text{ equivalent} = \sum_i (M_C \times \text{mass})_i \tag{20.3}$$

To evaluate human health damage at the midpoint level, the CO_2 equivalent is calculated and divided by the midpoint factor (M_P). This process determines the number of individuals affected, as shown in Eq. (20.4):

$$\text{Human health damage}_{\text{mid}} = \frac{CO_2 \text{ equivalent}}{M_P} \tag{20.4}$$

For endpoint human health damage, the ReCiPe method is used. This method first assesses the environmental impact of CO_2 at the midpoint, focusing on specific effects such as species loss and damage to human health. The intermediate impact is then translated into a final measure, providing a standardized assessment of the overall environmental damage caused by pollutants. This is calculated using Eq. (20.5):

$$\text{Human health damage}_{\text{end}} = \frac{CO_2 \text{ equivalent} \times E_c}{E_P} \tag{20.5}$$

Generally, the LCIA offers several advantages. First, it reduces the reliance on expert opinions in environmental impact assessments, promoting a more unbiased and standardized framework (Jamshidi et al., 2024b). Second, it unifies various environmental categories and aggregates their impacts, which typically cannot be combined due to differences in units, effective periods, and receiving bodies (e.g., air, marine, freshwater, agricultural, or industrial soil). Meanwhile, the adoption of enhanced agricultural practices is advised to alleviate long-term ecological risks and promote sustainability (Jamshidi et al., 2024a). Finally, it has the flexibility to include additional indices, such as the CF, in its calculations. Therefore it needs to be weighed along with other environmental indicators to ensure that the presented EPI is meaningful. This approach is crucial for a comprehensive assessment of GHG emissions, as it provides a balanced view of environmental impacts. Integrating various environmental indicators is essential for creating a meaningful EPI, which is crucial for a thorough GHG emission assessment (Jamous and Müller, 2013).

20.3 Findings and recommendations

SWAT-C and LCIA are both used to assess C equivalence, but they serve different purposes. SWAT-C focuses on simulating soil C dynamics, including the decomposition of residues and SOM, and the associated C transformation (Zhang et al., 2013). It includes detailed modeling of residue pools (lignin, nonlignin, and metabolic) and SOM pools (microbial, slow, and passive), with specific decomposition rates influenced by factors such as soil temperature, moisture, and tillage. SWAT-C is effective in predicting long-term average crop production and C fluxes, particularly in agricultural settings where detailed management data is available. On the other hand, LCIA is a third phase of LCA that evaluates the environmental impacts associated with all stages of a product's life cycle, from raw material extraction through production, use, and disposal (Hauschild, 2005). LCIA involves the global quantification of various environmental impacts, including GWP, which is used to assess C equivalence. It considers emissions and resource use throughout the entire life cycle of agricultural products. While SWAT-C provides detailed simulations of soil processes and C fluxes, LCIA offers a comprehensive evaluation of multiple environmental impacts, including C equivalence. SWAT-C is more suited for site-specific agricultural management and research, whereas LCIA is used for broader policy-making and sustainability assessments. Both tools are valuable for understanding and managing C dynamics in agriculture, but they provide different types of insights. These two indicators are detailed as follows:

Specific case studies: SWAT-C has been tested extensively in agricultural settings. For example, Zhang et al. (2013) used SWAT-C to simulate corn and soybean yields across the U.S. Midwest, demonstrating its effectiveness in predicting long-term average crop production where detailed management data were available. LCIA has been applied in various contexts, including a biorefinery case study where different LCIA methods were used to evaluate global warming potential, eutrophication potential, and water assessment (Wu and Su, 2020). This study highlighted the importance of selecting appropriate LCIA methods for accurate environmental impact assessments.

Quantitative data: SWAT-C studies often provide detailed quantitative data on soil C dynamics. For instance, a sensitivity analysis of a SWAT model identified key parameters influencing model outputs, such as soil temperature and moisture (Sala et al., 2020). LCIA studies also offer quantitative insights. A study on the sensitivity to weighting in LCIA showed how different weighting methods can influence the interpretation of environmental impacts, particularly in the context of photovoltaic technologies (Chatzisideris et al., 2016).

Limitations and challenges: SWAT-C faces challenges such as the need for detailed management data to achieve accurate simulations. Its performance can be reduced in areas where such data is sparse (Zhang et al., 2013). Specifically, for C_{seq} and related SOC analysis, the basic data for model calibration regarding eroded and transported sediments and soil nutrients are essential. Consequently, calibrating sediment transport in the basin by the SWAT model can be nominated as the bottleneck and most critical step of the proposed methodology. This calibration itself is very effective on soil nutrient transport and restoration, particularly for the TP parameter (Imani et al., 2017). It implies that a reliable SWAT-C is very dependent on the appropriate calibration of basin hydrology (water flow), nutrient transport (mainly TP), erosion, and sediment transport altogether. It emphasizes that these calibrations are required in advance to achieve a reliable record for eroded C yields and SOC for any practice. The complexity of the SWAT model requires significant computational resources, reliable data, and expertise for calibration and validation. LCIA also has its limitations. The choice of LCIA methods and the normalization step can significantly influence the results, leading to potential discrepancies in environmental impact assessments (Chatzisideris et al., 2016). Moreover, LCIA requires comprehensive life cycle inventory data, which can be resource-intensive to collect and manage.

Integration possibilities: Integrating the SWAT-C and LCIA can provide a more holistic environmental assessment. For example, SWAT-C's detailed soil C dynamics and results can be an input for LCIA's comprehensive impact assessment. This combination can help to evaluate the overall environmental impacts of agricultural practices. This integrated approach can involve using SWAT-C to model site-specific C fluxes at the basin scale, considering all influential farm practices and LULCs and then applying LCIA to assess the broader equivalent environmental impacts of these practices globally, including global warming potential and natural resource application (Van Haersolte-Van Hof et al., 2019).

Calculating agricultural GHG involves three comprehensive steps: Simulation, calibration, and the integrated application of the SWAT-C and LCIA models. These steps are detailed in the flowchart presented in (Fig. 20.11). The simulation phase encompasses data curation for basin modeling using SWAT. During calibration, hydro and nutrient

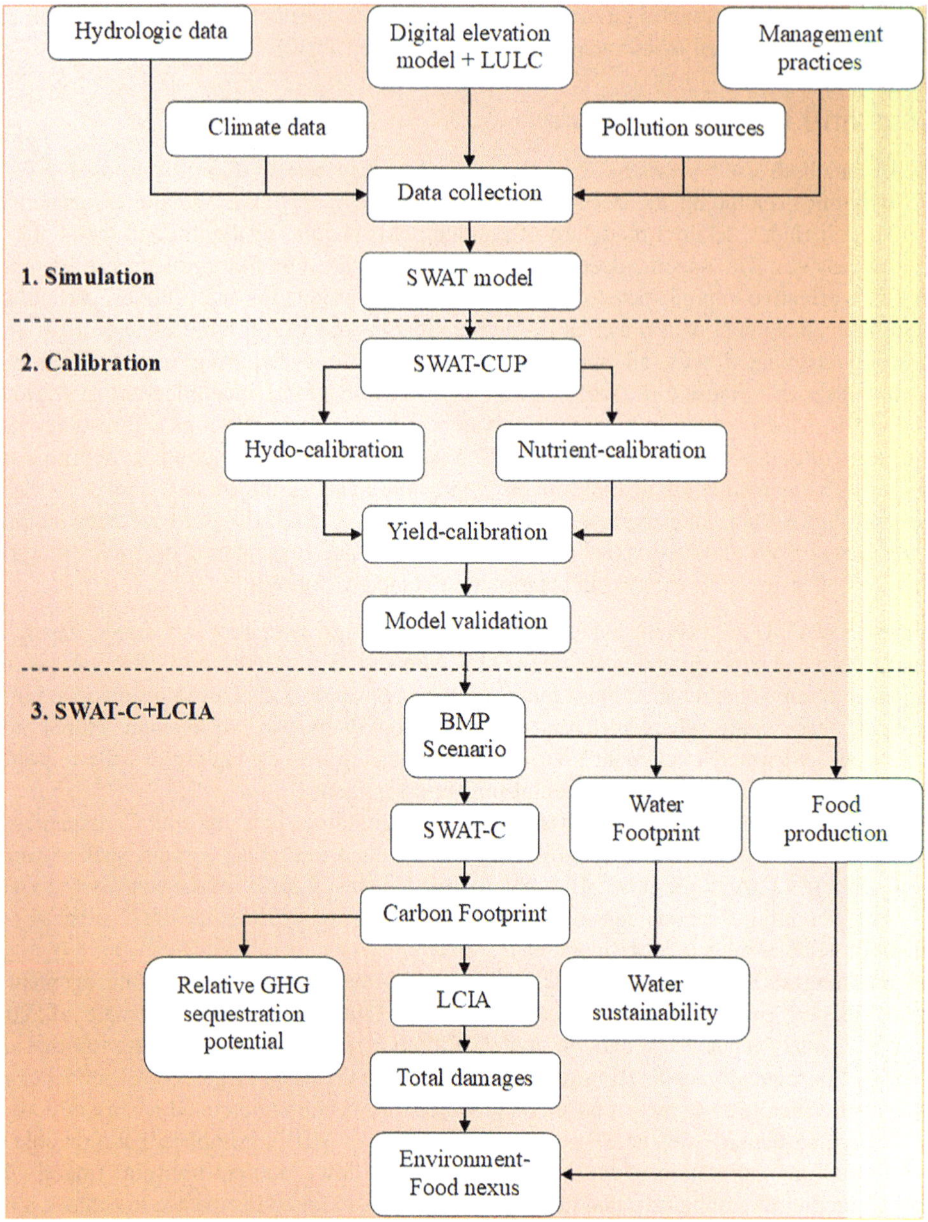

FIGURE 20.11 Overall recommended methodology based on the SWAT-C and LCIA for agricultural GHG calculation (by authors). *GHG*, Greenhouse gases; *LCIA*, life cycle impact assessment; *SWAT-C*, soil and water assessment tool for carbon.

data are utilized to validate the model. Finally, SWAT-C and LCIA are employed to assess environmental damages. The food EF is a metric developed to quantify the total environmental impacts associated with food production (Jamshidi et al., 2024a; Jamshidi and Naderi, 2023a; Tafazzoli and Jamshidi, 2024). It addresses the challenge of balancing the sustainability of BMPs with their effects on nutritional output. By integrating the SWAT with the LCIA method, the food EF evaluates the cumulative environmental impacts of BMPs on pollution loads, production yields, and water footprints. The food EF provides a quantitative framework for assessing the trade-offs between environmental sustainability and nutritional output, demonstrating that a decrease in food EF signifies a more environmentally sustainable food production method. By leveraging the strengths and capacities of the SWAT-C and LCIA tools, researchers and policymakers can gain a more comprehensive quantification method of the environmental impacts of agricultural practices, leading to more informed comparisons and decision-making for sustainable management strategies.

20.4 Policy perspectives

To address the challenges of GHG emissions and C_{seq} in agriculture, policy-makers require adopting a multifaceted approach that integrates scientific insights, economic incentives, and regulatory frameworks (Banerjee et al., 2020; Jhariya et al., 2022; Jhariya et al., 2023). Some key policy recommendations and perspectives based on the findings from the provided document are as follows:

1. *Promoting integrated assessment tools:* Policy-makers should encourage the adoption of advanced tools such as the SWAT-C model and LCIA. These tools provide a comprehensive framework for quantifying GHG emissions and C_{seq}, enabling evidence-based decision-making at all local, national, and global levels. By integrating these methodologies into environmental policies, governments can enhance their capacity for monitoring emissions and to employ targeted mitigation strategies.
2. *Motivating sustainable practices in agriculture:* Agriculture plays a dual role as a source of GHG emissions, and a potential C sink. Policy-makers should use incentives and fiscal policies to motivate farmers to implement BMPs such as:
   a. Reducing tillage and soil erosion during and after cultivation.
   b. Adopting vegetation filter strips, constructed wetlands, and hydroponic vegetation to reduce nutrient runoff and sedimentation.
   c. Composting and manure application to improve SOC while reducing reliance on chemical fertilizers. Obviously, economic incentives, such as subsidies, taxes, or C credits, can financially motivate farmers to adopt these sustainable practices.
3. *Establishing local carbon markets:* Developing local markets for trading C credits or other environmental services can encourage farmers to participate in emission reduction programs. These markets can also facilitate trading reclaimed water and nutrients from wastewater treatment plants, promoting resource efficiency while reducing environmental impacts.
4. *Strengthening international standards:* Harmonizing national policies with international standards, such as the GHG Protocol, ISO 14064, and IPCC guidelines, ensures consistency in GHG accounting and reporting. Policy-makers should also advocate for considering soil C changes in LCAs to provide a more holistic view of agricultural impacts.
5. *Supporting research and development:* Investment in research is critical for experimental and pilot studies, prior to refining models like the SWAT-C. This approach can refer to real and field-based data for accounting for the regional variations in farming practices, soil characteristics, and climate conditions. As a consequence, field-based experiments should be prioritized to validate these models and improve their accuracy.
6. *Addressing regional disparities:* Environmental policies should consider the regional variations in agricultural practices, nutrient requirements, and irrigation. A tailor-made approach that aligns with local traditions, economic conditions, and environmental factors can significantly enhance the effectiveness of mitigation strategies.
7. *Enhancing public awareness:* Educating farmers and stakeholders about the environmental impacts of agricultural practices is essential for altering environmentally disturbing behaviors. For instance, awareness-improving campaigns should emphasize the benefits of sustainable farming practices for both productivity and environmental health.

By implementing these policies, governments can create a more robust framework for reducing agricultural GHG emissions while enhancing C_{seq} capabilities. This integrated approach not only supports CCM but also promotes sustainable agricultural development.

20.5 Prospective shifts and innovations

The integration of SWAT-C with LCIA aims to create a new, comprehensive method for assessing C equivalents in agriculture. This innovative approach combines the detailed C fluxes measured by SWAT-C with the extensive impact assessments provided by LCIA. By leveraging the strengths and capacities of both methods, this new approach can offer a more accurate, inclusive, and quantitative measure of environmental impacts, including GHG emissions from the arable lands. Moreover, by integrating the global perspective of LCIA with the regional focus of SWAT-C, a worldwide applicable index can be developed to enhance environmental assessment and management efforts globally. This integration and development was similarly carried out by integrating SWAT-LCIA for translating the water consumption and pollution discharges of farmlands as equivalent environmental impacts on the basin scale (Jamshidi and Naderi, 2023a). This integrated approach could lead to further water-food assessments that even presented a new index for the quantification under the environment-food nexus. Likely, the SWAT-C-LCIA approach can also provide a model-based framework that can result in the quantification of the environment-food nexus index. However, this upgrading enables including C_{seq} and GHG emissions (N_2O or CO_2) as well in farm-based practices and management. This approach can improve the ability to develop effective strategies in both farm and basin scales for mitigating the origins of global warming in agriculture, using more adaptive local strategies against climate change, and putting one step toward more sustainable and green production. Additionally, by making the SWAT-C global through its combination with the LCIA, it would also be possible to account for C emissions from agriculture worldwide, providing a comprehensive tool for global environmental management through C offset programs and markets by involving new players: farmers. This combination will also make LCIA more precise in accounting for C equivalents by incorporating regional factors such as local temperature. This proposed approach not only supports better environmental decision-making but also helps create a more sustainable future (Raj et al., 2025) by accurately measuring and managing CF on a global scale.

20.6 Conclusion

This chapter provided an inclusive methodological framework for calculating and quantifying GHG emissions and C_{seq}, with a focus on the agricultural sector. The key findings and pivotal points of this chapter are summarized as follows:

- *Dual role of agriculture:* Agriculture is introduced as both a contributor to GHG emissions and a potential C sink. While activities like fertilizer application, deforestation, and livestock management release significant amounts of CO_2, CH_4, and N_2O, other practices such as soil C_{seq} can offset these emissions.
- *Advanced quantification tools:* This chapter introduced the SWAT-C model as an innovative quantification tool for estimating C losses and transformations on a basin scale. The SWAT-C can integrate soil erosion processes with C cycling, offering a more detailed simulation for CO_2 fluxes in large basins and ecosystems. Additionally, the LCIA method is recommended as a complementary tool for the SWAT-C to convert the simulated results into equivalent environmental impacts globally. It can provide a holistic evaluation of GHG emissions during farming or LULC changes.
- *Carbon footprint as a key metric:* CF is emphasized as a critical indicator for measuring GHG emissions in agriculture. It accounts for both direct and indirect emissions across the life cycle of agricultural products, providing practical insights for emission management.
- *International standards and protocols:* This chapter highlighted the importance of aligning GHG quantification efforts with international frameworks such as the GHG Protocol, ISO 14064, and IPCC guidelines. These standards ensure consistency, transparency, and accuracy in environmental reporting.
- *Best management practices:* Sustainable farming practices, for example, modifying tillage, cover cropping, using vegetated filter strips, and composting, are recommended to reduce emissions while enhancing SOC. These BMPs are regionally adaptable and can significantly mitigate the adverse environmental impacts.
- *Integrated environmental assessment:* By combining the basin scale of SWAT-C modeling capabilities with the LCIA, as a global impact assessment approach, this chapter presented an integrated methodology for evaluating the EF of agricultural practices. This approach supports more informed decision-making for policy-makers and stakeholders.
- *Future directions:* This chapter recommends field-based validations by experimental and pilot studies prior to the large-scale modeling. The developed SWAT-C can initially account for regional variations in soil characteristics, climate conditions, and farming practices. This chapter also advocated for the economic incentives, such as C markets, to encourage sustainable agricultural practices.

In conclusion, this chapter provided a robust framework for understanding and managing GHG emissions in agriculture through integrating advanced modeling tools, international standards, and sustainable practices. By addressing both local and global environmental impacts, this integrated approach can contribute to CCM while promoting sustainable agricultural development.

References

Abrol, Y., Raghuram, N., Sachdev, M., 2007. Agricultural nitrogen use and its environmental implications. IK International Pvt Ltd.

Acero, A.P., Rodríguez, C. and Ciroth, A. (2016). LCIA methods. Impact assessment methods in Life Cycle Assessment and their impact categories, 1–23.

Agency, U.S.E.P. (2024). Tool for Reduction and Assessment of Chemicals and Other Environmental Impacts (TRACI).

Arvanitoyannis, I.S., 2008. ISO 14040: Life Cycle Assessment (LCA) - Principles and Guidelines. Waste Management for the Food Industries. Elsevier Inc., Greece, pp. 97–132. Available at http://doi.org/10.1016/B978-012373654-3.50006-7.

Arzoumanidis, I., Fullana-I-Palmer, P., Raggi, A., Gazulla, C., Raugei, M., Benveniste, G., et al., 2014. Unresolved issues in the accounting of biogenic carbon exchanges in the wine sector. Journal of Cleaner Production 82, 16–22. Available at https://doi.org/10.1016/j.jclepro.2014.06.073.

Ayres, R.U., Schlesinger, W.H., Socolow, R.H., 1994. Human impacts on the carbon and nitrogen cycles. Industrial Ecology and Global Change 5, 121.

Babaeian, F., Delavar, M., Morid, S., Jamshidi, S., 2023. Designing climate change dynamic adaptive policy pathways for agricultural water management using a socio-hydrological modeling approach. Journal of Hydrology 627, 130398. Available at https://doi.org/10.1016/j.jhydrol.2023.130398.

Bachu, S., 2008. CO2 storage in geological media: Role, means, status and barriers to deployment. Progress in Energy and Combustion Science 34 (2), 254–273. Available at https://doi.org/10.1016/j.pecs.2007.10.001.

Banerjee, A., Meena, R.S., Jhariya, M.K., Yadav, D.K., 2020. Agroecological Footprints Management for Sustainable Food System. Springer, Singapore, pp. 1–514. Available at https://doi.org/10.1007/978-981-15-9496-0.

Bare, J., Young, D., Qam, S., Hopton, M., Chief, S., 2012. Tool for the Reduction and Assessment of Chemical and other Environmental Impacts (TRACI). US Environmental Protection Agency, Washington, DC, USA.

Benbi, D.K., 2013. Greenhouse gas emissions from agricultural soils: sources and mitigation potential. Journal of Crop Improvement 27 (6), 752–772. Available at https://doi.org/10.1080/15427528.2013.845054.

Bockheim, J., 2010. Cryosols as a test of our knowledge of Earth as a system. https://old.iuss.org/19th/WCSS/Symposium/pdf/0219.pdf.

Brandão, M., Levasseur, A., Kirschbaum, M.U.F., Weidema, B.P., Cowie, A.L., Jørgensen, S.V., et al., 2013. Key issues and options in accounting for carbon sequestration and temporary storage in life cycle assessment and carbon footprinting. International Journal of Life Cycle Assessment 18 (1), 230–240. Available at https://doi.org/10.1007/s11367-012-0451-6.

Broucek, J., 2014. Production of methane emissions from ruminant husbandry: a review. Journal of Environmental Protection 05 (15), 1482–1493. Available at https://doi.org/10.4236/jep.2014.515141.

Buder, I., 2020. Encyclopedia of Sustainable Management. Springer, pp. 1–8.

Cedrez, C.B., Andeweg, K., Casu, F., 2023. Circular Food Systems around the world: exploring concepts, ideas and opportunities. Wageningen Livestock Research.

Chadwick, D., Sommer, S., Thorman, R., Fangueiro, D., Cardenas, L., Amon, B., et al., 2011. Manure management: Implications for greenhouse gas emissions. Animal Feed Science and Technology 166-167, 514–531. Available at https://doi.org/10.1016/j.anifeedsci.2011.04.036.

Chandrasekharan, I., Chandrasekharan, B., Srinivasan, S., 2020. Environmental Performance Index 2020 and ranking of states of India. Paper submitted to IJEPDM.

Chatzisideris, M.D., Espinosa, N., Laurent, A., Krebs, F.C., 2016. Ecodesign perspectives of thin-film photovoltaic technologies: a review of life cycle assessment studies. Solar Energy Materials and Solar Cells 156, 2–10. Available at https://doi.org/10.1016/j.solmat.2016.05.048.

Cheng, K., Yan, M., Pan, G., Luo, T., Yue, Q., 2015b. Methodology for carbon footprint calculation in crop and livestock production. In: Muthu, S.S. (Ed.), The Carbon Footprint Handbook. CRC Press, Boca Raton, pp. 78–101. Available at https://doi.org/10.1201/b18929-7.

Cheng, K., Yan, M., Nayak, D., Pan, G.X., Smith, P., Zheng, J.F., et al., 2015a. Carbon footprint of crop production in China: an analysis of National Statistics data. The Journal of Agricultural Science 153 (3), 422–431. Available at https://doi.org/10.1017/s0021859614000665.

Ciais, P., Sabine, C., Bala, G., Bopp, L., Brovkin, V., House, J.I., 2014. Climate Change 2013: The Physical Science Basis. Contribution of Working Group I to the Fifth Assessment Report of the Intergovernmental Panel on Climate Change Change. Cambridge University Press, pp. 465–570.

Dalal, R.C., Allen, D.E., 2008. Greenhouse gas fluxes from natural ecosystems. Australian Journal of Botany 56 (5), 369–407. Available at https://doi.org/10.1071/bt07128.

Easterbrook, D.J., 2016. Chapter 9 - Greenhouse gases. In: Easterbrook, D.J. (Ed.), Evidence-Based Climate Science (Second Edition). Elsevier, pp. 163–173. Available at https://doi.org/10.1016/B978-0-12-804588-6.00009-4.

EPI, 2024. Environmental Performance Index, 2024 EPI Framework.

European Commission, 2024. European Platform on LCA | EPLCA. European Commission. Available at https://eplca.jrc.ec.europa.eu/.

Falkowski, P., Scholes, R.J., Boyle, E., Canadell, J., Canfield, D., Elser, J., et al., 2000. The global carbon cycle: a test of our knowledge of earth as a system. Science (New York, N.Y.) 290 (5490), 291–296. Available at https://doi.org/10.1126/science.290.5490.291.

FAO; IFAD; UNICEF; WFP; WHO, 2022. The State of Food Security and Nutrition in the World 2022: Repurposing food and agricultural policies to make healthy diets more affordable. FAO; IFAD; UNICEF; WFP; WHO, Rome. Available at https://doi.org/10.4060/cc0639en.

Finnveden, G., Hauschild, M.Z., Ekvall, T., Guinée, J., Heijungs, R., Hellweg, S., et al., 2009. Recent developments in life cycle assessment. Journal of Environmental Management 91 (1), 1–21. Available at https://doi.org/10.1016/j.jenvman.2009.06.018.

Fred Singer, S., 2008. Nature, Not Human Activity, Rules the Climate: Summary for Policymakers of the Report of the Nongovernmental International Panel on Climate Change. The Heartland Institute, Chicago, IL.

Frontera, P., Salieri, B., Righi, S., 2020. Comparison of the LCIA methods used for the evaluation of chemicals. In: Maranghi, S., Brondi, C. (Eds.), Life Cycle Assessment in the Chemical Product Chain: Challenges, Methodological Approaches and Applications. Springer International Publishing, Cham, pp. 33–51. Available at https://doi.org/10.1007/978-3-030-34424-5_2.

Gao, B., Ju, X., Su, F., Meng, Q., Oenema, O., Christie, P., et al., 2014a. Nitrous oxide and methane emissions from optimized and alternative cereal cropping systems on the North China Plain: A two-year field study. Science of the Total Environment 472, 112–124. Available at https://doi.org/10.1016/j.scitotenv.2013.11.003.

Gao, T., Liu, Q., Wang, J., 2014b. A comparative study of carbon footprint and assessment standards. International Journal of Low-Carbon Technologies 9 (3), 237–243. Available at https://doi.org/10.1093/ijlct/ctt041.

Goeppert, A., Czaun, M., Surya Prakash, G.K., Olah, G.A., 2012. Air as the renewable carbon source of the future: an overview of CO2 capture from the atmosphere. Energy and Environmental Science 5 (7), 7833–7853. Available at https://doi.org/10.1039/c2ee21586a.

Goldstein, B., Hansen, S.F., Gjerris, M., Laurent, A., Birkved, M., 2016. Ethical aspects of life cycle assessments of diets. Food Policy 59, 139–151. Available at https://doi.org/10.1016/j.foodpol.2016.01.006.

Golsteijn, L., Vieira, M., 2020. Applicability of the European Environmental Footprint (EF) methodology in Southern Mediterranean countries—learnings and recommendations for enabling EF-compliant studies in regions outside of Europe. The International Journal of Life Cycle Assessment 25 (12), 2407–2416. Available at https://doi.org/10.1007/s11367-019-01681-z.

Guinée, J.B., 2002. Handbook on Life Cycle Assessment: Operational Guide to the ISO Standards. Springer Science & Business Media.

Haney, E.B., Haney, R.L., White, M.J., Arnold, J.G., 2018. Comparison of Wheat Yield Simulated Using Three N Cycling Options in the SWAT Model. Open Journal of Soil Science 08 (08), 197–211. Available at https://doi.org/10.4236/ojss.2018.88016.

Hauschild, M.Z., 2005. Assessing environmental impacts in a life-cycle perspective. Environmental Science and Technology 39 (4). Available at https://doi.org/10.1021/es053190s.

Hauschild, M.Z., 2019. Life cycle impact assessment. In: Chatti, S., Laperrière, L., Reinhart, G., Tolio, T. (Eds.), CIRP Encyclopedia of Production Engineering. Springer, Berlin, Heidelberg, pp. 1053–1057. Available at https://doi.org/10.1007/978-3-662-53120-4_16861.

Heinze, C., Meyer, S., Goris, N., Anderson, L., Steinfeldt, R., Chang, N., et al., 2015. The ocean carbon sink – impacts, vulnerabilities and challenges. Earth System Dynamics 6 (1), 327–358. Available at https://doi.org/10.5194/esd-6-327-2015.

Horwath, W., 2007. 12 - Carbon cycling and formation of soil organic matter. In: Paul, E.A. (Ed.), Soil Microbiology, Ecology and Biochemistry (Third Edition). Elsevier, pp. 303–339. Available at https://doi.org/10.1016/B978-0-08-047514-1.50016-0.

Hsu, A., Esty, D., Levy, M., de Sherbinin, A., 2016. 2016 Environmental Performance Index (EPI). Available at https://doi.org/10.13140/RG.2.2.19868.90249.

Huijbregts, M.A., Steinmann, Z.J., Elshout, P.M., Stam, G., Verones, F., Vieira, M.D., et al., 2016. ReCiPe2016: a harmonised life cycle impact assessment method at midpoint and endpoint level. The International Journal of Life Cycle Assessment 22, 138–147. Available at https://doi.org/10.1007/s11367-016-1246-y.

Imani, S., Niksokhan, M.H., Jamshidi, S., Abbaspour, K.C., 2017. Discharge permit market and farm management nexus: an approach for eutrophication control in small basins with low-income farmers. Environmental Monitoring and Assessment 189 (7). Available at https://doi.org/10.1007/s10661-017-6066-4.

Intergovernmental Panel on Climate Change (IPCC), 2014. Chapter 2 - Observations: Atmosphere and Surface. Climate Change 2013 – The Physical Science Basis: Working Group I Contribution to the Fifth Assessment Report of the Intergovernmental Panel on Climate Change. Cambridge University Press, Cambridge, pp. 159–254. Available at https://doi.org/10.1017/CBO9781107415324.008.

IPCC, 2013. Climate Change 2013: The Physical Science Basis. In: Stocker, T.F., Qin, D., Plattner, G.K., Tignor, M., Allen, S.K., Boschung, J., Nauels, A., Xia, Y., Bex, V., Midgley, P.M. (Eds.), Contribution of Working Group I to the Fifth Assessment Report of the Intergovernmental Panel on Climate Change. Cambridge University Press, Cambridge, United Kingdom and New York, NY, pp. 1535. Available at https://www.ipcc.ch/site/assets/uploads/2017/09/WG1AR5_Frontmatter_FINAL.pdf.

ISO, 2006a. 14040 International standard. Environmental Management—Life Cycle Assessment—Principles and Framework.

ISO, 2006b. Greenhouse Gases-Part 1: Specification with Guidance at the Organization Level for Quantification and Reporting of Greenhouse Gas Emissions and Removals (ISO 14064-1: 2018); German and English version EN ISO 14064-1: 2018. International Organization for Standardization, Geneva, Switzerland.

ISO, 2006c. Greenhouse Gases-Part 2: Specification with Guidance at the Project Level for Quantification, Monitoring and Reporting of greenhouse Gas Emis-sion Reductions or Removal Enhancements (ISO 14064-2: 2019); German and English version EN ISO 14064-2: 2019. International Organization for Standard-ization, Geneva, Switzerland.

Jain, N., Bhatia, A., Pathak, H., Gupta, N., Sharma, D., Kaushik, R., 2015. Greenhouse gas emission and global warming. In: Khoiyangbam, R.S., Gupta, N. (Eds.), Introduction to Environmental Sciences. TERI Press, pp. 379–411.

Jaiswal, B., Agrawal, M., 2020. Carbon footprints of agriculture sector. In: Muthu, S. (Ed.), Carbon Footprints. Environmental Footprints and Eco-design of Products and Processes. Springer, Singapore, pp. 81–99. Available at https://doi.org/10.1007/978-981-13-7916-1_4.

Jamous, N., Müller, K., 2013. Environmental performance indicators. In: Dada, A., Stanoevska, K., Gómez, J.M. (Eds.), Organizations' Environmental Performance Indicators: Measuring, Monitoring, and Management. Springer, Berlin Heidelberg, pp. 3–18. Available at https://doi.org/10.1007/978-3-642-32720-9_1.

Jamshidi, S., Naderi, A., 2023a. A quantitative approach on environment-food nexus: integrated modeling and indices for cumulative impact assessment of farm management practices. PeerJ 11. Available at https://doi.org/10.7717/peerj.14816. e14816.

Jamshidi, S., Naderi, A., 2023b. Wetland Restoration Policies and the Sustainability of Agricultural Productions, Lessons Learnt from Zrebar Lake, Iran. In: Banerjee, A., Jhariya, M.K., Bargali, S.S., Palit, D. (Eds.), Ecorestoration for Sustainability. Wiley, pp. 113–166. Available at https://doi.org/10.1002/9781119879954.ch4.

Jamshidi, S., Niksokhan, M.H., Ardestani, M., 2014. Surface water quality management using an integrated discharge permit and the reclaimed water market. Water Science and Technology 70 (5), 917–924. Available at https://doi.org/10.2166/wst.2014.314.

Jamshidi, S., Imani, S., Delavar, M., 2020. Impact assessment of best management practices (bmps) on the water footprint of agricultural productions. International Journal of Environmental Research 14 (6), 641–652. Available at https://doi.org/10.1007/s41742-020-00285-y.

Jamshidi, S., Dehnavi, A., Roudbari, M.V., Yazdani, M., 2024a. An integrated approach through controlled experiment and LCIA to evaluate water quality and ecological impacts of irrigated paddy rice. Environmental Science and Pollution Research 31 (32), 45264–45279. Available at https://doi.org/10.1007/s11356-024-34188-8.

Jamshidi, S., Farsimadan, M., Mohammadi, H., 2024b. A holistic approach for performance evaluation of wastewater treatment plants: integrating grey water footprint and life cycle impact assessment. Water Science and Technology 89 (7), 1741–1756. Available at https://doi.org/10.2166/wst.2024.081.

Jhariya, M.K., Meena, R.S., Banerjee, A., Meena, S.N., 2022. Natural resources conservation and advances for sustainability. Elsevier. Available at https://doi.org/10.1016/C2019-0-03763-6 ISBN: 9780128229767.

Jhariya, M.K., Meena, R.S., Banerjee, A., Kumar, S., Raj, A., 2023. Agroforestry for Carbon and Ecosystem Management. Elsevier, India, pp. 1–434. Available at http://doi.org/10.1016/C2021-0-02916-3.

Kamyab, H., SaberiKamarposhti, M., Hashim, H., Yusuf, M., 2024. Carbon dynamics in agricultural greenhouse gas emissions and removals: a comprehensive review. Carbon Letters 34 (1), 265–289. Available at https://doi.org/10.1007/s42823-023-00647-4.

Kane, D., 2015. Carbon Sequestration Potential on Agricultural Lands: A Review of Current Science and Available Practices. National Sustainable Agriculture Coalition Breakthrough Strategies and Solutions, LLC, pp. 1–35. Available at https://sustainableagriculture.net/wp-content/uploads/2015/12/Soil_C_review_Kane_Dec_4-final-v4.pdf.

Keenan, T.F., Williams, C.A., 2018. The terrestrial carbon sink. Annual Review of Environment and Resources 43 (1), 219–243. Available at https://doi.org/10.1146/annurev-environ-102017-030204.

Kim, T., 2015. Applications of carbon footprint in urban planning and geography. In: Muthu, S.S. (Ed.), The Carbon Footprint Handbook. CRC Press, pp. 163–182. Available at https://doi.org/10.1201/b18929-10.

Kumar, S., Meena, R.S., Sheoran, P., Jhariya, M.K., 2024. Regenerative Agriculture for Sustainable Food Systems. Springer, Singapore, pp. 1–493. Available at https://doi.org/10.1007/978-981-97-6691-8.

Kweku, D., Bismark, O., Maxwell, A., Desmond, K., Danso, K., Oti-Mensah, E., et al., 2018. Greenhouse effect: greenhouse gases and their impact on global warming. Journal of Scientific Research and Reports 17 (6), 1–9. Available at https://doi.org/10.9734/jsrr/2017/39630.

Lal, R., 2004a. Agricultural activities and the global carbon cycle. Nutrient Cycling in Agroecosystems 70 (2), 103–116. Available at https://doi.org/10.1023/B:FRES.0000048480.24274.0f.

Lal, R., 2004b. Soil carbon sequestration impacts on global climate change and food security. Science (New York, N.Y.) 304 (5677), 1623–1627. Available at https://doi.org/10.1126/science.1097396.

Lal, R., 2008. Carbon sequestration. Philosophical Transactions of the Royal Society B: Biological Sciences 363 (1492), 815–830. Available at https://doi.org/10.1098/rstb.2007.2185.

Leahy, S., Clark, H., Reisinger, A., 2020. Challenges and prospects for agricultural greenhouse gas mitigation pathways consistent with the paris agreement. Frontiers in Sustainable Food Systems 4. Available at https://doi.org/10.3389/fsufs.2020.00069.

Levy, P.E., Cannell, M.G.R., Friend, A.D., 2004. Modelling the impact of future changes in climate, CO2 concentration and land use on natural ecosystems and the terrestrial carbon sink. Global Environmental Change 14 (1), 21–30. Available at https://doi.org/10.1016/j.gloenvcha.2003.10.005.

Luo, Y., 2007. Terrestrial carbon–cycle feedback to climate warming. Annual Review of Ecology, Evolution, and Systematics 38 (1), 683–712. Available at https://doi.org/10.1146/annurev.ecolsys.38.091206.095808.

Luyssaert, S., Jammet, M., Stoy, P.C., Estel, S., Pongratz, J., Ceschia, E., et al., 2014. Land management and land-cover change have impacts of similar magnitude on surface temperature. Nature Climate Change 4 (5), 389–393. Available at https://doi.org/10.1038/nclimate2196.

Lynch, J., Cain, M., Frame, D., Pierrehumbert, R., 2021. Agriculture's contribution to climate change and role in mitigation is distinct from predominantly fossil CO2-emitting sectors. Frontiers in Sustainable Food Systems 4. Available at https://doi.org/10.3389/fsufs.2020.518039.

Meena, R.S., Pradhan, G., 2023. Industrial garbage-derived biocompost enhances soil organic carbon fractions, CO2 biosequestration, potential carbon credits and sustainability index in a rice-wheat ecosystem. Environmental Research 235, 116525. Available at https://doi.org/10.1016/j.envres.2023.116525.

Meena, R.S., Pradhan, G., Kumar, S., 2025b. Energy flow, eco-efficiency, and economic circulation with recycled industrial waste compost application in wheat and subsequent rice farming. Science of the Total Environment 967. Available at https://doi.org/10.1016/j.scitotenv.2025.178779.

Meena, R.S., Pradhan, G., Kumar, S., Lal, R., 2023. Using industrial wastes for rice-wheat cropping and food-energy-carbon-water-economic nexus to the sustainable food system. Renewable and Sustainable Energy Reviews 187, 113756. Available at https://doi.org/10.1016/j.rser.2023.113756.

Meena, R.S., Pradhan, G.S., Kumar, S., Jhariya, M.K., Rao, C.S., 2025a. Carbon-Negative Agriculture. Springer, Singapore, pp. 1–500. Available at https://doi.org/10.1007/978-981-95-1483-0.

Meena, R.S., Lal, R., Kumar, S., Pradhan, G., Srinivasarao, C., Singh, A.K., et al., 2024a. Potential of Indian agriculture for capturing atmospheric CO2 and monetizing carbon credits to the farmers: an overview and policy framework. Advances in Agronomy 188, 101–206. Available at https://doi.org/10.1016/bs.agron.2024.06.001.

Meena, R.S., Pradhan, G., Singh, K., Kumar, S., Singh, A.K., Shashidhar, K.S., et al., 2024c. Agriculture models for restoring degraded land to enhance CO2 biosequestration and carbon credits in the Vindhyan region of India. Science of the Total Environment 929. Available at https://doi.org/10.1016/j.scitotenv.2024.172661.

Meena, R.S., Pradhan, G., Nalani, Kumar, S., Singh, A.K., Verma, S.K., et al., 2024b. Diversified cropping modules designed for soil restoration, CO2 sequestration, and generating carbon credits. Land Degradation and Development 35 (14), 4314–4329. Available at https://doi.org/10.1002/ldr.5224.

Minasny, B., Malone, B.P., McBratney, A.B., Angers, D.A., Arrouays, D., Chambers, A., et al., 2017. Soil carbon 4 per mille. Geoderma 292, 59–86. Available at https://doi.org/10.1016/j.geoderma.2017.01.002.

Müller-Nedebock, D., Chaplot, V., 2015. Soil carbon losses by sheet erosion: A potentially critical contribution to the global carbon cycle. Earth Surface Processes and Landforms 40 (13), 1803–1813. Available at https://doi.org/10.1002/esp.3758.

Nazir, M.J., Li, G., Nazir, M.M., Zulfiqar, F., Siddique, K.H.M., Iqbal, B., et al., 2024. Harnessing soil carbon sequestration to address climate change challenges in agriculture. Soil and Tillage Research 237. Available at https://doi.org/10.1016/j.still.2023.105959.

Ontl, T.A., Schulte, L.A., 2012. Soil carbon storage. Nature Education Knowledge 3 (10), 35. Available at https://www.nature.com/scitable/knowledge/library/soil-carbon-storage-84223790/.

Ourworldindata. (2024a). Global greenhouse gas emissions by sector. https://ourworldindata.org/grapher/ghg-emissions-by-sector.

Ourworldindata. (2024b). Greenhouse gas emissions by gas, World, 1850 to 2022. https://ourworldindata.org/grapher/ghg-emissions-by-gas.

Owusu, E., Ndunda, E., Koech, M., 2023. Carbon Footprint of Food Waste in Classified Star Hotels in Nairobi City County, Kenya. Adv Environ Stud 7 (1), 557–566.

Painkra, S., Singh, K.P., Singh, B., Jhariya, M.K., Banerjee, A., Sahu, K., et al., 2025. Energy budgeting and economic analysis of coriander crop production in central India. Environment, Development and Sustainability. Available at https://doi.org/10.1007/s10668-025-06217-0.

Pandey, D., Agrawal, M., 2014. Carbon footprint estimation in the agriculture sector. In: Muthu, S. (Ed.), Assessment of Carbon Footprint in Different Industrial Sectors, Volume 1. EcoProduction. Springer, Singapore, pp. 25–47. Available at https://doi.org/10.1007/978-981-4560-41-2_2.

Pandey, D., Agrawal, M., Pandey, J.S., 2011. Carbon footprint: current methods of estimation. Environmental Monitoring and Assessment 178 (1-4), 135–160. Available at https://doi.org/10.1007/s10661-010-1678-y.

Paustian, K., Babcock, B., Hatfield, J.L., Lal, R., McCarl, B.A., McLaughlin, S., et al., 2001. Agricultural Mitigation of Greenhouse Gases: Science and Policy Options. Conference on Carbon Sequestration, Washington, DC.

Petersen, B.M., Knudsen, M.T., Hermansen, J.E., Halberg, N., 2013. An approach to include soil carbon changes in life cycle assessments. Journal of Cleaner Production 52, 217–224. Available at https://doi.org/10.1016/j.jclepro.2013.03.007.

Ping, C.L., Jastrow, J.D., Jorgenson, M.T., Michaelson, G.J., Shur, Y.L., 2015. Permafrost soils and carbon cycling. SOIL 1 (1), 147–171. Available at https://doi.org/10.5194/soil-1-147-2015.

Pleijel, H., 2000. Ground-level ozone. A problem largely ignored in southern Europe, air pollution and climate series (12). https://www.airclim.org/sites/default/files/documents/APC12.pdf.

Pongratz, J., Schwingshackl, C., Bultan, S., Obermeier, W., Havermann, F., Guo, S., 2021. Land use effects on climate: current state, recent progress, and emerging topics. Current Climate Change Reports 7 (4), 99–120. Available at https://doi.org/10.1007/s40641-021-00178-y.

Pradhan, G., Meena, R.S., 2023b. Utilizing waste compost to improve the atmospheric CO2 capturing in the rice-wheat cropping system and energy-cum-carbon credit auditing for a circular economy. Science of the Total Environment 892. Available at https://doi.org/10.1016/j.scitotenv.2023.164572.

Pradhan, G., Meena, R.S., 2023a. Interaction impact of biocompost on nutrient dynamics and relations with soil biota, carbon fractions index, societal value of CO2 equivalent and ecosystem services in the wheat-rice farming. Chemosphere 339. Available at https://doi.org/10.1016/j.chemosphere.2023.139695.

Pradhan, G., Meena, R.S., Kumar, S., Lal, R., 2023. Utilizing industrial wastes as compost in wheat-rice production to improve the above and below-ground ecosystem services. Agriculture, Ecosystems & Environment 358, 108704. Available at https://doi.org/10.1016/j.agee.2023.108704.

Quinton, J.N., Govers, G., Van Oost, K., Bardgett, R.D., 2010. The impact of agricultural soil erosion on biogeochemical cycling. Nature Geoscience 3 (5), 311–314. Available at https://doi.org/10.1038/ngeo838.

Raihan, A., Begum, R.A., Nizam, M., Said, M., Pereira, J.J., 2022. Dynamic impacts of energy use, agricultural land expansion, and deforestation on CO_2 emissions in Malaysia. Environmental and Ecological Statistics 29 (3), 477–507. Available at https://doi.org/10.1007/s10651-022-00532-9.

Raj, A., Jhariya, M.K., Meena, R.S., 2025. Agroforestry for monetising carbon credits. Springer, Cham, pp. 1–609. Available at https://doi.org/10.1007/978-3-031-97151-8.

Ramachandra, T.V., Mahapatra, D.M., 2015. The science of carbon footprint assessment. In: Kannan, S.S. (Ed.), The Carbon Footprint Handbook. CRC Press, pp. 3–44. Available at https://doi.org/10.1201/b18929-3.

Ranaweera, D., Kumarasinghe, U., Senanayake, C., Dassanayake, R.S., Bandara, P.C., Koliyabandara, P.A., 2023. Biogeochemistry: essential link between geosphere and biosphere. Medical Geology: En route to One Health. Wiley, Sri Lanka, pp. 37–54. Available at http://doi.org/10.1002/9781119867371.ch2.

Roudbari, M.V., Dehnavi, A., Jamshidi, S., Yazdani, M., 2023. A multi-pollutant pilot study to evaluate the grey water footprint of irrigated paddy rice. Agricultural Water Management 282. Available at https://doi.org/10.1016/j.agwat.2023.108291.

Roy, O., Meena, R.S., Kumar, S., Jhariya, M.K., Pradhan, G., 2022. Assessment of land use systems for CO2 sequestration, carbon credit potential, and income security in Vindhyan region, India. Land Degradation and Development 33 (4), 670–682. Available at https://doi.org/10.1002/ldr.4181.

Sala, S., Laurent, A., Vieira, M., Van Hoof, G., 2020. Implications of LCA and LCIA choices on interpretation of results and on decision support. The International Journal of Life Cycle Assessment 25 (12), 2311–2314. Available at https://doi.org/10.1007/s11367-020-01845-2.

Sherif, S.E.A.E. (2023). Egypt's Policy Adjustment in International Climate Diplomacy and Its Reasons. Available at https://doi.org/10.13140/RG.2.2.16772.30082.

Shoji, S., Delgado, J., Mosier, A., Miura, Y., 2001. Use of controlled release fertilizers and nitrification inhibitors to increase nitrogen use efficiency and to conserve air and water quality. Communications in Soil Science and Plant Analysis 32 (7-8), 1051–1070. Available at https://doi.org/10.1081/CSS-100104103.

Shrestha, B.M., Desjardins, R.L., McConkey, B.G., Worth, D.E., Dyer, J.A., Cerkowniak, D.D., 2014. Change in carbon footprint of canola production in the Canadian Prairies from 1986 to 2006. Renewable Energy 63, 634–641. Available at https://doi.org/10.1016/j.renene.2013.10.022.

Sicard, A.J., Baker, R.T., 2020. Fluorocarbon refrigerants and their syntheses: past to present. Chemical Reviews 120 (17), 9164–9303. Available at https://doi.org/10.1021/acs.chemrev.9b00719.

Singh, H., Prasad, P.V.V., Northup, B.K., Ciampitti, I.A., Rice, C.W., 2023. Strategies for mitigating greenhouse gas emissions from agricultural ecosystems. Global Agricultural Production: Resilience to Climate Change. Springer International Publishing, Cham, pp. 409–440. Available at https://doi.org/10.1007/978-3-031-14973-3_16.

Smith, K., 1997. The potential for feedback effects induced by global warming on emissions of nitrous oxide by soils. Global Change Biology 3 (4), 327–338. Available at https://doi.org/10.1046/j.1365-2486.1997.00100.x.

Smith, S.V., Renwick, W.H., Buddemeier, R.W., Crossland, C.J., 2001. Budgets of soil erosion and deposition for sediments and sedimentary organic carbon across the conterminous United States. Global Biogeochemical Cycles 15 (3), 697–707. Available at https://doi.org/10.1029/2000gb001341.

Snyder, C.S., Bruulsema, T.W., Jensen, T.L., Fixen, P.E., 2009. Review of greenhouse gas emissions from crop production systems and fertilizer management effects. Agriculture, Ecosystems & Environment 133 (3-4), 247–266. Available at https://doi.org/10.1016/j.agee.2009.04.021.

Souri, M.A., Jamshidi, S., Moghaddam, H.K., 2025. Multi-pollutant water quality trading: a conditional approach for groundwater quality management. International Journal of Environmental Research 19 (2). Available at https://doi.org/10.1007/s41742-024-00714-2.

Sovacool, B.K., Griffiths, S., Kim, J., Bazilian, M., 2021. Climate change and industrial F-gases: A critical and systematic review of developments, sociotechnical systems and policy options for reducing synthetic greenhouse gas emissions. Renewable and Sustainable Energy Reviews 141. Available at https://doi.org/10.1016/j.rser.2021.110759.

Specification, P.A. (2008). Specification for the assessment of the life cycle greenhouse gas emissions of goods and services. Bsi Br. Stand. ISBN 978, 580.

Swamy, Y.V., Venkanna, R., Nikhil, G.N., Chitanya, D.N.S.K., Sinha, P.R., Ramakrishna, M., et al., 2012. Impact of nitrogen oxides, volatile organic compounds and black carbon on atmospheric ozone levels at a semi arid urban site in Hyderabad. Aerosol and Air Quality Research 12 (4), 662–671. Available at https://doi.org/10.4209/aaqr.2012.01.0019.

Swift, R.S., 2001. Sequestration of carbon by soil. Soil Science 166 (11), 858–871. Available at https://doi.org/10.1097/00010694-200111000-00010.

Tafazzoli, P., Jamshidi, S., 2024. Environment-food nexus in trout ponds: a developed index by combining grey water footprint and life cycle assessment. International Journal of Environmental Science and Technology 21 (14), 8885–8900. Available at https://doi.org/10.1007/s13762-024-05688-4.

Trabalka, J.R., 1985. Atmospheric carbon dioxide and the global carbon cycle. United States Dept. of Energy, Office of Energy Research, Office of Basic Energy Sciences, Carbon Dioxide Research Division, Washington, DC.

UNFCCC Secretariat, 2018. Report of the Conference of the Parties on its twenty-third session, held in Bonn from 6 to 18 November 2017. Part one: Proceedings. United Nations Climate Change, Bonn.

Van Haersolte-Van Hof, J., Greenway, F., Cho, A., 2019. LCIA approach to challenges to arbitrators. BCDR International Arbitration Review 6 (1), 191–216. Available at https://doi.org/10.54648/bcdr2021011.

Wakeland, W., Cholette, S., Venkat, K., 2012. Food transportation issues and reducing carbon footprint. In: Boye, J.I., Arcand, Y. (Eds.), Green Technologies in Food Production and Processing. Food Engineering Series. Springer US, Boston, MA, pp. 211–236. Available at https://doi.org/10.1007/978-1-4614-1587-9_9.

Weinstein, L.H., Davison, A., 2004. Fluorides in the Environment: Effect on Plants and Animals. CABI Press. Available at https://doi.org/10.1079/9780851996837.0000.

Weng, C.K., Boehmer, K., 2006. Launching of ISO 14064 for greenhouse gas accounting and verification. ISO Management Systems 15, 14–16.

West, T.O., Marland, G., 2002. A synthesis of carbon sequestration, carbon emissions, and net carbon flux in agriculture: comparing tillage practices in the United States. Agriculture, Ecosystems and Environment 91 (1-3), 217–232. Available at https://doi.org/10.1016/S0167-8809(01)00233-X.

Wintergreen, J., Delaney, T., 2007. ISO 14064, international standard for GHG emissions inventories and verification. https://gaftp.epa.gov/AIR/nei/ei_conference/EI16/session13/wintergreen.pdf.

Wu, Y., Su, D., 2020. Review of life cycle impact assessment (LCIA) methods and inventory databases. In: Su, D. (Ed.), Sustainable Product Development: Tools, Methods and Examples. Springer, Cham, pp. 39–55. Available at https://doi.org/10.1007/978-3-030-39149-2_3.

Yang, Q., Zhang, X., 2016. Improving SWAT for simulating water and carbon fluxes of forest ecosystems. Science of the Total Environment 569-570, 1478–1488. Available at https://doi.org/10.1016/j.scitotenv.2016.06.238.

Yao, Y., Huang, G., An, C., Chen, X., Zhang, P., Xin, X., et al., 2020. Anaerobic digestion of livestock manure in cold regions: technological advancements and global impacts. Renewable and Sustainable Energy Reviews 119. Available at https://doi.org/10.1016/j.rser.2019.109494.

Zhang, D., Shen, J., Zhang, F., Li, Y., Zhang, W., 2017. Carbon footprint of grain production in China. Scientific Reports 7 (1). Available at https://doi.org/10.1038/s41598-017-04182-x.

Zhang, L., Wylie, B.K., Ji, L., Gilmanov, T.G., Tieszen, L.L., 2010. Climate-driven interannual variability in net ecosystem exchange in the Northern Great Plains grasslands. Rangeland Ecology and Management 63 (1), 40–50. Available at https://doi.org/10.2111/08-232.1.

Zhang, X., 2018. Simulating eroded soil organic carbon with the SWAT-C model. Environmental Modelling & Software 102, 39–48. Available at https://doi.org/10.1016/j.envsoft.2018.01.005.

Zhang, X., Izaurralde, R.C., Arnold, J.G., Williams, J.R., Srinivasan, R., 2013. Modifying the Soil and Water Assessment Tool to simulate cropland carbon flux: Model development and initial evaluation. Science of the Total Environment 463-464, 810–822. Available at https://doi.org/10.1016/j.scitotenv.2013.06.056.

Zhao, J., Wang, T., Deng, J., Shu, C.M., Zeng, Q., Guo, T., et al., 2020. Microcharacteristic analysis of CH4 emissions under different conditions during coal spontaneous combustion with high-temperature oxidation and in situ FTIR. Energy 209. Available at https://doi.org/10.1016/j.energy.2020.118494.

Chapter 21

Measurement, reporting, and verification (MRV) of methodologies for carbon credits in cropping systems

Saikat Mondal[1] and Debnath Palit[2]

[1]Raghunathpur College, Raghunathpur, Purulia, West Bengal, India, [2]Durgapur Government College, Durgapur, West Bengal, India

Highlights

- Carbon credit assessment and valuation are gaining importance in agriculture.
- Proper MRV techniques for diverse agroecosystems are required.
- MRV helps generate a carbon market for agroecosystems.
- Adopting MRV and carbon credits leads to a sustainable farming system.

21.1 Introduction

Agricultural systems can both contribute to and mitigate the effects of climate change. Agriculture has a considerable potential to sequester carbon (C) through techniques like crop rotation, cover crops, no-till farming, and agroforestry (AF), even though it is responsible for about 24% of global greenhouse gas (GHG) emissions (Meena et al., 2025). Methodologies for measuring and verifying carbon sequestration (Cseq) in these systems are important due to the growing market for C credits (Raj et al., 2025; Meena, Lal, et al., 2024). Agricultural producers can sell C offsets based on verified decreases in GHG emissions or gains in Cseq through C credit systems such as the Voluntary Carbon Standard (VCS) and the Gold Standard. However, robust measurement, reporting, and verification (MRV) procedures that account for the intricacies of cropping systems are necessary to ensure the accuracy and reliability of these C credits. This chapter reviews the latest research on MRV frameworks for C credits in cropping systems, including advancements in C measurement techniques, data reporting, and the verification processes that underpin market integrity.

21.1.1 Overview of carbon credits in cropping systems

One metric ton of carbon dioxide (CO_2) emissions that have been cut down or eliminated from the atmosphere is represented as a C credit. A market-based strategy for encouraging sustainable farming methods, especially in cropping systems, is the use of C credits (Meena et al., 2025a). By using techniques such as cover crops, decreased tillage, and AF that improve soil C storage, farmers can receive C credits (Jhariya et al., 2022; Jhariya et al., 2023). After that, these credits can be traded on C markets, generating additional income and helping fight climate change (Fig. 21.1). Recent studies highlight the perspectives of farmers on C markets, emphasizing the importance of understanding their motivations and challenges in participating in these systems (Barbato and Strong, 2023). Additionally, research indicates that C farming can be profitable, although outcomes may vary geographically (Plastina et al., 2024). The evolution of C credit markets has also been examined, highlighting their significance in sustainable agricultural practices (Raina, 2024; Meena et al., 2024). Supportive policy frameworks are essential for the growth of C markets, including clear regulations, standards for C accounting, and incentives for participation. So, integrating C credits into cropping systems presents a promising opportunity for farmers to contribute to climate change mitigation (CCM) while enhancing their economic viability. Continued research and supportive policies will be crucial in maximizing the potential of C credits in agriculture (Kumar et al., 2024).

Carbon Incentives, Plans, and Policies in Agriculture. DOI: https://doi.org/10.1016/B978-0-443-33557-0.00027-6

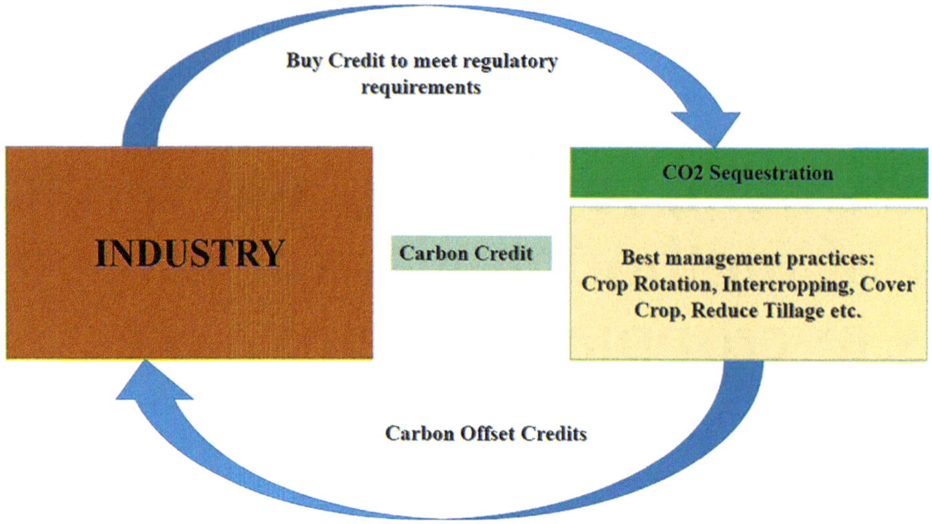

FIGURE 21.1 Carbon credit trading framework.

21.1.2 The role of cropping systems in carbon sequestration

Enhancing C_{seq} is crucial for reducing climate change, and cropping systems play a key role in this process. Recent research has highlighted numerous techniques that enhance soil C storage and overall soil health. One important study shows that, especially in semiarid areas, cover crops can raise soil organic carbon (SOC) levels (Meena et al., 2024). According to this study, adding cover crops to cropping systems enhances soil health and increases SOC, thereby promoting more sustainable farming practices (Acharya et al., 2024). The impact of different cropping methods on total organic carbon (TOC) and its fractions is evaluated in another significant study. According to research, different cropping systems may lead to notable variations in soil biological activity, which is essential for C_{seq} (Acharya et al., 2024). The annual C_{seq} rates in regenerative farming approaches throughout vineyards and arable crops are measured by Villat and Nicholas (2023). The potential of regenerative methods to greatly increase soil C reserves is highlighted by this study. The benefits of perennial farming systems over annual ones are also highlighted by Shang et al. (2024), who demonstrate that they can increase soil C stocks while preserving production. All things considered, these findings underscore the importance of cropping systems in Cseq and the need to adopt sustainable farming methods to enhance soil C storage and effectively address climate change (Banerjee et al., 2023; Banerjee et al., 2024; Meena et al., 2023). To improve C_{seq}, which is crucial for C mitigating the effects of climate change, cropping systems are necessary. Numerous agricultural techniques that enhance soil C storage and general soil health have been the subject of recent research (Meena and Pradhan, 2023). According to Acharya et al. (2024), the addition of cover crops to cropping systems significantly increases SOC levels, particularly in semiarid areas. By improving soil health and SOC, this integration encourages more environmentally friendly farming methods. Ankit et al. (2024) assess how various cropping systems influence TOC and its fractions. The findings indicate that diverse cropping systems can lead to substantial changes in soil biological activities, which are crucial for effective C_{seq}. Villat and Nicholas (2023) quantify annual C_{seq} rates in regenerative agricultural practices across arable cropland and vineyards. This study emphasizes the potential of regenerative practices to enhance soil C stocks significantly. Shang et al. (2024) highlight the benefits of perennial cropping systems over annual ones. It demonstrates that perennial systems can achieve higher soil C stocks while maintaining productivity, showcasing their effectiveness in C_{seq}.

21.2 The significance of measurement, reporting, and verification in carbon credit generation

C credits have emerged as a crucial tool for promoting sustainable behaviours and emissions reductions as international efforts to combat climate change intensify. Efficient MRV procedures are critical to the success of C credit schemes. To ensure that C credits accurately reflect actual, quantifiable, and verifiable emission reductions, MRV is essential.

21.2.1 Understanding measurement, reporting, and verification in carbon credit generation

Measurement entails quantifying the amount of CO_2 that is either sequestered or reduced by a particular project. Precise measurement is crucial for establishing baseline emissions and determining the actual reductions achieved. A variety of methodologies are utilized, including direct measurements, remote sensing technologies, and modeling techniques. The choice of methodology can significantly affect the credibility of the resulting C credits. Reporting involves the organized documentation of measurement outcomes, the methodologies employed, and any assumptions made throughout the process. Clear and transparent reporting is essential for stakeholders, including investors, regulatory agencies, and the general public, to understand the effectiveness of C credit initiatives. It also aids in meeting regulatory obligations and adhering to standards set by certification organizations. Verification is the independent evaluation of the reported data to confirm its accuracy and reliability. Third-party verifiers assess the methodologies, data, and processes utilized during the measurement and reporting stages. This verification process fosters trust in the C credit market by ensuring that the credits issued are valid and reflect actual emissions reductions.

21.2.2 Significance of measurement, reporting, and verification in carbon credit generation

Strong MRV processes enhance the credibility of C credits, which is crucial for establishing trust among stakeholders. This trust is crucial to the effective operation of C markets, as it encourages investment in projects that reduce C emissions. In many regions, verified C credits are required to meet emissions-reduction goals. Effective MRV ensures that projects adhere to regulatory standards, thereby facilitating participation in compliance markets. Accurate measurement and reporting simplify the C credit trading process, lowering transaction costs and improving market efficiency. This efficiency is essential for attracting investment in C credit initiatives. Furthermore, MRV processes guarantee that C credits genuinely represent emissions reductions, thereby enhancing the overall integrity of climate action efforts. In light of global climate agreements, such as the Paris Agreement, this is particularly important. Standardized MRV procedures are crucial for boosting the legitimacy of C credits in forestry projects, according to research by Houghton et al. (2023). The report promotes the use of uniform approaches to increase project comparability and transparency. The challenges and opportunities of MRV for agricultural C credits are examined by Popp et al. (2022). In order to increase measurement accuracy and save costs, the authors emphasize the use of cutting-edge technology like machine learning and remote sensing. The contribution of MRV to improving the efficiency of C credit schemes in low-income nations is examined by Zomer et al. (2021). The study highlights the need for technical assistance and capacity building to establish efficient MRV procedures in these areas. It is impossible to overestimate the importance of MRV in the creation of C credits. Strong MRV procedures are necessary to ensure the legitimacy, transparency, and efficiency of carbon credit schemes. Enhancing MRV in C markets requires the implementation of innovative technology, standardized protocols, and capacity-building initiatives, as a recent scientific study has shown. Stakeholders can promote trust, attract investment, and ultimately contribute to the fight against climate change by strengthening MRV procedures.

21.3 Carbon sequestration in soils

21.3.1 Soil carbon and the global carbon cycle

A significant fraction of the C in the planet's terrestrial ecosystems is stored in soil. Terrestrial environments contain roughly 3170 gigatons of C (GT; 1 GT=1 petagram=1 billion metric tons). Soil contains about 80% of this amount (2500 GT). There are two types of C in soil: inorganic (950 GT) and organic (1550 GT). Calcite, dolomite, and gypsum are examples of carbonate minerals, and elemental C makes up the latter (Lal, 2004). Compared to soil, which contains 560 GT of C, live plants and animals contain relatively less of it. Compared to the atmospheric pool of 800 GT, the soil C pool is roughly 3.1 times greater (Oelkers and Cole, 2008).

21.3.2 Soil carbon sequestration

The act of absorbing atmospheric CO_2 and storing it in the soil C pool is known as soil C_{seq}. This process is largely enabled by plants' ability to transform C into SOC through photosynthesis (Fig. 21.2) (Lal, 2011; Pradhan and Meena, 2023b; Majumder et al., 2008). Although it occurs much more slowly, CO_2 can also be converted into inorganic forms, such as secondary carbonates, in dry and semiarid locations. Between 50 and 100 gigatons of C have been released into the atmosphere since the Industrial Revolution as a result of the transition from natural ecosystems to agricultural activities, which has significantly reduced SOC levels. Reduced plant root biomass and residues returned to the soil, increased breakdown from soil tillage, and increased soil erosion are the causes of this depletion.

21.3.3 Carbon sequestration practices in cropping systems

A viable strategy for lowering atmospheric CO_2 concentrations to mitigate climate change has been made possible by the enormous potential of C_{seq} in farmland. Nevertheless, cropping techniques are necessary for this strategy.

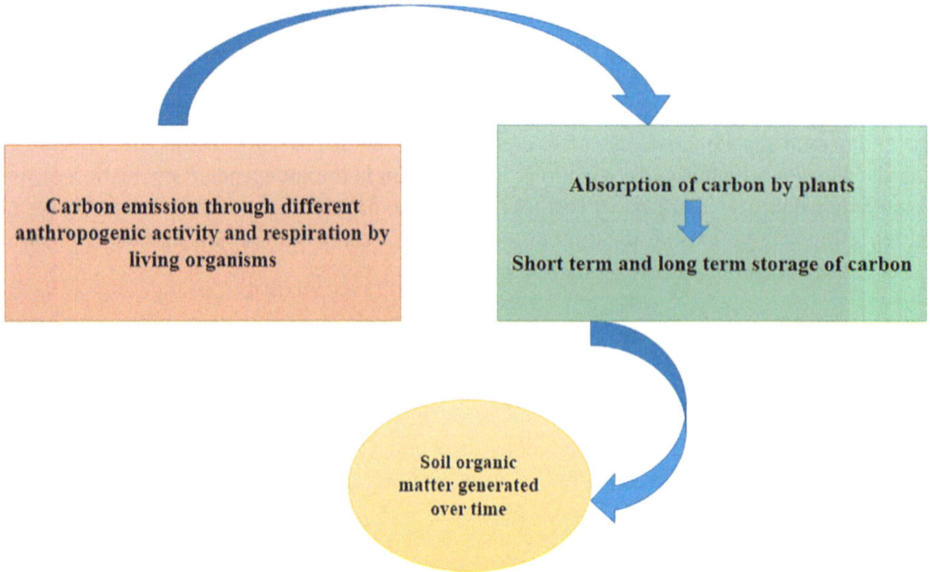

FIGURE 21.2 Capture of carbon and sequestration.

21.3.3.1 Agroforestry

It is well known that AFs can store C, which would greatly aid the fight against climate change. According to research, AF can improve a variety of regulating ecosystem services, increase C sinks, and lower CO_2 emissions (Meena et al., 2022). According to a study by Zomer et al. (2022), South America has the greatest potential to sequester more than 18 PgC if tree cover on agricultural land were to increase by 10% worldwide. AFs used by smallholder farmers in West Java Province also have substantial potential for C storage, according to a study by Siarudin et al. (2021). AF is a viable strategy for sustainable land use, with substantial potential for C_{seq}, according to another study by Nair et al. (2010). Stakeholders can encourage environmental sustainability and economic resilience among farmers, especially in developing nations, by improving knowledge of AFs and their management (Prasad et al., 2021; Jhariya et al., 2021; Jhariya et al., 2021). To fully reap the benefits of AF in the fight against climate change, more research and funding are needed. Additionally, AF was identified by Udawatta et al. (2022) as having the highest potential for C_{seq} among all land uses examined in the IPCC's LULUCF report (2000). These results underscore the significance of AF in C_{seq} and its potential to contribute to the fight against climate change.

21.3.3.2 Crop rotation

Crop rotation improves soil C_{seq} and biomass production, especially when it involves legumes and nonlegumes. By drastically reducing the need for chemical nitrogen fertilizers, legumes can help decrease the amount of fossil fuels used to produce fertilizer (Zentner et al., 2001; Jhariya et al., 2018; Zentner et al., 2004; Devi et al., 2024). On the other hand, improper crop rotation can lead to increased weed, disease, and insect infestations, which can reduce soil biomass and yields. Another good way to increase biomass output and soil C_{seq} is to increase cropping intensity by reducing bare land fallow periods (Pradhan and Meena, 2023a; Kumar et al., 2020). According to Dumanski et al. (1998), increased cropping intensity can also slow down the pace at which organic matter (OM) breaks down and the oxidation or mineralization of SOC. Low C-to-nitrogen residues could improve soil C_{seq} by 13–30 Tg yr^{-1} in the primary corn-soybean region of the United States, according to this study. This amounts to 1%–2% of the estimated annual C emissions from burning fossil fuels in the United States (1.4 Pg C yr^{-1}) (Marland and Boden, 1997).

21.3.3.3 Intercropping

By maximizing the use of sunlight through various plant designs and spatial arrangements, intercropping enhances crop yield. Different intercropping systems, including relay, mixed, row, and strip intercropping, are designed to meet particular cropping objectives and crop characteristics. When corn or sorghum is row interplanted with climbing crops, such as beans or sweet potatoes, for example, the latter can utilize the former's structure for improved access to sunlight, and the primary crops can still produce at their best even while supporting the vines. Choosing crops with staggered maturity dates can make harvesting easier, which is why strip cropping is frequently used, particularly when using combined harvesters. Permitting the later flowering and ripening of the slower-growing pigeonpea (Willy, 1983). Productivity can be greatly increased by intercropping; in South Carolina, for instance, intercropping southern peas and sweet corn required 30%–48% more area to provide the exact yield as monoculture systems (Francis and Decoteau, 1993). Compared with monoculture, a strip-cropping system consisting of oats, corn, and soybeans in Iowa enhanced net returns by 38% (Anon, 1990). Overall yields in the intercropping system were higher than in monoculture due to synergistic effects among crops, even though soybean yields on border rows decreased by 10% due to shading (Sullivan, 2003). These advantages result from reciprocal benefits, such as early-maturing oats using resources before corn produces shade and corn protecting oats from the wind. Additionally, soybeans can fix nitrogen, benefiting corn, and strip intercropping can reduce pest and pathogen infestations. For example, in Yunnan province, China, mixing two rice varieties effectively controlled blast disease, reducing the need for chemical fungicides (Wolf, 2000).

21.3.3.4 Mixed cropping

It is a successful intercropping technique that maximizes ecosystem productivity by planting two or more species in close proximity. With limited resources, this method helps control plant diseases, suppress weeds and pests, balance soil nutrient inputs and outputs, endure temperature extremes, and increase overall productivity (Hirst, 2009). The Native American "three sisters" system—corn, beans, and squash—is a prime example. In this arrangement, squash covers the ground to keep weeds out and minimize water evaporation, corn provides a support structure for the beans, and the beans add nutrients to the soil. This mutualistic interaction promotes the best biomass production and soil C_{seq}. Mixed cropping systems have been modified for simpler harvesting and management in contemporary agriculture. Peanut and sunflower, soybean and pigeon pea, wheat and chickpea, and other combinations have been shown to be successful. In nations like China and India, mixed cropping has a long history. For example, research has demonstrated that growing faba beans alongside cereals can improve disease management in Ethiopia (Sahile et al., 2008), increase yields of wheat and chickpea in Australia (Jahansooz et al., 2007), and benefit both parties in Europe (Gunes et al., 2007).

Relay intercropping, such as planting soybeans into standing winter wheat shortly before wheat harvest, maximizes spatial and temporal advantages, reduces fallow periods, and conserves soil moisture. Successful relay intercropping relies on proper timing and wheat row spacing; for example, a spacing of 25–38 cm is recommended for optimal soybean development in Ohio. This system effectively captures sunlight, increases net returns, mitigates risks associated with commodity price fluctuations, and ensures efficient use of natural resources, thereby improving biomass accumulation.

21.3.3.5 Cover cropping

An efficient way to improve SOC storage and C_{seq} is to grow cover crops. Winter cover crops, including rye, ryegrass, oats, peas, vetch, and clover, are usually planted in the fall in temperate countries. They survive the mild winter and grow again in the spring, covering bare ground during the off-season. When combined with crops like cotton and sorghum, the biomass production for vetch and rye in bi-culture can range from 5.7 to 8.2 Mg ha^{-1} above-ground and 372 to 880 kg ha^{-1} below-ground, resulting in total C inputs to the soil between 6.8 and 22.8 Mg ha^{-1} (Sainju et al., 2006). According to Sainju et al. (2006), using cover crops enhanced SOC by 6%–8% at depths of 0–10 cm. Summer cover crops, such as sun hemp, velvet bean, and sorghum sudan grass, are frequently grown in tropical and subtropical areas during hot, humid summers to conserve soil, retain moisture, and protect bare soil. In just three months, sun hemp, in particular, can yield up to 15 Mg ha^{-1} of above-ground biomass and 3.5 Mg ha^{-1} of below-ground biomass, adding about 8 Mg ha^{-1} of organic C input into the soil (Wang et al., 2005). All things considered, cover cropping systems are a useful tactic for enhancing C_{seq} and reducing climate change.

21.3.3.6 Companion cropping

Companion cropping, which involves growing companion crops alongside vegetables such as lettuce and cabbage, is a widely adopted approach in organic vegetable production. Because companion crops can enhance biodiversity and mitigate the impact of weeds and pests on vegetable crops, this approach is considered a viable alternative in organic farming. There are various ways that companion cropping might be advantageous. Pests may be drawn to certain companion crops instead of vegetable crops. For instance, because collards are more enticing to the pests, they can divert the diamondback moth (*Plutella xylostella*) from cabbage (Boucher, 2000). Certain plants emit compounds that can keep pests away or shield nearby plants. For example, the African marigold (*Tagetes erecta*) is a good companion for many garden crops because it releases thiopene, a nematode repellent. Allelochemicals, such as juglone from black walnut, can also help to manage weeds by preventing the growth of other plants. Mowed-down rye can be used as mulch to prevent weeds from growing without damaging transplanted vegetables like broccoli and tomatoes. More delicate species, like ginger, can benefit from the shade or wind protection that taller, denser plants can offer. For instance, alfalfa and other forage crops have historically been established using oats. In certain situations, the nursing effect may only involve physical contact that helps the primary crop. All things considered, through these numerous advantageous interactions, companion cropping in organic farming can increase crop resilience and productivity.

21.3.3.7 Ratoon cropping

A technique called "ratio cropping" enables a single crop to yield multiple harvests from a single planting. The crop needs to be perennial, have a well-established root system, and mature early for ratoon cropping to be successful. This method offers several benefits for soil C_{seq} and crop development. Ratoon cropping, for instance, reduces production costs by lowering expenditures on planting and site preparation. It optimizes the growth season, effectively captures sunlight energy, and can deliver higher yields and biomass per unit area within a certain timeframe. Because of its shorter growth cycle, it also uses less fertilizer and irrigation water than the original crop, helps to stop soil and water erosion, lessens nutrient leaching, and is typically more economically productive than conventional cropping systems. Sugarcane (*Saccharum officinarum* L.) and sorghum (*Sorghum bicolor* (L.) Moench) are two successful examples of ratooning (Rodolfo and Plucknett, 1975), in which the primary crop is harvested at a height of roughly 2.5–10 cm aboveground after maturity. Another vegetable that can be ratooned in tropical and subtropical climates is okra (*Abelmoschus esculentus*), which yields two or three harvests (Li et al., 2006).

Optimizing biomass output and enhancing crop development necessitate effective cropping techniques, including suitable fertilization and water management. However, traditional tillage techniques, especially moldboard plowing, can cause SOC to mineralize quickly, which depletes rather than sequesters it. Reducing soil disturbance and increasing the return of plant residues are essential for better C_{seq} in soil. Converting C_{seq} into stable forms, such as recalcitrant C or SOC, through carbonization or gradual humification is also crucial. Environmental factors such as soil type, temperature, and moisture significantly affect the stability of OC in plant residues and soil. Another important factor in determining OC's stability against degradation is plant makeup. According to McLauchlan and Hobbie's two-pool theories, OC in plants may generally be divided into two primary compartments: active and inert, which represent labile and recalcitrant pools, respectively (McLauchlan and Hobbie, 2004). Decomposable OC, resistant OC, microbial biomass OC, and humified OC are the four subcomponents of active OC (McLauchlan and Hobbie, 2004). The distribution of these OC compartments and their decomposition rates can be influenced by the physiological and chemical properties of plant wastes, including lignin content and the C-to-nitrogen (C:N) ratio. Through integrated cropping systems and techniques such as conservation tillage, cover crops, crop rotation, land use restoration, and fertilization, numerous studies have demonstrated that C_{seq} and SOC accumulation occur in croplands (Kuo and Jellum, 2002; West and Post, 2002; Sainju et al., 2002; Sainju et al., 2006). It is evident that the SOC pool has significant potential to store C, and when paired with efficient techniques, integrated cropping systems hold promise for removing C from the atmosphere and mitigating the effects of climate change.

21.4 Measurement of carbon sequestration

C_{seq} refers to the process of capturing and storing atmospheric CO_2 in CCM. Measurement of C_{seq} is crucial for understanding its effectiveness and involves various methods, including direct measurements from ecosystems, remote sensing, and modeling approaches. Key methods include the Eddy Covariance technique, which assesses gas exchange between the atmosphere and ecosystems, and loss-on-ignition (LOI) for soil C content analysis. Studies have shown that different land covers, such as forests and wetlands, have varying capacities for C storage, with forests generally being more effective. C_{seq} is a vital process in combating climate change, involving the capture and storage of atmospheric CO_2. Accurate measurement of C_{seq} is essential for evaluating the effectiveness of strategies to reduce atmospheric GHG concentrations (Pradhan et al., 2023).

21.4.1 Key measurement methods

Eddy covariance technique (Baldocchi, 2003): This method measures the exchange of gases between the atmosphere and ecosystems by analyzing fluctuations in wind speed and gas concentration. It provides real-time data on C fluxes, making it a valuable tool for understanding ecosystem-level C dynamics.

Soil carbon content analysis: Techniques such as LOI are employed to determine the OC content in soils. This method involves heating soil samples to combust OM, thereby allowing quantification of C stored in the soil (Heiri et al., 2001).

Remote sensing: Satellite and aerial imagery are utilized to assess vegetation cover and biomass, which are critical for estimating C storage in forests and other ecosystems. Remote sensing provides a broad-scale perspective on C_{seq} potential across different landscapes (Avitabile et al., 2011; Tian et al., 2023; Huang et al., 2022).

Modeling approaches: Various models, such as the Century model (Liu et al., 2003) or the C Budget Model (Pilli et al., 2018), simulate C dynamics in ecosystems based on input data like climate, soil type, and land use. These models help to predict future C_{seq} scenarios under different management practices.

Factors influencing carbon sequestration: The ability of various ecosystems to store C varies. For example, forests tend to store C more effectively than agricultural fields or grasslands. Due to their distinct hydrology and vegetation, wetlands also play a significant role in C storage. C_{seq} can be enhanced by employing sustainable land management techniques, such as conservation tillage, afforestation, and reforestation (Jhariya et al., 2019; Raj et al., 2023; Raj et al., 2024). On the other hand, methods such as intensive farming and deforestation can cause C emissions.

21.5 Reporting of carbon sequestration in agricultural systems

As CO_2 sources and sinks, agricultural systems are essential to the global C cycle. The significance of C_{seq} in agriculture as a method to reduce GHG emissions is becoming more and more recognized as a result of growing awareness of climate change. It is crucial to report C_{seq} in agricultural systems to monitor developments, inform policy, and encourage sustainable practices. By employing techniques such as cover crops, AF, reduced tillage, and improved grazing management, agriculture can help mitigate and possibly balance the C emissions from other sectors (Intergovernmental Panel on Climate Change IPCC, 2019).

21.5.1 Reporting frameworks

21.5.1.1 The greenhouse gas protocol

The GHG Protocol offers a detailed framework for measuring and reporting GHG emissions and removals, including C_{seq} in agriculture. It provides methodologies for quantifying emissions from agricultural practices and promotes transparency and consistency in reporting.

21.5.1.2 The Intergovernmental Panel on Climate Change guidelines

The Intergovernmental Panel on Climate Change provides guidelines for national GHG inventories, including methods for estimating C_{seq} in agricultural soils. These guidelines are commonly adopted by countries to report their emissions and removals, ensuring uniformity and comparability across various regions (Intergovernmental Panel on Climate Change IPCC, 2006).

21.5.1.3 Carbon credit programs

Numerous C credit programs, such as the Verified Carbon Standard (VCS) and the (Climate Action Reserve CAR, 2020), provide frameworks for quantifying and verifying C_{seq} in agricultural systems. These programs encourage farmers to implement sustainable practices by offering financial incentives for verified C credits (Verified Carbon Standard VCS, 2020). Despite the importance of reporting C_{seq}, several challenges exist: Accurate data on soil C stocks and changes are often limited, particularly in developing regions. C_{seq} rates can vary significantly based on soil type, climate, and management practices, making it difficult to generalize findings. Ensuring the accuracy and reliability of reported data is essential for maintaining credibility in C markets and policy frameworks.

21.6 Verification of carbon credits

C credits have become an essential instrument in the global effort to reduce GHG emissions as the world struggles to prevent climate change. By funding initiatives that absorb C or lower emissions elsewhere, businesses can use C credits as a unit of measurement to offset their emissions. However, the quality of the credits themselves determines how effective C credit schemes are; thus, strong verification procedures are required. Projects that either improve C_{seq} or lower emissions produce C credits. C credits, for example, can be generated through energy-efficiency upgrades, renewable-energy installations, and reforestation initiatives. One metric ton of CO_2 equivalent (CO_2e) that has been decreased or sequestered is usually represented by one C credit. Due to legal frameworks and voluntary pledges by governments and corporations to attain net-zero emissions, the market for C credits has expanded dramatically (World Bank, 2022; World Bank, 2024).

Verification is the process of ensuring that the C credits generated by a project are real, additional, and permanent. The integrity of C credits is crucial for several reasons: Verification builds trust among stakeholders, including investors, regulatory bodies, and the public, ensuring that C credits represent genuine emissions reductions. A reliable verification process is essential for the functioning of C markets. It ensures that credits can be traded with confidence, thereby promoting investment in C-reduction projects. Many jurisdictions require verified C credits to comply with emissions-reduction targets. Verification ensures that organizations meet their regulatory obligations. Verification helps to ensure that projects deliver the intended environmental benefits, contributing to global climate goals.

21.6.1 Verification methodologies

Verification of C credits typically involves several key steps (World Bank, 2022; World Bank, 2024):

21.6.1.1 Project design and baseline assessment

Before a project can generate C credits, it must be designed according to established methodologies that outline how emissions reductions will be achieved. A baseline assessment is conducted to determine the emissions that would have occurred in the absence of the project. This baseline serves as a reference point for measuring the project's impact.

21.6.1.2 Monitoring and data collection

Ongoing monitoring is essential to track project performance and measure actual emissions reductions. This may involve collecting data on factors such as biomass growth, energy production, or emissions from agricultural practices. Monitoring plans must be detailed and adhere to the methodologies established during project design.

21.6.1.3 Third-party verification

Impartial third-party verifiers evaluate the project's adherence to accepted standards and procedures. These verifiers visit the project site, review monitoring data, and assess whether the established procedures are being followed. Generally, the verification procedure adheres to accepted guidelines, such as those established by the Climate Action Reserve (CAR), the Gold Standard, or the Verified Carbon Standard (VCS).

21.6.1.4 Issuance of carbon credits

Once a project has been verified, C credits can be issued. The credits are then registered in a recognized registry, ensuring transparency and traceability. This process helps to prevent double-counting and ensures that each credit is unique.

Despite the importance of verification, several challenges persist, such as the complexity of measurement. Accurately measuring emissions reductions can be complex, particularly in projects involving biological processes, such as forestry or agriculture. Variability in factors like climate, soil type, and management practices can complicate assessments. The verification process can be costly, particularly for small-scale projects. High costs may deter participation in C markets, limiting the diversity of projects and the overall impact of C credit systems. The lack of universally accepted standards for verification can lead to inconsistencies in how projects are assessed. This can undermine confidence in C credits and create barriers to market participation. Instances of fraud or misrepresentation can occur, particularly in unregulated markets. Robust verification processes are essential to mitigate these risks and maintain the integrity of C credit systems (World Bank, 2022; World Bank, 2024).

21.7 Regulatory and market mechanisms for carbon credits in agriculture

C credits have emerged as a vital tool in the battle against climate change, especially within the agricultural sector (Boufous et al., 2024; Smith et al., 2019; Nayak et al., 2019). These credits enable farmers to capitalize on their efforts to sequester C and lower GHG emissions.

21.7.1 Regulatory mechanisms

Regulatory frameworks are essential for defining the rules and standards governing C credit markets. Key regulatory mechanisms include:

Cap-and-trade systems: These systems set a cap on total emissions and allow entities to buy and sell allowances. The agricultural sector can participate by generating credits through sustainable practices, which can then be sold to regulated industries that need to offset their emissions.

Carbon taxes: In some regions, C taxes are levied on emissions, providing an incentive for agricultural producers to adopt practices that minimize their C footprint. This creates a financial incentive for farmers to engage in C-credit generation.
Government programs: Various government initiatives support C farming through grants, technical assistance, and improved market access. Programs like the USDA's Conservation Stewardship Program promote practices that enhance C_{seq}.

21.7.2 Market mechanisms

Market mechanisms enable the trading of C credits, allowing farmers to monetize their sustainable practices. Key components include:

Voluntary carbon markets: These markets allow businesses and individuals to voluntarily purchase C credits to offset their emissions. Farmers can participate by selling credits generated from their sustainable practices.
Certification standards: To ensure the legitimacy of C credits, several certification standards were developed, such as the VCS and the Gold Standard. These standards provide guidelines for measuring, reporting, and verifying emissions reductions.
Blockchain technology: Emerging technologies like blockchain are being investigated to improve transparency and traceability in C credit transactions, ensuring that credits are not double-counted and that their origins can be verified (Zhang et al., 2024).
Recent studies have highlighted the effectiveness and challenges of C credit mechanisms in agriculture by Raina (2024) systematically reviews incentive mechanisms for C farming contracts, emphasizing the importance of clear guidelines and support for farmers to engage in C markets. Research by Barbato and Strong (2023) discusses farmer perspectives on C markets, noting the need for accurate measurement and verification of C_{seq} to build trust in these markets. Lokuge and Anders (2022) identify gaps in knowledge regarding C credit systems in agriculture, suggesting areas for further research and policy development.

21.8 Recent advancements in measurement, reporting, and verification in carbon credit generation

The field of C credit generation is undergoing a significant transformation, largely due to advancements in MRV systems. These innovations are vital for maintaining the integrity and efficiency of C markets, allowing stakeholders to invest in and trade C credits with confidence. Recent investigations underscore the need to incorporate digital technologies into MRV frameworks. To increase data accuracy and transparency, digital monitoring, reporting, and verification (D-MRV) systems use technologies such as blockchain, geospatial imagery, and geographic information systems. For example, a report from the World Bank (2024) highlights how D-MRV systems can lower costs and enhance the security of transactions within C markets, facilitating the generation of credible C credits (World Bank, 2024). The movement towards standardized MRV protocols has gained traction, especially within the forestry and agriculture sectors. Research conducted by Houghton et al. (2023) supports the need for consistent methodologies to improve comparability and transparency among C credit projects. Establishing such standards is crucial for fostering trust among stakeholders and ensuring that C credits reflect authentic emissions reduction (Houghton et al., 2023).

New measurement techniques are being introduced to enhance the precision and efficiency of MRV processes. A study by Popp et al. (2022) discusses how remote sensing and machine learning can significantly improve the accuracy of measurements for agricultural C credits. These technologies have the potential to lower costs and enhance the reliability of reported data, addressing some of the longstanding challenges in MRV (Popp et al., 2022). Recent research has also suggested developing modular, scalable MRV frameworks tailored to different ecosystems and project types. Batjes et al. (2024) created a framework that categorizes various MRV methodologies, allowing for adaptable and scalable implementation. This approach is particularly advantageous for projects in developing regions, where resources and technical skills may be limited (Batjes et al., 2024; Houghton et al., 2023). Stressing the importance of standardized MRV protocols in enhancing the credibility of C credits within forestry initiatives. Their study advocates for uniform methodologies to boost comparability and transparency across various projects (Houghton et al., 2023; Popp et al., 2022) examine the challenges and opportunities in MRV for agricultural C credits, emphasizing the necessity for innovative technologies to enhance measurement accuracy and reduce expenses (Popp et al., 2022). To increase implementation flexibility, Batjes et al. (2024) provide a framework for a modular and scalable MRV system that includes an inventory and classification of specific MRV approaches.

21.9 Future directions and research gaps in measurement, reporting, and verification methodologies

The future of MRV methodologies for C credits in cropping systems is evolving, with significant opportunities for research and development. Key areas for future exploration include the integration of advanced digital technologies, the establishment of standardized protocols across diverse agricultural practices, and addressing the complexities of soil C measurement. Using imagery from space and geographic information systems can improve crop growth and C_{seq} monitoring. D-MRV systems are expected to reduce costs and improve transparency in C market transactions (World Bank, 2022). Developing uniform MRV guidelines across various cropping systems is essential for consistency and credibility. Recent studies emphasize the need for best practices in data collection and reporting to facilitate global harmonization (Batjes et al., 2024). Research is needed to develop accurate methods for assessing soil C stocks, taking into account the variability in soil types and management practices. Innovative tools for farmers to measure soil C levels cost-effectively are also crucial (Smith et al., 2020).

Understanding how MRV systems influence smallholder farmers and their engagement in C markets is vital. Research should focus on the equitable distribution of benefits and barriers faced by marginalized communities (Fund, 2022). Ensuring data reliability and robust verification processes is essential. Promoting open data initiatives can enhance collaboration among stakeholders, including farmers and policymakers (Köhl et al., 2020). Analyzing the effectiveness of existing policies in supporting MRV practices is necessary. Identifying regulatory gaps that impede MRV implementation can help to advocate for policies that incentivize sustainable agricultural practices (Mungroo, 2023). Advancing MRV methodologies for C credits in cropping systems is crucial for meeting climate objectives. By addressing research gaps and focusing on innovative solutions, stakeholders can enhance the effectiveness of MRV systems, leading to sustainable agricultural practices and improved environmental outcomes.

21.10 Conclusion

MRV are essential components that underpin the credibility and effectiveness of C credit generation, acting as the foundation for C markets. As the global community increasingly relies on C credits to combat climate change, the significance of robust MRV systems becomes even more pronounced. These systems ensure that C credits accurately reflect genuine, measurable, and verifiable reductions in emissions, thereby fostering trust among various stakeholders, including investors, regulatory agencies, and the general public. Recent advancements in MRV technologies—such as digital monitoring, remote sensing, and machine learning—have a role in improving the precision and efficiency of C credit assessments. These technological innovations not only lower costs but also enhance data reliability, addressing many of the historical challenges linked to MRV processes. Additionally, the movement toward standardized protocols across different sectors, especially in forestry and agriculture, has played a crucial role in improving transparency and comparability. Such standardization fosters confidence in C markets, ensuring that credits from diverse projects can be reliably assessed and traded. The emergence of modular and scalable MRV frameworks also provides the necessary flexibility for implementation, catering to the varied requirements of projects across different ecosystems and regions. This adaptability is particularly vital for initiatives in developing nations, where resources and technical know-how may be scarce. By offering a structured yet flexible MRV approach, stakeholders can more effectively navigate the complexities of C-credit generation, ensuring their initiatives make a meaningful contribution to global climate objectives. Nonetheless, challenges persist. Accurately measuring emissions reductions, managing verification costs, and building capacity in less developed areas are significant obstacles that need to be addressed. Overall, effective MRV systems are not merely a regulatory necessity; they are vital to the success of C credit markets and the broader effort to combat climate change. MRV procedures can encourage investment in sustainable practices, preserve environmental integrity, and ultimately help achieve global climate goals by ensuring that C credits are reliable and accurately reflect the decrease in emissions. As C markets continue to evolve, the ongoing development of MRV will serve as a powerful instrument for climate action.

References

Acharya, P., Ghimire, R., Acosta-Martínez, V., 2024. Cover crop-mediated soil carbon storage and soil health in semi-arid irrigated cropping systems. Agriculture, Ecosystems & Environment 361, 108813. Available at https://doi.org/10.1016/j.agee.2023.108813.

Ankit, D., Prakash, S., Sheoran, P.K., Yadav, M., Kumari, S., Umar, K., et al., 2024. Different cropping systems impact soil health by improving soil biological activities and total organic carbon content. Archives of Agronomy and Soil Science 70 (1), 1–24. Available at https://doi.org/10.1080/03650340.2024.2419035.

Anon, 1990. Strip intercropping offers low-input way to boost yields. Sensible Agriculture, Mono Publication 7–8.

Avitabile, V., Herold, M., Henry, M., Schmullius, C., 2011. Mapping biomass with remote sensing: a comparison of methods for the case study of Uganda. Carbon Balance and Management 6 (1). Available at https://doi.org/10.1186/1750-0680-6-7.

Baldocchi, D.D., 2003. Assessing the eddy covariance technique for evaluating carbon dioxide exchange rates of ecosystems: past, present and future. Global Change Biology 9 (4), 479–492. Available at https://doi.org/10.1046/j.1365-2486.2003.00629.x.

Banerjee, A., Jhariya, M.K., Bargali, S.S., Palit, D., 2023. Ecorestoration for Sustainability. Wiley-Scrivener, USA, pp. 1–441 ISBN: 9781119879718.

Banerjee, A., Jhariya, M.K., Raj, A., Mechergui, T., 2024. Ecosystem Management: Climate Change and Global Sustainability—An Introduction. In: Banerjee, A., Jhariya, M.K., Raj, A., Mechergui, T. (Eds.), Ecosystem Management: Climate Change and Sustainability. Wiley, India, pp. 1–21. Available at https://doi.org/10.1002/9781394231249.ch1.

Barbato, C.T., Strong, A.L., 2023. Farmer perspectives on carbon markets incentivizing agricultural soil carbon sequestration. Npj Climate Action 2 (1). Available at https://doi.org/10.1038/s44168-023-00055-4.

Batjes, N.H., Ceschia, E., Heuvelink, G.B.M., Demenois, J., Le Maire, G., Cardinael, R., et al., 2024. Towards a modular, multi-ecosystem monitoring, reporting and verification (MRV) framework for soil organic carbon stock change assessment. Carbon Management 15 (1). Available at https://doi.org/10.1080/17583004.2024.2410812.

Boucher, J., 2000. Setting a trap. American Vegetable Grower 20–22.

Boufous, S., Wade, T., Chakravarty, S., Andreu, M., Bhadha, J.H., Her, Y.G., et al., 2024. Introduction to Carbon Credit Markets and their Potential for Florida Agricultural Producers [FE1154]. EDIS.

Climate Action Reserve (CAR), 2020. About the climate action reserve. Available from https://www.climateactionreserve.org/how/program-resources/programqandas/.

Devi, A., Jhariya, M.K., Raj, A., Banerjee, A., Singh, K.P., Singh, B., 2024. Acacia nilotica: a promising species for soil sustainability. In: Raj, A., Jhariya, M.K., Banerjee, A., Nema, S., Bargali, K. (Eds.), Land and Environmental Management through Forestry. Wiley, India, pp. 339–353. Available at https://doi.org/10.1002/9781119910527.ch14.

Dumanski, J., Desjardins, R.L., Tarnocai, C., Monreal, C., Gregorich, E.G., Kirkwood, V., et al., 1998. Possibilities for future carbon sequestration in canadian agriculture in relation to land use changes. Climatic Change 40 (1), 81–103. Available at https://doi.org/10.1023/a:1005390815340.

Francis, R., Decoteau, D.R., 1993. Developing an effective southernpea and sweet corn intercrop system. HortTechnology 3 (2), 178–184. Available at https://doi.org/10.21273/horttech.3.2.178.

Fund, B.E., 2022. Agricultural soil carbon credits. EDF. https://www.edf.org/sites/default/files/content/agricultural-soil-carbon-credits-protocol-synthesis.pdf.

Gunes, A., Inal, A., Adak, M.S., Alpaslan, M., Bagci, E.G., Erol, T., et al., 2007. Mineral nutrition of wheat, chickpea and lentil as affected by mixed cropping and soil moisture. Nutrient Cycling in Agroecosystems 78 (1), 83–96. Available at https://doi.org/10.1007/s10705-006-9075-1.

Heiri, O., Lotter, A.F., Lemcke, G., 2001. Loss on ignition as a method for estimating organic and carbonate content in sediments: reproducibility and comparability of results. Journal of Paleolimnology 25 (1), 101–110. Available at https://doi.org/10.1023/A:1008119611481.

Hirst, K.K., 2009. Mixed cropping, agricultural technique known as mixed cropping. http://archaeology.about.com/od/historyofagriculture/qt/mixed_cropping.html.

Houghton, R.A., Goodall, J.L., Houghton, J.T., 2023. Standardizing MRV protocols for carbon credits in forestry. Environmental Science & Policy 132, 1–10. Available at https://doi.org/10.1016/j.envsci.2022.11.005.

Huang, C., Gong, W., Pang, Y., 2022. Remote sensing and forest carbon monitoring—a review of recent progress, challenges, and opportunities. Journal of Geodesy and Geoinformation Science 5 (2), 124–147. Available at https://doi.org/10.11947/j.JGGS.2022.0212.

Intergovernmental Panel on Climate Change (IPCC), 2006. 2006 IPCC guidelines for national greenhouse gas inventories. Volume 4: Agriculture, forestry and other land use. Available from: https://www.ipcc-nggip.iges.or.jp/pu.

Intergovernmental Panel on Climate Change (IPCC), 2019. Climate change and land: an IPCC special Intergovernmental Panel on Climate Change (IPCC). (n.d.). Climate change 2007: the physical science basis. Available from: https://www.ipcc.ch/report/ar4/wg1/.

Jahansooz, M.R., Yunusa, I.A.M., Coventry, D.R., Palmer, A.R., Eamus, D., 2007. Radiation- and water-use associated with growth and yields of wheat and chickpea in sole and mixed crops. European Journal of Agronomy 26 (3), 275–282. Available at https://doi.org/10.1016/j.eja.2006.10.008.

Jhariya, M.K., Banerjee, A., Meena, R.S., Yadav, D.K., 2019. Sustainable Agriculture, Forest and Environmental Management. Springer, pp. 606. Available at https://doi.org/10.1007/978-981-13-6830-1.

Jhariya, M.K., Banerjee, A., Swaroop Meena, R., Kumar, S., Raj, A., 2021. Sustainable Intensification for Agroecosystem Services and Management. Springer, Singapore, pp. 1–870. Available at https://doi.org/10.1007/978-981-16-3207-5.

Jhariya, M.K., Banerjee, A., Yadav, D.K., Raj, A., Meena, R., Das, A., Yadav, G., Lal, R., 2018. Leguminous trees an innovative tool for soil sustainability. In: Meena, R., Das, A., Yadav, G., Lal, R. (Eds.), Legumes for Soil Health and Sustainable Management. Springer, Singapore, pp. 315–345. Available at https://doi.org/10.1007/978-981-13-0253-4_10.

Jhariya, M.K., Meena, R.S., Banerjee, A., Kumar, S., Raj, A., 2023. Agroforestry for Carbon and Ecosystem Management. Academic Press, pp. 1–434. Available at 10.1016/C2021-0-02916-3.

Jhariya, M.K., Meena, R.S., Banerjee, A., 2021. Ecological Intensification of Natural Resources for Sustainable Agriculture. Springer, Singapore, pp. 1–655. Available at https://doi.org/10.1007/978-981-33-4203-3.

Jhariya, M.K., Meena, R.S., Banerjee, A., Meena, S.N., 2022. Natural Resources Conservation and Advances for Sustainability. Elsevier. Available at https://doi.org/10.1016/C2019-0-03763-6.

Li, Y., Klassen, W., Lamberts, M., Olczyk, T., 2006. Okra production in Miami-Dade County, Florida. HS-875. http://edis.ifas.ufl.edu., no date)gl/vol4.html.

Kumar, S., Meena, R.S., Jhariya, M.K., 2020. Resources Use Efficiency in Agriculture. Springer Nature, Singapore, pp. 760. Available at https://doi.org/10.1007/978-981-15-6953-1.

Kumar, S., Meena, R.S., Sheoran, P., Jhariya, M.K., 2024. Regenerative Agriculture for Sustainable Food Systems. Springer, Singapore, pp. 1–493. Available at https://doi.org/10.1007/978-981-97-6691-8.

Kuo, S., Jellum, E.J., 2002. Influence of winter cover crop and residue management on soil nitrogen availability and corn. Agronomy Journal 94 (3), 501–508. Available at https://doi.org/10.2134/agronj2002.5010.

Köhl, M., Neupane, P.R., Mundhenk, P., 2020. REDD+ measurement, reporting and verification–a cost trap? Implications for financing REDD+MRV costs by result-based payments. Ecological Economics 168, 106513. Available at https://doi.org/10.1016/j.ecolecon.2019.106513.

Lal, R., 2011. Sequestering carbon in soils of agro-ecosystems. Food Policy 36 (1), S33. Available at https://doi.org/10.1016/j.foodpol.2010.12.001.

Lal, R., 2004. Soil carbon sequestration impacts on global climate change and food security. Science 304 (5677), 1623–1627. Available at https://doi.org/10.1126/science.1097396.

Liu, S., Bliss, N., Sundquist, E., Huntington, T.G., 2003. Modeling carbon dynamics in vegetation and soil under the impact of soil erosion and deposition. Global Biogeochemical Cycles 17 (2), 1074. Available at 10.1029/2002GB002010.

Lokuge, N., Anders, S., 2022. Carbon-credit systems in agriculture: a review of literature. Policy School 15, 12.

Majumder, B., Mandal, B., Bandyopadhyay, P.K., Gangopadhyay, A., Mani, P.K., Kundu, A.L., et al., 2008. Organic amendments influence soil organic carbon pools and rice–wheat productivity. Soil Science Society of America Journal 72 (3), 775–785. Available at https://doi.org/10.2136/sssaj2006.0378.

Marland, G., Boden, T.A., 1997. Trends: a compendium of data on global change. Carbon Dioxide Information Analysis Center, Oak Ridge National Laboratory.

McLauchlan, K.K., Hobbie, S.E., 2004. Comparison of labile soil organic matter fractionation techniques. Soil Science Society of America Journal 68 (5), 1616–1625. Available at https://doi.org/10.2136/sssaj2004.1616.

Meena, R.S., Lal, R., Kumar, S., Pradhan, G., Srinivasarao, C., Singh, A.K., et al., 2024. Potential of Indian agriculture for capturing atmospheric CO2 and monetizing carbon credits to the farmers: an overview and policy framework. Advances in Agronomy 188, 101–206. Available at https://doi.org/10.1016/bs.agron.2024.06.001.

Meena, R.S., Pradhan, G., Kumar, S., Lal, R., 2023. Using industrial wastes for rice-wheat cropping and food-energy-carbon-water-economic nexus to the sustainable food system. Renewable and Sustainable Energy Reviews 187, 113756. Available at https://doi.org/10.1016/j.rser.2023.113756.

Meena, R.S., Pradhan, G., 2023. Industrial garbage-derived biocompost enhances soil organic carbon fractions, CO2 biosequestration, potential carbon credits and sustainability index in a rice-wheat ecosystem. Environmental Research 235, 116525. Available at https://doi.org/10.1016/j.envres.2023.116525.

Meena, R.S., Pradhan, G., Kumar, S., 2025. Energy flow, eco-efficiency, and economic circulation with recycled industrial waste compost application in wheat and subsequent rice farming. Science of the Total Environment 967. Available at https://doi.org/10.1016/j.scitotenv.2025.178779.

Meena, R.S., Pradhan, G.S., Kumar, S., et al., 2025a. Carbon-Negative Agriculture. Springer Nature, Singapore.

Meena, R.S., Pradhan, G., Nalani, Kumar, S., Singh, A.K., Verma, S.K., et al., 2024. Diversified cropping modules designed for soil restoration, CO2 sequestration, and generating carbon credits. Land Degradation and Development 35 (14), 4314–4329. Available at https://doi.org/10.1002/ldr.5224.

Meena, R.S., Pradhan, G., Singh, K., Kumar, S., Singh, A.K., Shashidhar, K.S., et al., 2024. Agriculture models for restoring degraded land to enhance CO2 biosequestration and carbon credits in the Vindhyan region of India. Science of the Total Environment 929. Available at https://doi.org/10.1016/j.scitotenv.2024.172661.

Meena, R.S., Yadav, A., Kumar, S., Jhariya, M.K., Jatav, S.S., 2022. Agriculture ecosystem models for CO2 sequestration, improving soil physicochemical properties, and restoring degraded land. Ecological Engineering 176. Available at https://doi.org/10.1016/j.ecoleng.2022.106546.

Mungroo, A., 2023. Case study of technology adoption in carbon market. UNFCCC. Available from https://unfccc.int/sites/default/files/resource/MEP004_CfI_Addressing_Leakage_Mungroo.pdf.

Nair, P.K.R., Nair, V.D., Kumar, B.M., Showalter, J.M., 2010. Carbon sequestration in agroforestry systems. In: Sparks, D.L. (Ed.), Advances in Agronomy. Academic Press, pp. 237–307. Available at https://doi.org/10.1016/S0065-2113(10)08005-3.

Nayak, A.K., Rahman, M.M., Naidu, R., Dhal, B., Swain, C.K., Nayak, A.D., et al., 2019. Current and emerging methodologies for estimating carbon sequestration in agricultural soils: a review. Science of the Total Environment. 665, 890–912. Available at https://doi.org/10.1016/j.scitotenv.2019.02.125.

Oelkers, E.H., Cole, D.R., 2008. Carbon dioxide sequestration a solution to a global problem. Elements 4 (5), 305–310. Available at https://doi.org/10.2113/gselements.4.5.305.

Pilli, R., Kull, S.J., Blujdea, V.N.B., Grassi, G., 2018. The carbon budget model of the Canadian Forest Sector (CBM-CFS3): customization of the Archive Index Database for European Union countries. Annals of Forest Science 75 (3). Available at https://doi.org/10.1007/s13595-018-0743-5.

Plastina, A., Jo, H., Wongpiyabovorn, O., 2024. The business case for carbon farming in the USA. Carbon Balance and Management 19 (1). Available at https://doi.org/10.1186/s13021-024-00253-5.

Popp, A., et al., 2022. Innovative MRV approaches for agricultural carbon credits: challenges and opportunities. Global Change Biology 28 (4), 1234–1246. Available at https://doi.org/10.1111/gcb.15999.

Pradhan, G., Meena, R.S., 2023a. Interaction impact of biocompost on nutrient dynamics and relations with soil biota, carbon fractions index, societal value of CO2 equivalent and ecosystem services in the wheat-rice farming. Chemosphere 339. Available at https://doi.org/10.1016/j.chemosphere.2023.139695.

Pradhan, G., Meena, R.S., 2023b. Utilizing waste compost to improve the atmospheric CO2 capturing in the rice-wheat cropping system and energy-cum-carbon credit auditing for a circular economy. Science of the Total Environment 892. Available at https://doi.org/10.1016/j.scitotenv.2023.164572.

Pradhan, G., Meena, R.S., Kumar, S., Lal, R., 2023. Utilizing industrial wastes as compost in wheat-rice production to improve the above and below-ground ecosystem services. Agriculture, Ecosystems & Environment 358, 108704. Available at https://doi.org/10.1016/j.agee.2023.108704.

Prasad, R., Jhariya, M.K., Banerjee, A., 2021. Advances in Sustainable Development and Management of Environmental and Natural Resources: Economic Outlook and Opinions, 2-volume set. Apple Academic Press, New York. Available at https://doi.org/10.1201/9781003187455.

Raina, N., 2024. Incentive mechanisms of carbon farming contracts: A systematic mapping study. Journal of Environmental Management 352, 120126. Available at https://doi.org/10.1016/j.jenvman.2024.120126.

Raj, A., Jhariya, M.K., Banerjee, A., Jha, R.K., Singh, K.P., 2024. Agroforestry. Scrivener Publishing, Beverly, MA.

Raj, A., Jhariya, M.K., Banerjee, A., Nema, S., Bargali, K., 2023. Land and Environmental Management Through Forestry. Wiley-Scrivener, USA.

Raj, A., Jhariya, M.K., Meena, R.S., 2025. Agroforestry for Monetising Carbon Credits. Springer Nature, Switzerland.

Rodolfo, G.E., Plucknett, D.L., 1975. Ratoon cropping of sorghum: II. Effect of day length and temperature on tillering and plant development. Journal Series No. 1775, Hawaii Agricultural Station, University of Hawaii, Honolulu.

Sahile, S., Fininsa, C., Sakhuja, P.K., Ahmed, S., 2008. Effect of mixed cropping and fungicides on chocolate spot (*Botrytis fabae*) of faba bean (*Vicia faba*) in Ethiopia. Crop Protection 27 (2), 275–282. Available at https://doi.org/10.1016/j.cropro.2007.06.003.

Sainju, U.M., Singh, B.P., Whitehead, W.F., Wang, S., 2006. Carbon supply and storage in tilled and nontilled soils as influenced by cover crops and nitrogen fertilization. Journal of Environmental Quality 35 (4), 1507–1517. Available at 10.2134/jeq2005.0189.

Sainju, U.M., Singh, B.P., Whitehead, W.F., 2002. Long-term effects of tillage, cover crops, and nitrogen fertilization on organic carbon and nitrogen concentrations in sandy loam soils in Georgia, USA. Soil and Tillage Research. 63 (3-4), 167–179. Available at https://doi.org/10.1016/s0167-1987(01)00244-6.

Shang, Y., Olesen, J.E., Lærke, P.E., Manevski, K., Chen, J., 2024. Perennial cropping systems increased topsoil carbon and nitrogen stocks over annual systems—a nine-year field study. Agriculture, Ecosystems and Environment 365. Available at https://doi.org/10.1016/j.agee.2024.108925.

Siarudin, M., Rahman, S.A., Artati, Y., Indrajaya, Y., Narulita, S., Ardha, M.J., et al., 2021. Carbon sequestration potential of agroforestry systems in degraded landscapes in West Java, Indonesia. Forests 12 (6), 714. Available at https://doi.org/10.3390/f12060714.

Smith, P., Soussana, J.F., Angers, D., Schipper, L., Chenu, C., Rasse, D.P., et al., 2020. How to measure, report and verify soil carbon change to realize the potential of soil carbon sequestration for atmospheric greenhouse gas removal. Global Change Biology 26 (1), 219–241. Available at https://doi.org/10.1111/gcb.14815.

Smith, P., Soussana, J.F., Angers, D., Schipper, L., Chenu, C., Rasse, D.P., et al., 2019. How to measure, report and verify soil carbon change to realize the potential of soil carbon sequestration for atmospheric greenhouse gas removal. Global Change Biology.

Sullivan, P., 2003. Intercropping principles and production practices. Agronomy Systems Guide. Appropriate Technology Transfer for Rural Areas. Available at http://www.attra.ncat.org.

Tian, L., Wu, X., Tao, Y., Li, M., Qian, C., Liao, L., et al., 2023. Review of remote sensing-based methods for forest aboveground biomass estimation: progress, challenges, and prospects. Forests 14 (6), 1086. Available at https://doi.org/10.3390/f14061086.

Udawatta, R.P., Walter, D., Jose, S., 2022. Carbon sequestration by forests and agroforests: a reality check for the United States. Carbon Footprints 1 (1), 8. Available at https://doi.org/10.20517/cf.2022.06.

Verified Carbon Standard (VCS), 2020. VCS program overview. Available from: https://verra.org/project/vcs-program/.

Villat, J., Nicholas, K.A., 2023. Quantifying soil carbon sequestration from regenerative agricultural practices in crops and vineyards. Frontiers in Sustainable Food Systems 7. Available at https://doi.org/10.3389/fsufs.2023.1234108.

Wang, Q., Klassen, W., Li, Y., Codallo, M., Abdul-Baki, A.A., 2005. Influence of cover crops and irrigation rates on tomato yields and quality in a subtropical region. American Society for Horticultural Science, United States HortScience 40 (7), 2125–2131. Available at https://doi.org/10.21273/hortsci.40.7.2125.

West, T.O., Post, W.M., 2002. Soil organic carbon sequestration rates by tillage and crop rotation: a global data analysis. Soil Science Society of America Journal 66 (6), 1930–1946. Available at https://doi.org/10.2136/sssaj2002.1930.

Willy, R.W., 1983. Intercropping studies with annual crops. In: Nugent, J., O'Connor, M. (Eds.), Better Crops for Food (CIBA Foundation Symposium 97). Pitman, pp. 1–10.

Wolf, M.S., 2000. Crop strength through diversity. Nature 406, 681–682. Available at https://doi.org/10.1038/35021000.

World Bank, 2022. Digital monitoring, reporting, and verification systems and their application in future carbon markets. Washington, DC: World Bank. Available from: https://openknowledge.worldbank.org/bitstreams/b7272a89-6512-51e3-acd9-98ddf0a7f908/download.

World Bank, 2024. Digital monitoring, reporting, and verification systems and their impact on carbon markets.

Zentner, R.P., Campbell, C.A., Biederbeck, V.O., Miller, P.R., Selles, F., Fernandez, M.R., 2001. In search of a sustainable cropping system for the Semiarid Canadian Prairies. Journal of Sustainable Agriculture 18 (2-3), 117–136. Available at https://doi.org/10.1300/j064v18n02_10.

Zentner, R.P., Campbell, C.A., Biederbeck, V.O., Selles, F., Lemke, R., Jefferson, P.G., et al., 2004. Long-term assessment of management of an annual legume green manure crop for fallow replacement in the Brown soil zone. Canadian Journal of Plant Science 84 (1), 11–22. Available at https://doi.org/10.4141/p02-188.

Zhang, G., Chen, S.C.I., Yue, X., 2024. Blockchain technology in carbon trading markets: impacts, benefits, and challenges—a case study of the shanghai environment and energy exchange. Energies 17 (13). Available at https://doi.org/10.3390/en17133296.

Zomer, R.J., Bossio, D.A., Trabucco, A., Noordwijk, Mv, Xu, J., 2022. Global carbon sequestration potential of agroforestry and increased tree cover on agricultural land. Circular Agricultural Systems 2. Available at https://doi.org/10.48130/CAS-2022-0003.

Zomer, R.J., et al., 2021. Enhancing MRV in carbon credit systems in developing countries: a framework for action. Climate Policy 21 (5), 645–658. Available at https://doi.org/10.1080/14693062.2020.1861234.

Further Reading

Bernoux, M., Cerri, C.C., Cerri, C.E.P., Neto, M.S., Metay, A., Perrin, A.S., et al., 2006. Cropping systems, carbon sequestration and erosion in Brazil: a review. Agronomy for Sustainable Development 26 (1), 1–8. Available at https://doi.org/10.1051/agro:2006001.

Cullen, J.J., Boyd, P.W., 2008. Predicting and verifying the intended and unintended consequences of large-scale ocean iron fertilization. Marine Ecology Progress Series 364, 295–301. Available at https://doi.org/10.3354/meps07512.

Drake, B.G., Gonzalez-Meler, M.A., Long, S.P., 1997. More efficient plants: a consequence of rising atmospheric CO2? Annual Review of Plant Physiology 48, 608–637. Available at https://doi.org/10.1146/annurev.arplant.48.1.608.

Drinkwater, L.E., Wagoner, P., Sarrantonio, M., 1998. Legume-based cropping systems have reduced carbon and nitrogen losses. Nature 396 (6708), 262–265. Available at https://doi.org/10.1038/24376.

Ehrenfeld, J.G., Kaldor, E., Parmelee, R.W., 1992. Vertical distribution of roots along a soil toposequence in the New Jersey pinelands. Canadian Journal of Forest Research 22, 1929–1936. Available at https://doi.org/10.1139/x92-241.

Horrocks, M., Shane, P.A., Barber, I.G., D'Costa, D.M., Nichol, S.L., 2004. Microbotanical remains reveal Polynesian agriculture and mixed cropping in early New Zealand. Review of Palaeobotany and Palynology 131 (3-4), 147–157. Available at https://doi.org/10.1016/j.revpalbo.2004.03.002.

Hutchinson, J.J., Campbell, C.A., Desjardins, R.L., 2007. Some perspectives on carbon sequestration in agriculture. Agricultural and Forest Meteorology 142 (2-4), 288–302. Available at https://doi.org/10.1016/j.agrformet.2006.03.002.

Intergovernmental Panel on Climate Change (IPCC), 2007. Climate change 2007: The physical science basis. Cambridge University Press.

Janik, L., Spouncer, L., Correl, R., Skjemstad, J., 2002. Sensitivity analysis of the Roth-C soil carbon model. National Carbon Accounting System Technical Report No. 30, CSIRO Land and Water and Mathematical and Information Sciences, Australian Greenhouse Office.

Kuepper, G., Dodson, M., n.d. Companion planting: basic concepts and resources. Horticulture Technical Note. http://www.attra.ncat.org.

Li, Y., Klassen, W., Lamberts, M., Olczyk, T., 2006. Okra production in Miami-Dade County. Florida. HS 875. Available at http://edis.ifas.ufl.edu.

Maracchi, G., Sirotenko, O., Bindi, M., 2005. Impacts of present and future climate variability on agriculture and forestry in the temperate regions: Europe. Climate Change 70, 117–135. Available at https://doi.org/10.1007/s10584-005-5930-0.

McNeill, J.R., Winiwarter, V., 2004. Breaking the sod: humankind, history, and soil. Science 304, 1627–1629. Available at https://doi.org/10.1126/science.1099650.

Sainju, U.M., Whitehead, W.F., Singh, B.P., 2003. Agricultural management practices to sustain crop yields and improve soil and environmental qualities. The Science World 3, 768–789. Available at https://doi.org/10.1100/tsw.2003.100.

Sainju, U.M., Whitehead, W.F., Singh, B.P., 2005. Carbon accumulation in cotton, sorghum, and underlying soil as influenced by tillage, cover crops, and nitrogen fertilization. Plant and Soil 273 (1-2), 219–234. Available at https://doi.org/10.1007/s11104-004-1005-5.

Thompson, J.A., Kolka, R.K., 2005. Soil carbon storage estimation in a central hardwood forest watershed using quantitative soil-landscape modeling. Soil Science Society of America Journal 69, 1086–1093. Available at https://doi.org/10.2136/sssaj2004.0220.

World Meteorological Organization (WMO), 2006. Greenhouse gas bulletin. Geneva, Switzerland: World Meteorological Organization.

Chapter 22

Carbon incentives and quantification in agroforestry

Donald Mlambo and Edward Mufandaedza
Department of Forest Resources and Wildlife Management, National University of Science and Technology, Bulawayo, Zimbabwe

Highlights

- Agroforestry enhances carbon sequestration and mitigates climate change.
- Carbon incentives drive the adoption of agroforestry practices globally.
- Accurate carbon quantification is essential for carbon credit verification
- Agroforestry boosts carbon storage while improving farm productivity.
- Policies must align carbon incentives with agroforestry for long-term sustainability.

22.1 Introduction

The escalating climate crisis and the pressing requirement for green land-use practices have brought agroforestry systems (AFS) to the forefront of innovative environmental management solutions. Agroforestry (AF), which integrates trees into the farming system, presents a unique opportunity to enhance carbon sequestration (C_{seq}) while simultaneously improving agricultural productivity and biodiversity (Yasin et al., 2024; Quandt et al., 2023; Banerjee et al., 2024; Jhariya et al., 2022). AFS can sequester carbon (C), enhance biodiversity, improve soil health, and provide economic opportunities for rural communities (Dhyani et al., 2021; Rolo et al., 2023; Dobhal et al., 2024; Raj et al., 2024a; Raj et al., 2025). To address the climate problem, C incentives have emerged as a key strategy to incentivize landowners and farmers to implement C_{seq} practices and reduce greenhouse gas (GHG) emissions (Mbow et al., 2014; Terasaki Hart et al., 2023; Meena et al., 2025b). C incentives refer to financial or regulatory mechanisms designed to promote practices that reduce carbon dioxide (CO_2) emissions or enhance C_{seq}. In the context of AFS, these incentives can take various forms, including C credits, subsidies for green land management practices, and payments for ecosystem services (PES).

In addition to C_{seq}, AFS provides many other ecosystem services (ES), such as improved soil fertility, biodiversity conservation, and enhanced food and livelihood security (Mlambo et al., 2024; Raj et al., 2024b). These cobenefits can further support the case for C incentive programs that recognize the multifunctional nature of AF and its broader contribution to sustainable land management (Meena et al., 2022; Meena et al., 2025a). Despite its potential benefits, implementing C incentive programs in AFS faces several challenges. The challenges include a lack of accurate C measurement, accessibility of C markets by smallholder farmers, and the need for robust verification processes for C accounting (Wongpiyabovorn et al., 2022; Pan et al., 2022).

Quantifying the amount of C_{seq} using AFS is crucial for establishing credible C markets and ensuring that incentives are allocated appropriately (Kay et al., 2019; Singh et al., 2024; Meena, Lal, et al., 2024c). Various methodologies exist for measuring C stocks in trees, soils, and associated vegetation. These methods often involve complex calculations based on biomass measurements, assessments of soil organic matter (SOM), and remote sensing technologies. The complexity of AFS can complicate the quantification process, making it challenging to establish standardized methodologies for C accounting. Advancements in technology offer new avenues for improving C quantification methods (Yao et al., 2023; Liu et al., 2024; Manikandan et al., 2025).

With the right policies and support mechanisms in place, farmers can be motivated to adopt AF, making a substantial impact on C emissions. Several countries and regions have established C incentive programs that provide financial

Carbon Incentives, Plans, and Policies in Agriculture. DOI: https://doi.org/10.1016/B978-0-443-33557-0.00028-8

rewards or credits to landowners and farmers who adopt AF practices. Many governments worldwide are recognizing the potential of AFS to mitigate the climate crisis and are incorporating it into their national climate action strategies (Dhyani et al., 2021; Yasin et al., 2024). This can involve direct financial support, tax breaks, or other policy measures to encourage adoption. However, the level of policy support varies greatly between countries and regions. The objective of this chapter is to examine the interplay between C incentives and the accurate quantification of C stocks and fluxes within AFS.

22.2 Agroforestry and carbon pool build-up

22.2.1 Forms of agroforestry

The concept of AF has ancient roots, with indigenous communities practicing various forms of tree-crop integration for centuries (Dagar and Tewari, 2017). AF can be broadly categorized into several types, including alley cropping, silvopasture, forest farming, windbreaks and shelterbelts, riparian buffers, and home gardens (Fig. 22.1). The alley cropping system involves cultivating agricultural crops between tree rows. The trees provide shade, which can reduce soil temperature and moisture loss, creating a more benign microclimate for crops. Numerous tree species can fix nitrogen (N_2) in the soil, enhancing nutrient availability for companion crops. Tree leaf litter enriches the soil by adding organic matter (OM), which in turn enhances soil structure and improves fertility (Majaura et al., 2024; Meena, Pradhan, Nalani, et al., 2024a). Tree roots contribute to soil stabilization, thereby reducing the incidence of erosion, particularly on inclined land surfaces. The silvopasture system integrates trees, livestock, and forage into a single system. In such a system, trees provide shade for livestock, reducing heat stress and improving overall animal welfare (Jose and Dollinger, 2019; Kinneen et al., 2024). Trees also serve as windbreaks, protecting livestock from harsh winds and reducing stress, which can lead to improved growth rates and productivity. The presence of trees can enhance the quality of forage by promoting diverse plant growth, which in turn leads to better nutrition for livestock.

Forest farming is a type of AF that focuses on cultivating high-value crops under the protection and management of a forest canopy. Natural shade of trees protects sensitive crops (e.g., mushrooms, berries, and edible greens) from direct sunlight, helping to maintain optimal growing conditions and reduce heat stress. This practice helps to conserve forest resources while providing economic benefits (Chamberlain et al., 2015; Prasad et al., 2021).

Windbreaks and shelterbelts are agricultural practices designed to reduce wind speed and protect crops, livestock, and soil from the damaging effects of wind (Meena, Pradhan, Singh, et al., 2024b; Wang et al., 2024; Wen et al., 2024). By

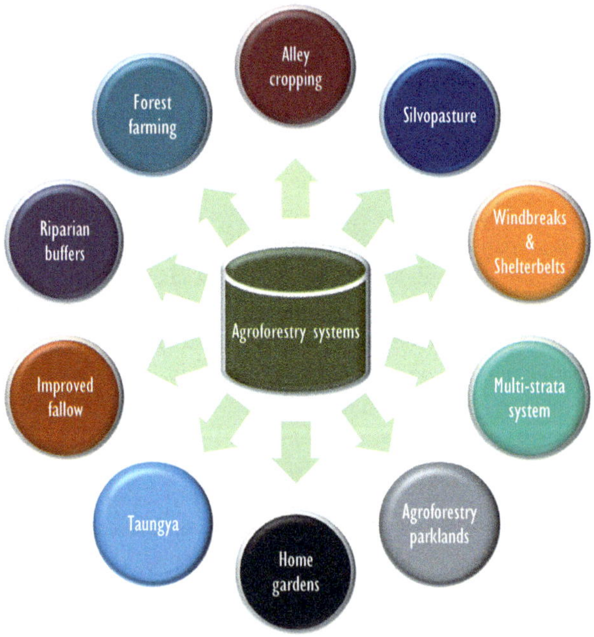

FIGURE 22.1 The most common agroforestry systems around the globe.

reducing wind speed, they create a more stable microclimate, which can enhance plant growth. Windbreaks and shelterbelts provide habitat and food for various wildlife species (Podhrázská et al., 2021). They reduce evaporation rates, helping to conserve soil moisture.

Home gardens are diverse, small-scale AFS typically found near homes in tropical regions. They combine fruit trees, vegetables, herbs, and sometimes livestock to provide food, fuel, and income from crops and livestock. The Taungya system is a traditional farming method practiced in some tropical regions where crops (usually annuals) are cultivated in newly established tree plantations. The farmers benefit from crop yields while the trees grow, promoting early income before forest establishment (Hemida et al., 2023; Isaac et al., 2024).

Riparian buffers are areas of trees and shrubs planted along waterways to help filter pollutants, reduce erosion, and enhance wildlife habitats. They protect aquatic ecosystems by intercepting sediment and other contaminants before they reach water bodies (kumwimba et al., 2024).

In a parkland AF, spaced-out trees are integrated into agricultural fields, allowing for significant areas of the land to be devoted to crop production. The trees often include both indigenous and native species. This system is common in savanna regions, offering a variety of ecological and economic benefits while maintaining a multifunctional land-use approach (Ljalem et al., 2024; Demise and Emire, 2024). A multistrata system involves planting multiple layers or strata of vegetation, including both woody plants and herbaceous plants, in a single area. This system mimics the structure of a natural forest, where different species occupy various vertical levels, thereby enhancing biodiversity and maximizing the use of space, light, and nutrients. Multistrata AF is particularly effective in the tropical regions where conditions are favorable for plant diversity.

Improved fallow integrates the use of fast-growing tree species or shrubs during the fallow period of traditional cropping systems. This approach aims to enhance soil fertility, improve land productivity, and increase ecosystem health (Musokwa and Mafongoya, 2021). Instead of leaving land idle with no vegetation, farmers plant specific leguminous or economically valuable trees or shrubs to benefit the soil and subsequent crops (Jhariya et al., 2018; Meena and Pradhan, 2023; Jhariya et al., 2019; Devi et al., 2024). Today, AF is increasingly recognized as a natural climate solution (Roy et al., 2022; Yasin et al., 2024; Quandt et al., 2023). AFS can yield multiple products while also improving soil health and reducing erosion. They contribute to C_{seq}, enhance wildlife habitats, and can increase farm resilience against climate change.

22.2.2 Carbon pool build-up in agroforestry systems

AF approaches increase biomass output by adding tree litter to soils, boosting OM content as it decomposes (Dollinger and Jose, 2018; Syano et al., 2023; Ngaba et al., 2024). AFS promotes C pool build-up through enhanced biomass production, improved soil health, reduced erosion, enhanced N_2 fixation, and sustainable management practices (Jhariya et al., 2021a,b). These mechanisms work synergistically to create resilient ecosystems that can sequester significant amounts of atmospheric CO_2 over time. Trees absorb atmospheric CO_2 through photosynthesis and can store between **0.3–15 tC ha^{-1} yr^{-1}** in aboveground biomass. The presence of trees in AFS leads to the continuous deposition of plant litter on the soil surface. This material gradually decomposes, creating a C-rich soil environment. In AFS, crop residues are often left on the soil surface or incorporated into the soil, which helps to build SOM and reduce soil erosion, thereby increasing C storage (Meena et al., 2023).

Tree roots and associated mycorrhizal fungi also play a crucial role in C_{seq} below the ground (Hawkins et al., 2023). Many tree species form symbiotic relationships with mycorrhizal fungi, which can further enhance C_{seq} (Luo et al., 2024). Mycorrhizal fungi can transport C from the roots to deeper soil layers and contribute to the formation of stable soil aggregates that protect C from decomposition. Root systems extend deep into the soil, where they store C in root tissues and OM. The C-rich root biomass is then incorporated into the soil as roots grow, die, and decompose. This is particularly important because C stored in deeper soil layers is less susceptible to rapid decomposition and mineralization, leading to long-term C_{seq} (Rolando et al., 2021; Pradhan and Meena, 2023b).

Root exudates, which are C-rich compounds released by plant roots, can interact with soil minerals and microorganisms to form stable SOM (Panchal et al., 2022). Tree roots positively impact soil structure, leading to better aggregation and creating stable pores and channels. This improved soil structure can protect SOM, including root-derived C, from decomposition and erosion. Soil erosion can lead to the loss of C in the topsoil (Mahmoodabadi et al., 2021; Holz and Augustin, 2021). AFS often incorporates tree plantings along waterbodies such as rivers, streams, or ponds. These vegetated riparian buffers help to control water flow, filter sediment, and trap nutrients from agricultural runoff. They act as buffers against water erosion, protecting the adjacent land from excessive soil loss (Prosser et al., 2020; Wu et al., 2023).

Some AFS incorporate the use of groundcover plants or cover crops between rows of trees or crops, which help to prevent soil erosion (Liu et al., 2021).

Effective management in AFS can optimize C_{seq} and contribute to climate stabilization strategies. The choice of tree species and their management can impact C_{seq}. Fast-growing high-biomass tree species tend to accumulate more C in their aboveground biomass and roots (Cuong et al., 2020). While pruning and thinning can enhance tree growth and C_{seq}, overthinning or excessive pruning can remove too much biomass, reducing the overall C storage capacity of the AFS (Pradhan and Meena, 2023a; Lin et al., 2024). Excessive grazing pressure can kill or suppress vegetation, reducing plant biomass. Since plants contribute to SOM through root systems and decaying plant material, loss of vegetation limits C inputs to the soil. Reduced plant cover leads to decreased OM input into the soil, impairing C_{seq}. This can lead to a decline in SOC levels. Overgrazing increases the risk of erosion, which can remove the topsoil where most OM is concentrated, further decreasing C storage.

Uncontrolled fires or improper fire management practices can release stored C back into the atmosphere, reducing the overall C_{seq} potential of AFS. Harvesting trees too frequently or at an early age can reduce the overall C storage capacity of the AFS, as older trees generally have more C stored in their biomass than younger trees. Short rotation cycles can also lead to soil degradation, further reducing C_{seq} potential. Reduced tillage practices disturb the soil less frequently than conventional tillage. This helps to maintain soil structure, which is essential for C storage (Krauss et al., 2022; Pradhan et al., 2023). Minimized disturbance facilitates the development of soil aggregates, and well-aggregated soils possess an enhanced ability to sequester C (Haddaway et al., 2017).

22.3 Carbon incentives through the agroforestry approach

C incentives can be financial, regulatory, or informational and are designed to promote activities that support climate action. These incentives can take the form of C credits, PES, and government incentives such as C taxes, subsidies, grants, low-interest loans, and tax breaks.

22.3.1 Carbon credits

C credits represent a permit to emit a certain amount of CO_2 or other GHGs. The goal of C credits is to reduce overall emissions by creating a financial incentive for companies and organizations to lower their C footprint (Banerjee et al., 2020, 2024). Each C credit typically represents one metric ton of CO_2 that has been prevented from entering the atmosphere. C credits can be generated through activities such as reforestation, green energy projects, and improvements in energy efficiency. AF projects can participate in C markets, where C credits are generated based on the amount of C_{seq} (Foster and Neufeldt, 2014). Platforms like Verra and Gold Standard allow AF projects to register and trade C credits. AF can also contribute to mandatory C offset programs like the EU Emissions Trading System. By engaging in C markets, AF practitioners can receive financial incentives for their C_{seq} efforts. Companies or countries that reduce their emissions can sell their surplus credits to those who exceed their emission limits. The idea is to establish a financial incentive for organizations to cut their emissions and invest in projects that mitigate the climate problem (Rivera-Niquepa et al., 2023).

22.3.2 Payments for ecosystem services

PES is a mechanism that incentivizes landowners to conserve and sustainably manage their land in ways that provide ecological benefits (Benjamin and Sauer, 2018). It can take various forms, such as C credits, soil conservation payments, water quality payments, biodiversity offsets, and AF premiums. In the context of AFS, landowners are compensated for the ES provided by integrating trees and crops on their land. Key ES in AFS that are commonly supported through PES schemes are presented in Table 22.1. PES can provide numerous benefits, including increased C_{seq}, enhanced soil health, biodiversity, and resilience to climate change (van Noordwijk, 2021). By rewarding landowners for these ES, PES can help promote environmentally friendly land management practices and support rural livelihoods.

The implementation of PES typically involves the establishment of payment schemes by governments, nongovernmental organizations, or private companies. Landowners who adopt AF practices can receive payments depending on the nature of ES they provide (Kay et al., 2019). Monitoring and verification mechanisms are often put in place to ensure that the desired outcomes are achieved. Despite the potential advantages of PES, there are significant obstacles to overcome. These include the design of effective payment schemes, the identification of suitable indicators for ES, the establishment of clear property rights, and the integration of local communities into decision-making processes.

TABLE 22.1 Key ecosystem services in AFS that are commonly supported through PES schemes.

Ecosystem services	Examples
C-related	C sequestration in trees and soil, C storage in biomass, and reduced emissions from deforestation.
Water-related	Water flow regulation, water quality improvement, watershed protection, and groundwater recharge.
Biodiversity	Habitat provision for wildlife, biological corridor creation, agricultural biodiversity conservation, and pollinator habitat protection.
Soil-related	Erosion control, soil fertility enhancement, and soil structure improvement
Additional services	Landscape beauty, cultural heritage preservation, climate regulation, and natural pest control.

AFS, agroforestry systems; *PES*, payments for ecosystem services.

TABLE 22.2 Examples of government incentives for agroforestry farmers.

Government incentives	Examples
1. Income tax benefits	Deductions for establishment costs, depreciation of agroforestry investments, tax credits for conservation practices, income averaging for long-term projects, reduced tax rates on timber income, and C credit income treatment
2. Property tax incentives	Lower assessment rates for agroforestry land, tax exemptions for conservation areas, reduced rates during the establishment period, special valuation for working lands, conservation easement tax benefits, and deferral programs during tree growth
3. Capital gains treatment	Preferential rates for timber sales, long-term capital gains qualification, basis allocation for
4. Grants, subsidies and other incentives	Grants, tax breaks, low-interest loans and free technical assistance

22.3.3 Government incentives

Governments can offer a wide variety of incentives to farmers and landowners for adopting AF C projects (Barbosa, 2024). The incentives include grants, subsidies, low-interest loans, and tax breaks to support the adoption of AF C-projects (Table 22.2). These incentives can help to cover establishment and planting costs; maintenance and management costs; training and capacity building; infrastructure development; marketing and value chain development; research and development; as well as other expenses.

22.4 Potential of agroforestry toward a greener economy

AF has great potential to contribute to a green economy (GE) by promoting sustainability, environmental conservation, and economic development (Pantera et al., 2021). A GE is one that aims to achieve sustainable development by promoting economic growth, social inclusion, and environmental protection (Chaaben et al., 2024). It involves transitioning to low-C, resource-efficient, and socially equitable economic systems that reduce environmental degradation and enhance resilience to climate change (Kumar et al., 2020; Kumar et al., 2024b).

22.4.1 Enhancing carbon sequestration and mitigating the climate crisis

One of AF's most significant contributions to a GE is its ability to C_{seq}. Trees play a crucial role in absorbing atmospheric CO_2, thereby minimizing the effects of climate variability. Research indicates that AFS can sequester significantly more C than conventional agriculture (Yasin et al., 2023; Ansari et al., 2024). The establishment of AFS provides an opportunity for landowners to participate in C markets, earning financial incentives for their C_{seq} efforts. This dual benefit (i.e., environmental impact and economic return) positions AF as a key strategy in achieving climate goals while promoting a GE.

22.4.2 Promoting biodiversity and ecosystem resilience

Biodiversity and ecosystem resilience are integral to a GE (Vargas-Hernández and Ali, 2022; Kopnina et al., 2024). Biodiversity provides essential resources, supports ES, increases resilience to environmental change, generates economic opportunities, and helps to mitigate climate impacts. Protecting and restoring biodiversity is not only an environmental imperative but also a pathway to sustainable economic growth. AFS systems are inherently designed to enhance biodiversity, a fundamental component of resilient ecosystems (Santos et al., 2022; Udawatta et al., 2019; Mlambo et al., 2024). Biodiversity is essential for the provision of ES, including food security, water provision, and the supply of raw materials. By incorporating diverse plant species, AF promotes habitat creation for a wide range of wildlife, including pollinators and beneficial insects. This biodiversity not only supports ecosystem health but also enhances agricultural productivity. Furthermore, AF practices contribute to soil conservation and water management, thereby reducing erosion and improving water quality (Kuyah et al., 2016).

N_2-fixing trees in AFS can increase soil N_2 availability, soil health, and crop yields (Rosenstock et al., 2014). The multifunctionality of AFS aligns with the principles of a CE, where resources are utilized efficiently, waste is minimized, and ecological balance is maintained. Biodiversity enhances ecosystem resilience, enabling ecosystems to better tolerate environmental changes and stresses, such as pollution and climate variability (Oliver et al., 2015). This resilience helps to maintain the stability of food systems and other economic activities, thereby supporting a sustainable economy. Biodiversity underpins rural livelihoods through agriculture, fisheries, and tourism (Wei et al., 2018). Conservation efforts can enhance these livelihoods by ensuring that natural resources are available for future generations, thus supporting a GE.

22.4.3 Soil health and water conservation

AF practices contribute to soil fertility improvement through the process of nutrient cycling facilitated by tree litter. Tree roots anchor soil, decreasing erosion and conserving topsoil quality. Roots also promote soil porosity, enabling more water to permeate and lowering runoff. Trees provide shade and reduce evaporation, which helps to maintain soil moisture levels. Soil health and water conservation are crucial components of promoting a GE. Enhanced soil contributes to increased crop yields, improved plant resilience, and decreased reliance on synthetic inputs such as fertilizers and pesticides. Reducing pollution from chemical runoff and ensuring access to clean water lead to better health outcomes. Soils with rich OM capture and store C, helping CCM (Don et al., 2024). Healthy soils support diverse ecosystems, promoting beneficial organisms that contribute to natural pest control and pollination.

Protecting water resources supports ecosystems that provide filtration, flood management, and wildlife habitat (Ferreira et al., 2023). Water conservation reduces the energy required for water treatment and distribution, thereby lowering CO_2 emissions. Efficient water management helps to adapt to changes in water availability due to climate change, ensuring stability for communities and industries. Practices that focus on soil and water conservation foster more resilient and sustainable communities, enhancing quality of life. This, in turn, can lead to economic growth while reducing the environmental impact of financial activities.

22.4.4 Creating economic opportunities and diversifying income streams

AF presents significant economic opportunities for farmers and rural communities. AF farmers can produce a variety of goods, including timber and nontimber products alongside traditional crops and livestock. This diversification helps to stabilize farmers' incomes and reduces their vulnerability to market fluctuations and climate-related risks (Kassie, 2018; Tebkew et al., 2024a). If one crop fails due to adverse weather or pests, farmers can still rely on tree products for income. This resilience is especially crucial as extreme weather events are becoming prevalent.

22.4.5 Job creation

AF practices require more hands-on management. Activities such as planting, pruning, harvesting, and maintaining diverse species can create a demand for seasonal and full-time labor (Mukhlis et al., 2022). Workers often require training to effectively manage the complexity of AFS, leading to skill development in areas such as sustainable farming practices, pest management, and resource conservation. The diverse nature of AFS creates a variety of jobs, from manual labor to skilled positions in management, research, and education. The seasonal nature of certain AF activities can provide temporary job opportunities, which is particularly beneficial in areas with high unemployment. The labor-intensive nature of AFS can help communities adapt to economic fluctuations by providing alternative livelihoods and income sources.

22.5 Quantification of incentives in agroforestry

Quantifying incentives in AFS is complex, as it involves assessing both the costs and benefits across various stakeholders and time scales. There is no single, universally accepted method, but several approaches can be used, often in combination. The quantification involves assessing the various forms of support, benefits, and regards that motivate farmers and landowners to adopt and sustain AF practices.

22.5.1 Cost–benefit analysis

Cost-benefit analysis (CBA) can be used to assess whether the benefits of implementing C projects outweigh costs, thereby guiding decision-making processes. In the context of AF, CBA can be particularly useful for quantifying both the economic and environmental benefits and costs of integrating trees into agricultural landscapes. The initial phase of performing a cost-benefit analysis for AF involves identifying all pertinent expenses and benefits linked to the adoption of AF methods. The costs of AF include establishment costs (planting materials, labor, land preparation), maintenance costs (pruning, pest control, fertilization), and potential opportunity costs (land that could have been used for other purposes). AF benefits include increased crop yields, income from product sales, improved soil fertility, C_{seq} (potentially generating C credits), and reduced erosion (Fig. 22.2). Yield measurements in AFS should not only include traditional crop yields but also tree products.

The second step is to quantify identified costs and benefits, which can be challenging due to the intangible nature of some of the benefits. Direct costs can often be easily quantified based on market prices for inputs such as seeds or labor.

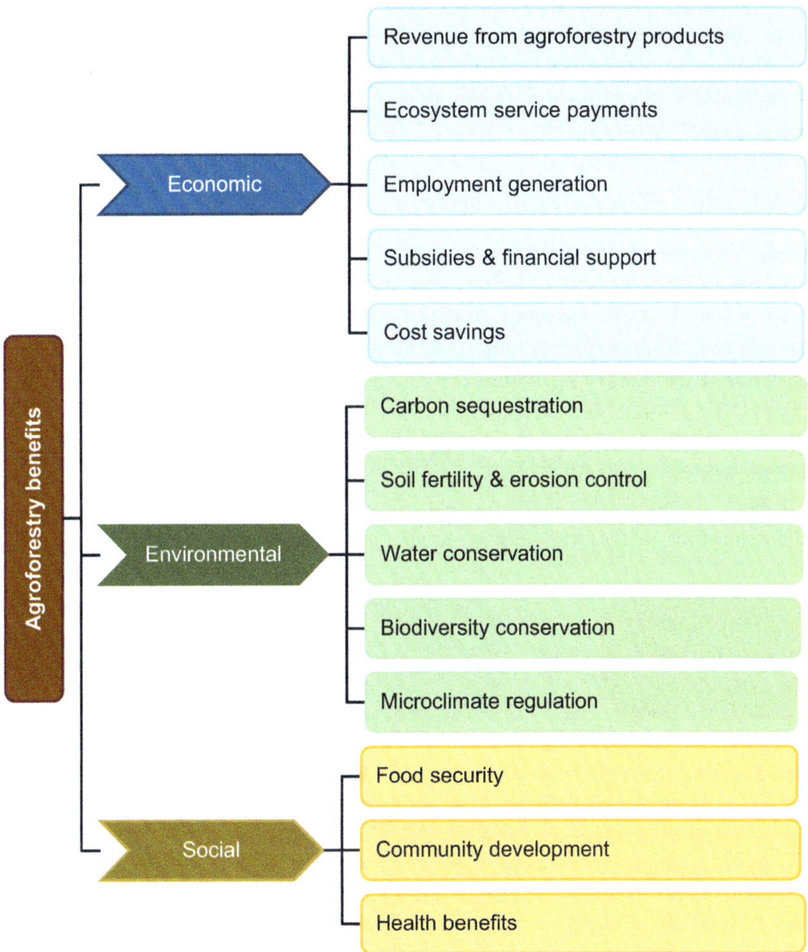

FIGURE 22.2 Agroforestry benefits.

Opportunity costs require an assessment of potential income from alternative land uses. Benefits need to be monetized using both market-based methods for direct products (e.g., selling timber) and nonmarket valuation methods for ecosystem services (e.g., increased crop yield). Benefits received in future years are worth less than those received today, due to time preference and the availability of investment opportunities. A discount rate is used to adjust future benefits to their present value. This step ensures that all future benefits from agroforestry practices are appropriately valued against current investments. If total benefits exceed total costs, then the AF project is considered economically viable. Conversely, if total expenses exceed total benefits, it may not be justified economically. This comparison helps stakeholders to make informed decisions about whether to implement AF practices based on their financial viability.

22.5.2 Contingent valuation methods

Contingent valuation methods (CVM) are survey-based economic techniques used to estimate the value of nonmarket resources, such as ES. These methods involve asking individuals how much they would be willing to pay (WTP) for specific environmental benefits or how much compensation they would require to accept a loss of these benefits. CVM is helpful in quantifying the value of ES provided by AFS. The first step in applying CVM is to identify the specific ES that AFS provides. These may include C_{seq}, biodiversity enhancement, soil health improvement, and water regulation.

Once the ES are identified, a survey instrument is designed to communicate the values of ES to respondents. This involves creating hypothetical scenarios where respondents can envision changes in the availability of ES. Questions for the survey are then formulated, asking respondents how much they would be willing to pay for maintaining or enhancing the identified ES. This could take the form of direct payment options or tax increases. Selecting an appropriate sample is critical for ensuring that results are representative of the broader population. Data collection can be conducted through interviews. Mean WTP is calculated for each respondent and aggregated to obtain a total WTP. Statistical methods (e.g., regression analysis) are used to identify factors that influence WTP, such as demographics (e.g., age, income), knowledge and awareness of ES, and perceived benefits and values associated with ES. WTP is translated into a monetary value per unit of service provided (e.g., per ton of C_{seq}).

22.5.3 Integrated valuation of ecosystem services and trade-offs

Integrated valuation of ES and trade-offs (InVEST) is a suite of modeling tools that can be effectively used in AF to value ES. It helps decision-makers to evaluate the trade-offs and synergies between the environmental, social, and economic benefits of AFS. InVEST provides models to assess how land-use and management decisions (e.g., implementing AFS) affect ES. These models can help to quantify benefits such as C storage, water regulation, soil retention, biodiversity, and agricultural production, while also identifying trade-offs between competing land uses.

Key steps for using the InVEST model are presented in (Fig. 22.3). InVEST can identify the ES provided by AFS, such as Cseq, soil conservation, water filtration, pollination, habitat provision, and climate regulation. InVEST uses spatial data (e.g., GIS, remote sensing) to map the AFS and model the ES. This involves creating spatial layers for each service, assigning values to each layer based on relevant data (e.g., C stocks, soil erosion rates), and modeling the relationships between ES and the characteristics of the AFS (e.g., tree species, canopy cover).

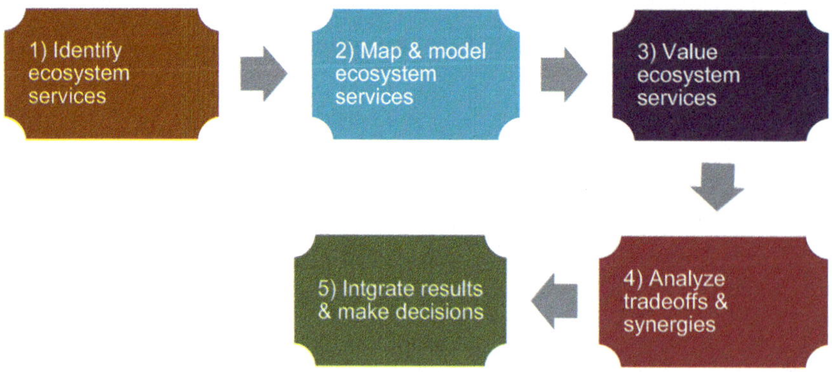

FIGURE 22.3 Key steps for using the InVEST model. *InVEST*, Integrated valuation of ES and Trade-offs.

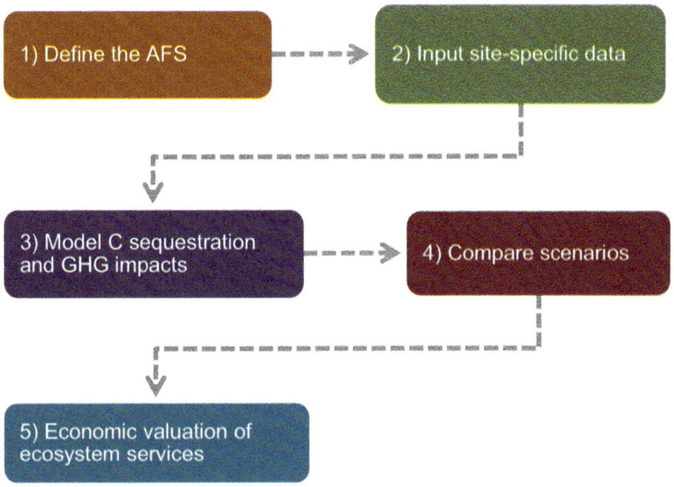

FIGURE 22.4 Key steps for using the COMET-Farm tool.

InVEST applies economic valuation techniques to estimate the financial benefit of ES. This can be done using market-based approaches (e.g., valuing C_{seq} based on C credit prices), nonmarket valuation methods (e.g., contingent valuation and choice experiments), and benefit transfer approaches (e.g., using existing valuation studies). InVEST helps to identify trade-offs and synergies between different ES. For example, assessing how increasing C_{seq} affects water filtration or how changes in tree species affect pollination and habitat provision. This information can be used to inform decision-making on AF management practices, evaluate the effectiveness of policy interventions or conservation efforts, and identify areas for improvement and opportunities for sustainable development.

22.5.4 COMET-Farm

COMET-Farm is a tool that can be used to quantify and value ES provided by AFS, particularly in terms of C_{seq} and GHG mitigation. This tool can quantify C_{seq} potential of trees, crops, and soils. It can estimate net GHG emissions, which are reduced through AF practices. Users can compare current land management practices with proposed AFS to assess the environmental benefits of transitioning to AF (Raj et al., 2023; Banerjee et al., 2023). While COMET-Farm does not directly assign monetary values to ES, the C_{seq} outputs can be linked to C credit pricing or the social cost of C to estimate economic benefits. The first step is to specify the type of AFS and then input details about land use, species composition, tree density, crop types, and management practices (Fig. 22.4). The second step involves providing location-specific data, including soil type, climate, and topography, as well as historical land-use data and current management practices. COMET-Farm uses this information, along with integrated databases, to model outcomes for ES. It then calculates aboveground and belowground C storage as well as GHG emissions reductions. Using this tool, C stored in tree biomass, roots, and SOM can be estimated. It also quantifies the long-term C_{seq} potential of AFS.

22.6 Prospects of agroforestry toward enhancing carbon incentives under changing climate

AF is poised to play a vital role in combating the climate crisis, as it aligns with goals for sustainability, biodiversity, and resilience in agricultural systems. With the urgent need to reduce GHG emissions, C incentives have emerged as a powerful tool to encourage sustainable land-use practices. In recent years, various governments and organizations have recognized the potential of AF in achieving climate goals (Dhyani et al., 2021).

22.6.1 Growing demand for carbon credits

As global efforts to mitigate the climate crisis intensify, the demand for C credits is expected to continue rising. Many governments are implementing stricter emissions regulations and targets for C neutrality, prompting companies to invest in C credits to offset their emissions. AFS can sequester substantial amounts of CO_2 through tree growth and soil enhancement. Estimates suggest that AFS can sequester 1–10 tC ha^{-1} yr^{-1}, depending on the system and local conditions. The market for C credits from AF is growing, with both compliance and voluntary markets recognizing the potential of these practices. C credits, which allow landowners to sell the C sequestered through AF practices, create a direct financial incentive for farmers. This mechanism not only encourages the adoption of sustainable practices but also aligns agricultural objectives with climate action goals. In regions where AF is embraced, farmers can earn additional income through C credits, thereby enhancing their overall financial stability while contributing to global C reduction efforts (Tamba et al., 2021)

Many forms of AF, such as alley cropping, silvopasture, and forest farming, can be developed into C credit projects, broadening market opportunities (Holderieath et al., 2012; Poudel et al., 2024). To participate in C markets, AF projects must adhere to specific standards and a verification process. AF projects, which combine C_{seq} with environmental and social benefits, are likely to be highly sought after in these markets. Organizations like Verra and the Gold Standard provide frameworks for certifying C credits from AFS. The effectiveness and participation of AF in C credit markets can depend on local regulations, market structures, and the recognition of AF practices within national C accounting frameworks. In developing regions, AF can provide additional economic incentives for farmers while contributing to global climate goals (Castle et al., 2021). AF practices need to be scaled up to achieve significant climate benefits, which will require policy support, farmer training, and market incentives.

22.6.2 Increasing support from governments

Governments worldwide are increasingly recognizing the importance of AF and are implementing various support mechanisms to encourage its adoption. Many governments provide financial support to farmers who adopt AF practices, such as subsidies for tree seedlings, fertilizers, and technical assistance. Some governments offer tax breaks or deductions to farmers who adopt AF practices. Several countries have developed national policies and strategies to promote AF, such as India, Nepal, Kenya, Ethiopia, Brazil, Costa Rica, and Indonesia. Countries such as Peru, Ghana, and Colombia are working to strengthen their land tenure systems to provide land rights to farmers.

22.6.3 Rising interest in nature-based solutions

The rising interest in nature-based solutions (NbS), like AF, reflects a shift toward more sustainable, integrated strategies to climate action (Stetter and Sauer, 2024; Corgo et al., 2024; Gebrael et al., 2024). The interest stems from the ability of AFS to address multiple climate, environmental, and social challenges synergistically, making them a compelling component of comprehensive climate action strategies. Investing in natural systems can often be a more cost-effective climate solution compared to engineered solutions, making AFS an attractive option, especially in resource-constrained contexts. Many nations are increasingly recognizing the value of NbS, such as AF, in their climate policies and frameworks (Turner et al., 2022; Davis et al., 2024). This is leading to increased incentives and support for adoption. AF can support climate adaptation at the local level by providing diversified income sources, improving food security, and drawing on traditional ecological knowledge. Beyond CCM and adaptation, AF provides numerous cobenefits, including biodiversity conservation, improved soil health, and livelihood support. This holistic approach is appealing to a wide range of stakeholders.

22.6.4 Advancement in monitoring, measurements, and market access

Developing robust methodologies for measuring and verifying C_{seq} in AFS can be complex, but it is essential for market participation. Technologies such as light detection and ranging (LiDAR), satellite imagery, and drones can accurately measure C stocks, biomass, and forest health (Sharma et al., 2022; Qin et al., 2022; Su et al., 2024; Borsah et al., 2023). These technologies are improving the accuracy and cost-effectiveness of monitoring AF C projects. Satellites and drones can monitor land-use changes, tree growth, and C_{seq} rates, providing accurate data for C credit verification. AI-powered tools can analyze satellite data, identify patterns, and predict C_{seq} potential. This helps to optimize AF practices and identify areas with high C_{seq} potential. Blockchain-based platforms can ensure transparency, security, and traceability of C credits (Zook, 2023; Šilenskytė et al., 2024). These build trust among stakeholders and facilitate the buying and selling of C credits. These technological advancements can enhance the efficiency, transparency, and scalability of agroforestry C projects, ultimately contributing to a more sustainable and climate-resilient future. Blockchain-based platforms can be used to connect farmers to credit markets, simplifying the process of selling credits generated from AF practices.

22.6.5 Integration with sustainable supply chains

There is a growing interest in integrating AF into the production of products such as coffee, cocoa, and palm oil to make supply chains more sustainable. C credits generated from such systems could attract premium buyers. AF practices that include shade trees alongside cocoa and coffee crops have been shown to lower C emissions while providing habitat for biodiversity (Somarriba et al., 2013; Niguse et al., 2022). Bamboo AFS can sequester a significant amount of C. Bamboo is a highly efficient C sink due to its fast growth rate, high biomass production, long lifespan (up to 100 years), and soil C_{seq} (Jember et al., 2023; Chiti et al., 2024). By improving the productivity of existing farmland through AF, farmers can meet their economic needs without further deforestation, thus preserving critical forest ecosystems. Integrating trees with supply chains not only addresses the climate crisis by enhancing C capture and promoting biodiversity but also supports economic resilience and improves soil health.

22.6.6 Scaling up smallholder participation

Scaling up AF practices in the smallholder farming sector is likely to be a more effective and sustainable approach for enhancing Cseq and combating the climate crisis compared to large-scale farming systems. Globally, small-scale farms represent 84% of all farms, and they occupy around 12% of the agricultural land (Lowder et al., 2021). Small-scale farmers are more likely to adopt AF practices due to their existing knowledge and experience with traditional farming methods. Additionally, it is easier to manage multiple species interactions on small farms where it is possible to provide detailed attention to individual trees.

The adoption of AF practices by small-scale farmers can significantly contribute to reduced input costs, improved livelihoods, and the development of a greener and resilient agricultural sector, alongside the numerous accompanying environmental and social benefits. For example, the use of N_2-fixing trees in AFS can reduce the need for synthetic fertilizers (Munroe and Isaac, 2014; Rosenstock et al., 2014). Furthermore, AFS can provide habitat for predators and parasites of crop pests, which help control pest populations and reduce the need for agrochemicals (Pumariño et al., 2015; Boinot et al., 2020; Mlambo, 2021). However, access to funding and resources is often a significant barrier for small-scale farmers seeking to adopt AF practices. Additionally, the lack of awareness or understanding of C markets can hinder farmers from participating in these initiatives. To overcome these challenges, targeted education and training programs are essential. Governments should be encouraged to develop policies and regulations that promote AF and provide incentives for small-scale farmers, including access to resources and markets.

22.7 R&D toward agroforestry and generation of carbon incentives

22.7.1 Research in carbon sequestration

While existing studies provide valuable insights, more research is needed to accurately quantify the C_{seq} potential of different AFS in various ecological and climatic conditions (Nath et al., 2021; Golicz et al., 2022). (Table 22.3) highlights some key research gaps that need to be addressed to advance our current understanding of C incentives and quantification in AFS. This includes refining methodologies for measuring C stocks in different components of the system (trees, soil, etc.) and understanding the factors influencing C_{seq} rates. Research can help to identify the most effective AF practices and species combinations for maximizing C_{seq} while also considering other ES and agricultural productivity. This includes investigating the effects of different tree species, planting densities, and management practices on C storage. C_{seq} is a long-term process, and more research is needed to understand how C stocks change over time in AFS. This includes studying the effects of age, succession, and disturbance on C storage and identifying potential tipping points or thresholds that could lead to C losses. For example, studies have shown that different tree species exhibit varying capacities for C capture, depending on factors such as growth rate, biomass production, and adaptability to local conditions (Jin et al., 2023; Ali et al., 2023).

Accurate and reliable monitoring and verification systems are crucial for tracking C_{seq} in AFS and ensuring the integrity of C markets. Further research is needed to develop cost-effective and efficient methods for monitoring C stocks and fluxes in these complex systems. Research can help to identify the barriers and opportunities for scaling up the adoption of C-sequestering AFS. This includes investigating the economic incentives and disincentives for farmers to adopt these practices, developing effective policies to support their implementation, and understanding the social and cultural factors that influence farmer decision-making. By addressing these knowledge gaps, further research can help unlock the full potential of AFS for mitigating climate change.

TABLE 22.3 Research focus areas and questions for advancing our current knowledge about C incentives and quantification in AFS.

Focus area	Key research questions
C sequestration and quantification	1. How can we improve the accuracy and precision of C sequestration measurement methods in AFS? 2. What are the most effective approaches for quantifying C sequestration in different AFS? 3. How can we better account for C losses due to deforestation, degradation, and other disturbances in AFS? 4. How can we develop standardized protocols for C accounting in AFS that are applicable across diverse geographical regions and management practices? 5. How can we improve the temporal resolution of C sequestration measurements to capture short-term fluctuations and long-term trends better? 6. What are the relative C sequestration potentials of different tree species and AFS under various environmental conditions? 7. How do different agroforestry management practices (e.g., thinning, pruning, harvesting) affect sequestration rates and storage? 8. How do soil properties and biodiversity influence C sequestration in AFS? 9. What is the long-term stability of C stored in AFS under different climate change scenarios? 10. How do AFS interact with other ecosystems (e.g., water cycling, biodiversity) to influence overall C sequestration?
Incentives and policy	1. What are the most effective types of financial incentives (e.g., C credits, tax breaks, cost-sharing) for promoting agroforestry adoption and C sequestration? 2. What are the optimal levels and durations of C incentives to maximize C sequestration while ensuring long-term sustainability? 3. How do different policy instruments (e.g., regulations, subsidies, market mechanisms) interact to influence C sequestration outcomes? 4. How can we design C incentives programs that are equitable and accessible to smallholder farmers and marginalized communities? 5. How do C incentive programs affect other agricultural practices and land use decisions? 6. How can we develop robust and transparent C markets that accurately reflect the value of C sequestration in agroforestry? 7. What are the potential risks and challenges associated with C market participation for smallholder farmers? 8. How can we integrate C markets with other markets for agroforestry products?

AFS, agroforestry systems.

22.7.2 Development of agroforestry carbon credits

One of the most prominent developments in recent years is the establishment of C credits that allow landowners to sell C credits generated through AFS (Kumar et al., 2024a). Research indicates that well-managed AFS can sequester significant amounts of CO_2, making them eligible for credits. Many countries and regions are developing C credit systems that reward farmers for adopting AF practices (Asadnabizadeh and Moe, 2024). These credits can be sold in C markets, providing financial incentives for sustainable practices. The generation of C incentives through AF hinges on the development of robust C market mechanisms. C credits, which represent verified reductions in GHG emissions, provide financial compensation to landowners who adopt C_{seq} practices. The establishment of clear standards for measuring and verifying C_{seq} in AFS is essential to ensure the credibility of these credits. As the market for C credits expands, researchers must contribute to the development of accurate C_{seq} methodologies, ensuring that AF practices are accurately represented and rewarded in C markets.

22.7.3 Technological advances in agroforestry

Recent technological advances have significantly improved the ability to quantify C_{seq} in AFS (Singh et al., 2024). Remote sensing technologies are being used to monitor changes in biomass and assess C stocks over time more accurately (Cheng et al., 2024). This data is essential for validating claims made by landowners regarding their contributions to C_{seq}. Remote sensing technologies, for example, can provide valuable data on tree growth, biomass accumulation, and C_{seq} rates in AFS (Leditznig and Klug, 2024; Dossa and Miassi, 2024). By utilizing satellite imagery and aerial surveys, researchers can monitor changes over time, enabling them to assess the effectiveness of different AFS and their contributions to C incentives. SOC is a critical component of total C_{seq} in AFS. Advances in soil sampling techniques, including portable X-ray fluorescence spectrometry and near-infrared spectroscopy, enable rapid assessment of soil properties without extensive laboratory analysis.

Researchers are also examining how AF can enhance resilience to climate change (Quandt et al., 2023). For instance, AFS can buffer crops against extreme weather events by improving soil structure and moisture retention. AF farms are likely to experience significantly lower yield losses during adverse weather conditions compared to conventional farms. The inclusion of AF in national and international climate policies can significantly enhance the generation of C incentives. Countries that commit to ambitious climate targets are increasingly looking to AF as a viable solution to meet their emissions reduction goals (Tebkew et al., 2024b; Bogale and Bekele, 2023). Governments that incorporate AF into their climate strategy might provide a favorable environment for farmers to engage in C markets, thereby boosting the adoption of sustainable techniques.

The future of AF research and development lies in innovative approaches that combine technology with traditional knowledge. Furthermore, participatory research models that involve farmers in the research process can lead to more relevant and practical solutions. By engaging local communities and farmers, researchers can gain insights into traditional practices and indigenous knowledge that may enhance AFS. This collaborative approach not only empowers farmers but also cultivates ownership and commitment to green land management practices.

Despite the promising prospects of AF and C incentives, several challenges remain. One significant hurdle is the accessibility of information and resources for smallholder farmers, who often lack the knowledge or financial means to implement AF practices (Tranchina et al., 2024; Houndjo Kpoviwanou et al., 2024). Targeted extension services and educational programs are essential to bridge this gap, equipping farmers with the knowledge and skills required to develop and implement green farming practices. Additionally, the fluctuating nature of C markets poses a risk for farmers investing in AFS. To mitigate this risk, policymakers need to establish stable and transparent C markets that ensure fair compensation for C_{seq} efforts. Long-term commitments from governments and organizations can help create a predictable environment that encourages investment in AF. Researchers have explored ways to scale up AF practices to achieve larger C_{seq} impacts, including the development of participatory approaches, farmer-led organizations, and mobile-based advisory services.

22.8 Policy perspectives for agroforestry and carbon incentive initiatives

As governments and organizations strive to meet ambitious climate targets, the development of robust policies that support AF and C incentives is crucial. This section examines the policy perspectives necessary to foster AF potential and maximize C incentives, ensuring a resilient agricultural future.

22.8.1 The need for comprehensive policy frameworks

To effectively harness the benefits of AF for C_{seq}, comprehensive policy frameworks must be established. These frameworks should encompass a range of elements, including financial incentives, regulatory measures, and capacity-building initiatives (Table 22.4). Creating a favorable environment for AF, policymakers can encourage farmers to embrace sustainable practices that help to climate mitigation (Menozzi et al., 2015). Financial incentives play a pivotal role in promoting AF. Governments can implement programs that offer subsidies, grants, or low-interest loans to farmers who integrate trees into their agricultural systems.

22.8.2 Integrating agroforestry into carbon markets

The integration of AF into existing C markets is another critical aspect of policy development (Table 22.4). Policymakers should work to establish clear standards for C accounting and verification specific to AF practices. This involves developing methodologies that accurately measure C_{seq} in different AFS, ensuring that farmers receive fair compensation for their efforts. International frameworks, such as the Paris Agreement, present opportunities for countries to include AF in their nationally determined contributions (Leiter, 2024). By explicitly recognizing AF as a means to achieve emissions reduction targets, governments can incentivize its adoption among farmers. Countries like Costa Rica have successfully integrated AF into their climate policies, resulting in increased C credits for landowners and enhanced forest cover. Such successful examples can inspire other nations to adopt similar approaches to promote C market mechanisms.

TABLE 22.4 Policy perspective for agroforestry and C incentives.

Policy	Mechanism	Examples
Regulatory framework	C protocols	Standardized measurement methodologies, third-party verification requirements, baseline establishment guidelines, and monitoring and reporting frameworks.
	Land use policies	Flexible zoning for AFS, protection for long-term land use rights, integration with conservation programs, and streamlined permitting processes
	Property rights	Clear ownership of C credits, transfer and trading rights, risk allocation mechanisms, and dispute resolution procedures.
	Policy integration	Alignment with agricultural subsidies, coordination with environmental regulations, compatibility with food security goals, and integration with climate action plans.
Financial incentives	C credit payments	Direct payments for verified C sequestration, performance-based rewards for maintaining tree cover, bundled ecosystem service payments, and premium rates for long-term commitments.
	Tax benefits	Property tax reductions for agroforestry land, income tax deductions for establishment costs, depreciation allowances for tree assets, and tax credits for sustainable agricultural practices
	Cost-share programs	Government matching funds for initial planting, subsidies or irrigation infrastructure, support for soil preparation and amendments, and equipment purchase assistance
	Premium product pricing	Certification premiums for sustainable products, market access to high-value organic markets, direct-to-consumer marketing support, and value-added product development assistance.
Technical support	Extension services	On-site technical assistance, design and planning support, problem-solving consultation, and regular system assessments.
	Research support	Species selection studies, system optimization research, economic analysis, and climate adaptation studies.
	Training programs	Hands-on implementation workshops, business planning assistance, marketing strategy development, and technical certification programs.
	Monitoring support	Data collection protocols, technology access and training, reporting assistance, and quality assurance programs.
Market development	C market access	Aggregation mechanisms for smallholders, access to trading platforms, price discovery tools, and risk management options.
	Supply chain development	Market linkage facilitation, quality control systems, storage and processing infrastructure, and transportation networks.
	Certification systems	Standards development, audit procedures, label recognition, and chain of custody tracking.
	Smallholder support	Cooperative development, group certification options, collective marketing programs, and financial services

AFS, agroforestry systems.

22.8.3 Capacity building and education

To facilitate the successful implementation of AF policies, capacity-building initiatives are essential. Policymakers should prioritize education and training programs that provide farmers with the information and skills required to employ AF technologies effectively (Table 22.4). This includes providing technical assistance on the selection of tree species, management practices, and integrating AFS into existing agricultural operations. Moreover, collaborative efforts between government agencies, research institutions, and NGOs can enhance capacity-building initiatives. By using local knowledge and expertise, these collaborations may build specialized training programs that address farmers' specific needs and concerns. Involving communities in policy formation cultivates ownership and promotes long-term commitment to green practices, eventually leading to more successful implementation of AF initiatives.

22.8.4 Policy incentives for research and innovation

Research and innovation are fundamental in advancing AF practices and maximizing C incentives (Table 22.4). Policymakers should allocate funding and resources to support research initiatives focused on AFS, C_{seq} methodologies, and sustainable land management practices. This investment in research will yield crucial insights into the efficacy of diverse AF approaches, thereby facilitating the formation of evidence-based policy decisions. Additionally, fostering innovation through public-private partnerships can lead to the development of novel technologies and practices that enhance AF's potential. For example, integrating digital tools and remote sensing technology can improve monitoring and verification processes for C credits. By investing in research and innovation, policymakers can create a dynamic environment where AF evolves in response to the changing climate.

22.8.5 Addressing challenges and ensuring equity

While the potential benefits of AF and C incentives are significant, policymakers must also address the challenges associated with implementing these initiatives. Ensuring equitable access to resources and support for smallholder farmers is paramount. Often, these farmers are constrained by a lack of information, funding, and technical assistance. Policies should prioritize inclusivity, ensuring that marginalized communities benefit from AF initiatives. Furthermore, policymakers need to establish robust systems for the continuous monitoring and evaluation of the effectiveness of AF policies to ensure their optimal impact. Regular assessments can provide insights into the effects of these initiatives on C_{seq}, biodiversity, and farmer livelihoods. This feedback loop enables adaptive management, ensuring that policies remain relevant and effective in achieving their intended objectives.

22.9 Future directives

The future directives for AF and C incentives must focus on enhancing policy integration, expanding financial access, fostering research and innovation, promoting education, and ensuring equity. As we move forward, it is essential to recognize that AF is not merely an agricultural practice but a vital strategy for achieving climate resilience and sustainability. By investing in AF and C incentives, the world can create a greener, more sustainable future—one where agricultural productivity coexists harmoniously with ecological health. Immediate climate action is essential, and the policies currently enacted by governments will shape the future agricultural landscape, benefiting both humanity and the environment for years to come. To maximize its potential in C_{seq}, clear directives must be established to guide the integration of AF into climate policies and C markets.

22.9.1 Enhancing policy integration and alignment

One of the primary directives for the future of AF and C incentives is to enhance policy integration across sectors. Governments must align AF initiatives with broader climate action plans, agricultural policies, and environmental regulations. This holistic approach enhances agricultural resilience and increases the chances of achieving global climate targets. For instance, countries can develop national AF strategies that outline specific goals for C_{seq}, biodiversity enhancement, and community engagement. These strategies should include measurable targets and timelines, allowing for accountability and progress tracking. By integrating AF into national climate commitments, countries can create a cohesive framework that incentivizes farmers to adopt sustainable practices.

22.9.2 Expanding access to financial incentives

To facilitate the widespread adoption of AF, it is essential to increase access to financial incentives. Policymakers should explore innovative funding mechanisms, such as green bonds, C credit trading systems, and results-based financing, to support AF initiatives. These financial tools can provide farmers with the necessary resources to establish and maintain AFS, ultimately enhancing their economic viability. In addition, governments should consider establishing targeted grant programs for smallholder farmers, who often face barriers to accessing funding. By providing financial support tailored to the specific needs of these farmers, policymakers can encourage the adoption of AF practices, thus contributing to both C_{seq} and rural development.

22.9.3 Fostering research and technological innovation

The future of AF and C incentives relies heavily on research and innovation. Policymakers should prioritize funding for research initiatives focused on AFS, C_{seq} methodologies, and the socio-economic impacts of these practices. Governments engaging in research can generate valuable data that informs evidence-based decision-making and policy formulation. Furthermore, technological advancements can play a pivotal role in enhancing AF practices. For instance, the integration of precision agriculture, remote sensing, and data analytics can improve monitoring and verification processes for C credits. These technologies enable more accurate assessments of C_{seq}, thereby enhancing the credibility of C markets and incentivizing farmer participation.

22.9.4 Promoting education and capacity building

To ensure the successful implementation of AF initiatives, education and capacity building must be prioritized. Policymakers should invest in comprehensive training programs designed to educate farmers on the establishment and sustainable management of AF projects. This includes providing technical support on tree selection, agro-ecological principles, and the integration of AF into existing farming systems. Moreover, collaborative efforts involving government agencies, NGOs, and research institutions can enhance capacity-building initiatives. By fostering partnerships that leverage local knowledge and expertise, policymakers can develop tailored training programs that focus on specific challenges encountered by farmers. Community-led education initiatives empower farmers, cultivate ownership, and promote long-term commitment to sustainable practices.

22.9.5 Ensuring equity and inclusivity

Future directives for AF and C incentives must prioritize equity and inclusivity. It is crucial to recognize that impoverished communities, including indigenous peoples and smallholder farmers, are disproportionately affected by the climate crisis. To ensure the success of AF initiatives, policymakers should prioritize equitable access to resources, support, and inclusive decision-making processes for the affected communities. This can be achieved by implementing participatory governance models that involve local communities in the development and implementation of policies. By incorporating the voices of those most experiencing the effects of the climate crisis, policymakers can create more effective and equitable AF programs that serve the needs of diverse populations.

22.10 Conclusion

The intersection of C incentives and quantification in AFS presents a promising avenue for addressing the dual challenges of the climate problem and the adoption of green agricultural practices. The incorporation of C credits and PES schemes has been shown to incentivize farmers to adopt AF practices, leading to increased C_{seq} and biodiversity conservation. However, the effectiveness of these incentives largely depends on the accuracy and transparency of C quantification methods. Recent studies have highlighted the importance of robust and reliable C quantification methods, such as LiDAR and hyperspectral remote sensing, to estimate C stocks and fluxes in AFS accurately. Despite the progress made, there are still challenges and limitations associated with C incentives and quantification in AF. These include issues related to data quality, methodology, and certification, as well as the need for more robust and inclusive governance frameworks to ensure that the benefits of C credits are equitably distributed among stakeholders.

Moving forward, collaboration among stakeholders will be essential to refine C incentive frameworks and enhance the effectiveness of AF as a natural climate solution. As the world advances into an era increasingly defined by climate challenges, the integration of C incentives and rigorous quantification methods in AF stands as a strategic imperative. This approach not only paves the way for a more resilient agricultural system but also underscores the vital role that sustainable practices play in securing a healthier planet for future generations. Harnessing the potential of AF through C management can create a sustainable and prosperous future that aligns environmental stewardship with agricultural innovation.

References

Ali, S., Khan, S.M., Ahmad, Z., Siddiq, Z., Ullah, A., Yoo, S., et al., 2023. Carbon sequestration potential of different forest types in Pakistan and its role in regulating services for public health. Frontiers in Public Health 10. Available at https://doi.org/10.3389/fpubh.2022.1064586.

Ansari, M.A., Ravisankar, N., Shamim, M., Rani, M., Prusty, A.K., Singh, R., et al., 2024. Accounting of carbon sequestration and tradeoff under various climatic scenarios in alternative agricultural system: a comprehensive framework toward carbon neutrality. Frontiers in Sustainable Food Systems 8. Available at https://doi.org/10.3389/fsufs.2024.1371255.

Asadnabizadeh, M., Moe, E., 2024. A review of Global Carbon Markets from Kyoto to Paris and beyond: the persistent failure of implementation. Frontiers in Environmental Science 12. Available at https://doi.org/10.3389/fenvs.2024.1368105.

Banerjee, A., Meena, R.S., Jhariya, M.K., Yadav, D.K., 2020. Agroecological Footprints Management for Sustainable Food System. Springer Singapore, India, pp. 1–514. Available at https://doi.org/10.1007/978-981-15-9496-0.

Banerjee, A., Jhariya, M.K., Bargali, S.S., Palit, D., 2023. Ecorestoration for Sustainability. Wiley-Scrivener, USA, pp. 1–441 ISBN: 9781119879718.

Banerjee, A., Jhariya, M.K., Raj, A., Mechergui, T., 2024. Ecosystem management: climate change and global sustainability—an introduction. Ecosystem Management: Climate Change and Sustainability. wiley, India, pp. 1–21. Available at https://doi.org/10.1002/9781394231249.ch1.

Barbosa, M.W., 2024. Government support mechanisms for sustainable agriculture: a systematic literature review and future research agenda. Sustainability 16 (5), 2185. Available at https://doi.org/10.3390/su16052185.

Benjamin, E.O., Sauer, J., 2018. The cost effectiveness of payments for ecosystem services—Smallholders and agroforestry in Africa. Land use Policy 71, 293–302. Available at https://doi.org/10.1016/j.landusepol.2017.12.001.

Bogale, G.A., Bekele, S.E., 2023. Sustainability of agroforestry practices and their resilience to climate change adaptation and mitigation in Sub-Saharan Africa: a review. Ekológia (Bratislava) 42 (2), 179–192. Available at https://doi.org/10.2478/eko-2023-0021.

Boinot, S., Mézière, D., Poulmarc'h, J., Saintilan, A., Lauri, P.-E., Sarthou, J.-P., 2020. Promoting generalist predators of crop pests in alley cropping agroforestry fields: farming system matters. Ecological Engineering 158, 106041. Available at https://doi.org/10.1016/j.ecoleng.2020.106041.

Borsah, A.A., Nazeer, M., Wong, M.S., 2023. LIDAR-based forest biomass remote sensing: a review of metrics, methods, and assessment criteria for the selection of allometric equations. Forests 14 (10). Available at https://doi.org/10.3390/f14102095.

Castle, S.E., Miller, D.C., Ordonez, P.J., Baylis, K., Hughes, K., 2021. The impacts of agroforestry interventions on agricultural productivity, ecosystem services, and human well-being in low- and middle-income countries: a systematic review. Campbell Systematic Reviews 17 (2). Available at https://doi.org/10.1002/cl2.1167.

Chaaben, N., Elleuch, Z., Hamdi, B., Kahouli, B., 2024. Green economy performance and sustainable development achievement: empirical evidence from Saudi Arabia. Environment, Development and Sustainability 26 (1), 549–564. Available at https://doi.org/10.1007/s10668-022-02722-8.

Chamberlain, J.L., Mitchell, D., Brigham, T., Hobby, T., Zabek, L.M., Davis, J., 2015. Forest farming practices. North American Agroforestry: An Integrated Science and Practice. wiley, United States, pp. 219–255. Available at https://doi.org/10.2134/2009.northamericanagroforestry.2ed.c9.

Cheng, F., Ou, G., Wang, M., Liu, C., 2024. Remote sensing estimation of forest carbon stock based on machine learning algorithms. Forests 15 (4), 681. Available at https://doi.org/10.3390/f15040681.

Chiti, T., Blasi, E., Chiriacò, M.V., 2024. Carbon sequestration in a bamboo plantation: a case study in a Mediterranean area. Journal of Forestry Research 35 (1). Available at https://doi.org/10.1007/s11676-024-01696-9.

Corgo, J., Cruz, S.S., Conceição, P., 2024. Nature-based solutions in spatial planning and policies for climate change adaptation: a literature review. Ambio 53 (11), 1599–1617. Available at https://doi.org/10.1007/s13280-024-02052-1.

Cuong, L., Hung, B., Bolanle-Ojo, O.T., Xu, X., Thanh, N., Chai, L., et al., 2020. Biomass and carbon storage in an age-sequence of acacia mangium plantation forests in southeastern region, vietnam. Ministerio de Agricultura Pesca y Alimentacion, China Forest Systems 29 (2), 1–16. Available at https://doi.org/10.5424/fs/2020292-16685.

Dagar, J.C., Tewari, V.P., 2017. Evolution of agroforestry as a modern science. In: Dagar, J., Tewari, V. (Eds.), Agroforestry. Springer, Singapore. Available at https://doi.org/10.1007/978-981-10-7650-3_2.

Davis, M., Cuevas, N.B., Gvein, M.H., 2024. Transforming ambition into action to catalyse nature-based solutions: insights from 250 good practice policy instruments. Nature-Based Solutions. 6. Available at https://doi.org/10.1016/j.nbsj.2024.100171.

Demise, S., Emire, A., 2024. Assessment of parkland tree species and their management practice in Ana Sora District of Guji Zone, Southern Ethiopia. International Journal of Plant & Soil Science 36 (5), 1033–1045. Available at https://doi.org/10.9734/ijpss/2024/v36i54669.

Devi, A., Jhariya, M.K., Raj, A., Banerjee, A., Singh, K.P., Singh, B., 2024. Acacia nilotica: a promising species for soil sustainability. Land and Environmental Management through Forestry. wiley, India, pp. 339–353. Available at https://doi.org/10.1002/9781119910527.ch14.

Dhyani, S., Murthy, I.K., Kadaverugu, R., Dasgupta, R., Kumar, M., Adesh Gadpayle, K., 2021. Agroforestry to achieve global climate adaptation and mitigation targets: are South Asian countries sufficiently prepared? Forests 12 (3), 303. Available at https://doi.org/10.3390/f12030303.

Dobhal, S., Kumar, R., Bhardwaj, A.K., Chavan, S.B., Uthappa, A.R., Kumar, M., et al., 2024. Global assessment of production benefits and risk reduction in agroforestry during extreme weather events under climate change scenarios. Frontiers in Forests and Global Change 7. Available at https://doi.org/10.3389/ffgc.2024.1379741.

Dollinger, J., Jose, S., 2018. Agroforestry for soil health. Agroforestry Systems 92 (2), 213–219. Available at https://doi.org/10.1007/s10457-018-0223-9.

Don, A., Seidel, F., Leifeld, J., Kätterer, T., Martin, M., Pellerin, S., et al., 2024. Carbon sequestration in soils and climate change mitigation—definitions and pitfalls. Global Change Biology 30 (1). Available at https://doi.org/10.1111/gcb.16983.

Dossa, K.F., Miassi, Y.E., 2024. Remote sensing methods and GIS approaches for carbon sequestration measurement: a general review. International Journal of Environment and Climate Change 14 (7), 222–233. Available at https://doi.org/10.9734/ijecc/2024/v14i74265.

Ferreira, C.S.S., Kašanin-Grubin, M., Solomun, M.K., Sushkova, S., Minkina, T., Zhao, W., et al., 2023. Wetlands as nature-based solutions for water management in different environments. Current Opinion in Environmental Science and Health 33. Available at https://doi.org/10.1016/j.coesh.2023.100476.

Foster, K., Neufeldt, H., 2014. Biocarbon projects in agroforestry: Lessons from the past for future development. Current Opinion in Environmental Sustainability 6 (1), 148–154. Available at https://doi.org/10.1016/j.cosust.2013.12.002.

Gebrael, K., Mitri, G., Kalantzi, O.-I., 2024. Overview of nature-based solutions for climate resilience in the MENA region. Nature-Based Solutions 6, 100159. Available at https://doi.org/10.1016/j.nbsj.2024.100159.

Golicz, K., Bellingrath-Kimura, S., Breuer, L., Wartenberg, A.C., 2022. Carbon accounting in European agroforestry systems – key research gaps and data needs. Current Research in Environmental Sustainability 4, 100134. Available at https://doi.org/10.1016/j.crsust.2022.100134.

Haddaway, N.R., Hedlund, K., Jackson, L.E., Kätterer, T., Lugato, E., Thomsen, I.K., et al., 2017. How does tillage intensity affect soil organic carbon? A systematic review. Environmental Evidence 6 (1). Available at https://doi.org/10.1186/s13750-017-0108-9.

Hawkins, H.J., Cargill, R.I.M., Van Nuland, M.E., Hagen, S.C., Field, K.J., Sheldrake, M., et al., 2023. Mycorrhizal mycelium as a global carbon pool. Current Biology 33 (11), R560. Available at https://doi.org/10.1016/j.cub.2023.02.027.

Hemida, M., Vityi, A., Hammad, Z.M., 2023. Socio-economic traits and constraints associated with smallholder farmers in Taungya agroforestry program in Sudan. Agroforestry Systems 97 (6), 1169–1184. Available at https://doi.org/10.1007/s10457-023-00855-x.

Holderieath, J., Valdivia, C., Godsey, L., Barbieri, C., 2012. The potential for carbon offset trading to provide added incentive to adopt silvopasture and alley cropping in Missouri. Agroforestry Systems 86 (3), 345–353. Available at https://doi.org/10.1007/s10457-012-9543-3.

Holz, M., Augustin, J., 2021. Erosion effects on soil carbon and nitrogen dynamics on cultivated slopes: a meta-analysis. Geoderma 397, 115045. Available at https://doi.org/10.1016/j.geoderma.2021.115045.

Houndjo Kpoviwanou, M.R.J., Sourou, B.N.K., Ouinsavi, C.A.I.N., 2024. Challenges in adoption and wide use of agroforestry technologies in Africa and pathways for improvement: a systematic review. Trees, Forests and People 17. Available at https://doi.org/10.1016/j.tfp.2024.100642.

Isaac, M., Muhammad, L., Joweria, N., 2024. Social and ecological contributions of the taungya agroforestry system in the restoration of Mount Elgon National Park, Uganda. East African Journal of Forestry and Agroforestry 7 (1). Available at https://doi.org/10.37284/eajfa.7.1.2307.

Jember, A.A., Taye, M.A., Gebeyehu, G., Mulu, G., Long, T.T., Jayaraman, D., et al., 2023. Carbon stock potential of highland bamboo plantations in northwestern Ethiopia. Carbon Balance and Management 18 (1). Available at https://doi.org/10.1186/s13021-023-00224-2.

Jhariya, M.K., Banerjee, A., Yadav, D.K., Raj, A., 2018. Leguminous trees an innovative tool for soil sustainability. In: Meena, R., Das, A., Yadav, G., Lal, R. (Eds.), Legumes for Soil Health and Sustainable Management. Springer, Singapore, pp. 315–345. Available at https://doi.org/10.1007/978-981-13-0253-4_10.

Jhariya, M.K., Banerjee, A., Meena, R.S., Yadav, D.K., 2019. Sustainable Agriculture, Forest and Environmental Management. Springer, pp. 606. Available at https://doi.org/10.1007/978-981-13-6830-1.

Jhariya, M.K., Banerjee, A., Swaroop Meena, R., Kumar, S., Raj, A., 2021a. Sustainable Intensification for Agroecosystem Services and Management. Springer, Singapore, pp. 1–870. Available at https://doi.org/10.1007/978-981-16-3207-5.

Jhariya, M.K., Meena, R.S., Banerjee, A., 2021b. Ecological Intensification of Natural Resources for Sustainable Agriculture. Springer Singapore, India, pp. 1–655. Available at https://doi.org/10.1007/978-981-33-4203-3.

Jhariya, M.K., Meena, R.S., Banerjee, A., Meena, S.N., 2022. Natural Resources Conservation and Advances for Sustainability. Elsevier. Available at https://doi.org/10.1016/C2019-0-03763-6.

Jin, S., Zhang, E., Guo, H., Hu, C., Zhang, Y., Yan, D., 2023. Comprehensive evaluation of carbon sequestration potential of landscape tree species and its influencing factors analysis: implications for urban green space management. Carbon Balance and Management 18 (1). Available at https://doi.org/10.1186/s13021-023-00238-w.

Jose, S., Dollinger, J., 2019. Silvopasture: a sustainable livestock production system. Agroforestry Systems 93 (1), 1–9. Available at https://doi.org/10.1007/s10457-019-00366-8.

Kassie, G.W., 2018. Agroforestry and farm income diversification: synergy or trade-off? The case of Ethiopia. Environmental Systems Research 6 (1). Available at https://doi.org/10.1186/s40068-017-0085-6.

Kay, S., Rega, C., Moreno, G., den Herder, M., Palma, J.H.N., Borek, R., et al., 2019. Agroforestry creates carbon sinks whilst enhancing the environment in agricultural landscapes in Europe. Land use Policy 83, 581–593. Available at https://doi.org/10.1016/j.landusepol.2019.02.025.

Kinneen, L., Escobar, M.P., Hernandez, L.M., Thompson, J., Ramos-Pastrana, Y., Córdoba-Suarez, E., et al., 2024. Silvopastoral systems benefit invertebrate biodiversity on tropical livestock farms in Caquetá, Colombia. Agricultural and Forest Entomology 26 (1), 126–134. Available at https://doi.org/10.1111/afe.12594.

Kopnina, H., Zhang, S.R., Anthony, S., Hassan, A., Maroun, W., 2024. The inclusion of biodiversity into environmental, social, and governance (ESG) framework: a strategic integration of ecocentric extinction accounting. Journal of Environmental Management 351, 119808. Available at https://doi.org/10.1016/j.jenvman.2023.119808.

Krauss, M., Wiesmeier, M., Don, A., Cuperus, F., Gattinger, A., Gruber, S., et al., 2022. Reduced tillage in organic farming affects soil organic carbon stocks in temperate Europe. Soil and Tillage Research 216, 105262. Available at https://doi.org/10.1016/j.still.2021.105262.

Kumar, S., Meena, R.S., Jhariya, M.K., 2020. Resources Use Efficiency in Agriculture Resources Use Efficiency in Agriculture. Springer, Singapore, pp. 1–760. Available at https://doi.org/10.1007/978-981-15-6953-1.

Kumar, A., Malik, M.S., Shabnam, S., Kumar, R., Karmakar, S., Das, S.S., et al., 2024a. Carbon sequestration and credit potential of gamhar (*Gmelina arborea* Roxb.) based agroforestry system for zero carbon emission of India. Scientific Reports 14 (1). Available at https://doi.org/10.1038/s41598-024-53162-5.

Kumar, S., Meena, R.S., Sheoran, P., Jhariya, M.K., 2024b. Regenerative Agriculture for Sustainable Food Systems. Springer, Singapore, pp. 1–493. Available at https://doi.org/10.1007/978-981-97-6691-8.

kumwimba, M.N., Akter, S., Li, X., Dzakpasu, M., Ifon, B.E., Manirakiza, B., et al., 2024. Nutrient and sediment retention by riparian vegetated buffer strips: impacts of buffer length, vegetation type, and season. Agriculture, Ecosystems and Environment 369. Available at https://doi.org/10.1016/j.agee.2024.109050.

Kuyah, S., Öborn, I., Jonsson, M., Dahlin, A.S., Barrios, E., Muthuri, C., et al., 2016. Trees in agricultural landscapes enhance provision of ecosystem services in Sub-Saharan Africa. International Journal of Biodiversity Science, Ecosystem Services and Management 12 (4), 255–273. Available at https://doi.org/10.1080/21513732.2016.1214178.

Leditznig, T., Klug, H., 2024. Estimating carbon stock in unmanaged forests using field data and remote sensing. Remote Sensing 16 (21), 3926. Available at https://doi.org/10.3390/rs16213926.

Leiter, T., 2024. Nationally determined contributions (NDCs) as a governance instrument – accounting for politics, negotiation progress, and related mechanisms under the Paris agreement. Environmental Politics 33 (3), 552–557. Available at https://doi.org/10.1080/09644016.2023.2252312.

Lin, N., Feng, M., Huang, H., Qiu, Z., Ma, T., Chen, S., 2024. Effects of thinning on carbon storage in a mixed broadleaved plantation in a subtropical area of China. Forests 15 (4), 638. Available at https://doi.org/10.3390/f15040638.

Liu, R., Thomas, B.W., Shi, X., Zhang, X., Wang, Z., Zhang, Y., 2021. Effects of ground cover management on improving water and soil conservation in tree crop systems: a meta-analysis. CATENA 199, 105085. Available at https://doi.org/10.1016/j.catena.2020.105085.

Liu, L., Zhou, W., Guan, K., Peng, B., Xu, S., Tang, J., et al., 2024. Knowledge-guided machine learning can improve carbon cycle quantification in agroecosystems. Nature Communications 15 (1). Available at https://doi.org/10.1038/s41467-023-43860-5.

Ljalem, S., Birhane, E., Teka, K., Berhe, D.H., 2024. Parkland trees on smallholder farms ameliorate soil physical-chemical properties in the semi-arid area of Tigray, Ethiopia. Journal of Arid Land 16 (1), 1–13. Available at https://doi.org/10.1007/s40333-024-0002-6.

Lowder, S.K., Sánchez, M.V., Bertini, R., 2021. Which farms feed the world and has farmland become more concentrated? World Development 142. Available at https://doi.org/10.1016/j.worlddev.2021.105455.

Luo, S., Schmid, B., Hector, A., Scherer-Lorenzen, M., Verheyen, K., Barsoum, N., et al., 2024. Mycorrhizal associations modify tree diversity–productivity relationships across experimental tree plantations. New Phytologist 243 (3), 1205–1219. Available at https://doi.org/10.1111/nph.19889.

Mahmoodabadi, M., Zamani, S., Yazdanpanah, N., 2021. Enrichment of soil organic carbon in the eroded sediments due to the simultaneous occurrence of rain and wind. Geoderma 386, 114926. Available at https://doi.org/10.1016/j.geoderma.2020.114926.

Majaura, M., Böhm, C., Freese, D., 2024. The influence of trees on crop yields in temperate zone alley cropping systems: a review. Sustainability 16 (8), 3301. Available at https://doi.org/10.3390/su16083301.

Manikandan, S., Kaviya, R.S., Shreeharan, D.H., Subbaiya, R., Vickram, S., Karmegam, N., et al., 2025. Artificial intelligence-driven sustainability: enhancing carbon capture for sustainable development goals– a review. Sustainable Development 33 (2), 2004–2029. Available at https://doi.org/10.1002/sd.3222.

Mbow, C., Van Noordwijk, M., Luedeling, E., Neufeldt, H., Minang, P.A., Kowero, G., 2014. Agroforestry solutions to address food security and climate change challenges in Africa. Current Opinion in Environmental Sustainability 6 (1), 61–67. Available at https://doi.org/10.1016/j.cosust.2013.10.014.

Meena, R.S., Pradhan, G., 2023. Industrial garbage-derived biocompost enhances soil organic carbon fractions, CO_2 biosequestration, potential carbon credits and sustainability index in a rice-wheat ecosystem. Environmental Research 235, 116525. Available at https://doi.org/10.1016/j.envres.2023.116525.

Meena, R.S., Yadav, A., Kumar, S., Jhariya, M.K., Jatav, S.S., 2022. Agriculture ecosystem models for CO_2 sequestration, improving soil physicochemical properties, and restoring degraded land. Ecological Engineering 176. Available at https://doi.org/10.1016/j.ecoleng.2022.106546.

Meena, R.S., Pradhan, G., Kumar, S., Lal, R., 2023. Using industrial wastes for rice-wheat cropping and food-energy-carbon-water-economic nexus to the sustainable food system. Renewable and Sustainable Energy Reviews 187, 113756. Available at https://doi.org/10.1016/j.rser.2023.113756.

Meena, R.S., Pradhan, G., Nalani, Kumar, S., Singh, A.K., Verma, S.K., et al., 2024a. Diversified cropping modules designed for soil restoration, CO_2 sequestration, and generating carbon credits. Land Degradation and Development 35 (14), 4314–4329. Available at https://doi.org/10.1002/ldr.5224.

Meena, R.S., Pradhan, G., Singh, K., Kumar, S., Singh, A.K., Shashidhar, K.S., et al., 2024b. Agriculture models for restoring degraded land to enhance CO_2 biosequestration and carbon credits in the Vindhyan region of India. Science of the Total Environment 929. Available at https://doi.org/10.1016/j.scitotenv.2024.172661.

Meena, R.S., Lal, R., Kumar, S., Pradhan, G., Srinivasarao, C., Singh, A.K., et al., 2024c. Potential of Indian agriculture for capturing atmospheric CO_2 and monetizing carbon credits to the farmers: an overview and policy framework. Advances in Agronomy 188, 101–206. Available at https://doi.org/10.1016/bs.agron.2024.06.001.

Meena, R.S., Pradhan, G.S., Kumar, S., Jhariya, M.K., Rao, C.S., 2025a. Carbon-Negative Agriculture. Springer Nature, Singapore.

Meena, R.S., Pradhan, G., Kumar, S., 2025b. Energy flow, eco-efficiency, and economic circulation with recycled industrial waste compost application in wheat and subsequent rice farming. Science of the Total Environment 967. Available at https://doi.org/10.1016/j.scitotenv.2025.178779.

Menozzi, D., Fioravanzi, M., Donati, M., 2015. Farmer's motivation to adopt sustainable agricultural practices. Bio-based and Applied Economics 4 (2), 125–147. Available at https://doi.org/10.13128/BAE-14776.

Mlambo, D., 2021. Ecological intensification: a step towards biodiversity conservation and management of terrestrial landscape. Ecological Intensification of Natural Resources for Sustainable Agriculture. Springer Singapore, Zimbabwe, pp. 77–102. Available at https://doi.org/10.1007/978-981-33-4203-3_3.

Mlambo, D., Sebata, A., Chichinye, A., Mabidi, A., 2024. Chapter 5 - Agroforestry and biodiversity conservation. In: Jhariya, M.K., Meena, R.S., Banerjee, A., Kumar, S., Raj, A. (Eds.), Agroforestry for Carbon and Ecosystem Management. Academic Press, pp. 63–78. Available at https://doi.org/10.1016/b978-0-323-95393-1.00008-7.

Mukhlis, I., Rizaludin, M.S., Hidayah, I., 2022. Understanding socio-economic and environmental impacts of agroforestry on rural communities. Forests 13 (4), 556. Available at https://doi.org/10.3390/f13040556.

Munroe, J.W., Isaac, M.E., 2014. N2-fixing trees and the transfer of fixed-N for sustainable agroforestry: a review. Agronomy for Sustainable Development 34 (2), 417–427. Available at https://doi.org/10.1007/s13593-013-0190-5.

Musokwa, M., Mafongoya, P.L., 2021. Effects of improved pigeonpea fallows on biological and physical soil properties and their relationship with maize yield. Agroforestry Systems 95 (2), 443–457. Available at https://doi.org/10.1007/s10457-021-00598-7.

Nath, A.J., Sileshi, G.W., Laskar, S.Y., Pathak, K., Reang, D., Nath, A., et al., 2021. Quantifying carbon stocks and sequestration potential in agroforestry systems under divergent management scenarios relevant to India's Nationally Determined Contribution. Journal of Cleaner Production 281. Available at https://doi.org/10.1016/j.jclepro.2020.124831.

Ngaba, M.J.Y., Mgelwa, A.S., Gurmesa, G.A., Uwiragiye, Y., Zhu, F., Qiu, Q., et al., 2024. Meta-analysis unveils differential effects of agroforestry on soil properties in different zonobiomes. Plant and Soil 496 (1-2), 589–607. Available at https://doi.org/10.1007/s11104-023-06385-w.

Niguse, G., Iticha, B., Kebede, G., Chimdi, A., 2022. Contribution of coffee plants to carbon sequestration in agroforestry systems of Southwestern Ethiopia. Journal of Agricultural Science 160 (6), 440–447. Available at https://doi.org/10.1017/S0021859622000624.

Oliver, T.H., Heard, M.S., Isaac, N.J.B., Roy, D.B., Procter, D., Eigenbrod, F., et al., 2015. Biodiversity and resilience of ecosystem functions. Trends in Ecology and Evolution 30 (11), 673–684. Available at https://doi.org/10.1016/j.tree.2015.08.009.

Panchal, P., Preece, C., Peñuelas, J., Giri, J., 2022. Soil carbon sequestration by root exudates. Trends in Plant Science 27 (8), 749–757. Available at https://doi.org/10.1016/j.tplants.2022.04.009.

Pantera, A, Mosquera-Losada, M.R., Herzog, F., den Herder, M., 2021. Agroforestry and the environment. Agroforestry Systems 95 (5), 767–774. Available at https://doi.org/10.1007/s10457-021-00640-8.

Pan, C., Shrestha, A., Innes, J.L., Zhou, G., Li, N., Li, J., et al., 2022. Key challenges and approaches to addressing barriers in forest carbon offset projects. Journal of Forestry Research 33 (4), 1109–1122. Available at https://doi.org/10.1007/s11676-022-01488-z.

Podhrázská, J., Kučera, J., Doubrava, D., Doležal, P., 2021. Functions of windbreaks in the landscape ecological network and methods of their evaluation. Forests 12 (1), 67. Available at https://doi.org/10.3390/f12010067.

Poudel, S., Pent, G., Fike, J., 2024. Silvopastures: benefits, past efforts, challenges, and future prospects in the United States. Agronomy 14 (7), 1369. Available at https://doi.org/10.3390/agronomy14071369.

Pradhan, G., Meena, R.S., 2023a. Interaction impact of biocompost on nutrient dynamics and relations with soil biota, carbon fractions index, societal value of CO_2 equivalent and ecosystem services in the wheat-rice farming. Chemosphere 339. Available at https://doi.org/10.1016/j.chemosphere.2023.139695.

Pradhan, G., Meena, R.S., 2023b. Utilizing waste compost to improve the atmospheric CO_2– capturing in the rice-wheat cropping system and energy-cum-carbon credit auditing for a circular economy. Science of the Total Environment 892. Available at https://doi.org/10.1016/j.scitotenv.2023.164572.

Pradhan, G., Meena, R.S., Kumar, S., Lal, R., 2023. Utilizing industrial wastes as compost in wheat-rice production to improve the above and below-ground ecosystem services. Agriculture, Ecosystems & Environment 358, 108704. Available at https://doi.org/10.1016/j.agee.2023.108704.

Prasad, R., Jhariya, M.K., Banerje, A., 2021. Advances in Sustainable Development and Management of Environmental and Natural Resources: Economic Outlook and Opinions, 2-volume set. Apple Academic Press, New York, pp. 1–428. Available at https://doi.org/10.1201/9781003187455.

Prosser, R.S., Hoekstra, P.F., Gene, S., Truman, C., White, M., Hanson, M.L., 2020. A review of the effectiveness of vegetated buffers to mitigate pesticide and nutrient transport into surface waters from agricultural areas. Journal of Environmental Management 261, 110210. Available at https://doi.org/10.1016/j.jenvman.2020.110210.

Pumariño, L., Sileshi, G.W., Gripenberg, S., Kaartinen, R., Barrios, E., Muchane, M.N., et al., 2015. Effects of agroforestry on pest, disease and weed control: a meta-analysis. Basic and Applied Ecology 16 (7), 573–582. Available at https://doi.org/10.1016/j.baae.2015.08.006.

Qin, S., Nie, S., Guan, Y., Zhang, D., Wang, C., Zhang, X., 2022. Forest emissions reduction assessment using airborne LiDAR for biomass estimation. Resources, Conservation and Recycling 181, 106224. Available at https://doi.org/10.1016/j.resconrec.2022.106224.

Quandt, A., Neufeldt, H., Gorman, K., 2023. Climate change adaptation through agroforestry: opportunities and gaps. Current Opinion in Environmental Sustainability 60, 101244. Available at https://doi.org/10.1016/j.cosust.2022.101244.

Raj, A., Jhariya, M.K., Banerjee, A., Meena, R.S., Kumar, S., Devi, A., et al., 2023. Agroforestry and ecosystem services. Agroforestry for Carbon and Ecosystem Management. Elsevier, India, pp. 205–221. Available at https://doi.org/10.1016/B978-0-323-95393-1.00027-0.

Raj, A., Jhariya, M.K., Banerjee, A., Jha, R.K., Singh, K.P., 2024a. Agroforestry. Wiley-Scrivener, USA.

Raj, A., Jhariya, M.K., Banerjee, A., Nema, S., Bargali, K., 2024b. Land and Environmental Management Through Forestry. wiley, India, pp. 1–375.

Raj, A., Jhariya, M.K., Meena, R.S., 2025. Agroforestry for Monetising Carbon Credits. Springer Nature, Switzerland.

Rivera-Niquepa, J.D., Zuluaga Muñoz, W., Rojas Hernández, A., 2023. Carbon credits: basics, the role of ICTs, and a bibliometric literature review. ParadigmPlus 4 (3), 1–16. Available at https://doi.org/10.55969/paradigmplus.v4n3a1.

Rolando, J.L., Dubeux, J.C.B., Souza, T.Cd, Mackowiak, C., Wright, D., George, S., et al., 2021. Organic carbon is mostly stored in deep soil and only affected by land use in its superficial layers: a case study. Agrosystems, Geosciences and Environment 4 (1). Available at https://doi.org/10.1002/agg2.20135.

Rolo, V., Rivest, D., Maillard, É., Moreno, G., 2023. Agroforestry potential for adaptation to climate change: a soil-based perspective. Soil Use and Management 39 (3), 1006–1032. Available at https://doi.org/10.1111/sum.12932.

Rosenstock, T.S., Tully, K.L., Arias-Navarro, C., Neufeldt, H., Butterbach-Bahl, K., Verchot, L.V., 2014. Agroforestry with N2-fixing trees: sustainable development's friend or foe? Current Opinion in Environmental Sustainability 6 (1), 15–21. Available at https://doi.org/10.1016/j.cosust.2013.09.001.

Roy, O., Meena, R.S., Kumar, S., Jhariya, M.K., Pradhan, G., 2022. Assessment of land use systems for CO_2 sequestration, carbon credit potential, and income security in Vindhyan region, India. Land Degradation and Development 33 (4), 670–682. Available at https://doi.org/10.1002/ldr.4181.

Santos, M., Cajaiba, R.L., Bastos, R., Gonzalez, D., Petrescu Bakış, A.L., Ferreira, D., et al., 2022. Why do agroforestry systems enhance biodiversity? Evidence from habitat amount hypothesis predictions. Frontiers in Ecology and Evolution 9. Available at https://doi.org/10.3389/fevo.2021.630151.

Sharma, S., Dhal, S., Rout, T., Acharya, B.S., 2022. Drones and machine learning for estimating forest carbon storage. Carbon Research 1 (1). Available at https://doi.org/10.1007/s44246-022-00021-5.

Šilenskytė, A., Butkevičienė, J., Bartminas, A., 2024. Blockchain-based connectivity within digital platforms and ecosystems in international business. Journal of International Management 30 (3), 101109. Available at https://doi.org/10.1016/j.intman.2023.101109.

Singh, R.K., Biradar, C.M., Behera, M.D., Prakash, A.J., Das, P., Mohanta, M.R., et al., 2024. Optimising carbon fixation through agroforestry: estimation of aboveground biomass using multi-sensor data synergy and machine learning. Ecological Informatics 79, 102408. Available at https://doi.org/10.1016/j.ecoinf.2023.102408.

Somarriba, E., Cerda, R., Orozco, L., Cifuentes, M., Dávila, H., Espin, T., et al., 2013. Carbon stocks and cocoa yields in agroforestry systems of Central America. Agriculture, Ecosystems and Environment 173, 46–57. Available at https://doi.org/10.1016/j.agee.2013.04.013.

Stetter, C., Sauer, J., 2024. Tackling climate change: agroforestry adoption in the face of regional weather extremes. Ecological Economics 224, 108266. Available at https://doi.org/10.1016/j.ecolecon.2024.108266.

Su, R., Du, W., Shan, Y., Ying, H., Rihan, W., Li, R., 2024. Aboveground carbon stock estimation based on backpack LiDAR and UAV multispectral imagery at the forest sample plot scale. Remote Sensing 16 (21), 3927. Available at https://doi.org/10.3390/rs16213927.

Syano, N.M., Nyangito, M.M., Kironchi, G., Wasonga, O.V., 2023. Agroforestry practices impacts on soil properties in the drylands of Eastern Kenya. Trees, Forests and People 14. Available at https://doi.org/10.1016/j.tfp.2023.100437.

Tamba, Y., Wafula, J., Magaju, C., St-Jacques, B., Stiem-Bhatia, L., Arias-Navarro, C., et al., 2021. A review of the participation of smallholder farmers in land-based carbon payment schemes. TMG and ICRAF Working Paper. Available at https://doi.org/10.35435/2.2021.4.

Tebkew, M., Asfaw, Z., Worku, A., Jacobson, M., 2024a. Contribution of agroforestry practices to income and poverty status of households in Northwestern Ethiopia. Discover Agriculture 2 (1). Available at https://doi.org/10.1007/s44279-024-00062-x.

Tebkew, M., Asfaw, Z., Worku, A., 2024b. The role of agroforestry systems for addressing climate change livelihood vulnerability of farmers of Northwestern Ethiopia. Heliyon 10 (17), e36196. Available at https://doi.org/10.1016/j.heliyon.2024.e36196.

Terasaki Hart, D.E., Yeo, S., Almaraz, M., Beillouin, D., Cardinael, R., Garcia, E., et al., 2023. Priority science can accelerate agroforestry as a natural climate solution. Nature Climate Change 13 (11), 1179–1190. Available at https://doi.org/10.1038/s41558-023-01810-5.

Tranchina, M., Reubens, B., Frey, M., Mele, M., Mantino, A., 2024. What challenges impede the adoption of agroforestry practices? A global perspective through a systematic literature review. Agroforestry Systems 98 (6), 1817–1837. Available at https://doi.org/10.1007/s10457-024-00993-w.

Turner, B., Devisscher, T., Chabaneix, N., Woroniecki, S., Messier, C., Seddon, N., 2022. The role of nature-based solutions in supporting social-ecological resilience for climate change adaptation. Annual Review of Environment and Resources 47, 123–148. Available at https://doi.org/10.1146/annurev-environ-012220-010017.

Udawatta, R.P., Rankoth, L., Jose, S., 2019. Agroforestry and biodiversity. Sustainability 11 (10), 2879. Available at https://doi.org/10.3390/su11102879.

van Noordwijk, M., 2021. Agroforestry-based ecosystem services. Land 10 (8), 770. Available at https://doi.org/10.3390/land10080770.

Vargas-Hernández, J.G., Ali, M.M., 2022. Sustainability and green socio-ecosystem resilience: a conceptual view. Int. J. Environ. Sustain. Green Technol 13 (1), 1–23. Available at https://doi.org/10.4018/IJESGT.292453.

Wang, J., Patruno, L., Chen, Z., Yang, Q., Tamura, Y., 2024. Windbreak effectiveness of single and double-arranged shelterbelts: a parametric study using large eddy simulation. Forests 15 (10), 1760. Available at https://doi.org/10.3390/f15101760.

Wei, F., Wang, S., Fu, B., Zhang, L., Fu, C., Kanga, E.M., 2018. Balancing community livelihoods and biodiversity conservation of protected areas in East Africa. Current Opinion in Environmental Sustainability 33, 26–33. Available at https://doi.org/10.1016/j.cosust.2018.03.013.

Wen, Y., Liu, B., Lin, L., Hu, M., Wen, X., Li, T.-Y., et al., 2024. Shelterbelt effects on soil redistribution on an arable slope by wind and water. CATENA 241, 108044. Available at https://doi.org/10.1016/j.catena.2024.108044.

Wongpiyabovorn, O., Plastina, A., Crespi, J.M., 2022. Challenges to voluntary ag carbon markets. Applied Economic Perspectives and Policy 45 (2), 1154–1167. Available at https://doi.org/10.1002/aepp.13254.

Wu, S., Bashir, M.A., Raza, Q.U.A., Rehim, A., Geng, Y., Cao, L., 2023. Application of riparian buffer zone in agricultural non-point source pollution control—a review. Frontiers in Sustainable Food Systems 7. Available at https://doi.org/10.3389/fsufs.2023.985870.

Yao, P., Yu, Z., Zhang, Y., Xu, T., 2023. Application of machine learning in carbon capture and storage: an in-depth insight from the perspective of geoscience. Fuel 333, 126296. Available at https://doi.org/10.1016/j.fuel.2022.126296.

Yasin, G., Nawaz, M.F., Zubair, M., Azhar, M.F., Mohsin Gilani, M., Ashraf, M.N., et al., 2023. Role of traditional agroforestry systems in climate change mitigation through carbon sequestration: an investigation from the semi-arid region of Pakistan. Land 12 (2). Available at https://doi.org/10.3390/land12020513.

Yasin, G., Nawaz, M.F., Sinha, D., Qadir, I., Altaf, M., Ashraf, M.N., et al., 2024. Agroforestry status, services, and its role in climate change mitigation through carbon sequestration under semi-arid conditions. Trees, Forests and People 17. Available at https://doi.org/10.1016/j.tfp.2024.100640.

Zook, M., 2023. Platforms, blockchains and the challenges of decentralization. Cambridge Journal of Regions, Economy and Society 16 (2), 367–372. Available at https://doi.org/10.1093/cjres/rsad008.

Section V

Plans and policies for carbon incentives

Chapter 23

Research, development, and future directives toward carbon financing

Donald Mlambo
Department of Forest Resources and Wildlife Management, National University of Science and Technology, Bulawayo, Zimbabwe

Highlights

- Research can boost investment in green technologies
- Development efforts must prioritize decarbonization projects
- Future directives should harmonize international carbon markets
- Blockchain technology can boost carbon market transparency
- Green growth promotes social equity and inclusion

23.1 Introduction

The interplay between carbon financing (CF) and a green economy (GE) is critical for addressing climate change (Li et al., 2024a; Yi and Zhang, 2024; Meena et al., 2025; Wu et al., 2024b; Raj et al., 2025). CF refers to financial resources allocated to projects that reduce, avoid, or sequester carbon (C) emissions. A GE is an economic system that has low-C, and is resource-efficient and socially inclusive (Eaton and Sheng, 2019; Meena et al., 2025). Sustained research, development, and innovation, backed by forward-thinking policies and strategic financing, are crucial for securing a sustainable future and environmental well-being in the decades to come. Numerous nations and corporations are committing to achieving C neutrality by the mid-21st century, thereby influencing future research and development (R&D) agendas (Elliott et al., 2023; Qadri et al., 2024). Amidst the intensifying climate crisis, the international community is progressively acknowledging the critical necessity for innovative financial mechanisms that promote green development (Song et al., 2023; Ma and Fei, 2024). A global momentum is increasingly building to encourage R&D in CF and the GE, as countries, businesses, and civil society organizations acknowledge the urgent need to address the climate problem, foster green development, and shift toward low-C, resource-efficient economies (Meena et al., 2022; Roy et al., 2022; Nguyen et al., 2024; Liu and Zhu, 2024; Daniya and Tang, 2024).

R&D in CF and the GE is focused on finding new ways to reduce greenhouse gas (GHG) emissions, increase energy efficiency, and promote green development. The future of CF and the GE is likely to be shaped by a number of factors, including government policies, technological advances, and changes, in consumer behavior. As more countries implement policies to reduce GHG emissions, the demand for CF is likely to increase, providing new opportunities for companies and organizations that can reduce their emissions. The GE is expected to continue to grow in the coming years, as more businesses and governments recognize the importance of sustainability and the need to transition to a GE. The shift to a GE hinges on harnessing innovation and technology to drive the development and adoption of eco-friendly solutions (Si et al., 2024; Liu et al., 2024; Teklie and Yağmur, 2024; Meena, Pradhan, Singh, et al., 2024a).

Effective policy frameworks, regulatory incentives, and government interventions are crucial for creating an enabling environment that supports CF and GE transitions. Many governments are leveraging a variety of climate-friendly technologies and policy instruments to accelerate investments in sustainable infrastructure, aiming to meet their climate goals and reduce GHG emissions (Dwivedi et al., 2022; Meena et al., 2024b). Recent reports suggest that trade in these technologies can further enhance decarbonization efforts (Miao et al., 2024; Zhu et al., 2024). Many countries face barriers that hinder their ability to achieve sustainability goals. Transitioning to a GE involves overcoming various barriers,

Carbon Incentives, Plans, and Policies in Agriculture. DOI: https://doi.org/10.1016/B978-0-443-33557-0.00029-X

including financial constraints, lack of awareness, and inadequate policies (Licastro and Sergi, 2021; Chukwu, 2020). These barriers limit efforts to balance social and environmental objectives with economic growth, often making it difficult for policies supporting sustainability to be effectively implemented and sustained (Jhariya et al., 2022; Jhariya et al., 2023). In order to effectively address the climate problem, countries must work together to reduce GHG emissions and promote green development. This will require cooperation on the development of international C markets and technology transfer.

R&D in CF and the GE are progressively emphasizing the need to address social equity, environmental justice, and inclusivity (Boyce et al., 2023). Building sustainable societies requires integrating environmental, economic, and social goals, ensuring a balanced relationship between humans and the natural world (Prasad et al., 2021; Meena, Lal, et al., 2024c). The objective of this chapter is to provide a comprehensive overview of the current R&D and future directives in the fields of CF and the GE. The goal is to equip readers with a deep understanding of the current state and future trajectory of CF and the GE, empowering them to make informed decisions and take meaningful action toward a more sustainable and equitable future.

23.2 Overview of carbon financing

Global CF reached new heights in 2022, with investments in C credits exceeding US$36 billion (Gareth and Sophie, 2024). Although there is significant funding available for decarbonization projects, challenges persist in the effectiveness, accessibility, and implementation of the funding mechanisms used to support these initiatives. CF attracts investments from private sectors, which are essential for scaling up sustainable projects. Countries can utilize CF to fulfill their commitments under international climate agreements, thereby enhancing their credibility and leadership on climate issues. Funding for green technologies can lead to advancements in energy efficiency, C capture, and other sustainable practices (Li et al., 2024a; Meena and Pradhan, 2023; Lee et al., 2023; Wu et al., 2024b). Financing climate action can help communities and ecosystems to withstand adverse weather events and changing climate conditions. By financing renewable energy and C reduction projects, CF enhances economic growth while reducing environmental impact.

CF often supports projects that contribute to a GE, such as green energy, energy efficiency, reforestation, and conservation. C markets (such as cap-and-trade systems) and other C pricing mechanisms create a financial incentive for entities to cut their C emissions (Xie, 2024; Fu et al., 2024; Si et al., 2024). This encourages green investments, which can drive the transition toward a GE. CF can help hedge against regulatory and market risks associated with the climate issue. By investing in low-C projects, businesses can position themselves favorably in a transitioning economy and safeguard against potential penalties related to C emissions.

Investments in C-reducing projects can result in job creation in emerging green industries, supporting economic growth while enhancing environmental sustainability (van der Ree, 2019; Meena et al., 2023). CF can spur innovation in clean technologies, leading to new solutions for reducing emissions and improving sustainability. This promotes the development of a resilient and adaptive economy that can meet both ecological and economic needs. CF mechanisms often facilitate international cooperation on climate change, allowing for cross-border investments in eco-friendly projects. The goal of CF is to provide funding for initiatives that decrease C emissions, thereby contributing to a low-C economy. The key components of CF are C markets, C credits, green finance, and climate funds (Fig. 23.1).

23.2.1 Carbon credits

C credits, also known as C offsets, are permits that allow the holder to emit a specific amount of a GHG. A credit typically represents one metric ton of carbon dioxide (CO_2) equivalent that can be emitted. The purpose of C credits is to provide a financial incentive for reducing emissions and to limit the total amount of GHGs released into the atmosphere. Projects that reduce emissions generate credits, which can be sold on C markets. C credits have several types or categories. These categories are based on the project type, verification standard, and location. Fig. 23.1 shows some of the main types of C credits.

23.2.2 Carbon markets

C markets are trading systems that allow the buying and selling of C credits or emission allowances. These markets are designed to incentivize the reduction of GHG emissions by putting a price on C (Asadnabizadeh and Moe, 2024; Pande, 2024; Pradhan and Meena, 2023). The basic premise is that companies or countries that emit more than their allocated amount of GHGs can purchase credits from those who have emitted less, thereby creating a financial incentive to reduce emissions. There are several types of C markets, which can be categorized based on their structure, scope, and regulatory framework (Fig. 23.1). The main types of C markets are compliance, voluntary, international, regional, and hybrid.

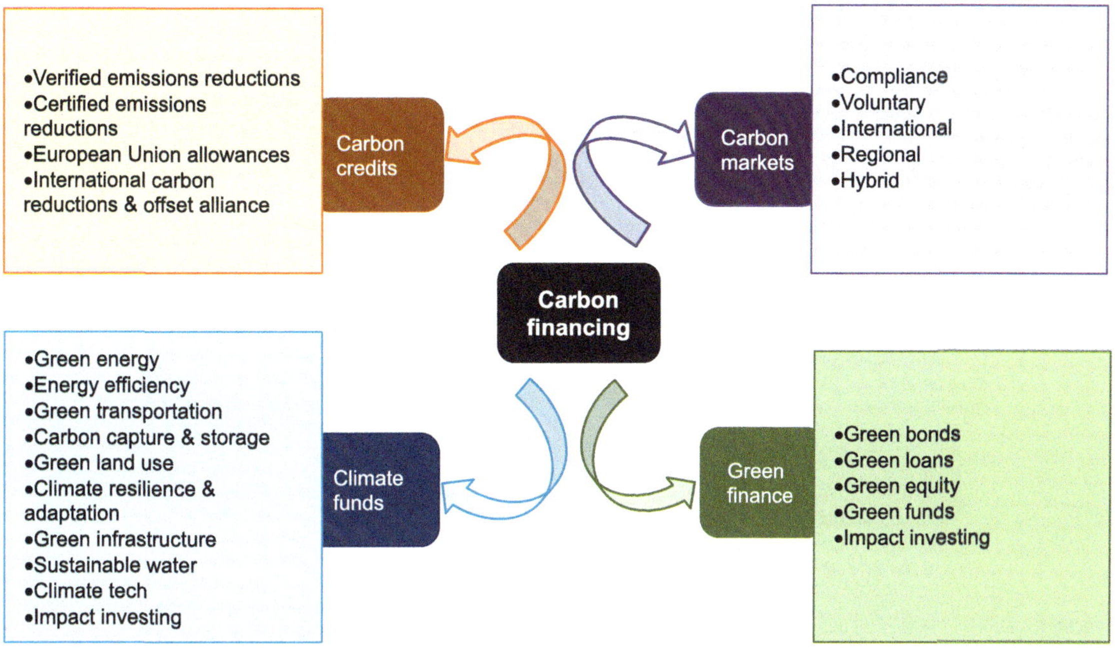

FIGURE 23.1 Global carbon financing mechanisms.
Compiled: Xie, 2024; Fu et al., 2024; Nasir and Ahmed, 2024.

Compliance C markets operate under mandatory emission reduction regimes (e.g., cap-and-trade and C taxes) that governments enforce. Cap-and-trade systems allocate emissions allowances to companies, which can be traded to meet compliance obligations (Michaelowa et al., 2023; Bryant and Webber, 2024). C taxes levy a direct tax on GHGs. Voluntary C markets operate outside of mandatory compliance regimes, allowing companies and individuals to purchase C credits voluntarily. International C markets facilitate the exchange of C credits between countries. Regional C markets operate with specific regions or sectors. Hybrid C markets combine elements of compliance and voluntary markets. These categories are not mutually exclusive, and some C markets may overlap or transition between categories as they evolve. The landscape is constantly changing, and new market mechanisms are emerging as countries and regions develop their climate policies and strategies (Wu, 2024; Xu et al., 2023).

23.2.3 Green financing

Green financing (GF) seeks to promote projects, initiatives, and investments that create beneficial environmental effects and support green development (Bryant and Webber, 2024). It seeks to offer financial assistance for projects that reduce GHG emissions and support the transition to a GE. This includes projects that enhance energy efficiency, promote green energy sources, and foster sustainable transportation systems (Fig. 23.2). GF promotes investments that align with global green development objectives, tackling environmental issues while also considering economic and social aspects. This includes projects that support biodiversity, conservation, and sustainable land use (Fig. 23.2). GF also encourages investment in green technologies and initiatives that minimize waste, improve recycling, and promote sustainable utilization of resources, thereby fostering a circular economy (CE) where materials are reused and repurposed. It supports projects that assist communities and ecosystems in adjusting to the effects of climate change.

GF supports R&D of innovative technologies and practices that can drive sustainability. This can include funding for green technology start-ups or projects aimed at developing green practices across various sectors. By integrating environmental considerations into finance, GF aims to promote long-term financial stability and risk management (Banerjee et al., 2020; Banerjee et al., 2023). This includes addressing environmental risks that could impact investment performance. GF typically supports a range of initiatives and activities aimed at promoting environmental sustainability (Mohd and Kaushal, 2018; Lazaro et al., 2023). The goal of CF is to support and facilitate the transition to a GE by providing financial resources for projects and initiatives that address the climate crisis. It encompasses a range of activities aimed at mitigating GHG emissions, adapting to climate disruption, and promoting Jhariya et al., 2021a,b). Various financial instruments fall under the umbrella of GF, such as green bonds, green loans, and sustainable investment funds

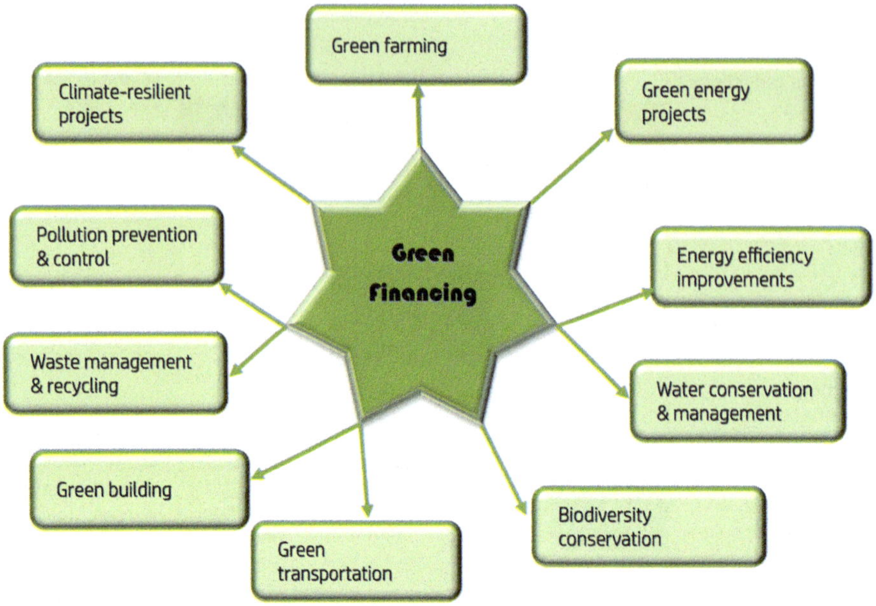

FIGURE 23.2 Examples of initiatives that can benefit from green financing.

(Fig. 23.1). These tools provide private investors with opportunities to support projects and companies that prioritize environmental sustainability while also generating financial returns. Governments, financial institutions, and businesses worldwide are increasingly recognizing the importance of green finance in addressing climate crisis and promoting green development (Akomea-Frimpong et al., 2022; Agrawal et al., 2024).

23.3 Green financing: concepts, principles, and applications

23.3.1 Green economy concept

A GE is characterized by low-C, resource-efficient, and socially inclusive (Table 23.1). It is an economy that represents a transformative shift toward green development by integrating environmental considerations into economic planning (Chaaben et al., 2024; Pradhan et al., 2023). In a GE, industries, transportation, buildings, and other sectors work to lower their GHG emissions, particularly CO_2 emissions, which contribute significantly to climate crisis (Eaton and Sheng, 2019; Li et al., 2024a). Emphasis is placed on using energy more efficiently, reducing waste, and improving overall energy productivity. The transition to a GE is anticipated to result in the generation of new employment opportunities, especially in green sectors of energy, agriculture, and other industries. A GE enhances resilience against climate-related impacts (Xiao and Fei, 2024). In a GE, communities can enjoy improved air and water quality, access to green spaces, and overall better living conditions.

23.3.2 Core principles of green financing

The GE is guided by five core principles that provide a comprehensive framework for aligning economic growth with green practices (Li et al., 2024a; Adamowicz, 2022). First, the well-being principle emphasizes enhancing the overall health and happiness of individuals and communities while ensuring the sustainable utilization of resources. Prioritizing well-being means ensuring fair access, protecting the environment, and supporting sustainable economic development to build strong, adaptable communities for generations to come. Second, the justice principle promotes equity within and between generations (Ehresman and Okereke, 2015). It seeks to ensure that benefits and costs are shared fairly among all members of society while empowering marginalized groups, including women. This principle emphasizes long-term perspectives on wealth creation that address current poverty and injustice while safeguarding opportunities for future generations.

Third, the planetary boundaries principle recognizes the ecological limits of our planet and advocates for the protection, restoration, and investment in natural systems (Li et al., 2021). It acknowledges the diverse values of nature (ecological, social, and economic) and employs precautionary measures to prevent critical loss of natural capital (Kosoy et al., 2012).

TABLE 23.1 The key characteristics of a GE.

Key characteristics	Description
Low-C	Shifting from high- to low-C energy sources.
Resource-efficient	Optimal use of resources, minimizing waste, and reducing environmental impact, while maintaining or increasing productivity and quality.
Socially inclusive	Policies, practices, or initiatives that promote equal opportunities, fairness, and participation for all people, irrespective of their background or circumstances.
Sustainable development	Fostering long-term ecological balance while promoting equitable resource distribution and economic opportunities.

GE, green economy.

Investments are directed toward conserving biodiversity, maintaining soil health, improving water quality, enhancing air purity, and protecting other vital ecosystems. Fourth, the efficiency and sufficiency principle supports sustainable consumption and production patterns. It encourages low-C economies that conserve resources through circular models of production where waste is minimized. This principle acknowledges the necessity for a substantial global shift in consumption patterns to ensure that resource use remains within sustainable limits while meeting the basic needs of all individuals. Fifth, the good governance principle requires institutions to be accountable, transparent, integrated across sectors, and responsive to community needs. Good governance involves public participation in the decision-making process, ensuring that policies are based on sound scientific evidence and local knowledge.

23.3.3 Application of green financing principles

GE principles can be applied in many sectors of the economy, including agriculture, energy production, transportation systems, and manufacturing (Georgeson et al., 2017; Zhu et al., 2023; Houssam et al., 2023; Meena et al., 2023). GE principles can be implemented through green farming practices that enhance productivity while minimizing environmental impact. Green farming often minimizes or eliminates synthetic fertilizers and pesticides, leading to healthier ecosystems and reduced pollution. These practices not only improve food security, but also contribute to carbon sequestration (C_{seq}) and resilience against climate changes. Transitioning to eco-friendly energy sources can help to reduce GHG emissions and dependence on fossil fuels (Liu and Han, 2024). Investing in green energy creates jobs at a much higher rate than the fossil fuel industries. Adopting green farming practices can help protect soil health, conserve water resources, reduce chemical inputs, and promote biodiversity (Mlambo et al., 2024). Promoting local food systems and reducing food waste are also important aspects of GE in the agriculture sector.

Green transportation systems are crucial for reducing C footprints (Rehman et al., 2023; Kwilinski et al., 2024). Enhancing public transportation options reduces individual car usage, resulting in lower emissions. The use of vehicles powered by green energy sources, such as electricity or biofuels, can help decrease our dependence on fossil fuels, which in turn helps lower GHG emissions and address climate issues. Additionally, green transportation options, such as public transit, biking, and walking, can help reduce congestion and improve air quality in urban areas. This can lead to cost savings and improved public health outcomes. Overall, investing in green transportation infrastructure and technology can create jobs, stimulate innovation, and drive sustainable economic growth (Acheampong et al., 2022).

Recycling, composting, and waste-to-energy programs are key to achieving a CE in waste management, reducing waste sent to landfills, and promoting sustainable resource management (Banerjee et al., 2024; Ibarra Vega and Bautista-Rodriguez, 2024). Several countries and organizations have adopted CE strategies and policies to promote resource efficiency, reduce waste, and facilitate material recovery (De Melo et al., 2022). Industrial symbiosis initiatives, where one company's waste becomes another's raw material, are gaining traction in various regions. Advocating for eco-friendly tourist practices, including sustainable lodging, wildlife preservation, and responsible travel, can help mitigate the adverse environmental and social impacts of tourism (Samal and Dash, 2023). Supporting cultural heritage preservation, and promoting awareness and sustainable tourism principles are also key components.

Transitioning to a GE has numerous obstacles that must be confronted (Söderholm, 2020). First, one of the primary challenges in transitioning to a GE is managing diffuse environmental risks. Unlike stationary pollution sources that can be easily monitored and regulated, diffuse emissions originate from scattered sources, such as transportation, agriculture, and industrial activities. These emissions are often challenging to track and regulate effectively, necessitating indirect strategies for monitoring and controlling them. Second, investment in green technologies, infrastructure upgrades, and sustainable practices often requires significant upfront costs (Agrawal et al., 2024). Businesses, governments, and individuals may face financial barriers in adopting these practices, especially in developing countries or for small businesses. Third, irregular or insufficient regulatory measures, financial support, and motivational mechanisms may impede the shift toward a GE (Howes et al., 2017). Explicit laws and regulations are crucial for establishing an equitable environment for green enterprises and promoting innovation in sustainable technology. Fourth, many people may not fully understand the benefits of a GE or how to incorporate sustainable practices into their daily lives (Bisschoff and Tshivshe, 2023). Education and awareness campaigns play a crucial role in promoting behavioral changes and fostering the widespread adoption of environmentally friendly initiatives. Fifth, developing and scaling up green technologies can be complex and require ongoing R&D (Nikyema and Blouin, 2020). Overcoming technical barriers and improving the efficiency of green technologies is crucial for a successful transition to a GE. Ultimately, attaining a GE necessitates global collaboration and coordination to tackle environmental issues, such as, climate change and biodiversity loss (Aisbett et al., 2023). Ensuring equity and inclusivity in the transition process is important to avoid widening social disparities between countries and communities.

23.4 Nexus between carbon financing and green financing

There is a substantial nexus between CF and the GE, as CF serves as a critical mechanism for promoting and enabling the transition toward a more environmentally sustainable and low-C economic model (Du, 2023; Li et al., 2024a). Research indicates an inverse relationship between CF and CO_2 emissions, suggesting that increased CF correlates with lower C emissions in certain economies. In other words, CF can directly drive decarbonization of the economy, as it enhances environmental performance while fostering financial development and addressing socio-economic needs. Fig. 23.3 shows the link between CF and GE.

23.4.1 Incentivizing emission reductions

CF, through C markets and pricing mechanisms, creates financial incentives for businesses, industries, and individuals to reduce their GHG (Cai et al., 2024). This promotes the implementation of greener technology, energy efficiency initiatives, and sustainable practices, which are fundamental components of the GE. Projects that reduce emissions can earn C credits, which represent a ton of CO_2 removed from the atmosphere or prevented from being emitted. These credits can be sold to companies or governments looking to offset their emissions. Companies can earn revenue by selling C credits, thus motivating them to invest in cleaner technologies. By internalizing the cost of C emissions, businesses are encouraged to adopt more sustainable practices (Santos, 2022). Government policies can link CF to a GE through the implementation of taxes or cap-and-trade systems that incentivize reductions in GHG emissions (Pan et al., 2024). Regulations that facilitate C credit trading can stimulate investment in green technologies. Various regulatory and voluntary markets exist that allow the trading of these credits, providing a financial incentive for green projects.

23.4.2 Funding green projects

The income obtained from the sale of C credits or offsets can be utilized to finance a range of eco-friendly initiatives and endeavors. These include green energy projects, reforestation efforts, sustainable agriculture techniques, and other environmental conservation and restoration programs that accord with the concepts of the GE. As the urgency to address the sustainability crisis grows, CF will likely play an increasingly vital role in funding green initiatives around the globe (Lee et al., 2022). GF facilitates the flow of capital to green sectors, scaling up sustainable investments, and accelerating the transition to a GE (Citil, 2024; Wang et al., 2024a). As CF supports the growth of green sectors, it contributes to green job creation. This aligns with the GE's goal of creating decent work opportunities in environmentally friendly industries (Fig. 23.4).

23.4.3 Promoting innovation

CF accelerates the transition to a GE by offering financial incentives for businesses and organizations to develop and adopt innovative, emission-reducing technologies and green practices (Jiang et al., 2023; Wang et al., 2024a). CF encourages firms to use cleaner, more efficient technology, such as green energy systems, energy-saving techniques, and C capture and storage solutions (Wang et al., 2024a). This drive for innovation not only helps companies achieve their emission

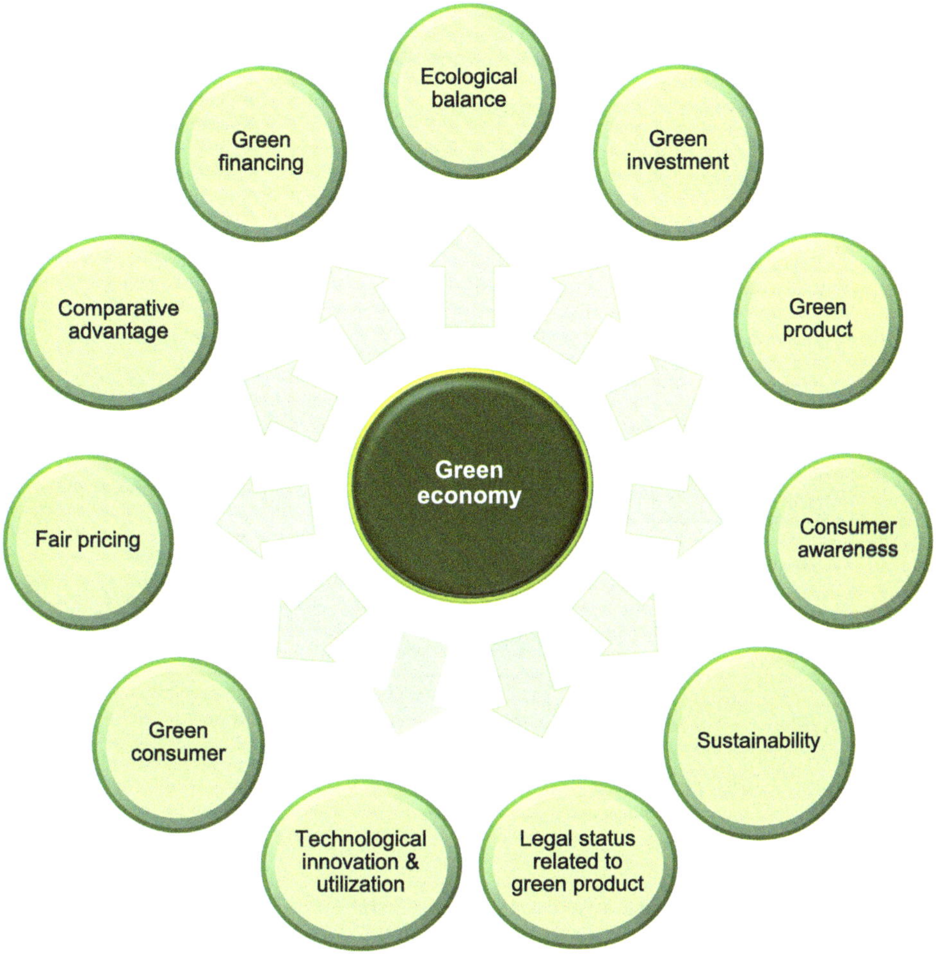

FIGURE 23.3 Elements of the GE. *GE*, green economy..
(Compiled: Ehresman and Okereke, 2015; Adamowicz, 2022; Lin et al., 2024).

reduction goals but also spurs the growth of GE by creating opportunities for new industries and job creation (Jiang et al., 2023). Furthermore, CF fosters R&D in green technologies, supports pilot projects for testing new green practices, and incentivizes collaboration among stakeholders to find creative solutions to environmental challenges.

CF creates economic incentives for businesses to adopt green technologies. Companies that invest in cleaner technologies can earn C credits by reducing their emissions below a certain threshold. This not only encourages firms to invest in green technologies but also fosters a competitive environment where innovation is rewarded financially (Zhao et al., 2024). CF also helps facilitate market access for innovative green solutions by lowering barriers to entry (Agrawal et al., 2024; Wu, 2024). For example, financing options, such as green bonds, specifically target environmentally friendly projects, making it easier for start-ups and small enterprises to focus on sustainability to secure funding. Additionally, as more investors seek out sustainable investment opportunities due to growing awareness of climate change issues, there is a growing demand for innovative products that contribute to a low-C future. CF often leads to increased collaboration among stakeholders in the GE, including governments, private sector players, non-governmental organizations, and academic institutions (Fontoura and Coelho, 2022). Such collaborations can result in knowledge sharing and joint ventures that drive innovation forward. For example, collaborations between technology innovators and financial institutions can result in the development of new funding models suited specifically to green innovations.

23.4.4 Enabling sustainable development

CF has the potential to stimulate sustainable development and build a more resilient and equitable future for everyone. By incentivizing emission reductions and promoting C-offset projects, CF contributes to environmental sustainability by

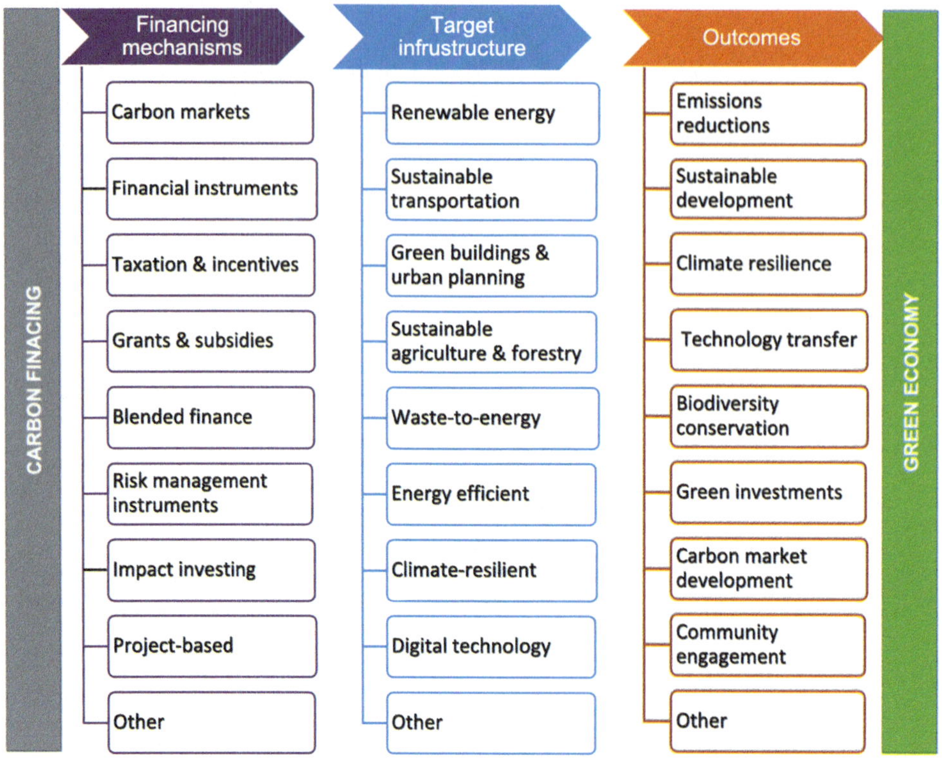

FIGURE 23.4 The nexus between CF. *CF*, carbon financing;*GE*, green economy. (Compiled: Jiang et al., 2023; Agrawal et al., 2024; Wang et al., 2024a).

mitigating climate change, preserving natural resources, and protecting ecosystems (Wu et al., 2024a). CF can support projects that benefit local communities, such as green energy installations in underserved areas or sustainable agriculture initiatives that provide livelihood opportunities. This can help promote social equity and empower marginalized populations. Investing in green technologies and sustainable practices through CF can drive green growth by creating new markets, supporting job creation in clean energy sectors, and fostering innovation and competitiveness (Tănasie et al., 2022).

23.4.5 Encouraging stakeholder participation

Stakeholder participation is crucial for the success of any initiative aimed at fostering a GE (Newig et al., 2023). Stakeholders can include governments, businesses, non-governmental organizations, local communities, and individuals. Their participation ensures that diverse opinions are acknowledged, and the benefits of green efforts are distributed equitably. CF provides financial incentives for stakeholders to invest in eco-friendly technologies and practices. By offering monetary rewards through mechanisms, such as C credits, stakeholders are more likely to engage in projects that contribute to emission reductions. For example, companies may invest in renewable energy sources or improve their operational efficiencies when they see a direct financial benefit from reduced emissions. Projects funded through CF mechanisms typically require partnerships between private sector entities, government agencies, and community organizations (Casady et al., 2024). This collaborative approach fosters dialogue and encourages stakeholder engagement as they work together toward common environmental goals.

23.4.6 Promoting resource efficiency

The principle of resource efficiency is intrinsically linked to the objective of optimizing resource use in the production of goods and services, thereby reducing resource consumption and minimizing waste and pollution (Yuyang, 2024). In a GE, optimizing resource efficiency is crucial, as it enables the attainment of green development goals by ensuring that economic growth is achieved without compromising environmental integrity. Many C-offset projects focus on improving resource efficiency, such as energy efficiency, waste management, and green land-use practices (Kumar et al., 2020;

Kumar et al., 2024). These initiatives directly contribute to the GE's goal of maximizing the efficient use of natural resources and minimizing waste. Financial incentives provided through CF can motivate businesses to implement green practices (Lou et al., 2023). Companies may receive funding or tax breaks for adopting energy-efficient processes or for reducing waste in their operations. This not only lowers their C emissions but also enhances their overall resource efficiency by optimizing the use of materials and energy.

CF often directs funds toward innovative technologies that enhance resource efficiency. For instance, investments in green energy sources reduce reliance on fossil fuels, which are typically more resource-intensive (Yadav and Mahalik, 2024). By promoting cleaner technologies, CF encourages industries to adopt practices that use fewer resources for energy production. CF can enable the shift toward a CE, in which resources are reused and recycled instead of being discarded after a single use. CF contributes to the creation of CE in which materials are kept in use for extended periods of time, thus boosting resource efficiency (Kirchherr et al., 2023).

23.4.7 Addressing social equity

While CF primarily focuses on reducing GHG emissions, it also has the potential to address social equity issues within the GE by promoting access to resources (Li et al., 2024b). Effective CF strategies prioritize the inclusion of vulnerable populations who are disproportionately affected by climate change. These populations often include low-income communities, indigenous peoples, and other marginalized groups. By ensuring that these communities have a voice in decision-making processes related to C finance projects, social equity can be promoted. For instance, community-based green energy projects can empower local residents while providing them with affordable energy solutions. The concept of environmental justice is closely tied to social equity within the framework of CF (Martin et al., 2016). Environmental justice necessitates fair representation and inclusive decision-making, ensuring that all voices are heard in the development of environmental policies. CF mechanisms that incorporate ecological justice principles ensure that benefits are equitably distributed among all segments of society, particularly those who have been historically marginalized.

23.4.8 Improving climate resilience

CF can support adaptation initiatives, including the construction of climate-resilient infrastructure, the implementation of early warning systems for extreme weather events, the enhancement of water resource management, and the development of sustainable agricultural practices that are more adaptable to shifting climatic conditions (Li et al., 2024a; Liu et al., 2024; Wu et al., 2024a,b). CF initiatives often support projects that focus on ecosystem restoration and conservation. Healthy ecosystems serve as natural buffers against climate-related hazards, thereby contributing to climate resilience. CF can help build the resilience of vulnerable communities through projects that focus on disaster risk reduction, livelihood diversification, capacity building, and community-based adaptation strategies. These initiatives empower communities to cope with climate impacts and reduce their vulnerability. CF supports green land management practices that improve soil health, increase agricultural productivity, conserve water resources, and reduce erosion. These practices enhance the resilience of ecosystems and communities to climate variability and extreme events. CF can be used to develop climate risk insurance mechanisms that provide financial protection to vulnerable populations against climate-related disasters. This helps communities to recover more quickly from climate shocks and enhance their resilience in the face of uncertainty.

23.5 Research on carbon financing and green financing

Financing a GE and reducing C emissions are essential for mitigating climate change and achieving green development goals. CF enables the transition to cleaner energy, supports innovation in green practices, and ensures that economic progress is achieved without compromising planetary health. Studies are increasingly examining how CF can facilitate the transition from fossil fuels to eco-friendly energy sources.

23.5.1 Does carbon financing reduce CO_2 emissions?

Numerous studies have shown that C trading has significantly reduced industrial CO_2 emissions, particularly in regions where these markets have been implemented (Lin and Huang, 2022; Weng et al., 2022; Han et al., 2022; Dechezleprêtre et al., 2023). However, there is also evidence suggesting that not all C-offset projects lead to actual emissions reductions due to leakage (Calel et al., 2021; Filewod and McCarney, 2023). Leakage refers to the unintended increase in C emissions in one area due to mitigation efforts implemented in another location. This can undermine the effectiveness of Coffsets, where the intended benefits of reduced emissions may be offset by increases elsewhere. It is essential for C-offset projects to accurately measure and address leakage to ensure that the credits generated represent real, additional, and permanent reductions in GHG emissions. Many certified green energy projects demonstrate evidence of emission reductions, casting doubt on the reliability of these offsets.

C markets have experienced significant price volatility, which can affect their effectiveness and investment certainty. High fluctuations in C prices can lead to uncertainty, making it challenging for businesses to plan investments and strategies related to emissions reductions (Wu et al., 2022; Ma et al., 2024). This unpredictability can undermine the market's stability and effectiveness in controlling emissions. The specific design of C markets has a significant impact on their effectiveness. Features, such as cap setting, allowance allocation methods, and price stabilization mechanisms, are crucial for market performance.

C pricing has had varying impacts across different sectors. Energy-intensive industries are generally more affected by C price volatility due to their high emissions and reliance on fossil fuels. Fluctuating C prices can significantly influence their operational costs and profitability (Lin et al., 2024). There is evidence that C markets can stimulate low-C innovation and technology adoption by providing economic incentives for companies to invest in R&D (Wang et al., 2024a). However, the strength of this effect depends on the C price level and policy certainty. C markets interact with other climate and energy policies, sometimes leading to unexpected outcomes. Complementary policies are often necessary to address market failures and achieve broader climate goals.

The voluntary C market has seen significant growth, driven by corporate net-zero commitments (World Economic Forum, 2023; Balderas et al., 2024). However, concerns about credit quality and additionality persist in voluntary markets. Engaging developing countries in C markets has been challenging but is seen as crucial for global emissions reduction. Capacity building and financial support are key to successful implementation in developing countries.

A significant body of research supports the notion that green bond issuance leads to lower C emissions (Alamgir and Cheng, 2023; Pang et al., 2024; Sharif et al., 2022; Xu and Li, 2023), with only a few exceptions finding no correlation (Si et al., 2024). Green bonds serve as a pivotal mechanism in channeling funds toward environmentally friendly projects, thus aiding the transition to a GE (Erb et al., 2018; Levi et al., 2019; Alamgir and Cheng, 2023; Pang et al., 2024). The growing demand for green bonds tends to lower the cost of capital for projects deemed environmentally friendly. As the financial landscape evolves, green bonds are likely to gain further significance, driving innovation, and fostering economic growth in alignment with climate goals.

23.5.2 Research directions on carbon financing and green financing

Research on CF and the GE has advanced significantly in recent years (Zhang et al., 2023; Xu and Liu, 2023; Liu and Zhu, 2024; Wu, 2024), but several gaps still exist that can be addressed to enhance understanding and effectiveness in these areas (Table 23.2). The impact of CF on emission reductions and other environmental benefits remains unclear due to inadequate standardized measurement frameworks. Developing robust frameworks that can accurately attribute emission reductions to specific CF mechanisms or projects is crucial. This includes differentiating between direct impacts and co-benefits. There is a lack of analysis on the long-term sustainability of green projects financed through CF mechanisms. Research is needed on the durability of benefits realized from these projects, including their resilience to changing climate conditions and their ability to continue providing emission reductions over time.

Investigating how different stakeholders perceive and respond to CF initiatives can help increase participation and improve outcomes. The effectiveness of various financial instruments (e.g., green bonds, C credits, and insurance products) in mobilizing capital for climate action is not well known. Research could explore innovations in financing mechanisms and their effectiveness at different scales and contexts, especially in developing countries. Much of the existing research focuses on mitigation, with less attention to how CF can facilitate adaptation. Exploring integrated strategies that simultaneously address mitigation and adaptation could enhance overall resilience and resource efficiency. Addressing these research gaps can provide critical insights that enhance the effectiveness of CF and support the transition to a GE. Such research will ultimately contribute to better policy-making, more impactful investments, and a more equitable and sustainable future.

23.6 Cutting-edge developments in global carbon financing and green financing

The global CF and the GE are experiencing rapid evolution, driven by increasing climate awareness, technological advancements, and evolving regulatory landscapes. There are many cutting-edge developments in the global CF and GE (Pan et al., 2019; Li et al., 2024a; Pereira et al., 2024). These developments include blockchain technology for transparency and traceability, remote sensing and the Internet of Things (IoT) for emission monitoring, C capture and storage, nature-based solutions (NbS), and the integration of C pricing mechanisms.

TABLE 23.2 Key research focus areas for CF and GE.

	Focus areas	Research needs
1	Sustainability of C projects and resilience	Investigating the long-term sustainability of CF projects and their resilience to changes in climate, policy, and market conditions.
2	Measurement and verification	Developing robust metrics and methodologies for measuring the impact of CF on emissions reductions and sustainability outcomes.
3	Quantification of CF benefits	Standardized metrics for quantifying social, economic, and environmental benefits of C projects.
4	Project validation and verification	Standardized methodologies for validating and verifying C-offset projects.
5	Pricing mechanisms	Evaluating the effectiveness and efficiency of various C pricing strategies and their impact on reducing emissions.
6	Voluntary vs. compliance markets	Assessing the differences in effectiveness and scalability between voluntary C markets and regulated markets.
7	Emerging technologies	Exploring how new technologies (e.g., blockchain and AI) can enhance CF and accounting.
8	Equitable distribution of benefits	Investigating how CF can address social inequalities.
9	Equity and inclusivity	Understanding the socio-economic impacts of CF on marginalized communities.
10	Public perception	Researching public attitudes toward CF and GE initiatives.
11	Cultural considerations	Understanding how cultural contexts influence the implementation and success of GE projects.
12	Technology transfer and innovation	Analyzing the processes and barriers to technology transfer in the context of CF and GE investments.
13	Regulatory challenges	Analyzing the barriers to implementing effective CFolicies.
14	Global vs. local perspectives	Researching how local environmental, social, and economic contexts influence the success of CF and the GE.
15	Stakeholder behavior	Assessing how psychological factors influence participation in C markets and green initiatives.

CF, carbon financing; *GE*, green economy

23.6.1 Carbon pricing mechanisms

Many countries are implementing C taxes and cap-and-trade programs to incentivize emission reductions and promote low-C investments (Pan et al., 2024). C taxes impose a direct tax on the C content of fossil fuels, while cap-and-trade is a system in which government sets a limit on the quantity of GHG emissions allowed from all participating entities. The government issues emission permits to companies, and these permits can be traded among companies. The C pricing mechanisms encourage companies to invest in green technologies, which helps to reduce their C emissions (Feng et al., 2024; Xu and Yang, 2024). By assigning a cost to C emissions, these mechanisms signal to investors that green technologies and practices can be financially beneficial, directing capital toward sustainable projects. C pricing mechanisms facilitate international cooperation on solutions to the climate problem, as countries can share best practices and potentially engage in C trading across borders. As a cutting-edge tool in addressing the climate problem, C pricing mechanisms not only help reduce GHG emissions but also play a pivotal role in shaping a sustainable and resilient global economy (Boroumand et al., 2022).

23.6.2 Blockchain-based carbon credits

A blockchain-based C credit is an innovative approach in combating climate instability and reducing GHG emissions (Pan et al., 2019; Swinkels, 2024). The concept involves leveraging blockchain technology to track, verify, and trade C credits transparently and securely (He et al., 2024; Xu et al., 2024). Blockchain technology enables the transparent tracking and traceability of C credits. Each credit is assigned a unique digital token or smart contract on the blockchain, containing information such as the project's origin, emission reduction methodology, and verification details. Blockchain-based platforms facilitate the trading and transfer of C credits between buyers and sellers. Participants can access a digital marketplace where they can buy, sell, or retire C credits. C credits can be represented as digital tokens on a blockchain. Smart contracts can automate transactions, ensuring secure and efficient transfers. Tokens are created to describe a specific quantity of C credits, and they can be bought, sold, and transferred on the blockchain. Tokenization enables fractional ownership, divisibility, and ease of trading, as tokens can be easily exchanged between participants without the need for intermediaries. Blockchain-based systems can provide a transparent audit trail, verify the legitimacy of emission reductions, and prevent double-counting. The immutability and transparency of blockchain records reduce the risk of fraud and provide stakeholders with confidence in the credibility and authenticity of C credits.

23.6.3 Artificial intelligence in carbon pricing

AI enhances the efficacy and accuracy of C pricing mechanisms by providing advanced data analytics, automation, and predictive capabilities. Policymakers and corporations may effectively tackle the climate problem by harnessing AI technologies and implementing more efficient and impactful C pricing systems. AI technologies can analyze large datasets, and accurately track and quantify C emissions from different industries and sectors. This data is essential for setting C prices based on actual emissions. This enables policymakers to make informed decisions when setting C prices and designing emission-reduction initiatives.

23.6.4 Nature-based solutions

NbS are initiatives that utilize natural ecosystems, processes, and materials to provide benefits for both the environment and human populations (Cassin, 2021; Sowińska-Świerkosz and García, 2022). Such initiatives include reforestation, afforestation, agroforestry, and ecosystem restoration that leverage the power of nature to C_{seq} and restore ecosystems (Table 23.3). NbS offer cost-effective and sustainable ways to tackle the climate problem, enhance biodiversity, and promote environmental resilience. By preserving and enhancing natural ecosystems, these initiatives help maintain biodiversity, ecosystem functions, and genetic diversity —essential components for the health of ecosystems and human well-being. NbS can generate C credits through various mechanisms, including biomass C_{seq}, soil C accumulation, and enhanced biodiversity value (Devi et al., 2024; Raj et al., 2024).

NbS often create new investment opportunities across sectors, including forestry, agriculture, and water management. Investors seeking sustainable or impact investments are increasingly looking for projects that provide environmental benefits alongside financial returns. NbS can improve the valuation of ecosystem services, such as C_{seq}, water filtration, and biodiversity enhancement, which can be monetized through mechanisms, such as C trading and payments for ecosystem services. By restoring ecosystems and maintaining healthy natural systems, NbS can reduce risks associated with the climate issue, which is more appealing to investors who seek to minimize potential losses (Debele et al., 2023). A variety of international funds, such as the Green Climate Fund, specifically finance nature-based projects aimed at climate adaptation and mitigation. Securing funding from these sources can enhance the financial viability of NbS initiatives. By attracting green financing through diverse funding avenues and enhancing economic resilience through sustainable practices and community engagement, NbS play an important role in facilitating a shift toward a GE.

23.6.5 Green bonds

Green bonds serve as a tool to facilitate investment in sustainable projects, helping to finance the transition to a low-C economy while meeting the investment needs of environmentally conscious investors (Sartzetakis, 2021; Zhang et al., 2024; Ahmed et al., 2024). They can be issued by various entities, including governments, municipalities, corporations, and financial institutions. Governments and municipalities can use green bonds to support their own environmental objectives and international commitments, such as the Paris Agreement. By funding projects that reduce emissions and enhance resilience to climate change, they can achieve policy goals while also attracting private investments. The green bond market has experienced significant growth over the past decade, driven by increasing investor demand for sustainable investment options and the global push for addressing climate change (Chen and Zhao, 2021; Adekoya et al., 2023). Green bonds may offer tax advantages, making them more appealing to investors compared to traditional bonds. The issuance of green bonds can attract a diverse range of investors who are interested in socially responsible investment (Sangiorgi and Schopohl, 2021). This influx of capital can help to scale up green projects and innovations, thereby fostering sustainable economic growth. Many green bonds are verified by third parties to ensure that the funds will be used for environmentally beneficial projects.

TABLE 23.3 Nature-based solutions (NbS) and their benefits to humans and the environment.

NbS solutions	Description	Benefits
Reforestation and afforestation	• Restoring degraded forests • Planting new forests in non-forested areas	• C_{seq} • Biodiversity enhancement • Soil protection
Wetland restoration	• Rehabilitating marshes, swamps, and other wetlands • Creating artificial wetlands for water purification	• Water filtration • Flood control • Wildlife habitat
Urban green spaces	• Developing parks and green roofs • Implementing urban gardens and green walls	• Air quality improvement • Heat island mitigation • Enhanced well-being
Agroforestry	• Integrating trees into agricultural landscapes	• Increased crop yields • Diversified income • Ecosystem services
Soil conservation	• Implementing cover crops and crop rotation • Employing no-till farming techniques	• Erosion prevention • Water retention • Nutrient cycling
Coastal and marine restoration	• Restoring mangroves, coral reefs, and seagrasses • Creating marine protected areas	• Coastal protection • Biodiversity support • C storage
River and watershed management	• Restoring natural riverbanks and floodplains • Implementing sustainable water management practices	• Flood mitigation • Water quality improvement • Habitat connectivity
Biodiversity corridors	• Creating connections between habitats to support wildlife movement • Protecting migration routes for species	• Species movement • Habitat preservation • Ecosystem resilience
Sustainable forestry practices	• Managing forests for timber while maintaining biodiversity • Implementing selective logging and reduced-impact logging techniques	• Long-term timber supply • Biodiversity conservation • Community livelihoods
Pollinator habitat creation	• Planting native flowers and plants to support bees and other pollinators • Reducing pesticide use in agricultural areas	• Increased crop pollination • Biodiversity support • Ecosystem stability

23.6.6 Technological innovations

Technological innovations, such as C capture and storage, efficiency improvements and renewable energy, precision farming, waste management technologies, transport innovations, smart grids, and energy storage create pathways for CF, attract investments, and promote GE (Si et al., 2024; Li et al., 2024a; Wang et al., 2024b). Investors are increasingly seeking chances to fund initiatives that have both positive environmental consequences and financial returns. R&D in C removal technologies, such as direct air capture and enhanced weathering, is expanding to provide scalable solutions for capturing and storing CO_2 from the atmosphere. CO_2 emissions from industrial processes can be captured and stored underground, creating opportunities for C credits under various C trading schemes.

The production of green hydrogen through renewable energy is being explored as a sustainable energy carrier and a potential replacement for fossil fuels in various applications. The development of electric and hydrogen fuel cell vehicles reduces emissions in the transportation sector, boosting interest from investors focused on sustainable mobility. Innovations, such as drones and IoT sensors, optimize resource use in agriculture, reducing emissions, and improving C_{seq} in soils, thereby attracting financing. Advanced recycling technologies can minimize waste and emissions, attracting investments aimed at sustainable waste management practices.

23.6.7 Circular economy initiatives

CE is an economic system designed to eliminate waste and promote continuous use of resources (Kirchherr et al., 2023; Souza Piao et al., 2024). In a CE, products are designed to be reused, repaired, and recycled as much as possible, rather than being disposed of after a single use (Fig. 23.5). This approach helps to minimize the extraction of raw materials, reduce energy consumption, and decrease the amount of waste generated (Hajoary et al., 2024). By closing the loop on resource use and waste production, a CE aims to create a more sustainable and environmentally friendly way of conducting economic activities. The CE complements and reinforces the GE by promoting sustainability, resource efficiency, and environmental stewardship, creating a holistic approach to economic and ecological resilience. Together, they foster a more sustainable future that benefits both people and the planet. Both the CE and GE promote shifts in consumer behavior toward sustainability (D'Amato and Korhonen, 2021). The CE encourages consumers to adopt practices, such as buying second-hand goods or supporting sustainable brands, which align with the GE principles.

23.7 Future directions for promoting carbon financing and the green economy

Advancing the GE through CF involves supporting environmentally friendly initiatives and projects by providing financial incentives based on the amount of C emissions reduced or sequestered. CF promotes a GE through policy frameworks, financial mechanisms, technological innovation, and public engagement.

23.7.1 Strengthening the government policy framework

Globally, governments are progressively acknowledging the significance of policies in advancing CF and a GE (Agrawal et al., 2024; Citil, 2024). Governments are responsible for developing and implementing supportive policies that incentivize C pricing, green investments, and sustainable practices. Strategic, vision-driven, and inclusive policies and frameworks can significantly enhance CF and foster the development of a GE (Wu et al., 2024a). A number of governments have established binding targets for C emissions reduction in line with international agreements. Such targets can stimulate the development of CF mechanisms as businesses and organizations seek to comply. Governments can set

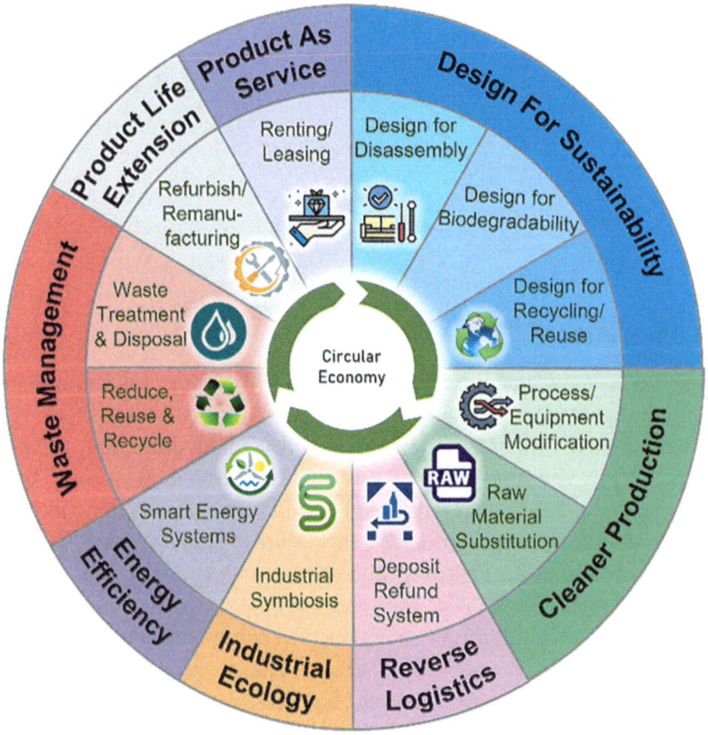

FIGURE 23.5 Circular economy.
Adapted: Srikanth and Visvanathan, 2023.

up regulated C trading markets where companies can buy and sell C credits, facilitating market-based solutions for emissions reductions. Stability and predictability in policy frameworks attract investment by reducing uncertainty (Kiptoo, 2024). Long-term commitments to sustainability from governments ensure that businesses understand the expected regulatory landscape. Governments can access international climate financing mechanisms (e.g., Green Climate Fund) to support domestic projects aimed at C reduction and sustainability.

23.7.2 Enhancing financial mechanisms

Financial mechanisms can be defined as tools or systems that facilitate the flow of funds toward specific goals. In the context of CF and the GE, these mechanisms include public funding, private investment, international climate finance, and C markets. Enhancing financial mechanisms is pivotal in advancing and promoting a GE. Green bonds have the potential to provide the necessary capital for various projects that contribute to reducing C emissions (Hassan et al., 2023; Bryant and Webber, 2024). The interest rates for green bonds are lower than those for traditional bonds due to their appeal to socially responsible investors. They can help transition economies shift toward low-C practices. Establishing or strengthening cap-and-trade systems allows companies to buy and sell emission permits. By enhancing these markets, countries can ensure more robust pricing and greater participation. There is a need to establish guarantees or insurance products that reduce the perceived risks associated with investing in green projects. For instance, credit enhancements can encourage banks to lend more freely to green energy projects. The public and private funding sources can be combined to leverage additional capital for sustainable development. This approach can help de-risk investments in emerging markets where traditional financing may be scarce. CF can be linked to successful achievement of developmental or environmental outcomes (i.e., pay-for-success), thus providing a strong incentive for performance (Saltman, 2017). Outcome-based financing that focuses on real-world impact must be encouraged. Enhancing access to finance for small-scale projects can significantly augment grassroots efforts toward achieving sustainability.

23.7.3 Promoting technological innovations

CF promotes technological innovations by providing financial incentives for investors to adopt green technologies that can help alleviate climate crisis (Dunbar et al., 2024; Agrawal et al., 2024). When companies receive funding or incentives for reducing their C emissions or implementing C-offset projects, they are encouraged to develop and adopt innovative technologies that can help them achieve their emission-reduction goals. By creating a market for C credits and incentivizing emission reductions, CF drives investment in new technologies, ultimately leading to technological advancements that stabilize the global climate system (Zhao and Hu, 2024). CF has supported the development of green infrastructure, which helps to reduce energy consumption and CO_2 emissions. CF has facilitated the development of C capture and storage technologies. This innovation helps reduce C emissions and address issues related to climate variability (Li et al., 2024a). CF has incentivized the development and adoption of electric vehicles and advanced battery technologies, making electric transportation more viable and environmentally friendly.

23.7.4 Capacity building and knowledge sharing

Capacity building for stakeholders enhances the skills necessary for effective participation in C markets, while knowledge sharing facilitates collaboration, innovation, and awareness. Capacity building includes training on how to develop projects that qualify for C credits, which enables stakeholders to design and implement projects that reduce GHG emissions. These projects can generate C credits that can be sold in C markets, providing financial resources for further environmental initiatives. Improved capacity allows entities to navigate better funding opportunities related to climate finance. This includes understanding how to apply for international grants or loans, which supports low-emission and climate-resilient development. Capacity building also aids governments in formulating effective policies that promote CF mechanisms (Robins, 2008). Well-informed policymakers can create regulations that encourage green investments (Farooq et al., 2024).

Knowledge sharing helps new project developers to avoid common pitfalls and adopt proven methodologies (Chisita and Fombad, 2020). Increased awareness can lead to greater public support for policies aimed at promoting a GE. As communities become more knowledgeable about environmental issues through capacity-building initiatives, they are better prepared to adapt to climate change impacts while simultaneously engaging in mitigation efforts through CF mechanisms. Knowledge-sharing initiatives help raise awareness about the importance of CF among different sectors of society. It can also lead to more innovative solutions for reducing emissions.

23.8 Conclusion

As we navigate the complexities of achieving net-zero emissions, innovative financial instruments, such as C credits, green bonds, and sustainability-linked loans, are essential in mobilizing capital toward projects that contribute to environmental sustainability. Continued investment in R&D not only enhances our understanding of effective C management practices, but also catalyzes technological advancements that can drive the transition toward a CE. Looking ahead, policymakers, businesses, and investors must collaborate on creating robust frameworks that ensure transparency, integrity, and accountability in CF initiatives. Emphasizing the importance of multi-stakeholder engagement and knowledge sharing will further strengthen the global response to climate challenges. As the urgency to combat climate change intensifies, future directives must focus on scaling up successful initiatives, increasing access to financing for sustainable projects, and integrating climate considerations into all facets of economic planning.

References

Acheampong, A.O., Dzator, J., Dzator, M., Salim, R., 2022. Unveiling the effect of transport infrastructure and technological innovation on economic growth, energy consumption and CO_2 emissions. Technological Forecasting and Social Change 182, 121843. Available at https://doi.org/10.1016/j.techfore.2022.121843.

Adamowicz, M., 2022. Green deal, green growth and green economy as a means of support for attaining the sustainable development goals. Sustainability 14 (10), 5901. Available at https://doi.org/10.3390/su14105901.

Adekoya, O.B., Abakah, E.J.A., Oliyide, J.A., Luis A, G.A., 2023. Factors behind the performance of green bond markets. International Review of Economics and Finance 88, 92–106. Available at https://doi.org/10.1016/j.iref.2023.06.015.

Agrawal, R., Agrawal, S., Samadhiya, A., Kumar, A., Luthra, S., Jain, V., 2024. Adoption of green finance and green innovation for achieving circularity: an exploratory review and future directions. Geoscience Frontiers 15 (4), 101669. Available at https://doi.org/10.1016/j.gsf.2023.101669.

Ahmed, R., Yusuf, F., Ishaque, M., 2024. Green bonds as a bridge to the UN sustainable development goals on environment: a climate change empirical investigation. International Journal of Finance and Economics 29 (2), 2428–2451. Available at https://doi.org/10.1002/ijfe.2787.

Aisbett, E., Raynal, W., Steinhauser, R., Jones, B., 2023. International green economy collaborations: chasing mutual gains in the energy transition. Energy Research & Social Science 104, 103249. Available at https://doi.org/10.1016/j.erss.2023.103249.

Akomea-Frimpong, I., Adeabah, D., Ofosu, D., Tenakwah, E.J., 2022. A review of studies on green finance of banks, research gaps and future directions. Journal of Sustainable Finance and Investment 12 (4), 1241–1264. Available at https://doi.org/10.1080/20430795.2020.1870202.

Alamgir, M., Cheng, M.-C., 2023. Do green bonds play a role in achieving sustainability? Sustainability 15 (13), 10177. Available at https://doi.org/10.3390/su151310177.

Asadnabizadeh, M., Moe, E., 2024. A review of global carbon markets from Kyoto to Paris and beyond: the persistent failure of implementation. Frontiers in Environmental Science 12. Available at https://doi.org/10.3389/fenvs.2024.1368105.

Balderas, A., Luna, E., Voora, V., Larrea, C., 2024. Voluntary Standards and Initiatives for Carbon Management: Navigating the Landscape. International Institute for Sustainable Development.

Banerjee, A., Meena, R.S., Jhariya, M.K., Yadav, D.K., 2020. Agroecological Footprints Management for Sustainable Food System. Springer, Singapore, pp. 1–514. Available at https://doi.org/10.1007/978-981-15-9496-0.

Banerjee, A., Jhariya, M.K., Bargali, S.S., Palit, D., 2023. Ecorestoration for Sustainability. Wiley-Scrivener, Beverly, MA, pp. 1–441.

Banerjee, A., Jhariya, M.K., Raj, A., Mechergui, T., 2024. Ecosystem management: climate change and global sustainability—an introduction. Ecosystem Management: Climate Change and Sustainability. wiley, India, pp. 1–21. Available at https://doi.org/10.1002/9781394231249.ch1.

Bisschoff, C., Tshivshe, L., 2023. Benefits and challenges of implementing green initiatives at public universities in South Africa. Journal of Economic Development, Environment and People 12 (2), 5–25. Available at https://doi.org/10.26458/jedep.v12i2.801.

Boroumand, R.H., Goutte, S., Porcher, T., Stocker, T.F., 2022. A fair and progressive carbon price for a sustainable economy. Journal of Environmental Management 303. Available at https://doi.org/10.1016/j.jenvman.2021.113935.

Boyce, J.K., Ash, M., Ranalli, B., 2023. Environmental justice and carbon pricing: can they be reconciled? Global Challenges 7 (4). Available at https://doi.org/10.1002/gch2.202200204.

Bryant, G., Webber, S., 2024. Climate Finance: Taking a Position on Climate Futures. Agenda Publishing, Newcastle upon Tyne, England. Available at https://www.agendapub.com/page/detail/climate-finance/?k=9781788214629.

Cai, D., Zhang, G., Lai, K.-hung, Guo, C., Su, B., 2024. Government incentive contract design for carbon reduction innovation considering market value under asymmetric information. Energy Policy 186, 114009. Available at https://doi.org/10.1016/j.enpol.2024.114009.

Calel, R., Colmer, J., Dechezleprêtre, A., Glachant, M., 2021. Do carbon offsets offset carbon? Centre for Climate Change Economics and Policy Working Paper 398/Grantham Research Institute on Climate Change and the Environment Working Paper371. London School of Economics and Political Science, London.

Casady, C.B., Cepparulo, A., Giuriato, L., 2024. Public-private partnerships for low-carbon, climate-resilient infrastructure: Insights from the literature. Journal of Cleaner Production 470, 143338. Available at https://doi.org/10.1016/j.jclepro.2024.143338.

Cassin, J., 2021. History and development of nature-based solutions: concepts and practice. In: Cassin, J., Matthews, J.H., Gunn, E.L. (Eds.), Nature-Based Solutions and Water Security. Elsevier, pp. 19–34. Available at https://doi.org/10.1016/b978-0-12-819871-1.00018-x.

Chaaben, N., Elleuch, Z., Hamdi, B., Kahouli, B., 2024. Green economy performance and sustainable development achievement: empirical evidence from Saudi Arabia. Environment, Development and Sustainability 26 (1), 549–564. Available at https://doi.org/10.1007/s10668-022-02722-8.

Chen, Y., Zhao, Z.J., 2021. The rise of green bonds for sustainable finance: global standards and issues with the expanding Chinese market. Current Opinion in Environmental Sustainability 52, 54–57. Available at https://doi.org/10.1016/j.cosust.2021.06.013.

Chisita, C.T., Fombad, M.C., 2020. Knowledge sharing to support climate change adaptation in Zimbabwe. VINE Journal of Information and Knowledge Management Systems 51 (2), 333–350. Available at https://doi.org/10.1108/vjikms-10-2019-0161.

Chukwu, V.E., 2020. Potentials, drivers and barriers to green economy transition: implications for Africa. Advanced Journal of Plant Biology 1 (1), 7–17. Available at https://doi.org/10.31248/ajpb2020.006.

Citil, M., 2024. Is green finance a prerequisite for green growth of G-20 economies. Innovation and Green Development 3 (4), 100170. Available at https://doi.org/10.1016/j.igd.2024.100170.

D'Amato, D., Korhonen, J., 2021. Integrating the green economy, circular economy and bioeconomy in a strategic sustainability framework. Ecological Economics 188, 107143. Available at https://doi.org/10.1016/j.ecolecon.2021.107143.

Daniya, G., Tang, D., 2024. Green finance and industrial low-carbon transition: a case study on green economy policy in Kazakhstan. Sustainability 16 (17), 7731. Available at https://doi.org/10.3390/su16177731.

De Melo, T.A.C., De Oliveira, M.A., De Sousa, S.R.G., Vieira, R.K., Amaral, T.S., 2022. Circular economy public policies: a systematic literature review. Procedia Computer Science 204, 652–662. Available at https://doi.org/10.1016/j.procs.2022.08.079.

Debele, S.E., Leo, L.S., Kumar, P., Sahani, J., Ommer, J., Bucchignani, E., et al., 2023. Nature-based solutions can help reduce the impact of natural hazards: a global analysis of NBS case studies. Science of The Total Environment 902, 165824. Available at https://doi.org/10.1016/j.scitotenv.2023.165824.

Dechezleprêtre, A., Nachtigall, D., Venmans, F., 2023. The joint impact of the European Union emissions trading system on carbon emissions and economic performance. Journal of Environmental Economics and Management 118, 102758. Available at https://doi.org/10.1016/j.jeem.2022.102758.

Devi, A., Jhariya, M.K., Raj, A., Banerjee, A., Singh, K.P., Singh, B., 2024. Acacia nilotica: a promising species for soil sustainability. Land and Environmental Management through Forestry. wiley, India, pp. 339–353. Available at https://doi.org/10.1002/9781119910527.ch14.

Du, G., 2023. Nexus between green finance, renewable energy, and carbon intensity in selected Asian countries. Journal of Cleaner Production 405, 136822. Available at https://doi.org/10.1016/j.jclepro.2023.136822.

Dunbar, K., Owusu-Amoako, J., Treku, D.N., 2024. Unveiling the nexus: carbon finance and climate technology advancements. International Review of Financial Analysis 96. Available at https://doi.org/10.1016/j.irfa.2024.103658.

Dwivedi, Y.K., Hughes, L., Kar, A.K., Baabdullah, A.M., Grover, P., Abbas, R., et al., 2022. Climate change and COP26: are digital technologies and information management part of the problem or the solution? An editorial reflection and call to action. International Journal of Information Management 63. Available at https://doi.org/10.1016/j.ijinfomgt.2021.102456.

Eaton, D., Sheng, F., 2019. Inclusive Green Economy: Policies and Practice. Zayed International Foundation for the Environment & Tongji University, Dubai, Shanghai.

Ehresman, T.G., Okereke, C., 2015. Environmental justice and conceptions of the green economy. International Environmental Agreements: Politics, Law and Economics 15 (1), 13–27. Available at https://doi.org/10.1007/s10784-014-9265-2.

Elliott, C., Schumer, C., Gasper, R., Ross, K., Singh, N., 2023. Realizing net-zero emissions: good practices in countries. World Resources Institute. Available at https://doi.org/10.46830/wriwp.22.00134.

Erb, C., Nelling, E., Zhang, Y., 2018. Green bonds: evidence on social and environmental impacts. J. Bus. Ethics 148, 355–370.

Farooq, U., Wen, J., Tabash, M.I., Fadoul, M., 2024. Environmental regulations and capital investment: does green innovation allow to grow? International Review of Economics and Finance 89, 878–893. Available at https://doi.org/10.1016/j.iref.2023.08.010.

Feng, T., Wang, X., Shi, Y., Tu, Q., 2024. The role of carbon price signal in green innovation: evidence from China. Journal of Environmental Management 370, 122787. Available at https://doi.org/10.1016/j.jenvman.2024.122787.

Filewod, B., McCarney, G., 2023. Avoiding carbon leakage from nature-based offsets by design. One Earth 6 (7), 790–802. Available at https://doi.org/10.1016/j.oneear.2023.05.024.

Fontoura, P., Coelho, A., 2022. How to boost green innovation and performance through collaboration in the supply chain: Insights into a more sustainable economy. Journal of Cleaner Production 359, 132005. Available at https://doi.org/10.1016/j.jclepro.2022.132005.

Fu, C., Lu, L., Pirabi, M., 2024. Advancing green finance: a review of climate change and decarbonization. Digital Economy and Sustainable Development 2 (1). Available at https://doi.org/10.1007/s44265-023-00026-x.

Gareth, B., Sophie, W., 2024. Climate Finance: Taking a Position on Climate Futures. Agenda Publishing, Newcastle upon Tyne. Available at https://hdl.handle.net/20.500.12657/88193.

Georgeson, L., Maslin, M., Poessinouw, M., 2017. The global green economy: a review of concepts, definitions, measurement methodologies and their interactions. Geo: Geography and Environment 4 (1). Available at https://doi.org/10.1002/geo2.36.

Hajoary, P.K., Ramani, V., Nuur, C., 2024. New for some, old for others: circular economy practices in ancient time. Circular Economy and Sustainability 4 (2), 815–825. Available at https://doi.org/10.1007/s43615-023-00323-9.

Han, Y., Tan, S., Zhu, C., Liu, Y., 2022. Research on the emission reduction effects of carbon trading mechanism on power industry: plant-level evidence from China. International Journal of Climate Change Strategies and Management 15 (2), 212–231. Available at https://doi.org/10.1108/ijccsm-06-2022-0074.

Hassan, T., Khan, Y., Safi, A., Chaolin, H., Wahab, S., Daud, A., et al., 2023. Green financing strategy for low-carbon economy: the role of high-technology imports and institutional strengths in China. Journal of Cleaner Production 415. Available at https://doi.org/10.1016/j.jclepro.2023.137859.

He, Y., Wang, S., Zhou, Z., Xiao, K., Xie, A., Wu, B., 2024. A Blockchain-based carbon emission security accounting scheme. Computer Networks 243, 110304. Available at https://doi.org/10.1016/j.comnet.2024.110304.

Houssam, N., Ibrahiem, D.M., Sucharita, S., El-Aasar, K.M., Esily, R.R., Sethi, N., 2023. Assessing the role of green economy on sustainable development in developing countries. Heliyon 9 (6). Available at https://doi.org/10.1016/j.heliyon.2023.e17306.

Howes, M., Wortley, L., Potts, R., Dedekorkut-Howes, A., Serrao-Neumann, S., Davidson, J., et al., 2017. Environmental sustainability: a case of policy implementation failure? Sustainability 9 (2), 165. Available at https://doi.org/10.3390/su9020165.

Ibarra Vega, D., Bautista-Rodriguez, S., 2024. The impact of circular economy strategies on municipal waste management: a system dynamics approach. Cleaner Engineering and Technology 21, 100761. Available at https://doi.org/10.1016/j.clet.2024.100761.

Jhariya, M.K., Meena, R.S., Banerjee, A., 2021b. Ecological Intensification of Natural Resources for Sustainable Agriculture. Springer, Singapore, pp. 1–655. Available at https://doi.org/10.1007/978-981-33-4203-3.

Jhariya, M.K., Meena, R.S., Banerjee, A., Meena, S.N., 2022. Natural Resources Conservation and Advances for Sustainability. Elsevier. Available at https://doi.org/10.1016/C2019-0-03763-6.

Jhariya, M.K., Banerjee, A., Swaroop Meena, R., Kumar, S., Raj, A., 2021a. Sustainable Intensification for Agroecosystem Services and Management. Springer, Singapore, pp. 1–870. Available at https://doi.org/10.1007/978-981-16-3207-5.

Jhariya, M.K., Meena, R.S., Banerjee, A., Kumar, S., Raj, A., 2023. Agroforestry for Carbon and Ecosystem Management. Elsevier, India, pp. 1–434. Available at https://doi.org/10.1016/C2021-0-02916-3.

Jiang, L., Niu, H., Ru, Y., Tong, A., Wang, Y., 2023. Can carbon finance promote high quality economic development: evidence from China. Heliyon 9 (12), e22698. Available at https://doi.org/10.1016/j.heliyon.2023.e22698.

Kiptoo, J., 2024. The influence of political stability on foreign direct investment (FDI). International Journal of Developing Country Studies 6 (1), 74–86. Available at https://doi.org/10.47941/ijdcs.2165.

Kirchherr, J., Yang, N.-H.N., Schulze-Spüntrup, F., Heerink, M.J., Hartley, K., 2023. Conceptualizing the circular economy (revisited): an analysis of 221 definitions. Resources, Conservation and Recycling 194, 107001. Available at https://doi.org/10.1016/j.resconrec.2023.107001.

Chapter 5: Circular economy for plastic waste management. In: Koottatep, T., Winijkul, E., Xue, W., Panuvatvanich, A., Visvanathan, C., Pussayanavin, T., Limphitakphong, N., Polprasert, C. (Eds.), Marine Plastics Abatement: Challenges, Implications, Assessments and Circularity (Volume 1). IWA Publishing, London, pp. 185–221. Available at https://doi.org/10.2166/9781789063202_0185.

Kosoy, N., Brown, P.G., Bosselmann, K., Duraiappah, A., Mackey, B., Martinez-Alier, J., et al., 2012. Pillars for a flourishing earth: planetary boundaries, economic growth delusion and green economy. Current Opinion in Environmental Sustainability 4 (1), 74–79. Available at https://doi.org/10.1016/j.cosust.2012.02.002.

Kumar, S., Meena, R.S., Jhariya, M.K., 2020. Resources use efficiency in agriculture. Springer Nature, Singapore, pp. 1–760. Available at https://doi.org/10.1007/978-981-15-6953-1.

Kumar, S., Meena, R.S., Sheoran, P., Jhariya, M.K., 2024. Regenerative Agriculture for Sustainable Food Systems. Springer, Singapore, pp. 1–493. Available at https://doi.org/10.1007/978-981-97-6691-8.

Kwilinski, A., Lyulyov, O., Pimonenko, T., 2024. Reducing transport sector CO_2 emissions patterns: environmental technologies and renewable energy. Journal of Open Innovation: Technology, Market, and Complexity 10 (1), 100217. Available at https://doi.org/10.1016/j.joitmc.2024.100217.

Lazaro, L.L.B., Grangeia, C.S., Santos, L., Giatti, L.L., 2023. What is green finance, after all? – Exploring definitions and their implications under the Brazilian biofuel policy (RenovaBio). Journal of Climate Finance 2, 100009. Available at https://doi.org/10.1016/j.jclimf.2023.100009.

Lee, C.-C., Li, X., Yu, C.-H., Zhao, J., 2022. The contribution of climate finance toward environmental sustainability: new global evidence. Energy Economics 111, 106072. Available at https://doi.org/10.1016/j.eneco.2022.106072.

Lee, C.-C., Wang, C.-song, He, Z., Xing, W.-wu, Wang, K., 2023. How does green finance affect energy efficiency? The role of green technology innovation and energy structure. Renewable Energy 219, 119417. Available at https://doi.org/10.1016/j.renene.2023.119417.

Levi, M., Arjomandi, A., Shams, S., Stewart, J., 2019. Green bonds and environmental performance: evidence from China. Journal of Corporate Finance 56, 430–449.

Li, H., Yang, Z., An, J., 2024a. Green finance, new infrastructure, and carbon emission performance in Chinese cities. Frontiers in Environmental Science 12. Available at https://doi.org/10.3389/fenvs.2024.1352869.

Li, K., Lin, W., Jiang, T., Mao, Y., Shi, W., 2024b. Driving carbon emission reduction in China through green finance and green innovation: an endogenous growth perspective. Environmental Science and Pollution Research 31 (9), 14318–14332. Available at https://doi.org/10.1007/s11356-024-32067-w.

Li, M., Wiedmann, T., Fang, K., Hadjikakou, M., 2021. The role of planetary boundaries in assessing absolute environmental sustainability across scales. Environment International 152, 106475. Available at https://doi.org/10.1016/j.envint.2021.106475.

Licastro, A., Sergi, B.S., 2021. Drivers and barriers to a green economy. A review of selected balkan countries. Cleaner Engineering and Technology 4. Available at https://doi.org/10.1016/j.clet.2021.100228.

Lin, B., Huang, C., 2022. Analysis of emission reduction effects of carbon trading: market mechanism or government intervention? Sustainable Production and Consumption 33, 28–37. Available at https://doi.org/10.1016/j.spc.2022.06.016.

Lin, Y., Mahmood, M.A., Meng, W., Ali, Q., 2024. Green economy transition in Asia Pacific: a holistic assessment of renewable energy production. Journal of Cleaner Production 437, 140648. Available at https://doi.org/10.1016/j.jclepro.2024.140648.

Liu, H., Han, P., 2024. Renewable energy development and carbon emissions: the role of electricity exchange. Journal of Cleaner Production 439, 140807. Available at https://doi.org/10.1016/j.jclepro.2024.140807.

Liu, W., Zhu, P., 2024. The impact of green finance on the intensity and efficiency of carbon emissions: the moderating effect of the digital economy. Frontiers in Environmental Science 12. Available at https://doi.org/10.3389/fenvs.2024.1362932.

Liu, X., Cui, J., Wu, Y., Yue, X., Shen, J., Guo, P., 2024. The nexus between green finance and carbon emissions: evidence from maturity mismatch in China. Sustainability 16 (10), 4319. Available at https://doi.org/10.3390/su16104319.

Lou, J., Hultman, N., Patwardhan, A., Mintzer, I., 2023. Corporate motivations and co-benefit valuation in private climate finance investments through voluntary carbon markets. npj Climate Action 2 (1). Available at https://doi.org/10.1038/s44168-023-00063-4.

Ma, J., Feng, J., Chen, J., Zhang, J., 2024. Volatility spillover from carbon prices to stock prices: evidence from China's carbon emission trading markets. Journal of Risk and Financial Management 17 (3), 123. Available at https://doi.org/10.3390/jrfm17030123.

Ma, Z., Fei, Z., 2024. Research on the mechanism of the carbon emission reduction effect of green finance. Sustainability 16 (7), 3087. Available at https://doi.org/10.3390/su16073087.

Martin, A., Coolsaet, B., Corbera, E., Dawson, N.M., Fraser, J.A., Lehman, I., et al., 2016. Justice and conservation: the need to incorporate recognition. Biological Conservation 197, 254–261. Available at https://doi.org/10.1016/j.biocon.2016.03.021.

Meena, R.S., Pradhan, G., 2023. Industrial garbage-derived biocompost enhances soil organic carbon fractions, CO_2 biosequestration, potential carbon credits and sustainability index in a rice-wheat ecosystem. Environmental Research 235, 116525. Available at https://doi.org/10.1016/j.envres.2023.116525.

Meena, R.S., Pradhan, G., Kumar, S., 2025. Energy flow, eco-efficiency, and economic circulation with recycled industrial waste compost application in wheat and subsequent rice farming. Science of the Total Environment 967. Available at https://doi.org/10.1016/j.scitotenv.2025.178779.

Meena, R.S., Pradhan, G., Kumar, S., Lal, R., 2023. Using industrial wastes for rice-wheat cropping and food-energy-carbon-water-economic nexus to the sustainable food system. Renewable and Sustainable Energy Reviews 187, 113756. Available at https://doi.org/10.1016/j.rser.2023.113756.

Meena, R.S., Yadav, A., Kumar, S., Jhariya, M.K., Jatav, S.S., 2022. Agriculture ecosystem models for CO_2 sequestration, improving soil physicochemical properties, and restoring degraded land. Ecological Engineering 176. Available at https://doi.org/10.1016/j.ecoleng.2022.106546.

Meena, R.S., Pradhan, G., Nalani, Kumar, S., Singh, A.K., Verma, S.K., et al., 2024b. Diversified cropping modules designed for soil restoration, CO_2 sequestration, and generating carbon credits. Land Degradation and Development 35 (14), 4314–4329. Available at https://doi.org/10.1002/ldr.5224.

Meena, R.S., Lal, R., Kumar, S., Pradhan, G., Srinivasarao, C., Singh, A.K., et al., 2024c. Potential of Indian agriculture for capturing atmospheric CO_2 and monetizing carbon credits to the farmers: an overview and policy framework. Advances in Agronomy 188, 101–206. Available at https://doi.org/10.1016/bs.agron.2024.06.001.

Meena, R.S., Pradhan, G., Singh, K., Kumar, S., Singh, A.K., Shashidhar, K.S., et al., 2024a. Agriculture models for restoring degraded land to enhance CO_2 biosequestration and carbon credits in the Vindhyan region of India. Science of the Total Environment 929. Available at https://doi.org/10.1016/j.scitotenv.2024.172661.

Miao, C., Chen, Z., Zhang, A., 2024. Green technology innovation and carbon emission efficiency: the moderating role of environmental uncertainty. Science of The Total Environment 938, 173551. Available at https://doi.org/10.1016/j.scitotenv.2024.173551.

Michaelowa, A., Honegger, M., Poralla, M., Winkler, M., Dalfiume, S., Nayak, A., et al., 2023. International carbon markets for carbon dioxide removal. PLOS Climate 2 (5), e0000118. Available at https://doi.org/10.1371/journal.pclm.0000118.

Mlambo, D., Sebata, A., Chichinye, A., Mabidi, A., 2024. Chapter 5 - Agroforestry and biodiversity conservation. In: Jhariya, M.K., Meena, R.S., Banerjee, A., Kumar, S., Raj, A. (Eds.), Agroforestry for Carbon and Ecosystem Management. Academic Press, pp. 63–73. Available at https://doi.org/10.1016/B978-0-323-95393-1.00008-7.

Mohd, S., Kaushal, V., 2018. Green finance: a step towards sustainable development. MUDRA: Journal of Finance Account. 5 (1). Available at https://doi.org/10.17492/mudra.v5i01.13036.

Nasir, N., Ahmed, W., 2024. Green finance initiatives and their potential to drive sustainable development. In: Naifar, N. (Ed.), Climate Change and Finance. Sustainable Finance. Springer, Cham. Available at https://doi.org/10.1007/978-3-031-56419-2_1.

Newig, J., Jager, N.W., Challies, E., Kochskämper, E., 2023. Does stakeholder participation improve environmental governance? Evidence from a meta-analysis of 305 case studies. Global Environmental Change 82, 102705. Available at https://doi.org/10.1016/j.gloenvcha.2023.102705.

Nguyen, D.T., Oanh, T.T.K., Bui, T.D., Dao, L.K.O., 2024. The impact of green finance on green growth: the role of green energy and green production. Heliyon 10 (16). Available at https://doi.org/10.1016/j.heliyon.2024.e36639.

Nikyema, G.A., Blouin, V.Y., 2020. Barriers to the adoption of green building materials and technologies in developing countries: the case of Burkina Faso. IOP Conference Series: Earth and Environmental Science 410 (1), 012079. Available at https://doi.org/10.1088/1755-1315/410/1/012079.

Pan, J., Cross, J.L., Zou, X., Zhang, B., 2024. To tax or to trade? A global review of carbon emissions reduction strategies. Energy Strategy Reviews 55, 101508. Available at https://doi.org/10.1016/j.esr.2024.101508.

Pan, Y., Zhang, X., Wang, Y., Yan, J., Zhou, S., Li, G., et al., 2019. Application of blockchain in carbon trading. Energy Procedia 158, 4286–4291. Available at https://doi.org/10.1016/j.egypro.2019.01.509.

Pande, R., 2024. Offsets, carbon markets, and climate and economic justice. Science 385 (6714). Available at https://doi.org/10.1126/science.ads1902.

Pang, H., Wu, C., Zhang, L., Haider Hashmi, S., 2024. The impact of green bond issuance on carbon emission intensity and path analysis. PLoS One 19 (6), e0304364. Available at https://doi.org/10.1371/journal.pone.0304364.

Pereira, P., Wang, F., Inacio, M., Kalinauskas, M., Bogdzevič, K., Bogunovic, I., et al., 2024. Nature-based solutions for carbon sequestration in urban environments. Current Opinion in Environmental Science & Health 37, 100536. Available at https://doi.org/10.1016/j.coesh.2024.100536.

Pradhan, G., Meena, R.S., 2023. Utilizing waste compost to improve the atmospheric CO_2 capturing in the rice-wheat cropping system and energy-cum-carbon credit auditing for a circular economy. Science of the Total Environment 892. Available at https://doi.org/10.1016/j.scitotenv.2023.164572.

Pradhan, G., Meena, R.S., Kumar, S., Lal, R., 2023. Utilizing industrial wastes as compost in wheat-rice production to improve the above and below-ground ecosystem services. Agriculture, Ecosystems & Environment 358, 108704. Available at https://doi.org/10.1016/j.agee.2023.108704.

Prasad, R., Jhariya, M.K., Banerje, A., 2021. Advances in Sustainable Development and Management of Environmental and Natural Resources: Economic Outlook and Opinions, 2-volume set. Apple Academic Press, New York, pp. 1–428. Available at https://doi.org/10.1201/9781003187455.

Qadri, H.Mu.D., Ali, H., Abideen, Zu, Jafar, A., 2024. Mapping the evolution of green finance research and development in emerging green economies. Resources Policy 91. Available at https://doi.org/10.1016/j.resourpol.2024.104943.

Raj, A., Jhariya, M.K., Meena, R.S., 2025. Agroforestry for Monetising Carbon Credits. Springer Nature, Cham.

Raj, A., Jhariya, M.K., Banerjee, A., Nema, S., Bargali, K., 2024. Land and Environmental Management Through Forestry. wiley, India, pp. 1–375.

Rehman, F.U., Islam, M.M., Miao, Q., 2023. Environmental sustainability via green transportation: a case of the top 10 energy transition nations. Transport Policy 137, 32–44. Available at https://doi.org/10.1016/j.tranpol.2023.04.013.

Robins, L., 2008. Perspectives on capacity building to guide policy and program development and delivery. Environmental Science & Policy 11 (8), 687–701. Available at https://doi.org/10.1016/j.envsci.2008.07.003.

Roy, O., Meena, R.S., Kumar, S., Jhariya, M.K., Pradhan, G., 2022. Assessment of land use systems for CO_2 sequestration, carbon credit potential, and income security in Vindhyan region, India. Land Degradation and Development 33 (4), 670–682. Available at https://doi.org/10.1002/ldr.4181.

Saltman, K.J., 2017. The promise and realities of pay for success/ social impact bonds. Education Policy Analysis Archives 25, 59. Available at https://doi.org/10.14507/epaa.25.2640.

Samal, R., Dash, M., 2023. Ecotourism, biodiversity conservation and livelihoods: understanding the convergence and divergence. International Journal of Geoheritage and Parks 11 (1), 1–20. Available at https://doi.org/10.1016/j.ijgeop.2022.11.001.

Sangiorgi, I., Schopohl, L., 2021. Why do institutional investors buy green bonds: evidence from a survey of European asset managers. International Review of Financial Analysis 75, 101738. Available at https://doi.org/10.1016/j.irfa.2021.101738.

Santos, G., 2022. Climate change policy and carbon pricing. Energy Policy 168, 112985. Available at https://doi.org/10.1016/j.enpol.2022.112985.

Sartzetakis, E.S., 2021. Green bonds as an instrument to finance low carbon transition. Springer, Greece Economic Change and Restructuring 54 (3), 755–779. Available at https://doi.org/10.1007/s10644-020-09266-9.

Sharif, A., Saqib, N., Dong, K., Khan, S.A.R., 2022. Nexus between green technology innovation, green financing, and CO_2 emissions in the G7 countries: the moderating role of social globalisation. Sustainable Development 30 (6), 1934–1946. Available at https://doi.org/10.1002/sd.2360.

Si, R., Wang, Y., Cao, M., Wen, H., 2024. Does green technology innovation promote green economic growth? –Examining regional heterogeneity between resource-based and non-resource-based cities. International Review of Economics & Finance 94, 103406. Available at https://doi.org/10.1016/j.iref.2024.103406.

Söderholm, P., 2020. The green economy transition: the challenges of technological change for sustainability. Sustainable Earth 3 (1). Available at https://doi.org/10.1186/s42055-020-00029-y.

Song, X., Yao, Y., Wu, X., 2023. Digital finance, technological innovation, and carbon dioxide emissions. Economic Analysis and Policy 80, 482–494. Available at https://doi.org/10.1016/j.eap.2023.09.005.

Souza Piao, R., Vincenzi, T.B., Vazquez-Brust, D.A., Yakovleva, N., Bonsu, S., Carvalho, M.M., 2024. Barriers toward circular economy transition: exploring different stakeholders' perspectives. Corporate Social Responsibility and Environmental Management 31 (1), 153–168. Available at https://doi.org/10.1002/csr.2558.

Sowińska-Świerkosz, B., García, J., 2022. What are nature-based solutions (NBS)? Setting core ideas for concept clarification. Nature-Based Solutions 2, 100009. Available at https://doi.org/10.1016/j.nbsj.2022.100009.

Swinkels, L., 2024. Trading carbon credit tokens on the blockchain. International Review of Economics & Finance 91, 720–733. Available at https://doi.org/10.1016/j.iref.2024.01.012.

Tănasie, A.V., Năstase, L.L., Vochiţa, L.L., Manda, A.M., Boţoteanu, G.I., Sitnikov, C.S., 2022. Green economy—green jobs in the context of sustainable development. Sustainability (Switzerland) 14 (8). Available at https://doi.org/10.3390/su14084796.

Teklie, D.K., Yağmur, M.H., 2024. The role of green innovation, renewable energy, and institutional quality in promoting green growth: evidence from african countries. Sustainability 16 (14), 6166. Available at https://doi.org/10.3390/su16146166.

van der Ree, K., 2019. Promoting green jobs: decent work in the transition to low-carbon, green economies. Revue internationale de politique de dévéloppement 11, 248–271. Available at https://doi.org/10.4000/poldev.3107.

Wang, C., Du, D., Liu, T., Zhu, Y., Yang, D., Huang, Y., et al., 2024a. Impact of green technology innovation on green economy: evidence from China. Sustainability 16 (19), 8557. Available at https://doi.org/10.3390/su16198557.

Wang, M., Wang, X., Liu, Z., Han, Z., 2024b. How can carbon trading promote the green innovation efficiency of manufacturing enterprises? Energy Strategy Reviews 53, 101420. Available at https://doi.org/10.1016/j.esr.2024.101420.

Weng, Z., Cheng, C., Xie, Y., Ma, H., 2022. Reduction effect of carbon emission trading policy in decreasing PM2.5 concentrations in China. International Journal of Environmental Research and Public Health 19 (23), 16208. Available at https://doi.org/10.3390/ijerph192316208.

World Economic Forum, 2023. Scaling Voluntary Carbon Markets: A Playbook for Corporate Action. White paper September 2023. https://www.weforum.org/publications/scaling-voluntary-carbon-markets-a-playbook-for-corporate-action/

Wu, G., Liu, X., Cai, Y., 2024b. The impact of green finance on carbon emission efficiency. Heliyon 10 (1), e23803. Available at https://doi.org/10.1016/j.heliyon.2023.e23803.

Wu, J., 2024. Role of green finance and carbon accounting in achieving sustainability. Humanities and Social Sciences Communications 11 (1). Available at https://doi.org/10.1057/s41599-023-02492-2.

Wu, L., Zhu, C., Wang, G., 2024a. The impact of green innovation resilience on energy efficiency: a perspective based on the development of the digital economy. Journal of Environmental Management 355, 120424. Available at https://doi.org/10.1016/j.jenvman.2024.120424.

Wu, X., Li, Z., Tang, F., 2022. The effect of carbon price volatility on firm green transitions: evidence from Chinese manufacturing listed firms. Energies 15 (20), 7456. Available at https://doi.org/10.3390/en15207456.

Xiao, Q., Fei, L., 2024. How does climate vulnerability impact green innovation? A hindrance to sustainable development. Innovation and Green Development 3 (4), 100169. Available at https://doi.org/10.1016/j.igd.2024.100169.

Xie, Y., 2024. The interactive impact of green finance, ESG performance, and carbon neutrality. Journal of Cleaner Production 456, 142269. Available at https://doi.org/10.1016/j.jclepro.2024.142269.

Xu, L., Yang, J., 2024. Carbon pricing policies and renewable energy development: analysis based on cross-country panel data. Journal of Environmental Management 366, 121784. Available at https://doi.org/10.1016/j.jenvman.2024.121784.

Xu, L., Solangi, Y.A., Wang, R., 2023. Evaluating and prioritizing the carbon credit financing risks and strategies for sustainable carbon markets in China. Journal of Cleaner Production 414, 137677. Available at https://doi.org/10.1016/j.jclepro.2023.137677.

Xu, S., Liu, Y., 2023. Research on the impact of carbon finance on the green transformation of China's marine industry. Journal of Cleaner Production 418, 138143. Available at https://doi.org/10.1016/j.jclepro.2023.138143.

Xu, X., Li, J., 2023. Can green bonds reduce the carbon emissions of cities in China? Economics Letters 226, 111099. Available at https://doi.org/10.1016/j.econlet.2023.111099.

Xu, Y., Tao, X., Das, M., Kwok, H.H.L., Liu, H., Kuan, K.K.L., et al., 2024. A blockchain-based framework for carbon management towards construction material and product certification. Advanced Engineering Informatics 61. Available at https://doi.org/10.1016/j.aei.2023.102242.

Yadav, A., Mahalik, M.K., 2024. Does renewable energy development reduce energy import dependency in emerging economies? Evidence from CS-ARDL and panel causality approach. Energy Economics 131. Available at https://doi.org/10.1016/j.eneco.2024.107356.

Yi, H., Zhang, Z., 2024. Green financial structure and carbon emissions: a structural matching perspective. Journal of Cleaner Production 434, 140103. Available at https://doi.org/10.1016/j.jclepro.2023.140103.

Yuyang, L., 2024. Natural resource efficiency and the road to a green economy: from scarcity to availability. Resources Policy 89, 104574. Available at https://doi.org/10.1016/j.resourpol.2023.104574.

Zhang, J., Yang, G., Ding, X., Qin, J., 2024. Can green bonds empower green technology innovation of enterprises? Environmental Science and Pollution Research 31 (7), 10032–10044. Available at https://doi.org/10.1007/s11356-022-23192-5.

Zhang, Z., Linghu, Y., Meng, X., Yi, H., 2023. Research on the carbon emission reduction effects of green finance in the context of environment regulations. Economic Research-Ekonomska Istrazivanja. 36 (1). Available at https://doi.org/10.1080/1331677X.2023.2179513.

Zhao, S., Hu, P., 2024. Analysis of the impact of carbon finance on digital green technology innovation of enterprises. Applied Mathematics and Nonlinear Sciences 9 (1). Available at https://doi.org/10.2478/amns.2023.2.01307.

Zhao, X., Shang, Y., Ma, X., Xia, P., Shahzad, U., 2024. Does carbon trading lead to green technology innovation: recent evidence from Chinese companies in resource-based industries. IEEE Transactions on Engineering Management 71, 2506–2523. Available at https://doi.org/10.1109/TEM.2022.3186905.

Zhu, H., Feng, T., Li, X., Ali, N., 2024. Green finance, green development and decarbonization of the energy consumption structure. PLOS ONE 19 (4), e0300579. Available at https://doi.org/10.1371/journal.pone.0300579.

Zhu, J.J., Zhang, R., Kanhalikham, K., Liu, Z., Shen, X., 2023. Green economy studies amongst the global climate change challenge between 2016 and 2022: a bibliometric review. Frontiers in Ecology and Evolution 11. Available at https://doi.org/10.3389/fevo.2023.1168437.

Chapter 24

Policy and legal frameworks for the successful implementation of green credits and financing

Arnab Banerjee[1], Manoj Kumar Jhariya[2], Ram Swaroop Meena[3], Abhishek Raj[4], Sharda Prasad Tripathi[5], Tamoghna Achharya[6], and Srimanta Gupta[7]

[1]*School of Studies in Environmental Science, Pt. Ravishankar Shukla University, Raipur, Chhattisgarh, India,* [2]*Department of Farm Forestry, Sant Gahira Guru Vishwavidyalaya, Ambikapur, Chhattisgarh, India,* [3]*Department of Agronomy, Institute of Agricultural Sciences, Banaras Hindu University, Varanasi, Uttar Pradesh, India,* [4]*Pt. Deendayal Upadhyay College of Horticulture and Forestry, Dr. Rajendra Prasad Central Agricultural University, Samastipur, Bihar, India,* [5]*Sant Gahira Guru Vishwavidyalya, Surguja, Ambikapur, Chhattisgarh, India,* [6]*Department of Marine Science, Berhampore University, Brahmapur, Odisha, India,* [7]*Department of Environmental Science, The University of Burdwan, Burdwan, West Bengal, India*

Highlights

- Emission reduction and combating climate change require a green economy.
- Green credit and finance are the pathway toward a green economy.
- Policy and legal framework are required to implement green financing.
- Adoption of green financing fulfills the targets of the SDGs.

24.1 Introduction

Poverty reduction is an event that can be achieved through a proper development process. However, the process of development is associated with economic gain and loss, which exerts its impact on the natural environment in negative ways, leading to unsustainable development. In the present context, changing climate and degradation of the environment are the main constraints for humankind (Sachs et al., 2019; Meena et al., 2025). Hence there is an urgent requirement to develop eco-friendly approaches that give proper social benefits toward protecting the environment, as well as develop a proper response to climate change problems (Devi et al., 2024). Monetization and marketing of the carbon (C) in the form of C credits under the banner of C trading, a low-C economy, is an innovative approach to combat the mega event of climate change (Meena et al., 2025; Raj et al., 2025). In the international context, developing a suitable financial market that would move toward a C-neutral economy requires policy formulation for proper identification and mobilization of funding to provide environmental benefits, along with support for corporate social responsibilities that would promote sustainable growth (Schaefer and Voss, 2011; Meena et al., 2024a,b). Such a wider dimension would aid in promoting green practices for economic growth that would lead to emission reduction as well as develop a resource-efficient economy for developed and developing countries (Prasad et al., 2021).

Literature review suggests that lack of proper funding hinders the effectiveness of such eco-friendly approaches of monetization of ecosystem services or promoting green economy (Schaefer and Voss, 2011). In the economic sector the primary drive is toward earning, and hence, such a type of thinking promotes unsustainable growth and development with little protection to the environment. When one considers combating climate change, the lack of proper financial tools has hindered the implementation of concepts of green credit and finance at the grassroots level. On the other hand, due to climatic changes, the prevalence of catastrophic events has increased considerably, besides setting the global warming level below 2°C. Under such a perplexed situation, the adoption of green finance would be a suitable move toward building a green economy, along with adaptation to a changing climate (Banerjee et al., 2023, 2024; Meena et al., 2024a).

Carbon Incentives, Plans, and Policies in Agriculture. DOI: https://doi.org/10.1016/B978-0-443-33557-0.00026-4

Now green financing is a step in the finance sector to move toward a low-C economy along with the development of resource efficiency for developing climate adaptability (Schaefer and Voss, 2011). In the upcoming times, as more and more countries move toward green financing through eco-friendly or green approaches, investment through green financing mechanisms would help to address the research gaps in the concerned area. It would also lead to the development of suitable policies and planning for addressing the research gaps. Now the problem is that there is no such development in these issues across the globe. For example, in green economics, there is a requirement for the creation of a hypothetical market for environmental goods, but very few countries have adopted such an approach. Evolution of green financing has necessitated the development of pension funds for green development projects (Schaefer and Voss, 2011). According to the report of the G20 Green Finance Study Group, there is a significant level of influence of bonds in promoting low-C emission and bringing the global emission of C below the 2°C target (Bergedieck et al., 2017; Meena et al., 2024b). Such a positive response to bonds was found to be highly promising across various continents and countries. The present chapter deals with the concept of green credit and finance and their functionality toward C emission reduction and moving toward a green economy.

24.2 Green credits and finance

At present times there is a new trend of gradual development of the concept of green funding, financing for promoting environmentally friendly projects. The primary motive behind such activities is to move toward a low-C economy to attain net-zero emissions. This has a significant contribution toward the green circular economy (Feng, 2016). Green financing has a long historical root, as China initiated the practice of green financing in 2007. Various forms of insurance, credit, and security systems were formulated in the economic sector for all-around sustainable development. In the G20 summit in 2016, principles for the formulation of a green financial system were jointly developed by the administration, the ministry, and the commission level of the People's Republic of China (Ruoyu, 2016). At the conference green financing was the central theme of discussion. Green finance is a broad concept that tends to regulate the economic and business activity for the protection of the environment, promote and divert the corporate funding toward sustainable development (SD), and bring an overall reform through green design, transportation, and other related financial fields. Practices such as exploring clean sources of energy, climate financing for GHG (greenhouse gas) emission reduction, reduction of energy consumption, technological intervention, and infrastructure reforms lead to the protection of the environment. All these policies need to be oriented toward a credit system for sustainable green finance development (Shaohong, 2013). Different forms of green financing and credit systems are shown in Fig 24.1.

Further development of the green industry, ecologically sound industrial civilization, and an effective legal system for green finance development are required for ecology-based green economy development and sustainability. Such movements of greening the economic sector would not only help to develop resiliency against climate change but would also create better employment opportunities along with the development of the economic sector. However, in funding from the private sector, there should be alignment of different forms of investment toward environmental protection and SD. In this way, there should be a steady preparation of the financial sector against climate risks and various threats to the environment (Berensman et al., 2020). Considering the case of climate change, there is a significant risk of loss associated with climate disasters. The transformation to a low-C market economy is a big challenge, and climate insurance should work effectively to compensate for such risk. The G20 summit should work toward the development of a green philosophy in the economic sector and funding schemes and screen out such activities and practices. There is a need for the development of green infrastructure and standards for the production of green products. Further, developing a green bond market economy, the establishment of an identical rating system through the application of green bond principles is very much required at the present time (Sachs et al., 2019). All these approaches would be successful if there were a continuous monitoring system.

Application of suitable policies rendering a net economic return at a higher level promotes the private sector to invest more in green development projects. For example, relief in taxes through the adoption of clean energy sources can be a suitable strategy. The United States has already implemented such a type of tax relief to promote wind and solar power. In this way, industrial firms or corporate sectors can enjoy tax deductions through investment in the nonconventional energy resources (Berensman et al., 2020). Such policies of tax relief can also address the spillover impact of one industry over the other sectors. In such a way, the return may be so high that a tax refund would promote encouraging more private sectors to invest more in such green energy resources (Sachs et al., 2019).

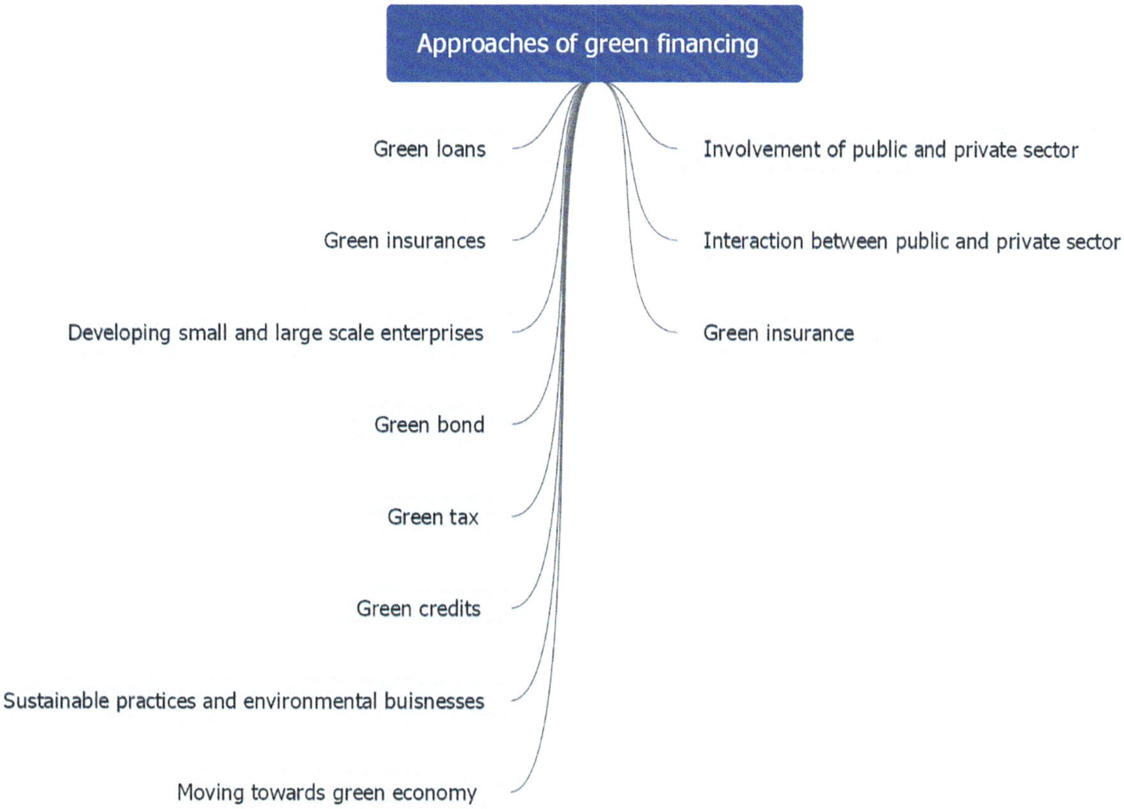

FIGURE 24.1 Different approaches to green financing.

TABLE 24.1 Green bond distribution by various sectors in India (Compiled: (Jha and Bakshi, 2019).

S. No.	Sector	% Contribution
1.	Private sector (nonfinancial corporations)	50%
2.	Private sector (financial corporation)	37%
3.	Government sector	16%
4.	Other financial organizations	37%

Another important policy includes village and community-based funds for various forms of medium- to small-sized green projects. One such example is the Hometown Investment Trust (HIT), which is a source of funding to promote and implement non-conventional sources of energy at the community level. In the process, HIT attracts investors to invest in eco-friendly projects according to their expertise. Investors tend to invest in such projects online. However, behind all these, there is an urgent need for proper development of the economy to promote the greening of economic activity. Such a thing can be achieved through some public policy, such as green bonds, which will serve the purpose of public banks. Table 24.1 presents the distribution of green bonds in India in 2019 (Jha and Bakshi, 2019).

Creation of a proper market with respect to sustainable infrastructure development in order to accommodate the risks associated with such types of approaches. Various examples in this regard include infrastructure venture trusts for G20 states and green fintech market for developed nations, followed by the development of Yieldcos or trust money for sustainable infrastructure development (Lee, 2020).

24.2.1 Challenges of implementing green finance at the root level

Although the scheme of green credit and finance is good and sustainable, its effective implementation faces some severe challenges. First, the cost associated with the environmental risk associated with such sustainable green infrastructure development projects remains hidden due to the uncertainty of occurrences of ecological disasters. Without considering the environmental risk cost associated with the project, the net economic gain would be reduced for the polluters. Uncertainty of estimation of environmental cost would also lead to miscalculation of the green economy of the project (Rashid and Uddin, 2018). Second, internalization of the environmental externality is a big issue, which may promote unsustainable behavior in place of sustainable behavior in the form of green financing. As a consequence, industrial pollution has become an inevitable fact, and hence, the problem of pollution persists within the economic activity. Third, all the investments in the banking sector cannot be driven toward green projects, as green projects have longer payback times than pollution-producing projects. Further, investing in green infrastructure projects, there is a probability of failure of new projects or technology, which would reduce the net economic return from the projects. To this, market risk is another crucial factor that increases the vulnerability of such projects in the economic sector. The transaction costs of such green projects are relatively high, and hence, the benefits from such projects become nullified. Most of the finance applications in this sector come from small-scale industries and hence are not very hospitable in financial terms. Lack of proper development of entrepreneurship in this aspect is also an important hindrance in implementing such projects (Khan et al, 2017).

24.3 Prospects for green financing

From a future perspective, green financing has immense potential for different sectors. Arrangement of funds through bond and equity in collaboration with the banking sector is an essential aspect for financing the private sector. The companies and organizations are required to disclose their socio-environmental perspective in their annual financial report, through which their performance will be evaluated through various forms of market indicators (Shakil et al., 2014). As a consequence, private sectoral companies would be responsible for such types of investments, followed by the adoption of a green finance system by various industries through the promotion of eco-friendly practices. Creation of demand for the adoption of eco-friendly technology would create the opportunity for cleaner and greener industries, which would be based on green financing. Further providing economic incentives for the effective implementation of green financing would help to promote green transformation and, hence, would help to reduce the environmental risk associated with such projects. This would frame the formation of the sustainable information sector.

Proper integration of short- and medium-sized enterprises with green financing is an important step for moving toward eco-friendly projects. One example of such a mechanism was observed in Bangladesh, where the Bangladesh Bank was providing such financial instruments as concessionary fund, credit guarantee, credit wholesale, etc. Such a financing system was aimed toward various forms of eco-friendly sectors, such as energy, green industry practices, food technology, and other allied fields. Identification of proper sectors and schemes for the success of the green finance system is very crucial for the greening of the economic system. It would vary on a case-by-case basis. Identification of potential sectors for effective fund utilization is the key to the success of the green finance system. Some of the possible sectors include energy, waste management, and developing green urbanization (Hossain et al., 2016). The effectiveness of the green financing system can be increased through regular monitoring of the banks and funds to ensure they are properly following the norms or standards. The guidelines related to these aspects should ensure that they are followed or not. Hence, a uniform reporting system should be mandatory for all the banks associated with such a type of green financing system. More investments and more supply of funds in the form of accessing green financing funds at the international level, and hence improve the effectiveness of different forms of green finance. Last but not least, the stakeholders from various sectors should be made well aware of the facts of green finance and their utility (Hoshen et al., 2017).

24.3.1 Green financing for sustainable landscapes

A joint project was initiated under the able leadership of CIFOR and UNEP. The project cost would be 1 billion dollars, funded by a global environment facility, aiming toward developing a green finance system for a sustainable landscape for a period of three years (2021–2024). The primary objective is to inhibit deforestation and promote various forms of sustainable land-use practices (Jhariya et al., 2022, 2023; Meena and Pradhan, 2023). The project would also help in boosting agribusiness activity along with a sustainable production system. The project would evaluate the social and environmental outcomes associated with investments made toward green credit, along with the objective of emission reduction, inhibition of deforestation, etc. The third outcome would be capacity building for the community forestry, such as social forestry, farm forestry, agroforestry, etc., as for the agribusiness sector. Proper evaluation of the said project would address the research gap of the utility of such commercial models, followed by an assessment of investment toward shifting the production model (https://www.fao.org/collaborative-partnership-on-forests/initiatives/green-finance-for-sustainable-landscapes/en).

24.4 Green credits

Agriculture is a miracle of development for the prosperity and well-being of humankind. It has always played a significant role in boosting the country's economy, especially for developing countries. Considering the improvement in productivity and quality of agricultural products, the process is very crucial in terms of boosting the rural economy along with ensuring food security for both urban and rural dwellers. Now the issue has become critical in the limelight of a changing climate, as new and novel methods are required to combat such a global problem. Hence there is an urgent need to develop practices toward the greening of the agriculture system through reformation in the practices of cultivation (Kumar et al., 2020; Kumar et al., 2024). Therefore, there is an urgent need to integrate the credit system with the production cycle of agriculture. It would provide an additional net economic return and offer an extra benefit to the farming community. As per one report, loans for green projects, such as green agriculture, have a one-third contribution of 500 trillion Vietnamese dong (Credit Department for Economic Sectors, 2023).

The concept of green lending involves promoting sustainable agriculture practices and environmental protection and fulfills the targets of SDG-2030. Under the scheme, loans are provided to the rural livelihood and farming community at a low interest rate, followed by flexible repayment mechanisms in comparison to other forms of financing. Such flexibility provides wider acceptance of the scheme and opens up more opportunities for investment in agro-based industries and enterprises, encouraging them to adopt innovative practices toward sustainable agriculture. Apart from having such benefits green lending suffers from some serious drawbacks, which include lacunae in the information regarding the product and inconsistency of standards across various countries worldwide. Hence, the general form of financing in the agricultural enterprises is more prevalent than such schemes across the globe. Research on green lending revealed a positive influence on the farm development and economy. It also works for achieving the SDG-2030 agenda for poverty alleviation (Xie et al., 2019). Further, comparison of green lending between India and China has also emphasized the positive impact of green lending on agricultural development and coping with climate change (Xiong et al., 2019; Meena et al., 2023). It was also reported to have a positive impact on sustainable agriculture practices (Li et al., 2019).

Green lending is a process that would help to improve production as well as develop a market for the completion of agroproducts for their export purposes (Al-Riffai et al., 2020). The success of such policy initiatives would be governed by policy support activities followed by an effective management schedule. Farmers who are reluctant to adopt such schemes due to financial crisis would be capable of doing so by obtaining green loans, which would act as a subsidy, supporting them in agricultural production. Hence, green loans would promote the productivity as well as the ecological health of the agroecosystem (Raza et al., 2018; Pradhan and Meena, 2023a). Further, the effectiveness of such green loans can be increased through a proper support policy in relation to agricultural extension products (Sharma and Saha, 2020). Many of the foreign studies have reflected positive outcomes in terms of adopting green loans in the agriculture sector. A similar observation was reported in Vietnam by various workers (Do et al., 2020). Similar positive results in Vietnam were reported by various workers in agriculture and livestock production (Tran et al., 2018). This has revolutionized the agriculture production sector, along with an increase in net income gain (Nguyen et al., 2019). Apart from gaining a positive impact, maximum efficiency in agricultural production is yet to be reached (Vu et al., 2021). Such initiatives would be highly fruitful for developing effective financial policies in the agricultural sector.

Apart from giving positive results, it has some inherent drawbacks also. Under the green lending scheme, small landholding farmers find it difficult to qualify for green loans. The major focus of green lending was given toward production and livelihood generation, and hence, the least importance was given to the other environmental components, such as the water resources. Hence, the impact appears to be negative from an environmental perspective due to a lack of proper monitoring and evaluation.

24.4.1 Green Credit Program

From an Indian perspective, the Green Credit Initiative was initiated by the Indian Prime Minister under the banner of the Green Credit Program (GCP) under the purview of the COP28 meeting held in the United Arab Emirates. It is actually a follow-up action of a government scheme named Lifestyle for Environment (LiFE) movement, which was launched by the Indian PM at the meeting of COP26 held at Glasgow in 2021. It was actually a movement toward commitment to the environment and developing a sustainable approach and attitude across the sectors of development. The major objective of the GCP was to stimulate eco-friendly approaches along with their proper incentivization. The underlying mechanism of the scheme includes earning the incentives as "Green credits." The program will be executed in separate phases, considering different sectors in the form of water management and afforestation, sustainable agriculture, mangrove ecosystem conservation, pollution control, and waste management, etc. It also emphasized developing green infrastructure (https://vajiramias.com/article/green-credit-programme-gcp/6620bc3e9c26110595046fee/).

Following these activities, the green credit rules were enacted on October 12, 2023, under the umbrella Act of 1986 of India. The primary purpose of framing such rules includes developing a voluntary principle toward a positive environmental attitude that would lead to the generation of green credit. At the beginning of the GCP, emphasis was given to massive plantation activity under different forms of land use under the regulation of the forest department. Economic incentives for such massive plantation activities would be provided by ICFRE (Indian Council of Forestry Research and Education) under the guidelines of MOEFCC (Ministry of Environment, Forest, and Climate Change). Individuals and various companies may come forward for such plantation activities and would be provided an economic incentive for such activity in the form of green credit under the scheme. The Forest Department of the State Government would be wholly responsible for undertaking the plantation activities. ICFRE would evaluate the effectiveness and success of the activity after two years of plantation, and for each tree, an individual green credit would be accorded to the participants of the plantation program. It means more the number of successes of the plantation of tree species, the more there will be earnings of green credit. The credits can be distributed in two ways: first, compliance with forest laws to obtain a significant number of green credits by the organization that is going for alternate land use practices other than the forest land uses, and second, compliance with specific norms of corporate social responsibility of the corporate sector (https://vajiramias.com/article/green-credit-programme-gcp/6620bc3e9c26110595046fee/.).

24.4.2 Concerns about the Green Credit Program and its changes

Initially, it was observed that green credits were not tradable. But with legal reforms, GCP has emphasized such trading programs to develop proper economic incentives through a market-based approach. Hence, such green credits would be tradable on the market at the domestic level. Such a type of trading would be acceptable by gaining C credits through emission reduction of C. Actually, the GCP provides a market-based approach to promote environmental activities that would help in the mitigation of the changing climate as well as the conservation of nature.

Under the present era of a changing climate, agriculture is the most severely affected sector. The developing countries are intensively dependent upon agricultural production, and hence stress has been placed on identifying proper cost-effective solutions in the form of organic farming, biofertilizer technology, crop rotation techniques, and C mitigatory agroecosystem-resilient agriculture practices. From the Indian perspective, major initiatives were taken to achieve the net-zero emission target by 2070. In this connection, the Indian Government has enacted the Energy Conservation Bill in 2022 to promote renewable energy as well as create a C market (https://vajiramias.com/article/green-credit-programme-gcp/6620bc3e9c26110595046fee/).

24.4.3 Prospects and concerns of green credits

From an Indian perspective, the GCP has significant prospects and challenges in terms of its effective implementation. Hence green credit has a significant impact on the diverse economic and business sectors. Activities that earn green credits open up the opportunities for green credits in the business activities. Further, this opens up opportunities for the agro-based industries and corporate sectors to participate in the C credit market, which may invest in projects that aim toward reducing C emissions, which would serve the purpose of revenue generation as well as fulfill the target of net-zero emissions by 2070 as a climate commitment. Further, for auditing of green credit, the corporate sector should have a robust platform for its analysis, as well as be capable of analyzing the environmental footprint of the business activity (Bellarby et al., 2008; Banerjee et al., 2020). (https://www.banqu.co/blog/indias-green-credit-programme-what-it-is-how-to-prepare).

Participating in the GCP may require investing in eco-friendly sustainable practices and developing green infrastructure in its initial phase, while there are some business activities that suffer from financial crisis and may be reluctant to participate under such schemes. The green credit approach needs to be applied in diverse sectors as well. There is a lacuna in proper evaluation in the value of green credits as well as their further transfer (FAO, 2009).

24.4.4 Market for green financing and credits

The reduction of C emission and improving C stock in soil, vegetation, and the oceanic environment has become a prime research areas and objectives. Accordingly, C has become an important economic resource for the growing civilization. The amount of GHG reduction is being converted into CO_2 equivalents, which are traded in the markets popularly known as C trading. Such C markets utilize these C credits in terms of currency, which helps the industrial and corporate sectors to develop trading of C emissions both in developed and developing nations. Within the C trading mechanisms, an agreement is made between the buyer and seller regarding the purchase of C credits. There are two conditions in this regard. First, those who are reducing C emissions and sequestering C would receive financial gain in terms of selling their C credits earned by them. Second, those who need to reduce their emissions and require optimum C credits would purchase the credits to balance their account toward C neutrality. C offsetting involves compensating for C emissions by purchasing the credits from those who have extra credits to sell. Such a condition arises when there is the least probability of reducing C emissions through any available process, technology, and methodology. Such C markets are voluntary in nature as the value of C credits varies depending upon the nature of the market and associated C offset projects (Hamilton et al., 2008; Pradhan and Meena, 2023b). According to one estimate, the CO-eq value in 2009 varies between €1.90 and €13. Further, there is also significant development and evolution in the C markets. From the C market point of view, there are two types of markets, *viz.*, the voluntary market and the regulatory compliance market. In the voluntary market, the trading of C is voluntary in nature. On the other hand, in the regulatory compliance market, there should be a GHG accounting for government and corporate institutions under legal compliance. Hence it comes under the regulatory framework under various C reduction regimes. There was significant variation in the C trading across the two types of C markets (Hamilton et al., 2009). The regulatory C market is associated with three types of Kyoto Protocol mechanisms, which include ETS (European Union Trading System), CDM (Clean Development Mechanism), and JI (Joint Implementation). The voluntary markets are very important for the forestry and agriculture sectors. The private sector usually purchases VER (Voluntary Carbon Credits). For buying the C credits, corporate social responsibility plays a significant role. There are some organizations that offer the C credits to become C neutral. For example, British Airways provides a C-neutral flight system, and Morgan Stanley provides equivalent C credits. Regarding the private sector, C credits can be purchased from green projects, C funds (The World Bank BioCarbon Fund), or from various companies (Baalman and Schlamadinger, 2008; Pradhan and Meena, 2023).

24.5 Legal framework for green financing and credits

To maximize the utilization of the green finance scheme, the proper development of a legal framework is necessary for the effective implementation of the green finance scheme. Therefore, proper development of an integrated legal system with regard to green finance, which should have a scientific approach with various forms of rules and regulations, both at the departmental level and the administrative level, is necessary. Such a legal framework needs to be aided by environmental regulations at the local level. Incorporating an action plan for environmental securities within the financial activities under the banner of green security is another important aspect of developing a legal framework for green finance. However, this process requires proper regulatory compliance, supervision, and monitoring, followed by a proper evaluation system. Further, reporting and sound development of an environmental information network are essential, as this would improve environmental protection and consciousness across various enterprises as well as guide the corporate sector in the decision-making process and have its own risk management policy (Feng, 2016). Under green finance, the legislation industry would itself act as a fund provider and promote various enterprises and entrepreneurship cells to undergo green infrastructure projects. Another milestone in this regard is the development of a green insurance legal system, which has already been implemented by China (Kang, 2017). From an industrial perspective, heterogeneous insurance schemes need to be designed under the green finance legislation framework. The legal framework for green finance is not as developed as in developed countries, as it is yet to be properly developed in our country. Most of the attention with respect to environmental protection is on the normative document, and the concept of a legal framework for green finance is a policy-oriented document, and hence is devoid of getting the prime focus for its proper development and its effective implementation. It was also observed that in most of the countries the legal framework is too fragile and hence their effective implementation appears to be too weak. Most of the larger enterprises across the various sectors ignore them and do not show compliance with the existing legal framework. Across the globe, the legislation related to green finance is very much limited to green credits of banks, and heterogeneous types of legislation persist for green bonds and insurance. Most of the existing legislation is focused toward pollution, and less attention is given to the energy sector and C finance mechanism (Guojun and Chao, 2017).

A Legal framework comprising soft and hard laws is usually formulated through councils and ministries of the respective countries, along with specific characteristic features and legal attributes. For the proper formulation of a regulatory framework, we must first look forward to the formulation of soft laws that focus on the formation of basic, normalized norms without the intervention of the government. It would be governed by public opinions as well as a social relevance perspective. Such formulation frames the basic framework for the formulation of regulated hard law legislation. Such a combination of soft laws and hard laws is very much essential in the field of green finance, which would increase the effectiveness of the implementation of such schemes. Further, at the institutional level, a synergistic combination of hard and soft laws would be highly fruitful for the effective implementation of green finance (EBF European Banking Federation, 2017).

24.6 Policy framework for green financing and credit development

For effective implementation of green finance at the global level, there should be a system of punishment for the violations of green finance and credit commitment, as well as existing rules and regulations. Generally, it was observed that financial institutions are recommended to issue loans following standard procedure and legal setup. However, there is no legal provision for punishment for violation of the regulation of environmental protection, as well as a lacuna for imparting an award for fulfilling the commitment. Further, policies could be made in terms of tax relaxation to promote the green finance-based activities (Wang and Zhu, 2018). For various industries that do not work in compliance with environmental protection, there should be a restriction on budget allocation as well as a proper punishment mechanism for their violation of the law. Proper economic incentives for encouraging the enterprises toward green activities should be there. There should be an economic incentive for every step of the activities during the effective implementation of green financing. It would also govern the decision-making processes of the various enterprises. Enterprises that are compatible with ecofriendly behavior and a green financing system may also benefit through policies such as low-interest loans. Such a type of promotion would help in the effective implementation of the green finance system. Uncertainty prevails in the technological upgradation of the processes under various sectors in terms of financial output than previously existing technology. Further, various green financing mechanisms across various levels and across various countries of the globe have been depicted in Table 24.2.

It also generates the issue of market risk. Financial corruption by the respective enterprises and misuse of green funds are other big problems of such schemes. Considering the aforesaid effects, there should be a proper evaluation of risk and negative impacts of the project. Inclusion of market risk is a major issue in this regard. Establishment of a proper task force to monitor the market mechanism and risks, as well as catalyzing activities for better market opportunity, can be done, which may culminate into a global network system of green market and financing. Proper development of technical terms associated with the green finance system is required in a transparent way so that overestimation of the social and environmental cost does not take place for a green project. It also requires participation from all parts of society. There should be an effective strategy for establishing private-public participation through the inclusion of policies related to public and finance in order to maximize the encouragement of better involvement in the green projects. From a policy perspective, there should be a long-term target for the green financing agenda for developing a sustainable finance system in the future. Providing direct subsidies for eco-friendly initiatives could be an effective strategy under a green finance system. The benefit could take the form of a tax reduction, subsidized funding for green projects, or other incentives. This has already been implemented in Italy in the form of a tax deduction associated with electricity consumption reduction by the consumers. Another important policy may include proper certification and validation of the green projects through third-party assessment in order to maintain transparency in the process. However, it may lead to additional costs for evaluation and hence may not be fruitful for small-scale projects. A level of awareness generation is required in both the public and banking sectors in order to make the green finance initiatives a part of the banking system. Such awareness generation campaigning can be achieved through various directories and guidelines prepared across the different countries of the globe. For example, the EIPP (the European Investment Project Portal) was launched to increase the visibility of such funded projects to global investors. This is also an essential part of policy and decision-making with regard to green finance and the credit system.

24.7 Conclusion

For the development of a country, economic development is the need of the hour. As growth and development are taking place at an unprecedented rate in the present era, it is giving rise to several forms of environmental perturbances. As a result, there is an urgent need for the transformation of the human-built environment. There is a widespread impact on

TABLE 24.2 Policy initiatives across various levels for green finance and credits.

S.N.	Level	Name	Details	Reference
1.	Global level	Paris Agreement and Climate Finance	Investing an amount of 100 billion dollars on an annual basis till 2025 to curb the CO_2 levels at the global level.	https://insideclimatenews.org/news/07112019/paris-climate-agreement-pledges-lack-urgency-ipcc-timeline-warning; https://www.theguardian.com/environment/climate-consensus-97-per-cent/2018/mar/28/climate-scientists-debate-a-flaw-in-the-paris-climateagreement
2.	Global level	UNEP Environment Fund	Flexible funding for UNEP members investing 70 million dollars during 2019.	https://www.unep.org/about-un-environment-programme/funding-and-partnerships/funding-facts/environment-fund. https://www.unep.org/about-un-environment-programme/funding-and-partnerships/funding-facts/environment-fund
3.	Global level	Social Investment Task Force	The United Kingdom Government initiated in 2013 for the improvisation of sustainable financing in the global market.	https://www.gov.uk/government/groups/social-impact-investment-taskforce#members-of-the-taskforce
4.	Regional level	European Green Deal	The European Union launched this scheme in 2019 for achieving net-zero emissions by 2050 and regulating resource use.	https://ec.europa.eu/info/strategy/priorities-2019-2024/european-green-deal_en
5.	Regional level	European Union Emission Trading System	The largest C trading system that works on the cap and trade principle, where there is capping on the greenhouse gas emissions by the industries.	https://ec.europa.eu/clima/policies/ets
6.	Regional level	European Budget Commitment	The European Union allocates 20% of the total budget and an investment of 864 million euros for climate-responsive projects and curbing climate change.	https://ec.europa.eu/clima/policies/budget/life_en
7.	Regional level	ASEAN Catalytic Green Finance Facility	Initiated by the Asian Development Bank in 2019 for promoting funding for developing green infrastructure.	ADBI (2020)
8.	Regional level	European Social Fund	Initiated by the European Union for the social and economic interrelationship among the members of the EU.	https://ec.europa.eu/esf/home.jsp
9.	National level	Green Finance Taskforce	The UK government initiated this scheme in 2017 to promote green financing and move toward a low-C economy.	Government of the United Kingdom (2019)
10.	National level	Green New Deal	The US initiated the scheme for developing a regulatory framework for the climate crisis.	https://www.reuters.com/world/us/ahead-bidens-climate-summit-lawmakers-relaunch-green-new-deal-2021-04-20
11.	National level	National Social Finance Policy	The UK government initiated the scheme to promote the social financial market.	Nicholls and Simon (2020)
12	Local level	Donut Economy	Amsterdam initiated the concept in 2020 toward emission reduction of air pollutants and greenhouse gas emissions across various cities in Europe.	https://www.theguardian.com/world/2020/apr/08/amsterdam-doughnut-model-mend-post-coronavirus-economy
13.	Local level	Fair Trade town	Initiated in the UK across the township to have an environment of fair trade.	http://www.fairtradetowns.org

agriculture, the production sector, forestry, the environment, the industrial sector, and both small- and large-scale enterprises under industrial activity. Hence, at the present time, we are thinking of greening our economy, which would bring some changes in the overall impact on the environment. In this way forward, there is an urgent need for the implementation of various schemes, such as green credit, finance, pollution tax, and money lending activities, to promote the development of green infrastructure, as well as more engagement of the different sectors in green projects in the form of eco-friendly technologies. However, such schemes face some challenges in their initiation as well as in their proper execution. Green finance is a broader concept that promotes the collection of proper funds that address environmental concerns as well as manage environmental gaps and associated financial risk. Green credit is a form of C financing in which credit is earned through the reduction of GHG emissions. Further, the earned credit undergoes economic exchange through the C trading market in the form of the purchase and sale of credits. Such a funding system would be very much applicable in terms of changing attitudes and developing the adoption of eco-friendly projects in various sectors of agriculture, forestry, and the environment. Such schemes would also aid in extra revenue generation and extra livelihoods for rural communities. On the other hand, schemes such as green taxes and pollution taxes would be fruitful for emission reduction in the industrial sector. Considering all such facts, a proper policy and legal framework formation is required through a proper decision-making process. Before going for proper policy making and legal reforms in this sector, an adequate understanding of the ifs and buts of various schemes and their associated problems needs to be dealt with lucidly in order to promote sustainable funding for green initiatives for achieving the goals of the 2030 Agenda for SD.

24.8 Future thrust of green financing and credits

From a future perspective, green finance and credit have immense potential, but they should be paved in a suitable direction and way. The foremost part includes promoting social funds with the targeting of the SDGs of the 2030 Agenda. Increasing investment is required to achieve the SDGs, as well as to address specific sectors such as education, employment, and health. Each of these sectors is an emerging sector that has immense scope for work. Various other sectors, such as energy, transport, and water, need to be focused on for the implementation of green financing to get a significant economic return. From a climate change perspective, green funding needs to be attributed toward mitigation, adaptation, and SD (Jhariya et al., 2021a,b). In all these aspects, the focus of green financing should be more on developing countries, as the climate crisis is more severe for them. Granting of funds should be given to innovativeness, promoting research, and exploration activities. An attempt should be made toward de-risking the later stage of development. Proper issue of green bonds would help to develop the green capital at the national level. The concept of impact bonds could be implemented from a socioeconomic and environmental perspective. Improvisation in the form of C taxes to move toward C neutrality by putting C taxes on the polluting industries or through tax relief for adopting new, greener technologies. New regulations need to be formulated for social financing as well as modifications of the banking sector to invest a portion of the assets in green financing. New schemes on result and success-based activities would also be beneficial for the future. Promote proper research and developmental activities in the available database to evaluate the outcome of various schemes under green financing and credit. Fund allocation can be done to develop industry-academic interlinkage and collaboration for better output through the implementation of green finance and credit. This would also pave the way for awareness generation and the formation of knowledge networks in this regard. Funding for promoting new research areas to develop a proper green market to stimulate more investors and participants to participate in the schemes.

References

ADBI, 2020. Green Finance Strategies for Post-COVID 19 Economic Recovery in South East Asia. Greening Recoveries for People and Planet. Asian Development Bank.

Al-Riffai, P., Breisinger, C., Raouf, M., 2020. Agricultural finance in MENA: needs and access constraints. Agricultural Finance Outlook 5 (1), 1–17. Available at https://doi.org/10.2478/afp-2020-0001.

Baalman, P., Schlamadinger, B., 2008. Scaling up AFOLU mitigation activities in non-annex-I countries. Working Paper. A report by Climate Strategies & GHG Offset Services for the Eliasch Review.

Banerjee, A., Meena, R.S., Jhariya, M.K., Yadav, D.K., 2020. Agroecological Footprints Management for Sustainable Food System. Springer, Singapore, pp. 1–514. Available at https://doi.org/10.1007/978-981-15-9496-0.

Banerjee, A., Jhariya, M.K., Bargali, S.S., Palit, D., 2023. Ecorestoration for Sustainability. Wiley-Scrivener, USA, pp. 1–441. Available at https://doi.org/10.1002/9781119879954.

Banerjee, A., Jhariya, M.K., Raj, A., Mechergui, T., 2024. Ecosystem management: climate change and global sustainability—an introduction. Ecosystem Management: Climate Change and Sustainability. wiley, India, pp. 1–21. Available at https://doi.org/10.1002/9781394231249.ch1.

Bellarby, J., Foereid, B., Hastings, A., Smith, P., 2008. Cool Farming: Climate Change Impacts of Agriculture and Mitigation Potential. Greenpeace International, Amsterdam.

Berensman, K., Volz, U., Bak, C., Bhattacharya, A., Huifang, T., Leipold, G. et al., 2020. Fostering sustainable global growth through green finance: what role for the G20? G20 Insights. Available from: https://www.g20-insights.org/policy_briefs/fostering-sustainable-global-growth-green-finance-role-g20/.

Bergedieck, L., Maheshwari, A., Ugaz, F.A., 2017. Green Finance: A Bottom-up Approach to Track Existing Flows. International Finance Corporation.

Credit Department for Economic Sectors, 2023. State Bank of Vietnam. Credit Department for Economic Sectors.

Devi, A., Jhariya, M.K., Raj, A., Banerjee, A., Singh, K.P., Singh, B., 2024. Acacia nilotica: a promising species for soil sustainability. Land and Environmental Management Through Forestry. wiley, India, pp. 339–353. Available at https://doi.org/10.1002/9781119910527.ch14.

Do, T.D., Duong, H.T.T., Le, N.H.A., Pham, T.P.T., 2020. Evaluating the impacts of funding forms on agriculture and agricultural exports in Vietnam. Journal of Economics Development 300 (10), 70–75.

EBF (European Banking Federation), 2017. Towards a Green Finance Framework Technical Report, pp. 1-45. Belgium, Germany.

FAO, 2009. Food Security and Agricultural Mitigation in Developing Countries: Options for Capturing Synergies. Rome, Italy. Available from: http://www.fao.org/docrep/012/i1318e/i1318e00.pdf.

Feng, Y., 2016. On the legal regulation of green Finance. People's Forum 14, 125–127. (In Chinese).

Government of the United Kingdom 2019. Green Finance Strategy. Transforming Finance for a Greener Future.

Guojun, A., Chao, C., 2017. International legislation and reference of Green Finance. China Financial Department 18, 74–76. (In Chinese).

Hamilton, K., Sjardin, M., Marcello, T., Xu, G., 2008. Forging a frontier: state of the voluntary carbon markets 2008. Ecosystem Marketplace & New Carbon Finance, New York.

Hamilton, K., Sjardin, M., Shapiro, A., Marcello, T., 2009. Fortifying the foundation: state of the voluntary carbon markets 2009. Ecosystem Marketplace & New Carbon Finance, New York.

Hoshen, M.S., Hasan, M.N., Hossain, S., Al Mamun, M.A., Mannan, A., 2017. Green financing: an emerging form of sustainable development in Bangladesh. Journal of Business and Management 19 (12), 24–30.

Hossain, D.M., Al Bir, A.T.S., Tarique, K.M., Momen, A., 2016. Disclosure of green banking issues in the annual reports: a study on Bangladeshi Banks. Middle East Journal of Business 11 (1), 19–30. Available at https://doi.org/10.5742/MEJB.2015.92758.

Jha, B., Bakshi, P., 2019. Green finance: fostering sustainable development in India. International Journal of Recent Technology and Engineering 8 (4), 3798–3801.

Jhariya, M.K., Banerjee, A., Swaroop Meena, R., Kumar, S., Raj, A., 2021a. Sustainable Intensification for Agroecosystem Services and Management. Springer, Singapore, pp. 1–870. Available at https://doi.org/10.1007/978-981-16-3207-5.

Jhariya, M.K., Meena, R.S., Banerjee, A., 2021b. Ecological Intensification of Natural Resources for Sustainable Agriculture. Springer, Singapore, pp. 1–655. Available at https://doi.org/10.1007/978-981-33-4203-3.

Jhariya, M.K., Meena, R.S., Banerjee, A., Meena, S.N., 2022. Natural Resources Conservation and Advances for Sustainability. Elsevier. Available at https://doi.org/10.1016/C2019-0-03763-6.

Jhariya, M.K., Meena, R.S., Banerjee, A., Kumar, S., Raj, A., 2023. Agroforestry for Carbon and Ecosystem Management. Academic Press, pp. 1–434. Available at https://doi.org/10.1016/C2021-0-02916-3.

Kang, Y., 2017. Green financial development and its legal system guarantee. Stock Market Bulletin 1, 4–11. (In Chinese).

Khan, A., Khan, B.F., Uddin, N., Azim, G.I.A., Islam, S., 2017. Financing green growth in bangladesh: challenges & opportunities. Economic Dialogue on Green Growth. UKaid.

Kumar, S., Meena, R.S., Jhariya, M.K., 2020. Resources Use Efficiency in Agriculture. Springer, Singapore, pp. 1–760. Available at https://doi.org/10.1007/978-981-15-6953-1.

Kumar, S., Meena, R.S., Sheoran, P., Jhariya, M.K., 2024. Regenerative Agriculture for Sustainable Food Systems. Springer, Singapore, pp. 1–493. Available at https://doi.org/10.1007/978-981-97-6691-8.

Lee, J.W., 2020. Green finance and sustainable development goals: the case of China. Journal of Asian Finance, Economics and Business 7 (7), 577–586. Available at https://doi.org/10.13106/jafeb.2020.

Li, L., Li, L., Li, W., 2019. Green agricultural finance and sustainable agriculture development in China: a study based on grey relational analysis. Sustainability 11 (17), 4702. Available at https://doi.org/10.3390/su11174702.

Meena, R.S., Pradhan, G., 2023. Industrial garbage-derived biocompost enhances soil organic carbon fractions, CO_2 biosequestration, potential carbon credits, and sustainability index in a rice-wheat ecosystem. Environmental Research, 116525. Available at https://doi.org/10.1016/j.envres.2023.116525.

Meena, R.S., Pradhan, G., Kumar, S., Lal, R., 2023. Using industrial wastes for rice-wheat cropping and food-energy-carbon-water-economic nexus to the sustainable food system. Renewable and Sustainable Energy Reviews 187 (November 2023), 113756. Available at https://doi.org/10.1016/j.rser.2023.113756.

Meena, R.S., Pradhan, G., Singh, K., Kumar, S., Singh, A.,K., Shashidhar, K.,S., et al., 2024a. Agriculture models for restoring degraded land to enhance CO_2 biosequestration and carbon credits in the Vindhyan region of India. Science of The Total Environment 929, 172661. Available at https://doi.org/10.1016/j.scitotenv.2024.172661.

Meena, R.S., Lal, R., Kumar, S., Pradhana, G., Rao, C.S., Singh, A.K., et al., 2024b. Potential of Indian Agriculture for Capturing Atmospheric CO_2 and Monetizing Carbon Credits to the Farmers: An Overview and Policy Framework. Advances in Agronomy 188, 101–206. Available at https://doi.org/10.1016/bs.agron.2024.06.001.

Meena, R.S., Pradhan, G.S., Kumar, S., Jhariya, M.K., Rao, C.S., 2025. Carbon-negative Agriculture. Springer, Singapore, pp. 1–500. Available at https://doi.org/10.1007/978-981-95-1483-0.

Nguyen, T.H., Tran, M.P., Le, H.P., 2019. Green credit and its impact on agriculture in Vietnam. Journal of Science Technology 57 (2), 40–45.

Nicholls, A., Simon, T., 2020. Dynamic persistence in UK policy making: the evolution of social investment ideas and policy instruments. Public Management Review 23 (6), 802–817. Available at https://doi.org/10.1080/14719037.2019.1699948.

Prasad, R., Jhariya, M.K., Banerje, A., 2021. Advances in Sustainable Development and Management of Environmental and Natural Resources: Economic Outlook and Opinions, 2-volume set. Apple Academic Press, New York, NY, pp. 1–428. Available at https://doi.org/10.1201/9781003187455.

Pradhan, G., Meena, R.S., 2023. Utilizing waste compost to improve the atmospheric CO2 capturing in the rice-wheat cropping system and energy-cum carbon credit auditing for a circular economy. Science of The Total Environment 892, 164572. Available at https://doi.org/10.1016/j.scitotenv.2023.164572.

Pradhan, G., Meena, R.S., 2023a. Interaction impacts of biocompost on nutrient dynamics and relations with soil biota, carbon fractions index, societal value of CO2 equivalent, and ecosystem services in the wheat-rice farming. Chemosphere, 139695. Available at https://doi.org/10.1016/j.chemosphere.2023.139695.

Pradhan, G., Meena, R.S., 2023b. Utilizing waste compost to improve the atmospheric CO2 capturing in the rice-wheat cropping system and energy-cum-carbon credit audit for a circular economy. Science of the Total Environment 164572. Available at https://doi.org/10.1016/j.scitotenv.2023.164572.

Raj, A., Jhariya, M.K., Meena, R.S., 2025. Agroforestry for Monetising Carbon Credits. Springer, Cham. Available at https://doi.org/10.1007/978-3-031-97151-8.

Rashid, M.H.U., Uddin, M.M., 2018. Green financing for sustainability: analysing the trends with challenges and prospects in the context of Bangladesh. International Journal of Green Economics 12 (3/4), 192. Available at https://doi.org/10.1504/ijge.2018.097876.

Raza, S.A., Ali, A., Hussain, A., Akhtar, M.A., 2018. Impact of microfinance on livelihood and environmental sustainability: evidence from Pakistan. Environmental Science and Pollution Research 25 (11), 10471–10483. Available at https://doi.org/10.1007/s11356-018-1667-y.

Ruoyu, L., 2016. China's green finance development status and policy recommendations. Macroeconomic Management 2016 (1), 58–60. (In Chinese).

Sachs, J.D., Woo, W.T., Yoshino, N., Taghizadeh-Hesary, F., 2019. Importance of Green Finance for Achieving Sustainable Development Goals and Energy Security. In: Sachs, J., Woo, W., Yoshino, N., Taghizadeh-Hesary, F. (Eds.), Handbook of Green Finance. Sustainable Development. Springer, Singapore, pp. 3–12. Available at https://doi.org/10.1007/978-981-13-0227-5_13.

Schaefer, J., 2011. Green Finance: An Innovative Approach to Fostering Sustainable Economic Development and Adaptation to Climate Change. Deutsche Gesellschaft für Internationale Zusammenarbeit (GIZ) GmbH. Available at https://www.greengrowthknowledge.org/sites/default/files/downloads/resource/Green_finance_GIZ.pdf.

Shakil, M.H., Azam, M.K.G., Raju, M., 2014. An evaluation of green banking practices in Bangladesh. European Journal of Business Management 6 (31), 8–16.

Shaohong, F., 2013. The legal system of Green Finance: the ought Choice and realistic construction from the perspective of sustainable development. Journal of Wuhan University 2013 (02), 66. (In Chinese).

Sharma, R., Saha, A., 2020. Promoting sustainable agriculture through green financing: the case of India. International Journal of Sustainable Development & World Ecology 27 (1), 61–70. Available at https://doi.org/10.1080/13504509.2019.1692573.

Tran, V.T., Phung, T.H., Nguyen, M.T., 2018. Green credit and its impact on business activities of farmers in Vietnam. Journal of Economics & Development 272 (4), 24–28.

Vu, V.H., Nguyen, V.T., Le, T.M.H., 2021. Comparing the impacts of green credit and other funding forms on agriculture and rural economic development in Vietnam. Journal of Agricultural and Rural Development 41 (2), 126–133.

Wang, H., Zhu, J., 2018. Analysis on the legal regulation of green finance development from the perspective of ecological civilization. Advances in Social Science, Education Humanities Research 181, 33–36.

Xie, Y., Sun, W., Huang, X., 2019. Green finance and agriculture: an empirical analysis of China's rural financial services. Journal of Cleaner Production 239, 118077. Available at https://doi.org/10.1016/j.jclepro.2019.118077.

Xiong, X., Zhang, Z., Ma, X., 2019. The role of green finance in sustainable agriculture: a comparative study of China and India. Sustainability 11 (10), 2938. Available at https://doi.org/10.3390/su11102938.

Chapter 25

Plans and policies for carbon financing in agroforestry

Sigit Andy Cahyono[1], Tyas Mutiara Basuki[1], Sumino[2], Leo Rio Ependi Malau[3], and Muhammad Rifqi Tirta Mudhofir[3]

[1]*Research Center for Ecology, National Research and Innovation Agency (BRIN), Cibinong, Indonesia,* [2]*Institute for Rural Technology Development (LPTP), Jl. Raya Palur, Karanganyar, Indonesia,* [3]*Research Center for Behavioral and Circular Economics, National Research and Innovation Agency (BRIN), Jakarta Selatan, Indonesia*

Highlights

- The potential of agroforestry (AF) in mitigating climate change (CC) through carbon sequestration (C_{seq}) and storage.
- Analysis of carbon (C) incentive policies and plans that support AF practices.
- An accessible and profitable C financing mechanism for farmers.
- Case studies of implementing C financing in AF in various countries.
- Identifying challenges and opportunities in the implementation of C finance policies.
- Strategic recommendations for the development of effective policies and plans.

25.1 Introduction

Agroforestry (AF), the practice of integrating trees, livestock, bees, fish, and insects into the agricultural landscape, plays an important role in mitigating climate change (CC), increasing biodiversity; improving soil health, food production, feed, fuel, and timber; and increasing farmers' incomes (Gemeda and Deresu, 2023; Prajapati et al., 2024; Satish et al., 2024; Tirkey et al., 2024; Meena et al., 2025), as well as encouraging a green economy. Agroforestry systems (AFs) are very effective in carbon sequestration (C_{seq}) and storing it in trees and above-ground biomass, underground root systems, and soil organic carbon (SOC) accumulation (Worku, 2024; Zomer et al., 2025; Meena et al., 2024; Dmuchowski et al., 2024; Jhariya et al., 2023), which can reach up to 100 T ha^{-1} SOC stock (Bajracharya et al., 2024; Karki et al., 2024). AF also helps in maintaining ecosystem services, such as pest control, pollination, nutrient cycling (Meena et al., 2024; Karki et al., 2024), and water regulation, as well as maintaining C stability below ground level (Kumar et al., 2023; Torquebiau, 2024). AF is one of the tools that facilitates both land restoration and CC mitigation simultaneously (Raj et al., 2023; Banerjee et al., 2023, 2024; Meena et al., 2024; Raj et al., 2024).

However, AF adoption faces challenges, including sociocultural resistance, knowledge gaps, policy support (Satish et al., 2024; VijayKumar et al., 2024), and financing. High AF initial financing, risks, and competition for resource inputs between trees and agricultural crops hinder large-scale implementation (VijayKumar et al., 2024).

One of the potential sources of financing for AF development is carbon (C) financing (Pradhan and Meena 2023a; Anderson and Zerriffi, 2012). However, farmers have limitations in accessing it. Without strong policy support and adequate accessibility to C markets, the potential of AF as a tool to mitigate CC has not been fully realized (Dumas et al., 2024). The problem of this research is how to design policies and incentive plans that can overcome these barriers and facilitate the adoption of AF based on C financing (Raj et al., 2025).

Currently, research focusing on C financing in AF is relatively limited (Zhou et al., 2024). Most existing studies place more emphasis on the technical aspects of AF or C markets in general. In contrast, studies that delve into how C policies and incentives can be practically implemented at the local level and for smallholders are still underexplored (Dumas et al., 2024). The research gap also includes a lack of comprehensive analysis of the economic, social, and environmental

Carbon Incentives, Plans, and Policies in Agriculture. DOI: https://doi.org/10.1016/B978-0-443-33557-0.00031-8

impacts of C finance in AF, as well as how such incentives can be designed to encourage farmer participation and improve their well-being. As such, there is a need to explore further policy approaches that can facilitate C financing in the AF sector, especially in developing countries.

This research will employ a qualitative approach, utilizing case studies from various countries that have successfully implemented AFC-finance policies. It focuses on how effective policies can enhance farmer participation, improve land productivity, and promote biodiversity restoration. This chapter is expected to guide policymakers in developing countries in adopting similar strategies.

25.2 Carbon financing in agroforestry

25.2.1 The concept of carbon financing

C finance is a mechanism linking the reduction of greenhouse gas (GHG) emissions with economic incentives, aiming to encourage the adoption of sustainable practices in various sectors, including agriculture and forestry (Prasad et al., 2021; Jiang et al., 2023; Meena et al., 2024). C financing mechanisms are conducted through emissions trading, C taxation, and C credits. The system is designed for emission reduction by setting a cost for C so that countries and businesses adopt cleaner technologies (Shapoval and Niedziółka, 2025). Blockchain and big data can also improve efficiency and transparency (Li, 2024; Shu, 2024). In China, "dual carbon targets" encourage high-C companies to leverage green payments supporting the transition of sustainable practices (Zhang et al., 2023; Meena et al., 2025), increased research, and reduction of financial constraints (Zhang and Wang, 2024).

In AF, C finance allows farmers and land managers to receive financial compensation based on the amount of C absorbed or stored through environmentally sound practices, such as tree planting, increased vegetation cover, and better land management (Jhariya et al., 2022). This mechanism is regulated through C markets, where C credits are traded between producers of emission reductions and buyers who want to offset their C footprint (Banerjee et al., 2021; Pradhan and Meena 2023b). One C credit is typically equivalent to one ton of carbon dioxide (CO_2) emission that is avoided or absorbed. C credits are verified by independent institutions that ensure C emission reductions or C_{seq} have occurred so that they can be traded in the C market (Woo et al., 2021). The C credits generated from these projects can then be sold to entities that want to offset their emissions, such as companies or governments that have an obligation to achieve emission-reduction targets. Farmers or AF project managers then receive income from the sale of these C credits, which serve as an economic incentive to encourage sustainable practices.

25.2.1.1 Relevant carbon finance models for agroforestry

Several C finance models are relevant and suitable for supporting AF, both at the local and global levels. One of the most common models is the voluntary C market, where unregulated economic actors, such as private companies, purchase C credits as part of their corporate social responsibility (CSR) initiatives or to enhance their environmental reputation (Spilker and Nugent, 2022). In this context, AF becomes a valid source of C credits, allowing smallholders or communities to participate in global markets.

Second, compliance C markets, such as the European Union Emissions Trading System (EU-ETS), also provide great opportunities. Under this scheme, entities regulated by emission regulations, such as companies in the energy and heavy industry sectors, are obliged to reduce their emissions or purchase C credits to meet emission targets (Li and Zhang, 2024). AF can be a provider of C credits resulting from sustainable practices implemented in developing countries, where the potential for C_{seq} is higher due to the availability of land.

In addition, there is a results-based financing (RBF) model, where payments are made to farmers or land managers based on verified C_{seq} results. This model is suitable for application in AF projects because accurate results in the form of C_{seq} or land-cover increase can be measured and reported periodically. This RBF allows farmers to receive direct income from their conservation efforts while incentivizing them to maintain sustainable practices in the future (Mintah et al., 2025).

Fourth, green bonds are also a growing model of C financing, where bonds issued by the government or financial institutions are used to fund environmentally friendly projects, including AF (Ning et al., 2023; Meena et al., 2024). Funds from these bonds can be used to support AF infrastructure, from providing tree seedlings to technical support for farmers. Bondholders get a return on capital from the proceeds from the sale of C credits generated by these projects.

25.2.1.2 Key mechanisms of carbon financing in agroforestry

C financing in AF involves creating and selling C credits derived from activities that reduce GHG emissions through specific practice mechanisms (Table 25.1). These C credits can be traded in compliance or voluntary C markets, carried

TABLE 25.1 Comparison of carbon financing mechanisms in agroforestry.

No	Mechanism	Purpose	Key features	Source
1	C offsetting	Generate credits for C sequestration through agroforestry practices	Involves MRV, sold in a compliance or voluntary marketplace	Anderson and Zerriffi (2012), Houghton (2020)
2	Clean development	Financing agroforestry projects in developing countries	Providing funding for projects that can reduce emissions and promote sustainable development	Cacho, et al. (2003), Murdiyarso and Herawati (2005)
3	Payment for environmental services	Encouraging sustainable land use practices	Direct payments to farmers for adopting agroforestry practices	Seeberg-Elverfeldt et al. (2008), Seeberg-Elverfeldt et al. (2009a)
4	C credit system	Create and trade credits based on sequestered C	Credits can be sold to companies or governments to offset emissions	Flugge and Abadi (2006), Kiyingi et al. (2019)
5	Voluntary C market	Allows companies and individuals to purchase credit voluntarily	Credits are traded on voluntary markets with a variety of standards and prices	Andela (2025), Honkomp and Schier (2024)
6	Vivo plans	Supporting community-based agroforestry projects	Focusing on equitable benefits and community engagement	Kiyingi et al. (2019), Murdiyarso and Herawati (2005)

MRV, monitoring, reporting, and verification.

out by the government or a company, to offset the C emissions it emits. AFs are suitable for C financing because of the dual benefits of C_{seq} and increased agricultural productivity (Seeberg-Elverfeldt et al., 2008, 2009a).

One of the main mechanisms of C financing is C offsets, where AF projects can generate C credits stored in trees and soil. For example, cocoa AF practices in Indonesia have been proven to absorb C. Paying for its environmental services with incentives to farmers to adopt more sustainable land use (Anderson and Zerriffi, 2012; Houghton, 2020). The case of Uganda with a *Grevillea robusta* AFs that can sequester C and increase profitability under a C offset scheme (Kiyingi et al., 2019).

Another mechanism used in C financing AF is C credits through the Clean Development Mechanism (CDM) scheme or Plan Vivo. This mechanism supports smallholders in implementing AF practices that sequester C and improve livelihoods (Cacho et al., 2003; Murdiyarso and Herawati, 2005). For example, in Western Australia, AF is used to overcome dryland salinity with C credits for landowners (Flugge and Abadi, 2006).

25.3 Prospects and constraints of carbon financing in agroforestry

25.3.1 Prospects

25.3.1.1 Environmental benefits

AFs preserve existing terrestrial C stocks and slow down deforestation by improving fallows and AF techniques to fulfill the need for food and fuelwood (Jose and Bardhan, 2012). Based on data adopted from the Intergovernmental Panel on Climate Change (Jose and Bardhan, 2012), it is estimated that by 2040, among various land uses, AF has the highest potential to sequester C due to the large areas (630 x10^6 ha) available worldwide.

At the aboveground AF, the system sequesters 0.29 Mg C/ha/year to 15.21 Mg C/ha/year and stores 30 Mg C/ha to 300 Mg C/ha up to 1 m soil layer depth (Nair, 2012). The capability of AFs to reduce GHG emissions, especially CO_2 emissions, depends on the applied method, in which the structure and function are affected by environmental, socioeconomic, tree species, and management conditions (Albrecht and Kandji, 2003). A study conducted by Udawatta and Jose (2012), has shown that the applied AF assists 33% offset of the United States (US) emission of 1600 Tg C/year from fossil fuel burning. These authors estimated C_{seq} potential for various AF types and found that silvopasture, alley cropping, and windbreaks in the US were 464, 52.4, and 8.6 Tg C/year, respectively. For riparian buffers, besides preserving water quality, they could also sequester C of 4.7 Tg C/year (Udawatta and Jose, 2012).

The AFs not only reduce C in the atmosphere through plant sequestration but also preserve SOC. A study conducted in humid tropical regions has found that vegetation in AFs has the potential to sequester C more than 70 Mg C/ha, and that the top 20 cm of soil, could sequester C up to 25 Mg/ha (Mutuo et al., 2005). Meanwhile, in the degraded lands of sub-humid tropics, improving fallow AF through maize cropping increased C stocks in the topsoil up to 1.6 Mg C/ha (Mutuo et al., 2005). A global study of SOC for mineral soil has been conducted by Jobbágy and Jackson (2000) using >2700 soil profiles based on three global datasets provided by data for climate, vegetation, and land use. This study revealed that at the top 20 cm of the first soil depth, the mean SOC for shrublands, grasslands, and forests was 33%, 42%, and 50%, respectively. Further, they found that at the global scale, the vertical distribution of SOC was somewhat more strongly related to vegetation than to climate; conversely, this was true for the absolute quantity of SOC. The total SOC content increases with higher precipitation and clay content, as well as with reduced temperature.

Biodiversity refers to the entirety of genetic, species, and ecological diversity that makes up life on the Earth, and it plays an essential role in conserving human life and the planet's health (Jose, 2012). However, this biodiversity has decreased rapidly because of the exploitation of natural resources, loss of species, invasion by alien species, impacts of environmental pollution, CC, ecosystem damage, and loss of habitats (Rands et al., 2010). AF is one of the alternatives that can be implemented to preserve biodiversity and fulfill human needs. AF can be applied to protect biodiversity loss because AF can provide 50%–80% variety compared to natural forests (Udawatta et al., 2019).

AFs have a higher potential to sequester atmospheric CO_2 than perennial crops, pastures, or natural grasslands. In these systems, the SOC is greatly affected by the biophysical and socioeconomic factors of the system parameters (Lorenz and Lal, 2014). The incorporation of trees in the systems improves soil properties, resulting in greater net C_{seq}. In sustainably managed AFs, a large amount of organic carbon (OC) returns to the soil as crop residues and tree litter (Jhariya et al., 2021a,b; Pradhan and Meena 2023a).

Those inputs can stabilize soil organic matter (SOM), decrease biomass decomposition rate, and preserve productivity (De Stefano and Jacobson, 2018; Oelbermann et al., 2004). The increase in SOM may also increase soil water retention; however, it is also affected by soil texture. The amount of OC in the soil and the proportion of textural components determine how variations in OC content affect soil water retention. In coarse soil with low C, an increase in C will cause an increase in water retention; conversely, it causes a reduction in water retention in fine-textured soils (Rawls et al., 2003). Further, Rawls et al. (2003) found that water retention of all textures increases in soil with high SOM.

25.3.1.2 Economic benefits

Subsistence farmers mainly apply AF in developing nations, and there is a compelling chance that these farmers will profit financially from AF if the C sequestered by AF practices is sold to developed countries. As a result, people expect AF to be used as a strategy for sequestering C, and it is pertinent that the current understanding of the subject be evaluated, as well as the realistic potential of AF as a biological method to C_{seq} (Ramachandran Nair et al., 2009). A study conducted in Java and Sumatra, Indonesia, has revealed that commercial and traditional AF contribute to food security. For instance, home gardens managed traditionally provide 20% greater food variations than commercial methods, while commercial home gardens could provide up to five times as much revenue as the traditional method (Duffy et al., 2021). Furthermore, they discovered that agro-silviculture presents a similar trade-off between diversity and income along a continuum of timber versus non-timber forest products. Traditional methods provide greater access to plants with therapeutic uses, but commercial systems derive 54% of their revenue from a single commodity crop, such as coffee (Duffy et al., 2021).

In addition, a study on the contribution of AFs to farmer income has been conducted in a state forest area in Parungpanjang, Indonesia. The research findings showed that the applied AF supported 15.8% of the farmer household income. The statistical analysis indicated that income from AF practices did not significantly affect total income due to variations in commercial crops, motivation, skill, and age-related physical capability (Desmiwati et al., 2021). Another study using data from 170 farmers in the semi-arid tropics of Central India has found that AF practices have positive impacts on food security and decrease income variation of farmers (Singh et al., 2023). The statistical analysis revealed that a 1% increase in tree density and tree diversity resulted in significant improvements in food security of 0.23% and 0.14% points, respectively. Consequently, the heterogeneity decreased by 0.38% and 0.16% (Singh et al., 2023).

Multi-strata AFs were studied in Ethiopia (Waldén et al., 2024). The research has found that the size of the farm and the applied fertilizer were the determining factors of the farm economy. Besides that, the net income positively correlated with the tree diversity; however, the number of legumes and large-diameter trees had a negative correlation with the income. This implies that the most economically feasible AF has a high potential for C_{seq} and thus serves as a C sink (Waldén et al., 2024; Devi et al., 2024; Pradhan and Meena 2023b). C finance offers additional income for farmers who adopt sustainable practices. In India, C income from AF increases farmers' incomes by 40%–50% (Singh et al., 2024), and in Ethiopia, the profitability of AF with C financing rises by 150% compared to monoculture agriculture (Waldén et al., 2020) and is more adaptive (Sima et al., 2024). In Southeast Asia, smallholder AFs contribute to C_{seq} and income from non-timber forest products (Roshetko et al., 2007). In Ghana, C payments can increase the profitability of organic cocoa AF (Agbotui et al., 2023). Additionally, Indonesia can contribute to reducing land degradation and promoting sustainable land management practices (Seeberg-Elverfeldt, 2009; Seeberg-Elverfeldt et al., 2009b). In Sulawesi, Indonesia, C payments are an incentive for farmers to adopt shade-intensive AF practices, increase household incomes, and reduce deforestation (Seeberg-Elverfeldt et al., 2008, 2009b). In addition, AF projects encourage sustainable agricultural practices, such as soil conservation and efficient water use, increasing soil productivity (Adhisankaran et al., 2024; Anderson and Zerriffi, 2012), and are included in policies, such as the Paris Agreement (Martins et al., 2024; Soloviy and Panchyshyn, 2024). These financial incentives motivate farmers to apply AF practices and sustainable land use.

25.3.1.3 Social benefits

Positive impacts of AFs on social aspects have been identified from numerous literatures. These include encouraging gender equality by empowering women, increasing food security at home by diversifying food sources, improving community-based cooperatives, preventing community exodus, and enhancing cultural engagement by involving the community in the creation of new ideas (Mukhlis et al., 2022). For the social aspect, AF practices create new off-farm jobs in rural areas, like drying harvested crops, harvesting wood, and producing furniture. These jobs provide women with an opportunity to participate in production activities, which will enhance gender equality and ultimately boost the rural economy. Additionally, AF can encourage the improvement of sociocultural aspects and participation of the community through group action, so that people can learn from one another, track down the strength of traditional wisdom and awareness, and then integrate these aspects with the creation of new awareness and technology (Octavia et al., 2022).

Although people have numerous benefits from AF practices, it is essential to rethink the interplay between nature and humans by stressing the common duty at landscape levels, especially for ecosystem services and biodiversity protection. Additionally, technology and community-based management are crucial factors in ensuring socioecological resilience and preserving biodiversity (Viñals et al., 2023). AF delivers numerous benefits for humans and the environment; however,

AF is not adjustable in every area. This is because AFs are influenced by soils, climate, crop types, markets, and local specificities, and therefore AF is not a universally applicable solution (Cialdella et al., 2023). However, when AFs are embraced and developed, these will offer numerous technical solutions that significantly affect farmers' earnings, national exports, national richness, and overall welfare (Cialdella et al., 2023).

25.3.1.4 *Institutional policies and support*

C finance drives policies and institutional reforms that encourage AF. For example, the integration of AF into the reduction of emissions from deforestation and forest degradation (REDD+) program that strengthens global climate action (Tirkey et al., 2024). In South Asia, regional cooperation and policy frameworks have promoted AF for sustainable development and climate mitigation (Dhyani et al., 2021).

25.3.2 Constraints

C financing in AF faces several constraints, including (1) high transaction cost, (2) insecure land ownership, (3) complexity and uncertainty, (4) dependence on external funding, and (5) limited awareness and capacity. The high transaction costs associated with C_{seq} measurement, monitoring, and verification burden smallholder farmers (Houghton, 2020; Cacho et al., 2003; Jindal et al., 2008). Intermediaries play an important role in connecting smallholder farmers with buyers of C credits (Clements and Moore, 2014). In Africa, smallholder projects in C trading schemes have low participation due to transaction costs (Jindal et al., 2008).

In Africa, insecure land ownership is a major obstacle to C investment due to the increased risk of local ownership displacement (Jindal et al., 2008). Likewise, in Indonesia, the lack of clarity on land ownership hinders C_{seq} projects (Seeberg-Elverfeldt et al., 2009a).

The complexity of C financing mechanisms hinders farmers from adopting them (Houghton, 2020). In addition, the uncertainty of C pricing and the legitimacy of C credits hinder farmers from participating (Awazi, 2025; Houghton, 2020; Pradhan et al., 2023). C financial markets are highly volatile due to institutional decisions, energy prices, and weather events, making it difficult to set a stable and predictable C price (El Amri et al., 2022). Studies in Germany and Australia show that the variability of C prices in the market will affect the profitability of AF practices (Flugge and Abadi, 2006; Plieninger, 2011). For this reason, the need for pre-financing and derivative C market agreements. Innovative solutions and external funding are needed to overcome these obstacles (Foster and Neufeldt, 2014).

C finance depends on sources of external financing that are difficult to predict and follow market fluctuations that create sustainability challenges if funds are in trouble (Awazi, 2025; Houghton, 2020). In Indonesia, this raises concerns about the long-term sustainability of AF projects (Seeberg-Elverfeldt, 2009).

Limited knowledge, awareness, and capacity gaps lead to low AF adoption despite the availability of financial incentives (Springgay and Pajel, 2024; Torquebiau, 2024). In Nepal, the adoption of AF is limited by the lack of knowledge and technical support among farmers (Dhakal, 2013).

25.4 Plans for implementing carbon financing approaches in agroforestry

25.4.1 Policy development strategies

Implementing the C finance approach in AF requires a multifaceted policy development strategy that integrates social, economic, and environmental considerations. The key to this strategy is to reduce transaction costs and increase farmers' participation in the C market (Cacho et al., 2003). The CDM Framework offers emission reduction and sustainable development that must also be designed to address social issues such as livelihoods and ecosystem restoration (Smith and Scherr, 2002). Credit-based policies can offset the costs of building AF and encourage farmers to adopt it (Awazi, 2025). Meanwhile, the sustainability mechanism finds that international platforms are crucial for attracting investment and that a policy framework is needed to promote low-C agricultural practices and integrate sustainability standards (Martins et al., 2024). In Indonesia, Seeberg-Elverfeldt (2009) showed that payments for environmental services (PES) can be an incentive for farmers to adopt shade-intensive AF. The Food and Agriculture Organization (FAO) assessment emphasizes the need for policy support, enhanced market access, and financial incentives to overcome barriers to AF adoption (Springgay and Pajel, 2024). In India, leveraging technologies and community approaches, such as remote sensing and the Farmer Producer Organization, can facilitate the monitoring and implementation of efficient AF projects, aligned with C neutrality goals (Ghosh and Sharma, 2024; Meena et al., 2025). India's National Action Plan on Climate Change (NAPCC) also underscores the potential of AF in CC adaptation and mitigation, although it highlights the need for more precise strategies and budget forecasts (Datta et al., 2024). In addition, C income from AF can significantly increase farmers' incomes, as seen in a study from India, where C finance projects can increase incomes by 40%–50% (Singh et al., 2024). However, challenges, such as high transaction costs, technical and financial barriers, and the need for a reliable C market framework, remain significant barriers (Houghton, 2020). Addressing these challenges through policy support, capacity building, and community engagement is critical to scaling up AF as a viable C financing strategy globally (Verma et al., 2024).

An important step toward developing a more accessible AF-based C market for local communities requires several strategies (Pancholi et al., 2023; Tavernier et al., 2024; Telwala, 2023). First, training local communities on AF and the importance of C markets, which includes planting techniques, plant maintenance, and how to calculate C emissions. Second, collaboration with universities or research institutions to conduct studies on the C_{seq} potential of AF practices applied by local communities. Third, the development of business models that enable local communities to engage in C market projects can be achieved by forming farmer groups that manage land together and share the proceeds. Fourth, consider seeking financing sources from governments, international institutions, or donors that support sustainability projects. Fifth, ensuring that AF practices are carried out appropriately to C certification standards so that communities can sell C credits at a reasonable price. Sixth, helping communities market C credits, including establishing partnerships with companies that want to offset their emissions. Seventh, strengthening local policies by encouraging policies that support the development of AF-based C markets, including incentives for the communities involved.

25.4.2 Operational measures

The development of AF-based C financing for a green economy requires planning that depends on the situation and conditions of the region, as well as the characteristics of the community. There are numerous guidelines, best practices, and directions for operational steps in C financing and trading in general that have been issued by credible institutions both nationally and internationally. Several operational steps that can be used as a reference, are (1) identification and selection of project locations for C financing for AF, which includes criteria for selecting target locations and communities, (2) capacity building and training, which includes educational programs to improve farmers' skills, understanding C financing, and increasing human and institutional capacity, (3) monitoring and evaluating mechanisms that are a system to measure project performance and impact on an ongoing basis (Dumas et al., 2024).

At the grassroots level, access to C finance for farmers is key to harnessing the potential of AF in mitigating CC and improving local economic well-being. However, to be able to access C finance, farmers must go through a series of procedures and meet various requirements designed to ensure the effectiveness of C projects, from implementation to verification, and C credit trading. Additionally, the roles of financial institutions and non-governmental organizations (NGOs) are vital in facilitating access and supporting farmers in implementing these projects.

To obtain C finance, farmers or farmer groups must first register their AF projects on available C schemes, including both voluntary C markets and regulated C markets. The first step that is usually required is the preparation of a project proposal that details an action plan to increase C_{seq} through AF, including the types of trees to be planted, the area to be restored, and the land management method to be employed. In many cases, farmers will require technical support from environmental experts or other relevant agencies to ensure that their projects are both scientifically and operationally feasible.

After that, the project must undergo a verification process by an accredited body to ensure that it will indeed absorb the estimated amount of C. The main requirements in this process include preliminary data on land conditions, an estimate of the amount of C that can be absorbed, and appropriate measurement methods. In addition, farmers or project managers must demonstrate their ability to maintain the project over the long term, given that C_{seq} typically occurs over many years and requires a commitment to sustainable management.

When verification is complete and C credits are generated, they can be traded in the C market. Farmers or project managers must ensure that they meet sustainability standards set by the market, such as the Verified Carbon Standard (VCS) or the Gold Standard, which often also entails the positive social impact of the project, including improvements in community well-being and biodiversity conservation.

Financial institutions and NGOs play a crucial role in facilitating access to finance for farmers, especially for smallholders who often face various technical and financial barriers. Financial institutions, such as development banks, microcredit institutions, or green investment institutions, can provide the initial financing that farmers need to start their AF projects. Many AF projects require initial investment in seedling purchases and land infrastructure improvements, as well as significant technical support before C_{seq} can be measured and verified (Shames et al., 2016).

NGOs often act as intermediaries between farmers and C markets, providing technical support and education to farmers on how to implement AF in accordance with global C standards. They also help farmers understand the complex procedures for accessing C finance, including navigating verification and certification requirements. Some NGOs are also collaborating with financial institutions to offer soft credit to farmers, enabling them to initiate their projects without incurring high financial risks.

In addition, NGOs can play a role in building a network of farmers and land managers to bring together small projects into one larger unit (Boyd et al., 2007). This is referred to as bundling or aggregation, where several small projects are combined to meet the economies of scale required to participate in the C market. This model is crucial for smallholders who may lack sufficient land or resources to generate C credits at scale, but can join groups to maximize their impact and gain access to global markets.

On the other hand, financial institutions also play a role in providing a secure and transparent payment mechanism (Campiglio, 2016). With transactions involving various parties, such as C credit buyers, farmers, and verifiers, a well-structured financial system is needed to ensure that farmers receive payments on time and in accordance with the market value of the C credits they generate.

The incentive mechanism in C financing in the field can be carried out with 3 systems that are adjusted to the conditions of farmers (Raina et al., 2024), which are (1) results-based payments. This involves payments that depend on the actual amount of C_{seq}, which requires strict monitoring and verification. They can be implemented through auctions, product label credits, or C credits. This mechanism encourages innovation and connects farmers to C markets, but it also poses challenges due to the need for strict monitoring standards. (2) Action-based payments; it is simpler and involves paying for specific actions taken by farmers, such as adopting certain sustainable practices. They require less monitoring and can be paid for upfront or after implementation, making them more accessible to farmers with limited resources. (3) Hybrid payments; combining elements of outcome-based payments and action-based payments, the hybrid mechanism offers low-risk and guaranteed payments while ensuring environmental benefits. This approach can motivate farmers to participate by providing a balance of security and incentives.

25.5 Policy arena for economic upliftment through carbon financing in agroforestry

25.5.1 Analysis of the current policy frameworks

The C trading market emerged as a historic moment for energy savings and emission reductions based on the framework of the Kyoto Protocol (Weng and Xu, 2018). Under the Kyoto Protocol, market mechanisms are GHG emissions by making CO_2 emissions a tradable commodity, viewed as a new approach to reducing CO_2 emissions. Several countries have issued policies to facilitate C trading with varying objectives, economic conditions, and environmental commitments. The main C trading markets are the EU-ETS, the Regional Greenhouse Gas Initiative (RGGI) of the US, California's Cap-and-Trade Program, the Tokyo Cap-and-Trade Program, New Zealand Emissions Trading Scheme, and South Korea's C trading program (Weng and Xu, 2018). The EU-ETS is the C trading scheme with the largest trading volume in the world, with a contribution of 90% of the global C trading scale (Dechezleprêtre et al., 2023). China has also developed its own C trading market since 2017, which is expected to contribute more to energy conservation and emission reduction (Weng and Xu, 2018).

The EU-ETS is one of the most established and ambitious C trading markets, covering more than 11,000 facilities in Europe and setting total GHG emission limits. The EU-ETS, launched in 2005, has allocated tradable emission permits to ±12,000 power plants and industrial plants across 31 EU countries and accounts for 40% of the EU's total GHG emissions (Dechezleprêtre et al., 2023). Some states in the US, such as California, also have their own C trading systems that are relatively similar to the EU-ETS, namely California's Cap-and-Trade Program, which covers the industrial and transportation sectors, to gradually reduce emissions.

China developed a national C trading system that initially included the power generation sector. China's C trading system has achieved remarkable results in volume and performance (Weng and Xu, 2018), as well as encouraged the development of C financing and reducing C emissions (Guo et al., 2022). However, the C trading system in China faces some significant gaps in terms of transaction efficiency, market supervision, development of relevant legislation, talent needs, and C finance product innovation (Zhou and Li, 2019). The development of a C trading system is also carried out by Canada by implementing a C pricing system that includes a C tax and an emissions trading system in several provinces. Japan has several C trading initiatives at the regional level, although there is no comprehensive national system. Some major cities, such as Tokyo, are implementing local emissions trading programs. Australia previously had a C trading system, but it was scrapped in 2014. Since then, there have been efforts to redevelop a policy framework that is more supportive of emissions reductions.

Indonesia has committed to C trading with an implementation mechanism using the cap-and-tax method combined with a C tax. Coal-fired steam power plants (PLTU) are the first subject to C tax because they are considered to produce the highest C emissions compared to other sectors. Indonesia has also developed policies to support C trading, including the National Action Plan for Reducing Greenhouse Gas Emissions (RAN-GRK), REDD+ scheme, and presidential regulation No. 98 of 2021. RAN-GHG aims to reduce GHG emissions across various sectors, including forestry, energy, and industry. REDD+ opensopportunities for C trading through the recognition of C credits from forest conservation projects. Meanwhile, Presidential Regulation No. 98 of 2021 establishes a framework for the implementation of C trading in Indonesia, which includes regulations on monitoring, reporting, and verification of emissions, as well as mechanisms for C trading. Indonesia's commitment to C trading is shown by increasing the use of renewable energy. Renewable energy investments can generate tradable C credits. Indonesia is also initiating a reduction in emissions from the transportation sector through the development of electric vehicles and a more efficient public transportation system. Indonesia is also active in various international forums, such as the Conference of the Parties (COP), to strengthen its position within the framework of global CC and C trading.

The CDM, as an implementation of the Kyoto Protocol, has encouraged developed countries to invest in C financing projects in developing countries, including AF initiatives. This framework enables farmers to receive compensation for the C stock sequestered in their plantations, providing financial incentives for sustainable practices (Agbotui et al., 2023; Luedeling and Neufeldt, 2012). In addition, the establishment of a Voluntary Carbon Offset (VCM) significantly increases the potential income by 40%–50% when engaging in C trading (Singh et al., 2024). However, the successful implementation of AF-based C finance is hampered by the lack of institutional frameworks, reliable sources of C finance, and the involvement of public and private-sector actors (Nyberg et al., 2020; Shames et al., 2016). Therefore, it is necessary to adopt a participatory approach that fosters social capital and encourages multistakeholder partnerships to enhance the inclusivity and effectiveness of C finance programs (Branca et al., 2021).

The economic viability of AFs is also related to the C income they can generate. C payoff can significantly increase AF profits compared to monoculture farming, making it an attractive option for smallholders (Waldén et al., 2020). However, to realize these benefits, effective management systems, high C prices, and strong institutional support are essential to reduce the transition and transaction costs associated with switching to AF practices (Verma et al., 2024; Waldén et al., 2020). In addition to economic incentives, policies to promote AF must also pay attention to social and environmental dimensions, such as community involvement in the design and implementation of AF projects to produce better benefits in C_{seq} and local livelihoods (Bhattarai et al., 2023; Sutrisno et al., 2024). In addition, integrating AF into broader CC strategies, such as REDD+ programs, can improve its effectiveness by aligning local practices with global C-reduction goals (Sutrisno et al., 2024).

25.5.2 Role of government and related institutions

25.5.2.1 *Supporting regulations and legislation*

The development of laws and regulations that facilitate C financing, particularly in the context of AF, is essential to improve C_{seq} and promote sustainable agricultural practices. The dual benefits of AF, which increase C_{seq} while providing economic benefits, emphasize the need for a regulatory framework that supports the provision of incentives for such practices. In terms of policy frameworks, C finance mechanisms, such as REDD+ programs, have demonstrated potential in promoting sustainable land-use practices, including AF. However, the effectiveness of this mechanism depends on its design and implementation. REDD+ needs to be developed into policy incentives that can effectively support AF initiatives (Loft et al., 2015). The tree planting program in Australia, which began in 2012, serves as a case study illustrating how a structured C credit system can incentivize land management practices that increase C_{seq} (Evans, 2018; Verschuuren, 2017). Ultimately, incentives for AF-based tree planting by communities must be supported by a combination of coherent and complementary policies to facilitate long-term participation and tangible environmental outcomes.

The integration of C finance into AF practices requires careful consideration of the key challenges in the C credit market, including additional properties, durability, and leakage. Therefore, effective policy design must be able to address these challenges to ensure that C credits generated through AF are credible and contribute meaningfully to emission-reduction targets (Thamo and Pannell, 2016; Pradhan et al., 2023). On the other hand, AF-based C financing requires robust monitoring and verification systems to track C stocks and ensure compliance with regulatory standards (England and Viscarra Rossel, 2018).

25.5.2.2 *Interinstitutional collaboration*

Interagency collaboration, including synergies between governments, research institutions, and international organizations, is critical in the development of AF-based C finance. Case studies on coffee AF in Indonesia show that farmers are often faced with limited capacity, capital, and networks, so collaboration is absolutely necessary to create an ecosystem that supports the success of AF-based C financing. Collaboration in AF-based C financing can be realized in several actions. First, integrated policies, namely cooperation between sectors in government, such as agriculture, forestry, and the environment, to formulate policies that support AF and C financing. Second, research and development, which is realized through collaboration between research institutions and the government to develop new technologies and methods related to efficient AF practices in storing C. The third relates to sources of financing, which can be realized through collaboration with international organizations that can provide funds and technical resources to support AF projects. Fourth, collaboration in training programs for farmers and other stakeholders to increase understanding of AF practices and the benefits of C financing. By creating strong synergies between various parties, AF-based C financing projects can run more effectively and sustainably.

25.5.3 Policy strategies for economic improvement

The development of a C market that is more accessible to local communities, especially those involved in AF, is crucial for enhancing environmental sustainability and socioeconomic resilience. However, the realization of these benefits depends on the establishment of clear land tenure rights and market infrastructure that facilitate participation in C trading (Verma et al., 2024).

Apart from AF making an essential contribution to improving the nutrition of local communities (Duffy et al., 2021), diversifying income sources and increasing resilience to the impacts of CC (Duffy et al., 2021; Ghimire et al., 2024). However, there are still challenges in implementing AFs and integrating them into C markets. Local communities often face obstacles, such as inadequate infrastructure, lack of access to information, and inadequate support to adopt new practices (Cole, 2010; Mathur et al., 2014). To address these challenges, a multi-faceted approach is needed that includes capacity building, community engagement, and the development of supporting policies that recognize the value of ecosystem services provided by AF (Mathur et al., 2014; Verma et al., 2024; Rahman et al., 2012).

25.5.4 Successful policy case studies

AF-based C financing policies demonstrate that this approach can enhance the economic and welfare of rural communities. AF provides access to communities around forests to engage in emission-reduction initiatives without losing access and benefits from forests (Holmes et al., 2017). In this section, various examples of implementation and successful integration of AF with C financing in various countries will be described. The practice of preserving AF has proven to be an effective C offset in Germany and suggests that measures to realize C_{seq} projects should include the collection of empirical evidence on the C_{seq} potential of AFs, as well as the development of localized demonstration projects to participate in existing C markets (Plieninger, 2011). The integration of AF practices with C financing in Germany has been shown to contribute to increased awareness of the environmental benefits of AF and provide economic incentives for farmers to invest in sustainable practices (Plieninger, 2011).

The next example of success comes from Lore Lindu National Park (LLNP), Indonesia. The rate of deforestation in LLNPs converted to smallholder cocoa plantations has been proven to be reduced by C financing policies with C prices received by farmers (Seeberg-Elverfeldt et al., 2009b). Furthermore, case studies on local communities in Panama offer a positive example of the role of AF in helping communities surrounding forest areas to engage in REDD+ (Holmes et al., 2017). AF-based C financing in Sub-Saharan Africa through REDD+ has been reported to benefit smallholders, with implementation having to be community-based (Thangata and Hildebrand, 2012). C financing with CO_2 payments has been shown to stimulate the attractiveness of farmers to develop cocoa AF in Ghana (Agbotui et al., 2023).

Based on a case study in Africa, Foster and Neufeldt (2014) identified the main challenges and opportunities in the AF-based C market in the form of a very high need for external funding at the beginning of a project and the need for innovative solutions to minimize the trade-off between livelihood goals and the environment. AF-based C financing should also emphasize the non-C benefits by using C income as a tool to help farmers transition to more sustainable and productive practices on their land (Foster and Neufeldt, 2014). Ultimately, achieving success in AF-based C financing requires community participation, financial incentives, international partnerships, and supportive policies (Campiglio, 2016; Dumas, 2023; Kiremu et al., 2022). By implementing the right policies and involving various stakeholders, these countries have succeeded in improving their economies while preserving the environment through C financing in AF.

25.6 Future roadmap

25.6.1 Future trends and opportunities

25.6.1.1 Technological innovation

Various approaches and methods have been developed by researchers for spatial C estimation with satellite imagery. Lifecycle and multicriteria analyses are good choices because they pay attention to the overall C cycle on various related elements (Grieco et al., 2024). The spatial decision support system is another approach that can be used to determine more comprehensive planning, especially for new areas developed for C trading (Keenan and Jankowski, 2019). This approach will support the perception of the development of new AF areas oriented toward potential buyers for C trading (Matilainen et al., 2023). Technological developments have encouraged the growth of new approaches to existing C estimation and its dynamic processes. Massive analysis through big data can also be involved, for example, in AgriBigCAT analysis, which allows decision-makers to include the wide availability of data in preparing future AF development plans (Kamilaris et al., 2018).

An accurate C estimation is key in the implementation of C trading between the two parties. Cutting-edge technology can reduce the information gap between providers and potential buyers, while also reducing calculation costs. The wider public access to the internet also has a positive impact on the emergence of various voluntary movements involved in MRV of C trading. Access to technology even allows people to have devices that can calculate emissions and C needs around their homes (Santier, 2023). Meanwhile, in the forestry sector, the use of satellite images is a method that contributes a lot to C estimation (Chen et al., 2022). However, satellite image analysis also requires a combination with an accurate and effective field calculation approach in estimating C storage and release in an area (Tabor and Connell, 2019). Digitalization is not only limited to the scope of spatial analysis, but also in a series of MRV (D-MRV) processes currently being developed by many parties, so that the transaction process becomes more efficient and encourages maximum implementation (World, 2022).

Certification of the use of clean energy or C_{seq} claims is an added value that buyers are aiming for. Certificates, as a form of recognition, continue to evolve in form, with the aim of facilitating public access and traceability. One form, which has been developing in recent years is smart contracts that use blockchain technology (Andela, 2025). Some countries have begun to examine the application of this form of contract, such as China and the EU, mainly because of the superior traceability of the certificate holder to the party that issued the certificate (Vilkov and Tian, 2023; Zhang et al., 2023). The integration of smart contracts also opens opportunities for easy C service transactions to be multiparty on the seller or buyer side of C services (He et al., 2024; Parhamfar et al., 2024). The decentralization of party involvement in this contract model presents a highly effective opportunity for C trading from forest areas, particularly in AF areas. The community, through farmer groups and the assistance of financial and banking institutions, can be directly involved with the buyers of C services. Apart from its advantages that tend to be efficient, the level of technology mastery is one of the aspects that need to be considered by service providers and buyers in the implementation of smart contracts (Vionis and Kotsilieris, 2024). For this reason, in the implementation carried out at the community level, it is necessary to have an intermediary institution in place that can be taken over by the government's technical implementation unit or technical implementer at the local government level. It is also necessary to establish a standard C pricing mechanism and a harmonized C standard market that can increase the credibility and effectiveness of C credits (Honkomp and Schier, 2024; Soloviy and Panchyshyn, 2024).

25.6.1.2 Changes in carbon market dynamics

The development of the C market is progressing dynamically, with some countries advancing at a significantly faster pace than others. Growth can be characterized by an increase in the number of countries that have expressed their desire to engage in C market trading (World, 2024). In addition to the state, various multinational companies have also begun to express interest in participating in the C market (Icap, 2024). Several international regions have begun to adopt C markets in their countries. Generally, the C market mechanism adopts two ways, which are through C taxes and emissions trading (Blanton et al., 2024). Another mechanism currently widely implemented is the use of credits for C storage and incentives provided by buyers to industries to adopt low-C development practices or implement emission reduction (Bherwani et al., 2024). Each mechanism has advantages and disadvantages and is greatly influenced by the agreement between the two parties. Its implementation has been carried out in various ways, starting from those that have been implemented a decade ago, for example, in Mexico (Blanton et al., 2024). Although some countries have implemented it for a long time, many other countries are still considering the policy (Pan et al., 2024). Various international institutions in each country also support initiatives to implement C markets.

From the perspective of engagement, C markets can be divided into two, namely mandatory and voluntary (Blanton et al., 2024). Mandatory C trading is present in countries with quite massive multinational company activities, so the company's response to C trading is positive. Strengthening market trust, especially for companies operating in many tropical forested countries, is a common motive. Voluntary C markets also exist in various forms, one of which is through the trade of environmental services, and have been implemented in several countries. The successful implementation of C market trading also involves several financial institutions, especially multinational banks, by providing schemes and encouraging initiatives from companies.

25.6.2 Recommendations for further development

25.6.2.1 Strengthening the policy framework

Weng and Xu (2018) argues that several steps need to be taken to strengthen the C trading system, including harmonization of C trading standards, increased openness and accessibility of information, and increased liquidity in the C market. Harmonization of trade standards, including accurate C estimates, requires a policy basis that can be carried out between countries within a geographical area, such as countries in Latin America, or between provinces that have been carried out by China (Blanton et al., 2024). The development of digitalization technology, as explained in the previous chapter, helps the distribution of information, as well as the liquidity of the C market. In addition to blockchain systems for smart contracts, C exchange-traded funds (ETFs) that have been tested in several countries are thought to be able to prevent the influence of policy uncertainty on C trading (Naysary and Shrestha, 2024). The application of this digitalization technology must be ensured in the policy of the C trading mechanism established between the two parties.

Steps to improve and strengthen policies in C trading, especially AF, are quite casuistic, depending on the systems or mechanisms that have been implemented to the extent that progress has been implemented. Until now, various countries have been debating about a more appropriate and efficient C trading mechanism between C taxes or emissions trading. The results of the study (Pan et al., 2024) state that there needs to be an integration of the two schemes so that they can be more adaptive to the conditions of the provider or buyer country. The subsidy mechanism over parties that can reduce C emissions, for example, by practicing environmentally friendly land cultivation or being able to reduce C emissions, also needs to be encouraged because it is easier to apply within a country's economic activities (Liu et al., 2024). Even the integration of the C trading mechanism with other policies related to CC mitigation is felt to be a focus for country leaders so that the policies taken become more efficient (Yu et al., 2024; Zhou et al., 2024).

Adaptivity is a feasible approach for every country to implement C trade. Initial policies or pilot projects can serve as the basis for developing larger, more comprehensive policies that impact all parties involved. From C trading projects that are innovative, the government can then implement strategic steps to accelerate policy implementation in larger aspects, such as across regions or fields. Even highly impactful policies, such as changing the design of renewable energy development plans, are felt to strengthen the implementation of C trading reciprocally between buyer and supplier (Wang et al., 2024). The state can also establish a separate institution that has a set of powers (control, allocation, and reporting and verification) or ensure that the function runs on an existing institution (Weng and Xu, 2018).

25.6.2.2 Increasing international cooperation

International cooperation in C trading is key to achieving a common target toward cleaner development, especially in the natural resource management and forestry sectors. The development of responses from each country and advancements in science have brought significant progress toward this step, but there are still challenges in ensuring an efficient and effective global C trading system. Forest ecosystems have their own unique characteristics and influencing factors in their C cycles. Unlike forests in developed countries, developing countries will be highly dependent on the ownership of natural resources, including forests, so decarbonization will be greatly affected by economic and social activities (Fu and Xu, 2023). Meanwhile, in the case of developed countries, decarbonization of forest areas tends to be dominated by natural factors, so preventive activities are very relevant (Dye et al., 2024). Differences in natural factors and knowledge are an opportunity for each country to be able to complement each other to realize fair C trading.

Collaboration between countries cannot develop perfectly when the institutions involved in C trading, both in the service provider system and the buyer entity, are not strong. Service providers or providers need a series of institutional functions ranging from those that maintain storage to institutions that have monitoring or evaluation functions (Santier, 2023). This position will be highly relevant to the buyer as it offers one of the advanced services needed. In some countries, this function is held by financial institutions operating in the country, such as national banks or private banks

(Babic, 2024). A series of institutional roles in realizing C trading are also dynamic, such as shifting from those initially held by national institutions to later being smaller and more efficient, like community groups.

Strengthening the role of community groups, including in forestry and AF activities, has been found to exist in various forms of resource management (Arifin et al., 2023; Kumar et al., 2020, 2024). Apart from the strong and long-standing interaction and trust between members, the group's ability is also supported by policies from the government in encouraging the strengthening of institutions so that they are able to adopt financial technology. AF, as a form of forest management, also has various models ranging from small-scale carried out around residences to large scales that are present in the form of combined groups (Muthuri et al., 2023). These various forms and uses result in policies made by the government that must be able to openopportunities for wide involvement.

Until now, several international institutions have made efforts to establish a common forum for the development of AF, including for C trading. The FAO has argued for the potential for small-scale forest resource management that needs to be involved in C trading since 2010 (FAO, 2010). To date, FAO has also facilitated several countries in developing various policies needed to realize C trading (Food and Agriculture, 2023). Another global collaboration that actively provides information and knowledge, regardless of administrative boundaries, is the Consultative Group on International Agricultural Research (CGIAR) through its Research Program on Forests, Trees, and Agroforestry (FTA). This forum continues to seek sustainable land management practices, especially to reduce GHG emissions in developing countries (Ngaiwi et al., 2024). The transfer of science and technology is always carried out in various forms of communities with the object of developing the skills of each individual involved in land management (Müller et al., 2024). In addition to the individual or community level, technology transfer is also carried out at a larger organizational level, for example, through group portfolio management so that environmental service rewards can be managed more effectively (Schut et al., 2024). These various multinational involvements are evidence that strengthens the need for collaboration between parties, with the advantages of their respective resources, so that AF management practices for C trading can be more easily implemented widely.

25.6.3 Future research agenda

The opportunity to adopt C trading for AF practices remains open for implementation. The potential for broad community involvement is one of the advantages, although the issue of skills and knowledge is one of the aspects that also needs to be considered (Raina et al., 2024). The following outlines some other aspects that are still a concern and can be the focus of future research.

Although it has a significant C storage function, forest management activities also experience C storage and release dynamics during their utilization. This is relevant to the pattern of AF utilization, although it is very rare to find a land-clearing method in practice. C dynamics is one aspect that still needs to be studied further, especially in AF-based land management practices. Several studies in other fields have been carried out with a risk assessment approach, namely, accurately measuring the C cycle at multiple levels (Grieco et al., 2024).

In addition to the C payment mechanism obtained by providers, one of the aspects that needs to be researched in the future is how to ensure that the benefits of C trading that have been obtained can be felt reciprocally in areas that are C receiver (Arifin et al., 2023). In addition, the value of C trading can also be felt by the community in the surrounding forest area or the area where forest development is possible. This is very relevant to AF practices that are very closely utilized with marginalized community groups, even though they provide ecosystem services for multiparties. There are still many studies related to communities or those around forest areas that play a role as providers. For example, the C trading mechanism for AF that guarantees its benefits can have a direct impact on the development of the community itself (van Noordwijk, 2021).

Another aspect that warrants further investigation is the community's perception and motivation. Especially in various scenarios that may be faced, such as limitations on land use or contract costs to ensure the sustainability of environmental services. According to Raina et al. (2024), high contract and financing mechanisms are still the main obstacles in the implementation of C trading globally. The business model that still needs to be studied is whether business entities in C trading can be carried out at the community level, and what needs to be prepared by the government so that later its role can be decentralized for more efficient transactions (Arifin et al., 2023). Existing models and mechanisms are still inefficient in paying existing C incentives, especially in involving farmers or communities directly (Raina et al., 2024).

One possible scheme to ensure the distribution of benefits for C trading services is to shorten the benefit distribution channel between providers and buyers. Like CSR, C trading in the future is felt to be directly carried out with a public–private partnership (PPP) scheme. Not only services for C_{seq}, but PPP also has the potential to be developed into various types of ecosystem services provided by public entities. Although the case in Italy highlights the need for government agencies to play a role in the early stages, especially in ensuring the ability of providers to also regulate when the recipient will be directly confronted with community groups (Gaglio et al., 2023). This assistance is primarily due to the potential costs associated with cooperation agreements and the lack of public awareness in conducting cooperation activities with external parties (Agarwal et al., 2023).

From the perspective of economic transactions, demand is a crucial aspect that needs to be considered in business, including for C trading. In addition to provider preferences, identifying potential buyers is crucial for further research, including their motives and the prices they are willing to pay. With the opening of buyer entities, such as a country or company, the diversity of purchase motives also has the opportunity to become higher. For this reason, various systems are needed for the purchase of C trading services that are fair so that it does not become a greenwashing action for companies (Blanton et al., 2024). This identification is also important to determine if there is a spatial relationship between providers and service buyers, as the C cycle is also influenced by the spatial dimension.

25.7 Conclusion

AF practices are recognized as playing a crucial role in addressing environmental challenges while enhancing agricultural productivity and ecosystem services. However, AF adoption faces various challenges, such as sociocultural resistance, knowledge gaps, policy support, and financing. High initial financing of AF, risks, and competition for resource inputs between tree and agricultural crops hinder large-scale implementation. One of the sources of financing for AF development is C financing. This chapter outlines the C finance mechanisms available to farmers, and the incentives designed to encourage participation in C schemes. There are several financing models relevant to AF in the context of C financing, including the voluntary C market, the compliance C market, RBF, and green bonds.

Mechanisms of C financing for AF include the use of C offsets, where AF projects can generate C credits that are stored in trees and soil. PES are made by providing incentives to farmers to adopt more sustainable land use practices. Another mechanism uses C credits to finance AF projects through the CDM scheme or Plan Vivo. This type of mechanism provides support to smallholders in implementing AF practices that sequester C and improve their livelihoods. Farmers must go through a series of procedures and meet various requirements to be able to access C finance designed to ensure the effectiveness of C projects, from implementation, verification, to C credit trading. Additionally, the roles of financial institutions and NGOs are vital in facilitating access and supporting farmers.

Implementing a C finance approach in AF requires a multifaceted policy development strategy that integrates social, economic, and environmental considerations. The key to this strategy is to reduce transaction costs and increase farmers' participation in the C market. In addition, the role of the government and institutions is related to the development of supportive regulations and legislation, and collaboration between institutions. Strategic options that can be implemented include integrating AF-based C financing with rural development and promoting local and regional C markets. This chapter also highlights several future trends and opportunities in the integration of C finance into AF, including technological innovation and shifting C market dynamics. Meanwhile, recommendations for further development that can be implemented include strengthening the policy framework and enhancing international cooperation.

References

Adhisankaran, K., Manivannan, V., Sakthivel, N., Balachandar, D., Ragunath, K.P., Vasumathi, V., 2024. Agricultural carbon credits: a pathway to environmental sustainability. Plant Science Today 11 (4), 1200–1211. Available at https://doi.org/10.14719/pst.3504.

Agarwal, V., Malhotra, S., Dagar, V., Pavithra, M.R., 2023. Coping with public-private partnership issues: a path forward to sustainable agriculture. Socio-economic Planning Sciences 89, 101703. Available at https://doi.org/10.1016/j.seps.2023.101703.

Agbotui, D.K., Ingold, M., Wiehle, M., Buerkert, A., 2023. Can carbon payments improve profitability of traditional conventional and organic cocoa agroforests? A case study in the Eastern Region of Ghana. Agroforestry Systems 97 (5), 813–831. Available at https://doi.org/10.1007/s10457-023-00828-0.

Albrecht, A., Kandji, S.T., 2003. Carbon sequestration in tropical agroforestry systems. Agriculture, Ecosystems and Environment 99 (1-3), 15–27. Available at https://doi.org/10.1016/S0167-8809(03)00138-5.

Andela, N., 2025. The role of carbon markets in reducing carbon emissions and strengthening the land sink at scale. EGU General Assembly 2024, Vienna. https://ui.adsabs.harvard.edu/abs/2024EGUGA..2612591A/abstract.

Anderson, E.K., Zerriffi, H., 2012. Seeing the trees for the carbon: agroforestry for development and carbon mitigation. Climatic Change 115 (3-4), 741–757. Available at https://doi.org/10.1007/s10584-012-0456-y.

Arifin, M.R., Raharja, B.S., Aligarh, F., Nugroho, A., 2023. A systematic literature review on community financing: avenues for further research. Cogent Business and Management 10 (2). Available at https://doi.org/10.1080/23311975.2023.2211208.

Awazi, N.P., 2025. Driving sustainable agroforestry through carbon credit-based policies: realities and perspectives. Natural Resources Conservation and Research 8 (1), 10184. Available at https://doi.org/10.24294/nrcr10184.

Babic, M., 2024. Green finance in the global energy transition: actors, instruments, and politics. Energy Research & Social Science 111, 103482. Available at https://doi.org/10.1016/j.erss.2024.103482.

Bajracharya, R.M., Gautam, K.D., Dahal, R.N., Lal Shrestha, H., 2024. Diversified agroforestry for climate change adaptation and mitigation in the himalayan region: potential for achieving multiple benefits. In: Zhang, L., Wang, S., Liu, L. (Eds.), Mitigating Global Climate Change - Enhancing Adaptation, Evaluation, and Restoration of Mountain Ecosystems. IntechOpen. Available at https://doi.org/10.5772/intechopen.113157.

Banerjee, A., Meena, R.S., Jhariya, M.K., Yadav, D.K., 2021. Agroecological footprints management for sustainable food system. Springer Nature Singapore 514. Available at https://doi.org/10.1007/978-981-15-9496-0.

Banerjee, A., Jhariya, M.K., Bargali, S.S., Palit, D., 2023. Ecorestoration for Sustainability. Wiley-Scrivener, USA, pp. 1–441.

Banerjee, A., Jhariya, M.K., Raj, A., Mechergui, T., 2024. Ecosystem Management: Climate Change and Sustainability. Wiley, India, pp. 1–669. Available at http://doi.org/10.1002/9781394231249.

Bhattarai, N., Watanabe, T., Avtar, R., Karky, B.S., Thapa, R.B., 2023. Harnessing REDD+ for community involvement and equitable benefit distribution: insights from Dhankuta District, Nepal. Journal of Green Economy and Low-Carbon Development 2 (2), 58–71. Available at https://doi.org/10.56578/jgelcd020202.

Bherwani, H., Banerji, T., Menon, R., 2024. Role and value of urban forests in carbon sequestration: review and assessment in Indian context. Environment, Development and Sustainability 26 (1), 603–626. Available at https://doi.org/10.1007/s10668-022-02725-5.

Blanton, A., Mohan, M., Galgamuwa, G.A.P., Watt, M.S., Montenegro, J.F., McTavish, F., et al., 2024. The status of forest carbon markets in Latin America. Journal of Environmental Management 352. Available at https://doi.org/10.1016/j.jenvman.2023.119921.

Boyd, E., Gutierrez, M., Chang, M., 2007. Small-scale forest carbon projects: adapting CDM to low-income communities. Global Environmental Change 17 (2), 250–259. Available at https://doi.org/10.1016/j.gloenvcha.2006.10.001.

Branca, G., Arslan, A., Paolantonio, A., Grewer, U., Cattaneo, A., Cavatassi, R., et al., 2021. Assessing the economic and mitigation benefits of climate-smart agriculture and its implications for political economy: a case study in Southern Africa. Journal of Cleaner Production 285, 125161. Available at https://doi.org/10.1016/j.jclepro.2020.125161.

Cacho, O.J., Marshall, G.R., Milne, M., 2003 Smallholder agroforestry projects: potential for carbon sequestration and poverty alleviation. https://www.fao.org/4/ae039e/ae039e00.pdf.

Campiglio, E., 2016. Beyond carbon pricing: the role of banking and monetary policy in financing the transition to a low-carbon economy. Ecological Economics 121, 220–230. Available at https://doi.org/10.1016/j.ecolecon.2015.03.020.

Chen, J., Gao, M., Cheng, S., Xu, Y., Song, M., Liu, Y., et al., 2022. Evaluation and drivers of global low-carbon economies based on satellite data. Humanities and Social Sciences Communications 9 (1). Available at https://doi.org/10.1057/s41599-022-01171-y.

Cialdella, N., Jacobson, M., Penot, E., 2023. Economics of agroforestry: links between nature and society. Agroforestry Systems 97 (3), 273–277. Available at https://doi.org/10.1007/s10457-023-00829-z.

Clements, C., Moore, K.M., 2014. Exploring the meso-level of agricultural carbon finance projects. Sustainable Intensification to Advance Food Security and Enhance Climate Resilience in Africa. Springer International Publishing, United States, pp. 489–513. Available at http://doi.org/10.1007/978-3-319-09360-4_26.

Cole, R.J., 2010. Social and environmental impacts of payments for environmental services for agroforestry on small-scale farms in southern Costa Rica. International Journal of Sustainable Development & World Ecology 17 (3), 208–216. Available at https://doi.org/10.1080/13504501003729085.

Datta, P., Behera, B., Rahut, D.B., 2024. India's approach to agroforestry as an effective strategy in the context of climate change: an evaluation of 28 state climate change action plans. Agricultural Systems 214. Available at https://doi.org/10.1016/j.agsy.2023.103840.

De Stefano, A., Jacobson, M.G., 2018. Soil carbon sequestration in agroforestry systems: a meta-analysis. Agroforestry Systems 92 (2), 285–299. Available at https://doi.org/10.1007/s10457-017-0147-9.

Dechezleprêtre, A., Nachtigall, D., Venmans, F., 2023. The joint impact of the European Union emissions trading system on carbon emissions and economic performance. Journal of Environmental Economics and Management 118, 102758. Available at https://doi.org/10.1016/j.jeem.2022.102758.

Desmiwati, D., Veriasa, T.O., Aminah, A., Safitri, A.D., Hendarto, K.A., Wisudayati, T.A., et al., 2021. Contribution of agroforestry systems to farmer income in state forest areas: a case study of Parungpanjang, Indonesia. Forest and Society 5 (1), 109–119. Available at https://doi.org/10.24259/fs.v5i1.11223.

Devi, A., Jhariya, M.K., Raj, A., Banerjee, A., Singh, K.P., Singh, B., 2024. Acacia nilotica: a promising species for soil sustainability. Land and Environmental Management through Forestry. Wiley, India, pp. 339–353. Available at http://doi.org/10.1002/9781119910527.ch14.

Dhakal, A., 2013. Evolution, adoption, and economic evaluation of an agroforestry-based farming system with and without carbon values: the case of Nepal. Available at https://doi.org/10.13140/RG.2.2.34527.61602.

Dhyani, S., Murthy, I.K., Kadaverugu, R., Dasgupta, R., Kumar, M., Adesh Gadpayle, K., 2021. Agroforestry to achieve global climate adaptation and mitigation targets: are South Asian Countries sufficiently prepared? Forests 12 (3), 303. Available at https://doi.org/10.3390/f12030303.

Dmuchowski, W., Baczewska-Dąbrowska, A.H., Gworek, B., 2024. The role of temperate agroforestry in mitigating climate change: a review. Forest Policy and Economics 159, 103136. Available at https://doi.org/10.1016/j.forpol.2023.103136.

Duffy, C., Toth, G.G., Hagan, R.P.O., McKeown, P.C., Rahman, S.A., Widyaningsih, Y., et al., 2021. Agroforestry contributions to smallholder farmer food security in Indonesia. Agroforestry Systems 95 (6), 1109–1124. Available at https://doi.org/10.1007/s10457-021-00632-8.

Dumas, Y., Ubacht, J., van Andel, E., Keledorme, L.A., 2024. Exploring the transition to agroforestry for smallholder farmers: a feasibility study for the Ashanti region of Ghana. Frontiers in Sustainable Food Systems 8. Available at https://doi.org/10.3389/fsufs.2024.1493753.

Dumas, Y.C.M., 2023. Carbon credit incentives for agroforestry: a feasibility study for smallholder farmers in Ghana's Ashanti region [Master's thesis]. University of Ghana.

Dye, A.W., Houtman, R.M., Gao, P., Anderegg, W.R.L., Fettig, C.J., Hicke, J.A., et al., 2024. Carbon, climate, and natural disturbance: a review of mechanisms, challenges, and tools for understanding forest carbon stability in an uncertain future. Carbon Balance and Management 19 (1). Available at https://doi.org/10.1186/s13021-024-00282-0.

El Amri, A., Oulfarsi, S., Sahib Eddine, A., El Khamlichi, A., Hilmi, Y., Ibenrissoul, A., et al., 2022. Carbon Financial Market. IGI Global, pp. 424–445. Available at http://doi.org/10.4018/978-1-7998-8210-7.ch017.

England, J.R., Viscarra Rossel, R.A., 2018. Proximal sensing for soil carbon accounting. SOIL 4 (2), 101–122. Available at https://doi.org/10.5194/soil-4-101-2018.

Evans, M.C., 2018. Effective incentives for reforestation: lessons from Australia's carbon farming policies. Current Opinion in Environmental Sustainability 32, 38–45. Available at https://doi.org/10.1016/j.cosust.2018.04.002.

FAO, 2010. Environment and natural resources management working paper carbon finance possibilities for agriculture, forestry, and other land use projects in a. Forestry, pp. 38–38. http://www.fao.org/docrep/012/i1632e/i1632e.pdf.

Flugge, F., Abadi, A., 2006. Farming carbon: an economic analysis of agroforestry for carbon sequestration and dryland salinity reduction in Western Australia. Agroforestry Systems 68 (3), 181–192. Available at https://doi.org/10.1007/s10457-006-9008-7.

Food and Agriculture, O., 2023. Carbon credits trading (general) regulations. Available from: https://www.fao.org/faolex/results/details/en/c/LEX-FAOC223506/.

Foster, K., Neufeldt, H., 2014. Biocarbon projects in agroforestry: lessons from the past for future development. Current Opinion in Environmental Sustainability 6 (1), 148–154. Available at https://doi.org/10.1016/j.cosust.2013.12.002.

Fu, C., Xu, M., 2023. Achieving carbon neutrality through ecological carbon sinks: a systems perspective. Green Carbon 1 (1), 43–46. Available at https://doi.org/10.1016/j.greenca.2023.08.005.

Gaglio, M., Lanzoni, M., Goggi, F., Fano, E.A., Castaldelli, G., 2023. Integrating payment for ecosystem services in protected areas governance: the case of the Po Delta Park. Ecosystem Services 60, 101516. Available at https://doi.org/10.1016/j.ecoser.2023.101516.

Gemeda, W.K., Deresu, Z., 2023. The role of agroforestry system in reducing farmers' exposure to climate change. Journal of the Selva Andina Biosphere 11 (2), 186–197. Available at https://doi.org/10.36610/j.jsab.2023.110200186.

Ghimire, M., Khanal, A., Bhatt, D., Dahal, D.D., Giri, S., 2024. Agroforestry systems in Nepal: enhancing food security and rural livelihoods – a comprehensive review. Food and Energy Security 13 (1). Available at https://doi.org/10.1002/fes3.524.

Ghosh, S., Sharma, J.V., 2024. Advancing India's agroforestry potential: leveraging technology and community approaches for sustainable development. Asian Research Journal of Agriculture 17 (4), 1053–1058. Available at https://doi.org/10.9734/arja/2024/v17i4619.

Grieco, R., Cervelli, E., Bovo, M., Pindozzi, S., Scotto di Perta, E., Tassinari, P., et al., 2024. The role of geospatial technologies for sustainable livestock manure management: a systematic review. Science of The Total Environment 954, 176687. Available at https://doi.org/10.1016/j.scitotenv.2024.176687.

Guo, Q., Su, Z., Chiao, C., 2022. Carbon emissions trading policy, carbon finance, and carbon emissions reduction: evidence from a quasi-natural experiment in China. Economic Change and Restructuring 55 (3), 1445–1480. Available at https://doi.org/10.1007/s10644-021-09353-5.

He, Q., Wang, J., Shi, R., He, Y., Wu, M., 2024. Enhancing renewable energy certificate transactions through reinforcement learning and smart contracts integration. Scientific Reports 14 (1). Available at https://doi.org/10.1038/s41598-024-60527-3.

Holmes, I., Kirby, K.R., Potvin, C., 2017. Agroforestry within REDD+: experiences of an indigenous Emberá community in Panama. Agroforestry Systems 91 (6), 1181–1197. Available at https://doi.org/10.1007/s10457-016-0003-3.

Honkomp, T., Schier, F., 2024. Scoping review of carbon pricing systems in forest sector models. Environmental Research Letters 19 (1). Available at https://doi.org/10.1088/1748-9326/ad101d.

Houghton, E., 2020. Carbon offsetting in agroforestry systems, Nicholas School of the Environment and Earth Sciences. https://hdl.handle.net/10161/33099.

Icap, 2024. Emissions trading worldwide: status report 2024. Available from: https://icapcarbonaction.com/system/files/document/240522_report_final.pdf.

Jhariya, M.K., Meena, R.S., Banerjee, A., 2021b. Ecological Intensification of Natural Resources for Sustainable Agriculture. Springer, Singapore, pp. 1–655. Available at https://doi.org/10.1007/978-981-33-4203-3.

Jhariya, M.K., Meena, R.S., Banerjee, A., Meena, S.N., 2022. Natural Resources Conservation and Advances For Sustainability. Elsevier. Available at https://doi.org/10.1016/C2019-0-03763-6.

Jhariya, M.K., Banerjee, A., Swaroop Meena, R., Kumar, S., Raj, A., 2021a. Sustainable Intensification for Agroecosystem Services and Management. Springer, Singapore, pp. 1–870. Available at https://doi.org/10.1007/978-981-16-3207-5.

Jhariya, M.K., Meena, R.S., Banerjee, A., Kumar, S., Raj, A., 2023. Agroforestry for Carbon and Ecosystem Management. Elsevier, India, pp. 1–434. Available at https://doi.org/10.1016/C2021-0-02916-3.

Jiang, L., Niu, H., Ru, Y., Tong, A., Wang, Y., 2023. Can carbon finance promote high quality economic development: evidence from China. Heliyon 9 (12), e22698. Available at https://doi.org/10.1016/j.heliyon.2023.e22698.

Jindal, R., Swallow, B., Kerr, J., 2008. Forestry-based carbon sequestration projects in Africa: potential benefits and challenges. Natural Resources Forum 32 (2), 116–130. Available at https://doi.org/10.1111/j.1477-8947.2008.00176.x.

Jobbágy, E.G., Jackson, R.B., 2000. The vertical distribution of soil organic carbon and its relation to climate and vegetation. Ecological Applications 10 (2), 423–436. Available at https://doi.org/10.1890/1051-0761(2000)010[0423:TVDOSO]2.0.CO;2.

Jose, S., 2012. Agroforestry for conserving and enhancing biodiversity. Agroforestry Systems 85 (1), 1–8. Available at https://doi.org/10.1007/s10457-012-9517-5.

Jose, S., Bardhan, S., 2012. Agroforestry for biomass production and carbon sequestration: an overview. Agroforestry Systems 86 (2), 105–111. Available at https://doi.org/10.1007/s10457-012-9573-x.

Kamilaris, A., Assumpcio, A., Blasi, A.B., Torrellas, M., Prenafeta-Boldú, F.X., 2018. Estimating the environmental impact of agriculture by means of geospatial and big data analysis: the case of catalonia. In: Otjacques, B., Hitzelberger, P., Naumann, S., Wohlgemuth, V. (Eds.), From Science to Society. Progress in IS. Springer, Cham, pp. 39–48. Available at https://doi.org/10.1007/978-3-319-65687-8_4.

Karki, H., Bargali, K., Bargali, S.S., 2024. Carbon sequestration potential of soils under different agroforestry systems of Kumaun Himalaya India and its implication for climate change mitigation. SSRN, India SSRN. Available at https://doi.org/10.2139/ssrn.4819865.

Keenan, P.B., Jankowski, P., 2019. Spatial decision support systems: three decades on. Decision Support Systems 116, 64–76. Available at https://doi.org/10.1016/j.dss.2018.10.010.

Kiremu, M., Scrimgeour, F., Mutegi, J., Mumo, R., 2022. Climate finance readiness: a review of institutional frameworks and policies in Kenya. Sustainable Environment 8 (1). Available at https://doi.org/10.1080/27658511.2021.2022569.

Kiyingi, I., Ocama, D., Mujuni, D., Nyombi, K., 2019. A bioeconomic analysis of the carbon sequestration potential of agroforestry systems: a case study of Grevillea robusta in South Western Uganda. Uganda Journal of Agricultural Sciences 17 (2), 219. Available at https://doi.org/10.4314/ujas.v17i2.7.

Kumar, R., Veeraragavan, M., Baral, K., Saikanth, D.R.K., Singh, V., Upadhyay, L., et al., 2023. Agroforestry and Its potential for sustainable land management and climate action: a review. International Journal of Environment and Climate Change 13 (12), 620–629. Available at https://doi.org/10.9734/ijecc/2023/v13i123722.

Kumar, S., Meena, R.S., Jhariya, M.K., 2020. Resources Use Efficiency in Agriculture Resources Use Efficiency in Agriculture. Springer, Singapore, pp. 1–760. Available at http://doi.org/10.1007/978-981-15-6953-1.

Kumar, S., Meena, R.S., Sheoran, P., Jhariya, M.K., 2024. Regenerative Agriculture for Sustainable Food Systems. Springer Science+Business Media, India, pp. 1–493. Available at http://doi.org/10.1007/978-981-97-6691-8.

Li, L., Zhang, D., 2024. Forest carbon offset protocols in compliance carbon markets. Forest Policy and Economics 165, 103253. Available at https://doi.org/10.1016/j.forpol.2024.103253.

Li, M., 2024. The application of financial technology in carbon finance. Highlights in Business, Economics and Management 40, 1066–1070. Available at https://doi.org/10.54097/38k7zf44.

Liu, Y., Zhao, W., Zhang, L., Li, X., Peng, L., Wang, Z., et al., 2024. An assessment framework for mapping the air purification service of vegetation at the regional scale. Forests 15 (2), 391. Available at https://doi.org/10.3390/f15020391.

Loft, L., Ravikumar, A., Gebara, M.F., Pham, T.T., Resosudarmo, I.A.P., Assembe, S., et al., 2015. Taking stock of carbon rights in REDD+ candidate countries: concept meets reality. Forests 6 (4), 1031–1060. Available at https://doi.org/10.3390/f6041031.

Lorenz, K., Lal, R., 2014. Soil organic carbon sequestration in agroforestry systems. A review. Agronomy for Sustainable Development 34 (2), 443–454. Available at https://doi.org/10.1007/s13593-014-0212-y.

Luedeling, E., Neufeldt, H., 2012. Carbon sequestration potential of parkland agroforestry in the Sahel. Climatic Change 115 (3-4), 443–461. Available at https://doi.org/10.1007/s10584-012-0438-0.

Martins, M.M.V., Bispo, S.Q.A., Silva, F.A., Burnquist, H.L., Rodriguez, A.M.B., 2024. Climate finance towards greener agriculture: a win-win strategy. Revista Tempo do Mundo (RTM) 34, 317–344. Available at https://doi.org/10.38116/rtm34art12.

Mathur, V.N., Afionis, S., Paavola, J., Dougill, A.J., Stringer, L.C., 2014. Experiences of host communities with carbon market projects: towards multi-level climate justice. Climate Policy 14 (1), 42–62. Available at https://doi.org/10.1080/14693062.2013.861728.

Matilainen, A., Andersson, E., Lähdesmäki, M., Lidestav, G., Kurki, S., 2023. Services for what and for whom? A literature review of private forest owners' decision-making in relation to forest-based services. Small-scale Forestry 22 (3), 511–535. Available at https://doi.org/10.1007/s11842-023-09541-3.

Meena, R.K., Kumari, T., Solanki, V.K., Partel, V., Singh, S.L., Sinha, R., 2024. Soil, water, and biodiversity conservation through agroforestry for crop production. In: Jatav, H.S., Rajput, V.D., Minkina, T., Van Hullebusch, E.D., Dutta, A. (Eds.), Agroforestry to Combat Global Challenges. Sustainable Development and Biodiversity, vol 36. Springer, Singapore, pp. 345–366. Available at https://doi.org/10.1007/978-981-99-7282-1_17.

Meena, R.S., Pradhan, G.S., Kumar, S., Jhariya, M.K., Rao, C.S., 2025. Carbon-Negative Agriculture. Springer, pp. 1–500.

Mintah, K.O., Nakouwo, S.N., Ahenkan, A., Bawole, J.N., 2025. Exploring climate change investment initiatives and funding challenges among faith-based organizations in Ghana. Development Policy Review 43 (2). Available at https://doi.org/10.1111/dpr.12826.

Mukhlis, I., Rizaludin, M.S., Hidayah, I., 2022. Understanding socio-economic and environmental impacts of agroforestry on rural communities. Forests 13 (4), 556. Available at https://doi.org/10.3390/f13040556.

Müller, A., Steinke, J., Dorado, H., Keller, S., Jiménez, D., Ortiz-Crespo, B., et al., 2024. Challenges and opportunities for human-centered design in CGIAR. Agricultural Systems 219. Available at https://doi.org/10.1016/j.agsy.2024.104005.

Murdiyarso, D., Herawati, H., 2005. Carbon forestry: who will benefit?: proceedings of Workshop on Carbon Sequestration and Sustainable Livelihoods, held in Bogor on 16-17 February 2005. Center for International Forestry Research (CIFOR). Available at https://doi.org/10.17528/cifor/001733.

Muthuri, C.W., Kuyah, S., Njenga, M., Kuria, A., Öborn, I., van Noordwijk, M., 2023. Agroforestry's contribution to livelihoods and carbon sequestration in East Africa: a systematic review. Trees, Forests and People 14. Available at https://doi.org/10.1016/j.tfp.2023.100432.

Mutuo, P.K., Cadisch, G., Albrecht, A., Palm, C.A., Verchot, L., 2005. Potential of agroforestry for carbon sequestration and mitigation of greenhouse gas emissions from soils in the tropics. Nutrient Cycling in Agroecosystems 71 (1), 43–54. Available at https://doi.org/10.1007/s10705-004-5285-6.

Nair, P.K.R., 2012. Carbon sequestration studies in agroforestry systems: a reality-check. Agroforestry Systems 86 (2), 243–253. Available at https://doi.org/10.1007/s10457-011-9434-z.

Naysary, B., Shrestha, K., 2024. Carbon exchange-traded funds market and economic policy uncertainty. Development and Sustainability in Economics and Finance 2-4, 100012. Available at https://doi.org/10.1016/j.dsef.2024.100012.

Ngaiwi, M.E., del Mar Esponda, M., Amanhui, G.A., Villarino, M.E., Andrade, R., Castro-Nunez, A., 2024. Exploring CGIAR's efforts towards achieving the paris agreement's climate-change targets. Journal of Agriculture and Food Research 18. Available at https://doi.org/10.1016/j.jafr.2024.101326.

Ning, Y., Cherian, J., Sial, M.S., Álvarez-Otero, S., Comite, U., Zia-Ud-Din, M., 2023. Green bond as a new determinant of sustainable green financing, energy efficiency investment, and economic growth: a global perspective. Environmental Science and Pollution Research 30 (22), 61324–61339. Available at https://doi.org/10.1007/s11356-021-18454-7.

Nyberg, Y., Musee, C., Wachiye, E., Jonsson, M., Wetterlind, J., Öborn, I., 2020. Effects of agroforestry and other sustainable practices in the Kenya agricultural carbon project (KACP). Land 9 (10), 1–22. Available at https://doi.org/10.3390/land9100389.

Octavia, D., Suharti, S., Murniati, Dharmawan, I.W.S., Nugroho, H.Y.S.H., Supriyanto, B., et al., 2022. Mainstreaming smart agroforestry for social forestry implementation to support sustainable development goals in indonesia: a review. Sustainability 14 (15), 9313. Available at https://doi.org/10.3390/su14159313.

Oelbermann, M., Paul Voroney, R., Gordon, A.M., 2004. Carbon sequestration in tropical and temperate agroforestry systems: a review with examples from Costa Rica and southern Canada. Agriculture, Ecosystems & Environment 104 (3), 359–377. Available at https://doi.org/10.1016/j.agee.2004.04.001.

Pan, J., Cross, J.L., Zou, X., Zhang, B., 2024. To tax or to trade? A global review of carbon emissions reduction strategies. Energy Strategy Reviews 55, 101508. Available at https://doi.org/10.1016/j.esr.2024.101508.

Pancholi, R., Yadav, R., Gupta, H., Vasure, N., Choudhary, S., Singh, M.N., et al., 2023. The role of agroforestry systems in enhancing climate resilience and sustainability- a review. International Journal of Environment and Climate Change 13 (11), 4342–4353. Available at https://doi.org/10.9734/ijecc/2023/v13i113615.

Parhamfar, M., Sadeghkhani, I., Adeli, A.M., 2024. Towards the net zero carbon future: a review of blockchain-enabled peer-to-peer carbon trading. Energy Science and Engineering 12 (3), 1242–1264. Available at https://doi.org/10.1002/ese3.1697.

Plieninger, T., 2011. Capitalizing on the carbon sequestration potential of agroforestry in germany's agricultural landscapes: realigning the climate change mitigation and landscape conservation agendas. Landscape Research 36 (4), 435–454. Available at https://doi.org/10.1080/01426397.2011.582943.

Pradhan, G., & Meena, R. S. (2023a). Interaction impact of biocompost on nutrient dynamics and relations with soil biota, carbon fractions index, societal value of CO2 equivalent and ecosystem services in the wheat-rice farming. Chemosphere, 339, 139695. https://doi.org/10.1016/j.chemosphere.2023.139695

Pradhan, G., & Meena, R. S. (2023b). Utilizing waste compost to improve the atmospheric CO2 capturing in the rice-wheat cropping system and energy-cum carbon credit auditing for a circular economy. Science of The Total Environment, 892, 164572. https://doi.org/10.1016/j.scitotenv.2023.164572.

Pradhan, G., Meena, R. S., Kumar, S., & Lal, R. (2023). Utilizing industrial wastes as compost in wheat-rice production to improve the above and below-ground ecosystem services. Agriculture, Ecosystems & Environment, 358, 108704. https://doi.org/10.1016/j.agee.2023.108704.

Prajapati, R., Singh, S., Anand, A., Mallesh, Y., Gulaiya, S., 2024. Green solutions: the impact of agroforestry on climate change resilience- a review. International Journal of Environment and Climate Change 14 (11), 1–13. Available at https://doi.org/10.9734/ijecc/2024/v14i114522.

Prasad, R., Jhariya, M.K., Banerje, A., 2021. Advances in Sustainable Development and Management of Environmental and Natural Resources: Economic Outlook and Opinions: Volume 1. Apple Academic Press, India, pp. 1–428. Available at http://doi.org/10.1201/9781003187455.

Rahman, S.A., Rahman, M.F., Sunderland, T., 2012. Causes and consequences of shifting cultivation and its alternative in the hill tracts of eastern Bangladesh. Agroforestry Systems 84 (2), 141–155. Available at https://doi.org/10.1007/s10457-011-9422-3.

Raina, N., Zavalloni, M., Viaggi, D., 2024. Incentive mechanisms of carbon farming contracts: a systematic mapping study. Journal of Environmental Management 352, 120126. Available at https://doi.org/10.1016/j.jenvman.2024.120126.

Raj, A., Jhariya, M.K., Meena, R.S., 2025. Agroforestry for Monetising Carbon Credits. Springer Nature, Switzerland.

Raj, A., Jhariya, M.K., Banerjee, A., Nema, S., Bargali, K., 2023. Land and Environmental Management Through Forestry. Wiley-Scrivener, USA.

Raj, A., Jhariya, M.K., Banerjee, A., Jha, R.K., Singh, K.P., 2024. Agroforestry. Wiley-Scrivener, USA.

Ramachandran Nair, P.K., Mohan Kumar, B., Nair, V.D., 2009. Agroforestry as a strategy for carbon sequestration. Journal of Plant Nutrition and Soil Science 172 (1), 10–23. Available at https://doi.org/10.1002/jpln.200800030.

Rands, M.R.W., Adams, W.M., Bennun, L., Butchart, S.H.M., Clements, A., Coomes, D., et al., 2010. Biodiversity conservation: challenges beyond 2010. Science (New York, N.Y.) 329 (5997), 1298–1303. Available at https://doi.org/10.1126/science.1189138.

Rawls, W.J., Pachepsky, Y.A., Ritchie, J.C., Sobecki, T.M., Bloodworth, H., 2003. Effect of soil organic carbon on soil water retention. Geoderma 116 (1-2), 61–76. Available at https://doi.org/10.1016/s0016-7061(03)00094-6.

Roshetko, J.M., Lasco, R.D., Angeles, M.S.D., 2007. Smallholder agroforestry systems for carbon storage. Mitigation and Adaptation Strategies for Global Change 12 (2), 219–242. Available at https://doi.org/10.1007/s11027-005-9010-9.

Santier, T., 2023. The next priorities of the Voluntary Carbon Market for mass adoption: the need for new technologies, carbon policy frameworks, and a meta registry. Available from: https://hal.science/hal-04301394.

Satish, P., Madiwalar, A.F., Lallawmkimi, M.C., Reddy, K.J., Parveen, S., Thejavath Laxman, A.P., et al., 2024. Agroforestry: multifunctional benefits and implementation strategies. Journal of Geography, Environment and Earth Science International 28 (10), 1–12. Available at https://doi.org/10.9734/jgeesi/2024/v28i10821.

Schut, M., Colomer, J., Proud, E., Bonaiuti, E., Dror, I., Kang'ethe, E., et al., 2024. Innovation portfolio management for responsible food systems transformation in the public sector: lessons, results and recommendations from CGIAR. Agricultural Systems 216, 103907. Available at https://doi.org/10.1016/j.agsy.2024.103907.

Seeberg-Elverfeldt, C., Schwarze, S., Zeller, M., 2008. Carbon finance schemes - incentives for forest and agroforestry systems. Nature Precedings. Available at https://doi.org/10.1038/npre.2008.2504.1.

Seeberg-Elverfeldt, C., Schwarze, S., Zeller, M., 2009a. Carbon finance options for smallholders' agroforestry in Indonesia. International Journal of the Commons 3 (1), 108–130.

Seeberg-Elverfeldt, C., Schwarze, S., Zeller, M., 2009b. Payments for environmental services – Carbon finance options for smallholders' agroforestry in Indonesia. International Journal of the Commons 3 (1), 108. Available at https://doi.org/10.18352/ijc.96.

Seeberg-Elverfeldt, C., 2009. Carbon Finance Schemes in Indonesia: Empirical Evidence of their Impact and Institutional Requirements.

Shames, S., Heiner, K., Kapukha, M., Kiguli, L., Masiga, M., Kalunda, P.N., et al., 2016. Building local institutional capacity to implement agricultural carbon projects: participatory action research with Vi Agroforestry in Kenya and ECOTRUST in Uganda. Agriculture & Food Security 5 (1). Available at https://doi.org/10.1186/s40066-016-0060-x.

Shapoval, Y., Niedziółka, P., 2025. Carbon Finance. Financial Innovation and Sustainability. Routledge, pp. 45–58. Available at https://doi.org/10.4324/9781003499138-4.

Shu, H., 2024. Research on innovation mechanism and realization path of carbon finance under the goal of "double carbon" Economics, Law and Policy 7 (2), p154. Available at https://doi.org/10.22158/elp.v7n2p154.

Sima, D.B., Estifanos, S., Asfaw, Z., 2024. Assessing the Role of Fruit Tree-Based Agroforestry in Climate Change Adaptation: A Case Study in Sofi District, Harari Regional State, Ethiopia. Research Square. Available at http://doi.org/10.21203/rs.3.rs-3872331/v1.

Singh, N., Biswas, D., Gokhale, Y., Kumar, K., 2024. Incentivising agroforestry through carbon revenue: augmenting farmers' income in India. Agricultural Science Digest - A Research Journal 44. Available at https://doi.org/10.18805/ag.d-6066 Of.

Singh, P., Choudhary, B.B., Dwivedi, R.P., Arunachalam, A., Kumar, S., Dev, I., 2023. Agroforestry improves food security and reduces income variability in semi-arid tropics of central India. Agroforestry Systems 97 (4), 509–518. Available at https://doi.org/10.1007/s10457-023-00806-6.

Smith, J., Scherr, S.J., 2002. Forest carbon and local livelihoods: assessment of opportunities and policy recommendations. Center for International Forestry Research (CIFOR). Available at https://doi.org/10.17528/cifor/001086.

Soloviy, I., Panchyshyn, Y., 2024. Climate and carbon markets: opportunities and barriers for merging instruments. Economic Scope. Available at https://doi.org/10.32782/2224-6282/191-66.

Spilker, G., Nugent, N., 2022. Voluntary carbon market derivatives: growth, innovation & usage. Borsa Istanbul Review 22, S109. Available at https://doi.org/10.1016/j.bir.2022.11.008.

Springgay, E., Pajel, P., 2024. Breaking barriers to agroforestry: FAO's global capacity needs assessment. Tropical Forest Issues(62). Available at https://doi.org/10.55515/tcdu9996.

Sutrisno, A., Agang, W., Hartono, T.T., Sayaza, M.D., 2024. Cultivating climate solutions: agroforestry's potentials and roles in North Kalimantan's REDD+ program. Ilomata International Journal of Social Science 5 (3), 743–753. Available at https://doi.org/10.61194/ijss.v5i3.1253.

Tabor, K., Connell, L., 2019. Applications of Forest Monitoring Tools for Development Projects. Bank Information Center (202), 28–28. www.bankinformationcenter.org.

Tavernier, H., Borremans, L., Bracke, J., Reubens, B., Wauters, E., 2024. Assessing the potential of different economic incentives for stimulating temperate agroforestry. A study in Flanders, Belgium. Agroforestry Systems 98 (6), 1873–1889. Available at https://doi.org/10.1007/s10457-024-00996-7.

Telwala, Y., 2023. Unlocking the potential of agroforestry as a nature-based solution for localizing sustainable development goals: a case study from a drought-prone region in rural India. Nature-Based Solutions 3, 100045. Available at https://doi.org/10.1016/j.nbsj.2022.100045.

Thamo, T., Pannell, D.J., 2016. Challenges in developing effective policy for soil carbon sequestration: perspectives on additionality, leakage, and permanence. Climate Policy 16 (8), 973–992. Available at https://doi.org/10.1080/14693062.2015.1075372.

Thangata, P.H., Hildebrand, P.E., 2012. Carbon stock and sequestration potential of agroforestry systems in smallholder agroecosystems of sub-Saharan Africa: mechanisms for 'reducing emissions from deforestation and forest degradation' (REDD+). Agriculture, Ecosystems & Environment 158, 172–183. Available at https://doi.org/10.1016/j.agee.2012.06.007.

Tirkey, A.I.P., Kumar, R., Prasad, R.S., Malakar, A., 2024. Agroforestry for carbon neutrality: an effective pathway to net zero. Journal of Advances in Biology & Biotechnology 27 (11), 591–602. Available at https://doi.org/10.9734/jabb/2024/v27i111643.

Torquebiau, E., 2024. The agroforestry-biodiversity-climate change nexus. Tropical Forest Issues(62). Available at https://doi.org/10.55515/fllz3215.

Udawatta, R.P., Jose, S., 2012. Agroforestry strategies to sequester carbon in temperate North America. Agroforestry Systems 86 (2), 225–242. Available at https://doi.org/10.1007/s10457-012-9561-1.

Udawatta, R.P., Rankoth, L., Jose, S., 2019. Agroforestry and biodiversity. Sustainability 11 (10), 2879. Available at https://doi.org/10.3390/su11102879.

van Noordwijk, M., 2021. Agroforestry-based ecosystem services: reconciling values of humans and nature in sustainable development. Land 10 (7), 699. Available at https://doi.org/10.3390/land10070699.

Verma, K., Sharma, P., Bhardwaj, D.R., Sharma, V., Thakur, P., Parmar, Y., 2024. Towards a greener future: scaling up agroforestry for global sustainability. Green and Low-Carbon Economy 1–5. Available at https://doi.org/10.47852/bonviewGLCE42022481.

Verschuuren, J., 2017. Towards a regulatory design for reducing emissions from agriculture: lessons from Australia's carbon farming initiative. Climate Law 7 (1), 1–51. Available at https://doi.org/10.1163/18786561-00701001.

VijayKumar, R., Tiwari, P., Daniel, S., Kumar, K.R., Mishra, I., Aneesh, K.S., et al., 2024. Agroforestry systems: a pathway to resilient and productive landscapes. International Journal of Environment and Climate Change 14 (12), 177–193. Available at https://doi.org/10.9734/ijecc/2024/v14i124617.

Vilkov, A., Tian, G., 2023. Blockchain's scope and purpose in carbon markets: a systematic literature review. Sustainability 15 (11), 8495. Available at https://doi.org/10.3390/su15118495.

Viñals, E., Maneja, R., Rufí-Salís, M., Martí, M., Puy, N., 2023. Reviewing social-ecological resilience for agroforestry systems under climate change conditions. Science of the total Environment 869, 161763. Available at https://doi.org/10.1016/j.scitotenv.2023.161763.

Vionis, P., Kotsilieris, T., 2024. The potential of blockchain technology and smart contracts in the energy sector: a review. Applied Sciences 14 (1), 253. Available at https://doi.org/10.3390/app14010253.

Waldén, P., Ollikainen, M., Kahiluoto, H., 2020. Carbon revenue in the profitability of agroforestry relative to monocultures. Agroforestry Systems 94 (1), 15–28. Available at https://doi.org/10.1007/s10457-019-00355-x.

Waldén, P., Eronen, M., Kaseva, J., Negash, M., Kahiluoto, H., 2024. Determinants of the economy in multistrata agroforestry in Ethiopia. Land use Policy 141, 107162. Available at https://doi.org/10.1016/j.landusepol.2024.107162.

Wang, S., Zhou, S., Wu, R., Feng, K., Hubacek, K., 2024. Interregional flows of embodied carbon storage associated with land-use change in China. Annals of the American Association of Geographers 114 (7), 1526–1545. Available at https://doi.org/10.1080/24694452.2024.2356849.

Weng, Q., Xu, H., 2018. A review of China's carbon trading market. Renewable and Sustainable Energy Reviews 91, 613–619. Available at https://doi.org/10.1016/j.rser.2018.04.026.

Woo, J., Fatima, R., Kibert, C.J., Newman, R.E., Tian, Y., Srinivasan, R.S., 2021. Applying blockchain technology for building energy performance measurement, reporting, and verification (MRV) and the carbon credit market: a review of the literature. Building and Environment 205. Available at https://doi.org/10.1016/j.buildenv.2021.108199.

Worku, A., 2024. The role of agroforestry in ecosystem services and mitigation of climate change. Vegetable crops of Russia (4), 111–119. Available at https://doi.org/10.18619/2072-9146-2024-4-111-119.

World, B., 2022. Digital monitoring, reporting, and verification systems and their application in future carbon markets. http://hdl.handle.net/10986/37622.

World, B., 2024. State and Trends of Carbon Pricing 2024 (9781802209204). Available from: https://hdl.handle.net/10986/41544.

Yu, C., Mao, J.-H., Huang, X., Zhang, Y., 2024. Carbon capture by phytomass storage and trading to mitigate climate change and preserve natural resources. Environmental and Sustainability Indicators 22, 100358. Available at https://doi.org/10.1016/j.indic.2024.100358.

Zhang, T.Y., Feng, T.T., Cui, M.L., 2023. Smart contract design and process optimization of carbon trading based on blockchain: the case of China's electric power sector. Journal of Cleaner Production 397. Available at https://doi.org/10.1016/j.jclepro.2023.136509.

Zhang, W., Wang, Z., 2024. Carbon finance and low-carbon technological change: evidence from China. Energy & Environment. Available at https://doi.org/10.1177/0958305x241230617.

Zhou, K., Li, Y., 2019. Carbon finance and carbon market in China: progress and challenges. Journal of Cleaner Production 214, 536–549. Available at https://doi.org/10.1016/j.jclepro.2018.12.298.

Zhou, K., Midkiff, D., Yin, R., Zhang, H., 2024. Carbon finance and funding for forest sector climate solutions: a review and synthesis of the principles, policies, and practices. Frontiers in Environmental Science 12. Available at https://doi.org/10.3389/fenvs.2024.1309885.

Zomer, R., Xu, J., Spano, D., Trabucco, A., 2025. Nature-based solutions: evaluating the global carbon sequestration potential of agroforestry and increased tree cover on agricultural land. EGU General Assembly 2024.

Chapter 26

Private sector contributions to carbon credit initiatives in Indian agriculture

Niladri Sekhar Mondal
Department of Environmental Science, School of Sciences, Netaji Subhas Open University, Kolkata, West Bengal, India

Highlights

- Carbon (C) credit has become a key component in emission reduction, climate action, and a low-C economy.
- C credit has shown a potential role and promise in the agricultural system.
- India has a rich heritage of agricultural production systems.
- The private sector plays a key role in implementing C credit in the Indian agricultural system.
- C credit has sensitized the farming community to C farming and a green economy.

26.1 Introduction

Carbon (C) credits are a key tool in the global strategy for climate change mitigation (CCM) by regulating and reducing greenhouse gas (GHG) emissions, providing economic incentives for emission reductions by creating a market for trading these credits (Raj et al., 2025; Meena et al., 2025a). The mechanism of C credits involves a cap-and-trade system, where C credits represent a permit or certificate allowing the holder to emit one ton of carbon dioxide (CO_2) or an equivalent amount of other GHG, and regulatory bodies set a cap on the total amount of GHGs that certain sectors or industries can emit. Suppose a company emits less than its allowance. In that case, they can sell the surplus credits to other companies that exceed their limits, creating a financial incentive for companies to reduce their emissions. This market-based approach encourages investment in sustainable practices and technologies, fostering a transition to a low-C economy (Meena et al., 2024; Meena et al., 2025b). Consequently, the entities step forward to offset projects, such as reforestation, agricultural integration, renewable energy projects, and methane capture from landfills, which are projects that reduce, remove, or avoid GHG emissions. Entities can invest in these projects to gain C credits, which can then be sold in the C market. C credits of variable nature of the projects involved, as well as a regulatory framework, exist across various countries of the globe.

Agriculture plays a crucial role in carbon sequestration (C_{seq}), capturing and storing atmospheric CO_2. Regenerative agricultural practices can significantly enhance the soil's ability to store C and reduce overall GHG emissions. This practice includes conservation tillage or no-till farming, growing crops to cover and protect soil; improving organic matter (OM), living roots (leguminous plants) in the soil all year round; integrating livestock to prevent overgrazing and enhance soil health; and agroforestry (AF) (Kumar et al., 2024; Meena et al., 2024; Devi et al., 2024; Jhariya et al., 2022; Jhariya et al., 2023). Such approaches would be highly helpful in serving multiple purposes toward combating climate change, biodiversity conservation, followed by improvement in the soil quality and health. On a global basis, the market for C credits is of a dual nature, which comprises voluntary and compliance markets. In the case of compliance markets, it comes under the action of C emission reduction at different scales, and the voluntary market is rather flexible and involves more people at the individual level, corporate level to participate in the C trading mechanism and move toward C neutrality.

26.1.1 Influence of the private sector on environmental sustainability

The private sector is playing a key role in the field of environmental sustainability through the promotion of C credit programs. Sustainability projects on a large-scale basis have been initiated through technical skill development

Carbon Incentives, Plans, and Policies in Agriculture. DOI: https://doi.org/10.1016/B978-0-443-33557-0.00030-6

infrastructure build-up, along with innovations in terms of startups and entrepreneurship. Under the corporate social responsibility of industries, various investments are being made in the promotion and adoption of green technologies. Industries are also making investments in the promotion and adoption of green technologies to increase their involvement in the C trading system. Simultaneously, it would also bring significant environmental benefits. Private sectors are also working toward optimum economic growth through startups, incubations, and entrepreneurship development, along with boosting agricultural productivity. Various forms of financial investments include funding ecorestoration projects, renewable energy promotion, and reforestation practices, as well as promoting sustainable agriculture practices. Development of technical skills for effective monitoring of the C_{seq} project that would lead to sound development of the C credit market. The private companies are investing more in innovations, and research and development activities and focusing more on emission reduction and C_{seq} that would lead to an effective C credit project.

Now, from the Indian perspective, agriculture is an all-round activity that contributes significantly to the country's gross domestic product (GDP) but has some inherent problems associated with it. One of the most significant issues includes small landholdings that hinder the adoption of sustainable approaches at the grassroots level. Another significant issue is the variability in the monsoonal rainfall pattern due to the uncertainties of climate change. Also, there is a practice of huge diversification of agricultural crop cultivation. Now, at present, context-intensive and industrialized agriculture practices lead to a higher level of soil fertility and quality loss (Jhariya et al., 2021a; Jhariya et al., 2021b). Further, depletion of water resources due to overexploitation for irrigation practices. Methane emissions from rice cultivation, nitrous oxide emissions from nitrogenous fertilizer are big issues that contribute significantly to the GHG emission profile of India. Now, various companies of the private sector are realizing the importance of the C credit initiative and are investing more in these aspects. Collaborative work of the NGOs with the farming community on C financing initiatives through C_{seq} and credits in the form of AF projects, sprinkler and drip irrigation projects, conservative agriculture practices, climate-resilient and eco-friendly agro-technologies, etc. All these would boost productivity as well as promote greater participation of the people in the C financing market.

26.2 Indian agriculture: an overview

India is a global agricultural powerhouse, acting as the most vital sector in the country, which contributes not only to the economy but also to food security, rural livelihoods, and cultural heritage. About 19.9% of India's GDP (gross domestic product) was contributed by agriculture in 2020–21, and it has been raised from 17.8% in 2019–20. Although the economic contribution has decreased over the last few decades, in the 1950s, agricultural production contributed about 50% of the Indian economy. Apart from these facts, Indian agriculture employs 58% of the population. Beyond its economic contribution, agriculture promotes the well-being of the rural population, which is about 63.64% of the total population of India, which means 909.12 million people are directly involved in agricultural practices. It is the foundation of India's socioeconomic system, directly employing millions of people and supporting related industries, such as retail, transportation, and agro-processing.

26.2.1 Salient features of Indian agriculture

India's agriculture is distinctive due to its diverse climate and topography and is also shaped by cultural and economic factors. India has a variety of climates and is categorized into 15 agro-climatic zones and 21 agroecological zones based on their various characteristics (Paramesh et al., 2022; Meena et al., 2024; Bhatt et al., 2019). These factors define its productivity and sustainability. Some of the characteristic features of Indian agriculture are mentioned below:

1. Diverse agro-climatic zones: India's vast and varied geography, spanning from the tropical south to the temperate north, has divided the country into 15 distinct agro-climatic zones. Tropical and temperate regions are due to the Tropic of Cancer running through the middle of the country. These agroclimatic zones are classified based on rainfall patterns, temperature ranges, soil types, and water availability. This diversity supports the cultivation of a wide array of crops. For instance, rice, sugarcane, cotton, and bananas are cultivated in tropical regions with ample rainfall and warm temperatures. Wheat, apples, and barley are suited to cooler, temperate regions of northern and mountainous states. Arid and semiarid areas with limited rainfall, such as Rajasthan and Gujarat, produce millet, pulses, and oilseeds. This diversification leads India to be a major producer of various food grains (rice and wheat), oilseeds, and horticultural products, helping to maintain the food security of the country and export earnings.
2. Dominance of small and marginal landholdings: Most of the farmers of India, about 85%, are operating on small and marginal landholdings of less than 2 hectares. The average farm size in India is approximately 1.08 hectares, which poses several challenges for modern large efficient instruments and large-scale agricultural practices. Climate-smart

technology is not or less feasible and increases the costs of production (Modi et al., 2020). Fragmented lands also increase the cost of high-quality seeds, fertilizers, and irrigation systems, posing an affordability issue for small farmers. The small and marginal nature of these farms makes them vulnerable to price volatility, climate risks, and input cost fluctuations, often leading to income instability among farmers (Azadi et al., 2021).

3. Dependence on monsoon rains: India's agricultural area (60%) is dependent on unreliable monsoon behavior. About 48.8% of 140 million hectares (Mha) of agricultural land is under coverage of an irrigation system, and the remaining 51.2% depends on rain (Sikka et al., 2022). As a result, crop cycles are highly dependent on the monsoon, which makes Indian agriculture vulnerable to climatic variations. The productivity of the rainfed area is approximately half in comparison with the irrigated area, which is about 1.1 t/ha and 2.8 t/ha, respectively (Singh et al., 2020).
4. Labor-intensive practices: Indian agriculture is mainly manual labor dependent (95% of the rice area) because of fragmented land ownership as well as for continuing with traditional practices for cultivation like manual planting, harvesting, and weeding, especially in rural and marginal areas (Balwinder-Singh et al., 2020). High labor costs and availability affect profitability, particularly for small-scale farmers who cannot afford mechanized equipment. It is estimated that during COVID-19, the breadbasket of India (Haryana and Punjab) accounted for economic losses of USD $674 m to $1.48 billion due to labor unavailability (Singh et al., 2020; Balwinder-Singh et al., 2020).
5. Wide variety of crops: India is a major global producer of a variety of crops, including staple grains like rice and wheat; pulses like chickpeas, pigeon peas, and lentils; oilseeds like groundnut, soybean, and mustard; and horticultural fruits like mangoes, bananas, and citrus (Kumar and Babu, 2018). These crops are critical for dietary protein, domestic oil production, and export revenue. India's diverse climate and soil contribute to its status as the world's leading producer of these crops, which are affected by changing seasons.

26.2.2 Key challenges in Indian agriculture

While agriculture is essential, it faces significant challenges that impact productivity, sustainability, and profitability. Since the green revolution, the cereal farming area has decreased from 38 Mha (in 1950–51) to 31 Mha (in 2003–04). However, production has increased more than twice as much, from 15 MT (in 1950–51) to 38 MT (in 2003–04) due to yield improvement from 408 to 1228 kg ha^{-1} (Ramakrishna and Rao, 2008). Simultaneously, consumption of chemical fertilizer and pesticides increased by 7 and 375-folds, respectively (Singh et al., 2016). So, although the green revolution increases food production in a shorter time, excessive inputs of agrochemicals lead to a decline in soil quality, productivity, and its multifunctionality, along with environmental pollution through the release of various chemical species, such as NO_3^-, PO_4^{3-}, and heavy metals. Such a type of negative impact of agrochemicals is prominent in the food bowl of India, Punjab and Haryana states (Kumari et al., 2019; Singh et al., 2019)—Indian agriculture is fighting against several other challenges, which are mentioned in Table 26.1.

26.2.3 Government initiatives and policies

The Indian government has introduced several key initiatives and policies to strengthen agriculture, promote farmer welfare, and implement different initiatives to boost the agricultural industry and improve economic wealth. Here are some highlights:

- Digital agriculture mission: Launched in 2024, it is designed as an umbrella scheme that aims to establish a comprehensive farmer-centric digital and space-tech ecosystem to enhance transparency and efficiency. It includes several tools like AgriStack, farmer IDs, and geo-referenced crop mapping to provide data-driven services to farmers and increase productivity.
- National Agriculture Market (e-NAM) Expansion: This e-marketing platform (One Nation, One Market) has been initiated by the Ministry of Agriculture and Farmers' Welfare, Government of India, on April 14, 2016. The government is integrating more mandis (markets) into the electronic National Agriculture Market (e-NAM) platform to create a unified market, increasing transparency and ease of trade for farmers. As of March 16, 2023, 1361 mandis of 23 States and 4 UTs have been integrated with the network.
- Pradhan Mantri Fasal Bima Yojana (PMFBY): Continued emphasis on crop insurance to protect farmers from climate risks, covering four crore farmers annually. It was implemented in 2016 to improve crop safety and guarantee that farmers receive the full benefit of crop insurance, with a maximum premium during 2% Kharif, 1.5% Rabi food and oilseed crops, and 5% annual commercial or horticultural crops, with the remaining actuarial or bid premium covered by the state and central govt. equally. The Government of India has earmarked Rs 16,000 crores for the financial year 2021–22 (https://www.indiabudget.gov.in/).

TABLE 26.1 Major key challenges in Indian agriculture.

Challenges	Description
Fragmented land holdings	The majority of farmers own small, fragmented plots, reducing economies of scale and productivity. About 86.2% of the operational land (146.45 Mha) is fragmented, creating difficulty in mechanization and promoting monocropping.
Water scarcity	Overexploitation of groundwater and uneven rainfall patterns affect irrigation and crop growth. Approximately 80% of India's water consumption is used for agricultural irrigation. Nowadays, 65% of irrigation is shifted to groundwater sources because of the steady decline of the water table.
Dependence on monsoons	A large percentage, about 51.2%, of agriculture is rainfed, making it vulnerable to erratic monsoon patterns.
Low Mechanization	Limited adoption of modern agricultural tools and technologies, especially in small-scale farms, and most of the agricultural operations are done by laborers. In India, mostly about 60%–70% of mechanization is done at the level of plowing, harvesting, threshing, and irrigation only, but there is much machinery available that can be used in seeding, weeding, etc.
Soil degradation	After the Green Revolution in the 1960s, overuse of fertilizers and pesticides has led to loss of soil fertility, reduced OM and humus content, and increased salinity.
Postharvest losses	India loses 10%–20% of its agricultural production because of a lack of adequate storage, processing, transportation, and inadequate market infrastructure, which leads to significant losses. Every year, approximately 16% of fruits and vegetables, 10% of oilseeds, 9% of pulses, and 6% of cereals are wasted due to a lack of storage facilities.
Market access	Farmers face challenges in accessing fair markets due to middlemen and a lack of price transparency because of the exploitation of local traders.
Climate change impact	Climate change can lead to extreme weather patterns, droughts, and floods that threaten crop yields and food security. Unpredictable rainfall or the time of rainfall affects several agricultural operations; for example, extreme rainfall during harvesting time causes total loss of crop production. In India between 2015–16 and 2021–22, hydro-meteorological calamities such as heavy rainfall and floods destroyed approximately 34 million hectares of cropped land
Credit availability	Many farmers lack access to formal credit systems, resulting in their dependence on informal moneylenders.
Price volatility	Fluctuating market prices of crops create economic instability for farmers. Sudden drops in prices reduce the income and profits of the farmers, especially small farmers, who are more vulnerable to these. This situation creates uncertainty for investing in farms, leading to a vicious cycle of poverty and low productivity.
Lack of diversification	Indian farmers are overdependent on a few crops (e.g., rice and wheat), which reduces resilience against market and climate risks.
Lack of training and research	Poor extension programs limit farmers' access to modern farming practices and research outputs. Farmers who do not receive training or extension services are more sensitive to pests and diseases, which can reduce productivity.

OM, organic matter.

- Pradhan Mantri Kisan Maan Dhan Yojana, 2019: It is mainly a government scheme meant for old age (after 60 years of age) protection and social security in terms of a pension of Rs. 3000/- for small and marginal farmers.
- Pradhan Mantri Kisan Samman Nidhi Yojana, 2018: The scheme is effective from 1.12.2018. It provided direct financial support to farmers with ₹ 6,000 per year, with ₹ 60,000 crore allocated in the 2024–25 budget to benefit small and marginal farmers.
- Soil health card scheme: Launched in 2015 to address soil degradation, this scheme provides farmers with soil health cards indicating the nutrient status of their soil and recommendations for balanced fertilizer use, aiming to make agriculture more productive and sustainable and increase crop yields (https://www.pib.gov.in/PressRelease).
- Atmanirbhar Bharat Initiatives: To make Indian agriculture future-ready, initiatives like the National Mission for Sustainable Agriculture, scientific warehousing, and drone technology adoption. The government has increased

investment in agriculture through the Agri-Tech Infrastructure Fund, promoted organic farming through the Paramparagat Krishi Vikas Yojana, reduced edible oil imports through the Atmanirbhar Oilseeds Abhiyan, and established Long-Term Irrigation Funds.

Apart from the above, several new initiatives have been introduced by the government, collectively aiming to enhance productivity, ensure food security, and improve farmers' incomes while addressing the challenges of climate change, market inefficiencies, and resource constraints (Kumar et al., 2020).

26.3 Carbon credit scenarios in Indian agriculture

The concept of C credits offers a market-based mechanism in global efforts for mitigating GHG emissions and combating climate change. They represent C credits, which are tradable certificates representing the reduction, removal, or avoidance of one metric tonne of carbon dioxide equivalent (tCO_2e) or an equivalent amount of other GHGs, such as methane. Although after the 1960s green revolution, nitrogen-based fertilizers increased food production, they also led to an increase in GHG emissions, primarily from methane (CH_4), nitrous oxide (N_2O), and CO_2. But in agriculture, C credits can be generated through the use of optimized fertilizer application or alternative rice cultivation methods that cut down methane and N_2O emissions and through C_{seq} and storing it in soil, biomass, or plants by AF, no-till farming, and cover cropping systems. In the country like India which belongs in two-thirds of the world's GHG emissions contributors, per capita emissions is 2.77 tCO_2e (Jones et al., 2023), the C credit framework presents a broader opportunity to incentivize sustainable farming practices, reduce emissions, and improve livelihoods of the farmers (Prasad et al., 2021; Ritchie et al., 2023; Meena and Pradhan, 2023).

There are two distinct categories of C markets: (1). Compliance C markets, established following the Paris Agreement, are used in regulatory cap-and-trade systems by governments or international organizations like the European Union Emission Trading Systems (EU ETS), the Regional Greenhouse Gas Initiative, or the California Cap-and-Trade Program. (2). Voluntary C credits are purchased by institutions to offset their C footprint outside of government-regulated markets voluntarily to make C-neutral. Companies in these markets must either meet emission reduction objectives or obtain C credits to offset their emissions (Meena et al., 2023). An institution or company may invest in reforestation or AF projects that absorb CO_2; the amount of CO_2 absorbed is awarded to that company as a C credit. Sometimes a company may invest in green energy generation.

Agriculture is a significant contributor to India's GHG emissions, accounting for about 17%–18% of total emissions. Among the total emissions, 74% is contributed from the agricultural emissions, caused by anaerobic decomposition in waterlogged fields (for e.g., traditional paddy cultivation) and CH_4 emissions from enteric fermentation in ruminant livestock farming. About 23% of GHG emissions in terms of N_2O happen due to excessive or inefficient application of nitrogen fertilizer. The remaining 3% of emissions occur due to open-field burning of crop residues, especially in northern India, and the fuel burnt in machinery like pumps, harvesters, tractors, etc. So, India's agricultural sector has the potential to take a major part in C-credit systems. In agriculture, C credits are earned by adopting sustainable and advanced agrotech that reduce GHG emissions or enhances the quantity of C_{seq} in soils and vegetation. Sustainable farming practices, such as no-till farming, cover cropping, AF, and improved manure management, are effectively increasing C_{seq} in soil and reducing the GHGs from agricultural fields, along with maintaining the integrity of the crop-soil ecosystem (Banerjee et al., 2023; Banerjee et al., 2024; Pradhan and Meena, 2023a). This approach not only contributes to climate goals but also provides economic benefits to farmers.

26.3.1 Current status of carbon credits in Indian agriculture

C farming promotes sustainable agricultural techniques that increase soil C uptake and reduce GHG emissions. Farmers can generate C credits utilizing sustainable farming techniques and sell them on the market. This advancement has the potential to address the fundamental issue of modernizing agricultural systems sustainably while providing additional support for farmers' economies. From this perspective, India is expected to see a significant growth in the C farming credits market, by taking the top position in the list of the highest estimated emission reductions. The number of agricultural projects that are currently active in India is more than 50, and is counted as the second highest after China, which is claimed to cut emissions of CO_2e by 37 MT every year. India's compliance and voluntary C markets and green credit trading scheme will focus on eight key sectors, with a focus on "Sustainable Agriculture-based Green Credit" to promote natural and regenerative farming, soil health, food nutrition, and agricultural productivity.

India, a country with a population of 142.86 crore, actively participates in C gain initiatives to combat the challenge of GHG emissions, along with maintaining a sufficient food supply for this large population. Conventional agricultural practices cause several issues, such as lowering of groundwater levels, land degradation, loss of soil rejuvenation capacity due to excessive use of chemical fertilizers, and, recently, for the last few decades, crop residue burning. C emissions due to crop residue burning increased to 33,834 Gigagram in 2020 from 19,340 Gigagram per year (Deshpande et al., 2023). India is actively working on emission reduction initiatives, including rice cultivation, nutrient optimization, livestock emissions management, and C_{seq} (Table 26.2). Rice cultivation methods like the system of rice intensification and alternate wetting and drying have been tested in Andhra Pradesh, Tamil Nadu, and Punjab to reduce 45%–90% CH_4 emissions, which is estimated to save 30–44 $MtCO_2e$/year (Ishfaq et al., 2020). Initiatives under the soil health card scheme aim to mitigate excessive application of nitrogen-based chemical fertilizers. It has been recorded that after optimized use of synthetic nitrogen fertilizer, that is, a reduction of 9.30 kg/ha, led to a loss of productivity of about to 0.42%, as it reduced the emission by 102 kg CO_2e (Rosas et al., 2015).

Livestock accounts for 14.5% of worldwide CH_4 emissions, with Asian countries being the significant contributors (Gerber et al., 2013). Enteric fermentation in ruminants, which possess four-compartment digestive tracts, is responsible for 44% of CH_4 emissions (Swamy and Bhattacharya, 2006). Ninety-five percent of methane is expelled via the oral cavity, whereas five percent is discharged through the anal canal. Other animals, including pigs, poultry, and small ruminants, contribute 7%–11% of overall methane emissions (FAO, 2021; Zhang et al., 2022). The Indian subcontinent, with 36 states and union territories, has the five states with the greatest emissions attributed to enteric fermentation and manure management, which are Uttar Pradesh (2550.92 Gg yr^{-1}), Rajasthan (1342.44 Gg yr^{-1}), Madhya Pradesh (1187.84 Gg yr^{-1}), Bihar (998.63 Gg yr^{-1}), Maharashtra (861.38 Gg yr^{-1}). Northeastern states such as Mizoram and Goa rank among the lowest in emissions, that is, 3.45 Gg yr^{-1} and 4.01 Gg yr^{-1} accordingly (Samal et al., 2024). These livestock emissions can be managed through enhanced feed additives, such as tannins or seaweed, for instance, Asparagopsis taxiformis (R) mixed with ruminant feed with 0.05%–0.20% helps to decrease CH_4 emissions up to 40% and 98% (Abbott et al., 2020; Kinley et al., 2020).

Agriculture practices in terms of AF, incorporating trees into agricultural areas, enhance biodiversity and significantly increase soil C_{seq} (Raj et al., 2023; Raj et al., 2024). That can be achieved through no-till agriculture, crop rotation, and

TABLE 26.2 Current status of carbon credits in Indian agriculture.

Aspect	Data/Details
Share in total Indian GHG emissions	Approx. 17.6% (agriculture sector), primarily CH_4, N_2O, and CO_2.
Potential C credit generation	Estimated to generate more than 100 million credits annually through various sustainable practices.
Active projects	More than 50 projects have been registered, covering rice cultivation, AF, soil C, and livestock.
Certified C credits issued	Approx. 15–20 million credits are issued annually from agriculture-related projects.
Key methodologies	VM0042 (Improved Agricultural Land Management), CDM AMS-III.A (methane reduction), Plan Vivo standards.
Primary practices for credits earned	• Rice CH_4 reduction: Alternate wetting and drying (AWD), SRI. • Soil C_{seq}: No-till farming, crop rotation, and cover cropping. • AF: Tree integration on farmlands. • Livestock management: CH_4 reduction through improved feed and manure practices.
Geographical focus areas	Punjab, Haryana, Tamil Nadu, Andhra Pradesh, Gujarat, Rajasthan, Madhya Pradesh, Uttar Pradesh, Karnataka.
C credit market participation	India contributes up to 30% of global VCM through forestry and agriculture projects.
Farmer participation	Only 15%–20% of eligible farmers engaged in C credit initiatives due to knowledge and awareness gaps.
Key challenges	• Lack of farmer knowledge on C credits. • Difficult to aggregate smallholder lands for projects. • Monitoring, reporting, and verification costs reduce profitability for small projects. • Absence of an efficient domestic C market for agriculture offsets.

AF, agroforestry; *GHG*, greenhouse gas; *SRI*, system of rice intensification; *VCM*, voluntary carbon market.

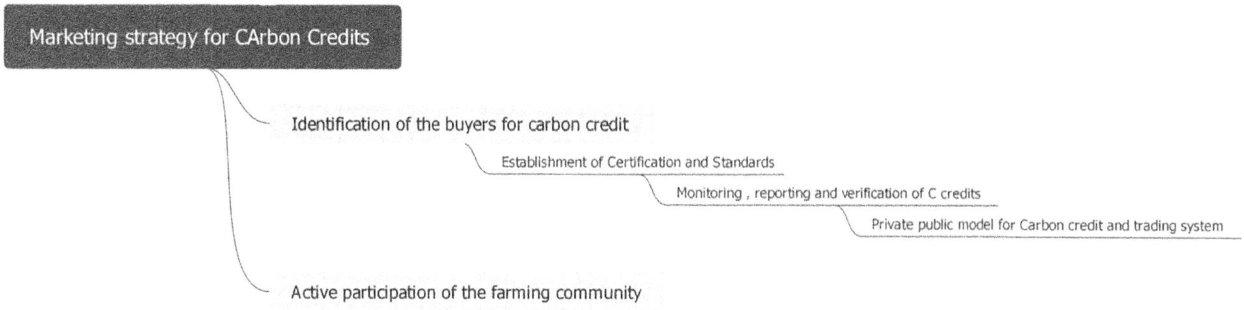

FIGURE 26.1 Mechanism for the proper marketing of carbon credits in the Indian agricultural system.

organic amendments. AFs in India have the capacity to sequester a rate of approx. 10 to 229.5 Mg C ha^{-1} (Kumar and Kunhamu, 2021). India's geographical condition offers a wide opportunity for expansion of AF under 328 Mha, including 79% of its current 13.7 Mha, which is highly or moderately suitable for tree incorporation on farmland (Panwar et al., 2022). Zero-Budget Natural Farming initiatives in Andhra Pradesh prioritize C_{seq} through natural agricultural methodologies with the capacity of 0.3–0.5 tCO_2e/ha/year across 50 Mha. India is also a major provider of C credits to voluntary carbon markets, with agricultural initiatives accredited by standards like the verified C standard or Gold Standard that have begun trading credits internationally.

Besides these wide opportunities, several impairments to widespread use of C credits in Indian agriculture exist, including knowledge and awareness gaps, high transaction costs, monitoring and verification issues, fragmented landholdings, and policy and regulatory uncertainties. Many farmers are unaware. They have limited access to technical information and capacity-building initiatives, which leave them a few steps behind for C credit mechanisms and the advantages of participating in C markets. They have limited access to technical information and capacity-building initiatives, which leaves them a few steps behind in further adapting conservation agriculture and C initiatives. Without aggregation, small-scale farmers find it unprofitable due to high transaction costs, including project registration, monitoring, verification, and certification costs. Monitoring and verification necessitate modern technology like remote sensing or on-ground sampling, and grouping projects poses logistical challenges for smallholders (Fig. 26.1). Apart from these, the absence of a robust domestic C market and clear guidelines for agricultural C credits adds uncertainty.

26.4 Carbon initiatives in India

India is at the forefront of global climate action, committed to reducing GHG emissions under its international obligations, including the Paris Agreement on Climate Change in 2016. The nation's commitment to reducing its C footprint is evident through a series of comprehensive initiatives, policies, and programs aimed at transitioning to a low-C economy while sustaining its economic growth (Banerjee et al., 2020; Pradhan and Meena, 2023b). As a signatory to the Paris Agreement, it has set ambitious targets under its Nationally Determined Contributions (NDCs) to decrease the emissions intensity of GDP by 33%–35% from 2005 levels by 2030 and 2.5–3 billion tons of CO_2e by creating a C sink through enhancing forest cover. Apart from that, at the Glasgow Climate Summit (Glasgow, Scotland) in 2021, also known as COP 26, India announced bold steps, termed the *Panchamrit*:

1. Achieving net-zero emissions by 2070
2. Increasing nonfossil energy capacity to 500 GW by 2030
3. Fulfill 50% of energy requirements from renewables by 2030
4. Reducing C emissions by 1 billion tons by 2030
5. Lowering the C intensity of the economy by 45% by 2030 compared to 2005 levels.

These goals underscore India's alignment with global efforts to limit the global temperature rise to below 2°C, with aspirations for a 1.5°C target. To achieve these commitments, the country is implementing a comprehensive framework that includes market-based mechanisms, regulatory reforms, and innovative programs aimed at reducing GHG emissions. India is developing a unified Indian Carbon Market (ICM) through the Carbon Credits Trading Scheme (CCTS). The CCTS is a market-based mechanism designed to limit emissions by setting targets for various emitters. Companies that exceed their targets can sell their surplus credits to those who fall short. This scheme, introduced through the Energy Conservation (Amendment) Act, 2022, is expected to be operational by 2026. The Bureau of Energy Efficiency (BEE) is responsible for implementing this scheme, which aims to cover a broad spectrum of industries while initially focusing on voluntary compliance before transitioning to mandatory requirements. The Indian C market is structured into two primary mechanisms: the compliance mechanism and the offset mechanism. The compliance mechanism focuses on energy-intensive businesses, establishing GHG emission intensity targets that these organizations are required to achieve, and it is expected to launch in 2025–26. The offset system incentivizes voluntary activities by those organizations not covered by compliance obligations, allowing them to earn C credits for additional GHG reductions that they achieve. It is anticipated that the ICM will become one of the largest C markets globally by 2030, targeting an estimated market value exceeding $1.2 billion. Complementing existing market-based emission reduction schemes like the Perform, Achieve, and Trade (PAT) scheme and Renewable Energy Certificates.

Among the various government initiatives, the PAT scheme is one of the important steps toward improving energy efficiency for energy-intensive industries. The major target area under the PAT scheme includes GHG emission reduction in various components of industrial sectors. It acts as creating markets for a higher energy-saving approach in the form of energy-saving certificates. Such certificates can be traded by the industries having excess savings of energy to industries falling short of achieving the target, along with proper compliance with rules and regulations and fulfilling norms and standards. Quantification is typically carried out in three phases, with the gradual inclusion of industrial sectors. However, such schemes are facing some lacunae in the form of a lack of awareness, ineffective implementation, and inefficiency in proper functioning. Now, to improve the efficiency of the PAT scheme, the government has initiated more stringent regulations, followed by the development of an effective network for monitoring.

26.5 Government plans, policies, and initiatives toward carbon credits

The Government of India has built up a robust legislative framework to promote credit and trading across the country, along with its effective implementation. The major task was the formulation of ICM under the leadership of the National Steering Committee for the Indian Carbon Market (NSCICM). Further, ICM would monitor the effective implementation of C credit across various sectors while fulfilling climate goals. The committee under NSCICM comprises members representing various organizations and ministries to have an interdisciplinary approach across various sectors toward the management of C. India has shown a climate commitment of net zero emissions by 2070 at the COP26 summit in 2021. Further, India has committed to emission reduction and promotion of sustainable development. More focus was given to renewable energy resources with a target of achieving 500 GW of nonfossil fuel capacity by 2030. More focus was given toward developing nonconventional sources of energy. Launching of the Green India movement for an increase in the forest cover as well as promotion of C_{seq} (Jhariya et al., 2019). India is also promoting the development and advancement of various technologies related to C capture, followed by emission reduction across various sectors.

26.6 Private sector contribution and initiatives for the carbon credit approaches in agriculture

In the context of India's agricultural landscape, the private sector is progressively playing a pivotal role in developing and promoting C credit initiatives in the agricultural system to combat climate change and enhance rural livelihoods. These initiatives aim to incentivize sustainable farming practices, improve farmers' incomes, and contribute to the overall reduction of GHG emissions through implementing various techniques to mitigate the reduction of CH_4 emissions during rice production and N_2O generated from synthetic fertilizers. In response, private enterprises have emissions and create economic value via C credits. These companies or entities are involved in different sectors of agricultural practices, depending on their needs and expertise; some of them are as follows.

Formulation of VM0042 methodology and quantification of GHG emission reduction, followed by soil C pool management through innovativeness in agricultural practices, has already been initiated. This would promote sustainable agriculture practices through participation in the C credit market. Various forms of agronomic transformations under the banner of VM0042 were implemented to promote sustainability in agricultural practices. The VM0042 methodology has an inherent advantage because of its flexibility, measuring and modeling approach for quantification of GHG emission reduction, and mathematical simulations for predicting future emission reduction based on new, innovative techniques. According to one estimate, 35 projects under the said methodology have given an annual turnover of 21.7 million tonnes of CO_2 equivalent. Under the methodology, specific criteria were set, and processes were designed to evaluate verified carbon units (VCUs) or credits.

26.6.1 Promotion of sustainable practices

Private entities are investing significantly in promoting sustainable agricultural practices. These include no-till farming, crop rotation, AF, organic farming, and the use of biofertilizers (Fig. 26.2).

For instance, Bayer, GenZero, Shell, and Mitsubishi aim to positively impact rice farmers in India through the Good Rice Alliance. The Good Rice Alliance (TGRA) is expanding its initiative from 25,000 hectares to encompass around 8500 hectares across the major rice-producing states in the country, including Andhra Pradesh, Bihar, Haryana, Karnataka, Odisha, Tamil Nadu, Telangana, Uttar Pradesh, and West Bengal. The alliance empowers farmers to implement sustainable climate-smart techniques such as Alternate Wetting and Drying and Direct Seeded Rice (DSR) to achieve C reductions in the maintained states. Since its inception, the initiative has benefited 10,000 farmers who have incorporated over 25,000 hectares. The initiative has reduced the agricultural expenses for the majority of participating farmers. At its current scale, TGRA produces approximately 100,000 tonnes of CO_2e methane emission reductions annually from agricultural activities. The need for this program is anticipated to increase, reaching between 330 million and 1.5 billion tCO_2e by 2030. The C offset platform sector in India is anticipated to expand at an annual rate of 28%, with a market valuation of US$ 68.5 million by 2033 (Ponce De León Baridó et al., 2023). ITC Limited, through its e-Choupal initiative, supports 4 million farmers across 10,000 villages, promoting techniques like conservation agriculture and precision farming. While the e-Choupal initiative itself does not directly generate C credits, it facilitates sustainable agricultural practices that can lead to C offset opportunities through promoting sustainable farming. Farmers are trained for reduced tillage, efficient use of fertilizers and water, and crop diversification and rotation practices. These practices enhance soil C_{seq} and reduce emissions, making farms eligible for C credit projects (Mukerji, 2020). Several studies indicate that conservation agriculture can reduce soil emissions by 20%–50%, while organic farming methods can decrease N_2O emissions by 17% per hectare.

PepsiCo has committed to implementing regenerative agricultural methods across 7 million acres by 2030 as part of its pep+ (PepsiCo Positive) strategy. This effort aims to improve the lives of over 250,000 people as part of its agricultural supply chain, while simultaneously sourcing 100% of its essential components sustainably. The Positive Agriculture Outcomes Fund, which provides funding for innovative farming techniques that enhance sustainability, is another way the corporation contributes to advancing initiatives aimed at addressing challenges in the agricultural sector (Kumari and Nair, 2024). Mahanadi Compressed Biogas Private Limited is also engaged in producing biogas from agricultural waste, which not only contributes to reducing emissions but also provides an alternative source of energy. It is anticipated that the activities it undertakes will result in a reduction of about 64,457 MT of CO_2 equivalent per year (Nozaki, 2023).

26.6.1.1 *Carbon credit projects*

The private sector is actively involved in creating C credit projects, often in partnership with global certification standards such as Verra and the Gold Standard. These initiatives encourage Indian farmers to rack up C credits in climate-conscious ways. There are many corporate players in this field, such as Grow Indigo, a joint venture between Mahyco (a domestic seed company) and Indigo Ag (a U.S.-based agricultural technology firm), which is at the forefront of implementing C credit programs in agriculture. The company has registered approx. 80,000 farmers across eight states, including Punjab, Haryana, and Maharashtra, who will use sustainable techniques such as DSR that reduces water usage by 12%–35%, along with labor costs and crop residue management that manage agricultural waste while improving soil health. The firm intends to distribute 75% of the earnings from C credits to farmers, offering them a significant financial incentive

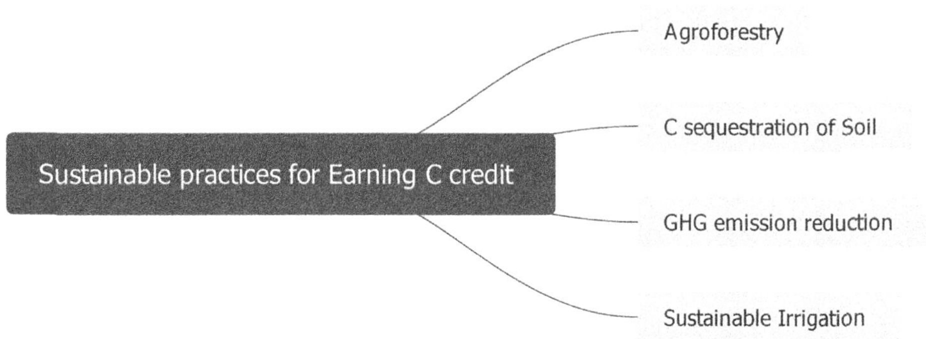

FIGURE 26.2 Sustainable practices for promoting C credit.

(Brummitt et al., 2024). Agri-tech firms, such as S4S Technologies, a nonprofit organization that aids female farmers and entrepreneurs by utilizing waste to produce and sell valuable food items, including ketchup. The organization links commercial purchasers to these items and allocates the majority of revenues to those small farmers. S4S supplies rural communities with solar-powered convection dryers and food processing machinery, therefore reducing expenses and facilitating on-site agricultural processing. By 2025, S4S intends to engage three million smallholder farmers and 30,000 enterprises, mitigating food waste by 1.2 MT and eliminating 10 MT of CO_2 from the environment.

26.6.2 Technology-driven interventions

Private companies are leveraging technology to streamline and scale C credit projects. They provide tools powered by artificial intelligence (AI) and remote sensing technology that can monitor soil C levels and emissions reductions across a large area with precision. For instance, Varaha ClimateAg Private Limited concentrates on extensive C farming initiatives aimed at substantial emission reductions. Varaha is projected to achieve roughly 4.8 million MT of CO_2e reductions yearly across an estimated two Mha of agricultural land. They utilize tech-powered evaluation (MRV) ARR (Afforestation, Reforestation, and Revegetation) and MRV (measurement, reporting and verification) regenerative agriculture, which use remote sensing and LiDAR-trained AI-ML satellite models to accurately estimate biomass and GHG emissions at scale, respectively. This initiative aims to enhance C_{seq} while providing financial incentives to farmers for adopting sustainable practices. The company applied for certification for a project with the highest projected emission reductions in India. They have secured 4 million US$ in seed-stage funding from Omnivore and others (Nozaki, 2023). Greneity Infocom Service Pvt. Ltd. is involved in initiatives that aim to generate around 173,250 MT of CO_2e reductions yearly, covering around 40,000 hectares of farmland. Greneity aims to utilize technology to enhance sustainable agriculture and facilitate farmers' access to C credit markets (Nozaki, 2023). Companies like *Crop in Climate* and Stellapps deploy IoT devices and satellite data to validate C_{seq} and track progress across 2 million Hectares under sustainable farming practices. These effective and accurate data-oriented technologies reduce agricultural input costs by 10%–20% and also cut the cost of MRV by up to **30%, finally** enhancing eligibility for C credit programs. Another example is Boomitra, an AI-driven leading international soil C marketplace, partnering with Kirishitan, a firm that advocates for sustainable agricultural methods. The startup is engaged in two initiatives encompassing 200,000 hectares of agricultural land to achieve yearly reductions of 750,000 MT of CO_2e. Boomitra focuses on integrating technology with agriculture to promote sustainable farming practices that qualify for C credits. In addition to India, the company is also engaged in initiatives throughout Latin America and East Africa and is collaborating with Kirishitan (formerly Carbon Farming India), which has had a customer base of farmers since 2018 (Meena et al., 2024; Nozaki, 2023).

26.6.3 Promotion of regenerative agriculture through investments

Private companies are directing substantial funds toward regenerative agricultural projects, focusing on improving soil health, biodiversity, and long-term C storage. General Mills aims to promote advanced regenerative agriculture across 1 million acres by 2030. The company involves farmers through workshops, technical support, and personalized coaching with the collaboration of the Noble Research Institute and the Soil Health Institute. They have developed a Regenerative Agriculture Self-Assessment Tool to help farmers align their practices with regenerative principles. A pilot project was run with wheat farmers in Kansas, showing increases in Soil Organic Matter (SOM) and C_{seq}. The One Planet Business for Biodiversity (OP2B) alliance, which includes major companies like Danone, McCain Foods, and others, is a business sector initiative that has achieved notable milestones in its fifth anniversary. The alliance, comprising 26 corporate members with a combined market value of $893 billion, has been promoting regenerative agriculture since its inception in 2019. From 2019 to 2023, the coalition has invested $3.6 billion in regenerative agriculture initiatives, aiming to impact over 12 million hectares by 2030. OP2B focuses on collective action across industries to drive change and support farmers in adopting sustainable practices. Many steel and energy companies of India are also investing in and supporting AF projects in Rajasthan and Gujarat, generating long-term C credit potential. Various studies estimated that widespread adoption of regenerative agricultural practices in India has the potential to sequester 0.3–0.6 gigatons of CO_2e annually, providing a significant opportunity for C credits. Not only are many big industrial giants like Nestlé and Unilever investing heavily in regenerative agriculture to achieve net-zero emissions. In France, Nestlé partnered with dairy farmers to adopt regenerative practices, including rotational grazing and natural soil enhancement, to achieve net-zero by 2050.

26.6.4 Public–private partnerships

A crucial element of the C credit ecosystem is the collaboration between commercial firms and governmental activities. The scope and impact of these efforts are enhanced by partnerships with initiatives like the National Mission for Sustainable Agriculture and PM-KUSUM. Solar irrigation pumps, funded by PM-KUSUM, reduce CO_2 emissions by 2–4 tonnes annually per pump (Terang and Baruah, 2023). The extensive use of these pumps is facilitated by funding from private groups. The implementation of sustainable practices over one Mha of agriculture has been facilitated via public-private partnerships. The private sector's involvement in India's agricultural C credit initiatives using its resources, creativity, and market access to promote sustainable practices represents a transformative approach to sustainable farming. Corporations are establishing systems by leveraging innovative practices that improve C_{seq}, diminish GHG emissions, and provide quantifiable, tradable C-offsets via investments in regenerative agriculture, public-private collaborations, and digital platforms such as e-Choupal. These projects are not only converting agriculture from a net emitter into a substantial solution for climate change but also promoting sustainable agricultural practices by incentivising farmers, facilitating access to new technology, and creating transparent carbon markets, thus aligning agriculture with global climate objectives and improving rural lives. These initiatives not only contribute to climate change mitigation but also enhance the livelihoods of farmers across the country. As the market matures and government support continues, India has the potential to become a leader in global agricultural C credits (Table 26.3). However, the path forward requires overcoming challenges such as standardizing measurement, ensuring equitable benefit distribution, and maintaining the integrity of C credits. Collaborative efforts among governments, corporations, and local communities are crucial to scaling these initiatives and maximizing their impact.

26.7 Policy perspectives on the carbon credit system in India

To fulfill the target of emission reduction, India is rapidly increasing its C credit system as per the Paris Agreement for developing climate resiliency across the country. The major policy of India in this direction includes moving toward SD along with greater participation of the corporate and industrial sectors under the C trading mechanism. One of the major initiatives includes C Credit Trading Scheme (CCTS) under the amended Energy Conservation Act of 2022. Such a policy provides a wider platform by providing a wider market of voluntary and compliance C trading systems that would focus on GHG emission reduction. The government of India is working across various interfaces and sectors for the generation of C credit and comes under the system of C trading. The government has also developed a proper monitoring network for both types of C markets that would lead to the build-up of ICM.

26.7.1 Institutional and regulatory framework

The Energy Conservation (Amendment) Act of 2022 authorizes the federal government to set up a C credit trading program. This includes procedures for issuing C credit certificates, which indicate one tonne of CO_2 equivalent that has been reduced or removed from the atmosphere. The Act includes Section 14(w), which grants the government the authority to develop a trading system and establish emission reduction objectives. However, the Act has been criticized for its lack of clarity over definitions of C credits and regulatory oversight, notably concerning the roles of several ministries and agencies engaged in implementation. The CCTS, introduced in 2023 under the Energy Conservation (Amendment) Act, aims to establish a regulated domestic C market in India. This initiative is crucial for reducing GHG emissions and facilitating the country's transition to a low-C economy. Below are the key features of the CCTS:

1. Regulatory framework: The Central Government is empowered to establish and implement a C trading scheme under Section 14(w) of The Energy Conservation (Amendment) Act of 2022, while a National Steering Committee oversees the governance of the C market, formulating procedures, rules, regulations, issuing C credit certificates; and monitoring market functions.
2. Compliance and voluntary markets: The CCTS has a dual market structure, with obligated entities meeting specific GHG emission intensity targets and nonobligated entities voluntarily purchasing C credits to offset their emissions. Entities that exceed their targets earn C credits, which can be sold or traded, while those that do not meet their targets must purchase credits to offset their excess emissions.
3. Carbon credit certificates: C credit certificates represent one tonne of CO_2e reduced or removed from the atmosphere; these certificates can be traded as tradable commodities in the C market. The BEE administers the issuance of C credit certificates based on recommendations from the National Steering Committee. In comparison, the Grid Controller of India maintains market records.

TABLE 26.3 Indian projects submitted for certification under the VM0042 methodology.

Sl. no.	Name project	Proponent	Estimated annual emission reductions (metric ton of CO_2 e)	Acres/ hectares
1	Reducing GHG emissions in irrigated crops through micro-irrigation in different agroecological regions of India	National Bank for Agriculture and Rural Development	100	12
2	Evaluation of SOC sequestration under natural farming conditions: Enhancing soil quality and capitalizing on C markets	National Bank for Agriculture and Rural Development	400	50
3	C credit farming through sustainable agriculture approaches in the upper Indo-Gangetic plains	National Bank for Agriculture and Rural Development	400	100
4	Harnessing the benefits of C farming *vis-Ã -vis* C credit for balancing economic development with climate change adaptation and mitigation	National Bank for Agriculture and Rural Development	350	52
5	C crediting in the rice-wheat system through crop residue management: a farmer participatory approach	National Bank for Agriculture and Rural Development	1295	100
6	Land management by organic farming practices in India	Qmark Foundation for Sustainable Livelihood Development	2,75,749	100,000
7	The Biwal C project	Multiple Proponents	92,377	30,000
8	Incentivising smallholder farmers to transition to low-emissions agriculture and AF	Climeverse Private Limited	4,63,650	133,603
9	Promoting regenerative agriculture and growth through income generation	Grow Indigo Private Ltd.	53,26,066	4,000,000
10	Regenerative agriculture in the Ganga-basin for farmer income and climate impact	Grow Indigo Private Ltd.	50,48,425	4,000,000
11	Maharashtra and Gujarat initiative for regenerative agriculture and income creation	Grow Indigo Private Ltd.	48,61,245	3,000,000
12	Developing the C project for farmers adopting regenerative agricultural practices in Gujarat state	Multiple Proponents	10,46,945	312,000
13	Green Shift: Catalyzing regenerative agriculture ecosystem for smallholder farmers in India, through C finance	Varaha ClimateAg Private Limited	45,05,732	1,658,000

14	Combating climate change with a sustainable agriculture and land use management project by Indian farmers	Himadri Energy International Private Limited	75,794	35,838
15	Sustainable agriculture project by Arvind in India	Multiple Proponents	79,875	98,223
16	GHG emission reduction through intermittent flooding in rice fields	Landmark Agri Exports Private Limited	58,722	8085
17	Optimization of the nitrogen fertilizer in India and Thailand by using satellite images analysis	Sagri Co., Ltd.	100	20
18	Developing C finance project for farmers adopting regenerative agriculture practices in the Indo-Gangetic plains	Varaha ClimateAg Private Limited	47,96,219	2,000,000
19	Waste management project-1 by BoxLNG	Mahanadi Compressed Biogas (CBG) Private Limited	64,457	64,457
20	Sustainable agriculture land management project by Biowin Agro Research	Biowin Agro Research	26,639	11,560
21	Sustainable agricultural land management project by Greneity	Greneity Infocom Service Pvt Ltd.	1,73,250	40,000
22	Boomitra C farming in the Indo-Gangetic plains	Boomitra Inc.	3,50,000	100,000
23	Boomitra C farming South of the Vindhyas	Boomitra Inc.	6,76,391	100,000
24	Regenerative agriculture in the rice-wheat-maize system for income generation	Grow Indigo Private Ltd.	42,67,622	1,534,000

4. Sectoral targets and benchmarks: India is set to implement emission intensity targets for energy-intensive sectors, aligning them with India's climate goals. This approach encourages effective emission reduction. Comprehensive MRV guidelines will be developed to ensure accurate tracking of emissions reductions and compliance with established targets.
5. Encouragement of clean technologies: The CCTS encourages clean technology adoption by assigning economic value to reduced CO_2e, promoting sustainable practices, and technological upliftment across sectors. It also supports emerging future generation energy sectors like green hydrogen, sustainable aviation fuel, and offshore wind energy, which highlights India's commitment toward innovative emission reduction solutions.
6. Future phases and development: The CCTS is set to transition from 2023 to 2025, with full implementation expected in subsequent years. The government plans to conduct capacity-building programs for stakeholders of the C market to improve understanding and operational efficiency in C credit trading.

India's institutional and regulatory framework for C credits is a blend of domestic and international mechanisms designed to support its climate goals and enable participation in global C markets. These systems focus on encouraging emissions reductions, C_{seq}, and sustainable practices across various sectors, including energy, industry, and agriculture. Various key institutions across India are working in this direction for the effective implementation of C credit in our agricultural practices. However, India has made significant contributions by sharing contributions in terms of having the second-largest CDM projects with more than 1000 projects in the sectors of renewable energy, industry, and management of waste. The CDM was framed in order to improve the larger participation of the developing world in the C trading system. The C market in India, known as ICM, is a unified market at the domestic level, which was defined in 2022, providing a broader platform for trading of CCCs to fulfill the target of emission reduction, followed by improved C trading and meeting international standards. From a regulatory framework point of view, various acts, statutes, and regulations were enacted for the effective implementation of the C trading system. However, there are diverse challenges in the form of a shortfall in participation, a low level of awareness level, and a lack of proper monitoring networks, which are big hindrances in implementing the schemes, which may be resolved through the PPP model framework.

26.8 Plan for enhancing carbon credits through smart agricultural practices

The integration of smart agricultural practices into C credit programs presents a significant opportunity for farmers to enhance profitability while contributing to CCM. Here's a comprehensive plan outlining how these practices can be implemented effectively.

26.8.1 Adoption of climate-smart agricultural practices

Climate-smart agriculture (CSA) encompasses a variety of practices aimed at increasing productivity while reducing GHG emissions. Techniques such as cover cropping, conservation tillage, AF, and organic fertilization not only improve soil health but also sequester CO_2 from the atmosphere.

Improved soil health: Practices that enhance SOC levels contribute to better soil fertility and water retention, which are crucial for sustainable farming.

Biodiversity enhancement: CSA encourages biodiversity, which can lead to more resilient agricultural systems.

26.8.1.1 Carbon credit generation and certification

C credit mechanism: Farmers can earn C credits by implementing practices that reduce emissions or sequester C (Pradhan et al., 2023). Each credit typically represents one metric ton of CO_2 equivalent that has been reduced or sequestered.

Certification process: To generate C credits, farmers must undergo a verification process that includes the following:

Data collection: Gathering baseline data on current farming practices.
Implementation of practices: Adopting additional sustainable practices.
Monitoring and reporting: Continuous tracking of emissions reductions using technologies like remote sensing and biogeochemical models.

26.8.1.2 Financial incentives and market access

Revenue streams: The sale of C credits provides farmers with an additional source of income. This can be facilitated through:

C markets: Farmers can sell their credits to businesses or individuals looking to offset their emissions, creating a market-driven approach to sustainability.

Investment in technology: Utilizing AGTECH solutions for precision agriculture can enhance the effectiveness of CSA practices and improve the quantification of C credits.

26.8.1.3 Government and institutional support

Policy frameworks: For C credit programs to succeed, robust policy support is essential. This includes:

Certification standards: Governments should establish clear standards for C credit certification to ensure quality and transparency.
Technical assistance: Providing farmers with guidance on best practices and access to resources can encourage participation in C credit programs.
MRV

Robust MRV systems: Implementing effective MRV systems is critical for ensuring the credibility of C credits. This involves:

Use of technology: Employing remote sensing and modeling tools to accurately measure changes in soil C levels and GHG emissions over time.
Third-party verification: Engaging independent verifiers to assess the accuracy of reported data ensures trust in the system.

26.9 Future of the carbon credit system in Indian agriculture

From a future perspective, C credit has significant promise under the Indian agriculture system with multifaceted benefits, starting from boosting agriculture production to livelihood generation to C offsetting to C trading. Hence, proper decision-making and capacity building, followed by awareness generation, is the need of the hour among the farming community. Further, the mechanism of the voluntary C market and regulatory compliance is working hard and fast to achieve the target of NDCs. The initiatives are also being taken at the local level for economic output generation through the C trading system. A prediction of generation of an amount of 3,94,405 lakh crores, which is equivalent to $480 billion, till 2070 is expected to be generated from the Indian agriculture system, which may act as a C-positive approach and can be adopted and integrated with C markets at the domestic level. Further, initiatives in the form of CCTS would be launched in 2026 as an acclimatization action policy.

26.10 Conclusion

Moving toward C neutralization and a low-C economy has become the fate of the future world. India is gradually putting its step forward in the field of C economy through sustainable agriculture practices. Developing voluntary C markets and strengthening the Indian C trading system are the major transformations under the Indian agriculture system. Both the government and private sectors are working hand in hand to develop the future potential of the C credit and trading system. Various schemes, economic subsidies, and projects are being implemented at a significant level for the better promotion of C credits in Indian agriculture practices. The major contribution and scope of C credit is immense for the small landholding farming community of the country, and due to its overall benefits, there is an enormous possibility for participating in the C market. Integration of sustainable practices and proper monitoring of C credits would aid in active participation of the farming community in the C market. Such policies aid in the participation of the farmer community in the C market and economy, as well as develop climate resiliency and economic crises.

References

Abbott, D.W., Aasen, I.M., Beauchemin, K.A., Grondahl, F., Gruninger, R., Hayes, M., et al., 2020. Seaweed and seaweed bioactives for mitigation of enteric methane: challenges and opportunities. Animals 10 (12), 1–28. Available at https://doi.org/10.3390/ani10122432.

Azadi, H., Movahhed Moghaddam, S., Burkart, S., Mahmoudi, H., Van Passel, S., Kurban, A., et al., 2021. Rethinking resilient agriculture: from climate-smart agriculture to vulnerable-smart agriculture. Journal of Cleaner Production 319. Available at https://doi.org/10.1016/j.jclepro.2021.128602.

Balwinder-Singh, P.B., Shirsath, M.L., Jat, A.J., McDonald, A.K., Srivastava, P., Craufurd, D.S., et al., 2020. Agricultural labor, COVID-19, and potential implications for food security and air quality in the breadbasket of India. Agricultural Systems 185. Available at https://doi.org/10.1016/j.agsy.2020.102954.

Banerjee, A., Meena, R.S., Jhariya, M.K., Yadav, D.K., 2020. Agroecological Footprints Management for Sustainable Food System. Springer, Singapore, pp. 1–514. Available at 10.1007/978–981-15–9496-0.

Banerjee, A., Jhariya, M.K., Raj, A., Mechergui, T., 2024. Ecosystem Management: Climate Change and Sustainability. Wiley, pp. 1–669. Available at 10.1002/9781394231249.

Banerjee, A., Jhariya, M.K., Bargali, S.S., Palit, D., 2023. Ecorestoration for Sustainability. Wiley-Scrivener, SA, pp. 1–441. Available at 10.1002/9781119879954.

Bhatt, D., Sonkar, G., Mall, R.K., 2019. Impact of climate variability on the rice yield in Uttar Pradesh: an agro-climatic zone based study. Environmental Processes 6 (1), 135–153. Available at https://doi.org/10.1007/s40710-019-00360-3.

Brummitt, C.D., Mathers, C.A., Keating, R.A., O'Leary, K., Easter, M., Friedl, M.A., et al., 2024. Solutions and insights for agricultural monitoring, reporting, and verification (MRV) from three consecutive issuances of soil carbon credits. Journal of Environmental Management 369. Available at https://doi.org/10.1016/j.jenvman.2024.122284.

Deshpande, M.V., Kumar, N., Pillai, D., Krishna, V.V., Jain, M., 2023. Greenhouse gas emissions from agricultural residue burning have increased by 75 % since 2011 across India. Science of the Total Environment 904. Available at https://doi.org/10.1016/j.scitotenv.2023.166944.

Devi, A., Jhariya, M.K., Raj, A., Banerjee, A., Singh, K.P., Singh, B., 2024. Acacia nilotica: a promising species for soil sustainability. Land and Environmental Management through Forestry. Wiley, India, pp. 339–353. Available at http://doi.org/10.1002/9781119910527.ch14.

FAO, 2021. Global livestock environmental assessment model (GLEAM). FAO, Rome, Italy. Available at https://www.fao.org/gleam/en.

Gerber, P.J., Steinfeld, H., Henderson, B., Mottet, A., Opio, C., Dijkman, J., et al., 2013. Tackling Climate Change Through Livestock – A Global Assessment of Emissions and Mitigation Opportunities. Food and Agriculture Organization of the United Nations (FAO), Rome. Available at https://www.fao.org/4/i3437e/i3437e.pdf.

Ishfaq, M., Farooq, M., Zulfiqar, U., Hussain, S., Akbar, N., Nawaz, A., et al., 2020. Alternate wetting and drying: a water-saving and ecofriendly rice production system. Agricultural Water Management 241. Available at https://doi.org/10.1016/j.agwat.2020.106363.

Jhariya, M.K., Banerjee, A., Meena, R.S., Yadav, D.K., 2019. Sustainable Agriculture, Forest and Environmental Management. Springer, Singapore, pp. 1–606. Available at https://doi.org/10.1007/978-981-13-6830-1.

Jhariya, M.K., Banerjee, A., Swaroop Meena, R., Kumar, S., Raj, A., 2021a. Sustainable Intensification for Agroecosystem Services and Management. Springer, Singapore, pp. 1–870. Available at https://doi.org/10.1007/978–981-16–3207-5.

Jhariya, M.K., Meena, R.S., Banerjee, A., 2021b. Ecological Intensification of Natural Resources for Sustainable Agriculture. Springer, Singapore, pp. 1–655. Available at https://doi.org/10.1007/978-981-33-4203-3.

Jhariya, M.K., Meena, R.S., Banerjee, A., Meena, S.N., 2022. Natural Resources Conservation and Advances for Sustainability. Elsevier. Available at https://doi.org/10.1016/C2019-0-03763-6.

Jhariya, M.K., Meena, R.S., Banerjee, A., Kumar, S., Raj, A., 2023. Agroforestry for Carbon and Ecosystem Management. Elsevier, India, pp. 1–434. Available at http://doi.org/10.1016/C2021-0-02916-3.

Jones, M.W., Peters, G.P., Gasser, T., Andrew, R.M., Schwingshackl, C., Gütschow, J., et al., 2023. National contributions to climate change due to historical emissions of carbon dioxide, methane, and nitrous oxide since 1850. Scientific Data 10 (1). Available at https://doi.org/10.1038/s41597-023-02041-1.

Kinley, R.D., Martinez-Fernandez, G., Matthews, M.K., de Nys, R., Magnusson, M., Tomkins, N.W., 2020. Mitigating the carbon footprint and improving productivity of ruminant livestock agriculture using a red seaweed. Journal of Cleaner Production 259. Available at https://doi.org/10.1016/j.jclepro.2020.120836.

Kumar, B.M., Kunhamu, T.K., 2021. Carbon sequestration potential of agroforestry systems in India: a synthesis. In: Udawatta, R.P., Jose, S. (Eds.), Agroforestry and Ecosystem Services. Springer, Cham, pp. 389–430. Available at https://doi.org/10.1007/978-3-030-80060-4_15.

Kumar, S., Meena, R.S., Sheoran, P., Jhariya, M.K., 2024. Regenerative Agriculture for Sustainable Food Systems. Springer, Singapore, pp. 1–493. Available at https://doi.org/10.1007/978-981-97-6691-8.

Kumar, S., Meena, R.S., Jhariya, M.K., 2020. Resources Use Efficiency in Agriculture. Springer, Singapore, pp. 1–760. Available at https://doi.org/10.1007/978-981-15-6953-1.

Kumar, K., Babu, S.C., 2018. Increasing Food System Resilience for Nutrition Sensitivity and Sustainability: A Decentralized Analysis for India. IFPRI Discussion Paper 01742. International Food Policy Research Institute. Available at https://cgspace.cgiar.org/server/api/core/bitstreams/24eaec30-d2b4-4c60-9de0-429154a21605/content.

Kumari, S., Nair, J., 2024. PepsiCo's sustainability strategy for building a positive value chain. IUP Journal of Supply Chain Management 21 (1), 63–78.

Kumari, T., Singh, R., Verma, P., 2019. Agriculture, forest and environmental management trinity: towards environmental sustainability and climate change mitigation. Tropical Ecology 60, 306–310. Available at https://doi.org/10.1007/s42965-019-00025-0.

Meena, R.S., Pradhan, G., Kumar, S., Lal, R., 2023. Using industrial wastes for rice-wheat cropping and food-energy-carbon-water-economic nexus to the sustainable food system. Renewable and Sustainable Energy Reviews 187, 113756. Available at https://doi.org/10.1016/j.rser.2023.113756.

Meena, R.S., Pradhan, G., 2023. Industrial garbage-derived biocompost enhances soil organic carbon fractions, CO_2 biosequestration, potential carbon credits and sustainability index in a rice-wheat ecosystem. Environmental Research 235, 116525. Available at https://doi.org/10.1016/j.envres.2023.116525.

Meena, R.S., Pradhan, G., Nalani, Kumar, S., Singh, A.K., Verma, S.K., et al., 2024. Diversified cropping modules designed for soil restoration, CO_2 sequestration, and generating carbon credits. Land Degradation and Development 35 (14), 4314–4329. Available at https://doi.org/10.1002/ldr.5224.

Meena, R.S., Pradhan, G., Kumar, S., 2025a. Energy flow, eco-efficiency, and economic circulation with recycled industrial waste compost application in wheat and subsequent rice farming. Science of the Total Environment 967. Available at https://doi.org/10.1016/j.scitotenv.2025.178779.

Meena, R.S., Pradhan, G.S., Kumar, S., Jhariya, M.K., Rao, C.S., 2025b. Carbon-Negative Agriculture. Springer, Singapore. Available at https://doi.org/10.1007/978-981-95-1483-0.

Modi, R.U., Manjunatha, K., Gauam, P.V., Nageshkumar, T., Sanodiya, R., Chaudhary, V., et al., 2020. Climate-smart technology based farm mechanization for enhanced input use efficiency. Climate Change and Indian Agriculture: Challenges and Adaption Strategies 325–358.

Mukerji, M., 2020. Re-examining strategic and developmental implications of e-Choupal, India. The Electronic Journal of Information Systems in Developing Countries 86 (4), e12132.

Nozaki, Y., 2023. Rise of carbon farming in India: world's large agrarian country expected to become the leading market for carbon farming credits. Mitsui and Co Global Strategic Studies Institute Monthly Report. Mitsui & Co. Global Strategic Studies Institute. Available at https://www.mitsui.com/mgssi/en/report/detail/__icsFiles/afieldfile/2023/03/28/2302i_nozaki_e.pdf.

Panwar, P., Mahalingappa, D.G., Kaushal, R., Bhardwaj, D.R., Chakravarty, S., Shukla, G., et al., 2022. Biomass production and carbon sequestration potential of different agroforestry systems in India: a critical review. Forests 13 (8). Available at https://doi.org/10.3390/f13081274.

Paramesh, V., Kumar, P., Shamim, M., Ravisankar, N., Arunachalam, V., Nath, A.J., et al., 2022. Integrated farming systems as an adaptation strategy to climate change: case studies from diverse agro-climatic zones of India. Sustainability 14 (18), 11629. Available at https://doi.org/10.3390/su141811629.

Ponce De León Baridó, P., Nielsen, J., Porsborg-Smith, A., Pineda, J., Owolabi, B., Gordon, M., 2023. In the Voluntary Carbon Market, Buyers Will Pay for Quality. https://web-assets.bcg.com/29/f5/3b36e7cb4ad49df092de00c9792d/bcg-in-the-voluntary-carbon-market-buyers-sept-2023-2.pdf.

Pradhan, G., Meena, R.S., 2023a. Interaction impact of biocompost on nutrient dynamics and relations with soil biota, carbon fractions index, societal value of CO_2 equivalent and ecosystem services in the wheat-rice farming. Chemosphere 339. Available at https://doi.org/10.1016/j.chemosphere.2023.139695.

Pradhan, G., Meena, R.S., 2023b. Utilizing waste compost to improve the atmospheric CO_2 capturing in the rice-wheat cropping system and energy-cum-carbon credit auditing for a circular economy. Science of the Total Environment 892. Available at https://doi.org/10.1016/j.scitotenv.2023.164572.

Pradhan, G., Meena, R.S., Kumar, S., Lal, R., 2023. Utilizing industrial wastes as compost in wheat-rice production to improve the above and below-ground ecosystem services. Agriculture, Ecosystems & Environment 358, 108704. Available at https://doi.org/10.1016/j.agee.2023.108704.

Prasad, R., Jhariya, M.K., Banerje, A., 2021. Advances in Sustainable Development and Management of Environmental and Natural Resources: Economic Outlook and Opinions, 2-volume set. Apple Academic Press, New York, pp. 1–428. Available at https://doi.org/10.1201/9781003187455.

Raj, A., Jhariya, M.K., Meena, R.S., 2025. Agroforestry for Monetising Carbon Credits. Springer, Cham. Available at https://doi.org/10.1007/978-3-031-97151-8.

Raj, A., Jhariya, M.K., Banerjee, A., Jha, R.K., Singh, K.P., 2024. Agroforestry. Wiley-Scrivener.

Raj, A., Jhariya, M.K., Banerjee, A., Nema, S., Bargali, K., 2023. Land and Environmental Management Through Forestry. Wiley-Scrivener. Available at 10.1002/9781119910527.

Ramakrishna, R., Rao, D.T., 2008. Strengthening Indian agriculture through dryland farming: need for reforms. Indian Journal of Agricultural Economics 63 (3), 460–476.

Ritchie, H., Roser, M., Rosado, P., 2023. CO_2 and greenhouse gas emissions, Our World in Data (2020). Published online at OurWorldinData.org. Retrieved from: 'https://ourworldindata.org/co2-and-greenhouse-gas-emissions' [Online Resource].

Rosas, F., Babcock, B.A., Hayes, D.J., 2015. Nitrous oxide emission reductions from cutting excessive nitrogen fertilizer applications. Climatic Change 132 (2), 353–367. Available at https://doi.org/10.1007/s10584-015-1426-y.

Samal, A., Sahu, S.K., Mishra, A., Mangaraj, P., Pani, S.K., Beig, G., 2024. Assessment and quantification of methane emission from Indian livestock and manure management. Aerosol and Air Quality Research 24 (6). Available at https://doi.org/10.4209/aaqr.230204.

Sikka, A.K., Alam, M.F., Mandave, V., 2022. Agricultural water management practices to improve the climate resilience of irrigated agriculture in India. Irrigation and Drainage 71 (1), 7–26. Available at https://doi.org/10.1002/ird.2696.

Singh, A., Upadhyaya, A., Kumari, S., Sundaram, P., Jeet, P., 2020. Role of agriculture in making India $5 trillion economy under corona pandemic circumstance. Journal of AgriSearch 6 (02). Available at https://doi.org/10.21921/jas.v6i02.18097.

Singh, R., Singh, H., Raghubanshi, A.S., 2019. Challenges and opportunities for agricultural sustainability in changing climate scenarios: a perspective on Indian agriculture. Tropical Ecology 60 (2), 167–185. Available at https://doi.org/10.1007/s42965-019-00029-w.

Singh, R., Srivastava, P., Singh, P., Raghubanshi, A.S., 2016. Sustainability indicators of the emergent agronomic practices: an Indian perspective. Environmental Science & Engineering: Climate Change and Sustainable Technology 12, 142–166.

Swamy, M., Bhattacharya, S., 2006. Budgeting anthropogenic greenhouse gas emission from Indian livestock using country-specific emission coefficients. Current Science 91 (10), 1340–1353.

Terang, B., Baruah, D.C., 2023. Techno-economic and environmental assessment of solar photovoltaic, diesel, and electric water pumps for irrigation in Assam, India. Energy Policy 183, 113807. Available at https://doi.org/10.1016/j.enpol.2023.113807.

Zhang, Z., Gao, X., Dong, W., Huang, B., Wang, Y., Zhu, M., et al., 2022. Plant cell wall breakdown by hindgut microorganisms: can we get scientific insights from rumen microorganisms? Journal of Equine Veterinary Science 115, 104027. Available at https://doi.org/10.1016/j.jevs.2022.104027.

Chapter 27

Action plans toward net-zero emissions

Krishna Kant Mina[1], Suchismita Satapathy[2], Satish Kumar Singh[1], Ramawatar Meena[1], Shashi Shekhar[3], Gourisankar Pradhan[2], Sanskriti[2], Sandeep Kumar[4], Manoj Kumar Jhariya[5], Arnab Banerjee[6], and Sanjeev Kumar[2]

[1]*Department of Soil Science and Agricultural Chemistry, Institute of Agricultural Sciences, Banaras Hindu University, Varanasi, Uttar Pradesh, India,* [2]*Department of Agronomy, Institute of Agricultural Sciences, Banaras Hindu University, Varanasi, Uttar Pradesh, India,* [3]*Department of Agricultural Engineering, Institute of Agricultural Sciences, Banaras Hindu University, Varanasi, Uttar Pradesh, India,* [4]*ICAR-Indian Agricultural Research Institute, Regional Station, Karnal, Haryana, India,* [5]*Department of Farm Forestry, Sant Gahira Guru Vishwavidyalaya, Ambikapur, Chhattisgarh, India,* [6]*School of Studies in Environmental Science, Pt. Ravishankar Shukla University, Raipur, Chhattisgarh, India*

Highlights

- Addressing the environmental impact of farming practices is required presently.
- Agricultural practices should strive toward net-zero emissions.
- Adoption of climate-smart practices for sustainable agriculture.
- Policy formulation and an action plan are required to fulfill the SDGs.

27.1 Introduction

The key aim of the "Green Revolution," which began in India in 1964–65, was to achieve food grain self-sufficiency. Over time, this revolution expanded to other regions of the country. In the period between 1960 and 1970, it caused the output of key grains like wheat (*Triticum aestivum*) and rice (*Oryza sativa*) to miraculously increase despite the decrease of coarse grain and legumes (Rajbhandari and Bhatta, 2008). Agricultural output increased dramatically during the Green Revolution due to the notable increase in yields per hectare. Between 1960 and 2000, yields enhanced for all rising nations by 208% for wheat, 109% for rice, 157% for maize (*Zea mays*), 78% for potatoes (*Solanum tuberosum*), and 36% for cassava (*Manihot esculenta*) (FAO, 2019). Beginning in the 1980s, Indian agriculture experienced problems because of the loss of natural and ecological resources, *viz.*, land and water; increased farming costs; and decreased productivity and farm production (Bezbaruah, 2014).

About 10%–20% of all human-induced greenhouse gas (GHG) output worldwide comes from agriculture, making it the world's largest contributor to GHG releases (Allen et al., 2019). By 2050, it is assumed that agricultural GHG emissions will reach 1.7 gigatons based on a baseline scenario. Approximately 13% of global anthropogenic GHG emissions are attributable to this sector (Wang et al., 2022). The top three countries in the world for carbon dioxide (CO_2) emissions are China, the USA, and the European Union (EU), with India coming in at number four (Jiang et al., 2019). The World Resources Institute Climate Analysis Indicators Tool estimated that India's total GHG emissions in 2014 were 3202 million Mt CO_2 or 6.55% of the world's total GHG emissions (Verma and Kumar Ghosh, 2022). Approximately 68.7% of GHG emissions in India are generated by the energy industry, followed by farming (19.6%), manufacturing (6%), land use change and forestry (3.8%), and waste (1.9%). Due to the burning of nonrenewable fuel sources and land-changing patterns, the gathering of GHGs, particularly CO_2, has grown by 40% since preindustrial times. This is mainly because agriculture not only exacerbates the effects of global warming but also plays a role in it by emitting large quantities of GHGs into the atmosphere. Current methods of cultivating arable land have raised serious worry about releasing more GHG into the atmosphere. Most GHG emissions in agriculture are due to the use and production of agricultural inputs, *viz.*, fertilizers, pesticides, farming equipment, excessive tillage, waste management, and water for agriculture (Meena et al., 2024a; Prasad et al., 2021).

Carbon Incentives, Plans, and Policies in Agriculture. DOI: https://doi.org/10.1016/B978-0-443-33557-0.00025-2

Additionally, important crops, *viz*., rice, maize, cotton (*Gossypium spp*.), wheat, and sugarcane (*Saccharum officinarum*) (which accounted for 80% of all crop emissions), and livestock like cattle (*Bos taurus)*, goats (*Capra aegagrus hircus*), buffalo (*Bubalus bubalis*), pigs (*Sus domesticus)*, and sheep (*Ovis aries*) (which accounted for 99% of all emissions related to livestock) contributed to India's overall CO_2 emissions (Garg et al., 2006). By 2030, these agricultural and animal species would have contributed 489 metric tonnes (Mt) of GHG release under the Business as Usual (BAU) scenario without reduction. Land consent, crop cultivation, irrigation, animal husbandry, fisheries, and aquaculture are agricultural practices that seriously affect GHG release (Pathak et al., 2022). However, 410 Mt CO_2 would be released under mitigation, suggesting a technical reduction possibility of about 78.67 Mt CO_2 annually. Technically possible mitigation methods might decrease agricultural emissions by about 18% overall (Sapkota et al., 2019).

Many world leaders gathered in Glasgow, UK, in 2021 for the United Nations (UN) Conference of the Parties (COP26) to seriously discuss the consequences of global warming. The conference intended to increase commitment to long-term progress regarding the Paris Agreement and UN Framework Convention on Climate Change by limiting the rise in global temperatures to 1.5°C over the preindustrial period (Michaelowa et al., 2018). Increased temperatures harm crops, leading to a significant decline in net growth and production. Heat stress conditions may impact agricultural productivity. Cropping practices and evapotranspiration are influenced by climate change, which in turn affects soil moisture retention, runoff, and plant water uptake. Moisture stress harms plants during their blooming, pollination, or grain-filling phases. The longer term concerns of poverty and the global food crisis are already a concern for emerging nations. Another concern is climate change, which could damage 11% of fertile land (Syers et al., 2008). Those in need of and rely on agriculture are most at risk from climate change. Crop failures and livestock losses are rising, causing significant economic harm and jeopardizing food security. Greater frequency of droughts and decreasing water availability could destroy significant portions of the tropics and impair water for drinking and irrigation in large communities of already vulnerable and underprivileged people (Rossing, 2010).

The significance of agriculture in India cannot be overstated; it is not merely an economic activity but a way of life for a significant portion of the population. However, the industry contributes significantly to GHG emissions from several sources, including CO_2 from shifts in land use, nitrous oxide (N_2O) from the use of fertilizers, and methane (CH_4) from the dairy sector (Meena et al., 2024a). As climate change poses increasingly severe threats to agriculture, including unpredictable weather patterns, water scarcity, and soil degradation, transformative management plans become imperative. Agriculture in India stands at the crossroads of progress and sustainability, grappling with the dual challenges of maintaining food security for an increasingly populous nation and mitigating the environmental impacts of its activities. India is just one of many countries that have embraced sustainable development (SD), which has led to the need to combat climate change and lower GHG releases, with its commitment to the Paris Agreement and the overarching aim of attaining net-zero emissions by 2070. To set the path for a more environmentally friendly and greener future, India needs to develop a thorough management strategy for the farming industry.

The management plans proposed here are designed to address the unique challenges of Indian agriculture, recognizing its diversity in terms of crops, regions, and socioeconomic conditions. The first pillar of the management plan focuses on mitigation strategies, seeking to reduce GHG emissions at the source. Precision farming techniques, such as sensor-based monitoring and variable rate technology (VRT), can optimize resource use, minimizing emissions from inputs like fertilizers and water. Organic farming practices, which exclude synthetic chemicals, not only contribute to emissions reduction but also enhance soil quality and ecosystem services (Meena and Pradhan, 2023; Meena and Pradhan, 2023). The second pillar revolves around adaptation strategies, acknowledging the inevitability of climate change impacts. Diversified cropping systems, integrated pest management (IPM), and rainwater harvesting are examples of climate-smart agriculture (CSA) methods that enable farmers to adapt to shifting environmental conditions while maintaining sustainable productivity. Research and development (R&D) efforts are crucial to breeding climate-resilient crop varieties that can thrive in the face of temperature extremes, erratic rainfall, and emerging pests and diseases. The third and perhaps most transformative pillar involves the overall transformation of agricultural systems toward sustainability. This entails a paradigm shift in energy use, embracing renewable sources for farm operations, and incorporating agroforestry (AF) practices for carbon sequestration (C_{seq}) (Jhariya et al., 2023; Jhariya et al., 2022). Technological advancements, including the adoption of precision agriculture (PA), artificial intelligence (AI), and blockchain for supply chain transparency, play a pivotal role in this transformation.

The suggested mitigation strategies for reaching net-zero emissions in Indian agriculture serve as a guide for protecting the environment and a means of establishing sustainable and resilient farming methods (Pradhan and Meena, 2023a). As India strives to meet its climate commitments, the agricultural sector stands as a linchpin in this endeavor, balancing the delicate need to feed a nation and safeguard the planet. Through collective action and innovative solutions, India can emerge as a global leader in sustainable agriculture, setting an example for the world to follow in the pursuit of a greener tomorrow. The agriculture sector adversely impacts climate change. Several agricultural activities, such as changes in cropping systems, reforming agricultural techniques for production, adaptation in natural farming (NF), reducing C footprint and water footprint, C stability in soil, and enhancing N use of efficient farming systems, may adopt precision input management (Meena and Pradhan, 2023). Net-zero emissions in agriculture are significantly influenced by farming practices that utilize smart tools for crop diversification, input management, and the adoption of renewable energy. Adoption of policies and action plans to reduce the greenhouse effect in India's farming sector is the aim of this paper. Crucially, effective policy frameworks are integral to the success of these management plans. Regulatory measures that incentivize sustainable practices, financial mechanisms that support the transition, and collaborative efforts involving government bodies, farmers, and the private sector are essential (Raj et al., 2025). These policies should promote the use of sustainable practices and ensure that the benefits and drawbacks of the shift are fairly distributed throughout the agricultural value chain.

27.2 Agriculture sector contributions to greenhouse gas emissions

Agricultural soils mainly contribute to the greenhouse effect by emitting and absorbing GHGs, *viz.*, CO_2, N_2O, and CH_4. In soil, CH_4 is generated when microbes break down organic matter (OM) under anaerobic circumstances. Therefore, an abundant spring of CH_4 may be present in rice fields. Continuous submersion, a multistory level of organic carbon (OC), and the addition of compost to wet soil all encourage CH_4 release. Additionally, the increase in global CH_4 budget is due to burning crop residue. CH_4 comes from a significant additional source, ruminant enteric fermentation. Nitrogenous fertilizers are a producer of N_2O in fertilized soils; however, in unfertilized soil, native N contributes to the emission of this GHG. As long as an appropriate nitrate source is available, soil moisture content and C content increase the formation of N_2O. Typically, after irrigation and precipitation, a spike in N_2O emissions is seen. The global N_2O budget is also increased by burning crop leftovers. Because they start the biological degradation of soil organic matter (SOM) and release CO_2, soil management methods like tillage are agriculture's key source of CO_2 emissions. To hasten the decomposition process, soil aggregates are broken up, oxygen delivery is increased, and the OM's surface area is exposed. Burning crop waste and utilizing fuel for various agricultural activities are two additional sources of CO_2 emissions. In 2005, the largest GHG in the agriculture sector was 38% from soil emission (CH_4+ N_2O), then enteric fermentation (CH_4) contributed 32% of the emission, and 12% was attributed to the burning of biomass (CH_4+ N_2O). In contrast, rice production CH_4 contributes 11%, the minimum contribution of GHG by 7% manure (CH_4 + N_2O) (Fig. 27.1).

27.2.1 Greenhouse gas emissions from Indian agriculture

27.2.1.1 Methane

One of the most complex alterations affecting CH_4 emissions is water management. A rice field's CH_4 emission effect shows how the soil's moisture content influences the peak flow. In comparison to intermittent soil conditions of wetting and drying, continuously saturated rice fields produced more CH_4 emissions (Arai, 2022; Liao et al., 2020; Meena et al., 2025). The mean release during alternate drying and wetting varied between 0.1 and 0.4 kg ha^{-1}; however, in saturated soil, it ranged from 0.3 to 0.6 kg ha^{-1}. Because methanogens are strictly anaerobes, soil saturation generates anaerobic

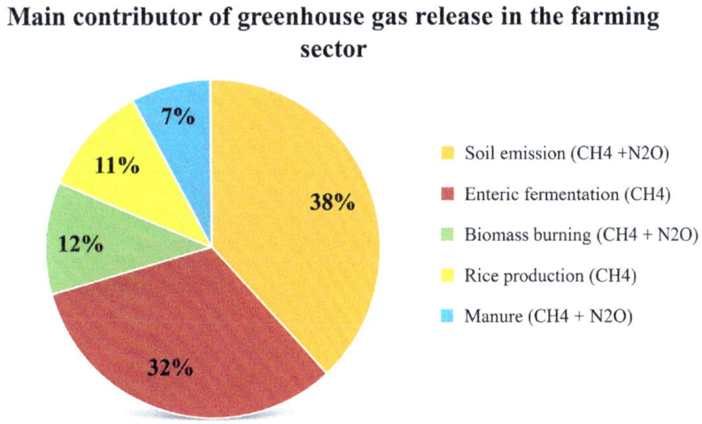

FIGURE 27.1 Sources of greenhouse gas (GHG) release from the farming industry.

conditions that are favorable for CH_4 generation. CH_4 generation nearly ceased when soggy soils were permitted to completely dry and become aquarobics. For instance, by providing the OC and N necessary for microbial growth and acting as an electron source, farmyard manure (FYM) increases CH_4 output. When 50% of mineral N was replaced with FYM instead of applying the complete amount of N through urea, the emission rate increased by 172% (Pathak et al., 2003). Another study found that using the full amount of N from organic sources resulted in the highest CH_4 emission (41.3 kg ha^{-1}) and the lowest (28.4 kg ha^{-1}) CH_4 emission rates (Bhatia et al., 2005). However, in contrast to FYM, the use of biogas slurry cuts emissions by 2.3 times, indicating that this approach is more appropriate for lowering CH_4 emissions (Lonappan, 2015). CH_4 emission is significantly influenced by the soil's organic carbon (SOC) concentration, texture, cation exchange capacity, and total nitrogen concentration (Adhya et al., 2009). Compared to the subsoil, the uppermost soil (0–15 cm depth) generates the majority (99%) of the CH_4 emission (Adhya et al., 2009). Different rice varieties emit CH_4 at significantly different rates (Jiang et al., 2017). Pusa 933 had the highest seasonal emission (27.2 kg ha^{-1}), and Pusa 169 had the minimum (15.6 kg ha^{-1}), with Pusa 1019, Pusa 834, Pusa Basmati, and Pusa 677 having intermediate values (Pathak et al., 2010). CH_4 emissions from soil can be decreased by 10–15% by using inhibitors of nitrification, such as dicyandiamide (DCD), neem oil, coated Ca-carbide, and neem-coated urea (Majumdar et al., 2002).

27.2.1.2 *Nitrous oxide*

By providing the bacteria with the OC and N they require, the use of manure plays a significant role in the production of N_2O. Denitrifiers use OC compounds as electron donors to produce biological components and energy. N_2O emissions increased when organic sources, such as FYM, OM, and agricultural waste, were applied to rice and wheat (Bhatia et al., 2005). N_2O emission increased as a result of the cultivation of legumes, however, to varying degrees depending on the crop (Ghosh et al., 2014). The type of fertilizer used and the soil's moisture content have the biggest effects on N_2O emissions. The nitrification of ammonium-N caused by soil applied with ammonium sulfate, urea, or ammonium nitrate resulted in the emission of more gases at field capacity than in soil fertilized with potassium nitrate. Nitrate-containing fertilizers emitted more N_2O while submerged. The findings indicated that improving irrigation techniques and using the appropriate fertilizer could help reduce emission levels (Pathak and Nedwell, 2001). By either preventing nitrification or reducing the amount of NO_3 available for denitrification, nitrification inhibitors decrease the amount of N_2O released. Examples of substances that function as nitrification inhibitors and reduce the formation of N_2O include neem cake, neem oil, and Karanja seed extract (karanjin) (Majumdar et al., 2002). The thiourea derivatives S-benzylisothiourea butanoate (SBT-butanoate) and S-benzylisothiourea furoate (SBT-furoate), which act as nitrification inhibitors, improved wheat output. Further, it also helps in reducing global warming potential by 8.9–19.5% (Bhatia et al., 2010). Tillage is essential for the soil's ability to emit N_2O. The total N_2O-N release increased by 12.2% when comparing conventional tillage to no-tillage (Bhatia et al., 2010).

27.2.1.3 *Carbon dioxide*

Tillage is the primary cause of CO_2 emissions from agricultural soil because it breaks up soil aggregates, boosts O_2 supply, and exposes OM to the surface, which encourages biological decomposition. Burning crop residue also raises CO_2 emissions, while off-site sources include fertilizer, herbicides, and agricultural machinery.

27.2.2 Greenhouse gas emissions from crop residue burning

In most Indian states, especially western Uttar Pradesh, Punjab, and Haryana, crop residue is burned in the fields, producing carbon monoxide (CO), CH_4, N_2O, SO_2 (sulfur dioxide), and many other gases. Rice, maize, wheat, jute (*Corchorus spp.*), cotton, sugarcane, rapeseed-groundnut (*Arachis hypogaea*), and mustard (*Brassica spp.*) are the nine crops whose remains are typically burned in fields. For rice, the percentages of stubble burned in the field were derived from (Gadde et al., 2009), while the percentages of residues oxidized were discovered in (Streets et al., 2003). The residue-burning release factors were extracted from (Andreae and Merlet, 2001). In Punjab, western Uttar Pradesh, and Haryana, burning rice residue on a large scale results in significant problems with pollution, health hazards, and nutritional loss in addition to the release of GHGs (Pathak et al., 2006) (Table 27.1).

27.2.3 Current status of greenhouse gases from a global perspective

The concentration of GHGs that trap heat in the air increased at a rate higher than normal between 2011 and 2020, setting a new record high for the amount of these gases in the atmosphere. In 2021, the pattern has remained consistent, according

TABLE 27.1 Sources of greenhouse gas (GHG) emission and adaptation methods to cut it off.

S. N.	Greenhouse gases	Source of greenhouse gas emission	Practices adopted to cut off greenhouse gases
1.	CO_2	Farm equipment is seen working across the field while tilling, planting, applying pesticides and fertilizers, and harvesting crops	**(a)** Sequestration of C (Küstermann et al., 2007) **(b)** Agriculture number of till (Berner et al., 2008; Teasdale et al., 2007) **(c)** The use of conservation farming (Pimentel and Burgess, 2014) **(d)** Mycorrhizal fungi (Mäder et al., 2000) **(e)** Compost (Hepperly et al., 2008) **(f)** The integrated crop-livestock system (Beauchemin and McGinn, 2005) **(g)** Ruminant (Beauchemin and McGinn, 2005) **(h)** Cover crops (Thorup-Kristensen et al., 2003) **(i)** Manure (Fließbach and Mäder, 2000) **(j)** Farming that is organic (Burger et al., 2008; Löschenberger et al., 2008)
2.	CH_4	**1.** Enteric fermentation, which causes ruminant animals like cows, buffaloes, goats, and sheep to release CH_4, is one source of manure **2.** Other sources include compost **3.** Deep animal bedding **4.** Livestock from pastures and liquid manure kept in lagoons	
3.	N_2O	**1.** Use of artificial fertilizer **2.** Animal waste from confinement operations	

to the World Meteorological Organization (WMO) GHG Bulletin. CO_2 is the primary GHG, and as of 2020, its content was 421 parts per million (ppm), which is 149% more than preindustrial levels. CH_4 and N_2O concentrations were 262 and 123% higher, respectively, in 1750, when human civilization first started disrupting the globe's natural balance. As long as emissions persist, global temperatures will rise. Due to the lengthy half-life of CO_2, even if emissions are swiftly brought to zero, the previous temperature will continue for several decades. This will lead to more severe weather, including intense heat and rain, ice melt, elevated sea level expansion, and acidification of the oceans, in addition to rising temperatures and widespread socioeconomic effects. Contribution to GHG emissions by different sectors play a different role in their source, and contributions are shown in Table 27.2.

27.2.4 Major cropping systems and their contributions to greenhouse gas emissions

The contribution of various field crops to the overall national GHG emissions varies greatly. 68% of field crop GHG emissions are attributable to the production of cereal crops, whereas only 7% are attributable to the cultivation of vegetables. The mean annual emissions from the utilization of land related to forestry and farming accounted for nearly 17% of GHG emissions from field crop cultivation between 2000 and 2010 (Meena et al., 2023).

27.2.4.1 Maize–groundnut cropping system

In the maize–groundnut system, groundnuts produced roughly 1.7 and 3.8 times more GHG emissions per tonne of the noble component than maize did under monospecific cropping, respectively. This might mean that the production of proteins and lipids has a higher environmental cost than carbs. This is not unexpected because protein and lipid synthesis require more complex mechanisms than carbohydrate synthesis, resulting in higher energy demands and higher CO_2 emissions.

27.2.4.2 Maize–soybean cropping system

The most significant reduction of GHG emissions is through the adaptation of several practices, such as the maize–soybean intercropping system, which shows higher sustainability than the maize–soybean monocropping system (Huang et al., 2013). Soybeans were chosen as the legume crop to preserve the soil because they are known to produce less C than other crops (Huang et al., 2013; Jensen et al., 2012). Peas share traits comparable to those of soy crops. According to Shen et al. (2018), compared to monoculture maize, intercropped maize's GHG emissions dropped by 7% during the growing season. The reason for this was that intercropped maize was better at recapturing NO_3, NH_4, and water. The

TABLE 27.2 Source and contribution to greenhouse gas (GHG) emission by different sectors (Gołasa et al., 2021).

S. N.	Greenhouse gas emissions in the sector	Total share (%)	Source of greenhouse gas emission
1.	Transportation	29	Fossil fuel burning for land, rail, cars, and maritime transportation
2.	Electricity and heat generation	25	Burning of coal, oil, and natural gas, heating homes, and lighting buildings
3.	Industry	23	Waste management practices (non-energy-related chemical, metallurgical, and mineral transformation processes)
4.	Commercial and residential	13	Fossil fuel burning for heat
5.	Land use change and forestry	12	Absorbing CO_2 from the air as a source of emission
6.	Agriculture **(a)** Rice CH_4 (2%) **(b)** Energy (agriculture 6%) **(c)** Soil fertilization (51%) **(d)** Ruminant wastes on pastures (2%) **(e)** Manure management (12%) **(f)** Ruminant enteric fermentation (27%)	10	Raising of crops and livestock, deforestation

maize–soybean intercropping also improved N use efficiency, land productivity, and length of root, which would lead to greater land production while using less fertilizer (Chen et al., 2017; Ren et al., 2017).

27.2.4.3 Maize–forage (ryegrass) cropping system

Ryegrass was cited by Parajuli et al. (2017) as a crop with a high potential for GHG emissions. With the global rise in protein demand, there is an increased need for animal feed and fodder. Ryegrass is a good quality crop that may be cultivated for forage in winter. Through technology like nitrification inhibitors, which have been demonstrated to reduce N_2O emissions from unfertilized soil, research is being conducted to mitigate the C emissions of ryegrass.

27.3 Reforming agricultural techniques for sustainable production

To evaluate climate sustainability, scientists and directors have resorted to life cycle analysis (LCA), which takes into account all GHG causes and sinks that are directly or indirectly related to agriculture (Sieverding et al., 2020). A typical LCA for agriculture encompasses all functions related to creating agricultural inputs, from the farm gate to the end user, and evaluates crops for their energy use, GHG emissions, and other environmental impacts. To evaluate biofuels based on agriculture, the LCA has been employed extensively. According to studies that evaluated the sustainability of biofuel production using the LCA method, farming operations are largely responsible for the GHG emissions of several biofuels, including 36% of bioethanol made from corn (Wang et al., 2021). In the modern day, LCAs are used to evaluate agricultural sustainability on a nationwide level (Sieverding et al., 2020). To establish priorities and focus on future GHG mitigation initiatives, LCA studies in the agriculture sector should not only compute GHG release levels but also pinpoint ecological hotspots. For instance, one of the leading causes of GHG emissions in agriculture is the N_2O emissions from chemical N fertilizers. Goals being worked toward include decreased N fertilizer consumption and N_2O emissions from N fertilizers.

The following practices were applied to reform the agricultural techniques.

27.3.1 Manufacturing low-carbon fertilizers and chemicals

These "greener" methods can considerably lessen agricultural GHG emissions and the sector's reliance on fossil fuels. Ammonia, the main component of all N fertilizers, can be created from water, air, and renewable electricity generated by solar or wind energy (Liu et al., 2020). Ammonia GHG emissions can be reduced by 91%, from 2.55 to 0.22 g CO_2 per gram of ammonia, when manufactured from N_2 and H_2 utilizing cryogenic distillation and low-temperature electrolysis, respectively. If all N fertilizers used for corn growth are made from green ammonia instead of the traditional method of manufacturing ammonia from natural gas using steam CH_4 reforming, the beginning to farm-gate GHG release of maize grown in the US can be decreased by 18% (Wang et al., 2020).

27.3.2 Reducing farm energy and fuel consumption

Petroleum products, natural gas, and electricity are all required fuels for agricultural equipment (such as trackers and pumps) that are used in several farming operations. Since the early 1990s, some initiatives, including the USDA's Agricultural Resource Management Survey (ARMS), have regularly evaluated on-farm energy use in the United States. In a current publication, the USDA's Economic Research Service provides a summary of the ARMS statistics. (Hitaj and Suttles, 2016), which also provides recommendations on how to reduce GHG emissions associated with farm energy use. These suggestions include using conservation tillage practices, switching to renewable energy produced on-farm rather than relying on the grid, and employing anaerobic digesters to convert animal waste into biogas as a secondary energy source. Tillage that is decreased through the use of conservation techniques is likely to minimize GHG emissions by synergistically increasing soil C absorption while also reducing on-farm fuel use. A different approach to lower the GHG emissions related to farming is to switch from natural gas to renewable energy, particularly bioenergy (Pradhan and Meena, 2023b). When compared to fossil fuels, biodiesel made from animal fats and vegetable oils can reduce GHG emissions by 24–73% during its whole life cycle (Wang et al., 2020). Anaerobic digestion of biomass can also produce biogas, which can be exported to other industries to meet the demand for renewable energy in various sectors (e.g., transportation, power). Biogas, which is primarily composed of CH_4 and CO_2, is produced through anaerobic digestion when bacteria break down agricultural waste, such as compost or crop leftovers. Additionally, biogas can be processed into renewable natural gas (RNG), which can be used as a substitute for regular natural gas. When compared to fossil natural gas, RNG can cut GHG release by 136% (in cattle manure) and 317% (in swine manure) throughout a lifetime (Wang et al., 2020). Due to RNG production there has been a significant reduction in CH_4 emissions from animal manure.

27.3.3 Enhancing the soil organic carbon pool

There is a possibility of storing soil C during farming by converting crop residue after harvest into SOC, even though some land-use shifts, such as converting native grassland or woodland to agricultural land, would cause soil C to be emitted as CO_2. Conservation measures often increase soil C inputs and decrease soil C losses to encourage soil C_{seq} (Paustian et al., 2016; Roy et al., 2022). GHG emission reduction through regenerative agriculture is urgently needed.

Tillage is essential for managing crop waste and preparing a seedbed for succeeding crops; however, excessive tillage might be the primary cause of soil disturbance that enhances both CO_2 emissions through increased microbial activity and physical soil C loss due to soil erosion. Farmers can now apply conservation tillage techniques that result in lower farmed energy consumption and higher soil C_{seq} because of advancements in tillage technologies and agronomic techniques (Paustian et al., 2016; Raj et al., 2021). Conservation agriculture has been found to improve soil C_{seq} in some regions when combined with crop rotations that incorporate cover crops during the fallow and off-seasons (Jian et al., 2020). C can be stored in soil when livestock dung is applied. However, the net GHG removal from compost use should take into account the fuels used to transport manure from animal feeding operations to farms for application in soil (Qin et al., 2018). Additionally, innovative methods are being executed, like developing crop varieties with deep roots that retain more C in the roots' dry matter (Paustian et al., 2016) (Fig. 27.2). To balance soil health and long-run farmland productivity while creating cellulosic biofuels, research has examined how to use biomass wastes, such as stover residue, in high-biomass production scenarios (Wilhelm et al., 2004). Crop leftovers can be used as a bioenergy fuel while retaining the soil's C pool if they are properly handled. Early research raises questions about how much stover is being removed, which could affect soil C (Xu et al., 2019).

27.3.4 Natural farming

NF is a set of practices that aims at agroecologically based, varied agricultural systems that integrate crops and trees with animals that embrace functional biodiversity while using no chemicals (Rosset and Martínez-Torres, 2012). NF is another name for "Zero Budget Natural Farming" (ZBNF). They have been widely adopted in numerous states, including Himachal Pradesh, Andhra Pradesh, Karnataka, and Maharashtra. By using natural fertilizers like Jeevamritha, Beejamritha, and Neemastra instead of chemical fertilizers, it has reduced production costs. Intercropping and mulching

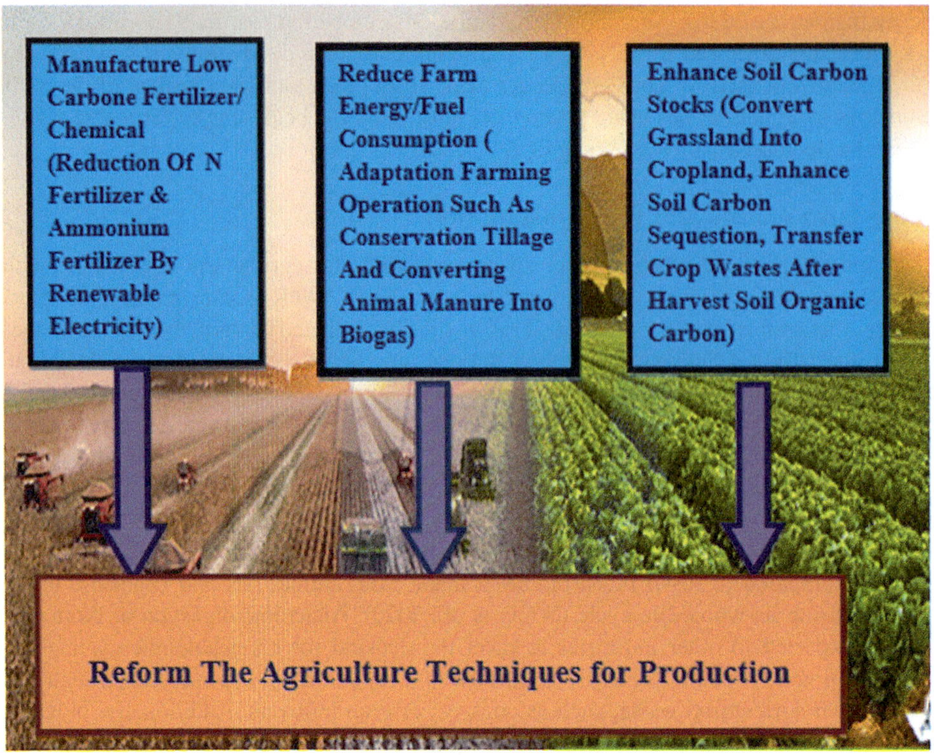

FIGURE 27.2 Techniques to enhance soil organic carbon stocks.

have also been utilized (Palekar, 2006). To attain agricultural sustainability and lower the C footprint, a sizable portion of FYM is added to the soil to improve soil health, SOC, and organic farming practices (Banerjee et al., 2020). Highlights ZBNF as one of the most promising agricultural policies for increasing farmers' incomes while also reducing fertilizer use and improving farm income. The alternative cultivation methods, like NF and ZBNF, are classified as organic farming models. ZBNF's primary goal is to reduce the use of agrochemicals and preserve agricultural productivity in a sustainable environment. Additionally, it supports climate-friendly agriculture practices, soil fertility, and the restoration of SOM.

27.3.5 The zero-budget natural farming method of Subhash Palekar

The ZBNF seeks to enhance soil structure by incorporating OM and microbial inoculants. One NF technique that can promote nutrient decomposition and recycling is microbial culture. Other NF practices include using local seeds, integrating crops with trees and livestock, properly spacing crops, conserving water, mulching heavily, planting many intercrops, and rotating crops (Fig. 27.3).

27.3.5.1 Natural farming practice

There are primarily four crucial elements:

27.3.5.1.1 Jeevamritha/Jeevamrutha

This fermented microbial culture serves as a promoting factor to enhance soil microbial activity and promote the growth of earthworm populations.

Jeevamritha manufacturing procedure

200 L of water should be placed in the barrel.
Add 10 kg of recent, local cow manure.
Add 5–10 L of old cow urine.
Include 2 kg of jaggery.

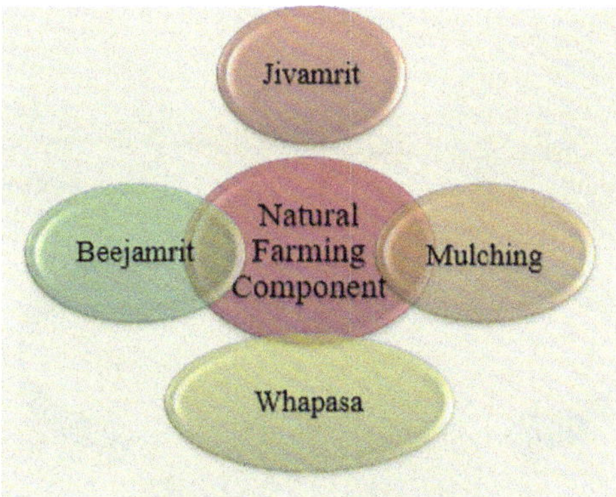

FIGURE 27.3 Components of natural farming.

Add 2 kg of flour made from pulses.

Mix the solution thoroughly and add a small amount of farm soil. Ferment the mixture for 48 hours in the shade. Cow dung and urine contain both aerobic and anaerobic bacteria, which grow and consume OM. They aid in preventing bacterial and fungal plant diseases as well.

How to use Jeevamritha is applied to crops as a 10% foliar spray or irrigation water, and it can be kept for up to 15 days. Applying drip irrigation is Jeevamritha in Maharashtra.

27.3.5.1.2 Beejamritha

They were also employed for treating seeds and other planting-related materials. The prevention of fungal, soil-borne, and seed-borne diseases in young roots is another benefit.

Beejamritha production process Combine the neighborhood cow dung, natural fungicide, cow urine (which has antibacterial properties), lime, and seed. For a 12-hour urine-leaving period, the excrement is bound in a cloth. Cow dung is pressed, cow urine is removed, and 50 grams of lime is combined.

How to use As a sowing agent, Beejamritha is applied to seeds, coated, and mixed well by hand. Just quickly dip the legumes in the crop, then let them dry.

27.3.5.1.3 Acchadana: Mulching

Mulching is the use of any substance to cover the soil in the ZBNF through the practices adopted. Mulch the soil to protect the upper layer, promoting aeration and conserving moisture while avoiding the need for deep tillage. Straw material may be used as dried biomass of crop residue from previous crop activity, of soil biota, of OM decomposition, and humus production. In multiple cropping systems, monocotyledons and dicotyledons are cultivated on the same field to supply every necessary nutrient for the crop, and soil dicotyledons like pulses fix atmospheric N, and cereals supply nutrients like potash, phosphate, and sulfur.

27.3.5.1.4 Whapasa: Moisture

It is a state in which soil contains both air and water molecules.

27.4 Reducing the footprint of water use

Water must be used as a preceding resource in a manner helpful for humanity (Bhatt and Kukal, 2017). Large amounts of water are utilized globally, especially the subterranean water table, to meet the demands of the continuously increasing population, according to (Bruinsma, 2009). Additionally, it is necessary to transition from flood irrigation to alternative wetting or drying methods of irrigation to reduce global water demand, particularly in rice-based farming. This is because flood irrigation and excessive fertilization will enhance insect-pest and disease attacks, lower yield levels, and ultimately reduce the region's water and land productivity. Flood irrigation is needed for the growth of traditional crops, particularly rice, and it contributes to the release of GHG like CO_2, CH_4, and others that cause global warming. In any event, employing efficient GHG control technologies will enable the reduction of water footprints (WFs) (Fig. 27.4).

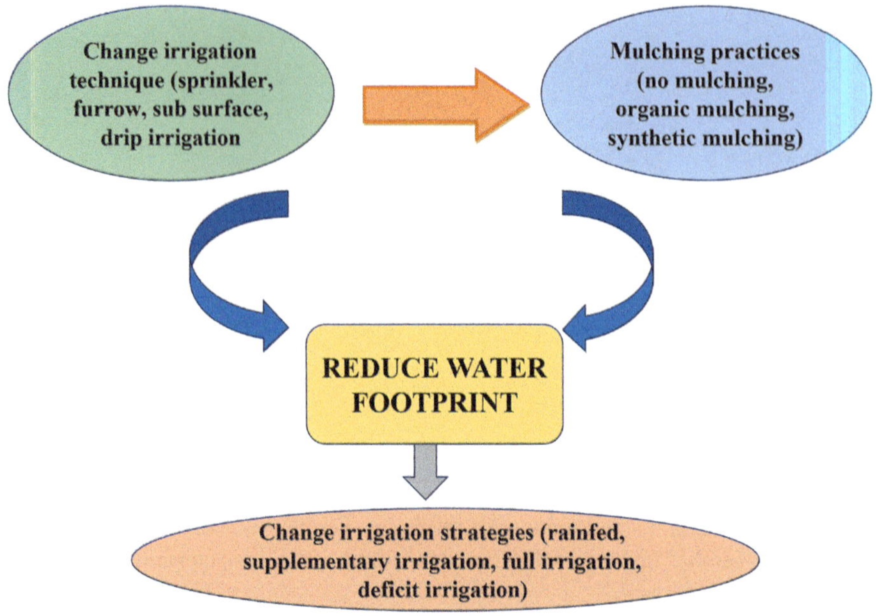

FIGURE 27.4 Techniques to reduce water footprint.

World GHG emissions grew to 49.5 megatons of CO_2 equivalents in 2010, the highest amount ever, as per the IPCC's Fourth Assessment Report from 2007 (Pachauri, et al., 2014). By lowering the WFs in rice-based farming, GHG emissions must be minimized. Additionally, due to the rice's puddle-flooded conditions, GHG emissions are to blame for the release of NO_2, CO, CO_2, CH_4, etc., which improve irrigation demand (Simpson et al., 2014). As a result, there is a connection between the flooded irrigation conditions for rice, GHG emissions, longer dry spells, greater irrigation inputs, and WFs. Things became more difficult due to the effects of climate change, which include relatively higher CO_2 and general temperatures, especially in the rainfed areas. A coordinated strategy for lowering WFs in South Asia must involve agronomists (for new planting techniques), soil scientists (for innovative approaches to irrigation under texturally distinct soils), entomologists or pathologists (for reducing insect, pest, or disease infestation); political leaders (for framing after keeping in mind the socioeconomic circumstances), and breeding experts (for new plant cultivars with lower WFs and that might tolerate stressed situations). Field crops can make better use of water by keeping crop residues on the outermost layer, cementing the water channels, or installing irrigation pipelines below ground. Additionally, under single-crop systems, claimed lower water inputs due to the adoption of any technique may result in higher water inputs in the following crop, yielding nearly the same water inputs for the entire system. For the past ten years, most research on a single crop—mainly rice—has been published in prestigious journals like *Field Crop Research* and *Agricultural Water Management*. These studies have often included complex conclusions or suggestions for lowering WFs without accounting for the time that has passed or the next upland crop. As a result, the importance of the latter two must be understood. To produce irrigated crops systematically and sustainably, the agricultural industry must reduce its WF. This involves altering management methods, such as mulching; irrigation methods (drip/subsurface drip irrigation, furrow, and sprinkler), and irrigation techniques (rainfed, supplemental irrigation, and deficit irrigation). Using drip or subsurface drip irrigation systems and systemic mulching techniques to lower the consumptive WFs for tomato production.

27.5 Carbon footprint

In an ambient condition, other GHGs are created as a byproduct or as part of a process. Recent investigations have shown that the C footprint is one of the key environmentally friendly indicators linked to global warming (Meena et al., 2024b). The biggest threat to climate change and the continuous increase in ocean and atmospheric temperatures in the twenty-first century is also thought to be the C footprint (Abbott, 2008). The rise in GHG content in the air was widely believed to be caused by deforestation, altered land use patterns, soil burning, etc. Some GHGs, including CO_2, N_2O, CH_4, HCFCs (Hydrochlorofluorocarbons), and ozone (O_3), have been recognized as key sources of atmospheric effluence. According to

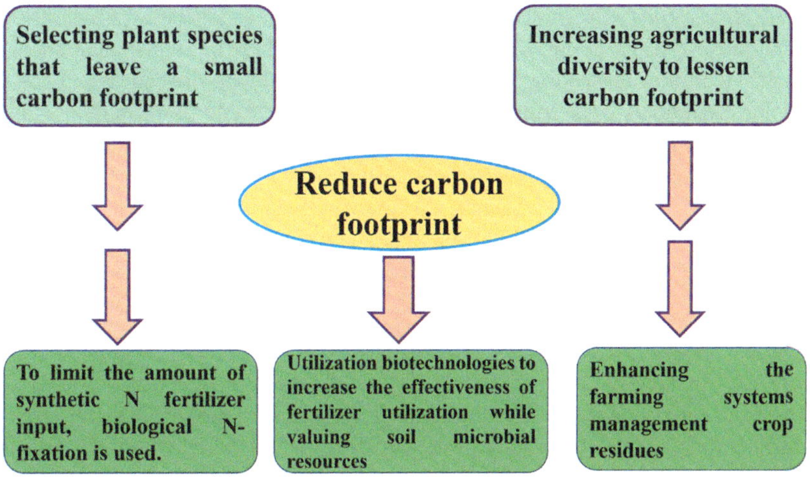

FIGURE 27.5 Techniques to reduce carbon footprint.

calculations, CO_2 emits the majority of GHGs caused by human activity, accounting for 63% of all GHGs released into the atmosphere (Bhatt, 2023). It may reduce as presented in Fig. 27.5.

27.5.1 Reducing carbon footprint

27.5.1.1 Selecting plant species that leave a small carbon footprint

A grain product's C footprint relies on plant species, agriculture technology, and environmental factors. The primary sources of emissions are transportation, farming, storage, and the use of N fertilizers, among other things. Breakdown of leftovers in both production N fertilizer applications makes up between 57 and 65% of the total reduction of crop waste, and the total emission of another 16–30%, depending on the percentages. The manufacture of phosphorus fertilizers, herbicides, and fungicides accounts for more than 13–18% of the overall emissions, compared to other field operations. Denitrification primarily produces N_2O and is heavily impacted by soil conditions and its moisture content (Flynn et al., 2005). Rainfall increases N_2O emissions under moisture-limited circumstances (Flynn et al., 2005). The three N-fixing crops, also known as pulse crops, lentils (*Lens culinaris*), chickpeas (*Cicer arietinum*), and dry peas (*Pisum sativum*), had a mean emission of 339 kg CO_2e ha^{-1}, which was 65% lower than the emissions of spring wheat and canola. Overall crop emissions are typically closely linked to the quantities of N fertilizer used, with crop production and N levels in different crop components, such as roots and straw, acting as modifiers. (Drinkwater et al., 1998) observe that legume-based cropping systems improve SOC and reduce N emission compared to cereal-based systems. Because of this, the impact of crop type on SOC is considered to be fairly minor in comparison to the impact of soil N and other parameters in our assessment of the C footprint.

27.5.1.2 Increasing agricultural diversity to lessen carbon footprints

The C footprint of crop products can be decreased by using varied cropping techniques. In a field study in southern Saskatchewan, (Gan et al., 2003) observed that, compared to monoculture systems, diverse cropping systems significantly reduced production inputs and increased durum wheat grain and straw yields. Improved agronomic techniques, such as early planting, optimal plant population density, and suitable crop rotation sequences, have been shown in multiple studies to increase crop yields without relying on rice as an agricultural input (Gan et al., 2010; Kirkegaard et al., 2008). It has been recognized that controlling weeds with a range of cropping patterns serves as one of the most crucial elements in the development of ecologically sound agricultural systems (Menalled et al., 2001). Weeds are susceptible to a variety of factors that can cause their death in crop rotations. Weeds are subject to a range of pressures when crops are grown with various planting and harvesting dates, morphologies, and phenologies, as well as competitive traits and crop leftovers. Increasing crop diversity in a rotation exposes weeds to a wider variety of cultivation techniques. These crop species transitions produce microenvironments that are unsuitable for the growth and spread of any specific weed species. Thus, crop variety reduces the amount of weeds while increasing the production of agriculture (Westerman et al., 2005) and pesticide input (Menalled et al., 2001). Herbicide use in crop production can be significantly reduced by employing alternative cultural techniques (Harker et al., 2009). For instance, when a high-management technique (i.e., more competitive varieties, higher-than-normal seeding rates, and rotating cereal with oilseed crops) was utilized together with a half rate (1/2%) of herbicides, wild oats were controlled to a level similar to that of a low-management approach (i.e., less competitive cultivars, normal to lower seeding rates, and cereal monoculture) combined with a full rate (1%) of herbicides (Harker et al., 2009). These findings suggest that varied cropping systems can significantly lower the number of pesticides used during crop production, lowering the food products' C footprint. Even though pesticides only make up a small portion of the estimated C footprint, improving crop health through better agronomic management offers opportunities to increase crop productivity while also lowering the C footprints associated with the cultivation of field crops.

27.5.1.3 To limit the amount of synthetic nitrogen fertilizer input, biological N fixation is used

Agriculture can decrease its dependency on synthetic N fertilizers and, consequently, its C footprint by incorporating legumes into crop rotations to utilize biological N-fixation (Crews and Peoples, 2004). Research has demonstrated that better N use and its application techniques, such as side-dressing, site-specific methods, and timely application, can significantly boost N use efficiency (Malhi et al., 2001; Peng et al., 2010). Effective solar-driven N_2-fixing systems, such as legume–rhizobial associations, convert atmospheric N_2 into ammonia without causing net CO_2 emissions. Pulse crops fix their own N needs and also leave some of their biologically fixed N in the soil. The genuine share of biologically fixed N by pulses has frequently been underestimated due to the neglect of the rhizodeposition, nodules, and roots that are left in the soil after crop cultivation. Chickpeas, for example, contribute 58 kg ha^{-1} yearly to the soil N pool, despite being previously considered a poor source of soil N with an N balance near zero. Taking into account the release of mineralized N from roots and nodules, this quantity could be considerably greater (Herridge et al., 2008). From agricultural areas, a significant quantity of N is exported as legume grain. N fixation is a mechanism by which pulse plants fix a significant amount of atmospheric N to fulfill their demands for N for growth and development and to add to soil N. Growing conditions like dryness and high soil nitrate levels can harm nitrogenase activity in pulse nodules, which lowers N fixation (Marino et al., 2009; Walley et al., 2007). By biologically fixing N, pulses not only help to cut the amount of N fertilizers used in farming operations but also, through as-yet-unidentified processes, improve the yield of subsequent crops (Gan et al., 2003; Kirkegaard et al., 2008). Pulses generate around 21 million tonnes of fixed N annually, which makes up one-third of all biological N_2 fixation in agricultural methods (Herridge et al., 2008). Increasing the frequency of pulses in crop rotations, reducing the adverse impacts of high soil N residual on N_2-fixation, and enhancing the occurrence of N mineralization from pulse leftovers and the peak N needs of the subsequent crop are all feasible. (Marino et al., 2009).

27.5.1.4 Utilizing biotechnologies to increase the effectiveness of fertilizer utilization while valuing soil microbial resources

Because several bacteria and fungi affect how plant roots grow, they can act as biofertilizers and biopesticides, and soil microbes can be used to enhance the effectiveness of these nutrients (Hynes et al., 2008). By altering the metabolic processes of plants, soil microorganisms such as arbuscular mycorrhizal fungi (AM fungi), P-solubilizing fungi and bacteria, and N_2-fixing bacteria can improve plant growth and nutrition. Numerous bacteria generate enough auxins, gibberellins, cytokinins, and ethylene to promote the growth of plants, improve root branching, or promote shoot progress (Van Loon, 2007). Field crops in the prairie region can yield various rhizobacteria that encourage plant growth (Hynes et al., 2008). Numerous bacteria are equipped with the enzyme 1-aminocyclopropane-1-carboxylic acid (ACC) deaminase and feed on ACC, an ethylene precursor. They eliminate ACC from the rhizosphere, promoting plant development and root elongation. To mitigate the adverse consequences of environmental stressors caused by toxic metals, the AM fungi are also known to increase hormone production, bind metals, change the concentration of electrolytes in plant cytoplasm, and draw water from the soil (Audet and Charest, 2008), high temperatures (Paradis et al., 1995), and severe drought on their host plants (Hynes et al., 2008). It is becoming clear that some endophytic fungi, such as dark septate endophytes, are significant plant allies. In challenging conditions, the prevalence of these endophytes can improve the host plants' ability to withstand drought (Yang et al., 2020). Biopesticides encompass bacteria as well as fungi. While *Trichoderma* and other species like AM fungi are antagonists that produce toxins, AM fungi can "sanitize" the rooting soil via baiting and nutritional interactions (Reino et al., 2008). Through siderophore-related Fe sequestration and starvation, *Pseudomonas fluorescence* also inhibits soil-borne pathogens (Choudhary et al., 2009). Systemically acquired resistance (SAR) and systemically induced resistance (SIR) to infections, which microorganisms might use to activate or "prime" plant defense mechanisms, can also affect gene expression (Van Loon, 2007). Rhizobacteria that support plant growth generate SAR and SIR in a two-way process that resembles legume–rhizobia cross-talk. Plants may also synthesize bioactive chemicals, which opens the door to regulating plant disease resistance with synthetic signal molecules. To regulate the agricultural rhizosphere and reduce the number of agrochemicals used on farms, compounds that target the microbial companions of plants could be utilized. This would reduce the C footprint of crops because bioactivity requires only small quantities of these signal molecules.

27.5.1.5 Enhancing the farming system's management of crop residue

Through conservation tillage, stubble that is kept on the soil's surface can offer the following advantages: increase soils' ability to sequester CO_2 from the atmosphere; reduce fuel consumption and GHG emissions; save soil surfaces from erosion (wind and water), which reduces the need for fertilizer (Malhi et al., 2001). Improved plant-mycorrhizal relationships, enhanced soil aggregate stability, rainfall capture effectiveness, and water-holding capacity boost biodiversity both above and below ground, and lower N_2O release (Ussiri et al., 2009). Additionally, conventional or strip tillage techniques can be used to incorporate agricultural waste into the soil near the surface (Al-Kaisi and Licht, 2004). By heating plant material in an oxygen-limited atmosphere (such as slow pyrolysis), biochar is created. C can be locked in the soil by using charcoal, a stable material that is high in C. Biochar has attracted a lot of interest as a potential technique to sequester atmospheric C because it can store C in the soil for hundreds to thousands of years. Furthermore, according to recent studies, using biochar can reduce leaching, enhance the availability of nutrients for plant growth, decrease the amount of fertilizer required for cultivation, and lower the emission of CH_4 and N_2O. (Laird et al., 2009).

27.6 Carbon stability in soil

Due to their ample C pool, soils can reduce GHG emissions by storing C in either aboveground biomass or soils (Fig. 27.6). By lowering emissions, managing the physicochemical characteristics of soil and vegetation also aids in the fight against climate change. Trapping C in agricultural soils has generated a lot of interest globally, both to enhance soil quality and lower the possibility of global warming (Bernoux et al., 2009; Paustian et al., 2016; Smith, 2008). Trapping C

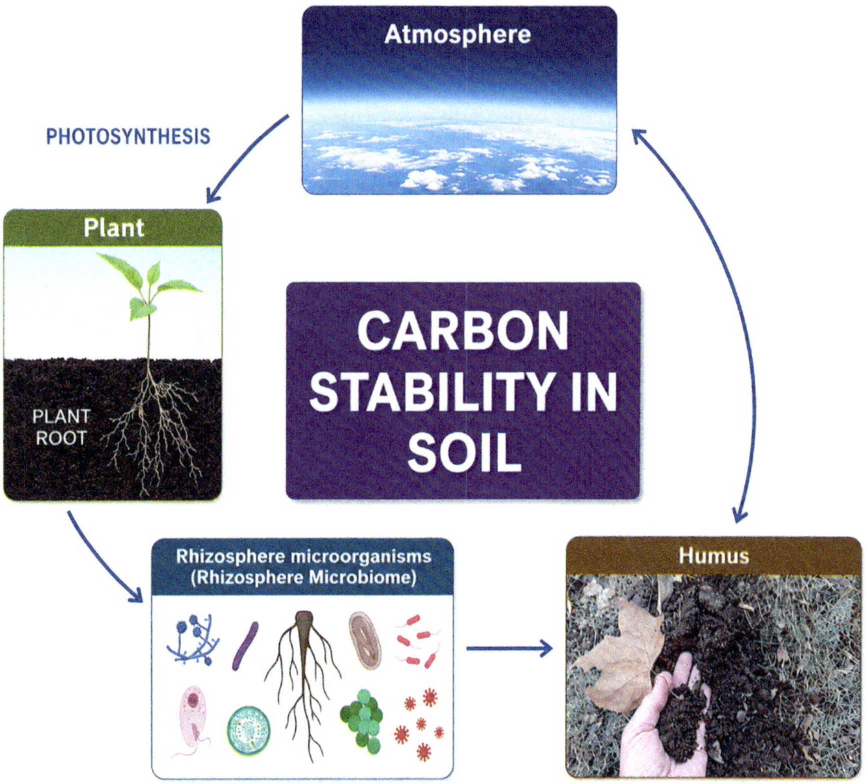

FIGURE 27.6 Mechanism of carbon stabilization in soil.

in agricultural soils around the world has been linked to improving soil quality and lowering the chance of global warming (Bernoux et al., 2009; Lal, 2011; Smith, 2008).

The expression C_{seq} has many different definitions Bernoux et al. (2005). Overall, though, it relates to both natural and intentional methods that either separate CO_2 from the environment or shift it from sources of emission, keeping it in the earth's trees and shrubs, soils, and sediments, as well as in the oceans and geological structures. This process is used to arrest and charge CO_2 in a stable form over an extended period. This process may be indirect, geological, chemical, biological, or physical. It is commonly referred to as "C fixation" when plants directly take up inorganic CO_2 through soil chemical reactions or photosynthesis. Biological processes known as "C sinks" take place in forests, oceans, soils, marshes, and other ecological systems to keep CO_2. The idea of "soil C_{seq}" should not be restricted to problems of CO_2 balancing or C storage because soils are connected to CH_4 and N_2O as well as CO_2 emissions. According to Bernoux et al. (2005), the overall GHG discharge must be computed at the plot level, or ideally at the level of the entire soil–plant pools of agricultural systems in C-CO_2 or CO_2-e, taking into account as many emission sources and sinks for the whole soil–plant system as possible. These can be solid, dissolved, organic, or mineral-based, and they can originate from various pools within the ecosystem. According to Bernoux et al., 2005, "soil C_{seq}" or, more accurately, "soil-plant C_{seq}" should be seen as the outcome of comparing all the sources and sinks of emissions from a particular agro-ecosystem to a recommendation for a given period and computing the net balance of all GHGs, indicated as C-CO_2 or CO_2-e. In addition to slowing down climate change, SOM is crucial for nutrient cycling, provides energy for soil microorganisms, serves as a biological control agent for plant, animal, and even human diseases, and influences several ecosystem services, such as the accessibility and integrity of water (Swift et al., 2004). SOM conversion is more significant than the accumulation of deposits of inactive OM since SOM is most advantageous when it breaks down and introduces nutrients and energy (Lehmann and Kleber, 2015). Furthermore, to the elements of SOM that support C storage (or soil C_{seq}) and the reduction of GHG emissions, we suggest that a definition of C_{seq} ought to encompass the special dynamic turnover that creates the labile pools necessary to maintain soil health.

27.7 Enhancing nitrogen use efficiency

A part of the chemical N fertilizer used for crop growing is immediately transformed to N_2O by soil microorganism action called nitrification and denitrification and discharged into the atmosphere. Additionally, nitrate leaching, runoff, and the volatilization of N fertilizers cause indirect N_2O emissions. Because GHG emissions associated with the use (and production) of N fertilizers make up a sizable portion of the total GHG emissions of agricultural commodities from cradle-to-farm-gate activities, there is a need to improve N use efficiency (for example, about 60% for U.S. corn farming). These emissions have a greater global warming potential (GWP) of 265 for N_2O emissions. To maximize N fertilizer consumption, numerous technologies are being developed and used. VRT-based precision agriculture practices are now being adopted at a rate of 8–35% for maize farming in the Midwest of the United States between 2010 and 2016 (Baranski et al., 2018). Since soil properties differ between fields, farmers use VRT techniques to segment their fields and choose the appropriate fertilizer application rates depending on the findings of soil tests. The concept of "proper rate" has been recently incorporated into the 4 R framework to manage N fertilizer in farms in a manner that reduces both direct and indirect on-field N_2O emissions (right time, right rate, right form, and right place). Research also showed that nitrification inhibitors, a form of increased efficiency fertilizer (IEF), may cut N_2O emissions and enhance N use efficiency (Thapa et al., 2016). Nitrification inhibitors, for example, can be used to stop nitrification in the soil from working and reduce N_2O emissions by 33–44% than traditional fertilizers (Thapa et al., 2016) (Table 27.3).

27.8 Precision input management

Adaptation of precision input management in the agricultural field reduces GHG by enhancing the capacity of soil to utilize the C stock, PA requires minimum tillage operation (Angers and Eriksen-Hamel, 2008), less N fertilization (Waldrop et al., 2004), lesser in direct GHG emission is caused by the reduction in fuel consumption from less tractor-related field operations, and during agricultural field operations, reducing inputs. Despite this, these strategies affect farm production by optimizing agricultural inputs to generate higher or comparable yields at a lower cost than traditional methods. While considering that crop production should be preserved or even made more difficult, GHG mitigation methods can significantly reduce the number of inputs required for new technology and procedures for all farming activities. This will help in the effort to limit agriculture's influence on climate change. In the first category, PA for crop farming is enclosed with a focus on fertilizer and water management. Agricultural GHG emission mitigation measures should focus on reducing N_2O emissions from soil N handling and increasing crop productivity to avoid future land

TABLE 27.3 Practices adopted to enhance nitrogen use efficiency.

S. N.	Practices	Effect	References
1.	Fertilizer application in wetland conditions	In rice, crops enhance crop growth, yield N utilization, and glutamate synthetase activity in the root	Sun et al. (2012)
2.	Management of N and water irrigation	Reduce the N application and the water application	Yang et al. (2020)
3.	Soil application with Zeolite + N fertilizer (Urea)	Enhance N use efficiency in rice	Kavoosi (2007)
4.	Application of Zeolite and nitrifying bacteria	Reduce ammonia loss and improve plant matter	Motesharezadeh et al. (2015)
5.	Use of Azolla compost and rice straw	In the paddy field, improve the total concentration of N	Novair et al. (2020)
6.	Continuous no-till and covering	Enhance N absorption	Habbib et al. (2020)
7.	Apply nitrification inhibitor (3,4-dimethyl pyrazole phosphate, DMPP)	Increase recovery of N and reduce N loss	Wallace et al. (2020)
8.	Drying –rewetting technique	Enhance N use efficiency and crop productivity	Zhang et al. (2018)

transformation. Hardware and sensors (such as positioning and guiding, crop sensing for water stress, nutrients, and yield sensing; ecological sensing; seedbed preparation; and fertilizer in addition to the soil profile); evaluating data and decision-support tools (which include protocols and standards for field data layer output, data analysis methods for identifying leadership zones, and user-friendly software); and commercial technology for PA (Fig. 27.7).

Loudjani (2014) categorized PATs (Precision Agriculture Technologies) for crop and livestock farming into three divisions: remote sensing, variable rate applications, and guidance systems. The division was made linearly to match the timeframe of the technologies' adoption. According to Fountas et al. (2015), innovations such as variable rate irrigation, weeding, and application are examples of reactive methods, which are also referred to as implements, hardware, and software that work together to alter the location of agricultural inputs in the field. All types of automatic steering and guidance for tractors and self-propelled farm machinery, including controlled traffic farming, machine guidance, and driver support, are included in guidance systems (i.e., hardware and software that guide tractors and implements across a field). technology for recording (e.g., sensors mounted on satellite platforms, acquiring spatial data, and rolling airborne or ground-based stations). Any farming practice, including traditional ones, can benefit from guidance methods, which concentrate on precise machinery movement within and between fields. These technologies have been shown to reduce overlap, which in turn lowers the use of inputs such as seeds, fertilizer, and chemical pesticides, as well as fuel usage for self-propelled machinery. To collect data in the field (during and after the crop season), storage technologies are necessary. Any kind of PA usage can use the information after it has been obtained. Responding technologies minimize the amount of water, seeds, fertilizer, and pesticides required for crop growth by utilizing the data produced by recording systems. Because of the benefits of better quality and fewer adverse environmental effects, it is expected that the right combination of these three categories will increase or at least keep output.

27.9 India's action plan for net-zero emissions

Most of the GHG-emitting countries are the USA, China, and India. Approximately 96% of GHG-emitting sectors are industries, agriculture, transportation, and infrastructure/building.

Following the plan adopted to achieve net-zero emissions:

- Making low C energy sources more adaptable
- Green mobility adaptation

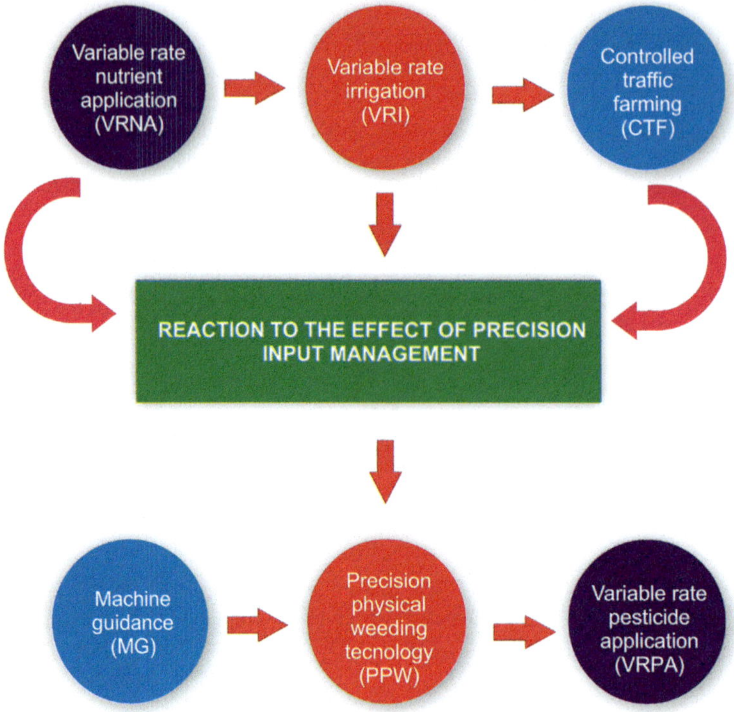

FIGURE 27.7 Reaction to the effect of precision input management.

- Energy source decarbonization
- Advanced green construction and building technology
- Promote sustainable farming

27.9.1 Using low-carbon sources of energy

In terms of global GHG production, India ranks third (Bhatt, 2023). Energy interventions must be comprehensive in India's energy sector and use fossil fuels and renewable energy sources (coal, gas, and oil). Creative solutions are needed to cut emissions in the energy transportation sector for consumers. India's expanding energy needs to construct new coal-fired power plants, which frequently use green energy that can be refined using renewable solutions (bundled solutions, including hybrid + thermal + storage). India must set a new 2070 GHG emission target. By 2060, all energy currently consumed from coal will be phased out, and by 2070, solar and wind energy will have increased from their current levels of 100 GW to more than 7400 GW (Bhatt, 2023). Coal-fired power stations, oil refineries, and other sources account for the majority of India's emissions (Bhatt, 2023). Maintaining and operating parameter optimization of artificial intelligence/machine learning is also effective for reducing emissions and increasing output. Continuous use of fossil fuel energy is most important for reducing the GHG footprint. Continued use of fusion fuel plants also reduces the most significant incremental level for fuel efficiency. Reduce energy waste in distribution work by implementing smart grids, which are beneficial for reducing power flows and enabling two-way communication, thereby helping consumers reduce their emissions. The grid can permit more gas plants and cut coal's CO_2 emissions by compressing peaking reserves. Coal-fired power plant C stores, commercial green hydrogen (H) incentives, and business C pricing are all speeding up the journey. Redirecting investment, increasing additional revenue, reforming the power distribution system, promoting market (virtual product) derivative markets for energy, and guaranteeing equitable and open third-party access to the transmission and distribution network will all require government policy assistance. By taking these steps, India can lower its emissions.

27.9.2 Use of clean energy and eco-friendly transportation

India's transportation industry accounts for half of the nation's overall oil demand. Because more people own cars and use the roads more frequently, the demand for oil has nearly doubled over the last 20 years. India may need to adjust its fuel efficiency standards to meet the Global Standards Demand (GSD) for the current BS-VI requirements for cars, as well as the TREM and CEV criteria for tractors and construction equipment. Nowadays, gas-based fuels (CNG and LNG) and biofuels like ethanol and methanol burn cleaner than gasoline and diesel. Commercially viable sustainable aviation fuel is made from waste and agricultural byproducts, and it uses power-to-liquid hydrogen technology, which has a 100% lower C footprint than conventional jet fuel over its entire life cycle. In India, where H mobility technology is most needed, there is a chance that fossil fuel-powered LCVs, trucks, and buses will be replaced by H vehicles over the next few decades.

27.9.3 Decarbonization of the energy sector

The steel, iron, cement, chemical, and fertilizer sectors in India's manufacturing economy may have the country's highest GHG emission levels (Bhatt, 2023). Enhance this sector's potential for economic growth and urbanization to minimize GHG emissions. For India to continue its manufacturing expansion under the current circumstances, a C-neutral industrial growth model will be essential. GHG emissions in this industry are significantly more difficult to reduce due to such power transport, as the fossil fuel feedstock is responsible for about 40% of emissions, necessitating changes to the production process and technologies. We need electric furnaces to change how fossil fuels are used because there are approximately 30% more GHG emissions when burned at high temperatures (800°C). The viability of technology and cost both fluctuate. Renovation of new technology and manufacturing processes that have a smaller C footprint is difficult to finance, given the sector's current economics. Regulations are needed to minimize GHG emissions. The Perform, Achieve, and Trade (PAT) strategy is useful for driving adaptation of certain early-win measures when GHG emissions decrease through EBITDA positive, improves energy and thermal efficiency, waste heat recovery systems, and initially substitutes high GHG feedstock (e.g., clinker reduction in cement). In this industry, decarbonization is needed to measure the regulation of demand, such as the circular economy, constantly improving energy efficiency, heat is electrically powered where applicable, using zero-C electricity, technology that stores C, consumes, and captures, utilizing C-free fuels like biomass and green or blue H, and creation of new technologies using non-fossil fuels.

27.10 Pathways for decarbonizing the industrial sector

SBTi (Science Based Targets Initiative)-based emission reduction goals have been volunteered by roughly 60 of the most prominent Indian companies, as well as a number of international businesses operating in India. Commercially viable technology adaptation is beneficial for the GHG emission industry only between the fifth and the quarter of the required reduction for the 1.5°C objectives archive. International cooperation will be enhanced by relatively lower prices for decarbonized goods when compared to conventional commodities, the accessibility of natural resources (like biomass or geological storage space for CO_2 capture), the implementation of new facilities, the affordability of novel innovations like hydrogen, and government subsidies and assistance for the steel industry to embrace environmentally friendly technologies.

27.11 Advanced green construction and building technology

25% of India's total GHG emissions come from its cities. India's per-capita consumption of 1.8 tons of CO_2 equivalent is 40% greater than the predicted total city per-capita emissions of 2.6 tons of CO_2 equivalent. India intends to create low C infrastructure more effectively if cities become involved. India may be at two crucial junctures in its shift to greener cities, structures, and infrastructure: To minimize city water reliance and increase energy efficiency, India's proposal to construct a transit-oriented, water-efficient metropolis calls for a decentralized method to gather, treat, and monitor water supply use. Appropriate waste management, such as optimizing waste creation, collection, disposal facilities, and gasification through generating energy from most of the large, open sink sources. Green places can aid the localization of C sequestration. Creating an energy-efficient new building, such as one that incorporates modular construction, retrofitting existing buildings, and using the best possible cross-ventilation and thermal insulation, will help minimize urban CO_2 emissions. Creating an energy-efficient new building, such as one incorporating modular construction, retrofitting existing buildings, and using the best possible cross-ventilation and thermal insulation, will help minimize urban CO_2 emissions. Agriculture and the livestock industry account for 18% of all national GHG emissions. Agriculture is responsible for 80% of global GHG emissions, of which 48% are GHGs like N_2O, due to the use of synthetic N fertilizers and the mixing of manure with soil. Approximately 80% of the CH_4 emissions in the agricultural sector are attributed to the fermentation process, livestock farming, and manure management. Burning savannas and using crop residue are the remaining sources of GHG emissions, with rice production being the second-largest source of CH_4. Due to its higher GHG emissions, the agriculture sector is the main cause of climate change. Frequent heat waves, dry spells, and erratic, abnormal rainfall are the main effects of climate change on India's agricultural output. The increased need for food production must be satisfied while reducing agricultural GHG emissions, which is a challenge as population growth continues. Since more than 100 million farmers in India are currently authorized to adopt PA (mainly modern rice cultivation practices that reduce N/urea utilization and increase N use efficiency, grouping the low and no-tillage practices), sustainable agriculture will be one of the most multiplex transformations for India shortly. By raising cattle and utilizing renewable energy sources, sustainable animal husbandry can help reduce the amount of CH_4 released into the atmosphere. It can also help with adopting electric farm equipment and converting diesel pumps to renewable energy sources.

27.12 Conclusion

Even developed nations are reluctant to initiate such programs because agricultural mitigation is costly and challenging, including the impact on farmers and the associated monitoring costs. One of the most vulnerable populations in India's agricultural sector, which has historically borne little responsibility for climate change, will bear the majority of the financial cost of mitigation. They will also facilitate the adoption of C space through mitigation in Indian agriculture, particularly through C trading in the cap-and-trade sense or C offsets, where advanced nations are the main buyers. Although these arguments hold for all C trading, emissions from the agriculture sector must be viewed as an essential component of guaranteeing future food security. This needs to be taken into account until agricultural decarbonization is accomplished without endangering food security and productivity. There is no reason to rush the development of C markets in agriculture. Furthermore, there is no strong argument in favor of mitigation being a top priority for Indian agriculture. Mitigation is a cost-benefit that significantly increases agricultural input efficiency, but adaptation should be the primary focus when it comes to climate change and agriculture in India.

References

Abbott, J., 2008. What is a carbon footprint? The Edinburgh Centre for carbon management. Report ECCM-EM-483-2007.

Adhya, T.K., Sharma, P.D., Kumar Gogoi, A., 2009. Mitigating greenhouse gas emission from agriculture. In: Singh, S.N. (Ed.), Climate Change and Crops. Environmental Science and Engineering. Springer, Berlin, Heidelberg, pp. 329–344. Available at 10.1007/978-3-540-88246-6_15.

Al-Kaisi, M., Licht, M.A., 2004. Effect of strip tillage on corn nitrogen uptake and residual soil nitrate accumulation compared with no-tillage and chisel plow. Agronomy Journal 96 (4), 1164–1171. Available at https://doi.org/10.2134/agronj2004.1164.

Allen, J., Pascual, K.S., Romasanta, R.R., Van Trinh, M., Van Thach, T., Van Hung, N., et al., 2019. Rice straw management effects on greenhouse gas emissions and mitigation options. Sustainable Rice Straw Management. Springer International Publishing, Philippines, pp. 145–159. Available at http://doi.org/10.1007/978-3-030-32373-8_9.

Andreae, M.O., Merlet, P., 2001. Emission of trace gases and aerosols from biomass burning. Global Biogeochemical Cycles 15 (4), 955–966. Available at https://doi.org/10.1029/2000gb001382.

Angers, D.A., Eriksen-Hamel, N.S., 2008. Full-inversion tillage and organic carbon distribution in soil profiles: a meta-analysis. Soil Science Society of America Journal 72 (5), 1370–1374. Available at https://doi.org/10.2136/sssaj2007.0342.

Arai, H., 2022. Increased rice yield and reduced greenhouse gas emissions through alternate wetting and drying in a triple-cropped rice field in the Mekong Delta. Science of the Total Environment 842, 156958. Available at https://doi.org/10.1016/j.scitotenv.2022.156958.

Audet, P., Charest, C., 2008. Allocation plasticity and plant-metal partitioning: meta-analytical perspectives in phytoremediation. Environmental Pollution 156 (2), 290–296. Available at https://doi.org/10.1016/j.envpol.2008.02.010.

Banerjee, A., Meena, R.S., Jhariya, M.K., Yadav, D.K., 2020. Agroecological Footprints Management for Sustainable Food System. Springer, Singapore, pp. 1–514. Available at 10.1007/978-981-15-9496-0.

Baranski, M., Caswell, H., Claassen, R., Cherry, C., Jaglo, K., Lataille, A., et al., 2018. Agricultural conservation on working lands: trends from 2004 to present. Technical Bulletin Number 1950. Office of the Chief Economist, U.S. Department of Agriculture, Washington, DC, pp. 127. Available at https://www.usda.gov/sites/default/files/documents/USDA_Conservation_Trends.pdf.

Beauchemin, K.A., McGinn, S.M., 2005. Methane emissions from feedlot cattle fed barley or corn diets1. Journal of Animal Science 83 (3), 653–661. Available at https://doi.org/10.2527/2005.833653x.

Berner, A., Hildermann, I., Fließbach, A., Pfiffner, L., Niggli, U., Mäder, P., 2008. Crop yield and soil fertility response to reduced tillage under organic management. Soil and Tillage Research 101 (1-2), 89–96. Available at https://doi.org/10.1016/j.still.2008.07.012.

Bernoux, M., Feller, C., Cerri, C.C., Eschenbrenner, V., Cerri, C.E.P., 2005. Soil carbon sequestration. In: Roose, E.J., Lal, R., Feller, C., Barthes, B., Stewart, B.A. (Eds.), Soil Erosion and Carbon Dynamics. CRC Press, Boca Raton, pp. 13–22.

Bernoux, M., Cerri, C.C., Cerri, C.E.P., Neto, M.S., Metay, A., Perrin, A.S., et al., 2009. Cropping systems, carbon sequestration and erosion in Brazil: a review. Sustainable Agriculture. Springer Netherlands, France, pp. 75–85. Available at 10.1007/978-90-481-2666-8_7.

Bezbaruah, M.P., 2014. Agricultural development in India: post-reform experience. Arthabeekshan 22, 159–172.

Bhatia, A., Pathak, H., Jain, N., Singh, P.K., Singh, A.K., 2005. Global warming potential of manure amended soils under rice–wheat system in the Indo-Gangetic plains. Atmospheric Environment 39 (37), 6976–6984. Available at https://doi.org/10.1016/j.atmosenv.2005.07.052.

Bhatia, A., Sasmal, S., Jain, N., Pathak, H., Kumar, R., Singh, A., 2010. Mitigating nitrous oxide emission from soil under conventional and no-tillage in wheat using nitrification inhibitors. Agriculture, Ecosystems & Environment 136 (3-4), 247–253. Available at https://doi.org/10.1016/j.agee.2010.01.004.

Bhatt, R., Kukal, S.S., 2017. Soil evaporation studies using mini-lysimeters under differently established rice-wheat cropping sequence in Punjab, India. Journal of Applied and Natural Science 9 (1), 222–229. Available at https://doi.org/10.31018/jans.v9i1.1177.

Bhatt, J.R., 2023. Debunking the narrative of india as a large greenhouse gas emitter. Current Science 124 (12), 1378. Available at https://doi.org/10.18520/cs/v124/i12/1378-1383.

Bruinsma, J., 2009. The resource outlook to 2050, in: Expert Meeting on How to Feed the World. pp. 1–33. https://www.fao.org/4/ak542e/ak542e06.pdf

Burger, H., Schloen, M., Schmidt, W., Geiger, H.H., 2008. Quantitative genetic studies on breeding maize for adaptation to organic farming. Euphytica 163 (3), 501–510. Available at https://doi.org/10.1007/s10681-008-9723-4.

Chen, P., Du, Q., Liu, X., Zhou, L., Hussain, S., Lei, L., et al., 2017. Effects of reduced nitrogen inputs on crop yield and nitrogen use efficiency in a long-term maize-soybean relay strip intercropping system. PLoS One 12 (9). Available at https://doi.org/10.1371/journal.pone.0184503 e0184503.

Choudhary, D.K., Prakash, A., Wray, V., Johri, B.N., 2009. Insights of the fluorescent pseudomonads in plant growth regulation. Current Science 97 (2), 170–179.

Crews, T.E., Peoples, M.B., 2004. Legume versus fertilizer sources of nitrogen: ecological tradeoffs and human needs. Agriculture, Ecosystems & Environment 102 (3), 279–297. Available at https://doi.org/10.1016/j.agee.2003.09.018.

Drinkwater, L.E., Wagoner, P., Sarrantonio, M., 1998. Legume-based cropping systems have reduced carbon and nitrogen losses. Nature 396 (6708), 262–265. Available at https://doi.org/10.1038/24376.

FAO, 2019. The State of Food and Agriculture 2019. Moving forward on food loss and waste reduction. FAO, Rome, pp. 1–23. Available at https://openknowledge.fao.org/server/api/core/bitstreams/11f9288f-dc78-4171-8d02-92235b8d7dc7/content.

Fließbach, A., Mäder, P., 2000. Microbial biomass and size-density fractions differ between soils of organic and conventional agricultural systems. Soil Biology and Biochemistry 32 (6), 757–768. Available at https://doi.org/10.1016/S0038-0717(99)00197-2.

Flynn, H.C., Smith, J., Smith, K.A., Wright, J., Smith, P., Massheder, J., 2005. Climate- and crop-responsive emission factors significantly alter estimates of current and future nitrous oxide emissions from fertilizer use. Global Change Biology 11 (9), 1522–1536. Available at https://doi.org/10.1111/j.1365-2486.2005.00998.x.

Fountas, S., Carli, G., Sørensen, C.G., Tsiropoulos, Z., Cavalaris, C., Vatsanidou, A., et al., 2015. Farm management information systems: current situation and future perspectives. Computers and Electronics in Agriculture 115, 40–50. Available at https://doi.org/10.1016/j.compag.2015.05.011.

Gadde, B., Bonnet, S., Menke, C., Garivait, S., 2009. Air pollutant emissions from rice straw open field burning in India, Thailand and the Philippines. Environmental Pollution 157 (5), 1554–1558. Available at https://doi.org/10.1016/j.envpol.2009.01.004.

Gan, Y.T., Miller, P.R., McConkey, B.G., Zentner, R.P., Stevenson, F.C., McDonald, C.L., 2003. Influence of diverse cropping sequences on durum wheat yield and protein in the Semiarid Northern great plains. Agronomy Journal 95 (2), 245–252. Available at https://doi.org/10.2134/agronj2003.2450.

Gan, Y.T., Kutcher, R., Menalled, F., Lafond, L., Brandt, S., 2010. Crop diversification and intensification with broadleaf crops in cereal-based cropping systems in the Northern Great Plains of North America. Recent trends in soil science and agronomy research in the Northern Great Plains of North America, pp. 277–301.

Garg, A., Shukla, P.R., Kapshe, M., 2006. The sectoral trends of multigas emissions inventory of India. Atmospheric Environment 40 (24), 4608–4620. Available at https://doi.org/10.1016/j.atmosenv.2006.03.045.

Ghosh, P.K., Kumar, N., Yadav, A., 2014. Role of resource conservation technologies in mitigating climate change. Resource Conservation Technology in Pulses 312.

Gołasa, P., Wysokiński, M., Bieńkowska-Gołasa, W., Gradziuk, P., Golonko, M., Gradziuk, B., et al., 2021. Sources of greenhouse gas emissions in agriculture, with particular emphasis on emissions from energy used. Energies 14 (13), 3784. Available at https://doi.org/10.3390/en14133784.

Habbib, H., Hirel, B., Spicher, F., Dubois, F., Tétu, T., 2020. In winter wheat (*Triticum aestivum* L.), No-till improves photosynthetic nitrogen and water-use efficiency. Journal of Crop Science and Biotechnology 23 (1), 39–46. Available at https://doi.org/10.1007/s12892-019-0122-0.

Harker, K.N., O'Donovan, J.T., Irvine, R.B., Turkington, T.K., Clayton, G.W., 2009. Integrating cropping systems with cultural techniques augments wild oat (*Avena fatua*) management in barley. Weed Science 57 (3), 326–337. Available at https://doi.org/10.1614/WS-08-165.1.

Hepperly, P., Moyer, J., Pimentel, D., Douds Jr, D., Nichols, K., Seidel, R., 2008. Organic maize/soybean cropping systems significantly sequester carbon and reduce energy use. Cultivating the Future Based on Science 2.

Herridge, D.F., Peoples, M.B., Boddey, R.M., 2008. Global inputs of biological nitrogen fixation in agricultural systems. Plant and Soil 311 (1-2), 1–18. Available at https://doi.org/10.1007/s11104-008-9668-3.

Hitaj, C., Suttles, S., 2016. Trends in U.S. Agriculture's Consumption and Production of Energy: Renewable Power, Shale Energy, and Cellulosic Biomass. U.S. Department of Agriculture, Economic Research Service. Available at https://www.ers.usda.gov/publications/pub-details?pubid=74661.

Huang, J., Chen, Y., Sui, P., Gao, W., 2013. Estimation of net greenhouse gas balance using crop- and soil-based approaches: two case studies. Science of the Total Environment 456-457, 299–306. Available at https://doi.org/10.1016/j.scitotenv.2013.03.035.

Hynes, R.K., Leung, G.C.Y., Hirkala, D.L.M., Nelson, L.M., 2008. Isolation, selection, and characterization of beneficial rhizobacteria from pea, lentil, and chickpea grown in western Canada. Canadian Journal of Microbiology 54 (4), 248–258. Available at https://doi.org/10.1139/W08-008.

Jensen, E.S., Peoples, M.B., Boddey, R.M., Gresshoff, P.M., Henrik, H.N., J.R. Alves, B., et al., 2012. Legumes for mitigation of climate change and the provision of feedstock for biofuels and biorefineries. A review. Agronomy for Sustainable Development 32 (2), 329–364. Available at https://doi.org/10.1007/s13593-011-0056-7.

Jhariya, M.K., Meena, R.S., Banerjee, A., Meena, S.N., 2022. Natural Resources Conservation and Advances for Sustainability. Elsevier. Available at https://doi.org/10.1016/C2019-0-03763-6.

Jhariya, M.K., Meena, R.S., Banerjee, A., Kumar, S., Raj, A., 2023. Agroforestry for Carbon and Ecosystem Management. Elsevier, India, pp. 1–434. Available at 10.1016/C2021-0-02916-3.

Jian, J., Du, X., Reiter, M.S., Stewart, R.D., 2020. A meta-analysis of global cropland soil carbon changes due to cover cropping. Soil Biology and Biochemistry 143. Available at https://doi.org/10.1016/j.soilbio.2020.107735.

Jiang, Y., van Groenigen, K.J., Huang, S., Hungate, B.A., van Kessel, C., Hu, S., et al., 2017. Higher yields and lower methane emissions with new rice cultivars. Global Change Biology 23 (11), 4728–4738. Available at https://doi.org/10.1111/gcb.13737.

Jiang, J., Ye, B., Liu, J., 2019. Research on the peak of CO_2 emissions in the developing world: current progress and future prospect. Applied Energy 235, 186–203. Available at https://doi.org/10.1016/j.apenergy.2018.10.089.

Kavoosi, M., 2007. Effects of zeolite application on rice yield, nitrogen recovery, and nitrogen use efficiency. Communications in Soil Science and Plant Analysis 38 (1), 69–76. Available at https://doi.org/10.1080/00103620601093652.

Kirkegaard, J., Christen, O., Krupinsky, J., Layzell, D., 2008. Break crop benefits in temperate wheat production. Field Crops Research 107 (3), 185–195. Available at https://doi.org/10.1016/j.fcr.2008.02.010.

Küstermann, B., Wenske, K., Hülsbergen, K.-J., 2007. Modellierung betrieblicher C-und N-Flüsse als Grundlage einer Emissionsin-ventur. https://orgprints.org/id/eprint/9654/

Laird, D., Fleming, P., Wang, B., Karlen, D., 2009. Impact of biochar amendments on soil quality for a typical midwestern agricultural soil. In: Presentation Made at the North American Biochar Conference. pp. 9–12.

Lal, R., 2011. Sequestering carbon in soils of agro-ecosystems. Food Policy 36 (1), S33. Available at https://doi.org/10.1016/j.foodpol.2010.12.001.

Lehmann, J., Kleber, M., 2015. The contentious nature of soil organic matter. Nature 528 (7580), 60–68. Available at https://doi.org/10.1038/nature16069.

Liao, B., Wu, X., Yu, Y., Luo, S., Hu, R., Lu, G., 2020. Effects of mild alternate wetting and drying irrigation and mid-season drainage on CH4 and N_2O emissions in rice cultivation. Science of the Total Environment 698, 134212. Available at https://doi.org/10.1016/j.scitotenv.2019.134212.

Liu, X., Elgowainy, A., Wang, M., 2020. Life cycle energy use and greenhouse gas emissions of ammonia production from renewable resources and industrial by-products. Green Chemistry 22 (17), 5751–5761. Available at https://doi.org/10.1039/d0gc02301a.

Lonappan, A.C., 2015. Substrate impact on biogas production and manurial value of slurry.

Löschenberger, F., Fleck, A., Grausgruber, H., Hetzendorfer, H., Hof, G., Lafferty, J., et al., 2008. Breeding for organic agriculture: the example of winter wheat in Austria. Euphytica 163 (3), 469–480. Available at https://doi.org/10.1007/s10681-008-9709-2.

Loudjani, P., 2014. Precision Agriculture: An Opportunity for EU-Farmers—Potential Support with the CAP 2014-2020. EPRS: European Parliamentary Research Service, Belgium. Available at https://policycommons.net/artifacts/1339069/precision-agriculture/1948411/.

Mäder, P., Edenhofer, S., Boller, T., Wiemken, A., Niggli, U., 2000. Arbuscular mycorrhizae in a long-term field trial comparing low-input (organic, biological) and high-input (conventional) farming systems in a crop rotation. Biology and Fertility of Soils 31 (2), 150–156. Available at https://doi.org/10.1007/s003740050638.

Majumdar, D., Pathak, H., Kumar, S., Jain, M.C., 2002. Nitrous oxide emission from a sandy loam Inceptisol under irrigated wheat in India as influenced by different nitrification inhibitors. Agriculture, Ecosystems and Environment 91 (1-3), 283–293. Available at https://doi.org/10.1016/S0167-8809(01)00223-7.

Malhi, S.S., Grant, C.A., Johnston, A.M., Gill, K.S., 2001. Nitrogen fertilization management for no-till cereal production in the Canadian great plains: a review. Soil and Tillage Research 60 (3-4), 101–122. Available at https://doi.org/10.1016/s0167-1987(01)00176-3.

Marino, D., Pucciariello, C., Puppo, A., Frendo, P., 2009. Chapter 5 the redox state, a referee of the legume-rhizobia symbiotic game. Advances in Botanical Research 52 (C), 115–151. Available at https://doi.org/10.1016/S0065-2296(10)52005-6.

Meena, R.S., Pradhan, G., 2023. Industrial garbage-derived biocompost enhances soil organic carbon fractions, CO_2 biosequestration, potential carbon credits and sustainability index in a rice-wheat ecosystem. Environmental Research 235, 116525. Available at https://doi.org/10.1016/j.envres.2023.116525.

Meena, R.S., Pradhan, G., Kumar, S., Lal, R., 2023. Using industrial wastes for rice-wheat cropping and food-energy-carbon-water-economic nexus to the sustainable food system. Renewable and Sustainable Energy Reviews 187. Available at https://doi.org/10.1016/j.rser.2023.113756.

Meena, R.S., Pradhan, G., Singh, K., Kumar, S., Singh, A.K., Shashidhar, K.S., et al., 2024a. Agriculture models for restoring degraded land to enhance CO_2 biosequestration and carbon credits in the Vindhyan region of India. Science of the Total Environment 929. Available at https://doi.org/10.1016/j.scitotenv.2024.172661.

Meena, R.S., Pradhan, G., Nalani, Kumar, S., Singh, A.K., Verma, S.K., et al., 2024b. Diversified cropping modules designed for soil restoration, CO_2 sequestration, and generating carbon credits. Land Degradation and Development 35 (14), 4314–4329. Available at https://doi.org/10.1002/ldr.5224.

Meena, R.S., Pradhan, G., Kumar, S., 2025. Energy flow, eco-efficiency, and economic circulation with recycled industrial waste compost application in wheat and subsequent rice farming. Science of the Total Environment 967. Available at https://doi.org/10.1016/j.scitotenv.2025.178779.

Menalled, F.D., Gross, K.L., Hammond, M., 2001. Weed aboveground and seedbank community responses to agricultural management systems. Ecological Applications 11 (6), 1586–1601. Available at https://doi.org/10.1890/1051-0761(2001)011[1586:WAASCR]2.0.CO;2.

Michaelowa, A., Allen, M., Sha, F., 2018. Policy instruments for limiting global temperature rise to 1.5°C–can humanity rise to the challenge? Climate Policy 18 (3), 275–286. Available at https://doi.org/10.1080/14693062.2018.1426977.

Motesharezadeh, B., Arasteh, A., Pourbabaee, A.A., Rafiee, G.R., 2015. The effect of zeolite and nitrifying bacteria on remediation of nitrogenous wastewater substances derived fromcarp breeding farm. International Journal of Environmental Research 9 (2), 553–560.

Novair, S.B., Hosseini, H.M., Etesami, H., Razavipour, T., 2020. Rice straw and composted azolla alter carbon and nitrogen mineralization and microbial activity of a paddy soil under drying–rewetting cycles. Applied Soil Ecology 154, 103638.

Pachauri, R.K., Allen, M.R., Barros, V.R., Broome, J., Cramer, W., Christ, R., et al., 2014. Climate Change 2014: Synthesis Report. In: Pachauri, R., Meyer, L. (Eds.), Contribution of Working Groups I, II and III to the Fifth Assessment Report of the Intergovernmental Panel on Climate Change. IPCC, Geneva, pp. 151. Available at https://epic.awi.de/id/eprint/37530/.

Palekar, S., 2006. The philosophy of spiritual farming.

Paradis, R., Dalpé, Y., Charest, C., 1995. The combined effect of arbuscular mycorrhizas and short-term cold exposure on wheat. New Phytologist 129 (4), 637–642. Available at https://doi.org/10.1111/j.1469-8137.1995.tb03032.x.

Parajuli, R., Kristensen, I.S., Knudsen, M.T., Mogensen, L., Corona, A., Birkved, M., et al., 2017. Environmental life cycle assessments of producing maize, grass-clover, ryegrass and winter wheat straw for biorefinery. Journal of Cleaner Production 142, 3859–3871. Available at https://doi.org/10.1016/j.jclepro.2016.10.076.

Pathak, H., Nedwell, D.B., 2001. Nitrous oxide emission from soil with different fertilizers, water levels and nitrification inhibitors. Water, Air, and Soil Pollution 129 (1-4), 217–228. Available at https://doi.org/10.1023/a:1010316215596.

Pathak, H., Prasad, S., Bhatia, A., Singh, S., Kumar, S., Singh, J., et al., 2003. Methane emission from rice–wheat cropping system in the Indo-Gangetic plain in relation to irrigation, farmyard manure and dicyandiamide application. Agriculture, Ecosystems & Environment 97 (1-3), 309–316. Available at https://doi.org/10.1016/s0167-8809(03)00033-1.

Pathak, H., Singh, R., Bhatia, A., Jain, N., 2006. Recycling of rice straw to improve wheat yield and soil fertility and reduce atmospheric pollution. Paddy and Water Environment 4 (2), 111–117. Available at https://doi.org/10.1007/s10333-006-0038-6.

Pathak, H., Bhatia, A., Jain, N., Aggarwal, P.K., 2010. Greenhouse gas emission and mitigation in Indian agriculture – a review. In: Singh, B. (Ed.), ING Bulletins on Regional Assessment of Reactive Nitrogen, Bulletin No. 19. SCON-ING, New Delhi, pp. 1–34.

Pathak, M., Slade, R., Pichs-Madruga, R., Ürge-Vorsatz, D., Shukla, R., Skea, J., 2022. Climate Change 2022 Mitigation of Climate Change: Technical Summary.

Paustian, K., Lehmann, J., Ogle, S., Reay, D., Robertson, G.P., Smith, P., 2016. Climate-smart soils. Nature 532 (7597), 49–57. Available at https://doi.org/10.1038/nature17174.

Peng, S., Buresh, R.J., Huang, J., Zhong, X., Zou, Y., Yang, J., et al., 2010. Improving nitrogen fertilization in rice by site-specific N management. A review. Agronomy for Sustainable Development 30 (3), 649–656. Available at https://doi.org/10.1051/agro/2010002.

Pimentel, D., Burgess, M., 2014. An environmental, energetic and economic comparison of organic and conventional farming systems. In: Pimentel, D., Peshin, R. (Eds.), Integrated Pest Management. Springer, Dordrecht, pp. 141–166. Available at https://doi.org/10.1007/978-94-007-7796-5_6.

Pradhan, G., Meena, R.S., 2023. Interaction impacts of biocompost on nutrient dynamics and relations with soil biota, carbon fractions index, societal value of CO_2 equivalent and ecosystem services in the wheat-rice farming. Chemosphere 339, 139695. Available at https://doi.org/10.1016/j.chemosphere.2023.139695.

Pradhan, G., Meena, R.S., 2023b. Utilizing waste compost to improve the atmospheric CO_2 capturing in the rice-wheat cropping system and energy-cum-carbon credit auditing for a circular economy. Science of The Total Environment 892, 164572. Available at https://doi.org/10.1016/j.scitotenv.2023.164572.

Prasad, M., Mahawer, S.K., Govindasamy, P., Srinivasan, R., Kumar, S., 2021. Does agriculture can be carbon neutral. AgroScience Today 2, 115–117.

Qin, Z., Canter, C.E., Dunn, J.B., Mueller, S., Kwon, H., Han, J., et al., 2018. Land management change greatly impacts biofuels' greenhouse gas emissions. GCB Bioenergy 10 (6), 370–381. Available at https://doi.org/10.1111/gcbb.12500.

Raj, A., Jhariya, M.K., Banerjee, A., Meena, R.S., Nema, S., Khan, N., et al., 2021. Chapter 13 - Agroforestry a model for ecological sustainability. Natural Resources Conservation and Advances for Sustainability. Elsevier, India, pp. 289–307. Available at 10.1016/B978-0-12-822976-7.00002-8.

Raj, A., Jhariya, M.K., Meena, R.S., 2025. Agroforestry for Monetising Carbon Credits. Springer, Cham. Available at https://doi.org/10.1007/978-3-031-97151-8.

Rajbhandari, B.P., Bhatta, G.D., 2008. Food crops: Agro-ecology and Modern Agro-techniques. Kathmandu, Nepal. Himalayan College of Agricultural Sciences and Technology, pp. 72–75.

Reino, J.L., Guerrero, R.F., Hernández-Galán, R., Collado, I.G., 2008. Secondary metabolites from species of the biocontrol agent Trichoderma. Phytochemistry Reviews 7 (1), 89–123. Available at https://doi.org/10.1007/s11101-006-9032-2.

Ren, Y.Y., Wang, X.L., Zhang, S.Q., Palta, J.A., Chen, Y.L., 2017. Influence of spatial arrangement in maize-soybean intercropping on root growth and water use efficiency. Plant and Soil 415 (1-2), 131–144. Available at https://doi.org/10.1007/s11104-016-3143-3.

Rosset, P.M., Martínez-Torres, M.E., 2012. Rural social movements and agroecology: context, theory, and process. Ecology and Society 17 (3). Available at https://doi.org/10.5751/ES-05000-170317.

Rossing, T., 2010. Water scarcity, climate change, and the poor. Reducing Poverty, Protecting Livelihoods, and Building Assets in a Changing Climate 21.

Roy, O., Meena, R.S., Kumar, S., Jhariya, M.K., Pradhan, G., 2022. Assessment of land use systems for CO_2 sequestration, carbon credit potential, and income security in Vindhyan region, India. Land Degradation and Development 33 (4), 670–682. Available at https://doi.org/10.1002/ldr.4181.

Sapkota, T.B., Vetter, S.H., Jat, M.L., Sirohi, S., Shirsath, P.B., Singh, R., et al., 2019. Cost-effective opportunities for climate change mitigation in Indian agriculture. Science of the Total Environment 655, 1342–1354. Available at https://doi.org/10.1016/j.scitotenv.2018.11.225.

Shen, Y., Sui, P., Huang, J., Wang, D., Whalen, J.K., Chen, Y., 2018. Greenhouse gas emissions from soil under maize–soybean intercrop in the North China Plain. Nutrient Cycling in Agroecosystems 110 (3), 451–465. Available at https://doi.org/10.1007/s10705-018-9908-8.

Sieverding, H., Kebreab, E., Johnson, J.M.F., Xu, H., Wang, M., Grosso, S.J.D., et al., 2020. A life cycle analysis (LCA) primer for the agricultural community. Agronomy Journal 112 (5), 3788–3807. Available at https://doi.org/10.1002/agj2.20279.

Simpson, M.S., Voorhees, E.M., Hersh, W., 2014. 23rd Text REtrieval Conference, TREC 2014 - Proceedings National Institute of Standards and Technology (NIST) United States Overview of the TREC 2014 Clinical Decision Support Track.

Smith, P., 2008. Land use change and soil organic carbon dynamics. Nutrient Cycling in Agroecosystems 81 (2), 169–178. Available at https://doi.org/10.1007/s10705-007-9138-y.

Streets, D.G., Bond, T.C., Carmichael, G.R., Fernandes, S.D., Fu, Q., He, D., et al., 2003. An inventory of gaseous and primary aerosol emissions in Asia in the year 2000. Journal of Geophysical Research: Atmospheres 108 (D21). Available at https://doi.org/10.1029/2002jd003093.

Sun, Y., Ma, J., Sun, Y., Xu, H., Yang, Z., Liu, S., et al., 2012. The effects of different water and nitrogen managements on yield and nitrogen use efficiency in hybrid rice of China. Field Crops Research 127, 85–98. Available at https://doi.org/10.1016/j.fcr.2011.11.015.

Swift, M.J., Izac, A.-M.N., van Noordwijk, M., 2004. Biodiversity and ecosystem services in agricultural landscapes—are we asking the right questions? Agriculture, Ecosystems & Environment 104 (1), 113–134. Available at https://doi.org/10.1016/j.agee.2004.01.013.

Syers, J.K., Johnston, A.E., Curtin, D., 2008. Efficiency of soil and fertilizer phosphorus use. FAO Fertilizer and Plant Nutrition Bulletin 18, 5–50.

Teasdale, J.R., Coffman, C.B., Mangum, R.W., 2007. Potential long-term benefits of no-tillage and organic cropping systems for grain production and soil improvement. Agronomy Journal 99 (5), 1297–1305. Available at https://doi.org/10.2134/agronj2006.0362.

Thapa, R., Chatterjee, A., Awale, R., McGranahan, D.A., Daigh, A., 2016. Effect of enhanced efficiency fertilizers on nitrous oxide emissions and crop yields: a meta-analysis. Soil Science Society of America Journal 80 (5), 1121–1134. Available at https://doi.org/10.2136/sssaj2016.06.0179.

Thorup-Kristensen, K., Magid, J., Jensen, L.S., 2003. Catch crops and green manures as biological tools in nitrogen management in temperate zones. Advances in Agronomy 79, 227–302. Available at https://doi.org/10.1016/S0065-2113(02)79005-6.

Ussiri, D.A.N., Lal, R., Jarecki, M.K., 2009. Nitrous oxide and methane emissions from long-term tillage under a continuous corn cropping system in Ohio. Soil and Tillage Research 104 (2), 247–255. Available at https://doi.org/10.1016/j.still.2009.03.001.

Van Loon, L.C., 2007. Plant responses to plant growth-promoting rhizobacteria. In: Bakker, P.A.H.M., Raaijmakers, J.M., Bloemberg, G., Höfte, M., Lemanceau, P., Cooke, B.M. (Eds.), New Perspectives and Approaches in Plant Growth-Promoting Rhizobacteria Research. Springer, Dordrecht, pp. 243–254. Available at https://doi.org/10.1007/978-1-4020-6776-1_2.

Verma, P., Kumar Ghosh, P., 2022. REDD+ Strategy for forest carbon sequestration in India. The Holistic Approach to Environment 12 (3), 117–130. Available at https://doi.org/10.33765/thate.12.3.4.

Waldrop, M.P., Zak, D.R., Sinsabaugh, R.L., Gallo, M., Lauber, C., 2004. Nitrogen deposition modifies soil carbon storage through changes in microbial enzymatic activity. Ecological Applications 14 (4), 1172–1177. Available at https://doi.org/10.1890/03-5120.

Wallace, A.J., Armstrong, R.D., Grace, P.R., Scheer, C., Partington, D.L., 2020. Nitrogen use efficiency of 15N urea applied to wheat based on fertiliser timing and use of inhibitors. Nutrient Cycling in Agroecosystems 116 (1), 41–56. Available at https://doi.org/10.1007/s10705-019-10028-x.

Walley, F.L., Clayton, G.W., Miller, P.R., Carr, P.M., Lafond, G.P., 2007. Nitrogen economy of pulse crop production in the Northern Great Plains. Agronomy Journal 99 (6), 1710–1718. Available at https://doi.org/10.2134/agronj2006.0314s.

Wang, F., Mu, Z., Guo, T., Huang, A., Lin, X., Shi, X., et al., 2020. Effect of long-term differentiated fertilisation regimes on greenhouse gas emissions from a subtropical rice-wheat cropping system. Plant, Soil and Environment 66 (4), 167–174. Available at https://doi.org/10.17221/693/2019-PSE.

Wang, M., Elgowainy, A., Lee, U., Bafana, A., Banerjee, S., Benavides, P.T., et al., 2021. Summary of Expansions and Updates in GREET® 2021. Argonne National Lab. (ANL), Argonne, IL (United States).

Wang, S., Wen, Y., Shi, Z., Nuran Zaini, I., Göran Jönsson, P., Yang, W., 2022. Novel carbon-negative methane production via integrating anaerobic digestion and pyrolysis of organic fraction of municipal solid waste. Energy Conversion and Management 252, 115042. Available at https://doi.org/10.1016/j.enconman.2021.115042.

Westerman, P.R., Liebman, M., Menalled, F.D., Heggenstaller, A.H., Hartzler, R.G., Dixon, P.M., 2005. Are many little hammers effective? Velvetleaf (*Abutilon theophrasti*) population dynamics in two- and four-year crop rotation systems. Weed Science 53 (3), 382–392. Available at https://doi.org/10.1614/WS-04-130R.

Wilhelm, W.W., Johnson, J.M.F., Hatfield, J.L., Voorhees, W.B., Linden, D.R., 2004. Crop and soil productivity response to corn residue removal: a literature review. Agronomy Journal 96 (1), 1–17. Available at https://doi.org/10.2134/agronj2004.1000a.

Xu, H., Sieverding, H., Kwon, H., Clay, D., Stewart, C., Johnson, J.M.F., et al., 2019. A global meta-analysis of soil organic carbon response to corn stover removal. GCB Bioenergy 11 (10), 1215–1233. Available at https://doi.org/10.1111/gcbb.12631.

Yang, Z., Li, N., Ma, P., Li, Y., Zhang, R., Song, Q., et al., 2020. Improving nitrogen and water use efficiencies of hybrid rice through methodical nitrogen–water distribution management. Field Crops Research 246, 107698. Available at https://doi.org/10.1016/j.fcr.2019.107698.

Zhang, H., Yu, C., Kong, X., Hou, D., Gu, J., Liu, L., et al., 2018. Progressive integrative crop managements increase grain yield, nitrogen use efficiency and irrigation water productivity in rice. Field Crops Research 215, 1–11. Available at https://doi.org/10.1016/j.fcr.2017.09.034.

Index

Note: Page numbers followed by "*f*" and "*t*" refer to figures and tables, respectively.

B

C

D

E

O

P

Q

R

S

Printed and bound by CPI Group (UK) Ltd, Croydon, CR0 4YY
14/09/2026
02216388-0010